Palaeopathology and Evolutionary Medicine

Palaeopathology and Evolutionary Medicine

An Integrated Approach

Kimberly A. Plomp

Associate Professorial Fellow, Archaeological Studies Program, University of the Philippines, Philippines; and Adjunct Professor, Department of Archaeology, Simon Fraser University, Canada

Charlotte A. Roberts

Emeritus Professor, Department of Archaeology, University of Durham, UK

Sarah Elton

Professor, Department of Anthropology, University of Durham, UK

Gillian R. Bentley

Professor, Department of Anthropology, University of Durham, UK

OXFORD
UNIVERSITY PRESS

OXFORD
UNIVERSITY PRESS

Great Clarendon Street, Oxford, OX2 6DP,
United Kingdom

Oxford University Press is a department of the University of Oxford.
It furthers the University's objective of excellence in research, scholarship,
and education by publishing worldwide. Oxford is a registered trade mark of
Oxford University Press in the UK and in certain other countries

Published in the United States of America by Oxford University Press
198 Madison Avenue, New York, NY 10016, United States of America

British Library Cataloguing in Publication Data
Data available

Library of Congress Control Number: 2021950129

ISBN 978–0–19–884971–1 (hbk)
ISBN 978–0–19–884972–8 (pbk)

DOI: 10.1093/oso/9780198849711.001.0001

Printed and bound by
CPI Group (UK) Ltd, Croydon, CR0 4YY

Foreword

Studying the evolution of human health could not be more relevant than during the current time of a global pandemic crisis! The investigation of the why and how human health changes in prehistoric and historic times is important today, not only to scientists but also to many societal stakeholders. The fact that this topic of research is explicitly addressed in this impressive book make it beneficial beyond the usual expectations for such compilations.

The title of this book specifically mentions the **integration** of the two fields of paleopathology and evolutionary medicine. This is a rather new perspective, as hitherto the two fields have been addressed individually—which makes the book particularly praiseworthy. These two fields overlap more than is usually thought. In fact, one could argue that certain perspectives of evolutionary medicine are key parts of palaeopathology, and vice versa. Stressing this important point is the worthy goal of this compilation.

Palaeopathology is not only the study of ancient diseases in humans, but also in animals, reflecting a holistic 'One Health' approach. Examining the multifactorial ecological niche of health has been promoted since ancient times, for example, by the medical doctor Hippocrates (*c.* 460 BCE–*c.* 370 BCE). This promotion has continued and is also the basis of an evolutionary understanding of medical issues. Thus, integrating the two fields of palaeopathology and evolutionary medicine should lead to a greatly needed **'Paleo-One-Health' approach** of medicine. It should also lay the groundwork for current and future One Health scientific work.

Both evolutionary medicine and palaeopathology have their own traditions and methodologies. Palaeopathological research can often addresses medical cases of the past (and, for good reason, it is often criticised for its focus on single case studies).

On the other hand, evolutionary medicine often provides more general theoretical perspectives that focus on the ultimate cause of a particular disease. Integrating these two approaches is mutually beneficial. Since both fields independently suffer from a partial lack of recognition in (medical) science, by integrating the knowledge and understanding from the two fields we might overcome this 'medical hesitancy' in the future, thereby strengthening both fields.

I fully thank all the authors—and in particular the editors–for compiling such a wonderful treatise. Not only are all of the diverse topics highly relevant, but also the chapters provide various perspectives on how the combination of the two fields will benefit science, and eventually, society as a whole.

Finally, can we *all* better help to promote the integration of the two fields? For my part, I always emphasise the crucial need to bring these scientific issues into the mindset of medical doctors, especially clinical researchers. However, it is equally important to integrate these same issues into the toolkits of those who are responsible for designing the contents of medical curricula used by the future generations of medical doctors who will contribute to the effective care of their patients.

To me, medicine without perspectives from evolution and history is not optimal medicine. For that reason, this volume clearly moves us forwards to a better medical future.

Frank Rühli
Dean, Medical Faculty, University of Zurich, Switzerland;
Founding Chair and Director, Institute of Evolutionary Medicine, University of Zurich;
President, International Society of Evolution, Medicine and Public Health

Acknowledgements

Preparing a book such as this is truly a team effort, and each co-editor acknowledges the collaboration and comradeship of each other during all our endeavours. Our utmost thanks to the authors for preparing their contributions during the very difficult time of the Covid-19 pandemic. The reviewers also pulled out all the stops to give us their views in a timely fashion, and we thank them from the bottom of our hearts. We are grateful to Frank Rühli and Jane Buikstra for their thoughtful and insightful Foreword and Afterword, respectively. We thank Ian Sherman and Charles Bath at OUP for their unflagging enthusiasm and helpful suggestions.

SE thanks John Russell for his support during the preparation of this volume, and CR thanks Stewart Gardner for his understanding and patience over the last few years as we have prepared this book. CR would also like to acknowledge how Nesse and Williams' book (*Why We Get Sick*, 1994) sparked her initial interest in evolutionary medicine.

Table of Contents

15   Metabolic diseases in bioarchaeology: an evolutionary medicine approach   **284**

Jonathan C. Wells, Nelissa Ling, Jay T. Stock, Hallie Buckley, and William R. Leonard

16   The palaeopathology of traumatic injuries: an evolutionary medicine perspective   **303**

Ryan P. Harrod and Anna J. Osterholtz

List of Abbreviations

2D	two-dimensional
3D	three-dimensional
ACL	anterior cruciate ligament
ACTH	adrenocorticotropin-releasing hormone
ADH	antidiuretic hormone
aDNA	ancient DNA
AEP	acquired enamel pellicle
ALL	acute lymphoblastic leukaemia
ALSPAC	Avon Longitudinal Study of Parents and Children
AS	ankylosing spondylitis
ASD	acquired spinal disease
BABAO	British Association of Biological Anthropology and Osteoarchaeology
BCG	Bacillus Calmette–Guérin
BCE	before common era
BCS70	1970 British Cohort Study
BIA	bioelectrical impedance analysis
BMI	body mass index
BMR	basal metabolic rate
BP	before present
CAC	coronary artery calcium
CAD	coronary artery disease
CCP	cancer crude prevalence
CDC	Centers for Disease Control and Prevention
CDV	canine distemper virus
CMI	cell-mediated immunity
CMV	cytomegalovirus
CRH	corticotropin-releasing hormone
CRP	C-reactive protein
CT	computed tomography
CVD	cardiovascular disease
DALY	disability-adjusted life years
DARC	Duffy antigen receptor for chemokines
DED	dental enamel defects
DHEA-S	dehydroepiandrosterone-sulfate
DISH	diffuse idiopathic skeletal hyperostosis
DNA	deoxyribonucleic acid
DOHaD	developmental origins of health and disease
DXA	dual-energy X-ray absorptiometry
EM	Evolutionary Medicine
EnA	enteropathic arthropathy
EWAS	epigenome-wide association studies
FGF2	fibroblast growth factor receptor 2
G6PD	glucose-6-phosphate dehydrogenase
GCRF	Global Challenges Research Fund
GDP	gross domestic product
GHHP	Global History of Health Project
GIT	gastrointestinal tract
HCO	Holocene Climatic Optimum
HDL	high-density lipoprotein
HFCS	high fructose corn syrup
HIPAA	Health Insurance Portability and Accountability Act
HIV	human immunodeficiency virus
HL	Harris lines
HLA	human leukocyte antigen
HLHS	hypoplastic left heart syndrome
HMV	human measles virus
HPA	hypothalamic-pituitary-adrenal
IDH	intervertebral disc height
IL-6	interleukin-6
IRB	Internal Review Board
IRMS	isotope ratio mass spectrometry
ISEMPH	International Society for Evolution, Medicine and Public Health
ky BP	thousands of years before present
LDL	low-density lipoprotein
LEH	linear enamel hypoplasia
LHT	life history theory
LMIC	lower- and middle-income country
MCA	Medieval Climate Anomaly
MCS	Millennium Cohort Study
MDRTB	multi-drug-resistant TB
MDT	multiple drug treatment
MERS	Middle East respiratory syndrome
MSU	monosodium urate

MTBC	*Mycobacterium tuberculosis* complex
NAGPRA	Native American Graves Protection and Repatriation Act
NCD	non-communicable (chronic) disease
NCDS	National Child Development Study
NCHS	National Center for Health Statistics
NHS	National Health Service
NSHD	National Survey of Health and Development
OD	obstetric dilemma
PAR	predictive adaptive response
PCR	polymerase chain reaction
PE	proline-glutamate
PPE	proline-proline-glutamate
PsA	psoriatic arthropathy
PTSD	post-traumatic stress disorder
ReA	reactive spondyloarthropathy/Reiter's syndrome
RPV	rinderpest virus
RR-TB	rifampicin-resistant TB
SAM	sympathetic-adrenal-medullary
SARS	severe acute respiratory syndrome
SES	socio-economic status
SFI	skeletal frailty index
SI-NMNH	National Museum of Natural History, Smithsonian Institution
SNP	single nucleotide polymorphism
T2DM	Type 2 diabetes mellitus
TB	tuberculosis
THLHP	Tsimane Health and Life History Project
TMRCA	the most recent common ancestor
TNF-α	tumour necrosis factor-α
TRF	thyrotropic-releasing factor
TSH	thyroid-stimulating hormone
USDA	United States Department of Agriculture
WEIRD	Western, Educated, Industrialized, Rich, and Democratic
XDRTB	extreme drug-resistant TB

List of Contributors

Ella Been, Head of Motion Analysis, Ono Academic College, Israel

Gillian R. Bentley, Professor, Durham University, Department of Anthropology, UK

Michaela Binder, Research Associate, Austrian Archaeological Institute, Austria

Kelly E. Blevins, Graduate Assistant, Arizona State University, School of Human Evolution and Social Change, USA

Amy M. Boddy, Assistant Professor, University of California Santa Barbara, Department of Anthropology, USA

Kirsten Bos, Group Leader of Molecular Palaeopathology, Max Planck Institute for the Science of Human History, Germany

Hallie Buckley, Professor, University of Otago, Department of Anatomy, New Zealand

Jane Buikstra, Professor, Arizona State University, School of Human Evolution and Social Change, USA

Nicole Burt, Curator, Cleveland Museum of Natural History, Department of Human Health & Evolutionary Medicine, USA

Mark Collard, Professor, Simon Fraser University, Department of Archaeology, Canada

Zachary Compton, Grad Service Assistant, Arizona State University, Arizona Cancer and Evolution Center, USA

Peter D. O. Davies, Retired Professor, University of Liverpool, School of Medicine, UK

Sarah-Louise Decrausaz, Researcher and University Lecturer, University of Victoria, Department of Anthropology, Canada

Sharon N. DeWitte, Professor, University of South Carolina, Department of Anthropology, USA

Ashley N. Edes, Animal Welfare Scientist, Saint Louis Zoo, Department of Reproductive and Behavioral Sciences, USA

Sarah Elton, Professor, Durham University, Department of Anthropology, UK

Vinicius M. Fava, Research Associate, Research Institute of the McGill University Health Centre, Canada.

Caleb E. Finch, Professor, University of Southern California, Leonard Davis School of Gerontology, USA

Frances Galloway, Anglia Ruskin University, School of Nursing and Midwifery, UK

Julia Gamble, Assistant Professor, University of Manitoba, Department of Anthropology, Canada

Alexandra M. Greenwald, Assistant Professor, University of Utah; and Faculty Affiliate, Arizona State University, USA

Ryan P. Harrod, Affiliate Professor University of Alaska Anchorage; and Dean of Academic Affairs, Garrett College, Department of Anthropology, USA

Marissa L. Ledger, University of Cambridge, UK; and University of Alberta, Canada

William R. Leonard, Professor and Department Chair, Northwestern University, Department of Anthropology, USA

Malcolm C. Lillie, Professor, Umeå University, Sweden; and Visiting Professor, University of Hull, UK

Nellissa Ling, University of Otago, Department of Anatomy, New Zealand

Carina Marques, Department of Anthropology. The University of Texas, Rio Grande Valley, Texas, USA; and University of Coimbra, Professor, Research Centre for Anthropology and Health, Portugal

Piers D. Mitchell, Senior Research Associate, University of Cambridge, Department of Archaeology, UK

Anna J. Osterholtz, Assistant Professor, Mississippi State University, Department of Anthropology and Middle Eastern Cultures, USA

Kimberly A. Plomp, Associate Professorial Fellow, Archaeological Studies Program, University of the Philippines, Philippines; and Simon Fraser University, Department of Archaeology, Canada

Charlotte A. Roberts, Emeritus Professor, Durham University, Department of Archaeology, UK

Chris J. Rowan, Director of Inpatient Heart Failure, Renown Regional Medical Center, USA

Frank Rühli, Director, University of Zurich Institute of Evolutionary Medicine, Switzerland

David M. Scollard, Retired, U.S. Department of Health and Human Services, USA

Tanya M. Smith, Professor, Griffith University, Australian Research Centre for Human Evolution & Griffith Centre for Social and Cultural Research, Australia

Jay T. Stock, Professor, University of Western Ontario, Department of Anthropology, Canada

Anne C. Stone, Professor, Arizona State University, School of Human Evolution and Social Change, USA

M. Linda Sutherland, MemorialCare Health System, USA

Daniel H. Temple, Director of Graduate Programs, George Mason University, Department of Sociology and Anthropology, USA

Gregory S. Thomas, Medical Director, MemorialCare Health System; and Clinical Professor, University of California, Irvine, USA

Richard Thomas, Professor and Dean of Research, University of Leicester, School of Archaeology and Ancient History, UK

Randall C. Thompson, Professor of Medicine, University of Missouri-Kansas City School of Medicine, USA

Christina Warinner, Assistant Professor, Harvard University, USA; and Group Leader, Max Planck Institute for the Science of Human History, Germany

Nicholas W. Weis, Kansas State University, USA

Jonathan C. Wells, Professor, UCL Great Ormond Street Institute of Child Health, Childhood Nutrition Research Centre, Population, Policy and Practice Programme, UK.

Elizabeth W. Uhl, Assistant Professor, University of Georgia, Department of Pathology, USA

What's it all about? A legacy for the next generation of scholars in evolutionary medicine and palaeopathology.

Kimberly A. Plomp, Charlotte A. Roberts, Sarah Elton, and Gillian R. Bentley

The great mystery of medicine is the presence, in a machine of exquisite design, of what seem to be flaws, frailties, and makeshift mechanisms that give rise to most disease.

Nesse and Williams 1994, p. 3

1.1 Background

Evolutionary medicine (EM) has sought to describe why we, as a species, are vulnerable to diseases (Ewald, 1994; Nesse & Stearns, 2008; Williams & Nesse, 1991), in contrast to focusing on treatments, the domain of more traditional clinical approaches to medicine. Evolutionary approaches to health have already impacted many areas of clinical study, leading, for example, to new understandings of antibiotic resistance (Morley et al., 2020), cancer (Gatenby & Brown, 2020; Staňková et al., 2019), inflammatory bowel disease (Axelrad et al., 2020), and myopia (Long, 2018), as well as different ways of approaching life-history events like childbirth (Ball, 2017; Trevathan, 2011). These successes have led to the recognition of EM's value across many in the medical community (Alcock, 2012; Stearns et al., 2010), as well as the adoption of evolutionary theory as a core concept in pre-medical education, including by the Association of American Medical Colleges and the Howard Hughes Medical Institute (Grunspan et al., 2018). Some researchers are actively incorporating evolutionary models in their approaches to curing diseases such as cancer,

for example, at the H. Lee Moffitt Cancer Center and Research Institute in Tampa, Florida (Gatenby & Brown, 2020), and also including them in clinical rounds in major US hospitals, for example, the UCLA, Department of Medicine Grand Rounds in Evolutionary Medicine, and at the University of Michigan (Dolgin, 2015).

Even with these recent advances, the primary goal of EM has been to provide what are termed 'ultimate' explanations for human vulnerabilities rather than the 'proximate' ones more common to a doctor's consultation. Proximate and ultimate are important concepts in biology, with Mayr (1961) among the first to formalise them. Probably the most prominent elaboration is by Tinbergen (1963) in his 'Four Questions' (causation [mechanism], ontogeny, evolution, survival value), usefully employed in EM (Nesse, 2019). Ultimate, or 'why', explanations usually seek to understand the evolutionary basis for a trait or behaviour, often considering fitness benefits and natural selection (Laland et al., 2011; Mayr, 1961). Proximate, or 'how', explanations are much more focused on the 'here and now'— the immediate influences (Laland et al., 2011; Mayr,

Kimberly A. Plomp et al., *What's it all about?* In: *Palaeopathology and Evolutionary Medicine*. Edited by Kimberly A. Plomp et al., Oxford University Press.
© Oxford University Press (2022). DOI: 10.1093/oso/9780198849711.003.0001

1961). Thus, to use an example from Nesse (2019), the ultimate ('why') explanation for fever is the selective advantage that it brings and the evolution of the response, with the proximate considerations being how the body instigates the fever (i.e., through detection of a pathogen), regulates its thermostat, and adjusts body temperature (see also Best & Schwartz, 2014). However, although they are complementary, proximate and ultimate explanations are often treated as dichotomous (Laland et al., 2011). Nevertheless, with greater understanding of plasticity, epigenetics, and evolutionary developmental biology, now incorporated specifically into evolutionary biology via the 'Extended Synthesis' (Pigliucci & Müller, 2010), plus cultural evolution, it is increasingly clear that 'proximate mechanisms contribute to the dynamics of selection' (Laland et al., 2011, p. 1515). Thus, although proximate influences are the primary consideration of clinical mechanism, a major goal of EM should be to integrate ultimate explanations to create a broader, synergistic framework that seeks to improve health and well-being in contemporary populations.

Understanding and improving health and well-being in contemporary populations cannot be accomplished through the narrow lens of the present. Evolution occurs at different temporal scales, depending on circumstances and, as several chapters in this book discuss, humans are still evolving. Gaining detailed insights from the past is therefore a fundamental element of the study of human evolution, including health and disease. With that in mind and given the depth of time needed to study the co-evolution of diseases and humans, it is surprising that bioarchaeology (the study of archaeological human remains) has not been placed more prominently in the development of EM as a discipline. In particular, the subdiscipline of palaeopathology (the study of pathological conditions/abnormal variation evident in archaeological remains) has much to offer EM. The Paleopathology Association's motto is 'Let the dead teach the living', and the insights afforded by the detailed study of skeletal remains, both on an individual and a population basis, are profoundly helpful in understanding processes of health and disease in recent times, from both proximate and ultimate

perspectives. A corollary of this is that palaeopathological research itself could benefit from greater incorporation of evolutionary theory within it. It was from these starting points that we developed the present book.

1.2 Palaeopathology and medicine

The earliest work that can be recognised as part of the discipline now known as palaeopathology is documented from the seventeenth century in Switzerland (Roberts & Buikstra, 2012). Once the discipline was established more formally, it is notable that many palaeopathology scholars born in the early part of the twentieth century were often physicians (e.g. Art Aufderheide, Aidan Cockburn, Jens Peder Hart Hansen, Joseph Jones, George Daniel Morse, Walter Putschar, Marc Armand Ruffer, Andrew Sandison, Eugen Strouhal, and Calvin Wells). Indeed, it was recommended that palaeopathologists should have a background in medicine. This is compared to the late twentieth century, when many palaeopathologists tended to emerge from anthropology and archaeology degrees, a tradition that continues today. One could argue that this is because more specific training in bioarchaeology, including palaeopathology, also developed from the 1990s onwards, especially in Europe, albeit more slowly in some parts of the world.

Although there was a flourishing of evolutionary thinking applied to disease shortly after Darwin's publications, referred to as Medical Darwinism (Zampieri, 2009), EM is a field of study that emerged in the early 1990s. Thus, one might forgive earlier scholars (i.e., physicians) for not integrating evolutionary theory into their palaeopathological work, notwithstanding the emphasis that Hart Hansen seemed to place on palaeopathology's relevance to present and future health (Lynnerup & Frohlich, 2012). It is less clear why scholars practicing palaeopathology from the 1990s onwards have tended not to adopt a more evolutionary approach to the origins and presence of disease in archaeological human remains. In addition, one wonders why EM has, relatively speaking, neglected evidence and perspectives from palaeopathology. One explanation may be because of the influence of disciplinary boundaries, with palaeopathology focusing on its 'standard' canon of literature (e.g. *American Journal of Physical Anthropology*, *International Journal of Osteoarchaeology*, *Journal of Paleopathology*, *Journal of Archaeological Science*, and latterly, the *International Journal of Paleopathology*) and EM looking more to publications in evolutionary biology.

Human remains represent a key source of direct evidence for individual life experiences in the past. Through their study, it is possible to gain insights into mortality and fertility patterns, diet, a range of diseases, growth, and episodes of growth arrest. This information can be captured for individuals across potentially all cohorts of a population (young, old, different sexes, and socioeconomic settings), and thus can both complement and test understandings drawn from historical documents (Herring & Swedlund, 2002; Mitchell, 1999, 2011, 2017) and be used to explore the long history of disease in humans. Furthermore, bioarchaeology is a rapidly developing, holistic field (Buikstra, 1977; Buikstra & Roberts, 2012; DeWitte & Stojanowski, 2015; Roberts & Manchester, 2005) that encompasses palaeopathology (Roberts, 2016), palaeodemography (Chamberlain, 2006; Milner et al., 2018), and palaeoepidemiology (Milner & Boldsen, 2017; DeWitte & Stojanowski, 2015; Waldron, 1994). Bioarchaeology and its subdisciplines recognise that the information 'read' from human remains has been influenced by many factors, and that it is imperative to contextualise such remains to understand any patterns observed (DeWitte & Stojanowski, 2015).

This is an ideal time to integrate evolutionary approaches into palaeopathology. Over the past twenty years, there have been more 'big picture' approaches to health in the past using, for example, greater sample sizes and datasets from larger geographic regions to explore disease frequency over deep time ranges (e.g., Steckel & Rose, 2002; Steckel et al., 2019). The emergence of research into pathogen DNA, both modern and ancient, since the early 1990s has further facilitated comparative approaches and enabled time frames for pathogen evolution to be established. These developments provide the context that has allowed more scholars to recognise the value of EM in palaeopathology.

1.3 Aims of the book

Palaeopathology and Evolutionary Medicine: An Integrated Approach seeks to fill the identified gap between scholars working in EM and palaeopathology, presenting novel, interdisciplinary research to illustrate how palaeopathology is currently being blended with evolutionary theory to address questions about human health. It also highlights potential avenues of future research to promote collaboration between palaeopathology and EM. Discussions about this book started in September 2016. By February 2017, a proposal for Oxford University Press (OUP) was being drafted and by October 2017 all the editors were 'on board'. In February 2018, invitations were sent to authors, and by March 2019, a contract was signed for the volume with OUP. The contributors to the volume are experts in many different fields, including medicine and other health sciences, biological anthropology and bioarchaeology, and each chapter presents innovative research using varied approaches. There is no geographical focus, thus ensuring that the research presented is relevant in a global context. In many cases, authors with a more biological or clinical background are teamed with bioarchaeologists, and we have tried to achieve a mix of backgrounds and experiences. For example, Chapter 9 is written by Charlotte Roberts (a palaeopathologist), David Scollard (a clinical leprologist), and Vinicius Fava (a geneticist). Each submitted chapter was initially subject to two external reviews and in subsequent versions was reviewed by each of the co-editors until a final version was achieved.

Three questions were central to the development of the volume. The first was how palaeopathological research on specific topics contributes to the understanding of the evolution of health and disease. The second was how palaeopathology offers a unique perspective and new understanding of those topics. The third was how this new understanding contributes to modern clinical research and medicine. Authors were asked to consider these questions as they wrote their chapters. The topics studied represent the key themes within EM that emerged from George Williams's and Randy Nesse's seminal paper 'The Dawn of Darwinian Medicine' in 1991 (Table 1.1). While these themes have

Table 1.1 Suggested categories for study within Evolutionary Medicine with commonly used examples (Williams & Nesse, 1990). The chapters in this volume are listed here according to these categories.

Categories within EM	Examples	Chapter
Co-evolution of host and pathogen	The rapid evolution of bacteria has led to the emergence of multi-drug-resistant pathogens that are becoming increasingly difficult to treat.	Chapter 8: Human resistance and the evolution of plague in Medieval Europe Chapter 9: Leprosy is down but not yet out: new insights shed light on its origin and evolution Chapter 10: Preventable and curable, but still a global problem: tuberculosis from an evolutionary perspective Chapter 11: Evolutionary perspectives on human parasitic infection: from ancient parasites to modern medicine Chapter 17: Uncovering tales of transmission: an integrated palaeopathological perspective on the evolution of shared human and animal pathogens
Constraints on selection	Adaptations can only arise based on existing characteristics producing limitations on these adaptations.	Chapter 3: Acquired spinal conditions in humans: the roles of spinal curvature, the shape of the lumbar vertebrae, and evolutionary history Chapter 4: Birthing humans in the past, the present and future: how birth can be approached holistically through an evolutionary medicine lens
Mismatch with current industrialised societies	Current lifeways that are at odds with our past have led to an epidemic of chronic conditions such as obesity, type 2 diabetes, and CVDs.	Chapter 5: Isotopic reconstruction of ancient human diet and health: implications for evolutionary medicine Chapter 6: Developmental, evolutionary and behavioural perspectives on oral health Chapter 7: Palaeoecology: considering proximate and ultimate influences in human diets and environmental responses in the early Holocene Dnieper River region of Ukraine Chapter 11: Evolutionary perspectives on human parasitic infection: from ancient parasites to modern medicine Chapter 12: Cardiovascular disease (CVD) in ancient people and contemporary implications Chapter 13: Connecting palaeopathology and evolutionary medicine to cancer research: past and present Chapter 15: Metabolic diseases in bioarchaeology: an evolutionary medicine approach
Physiological defences	Traits such as fever and coughs are debilitating and unpleasant but are designed to reduce the effects of a pathogen on the host.	Chapter 12: Cardiovascular disease (CVD) in ancient people and contemporary implications Chapter 16: The palaeopathology of traumatic injuries: an evolutionary medicine perspective
Reproduction at the expense of health	Selection does not operate to improve health and well-being but to improve 'fitness' and reproductive success.	Chapter 4: Birthing humans in the past, the present and future: how birth can be approached holistically through an evolutionary medicine lens
Trade-offs	Studies of different environments experienced during early life development and subsequent health of individuals across the life course have uncovered evidence of trade-offs between the two.	Chapter 2: Developmental origins of health and disease (DOHaD): perspectives from bioarchaeology Chapter 11: Evolutionary perspectives on human parasitic infection: from ancient parasites to modern medicine Chapter 14: Stress in bioarchaeology, epidemiology and evolutionary medicine: an integrated conceptual model of shared history from the descriptive to the developmental

evolved (Nesse, 2005; Nesse & Stearns, 2008; Nesse & Williams, 1994; Williams & Nesse, 1991), they remain useful for organising the domains within which EM is incorporated. This book also covers areas/diseases that are less well studied in EM. One example is leprosy, which has had a long (but often forgotten) impact on human health across the ages. Another example, examined in several chapters, is the One Health Initiative, a movement first described in the early 2000s in relation to SARS (severe acute respiratory disease) (MacKenzie & Jeggo, 2019) that combines the study of human and animal health and their shared environments. The next section outlines how the chapters in the volume have addressed the overarching categories in EM (Williams & Nesse, 1991; also see Table 1.1), including newer perspectives such as One Health.

1.4 Thematic overview

1.4.1 Co-evolution of host and pathogen

The co-evolution of hosts and pathogens has been likened to an arms race (Capewell et al., 2015; Pays et al., 2014; Safari et al., 2020), where each attempt to outwit the other with micro-evolutionary changes conferred either greater contagion or increased immunological protection. Some contributions to EM, such as Brockhurst and colleagues (2014) and MacPherson and colleagues (2018), have invoked van Valen's (1973) Red Queen Hypothesis, based on the character in the Alice in Wonderland story who states that, although she runs as much as she can, she remains in one place. Several chapters, covering a variety of conditions (Chapter 8 on plague, Chapter 9 on leprosy, and Chapter 10 on tuberculosis), relate to this theme. Two of these discuss diseases (leprosy and plague) that are more commonly limited to developing countries today and have received less attention in recent years from clinical research. It is also argued that what attention has been paid to them arose in the colonial context of treating tropical diseases and other maladies that either might have affected white colonialist administrators or their labour force (e.g., leprosy: Buckingham, 2002; Edmond, 2006; Zaman, 2005). This neglect by contemporary Western medicine and the comparatively greater attention by palaeopathologists reminds us of the potential utility of using palaeopathology to inform current knowledge of these neglected diseases, particularly where there are concerted efforts on the part of some funders to tackle them anew (e.g., Gates Foundation, and the Global Challenges Research Fund (GCRF) initiative by the UK Government).

In Chapter 8—*Human resistance and the evolution of plague in Medieval Europe*, Kirsten Bos and Sharon N. DeWitte take us on a tour of (mostly) medieval Europe to describe episodes of plague that affected the continent for hundreds of years as waves of the disease advanced and receded. They outline how bioarchaeological data, including information derived from ancient DNA, has assisted in the reconstruction of plague lineages, while analyses of palaeopathological and palaeodemographic data further contextualise how and why certain populations may have become more susceptible to plague pandemics. While currently having the means to control plague outbreaks with antibiotics and existing vaccines, we are reminded that, only in the last decade, several outbreaks killed almost 300 individuals in Madagascar (Rakotosamimanana et al., 2021).

When reading Chapter 8, one cannot but help raise comparisons to the current and unequal world experience of the COVID-19 pandemic, even though the political, socio-economic, and scientific situation of contemporary human populations gives us opportunities to tackle diseases with a much-enhanced repertoire of defences. Still, Bos and DeWitte point to the conditions in medieval Europe during the Second Pandemic of the plague that echo current socio-economic inequalities today, in that individuals with fewer resources who suffer from malnutrition and co-morbidities would have become far more susceptible to both morbidity and mortality (see, for example, Abrams et al., 2020; and Bentley, 2020; on COVID-19).

Similar themes emerge from Chapter 9—*Leprosy is down but not yet out: new insights shed light on its origin and evolution*, written by Charlotte A. Roberts, David M. Scollard, and Vinicius M. Fava. In it, the authors outline how the main organism that causes leprosy, *Mycobacterium leprae*, lost its genetic diversity after thousands of years of host/pathogen co-evolution with humans. This then led to much greater resistance in human populations so that, from about the fourteenth century CE onwards, barring selectivity in the palaeopathological record, leprosy in Europe became less common. In addition, the authors discuss the possibility that our species' long relationship with another mycobacterium, *Mycobacterium tuberculosis*, may have given humans a leading edge in the potential arms-race with leprosy, resulting in the eventual end of the medieval leprosy pandemic in Europe. However, the data also show that evolving resistance to *M. leprae* has involved immunological trade-offs, where relevant mutations conferring resistance to leprosy are associated with increased susceptibility to inflammatory bowel disease and Parkinson's Disease (Kaushansky & Ben-Nun, 2014; Sulzer et al., 2017; Wang et al., 2016). Like the authors of Chapter 8, Roberts, Scollard and Fava consider socio-economic and political contexts, examining when and where populations affected by leprosy resided, making note of the fact that some high-status burials are associated with the disease, and downplaying the idea that leprosy would always have carried stigma or that it specifically targeted the poor.

In Chapter 10—*Preventable and curable, but still a global problem: tuberculosis from an evolutionary perspective*, authors Charlotte A. Roberts, Peter D. O. Davies, Kelly E. Blevins, and Anne C. Stone provide a detailed overview and unique perspective on one of today's most deadly killers, tuberculosis (TB). Modern genetic work investigating the diversity and evolution of *M. tuberculosis* has indicated that, instead of being a zoonotic disease as previously thought, TB was spread from humans to other animals (Brosch et al., 2002; Gordon et al., 1999). These genetic studies are not limited to modern pathogen DNA, as ancient microbial DNA from bioarchaeological remains have provided vital insight into the selection pressures acting on *M. tuberculosis* as it evolved to infect new hosts, with ancient genomic epidemiology being used to trace the spread of outbreaks through time. When integrated with palaeopathological evidence, we can piece together the long co-evolutionary history of humans and TB. In particular, palaeopathological evidence, along with ancient and modern genomes, have indicated that human migrations, population interactions, and social upheavals have affected the distribution of TB throughout the world (Brynildsrud et al., 2018; Pepperell et al., 2011), as they do today.

The co-evolution of humans and TB is currently a particularly important topic due to the ongoing evolutionary arms race between our species and *M. tuberculosis*. TB is one of the top ten causes of death worldwide (WHO, 2019, 1), and we are starting to

see an increase in infection rates in some areas of the world, especially in lower- and middle-income countries (LMICs) (Macedo Couto et al., 2019). One reason for this re-emergence is that, through co-evolution and the evolutionary arms race, certain strains of the mycobacterium have evolved resistance to antibiotics (WHO, 2020). As such, TB is a prime example of how understanding the evolution and history of a disease can have critical importance for the health of people today.

In Chapter 11—*Evolutionary perspectives on human parasitic infection: from ancient parasites to modern medicine*, Marissa L. Ledger and Piers D. Mitchell tackle the topic of parasitic infections from the human past, and how humans have coevolved to tolerate and combat their effects. This topic inevitably links to the EM theme of 'trade-offs', since co-existing with helminths or other parasites necessarily involves some health compromises. Infections with soil-transmitted helminths have become some of the best examples of areas within EM that have potential clinical applications. It is increasingly appreciated that helminth (and other) infections serve to educate the immune system during development and that our modern sanitised environments may rob us of the opportunity to acquire this education (Rook, 2012; Rook et al., 2004). These ideas were first formulated in the Hygiene and Old Friends Hypotheses, where epidemiological studies first pointed to an apparent immunological protection against allergies for children who grew up on farms or in more rural areas.

Chapter 11 also fits the EM theme of mismatch, with the Hygiene Hypothesis highlighting the disjunction between our contemporary, clean environments that have made us more prone to auto-immune disorders and infection. There have since been numerous publications providing evidence for these associations in various geographic regions (Hoberg et al., 2001; Rassi et al., 2010; Zhu et al., 2015), while clinical trials have been conducted to see if infections with pinworms might alleviate other autoimmune conditions such as Crohn's disease or irritable bowel syndrome (Rook et al., 2004). As pointed out by Ledger and Mitchell, these trials have not been wholly successful, but this could be explained by the use of pinworms derived from pigs rather than humans. They rightfully note that palaeopathology and other evidence from the archaeological record could greatly assist our understanding of our co-evolutionary past with various parasites.

In Chapter 17—*Uncovering tales of transmission: an integrated palaeopathological perspective on the evolution of shared human and animal pathogens*, Elizabeth W. Uhl and Richard Thomas provide an extensive overview of some of the more common pathogens that either have a zoonotic origin or that humans have transmitted to animals. These include tapeworms, *Mycobacterium tuberculosis* and *Mycobacterium bovis* that mainly cause TB in humans, *Plasmodium falciparum* causing malaria, *Brucella* causing brucellosis, *Burkholderia mallei* causing the disease 'glanders' that mostly affects equines but can also infect humans, and the zoonotic *Morbillivirus* virus that causes measles. The chapter also integrates materials from palaeopathology, history, genetics, palaeoecology, and medicine to stress the importance of these pathogens during human evolutionary prehistory, drawing attention to the major transitions such as the development of farming when new opportunities arose for zoonotic and reverse zoonotic disease transmission as human populations started to live in permanent settlements.

1.4.2 Constraints on selection

Two chapters address the issue of constraints on selection. This key theme in EM argues that existing features within an organism limit how selective pressures can modify those features to accommodate or adapt to new environments (Nesse and Williams, 1994). One of the common examples used to illustrate this theme has been our transition from a quadrupedal primate to a bipedal one, involving not only changes to our spinal column but to many other aspects of our skeletal and muscular architecture. Bipedality introduced new challenges and risks for lower back pain and associated pathologies not often experienced (as far as we can tell) by our closest non-bipedal primate relatives, who have a low documented frequency of such conditions (Anderson, 1999; Castillo & Lieberman, 2015).

In Chapter 3—*Acquired spinal conditions in humans: the roles of spinal curvature, the shape of the*

lumbar vertebrae and evolutionary history, Kimberly A. Plomp, Ella Been, and Mark Collard discuss why certain acquired spinal diseases (ASDs) like arthritis, herniated discs, and spondylolysis, afflict humans. As well as providing a wealth of clinical and palaeopathological information about these conditions, the authors present evolutionary ideas to explain why humans are vulnerable to ASDs. Using their earlier work, they present the 'Ancestral Shape Hypothesis' that draws on the range of variation in three-dimensional shape of human vertebrae to suggest that those with characteristics closely matched to our closest primate relatives and hominin ancestors are more susceptible to Schmorl's nodes (a type of disk herniation where the interior fluid from one vertebral disc can obtrude into adjacent vertebral endplates, causing inflammation). Similarly, Plomp and colleagues argue that humans with vertebrae at the more extreme end of characteristics associated with bipedalism are more vulnerable to spondylolysis (a stress fracture in part of the vertebral arch often associated with an underlying weakness). They suggest that these hypotheses could have predictive value in allowing clinicians to assess individual risks for ASDs.

Chapter 4—*Birthing humans in the past, the present, and future: how birth can be approached holistically through an evolutionary medicine lens*, written by Sarah-Louise Decrausaz and Frances Galloway, deals with another anatomical constraint associated with bipedalism, namely the difficulties of human childbirth where a relatively large-headed infant must pass through a narrow, bipedal pelvis. Nesse and Williams (1994) described the difficulties of human childbirth as a 'legacy' of evolution. However, Decrausaz and Galloway examine how this 'legacy' was conceived as an 'Obstetrical Dilemma' (Washburn, 1960) by reviewing several criticisms of this concept. They discuss that more recent research considers how other factors, such as access to medical care, discrimination, and socio-economic status are all important variables in determining the health and morbidity of mothers and infants during birth. Using the bioarchaeological record, the authors also outline the difficulties in identifying parturition ('scars' on the pelvic bones) in the archaeological record, and its potential relationship to maternal and infant mortality. They suggest that comparing the prevalence of birth-related deaths between populations, both in the past through bioarcheological evidence and in today's modern world, can help us understand the implications of evolutionary, biocultural, environmental, and medical adaptations for the human birth process.

1.4.3 Mismatch with current industrialised societies

Many approaches using EM relate to this third theme, where chronic conditions observed within contemporary, (mostly) industrialised societies are thought to reflect disjunctions between our current and ancient lifeways. Increasing technologies, globalisation, and wealth, along with sedentary jobs, have led to a much less active lifestyle for many humans, with consequences for general health and increasing rates of overweight and obesity (Agha & Agha, 2017; Hamilton et al., 2018). This lack of exercise coupled with malnutrition, specifically referring here to poor-quality, obesogenic diets, are thought to be responsible for epidemics of Type 2 diabetes mellitus (T2DM), hypertension, and cardiovascular diseases (CVDs). As LMICs are becoming more exposed to globalisation and Westernised diets, these chronic conditions are increasing. A common narrative in some of the earlier EM literature is that contemporary, poor-quality diets are at odds with those of our hunting and gathering forebears, whom, it is argued, ate fewer carbohydrates and grains, more meat protein (with the meat consumed being lean and derived from wild animals) and, overall, a diet enriched in gathered fruits and vegetables with high fibre intakes.

There is now greater appreciation in the EM literature that our prehistoric and pre-agricultural ancestors, depending on their local ecologies, experienced a wide diversity of diets. In Chapter 5—*Isotopic reconstruction of ancient human diet and health: implications for evolutionary medicine*, Nicole Burt and Alexandra M. Greenwald discuss techniques of stable isotope analyses, which allow researchers to reconstruct major dietary elements from the bones and teeth of skeletal remains. By integrating palaeopathological and isotopic data with an evolutionary and biocultural framework, they illustrate

the diversity of human diets throughout our evolution and argue that, within a diverse diet, domesticated agricultural products play an important, nutritionally-positive role. Using evidence from isotope studies they argue that poor health in some populations is because of a switch from a traditional diet, whether that was based on agriculture or foraging. As in several of the previous chapters, Burt and Greenwald make a compelling case that this mismatch, with a paucity of nutritious food and the prevalence of associated health problems, is the direct result of the legacy of colonialism. They also call for a decolonisation of nutritional recommendations to help tackle diseases related to poor nutrition today.

Chapter 6—*Developmental, evolutionary, and behavioural perspectives on oral health*, by Tanya M. Smith and Christina Warriner, complements Chapter 5. Smith and Warriner focus on the dentition, outlining in detail its pattern of growth, various dental pathologies, and changes that have occurred as human diet and lifestyles have changed across millennia. This chapter has particular resonance for understanding early-life development and, in that sense, complements Chapter 2 by Gamble and Bentley, which also provides evidence for how dental disruptions can shed light on early-life history. In addition, Smith and Warriner address several facets of both the agricultural and industrial revolutions and how changes in dental wear and diets have altered the dentition's structure and vulnerability to caries and calculus accretions. They also outline important considerations about how these pathologies can affect human health more broadly.

Chapter 7—*Palaeoecology: considering proximate and ultimate influences on human diets and environmental responses in the early Holocene Dnieper River region of Ukraine*, by Malcolm C. Lillie and Sarah Elton, integrates palaeopathology and isotope data in a case study of the early Holocene of eastern Europe, a time of dietary transition for humans. They too emphasise the importance of palaeoenvironmental and dietary diversity in human evolution, critiquing the notion of the 'Stone Age diet' and arguing that, on top of developing ultimate explanations for human health and well-being, we need to be aware of how proximate factors, such as small ecological changes between groups, may impact health status. By comparing groups in a small geographic area over a few thousand years, they describe palaeopathological evidence for small, transient changes in behaviour—rather than large-scale revolutions—that have adverse health outcomes. They suggest that one key challenge for EM in the future is to combine proximate and evolutionary perspectives to improve our understanding of health in the past, as well as make a positive impact on health status in living populations.

Many publications in EM have underscored the seriousness of CVDs, including atherosclerotic plaques that build up in the major blood vessels, as one of the more common chronic conditions to afflict contemporary human populations. However, as is clear from Chapter 12—*Cardiovascular disease (CVD) in ancient people and contemporary implications* by Randall C. Thompson, Chris J. Rowan, Nicolas W. Weis, M. Linda Sutherland, Caleb E. Finch, Michaela Binder, Charlotte A. Roberts, and Gregory S. Thomas, there is abundant evidence to show that our prehistoric and historic ancestors also suffered from atherosclerosis. Evidence often arises from mummies or other preserved bodies, such as Ötzi, the Iceman, where scans have clearly shown the build-up of arterial calcified plaques. At times, calcified remains of plaque can actually be found alongside skeletal remains in graves in anatomical positions that can inform us which arteries were afflicted. Therefore, it appears that atherosclerosis is not exclusive to contemporary humans.

In cases of elite and wealthy individuals (a category into which some mummies from Egypt would fall), one might speculate that rich diets and an indolent lifestyle would predispose the individual to CVDs. However, there is no basis to believe that Ötzi (Neolithic European Alps) or people from ancient Peru would have fallen into these categories and the majority of such discoveries from older individuals suggest that atherosclerosis has been a common feature of human ageing. Therefore, if not derived from over-rich diets and lack of physical activity in these prehistoric individuals, what then might explain the existence of this disease in the past? One possibility is related to the Antagonistic Pleiotropy hypothesis, which states that some genes that provide benefits for early-life

survival and reproduction can have harmful effects later in life (Williams, 1966). Specifically, Thompson and colleagues argue that certain genes coding for inflammatory responses in young people may cause atherosclerosis in older adults. Higher exposure to infectious diseases and perhaps poor air quality throughout human history (mostly derived from indoor smoke from fires) would have placed selective pressure on genes that code for inflammatory responses at the cost of a trade-off for CVD in later life.

Metabolic disorders are also thought to be primarily a mismatched feature of our current lifeways, but Chapter 15—*Metabolic diseases in bioarchaeology: an evolutionary medicine approach* by Jonathan C. Wells, Nelissa Ling, Jay T. Stock, Hallie Buckley, and William R. Leonard contradicts this association. While evidence for T2DM is almost totally absent in skeletal remains (making definitive reconstructions very difficult), the authors argue for an integrated approach that could permit inferential diagnoses. Related metabolic conditions of gout and diffuse idiopathic skeletal hyperostosis (DISH) can be diagnosed from skeletal changes and have been argued to be, at times, associated with other metabolic disorders, including T2DM, and CVDs. While gout, DISH, and T2DM are commonly linked to more affluent populations who could gain access to diets rich in meat and even alcohol, there are examples of hunter–gatherer and other subsistence groups who appear to have had high rates of both gout and DISH, which may be partly related to genetic risk. These bioarchaeological data again warn us that assumptions of mismatch made in EM may not always be correct. This chapter also provides both ecological and life history models within which to contextualise the potential presence of metabolic conditions in past environments.

Certain cancers are also thought to be the result of mismatches between our modern and ancestral lifeways. One such neoplasm is lung cancer, high rates of which have arisen because of the cultural practice of smoking. Reproductive cancers, such as those of the breast and uterus, are linked to the recent dramatic declines in fertility for women, as well as a rise in the age at first birth. These factors have led to significantly greater exposure to high levels of reproductive hormones, including oestrogen and progesterone, during many more menstrual cycles, the cyclical nature of which repeatedly expose reproductive tissues to proliferation and replication during which errors in cell turnover could lead to cancer. It has been estimated, for example, that our female ancestors would perhaps only have menstruated around forty-five times in their lifetimes due to their higher fertility and repeated long bouts of breastfeeding that suppressed menstrual cyclicity (Strassman, 1999). In contrast, contemporary women who use effective contraceptives now menstruate about 450 times across their reproductive lives and frequently have a longer reproductive lifespan, increasing the likelihood of replication errors and neoplasms (Short, 1976; Strassman, 1999).

In Chapter 13—*Connecting palaeopathology and evolutionary medicine to cancer research: past and present*, Carina Marques, Zachary Compton, and Amy M. Boddy outline cases where mismatch has increased the risk for developing cancers in contemporary human populations and consider proximate and ultimate 'distinctions' in some detail, based on the framework presented by Tinbergen (1963). Using a compilation of palaeopathological data, Marques and colleagues suggest cancer likely occurred more frequently in the past than previously thought. Furthermore, using palaeopathological evidence dating from 5000 BCE to 1900 CE, they posit that cancer rates have not been increasing across time, contrary to ideas commonly stated in the EM literature (Gluckman et al., 2016; Greaves & Aktipis, 2016; Hochberg & Nobel, 2017). In Chapter 16, it is also stressed that many cancers originating from infectious diseases would have been much more prevalent in the past and are likely to be invisible in the palaeopathological record. In that case, reconstructions of past cancers rates from existing evidence are likely underestimated.

1.4.4 Physiological defences

Although physiological defences are a theme in EM, trauma is one area of pathology that has received little attention from EM scholars. Nesse and Williams had one chapter on pain as a 'defence' in their 1994 book *Why We Get Sick*, and there has been some consideration of an evolved human capacity to survive major trauma (Singer, 2017). Warfare and

inter-group or inter-individual competition often result in debilitating or fatal trauma that can leave a direct signature in the bioarchaeological record. For example, many skeletons have been found with evidence of severe injuries, especially to the skull, and sometimes with weapons still embedded in skeletal parts. Clearly, competition and warfare have been a major part of human history and prehistory and have contributed to overall ill-health of individuals and sometimes populations. During the daily course of life, our human ancestors must also have experienced a range of injuries beyond those that were the result of interpersonal violence, whether from hunting, making flint tools, or tending fires, among other daily activities.

In Chapter 16—*The palaeopathology of traumatic injuries: an evolutionary medicine perspective*, Ryan P. Harrod and Anna J. Osterholtz discuss how evidence of trauma in the archaeological record can provide a wealth of biocultural information about how humans have dealt with trauma throughout our history. They propose that humans have evolved both biological and social mechanisms to survive and heal from trauma, and that identifying these processes in the archaeological record can help us understand the role that interpersonal violence, warfare, and injury have played in the development of healing practices and social organisation. In addition to trauma, CVD (as discussed in Chapter 12) also has relevance for an EM perspective on physiological defences. Specifically, the ability of our body to respond to infection with an inflammatory response can be classified as a physiological defence mechanism. Chapter 12 expertly aligns this defence mechanism with the propensity for humans to develop CVD in later life.

1.4.5 Reproduction at the expense of health

Evolutionary success depends on the transmission of one's genes to the next generation, regardless of the impact that it has on the overall health of the parent. This concept is perfectly summarised and discussed in Sarah-Louise Decrausaz and Frances Galloway's contribution on childbirth and maternal health (Chapter 4). On top of outlining ultimate and proximate reasons for difficulty in childbirth, the authors also provide staggering statistics on how many women throughout human history have suffered health issues and mortality due to complications from giving birth. Maternal death during childbirth is not unique to humans, although it has long been accepted that humans have a somewhat unique situation of having to fit a large-headed infant through a narrow pelvis. This biological constraint, along with socio-economic and political issues, makes childbirth (except in the case of surrogacy) a potentially dangerous event that women must face if they wish to be evolutionarily successful.

1.4.6 Trade-offs

Evolutionary trade-offs occur when adaptations that optimise fitness also have negative repercussions for health or longevity (Nesse & Stearns, 2008). This concept has already been discussed in relation to several chapters in this volume. For example, to reiterate, Chapter 3 on spinal disease and bipedalism provides compelling evidence that humans may be prone to back problems due to a trade-off for the ability to walk and stand bipedally. Similarly, Julia Gamble and Gillian Bentley, in Chapter 2—*Developmental origins of health and disease (DOHaD): perspectives from bioarchaeology*, provide a comprehensive discussion on how adaptations to survive stressors early in life may lead to poor health. Relevant skeletal stress indicators that can be identified in the archaeological record include a relatively short stature and developmental vertebral stenosis, both of which indicate that during growth, the child went through a period of stress that was severe enough to arrest skeletal development. Although short stature is unlikely to cause health issues later in life, spinal stenosis (a smaller vertebral neural canal), can cause pressure on the spinal cord that can be painful and debilitating as an adult (Sekiguchi et al., 2012), making it a 'trade-off' for survival during childhood. Gamble and Bentley use examples such as this to highlight how palaeopathology enables investigations into stress responses across all ages, similar to longitudinal analyses, and how continued integration of palaeopathology with EM will help develop the DOHaD model.

Chapter 14—*Stress in bioarchaeology, epidemiology, and evolutionary medicine: an integrated conceptual model of shared history from the descriptive to the developmental*, by Daniel H. Temple and Ashley N. Edes, tackles a topic that aligns closely with DOHaD, namely life history theory. The authors outline a historical framework for the study of stress in bioarchaeology, epidemiology and EM to develop a better understanding of stress in the past. They use life history theory to demonstrate that surviving times of stress often requires a trade-off between immediate health and survival and later-life morbidity. These trade-offs, along with socio-economic and cultural flexibility, allow humans to weather episodes of stress, a trend that can be demonstrated by examining the skeletal health of populations who lived during arduous, traumatic, and difficult times. The authors make a poignant argument that humans will likely face unprecedented environmental challenges as the climate crisis continues. Combining a multidisciplinary approach, including palaeopathology and EM, to develop a holistic understanding of how humans adapt to survive stress will likely be a critical consideration today and in the future.

1.5 The One Health Initiative and future directions

One area of interest that has increased in visibility in EM is the One Health Initiative. This is a movement that started in 2011, and while it was not considered in the initial EM literature, it has since been gaining prominence (e.g. Natterson-Horowitz & Bowers, 2012). One Health's approach to improving health and well-being focuses on preventing risks and mitigating the effects of crises that originate when humans, animals, and their environments come together (MacKenzie & Jeggo, 2019; OIE, no date). According to Iossa and White (2018, 858), it 'seeks to improve health and well-being through the integrated management of disease risks at the interface between humans, animals, and the natural environment'. We highlight One Health in this volume as an essential area of study in EM and palaeopathology, now and in the future. Chapters that have relevance to One Health include Chapter 8 on plague, Chapter 9 on leprosy, Chapter 10 on TB and Chapter 12 on parasites. More specifically, Chapter 17 explores zoonotic and associated diseases and pathogens. Many of the diseases discussed in Chapter 17 would be hard to trace in the palaeopathological record without the advancement of newer genetic-based analytical techniques that allow for identification of DNA and RNA from ancient human and other animal remains. Of course, one of the constraints of the bioarchaeological, and specifically palaeopathological, record is the selectivity of evidence that prevents fuller reconstructions of past health, and the many caveats inherent when using skeletal data for disease, even in context (Wood et al., 1992; Wright & Yoder, 2003; DeWitte & Stojanowski, 2015). Finally, some important methodological considerations in reconstructing human health in the past are discussed in Chapter 18, the concluding chapter of this volume.

1.6 Conclusion

Although the main aim of this book is to illustrate how palaeopathology is currently being integrated with evolutionary theory to answer questions about human health, this relationship is symbiotic in that immersing an evolutionary framework into many palaeopathological studies would also benefit bioarchaeology as a field. Despite palaeopathology's ability to provide an immense amount of information and insight into the health of past peoples, there are a few issues that, historically, have decreased its perceived relevance to medical professionals. One criticism is its significant focus on individual and unique 'case studies' instead of hypothesis-driven, population-based research (Rühli et al., 2016). Nevertheless, the publication of 'case studies' is an essential practice for palaeopathology, as are population-based palaeoepidemiological approaches, and ultimately the accumulation of the former ultimately helps bring together dispersed findings from different populations and time periods for synthetic studies (e.g. Roberts & Cox, 2003).

In essence, medicine also takes the same approaches, with a considerable part of the clinical literature devoted to case studies and reports of disease presentation in individuals, in contrast to the more population-based emphasis of epidemiology and public health. This highlights that palaeopathology and medicine have synergies in the approaches used to explore health at individual and population levels. Incorporating evolutionary theory into palaeopathological research will provide a useful, and at times essential, framework with which to analyse and interpret disease and disease processes throughout human history. In tandem with the use of new technologies and types of data, including ancient DNA, biochemical, and histological analyses (Bouwman & Brown, 2005; Monot et al., 2005; Roberts & Ingham, 2008; Von Hunnius, 2009), incorporating evolutionary theory into palaeopathological studies will ultimately strengthen palaeopathology's relationship with both modern medicine and evolutionary biology. The future looks bright for further synergies between EM and palaeopathology that go well beyond this book.

References

Abrams, D., Hand, D. J., Heath, A., Nazroo, J., Richards, L., Karlsen, S., Mills, M. & Roberts, C.; Centre for Homelessness Impact. (2020). *What makes a community more vulnerable to COVID-19?* British Academy, London.

Agha, M. & Agha, R. (2017). The rising prevalence of obesity–part A: impact on public health. *International Journal of Surgery: Oncology*, 2(7), e17. doi: 10.1097/IJ9.0000000000000017.

Alcock, J. (2012). Emergence of evolutionary medicine: publication trends from 1991-2010. Journal of Evolutionary Medicine, 1(2), 1–12. doi:10.4303/jem/235572

Anderson, R. (1999). 'Human evolution, low back pain, and dual-level control', in W. R. Trevathan, E. O. Smith & J. J. McKenna (eds), *Evolutionary Medicine*, pp. 333–349. Oxford University Press, New York.

Axelrad, J. E., Cadwell, K. H., Colombel, J. F. & Shah, S. C. (2020). Systematic review: gastrointestinal infection and incident inflammatory bowel disease. *Alimentary Pharmacology & Therapeutics*, 51(12), 1222–1232. doi: 10.1111/apt.15770

Ball, H. L. (2017). Evolution-informed maternal–infant health. *Nature Ecology & Evolution*, 1(3), 1–3. doi: 10.1038/s41559-017-0073

Bentley, G. R. (2020). Don't blame the BAME: ethnic and structural inequalities in susceptibilities to COVID-19. *American Journal of Human Biology* 32(5), e23478. doi: 10.1002/ajhb.23478

Best, E.V. & Schwartz, M. D. (2014) Fever. *Evolution, Medicine and Public Health*, 2014(1), 92. https://doi.org/10.1093/emph/eou014

Bouwman, A. & Brown, T. A. 2005 The limits of biomolecular palaeopathology: ancient DNA cannot be used to study venereal syphilis. *Journal of Archaeological Science*, 32, 703–713. doi:10.1016/j.jas.2004.11.014

Brockhurst, M. A., Chapman, T., King, K. C., Mank, J. E., Paterson, S. & Hurst, G. D. D. (2014). Running with the Red Queen: the role of biotic conflicts in evolution. *Proceedings of the Royal Society B: Biological Sciences*, 281(197), article 20141382. http://doi.org/10.1098/rspb.2014.1382

Brosch, R., Gordon, S. V., Marmiesse, M., Brodin, P., Buchrieser, C., Eiglmeier, K., Garnier, T., Gutierrez, C., Hewinson, G., Kremer, K., Parsons, L. M., Pym, A. S., Samper, S., van Soolingen, D. & Cole, S. T. (2002). A new evolutionary scenario for the *Mycobacterium tuberculosis* complex. *Proceedings of the National Academy of Sciences of the United States of America*, 99(6), 3684–3689. doi: 10.1073/pnas.052548299

Brynildsrud, O. B., Pepperell, C. S., Suffys, P., Grandjean, L., Monteserin, J., Debech, N., Bohlin, J., Alfsnes, K., Pettersson, J. O.-H., Kirkeleite, I., Fandinho, F., Aparecida da Silva, M., Perdigao, J., Portugal, I., Viveiros, M., Clark, T., Caws, M., Dunstan, S., Thai, P. V. K., Lopez, B., Ritacco, V., Kitchen, A., Brown, T. S., van Soolingen, D., O'Neill, M. B., Holt, K. E., Feil, E. J., Mathema, B., Balloux, F. & Eldholm, V. (2018). Global expansion of *Mycobacterium tuberculosis* lineage 4 shaped by colonial migration and local adaptation. *Science Advances*, 4(10), eaat5869.

Buckingham, J. (2002). *Leprosy in colonial South India: medicine and confinement*. Palgrave, Basingstoke.

Buikstra, J. E. (1977). 'Biocultural dimensions of archaeological study: a regional perspective', in R. Blakely (ed), *Biocultural adaptation in prehistoric America*, pp. 67–84. University of Georgia Press, Athens, GA.

Buikstra, J. E. & Roberts, C. A. (eds). (2012). *A global history of paleopathology: pioneers and prospects*. Oxford University Press, New York.

Capewell, P., Cooper, A., Clucas, C., Weir, W. & Macleod, A. (2015). A co-evolutionary arms race: trypanosomes shaping the human genome, humans shaping the trypanosome genome. *Parasitology*, 142(S1), S108–S119. doi: 10.1017/S0031182014000602

Castillo, E. R. & Lieberman, D. E. (2015). Lower back pain. *Evolution, Medicine, and Public Health*, 2015(1), 2–3. https://doi.org/10.1093/emph/eou034

Chamberlain, A. (2006). *Demography in archaeology*. Cambridge University Press, Cambridge.

DeWitte, S.N. & Stojanowski, C.M. (2015). The osteological paradox 20 years later: past perspectives, future directions. *Journal of Archaeological Research*, 23, 397–450. https://doi.org/10.1007/s10814-015-9084-1

Dolgin, E. (2015). Inner workings: taking evolution to the clinic. *Proceedings of the National Academy of Sciences*, 112(44), 13421–13422. doi: 10.1073/pnas.1516954112

Edmond, R. (2006). *Leprosy and empire: a medical and cultural history*. Cambridge University Press, Cambridge.

Ewald, P. W. (1994). *Evolution of infectious disease*. Oxford University Press, New York.

Gatenby, R.A., Brown, J.S. (2020). Integrating evolutionary dynamics into cancer therapy. *Nature Reviews Clinical Oncology*, **17**, 675–686 https://doi.org/10.1038/s41571-020-0411-1

Gluckman, P., Beedle, A., Buklijas, T., Low, F. & Hanson, M. (2016). *Principles of evolutionary medicine*. 2nd edn. Oxford University Press, Oxford.

Gordon, S. V., Heym, B., Parkhill, J., Barrell, B. & Cole, S. T. (1999). New insertion sequences and a novel repeated sequence in the genome of *Mycobacterium tuberculosis* H37Rv. *Microbiology*, 145(4), 881–892. doi: 10.1099/13500872-145-4-881

Greaves, M. & Aktipis, A. (2016). 'Mismatches with our ancestral environments and cancer risk', in C. C. Maley & M. Greaves (eds), *Frontiers in cancer research: evolutionary foundations, revolutionary directions*, pp. 195–215. Springer, New York,

Grunspan, D. Z., Nesse, R. M., Barnes, M. E. & Brownell, S. E. (2018). Core principles of evolutionary medicine: a Delphi study. *Evolution, Medicine, and Public Health*, 2018(1), 13–23. doi: 10.1093/emph/eox025

Hamilton, D., Dee, A. & Perry, I. J. (2018). The lifetime costs of overweight and obesity in childhood and adolescence: a systematic review. *Obesity Reviews*, 19(4), 452–463. https://doi.org/10.1111/obr.12649

Herring. A. & Swedlund, A. (2002). *Human biologists in the archives. Demography, health, nutrition and genetics in historical populations*. Cambridge University Press, Cambridge.

Hoberg, E. P., Alkire, N. L., Queiroz, A. D. & Jones, A. (2001). Out of Africa: origins of the Taenia tapeworms in humans. *Proceedings of the Royal Society of London. Series B: Biological Sciences*, 268(1469), 781–787. doi: 10.1098/rspb.2000.1579

Hochberg, M. E. & Noble, R. J. (2017). A framework for how environment contributes to cancer risk. *Ecology Letters*, 20 (2), 1–18. doi: 10.1111/ele.12726

Iossa, G. & White, P.C.L. (2018). The natural environment: a critical missing link in national action plans on antimicrobial resistance. *Bulletin of the World Health Organization*, 96, 858–860. doi: 10.2471/BLT.18.210898

Kaushansky, N. & Ben-Nun, A. (2014). DQB1*06:02-associated pathogenic anti-myelin autoimmunity in multiple sclerosis-like disease: potential function of DQB1*06:02 as a disease-predisposing allele. *Frontiers in Oncology*, 4, 280. doi: 10.3389/fonc.2014.00280

Laland, K. N., Sterelny, K., Odling-Smee, J., Hoppitt, W. & Uller, T. (2011). Cause and effect in biology revisited:

is Mayr's proximate-ultimate dichotomy still useful? *Science* 334: 1512–1516. doi: 10.1126/science.1210879

Long, E. (2018). Evolutionary medicine: why does prevalence of myopia significantly increase? *Evolution, Medicine, and Public Health*, 2018(1), 151–152. doi: 10.1093/emph/eoy017

Lynnerup, N. & Frohlich, B. (2012). 'Jans Peder Hart Hansen and the paleopathology of the Greenland mummies', in J. E. Buikstra & C. A. Roberts (eds), *The global history of paleopathology: pioneers and prospects*, pp. 40–43. Oxford University Press, New York.

Macedo Couto, R., Ranzani, O. T. & Waldman, E. A. (2019). Zoonotic tuberculosis in humans: control, surveillance, and the one health approach. *Epidemiologic Reviews*, 41(1), 130–144. doi: 10.1093/epirev/mxz002

MacKenzie, J. S. & Jeggo, M. (2019). The One Health Approach—why is it so important? *Tropical Medicine and Infectious Disease*, 4(8), 889. doi: 10.3390/tropicalmed4020088

MacPherson, A., Otto, S. P., Nuismer, S. L. (2018). Keeping pace with the Red Queen: identifying the genetic basis of susceptibility to infectious disease, *Genetics*, 208(2), 779–789. https://doi.org/10.1534/genetics.117.300481

Mayr, E. (1961). Cause and effect in biology. *Science*, 134, 1501–1506. doi:10.1126/science.134.3489.1501

Milner, G. R. & Boldsen, J. L. (2017). Life not death: epidemiology from skeletons. *International Journal of Paleopathology*, 17, 26–39. https://doi.org/10.1002/(SICI)1099-1212(199606)6:3<221::AID-OA267>3.0.CO;2-2

Milner, G. R., Wood, J. W. & Boldsen, J. L. (2018). 'Paleodemography: Problems, progress, and potential', in M. A. Katzenberg & A. L. Grauer (eds), *Biological anthropology of the human skeleton*, 3rd edn, pp. 593–634. Wiley-Blackwell, New York.

Mitchell, P. D. (1999). The integration of the palaeopathology and medical history of the crusades. *International Journal of Osteoarchaeology*, 9(5), 333–343. https://doi.org/10.1002/(SICI)1099-1212(199909/10)9:5<333::AID-OA493>3.0.CO;2-W

Mitchell, P. D. (2011). 'Integrating historical sources with paleopathology', in A. Grauer (ed), *A companion to paleopathology*, pp. 310–323. Wiley-Blackwell, New York.

Mitchell, P. D. (2017). Improving the use of historical written sources in paleopathology. *International Journal of Paleopathology*, 19, 88–95. doi: 10.1016/j.ijpp.2016.02.005

Monot, M., Honoré, N., Garnier, M. T. Araoz, R., Coppée, J-Y., Lacroix, C., Sow, S., Spencer, J. S., Truman, R. W., Williams, D. L., Gelber, R., Virmond, M., Flageul, B., Cho, S.-N., Ji, B., Paniz-Mondolfi, A., Convit, J., Young,

S., Fine, P. E., Rasolofo, V., Brennan, P. J., & Cole, S. T. (2005). On the origin of leprosy. *Science*, 308, 1040–1042.

Morley, V. J., Kinnear, C. L., Sim, D. G., Olson, S. N., Jackson, L. M., Hansen, E., Usher, G. A., Showalter, S. A., Pai, M. P., Woods, R. J. & Read, A. F. (2020). An adjunctive therapy administered with an antibiotic prevents enrichment of antibiotic-resistant clones of a colonizing opportunistic pathogen. *Elife*, 9, e58147.

Natterson-Horowitz, B. & Bowers, K. (2012). *Zoobiquity: what animals can teach us about being human*. Virgin Books, London.

Nesse, R. M. (2005). Maladaptation and natural selection. *The Quarterly Review of Biology*, 80, 62–67

Nesse, R. M. (2019). Tinbergen's four questions: two proximate, two evolutionary. *Evolution, Medicine, and Public Health*, 2019(1), 2. https://doi.org/10.1093/emph/eoy035

Nesse, R. M. & Stearns, S. C. (2008). The great opportunity: evolutionary applications to medicine and public health. *Evolutionary Applications*, 1(1), 28–48.

Nesse, R. M. & Williams, G. C. (1994). *Why we get sick: the new science of Darwinian medicine*. Times Books, New York.

Pays, E., Vanhollebeke, B., Uzureau, P., Lecordier, L. & Pérez-Morga, D. (2014). The molecular arms race between African trypanosomes and humans. *Nature Reviews Microbiology* **12**, 575–584 https://doi.org/10.1038/nrmicro3298

Pepperell, C. S., Granka, J. M., Alexander, D. C., Behr, M. A., Chui, L., Gordon, J., Guthrie, J. L., Jamieson, F. B., Langlois-Klassen, D., Long, R., Nguyen, D., Wobeser, W. & Feldman, M. W. (2011). Dispersal of *Mycobacterium tuberculosis* via the Canadian fur trade. *Proceedings of the National Academy of Sciences*, 108(16), 6526–6531.

Pigliucci, M. & Müller, G. B. (2010). *Evolution—the extended synthesis*. MIT Press, Cambridge, MA.

Rakotosamimanana, S., Kassie, D., Taglioni, F., Ramamonjisoa, J., Rakotomanana, F. & Rajerison, M. (2021). A decade of plague in Madagascar: a description of two hotspot districts. *BMC Public Health*, 21, 1112. https://doi.org/10.1186/s12889-021-11061-8

Rassi Jr, A., Rassi, A. & Marin-Neto, J. A. (2010). Chagas disease. *The Lancet*, 375(9723), 1388–1402.

Roberts, C. A. (2016). Palaeopathology and its relevance to understanding health and disease today: the impact of environment on health, past and present. *Anthropological Review*, 79, 1–16.

Roberts, C. A. & Buikstra, J. E. (2012). 'Conclusions', in J. E. Buikstra & C. A. Roberts (eds), *A global history of paleopathology: pioneers and prospects*, 765–777. Oxford University Press, New York.

Roberts, C. A. & Cox, M. (2003). *Health and disease in Britain: from prehistory to the present day*. Sutton Publishing, Stroud.

Roberts, C. A. & Ingham, S. (2008). Using ancient DNA analysis in palaeopathology: a critical analysis of published papers with recommendations for future work. *International Journal of Osteoarchaeology*, 18, 600–613.

Roberts, C. A. & Manchester, K. (2005). *The archaeology of disease*. The History Press, Stroud.

Rook, G. A. W. (2012). Hygiene hypothesis and autoimmune diseases. *Clinical Reviews in Allergy & Immunology*, 42, 5–15. https://doi.org/10.1007/s12016-011-8285-8

Rook, G. A. W., Adams, V., Hunt, J., Palmer, R., Martinelli, R. & Rosa Brunet, L. (2004). Mycobacteria and other environmental organisms as immunomodulators for immunoregulatory disorders. *Springer Seminars in Immunology*, **25**, 237–255. https://doi.org/10.1007/s00281-003-0148-9

Rühli, F. J., Galassi, F. M. & Haeusler, M. (2016). Palaeopathology: current challenges and medical impact. *Clinical Anatomy*, 29(7), 816–822.

Safari, F., Sharifi, M., Farajnia, S., Akbari, B., Baba Ahmadi, M. K., Negahdaripour, M. & Ghasemi, Y. (2020) The interaction of phages and bacteria: the co-evolutionary arms race. *Critical Reviews in Biotechnology*, 40(2), 119–137, doi: 10.1080/07388551.2019.1674774

Sekiguchi, M., Wakita, T., Otani, K., Onishi, Y., Fukuhara, S., Kikuchi, S. & Konno, S. (2012). Development and validation of a symptom scale for lumbar spinal stenosis. *Spine*, 37(3), 232–239.

Short, R. V. (1976). The evolution of human reproduction. *Proceedings of the Royal Society of London. Series B. Biological Sciences*, 195(1118), 3–24.

Singer, M. (2017). Critical illness and flat batteries. *Critical Care*, 21(3), 69–73.

Staňková, K., Brown, J. S., Dalton, W. S. & Gatenby, R. A. (2019). Optimizing cancer treatment using game theory: a review. *JAMA Oncology*, 5(1), 96–103.

Stearns, S. C., Nesse, R. M., Govindaraju, D. R. & Ellison, P. T. (2010). Evolutionary perspectives on health and medicine. *Proceedings of the National Academy of Sciences*, 107(Suppl 1), 1691–1695.

Steckel, R. H., Larsen, C. S., Roberts, C. A. & Baten, J. (eds). (2019). *The backbone of Europe: health, diet, work and violence over two millennia*. Cambridge University Press, Cambridge.

Steckel, R. H. & Rose, J. C. (eds). (2002). *The backbone of history: health and nutrition in the western hemisphere*. Cambridge University Press, Cambridge.

Strassmann, B. I. (1999). Menstrual cycling and breast cancer: an evolutionary perspective. *Journal of Women's Health*, 8(2), 193–202. doi: 10.1089/jwh.1999.8.193

Sulzer, D., Alcalay, R. N., Garretti, F., Cote, L., Kanter, E., Agin-Liebes, J., Liong, C., McMurtrey, C., Hildebrand, W. H., Mao, X., Dawson, V. L., Dawson, T. M., Oseroff, C., Pham, J., Sidney, J., Dillon, M. B., Carpenter, C., Weiskopf, D., Phillips, E., Mallal, S., Peters, B., Frazier, A., Lindestam Arlehamn, C. S. & Sette, A. (2017). T cells of Parkinson's disease recognize α-synuclein peptides. *Nature*, 546(7660), 656–661.

Tinbergen, N. (1963). On aims and methods of ethology. *Zeitschrift für tierpsychologie*, 20(4), 410–433.

Trevathan, W. R. (2011). *Human birth: an evolutionary perspective*. Routledge, Abingdon.

Van Valen, L. (1973). A new evolutionary law. *Evolutionary Theory*, 1, 1–30.

Von Hunnius, T. (2009). Using microscopy to improve a diagnosis: an isolated case of tuberculosis-induced hypertrophic osteopathy in archaeological dog remains. *International Journal of Osteoarchaeology*, 19(3), 397–405.

Waldron, T. (1994). *Counting the dead: the epidemiology of skeletal populations*. John Wiley and Sons. Chichester.

Wang, Z., Sun, Y., Fu, X. A., Yu, G., Wang, C., Bao, F., Yue, Z., Li, J., Sun, L., Irwanto, A., Yu, Y., Chen, M., Mi, Z., Wang, H., Huai, P., Li, Y., Du, T., Yu, W., Xia, Y., Xiao, H., You, J., Li, J., Yang, Q., Wang, N., Shang, P., Niu, G., Chi, X., Wang, X., Cao, J., Cheng, X., Liu, H., Liu, J. & Zhang, F. (2016). A large-scale genome-wide association and meta-analysis identified four novel susceptibility loci for leprosy. *Nature Communications*, 7(1), 1–8.

Washburn, S. L. (1960). Tools and human evolution. *Scientific American*, 203(3), 62–75.

Williams, G. C. (1966). *Adaptation and natural selection: a critique of some current evolutionary thought*. Princeton University Press, Princeton.

Williams, G. C. & Nesse, R. M. (1991). The dawn of Darwinian medicine. *The Quarterly Review of Biology*, 66(1), 1–22.

Wood, J. W., Milner, G. R., Harpending, H. C. & Weiss, K. M. (1992). The osteological paradox. Problems of inferring health from skeletal samples. *Current Anthropology*, 33(4), 343–370

World Health Organization (WHO). (2019). *Global tuberculosis report 2019*. World Health Organization, Geneva.

World Health Organization (WHO). (2020). *Global tuberculosis report 2020*. World Health Organization, Geneva.

Wright, L. E. & Yoder, C. J. (2003). Recent progress in bioarchaeology: approaches to the osteological paradox. *Journal of Archaeological Research* 11(1):43–70

Zaman, S. (2005). *Broken limbs, broken lives: ethnography of a hospital ward in Bangladesh*. Het Spinhuis, Amsterdam.

Zampieri, F. (2009). Medicine, evolution, and natural selection: an historical overview. *Quarterly Review of Biology*, 84(4), 333–355. doi: 10.1086/648122.

Zhu, E. F., Gai, S. A., Opel, C. F., Kwan, B. H., Surana, R., Mihm, M. C., Kauke, M. K., Moynihan, K. D., Angelini, A., Williams, R. T., Stephan, M. T., Kim, J. S., Yaffe, M. B., Irvine, D. J., Weiner, L. M., Dranoff, G. & Wittrup, K. D. (2015). Synergistic innate and adaptive immune response to combination immunotherapy with anti-tumor antigen antibodies and extended serum half-life IL-2. *Cancer Cell*, 27(4), 489–501. doi: 10.1016/j.ccell.2015.03.004

Developmental origins of health and disease (DOHaD): perspectives from bioarchaeology

Julia Gamble and Gillian Bentley

2.1 The developmental origins model in contemporary and past populations

The developmental origins of adult health and disease (DOHaD) model is a well-established paradigm concerning plasticity in early life and has been adopted into Evolutionary Medicine (EM) (Gluckman et al., 2016; Rickard, 2016). It relates primarily to the uterine and infant environments (up to age 2), and to nutritional deficits experienced during this time that appear to shape health in later adulthood. DOHaD practitioners have conducted numerous observational and experimental studies in contemporary human populations and animal models to examine how early life stressors might influence susceptibility to later life chronic, or non-communicable, pathologies including cardiovascular diseases (CVDs), hypertension, stroke, type 2 diabetes (T2DM) and obesity (see Gluckman & Hanson, 2006; Newnham & Ross, 2009). For example, systematic reviews and meta-analyses of human studies have consistently shown a relationship between birth weight and risk for T2DM in later life, where low birth weights (<2500 g) increase the odds of susceptibility to this disease (e.g. Harder et al., 2007; Tian et al., 2019; Zhao et al., 2018).

The DOHaD framework has been applied primarily to contemporary populations, but the effects of developmental environments were presumably also operating in recent history. Taking a more bioarchaeological approach to DOHaD, the deceased, whether ancient or recent, are essentially 'witnesses' to the past. They potentially record permanent markers of early life disruption that enable reconstructions of historic and prehistoric developmental environments and their impact on adult health, thus facilitating a much broader DOHaD picture. In this chapter, therefore, we underscore the utility of integrating bioarchaeological information with that drawn from comparative studies of contemporary individuals to gain insight into the DOHaD model.

We introduce a brief caveat here in relation to bioarchaeological reconstructions and the debate surrounding the terms 'stress' and 'health' (e.g. Reitsema & McIlvaine, 2014; Temple & Goodman, 2014; see Chapter 14 in this volume). For purposes of this chapter, we refer to indicators of growth disruption in skeletal remains as 'stress' experiences and use the term 'health' generically to refer to the absence of visible skeletal indicators of disrupted growth or disease. However, while simplified here, any bioarchaeological study should consider the complexities of interpreting individual human experiences from skeletal remains carefully, as outlined further in Chapter 1. With these caveats in mind, we provide examples in this chapter where approaches derived from both archaeological and contemporary populations can spur greater collaboration between human biologists and bioarchaeologists. As a discipline, EM offers great potential in fostering better dialogue between these complementary fields.

Julia Gamble and Gillian Bentley, *Developmental origins of health and disease (DOHaD)*. In: *Palaeopathology and Evolutionary Medicine*. Edited by Kimberly A. Plomp et al., Oxford University Press. © Oxford University Press (2022). DOI: 10.1093/oso/9780198849711.003.0002

2.1.1 Bioarchaeological evidence

To date, the DOHaD paradigm has made a relatively recent appearance in bioarchaeology (Armelagos et al., 2009; Gowland, 2015; Humphrey & King, 2000; Klaus, 2014; Roberts & Steckel, 2019; Temple, 2014, 2019; see also Chapter 14) while studies of foetal bioarchaeology are still, should we say, in their infancy (e.g. Blake, 2017; Halcrow et al., 2017; Lewis, 2017). It is tied heavily to the concept of frailty, which considers individuals in mortality samples to be at differential risk of succumbing to morbidity and mortality at any given age (Wood et al., 1992). Of particular interest from the DOHaD perspective are those conditions reflecting disruptions to early life growth and development classified according to skeletal and dental indicators (Table 2.1). These include Harris lines (Box 2.1), cribra orbitalia and porotic hyperostosis (Box 2.2), reduced vertebral neural canal dimensions and other growth dimensions (stature, body proportions, fluctuating asymmetry). Dental indicators include enamel hypoplasia and associated conditions (i.e. accentuated striae of Retzius, enamel opacities and disruptions in dentine and cementum).

Box 2.1   Harris lines

Harris lines (HL) are horizontally expressed increases in bone density appearing in the long bone shaft created by changes in mineralisation along the growth plate during long bone formation and development (Figure 2.1) (Alfonso et al., 2005; Elliot et al., 1927; Harris, 1933). These are frequently interpreted as pathological, induced, arrested growth caused by dietary deficiency or disease, although they may also form in response to normal non-linear physiological changes in growth rates (Alfonso et al., 2005; Gindhart, 1969; Kulus & Dąbrowski, 2019; Lewis & Roberts, 1997; Papageorgopoulou et al., 2011; Scott & Hoppa, 2015), such as those caused by saltatory[1] growth (Lampl et al., 1992). While more work is needed to elucidate their aetiology, HL are broadly recognised to form in response to a range of childhood experiences, and are referred to as non-specific stress markers (Grolleau-Raoux et al., 1997). Given that HL form in relation to the ossification front (i.e. the area of active bone formation at the ends of long bones), it is also possible to estimate age at their formation (Kulus & Dąbrowski, 2019). However,

---

bone remodelling during adulthood can obliterate them, leading to increased and unknown errors in scoring for older individuals (Kulus & Dąbrowski, 2019).

[1] Saltatory describes a 'stepwise' pattern of normal growth fluctuating between more rapid periods and periods of stasis (Lampl et al., 1992).

Box 2.2   Cribra orbitalia and porotic hyperostosis

Cribra orbitalia and porotic hyperostosis are resorptive, porous lesions formed during childhood (McFadden & Oxenham, 2020; Stuart-Macadam, 1985), and are related to red blood cell production. The former forms in the orbital roof of the skull while the latter forms on the cranial vault (Figure 2.1). While frequently considered together, the aetiology of these two conditions and the extent to which this is shared is problematic (Rivera & Mirazón Lahr, 2017; Stuart-Macadam, 1989; Walker et al., 2009). Traditionally interpreted as evidence for dietary iron deficiency anaemia (Oxenham & Cavill, 2010; Stuart-Macadam, 1989, 1992a, 1992b), both cribra orbitalia and porotic hyperostosis appear to reflect a more diverse origin, including iron deficiency anaemia due to disease (in particular, helminth infections), deficiencies in vitamins B_{12} and B_9 (folate) and haemolytic and megaloblastic anaemia caused by blood disorders (McFadden & Oxenham, 2020; Rothschild, 2012; Stuart-Macadam, 1992a; Walker et al., 2009). More recently, a potential association with respiratory infections has been identified for porotic hyperostosis (O'Donnell et al., 2020).

Measurement of various body metrics in bioarchaeology can provide insight into different developmental periods in the context of past populations. Stature (seen as capturing cumulative growth) has long been evaluated as a proxy for individual and population health (Bogin, 1995, 1999, 2021; Stinson, 2012) and has been considered in relation to other markers of stress and health (Gamble, 2020; Kemkes-Grottenthaler, 2005; Roberts & Steckel, 2019; Steckel, 2005). For more details on stature estimation for past populations, see Table 2.1. Long bone growth also occurs at different times, depending on the specific bone, and has been used to evaluate shorter periods of growth, as well as changes in body proportions potentially associated

Table 2.1 Brief descriptions of skeletal indicators.
This table provides brief descriptions of each of the skeletal indicators of stress, growth, and health as outlined in this Chapter, tying each in with its relevance to the DOHaD framework.

Key Conditions	Description
Dental developmental defects	Dental developmental defects are caused by disruption in the development of incrementally forming dental tissues due to physiological disturbance (from infection, malnutrition, etc.). While dental enamel has been the most heavily studied tissue due to its resilience to degradation in the burial environment (compared to bone) and lack of remodelling once formed, dentine and more recently cementum (Mani-Caplazi et al., 2017) have also been studied. This extends the period captured by dental tissues from the neonatal (and even foetal) period into adulthood. As incrementally forming tissues, it is also possible to link these episodes of growth disruption to particular ages during development using these tissues. However, enamel and dentine can be obliterated through life from dental wear, usually impacting the earlier forming cuspal portions of the crown most heavily.
Harris lines	Harris lines are lines of increased bone density, visible through radiographic methods, and formed in relation to a change in the rate of endochondral growth (i.e. growth in length) (Alfonso et al., 2005). They act as non-specific markers of arrested or catch-up growth that may form in relation to stress episodes during early life, and can potentially be associated with age at formation (Kulus & Dąbrowski, 2019). They are best evaluated in non-adults, as bone remodelling leads to increased obliteration over the life course.
Stature	Stature is most commonly estimated from the skeleton using regression formulae applied to long bone lengths (Dupertuis & Hadden 1951; Formicola & Franceschi 1996; Genovés 1967; Trotter 1970; Trotter & Gleser 1958), but other techniques include the anatomical method (which measures every bone that contributes to stature and adds them together with a soft tissue correction) (Fully, 1956; Raxter et al., 2006) or by measuring stature *in situ* in the grave (Boldsen, 1984; Petersen, 2005, 2011). Raw long bone lengths are also increasingly used as proxies for stature to avoid errors encountered through the application of regression formulae (e.g. Boldsen, 1984; Boldsen, 1990; DeWitte & Hughes-Morey, 2012; DeWitte & Slavin, 2013; DeWitte & Wood, 2008; Gamble, 2020; Gunnell et al., 2001; Kemkes-Grottenthaler, 2005; Maijanen, 2009; Maijanen & Niskanen, 2006; Meiklejohn & Babb 2011). Stature is a frequently used indicator of population health and reflects the cumulative effect of growth across the skeleton (including any episodes of growth arrest and catch-up growth). It is often employed to evaluate broad population changes in health, including in periods of transition (Lewis, 2002) and in relation to socio-economic status (see, for example, Boix and Rosenbluth, 2017; Hughes-Morey, 2015; Ives and Humphrey, 2017; Robb et al., 2001; Weiss et al., 2019).
Growth metrics (e.g. long bone lengths, vertebral neural canal dimensions, vertebral body heights)	Since separate osteological elements undergo various patterns and timings of growth, measurements taken from one element may reflect a different period of growth than those taken from another (Bogin, 2021; Newman & Gowland, 2015). Measurement of the anteroposterior and transverse dimensions of the vertebral neural canals, for example, can capture early life growth disruption as formation should be largely complete by 5 years of age (in the lumbar region) (Clark et al., 1986; Newman & Gowland, 2015; Watts, 2013, 2015). Long bones reflect longer growth spans, with different periods being involved for growth completion across elements. Reconstruction of growth profiles through examination of diaphyseal length has potential to highlight patterns of growth stunting (Gowland & Newman, 2018). The early forming nature of vertebral neural canals theoretically means they will capture a more constrained and earlier window of growth than long bones and be less subject to catch-up growth. They have consequently been used as early life indicators of growth disruption (Amoroso & Garcia, 2018; Clark et al., 1986; Newman & Gowland, 2015; Primeau et al., 2018; Watts, 2011, 2013, 2015).
Developmental anomalies (e.g. scurvy, rickets)	Developmental anomalies represent a suite of changes that can be linked to a specific condition occurring during development, often having environmental implications. Included here are metabolic conditions. Scurvy is usually linked to vitamin C deficiency and is visible through new bone formation on the orbital roofs (distinguishable from cribra orbitalia) and porotic changes particularly towards the posterior maxillary and inferior palatine surfaces and around the greater wing of the sphenoid (Brickley et al., 2020; Brickley & Ives, 2008; Mays, 2014; Ortner & Ericksen, 1997). Rickets is most commonly linked to insufficient vitamin D synthesis and characteristic changes include bowing of the lower (weight-bearing) limbs and the formation of interglobular dentine (i.e. spaces in dentine resulting from disrupted dentine mineralisation during formation) (D'Ortenzio et al., 2016, 2017; Mays et al., 2006). Scurvy is frequently examined in relation to dietary and socio-economic influences (Mays, 2014). Rickets is also often examined in relation to socio-economic conditions and environment (for example, industrialisation and urbanisation that led to less exposure to sunlight) (Mays et al., 2006; Schattmann et al., 2016). The presence of these conditions can speak to population health and to conditions that influenced individuals during early life with projected later life repercussions.

continued

Table 2.1 *Continued*

Key Conditions	Description
Cribra orbitalia and porotic hyperostosis	Cribra orbitalia and porotic hyperostosis are, respectively, porous lesions in the orbits and in the bones of the cranial vault caused by expansion of the bone marrow (Brickley, 2018). While considered together here and similar in formation process, the relationship between the two has been debated (Oxenham & Cavill, 2010; Rivera & Mirazón Lahr, 2017; Walker et al., 2009). The aetiology of both is complex, and commonly linked to anaemia (either congenital or acquired) (Brickley, 2018; Oxenham & Cavill, 2010; Rivera & Mirazón Lahr, 2017; Walker et al., 2009). Differential diagnosis includes nutritional deficiencies (iron, ascorbic acid or vitamins B_9 or B_{12}), parasitic infections and haemolytic and megaloblastic anaemias (Novak et al., 2018; Rivera & Mirazón Lahr, 2017; Walker et al., 2009). Most recently, respiratory infection has been implicated (O'Donnell et al., 2020). The changing distribution of red versus yellow marrow through life (i.e. marrow linked to red blood cell production versus fat storage) and thickness of the cranial bones likely contribute to the primary formation of these lesions in childhood as opposed to adulthood and remodelling through life can further influence visibility in adult remains (Brickley, 2018). Osteological evaluation of these lesions should take both severity and degree of healing into account (see, for example, Rinaldo et al. 2019).
Periosteal new bone formation/periosteal reaction	Periosteal new bone formation represents a rapidly formed new (woven) bone layer produced by the periosteum in response to a range of conditions, including trauma, metabolic disease and infection (Weston, 2012). This condition has often been referred to as periostitis, but this latter term should be considered more specific to an inflammatory/infectious response (Weston, 2012). Most commonly found in the tibia, this condition can appear at any point over the life course. However, evaluation in infant remains can be particularly problematic (Weston, 2012). In the absence of associated circumstances or pathological conditions, it is often considered a non-specific stress indicator (Goodman et al., 1988) and, more recently, an indicator of overall health and mortality (DeWitte, 2014; DeWitte & Wood, 2008). The drawbacks of such an approach, and the restrictions of generalising periosteal new bone formation to a single non-specific phenomenon, have been the subject of some debate (Klaus, 2014; Weston, 2012).

with periods of growth disruption (Bogin et al., 2002; Boldsen, 1998; Gamble, 2020; Jantz & Jantz, 1999). Vertebral neural canals are potentially valuable in capturing shorter periods of time, with approximately 95% of their growth completed by age 5 (DiMeglio, 1992; DiMeglio & Bonnel, 1989; Newman & Gowland, 2015). Work on the use of these vertebral neural canal dimensions (measured transversely and antero-posteriorly across the canal) as evidence of early life growth disruption in the past was undertaken by Clark et al. (1985, 1986), was further recognised by Goodman and colleagues (1988), and has recently seen increased attention in bioarchaeology (Amoroso & Garcia, 2018; Newman & Gowland, 2017; Watts, 2011, 2013, 2015).

2.1.2 Insights from teeth

The incremental nature of dental structures and their early formation from foetal life to adolescence makes them particularly relevant for the study of childhood stress and DOHaD. Dental enamel is also permanent once formed and, unlike bone, does not remodel.

Dental enamel has been studied extensively in bioarchaeology both as macroscopic (see, for example, Boldsen, 2007; Guatelli-Steinberg, 2015; Palubeckaitė et al., 2002) and microscopic surface manifestations of growth disruption, known as dental enamel hypoplasia (Cares Henriquez & Oxenham, 2017, 2019; Gamble, 2020; Temple, 2014, 2019; Temple et al., 2012). Enamel can also contain internal disruptions, known as accentuated striae of Retzius or Wilson Bands (Gamble, 2014; Gamble et al., 2017; Garland, 2020; Goodman & Rose, 1990; Lorentz et al., 2019; Thomas, 2003). We refer in general to dental enamel defects (DED) here for simplicity, encompassing both internal and surface markers. Parallel lesions can also be seen in the softer dental tissue underlying enamel, called dentine, which is subject to some remodelling (referred to as the secondary dentine) (see Hillson, 2014 for an overview). More recent work has explored the utility of growth arrest lines in cementum (the mineralised tissue covering the tooth root) to consider experiences of physiological disruption occurring into adulthood (Mani-Caplazi et al., 2017).

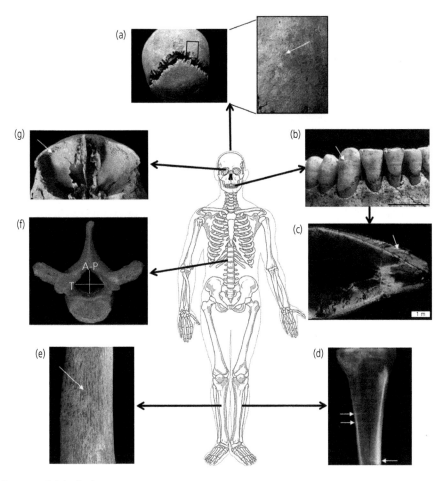

Figure 2.1 Illustration of skeletal indicators.
This figure shows examples of skeletal indicators (traits marked by white arrows) used by bioarchaeologists to reconstruct health, growth and stress from skeletal remains of past populations, alongside location of occurrence. **a:** a moderate example of healing porotic hyperostosis with an inset image showing the porotic nature of the bony change in closer view; **b:** dental enamel hypoplasia with a linear defect; **c:** accentuated striae of Retzius in the same canine highlighted in image b; **d:** Harris lines in an adult tibia; **e:** periosteal new bone formation on an adult tibia; **f:** a thoracic vertebra with the antero-posterior and transverse dimensions of the vertebral neural canal commonly measured for signs of growth disruption highlighted; **g:** porotic changes (known as CO) in the orbits.
Image credits: **a–c, e–d** (J. Gamble); **d** (courtesy of C. Roberts); skeleton schematic (public domain image from https://www.needpix.com/photo/32558/).

DEDs have significant potential to inform the DOHaD framework through life course considerations and have been explored in a number of studies with a heavy focus on the permanent dentition (Amoroso et al., 2014; Armelagos et al., 2009; Boldsen, 2007; Gamble et al., 2017; Garland, 2020; Lorentz et al., 2019; Roberts & Steckel, 2019; Steckel, 2005; Temple, 2014, 2019; Watts, 2013; Yaussy et al., 2016). Enamel on some of the permanent teeth in adults, such as the first molar, can also provide evidence of insults from birth, provided there has been only limited enamel wear. In these cases, the neonatal line (a pronounced internal enamel defect linked to the birth event) can be used to anchor post-natal developmental sequences. More commonly, DED studies cover a slightly later period than that frequently emphasised in DOHaD. This is because dental wear on permanent teeth obliterates the earliest growth, and studies of deciduous teeth cannot demonstrate their potential influences on later life health (i.e. being lost in childhood).

Fewer studies have considered deciduous enamel defects, partly because deciduous teeth are harder to recover and observe archaeologically if they are not fully erupted (Fitzgerald & Saunders, 2005). Their thinner enamel also makes them friable and subject to damage (Fitzgerald & Saunders, 2005). Nevertheless, a few studies have investigated deciduous DED, both macroscopically (Cook & Buikstra, 1979) and microscopically (FitzGerald et al., 2006; Fitzgerald & Rose, 2008; Fitzgerald & Saunders, 2005; Simpson et al., 1999; Ządzińska et al., 2015). Ządzińska and colleagues (2015) looked at deciduous incisors from eighty-three individuals from the late mediaeval Polish cemetery of Brześć Kujawski. They found that individuals who died younger tended to have both more accentuated striae of Retzius and a thicker neonatal line. These results suggest that the width of this line may relate in some way to child frailty and later life stress episodes. Alternatively, differences in infant frailty might have influenced both the width of the neonatal line and infant mortality (Ządzińska et al., 2015). Lorentz and colleagues (2019) considered enamel defects in the deciduous mandibular canine for a small sample of individuals (n = 15) from the Bronze Age Iranian site of Shahr-i Sokhta. They anchored all developmental sequences in the teeth to the neonatal line and then examined pre-natal and post-natal accentuated striae of Retzius in relation to age at death, identifying possible reductions connected to pre-natal but not post-natal accentuated striae of Retzius. Their results remain preliminary because of small sample sizes and low levels of significance. Nevertheless, the suggestion of different mortality patterns relating to pre-natal stress points to a useful avenue of future exploration in relation to DOHaD.

2.2 History of development of the DOHaD model

Work on DOHaD in living populations was already being initiated in the early to mid-twentieth century by many eminent nutritionists, including Elsie Widdowson and Robert McCance. These scholars, using primarily animal models and some human cadavers, conducted studies of body composition during foetal and infant life, and investigated the impact of early life nutritional deficits on later life growth

and metabolism (e.g. Widdowson & McCance, 1975; Widdowson & Spray, 1951; see Buklijas, 2014). Widdowson and McCance, to some extent, built on the work of forebears such as Joseph Barcroft (1947), while ideas about critical windows of development were already in circulation earlier in the twentieth century (e.g. Kermack et al., 2001; Stockard, 1921).

2.2.1 Famine studies relating to DOHaD and bioarchaeology

Detrimental consequences arising from the Second World War permitted researchers during the mid-part of the twentieth century to explore the effects of nutritional deprivation during different parts of foetal and infant life. For example, the Dutch Famine or Hunger Winter, which occurred when the Germans blockaded Amsterdam in 1944–45, exposed pregnant women and their foetuses to the effects of severe nutritional restriction. Smith (1947) recorded the decline in birth weights obtained from clinics in Rotterdam at the time, but Barcroft (1947) also noted the effects of the famine by gestational trimester.

The Siege of Leningrad by the German forces that lasted for 872 days from 1941–44 resulted in an even more severe, wartime nutritional deprivation affecting foetal health (Vågerö et al., 2013). Antonov (1947) recorded that birth weights for babies born following the Leningrad Siege fell by 500–600 g, while later studies from the 1970s–1990s recorded the effects of this on adult health of the survivors. Sparen and colleagues (2004) concluded that those exposed to famine during puberty (i.e. not during gestation or infancy) were more at risk for CVDs.

Another source of famine data that has proved useful for studies related both to bioarchaeology and DOHaD is the Great Chinese Famine imposed by Chairman Mao's agricultural policies in 1959–61 that affected millions of people across China. During this time, energy and protein intake decreased by between 25–40%, but there was significant variation between provinces (Wang et al., 2012; Zhou & Corruccini, 1998). This is one of the few cases of relatively contemporary famines where bioanthropologists have examined the evidence for DED in relation to famine exposure (Zhou & Corruccini, 1998) and found that the proportion of individuals with teeth affected by DED increased significantly during the famine compared to either individuals born before or afterwards. Huang and colleagues (2010) also recorded reduced stature and body mass in individuals exposed to the famine as well as an increase in rates of hypertension among the older survivors. However, a recent systematic review and meta-analysis of literature about the potential adverse effects of the Great Chinese Famine on health in later life failed to find any, except for an increased prevalence of schizophrenia among survivors who were gestated and born during this time (Li & Lumey, 2017).

Other hunger periods that have been studied extensively by bioarchaeologists date to a momentous time during the fourteenth century AD (from 1315 to 1317) when many parts of Europe experienced the Great Famine, the Great Bovine Pestilence of 1317–20 in Northern Europe (DeWitte & Slavin, 2013; Slavin, 2016), and the Black Death epidemic stretching across Europe, the Middle East, East Asia, and Africa in the mid-fourteenth century (Green, 2014a, 2014b). The preservation of large 'plague pits' in London (England) dating to the Black Death has permitted extensive research into frailty and selective mortality in a sample of people who died within a tightly constrained period (DeWitte & Hughes-Morey, 2012; DeWitte & Wood, 2008). Individuals (particularly females) of shorter stature appeared to be at increased risk of mortality during both the Black Death and the Great Famine (DeWitte & Hughes-Morey, 2012; DeWitte & Slavin, 2013). Insights from other mediaeval famine contexts suggest that mortality during famine periods differentially impacted individuals who had experienced childhood stress (measured through DEDs and reduced femoral length) (DeWitte & Yaussy, 2017; Yaussy et al., 2016; Yaussy & DeWitte, 2018).

The Irish Famine marks a more recent historical period studied by bioarchaeologists (Geber, 2014, 2015; Geber & Murphy, 2012). Using the Kilkenny Union Workhouse cemetery sample (dating to 1847–51), Geber (2014) examined DED, Harris lines and growth arrest (measured through femoral lengths), in 545 juveniles. Here, fewer defects developed during the famine years, while individuals with more DED appear to have lived longer than those with fewer, suggesting that they may

have been more resilient in the face of famine. Geber (2014) also suggested that these findings may reflect earlier exposure to infections, particularly in a workhouse context, providing protective immunity. Harris lines were also ubiquitous in the Kilkenny sample, with an expected growth-spurt associated peak in formation between 8 and 9 years of age, and also more Harris lines than expected in individuals aged 13–17 (Geber, 2014). When combined with femoral evidence for stunting, this was interpreted as delayed or reduced adolescent growth due to stress and insufficient physiological resources (resulting from nutritional deficiency and possibly infectious diseases prevalent in the workhouse population). Slightly reduced stature in individuals with DED also suggested a connection between earlier physiological disruption and growth stunting. Evidence for rickets, scurvy and infectious diseases has also been identified through historical records and, more specifically, bioarchaeological analyses of the Kilkenny workhouse sample (Geber, 2016; Geber & Murphy, 2012).

2.2.2. Development of DOHaD hypotheses

Following observational studies in the twentieth century from famines associated with the Second World War among living human beings, studies in northern Europe during the 1970s began to refer to correlations between early life environmental conditions and risks for CVDs (e.g. Forsdahl, 1977). The field of early life development and later life health advanced considerably, however, under the influence of David Barker and his colleagues at the University of Southampton from the 1980s onwards (see also Chapter 14). Barker and his teams were able to access midwifery records from around England, the most exceptional of which dated from 1911–30 and originated in Hertfordshire, a county just north of London (Barker, 1994; Barker & Osmond, 1986; Dickinson et al., 2016). Individuals listed in these records were traced through the UK National Health Service (NHS) to follow their patterns of morbidity and recorded causes of death. Barker (1994) found that individuals born at low or very high birth weights were significantly more likely to have experienced CVDs, circulatory disorders such as strokes, hypertension and T2DM in later life. Other records of neonates from different parts of the UK and elsewhere also showed associations between morbidity from CVDs and being stunted or thin at birth (Barker, 1994). These associations have since been confirmed by numerous studies (e.g. Barker, 1992; Knop et al., 2018; Whincup et al., 2008).

As work on developmental origins was ongoing, Barker and his colleagues formulated the 'Foetal Programming' (Lucas, 1991) and 'Thrifty Phenotype' hypotheses (Hales & Barker, 1992; Hales & Barker, 2001), arguing that environmental challenges experienced in utero would have lasting consequences for phenotypic health in later life (Barker, 1994, 2002). Barker and others argued, for example, that nutritional constraints in utero would affect particular organs while sparing the brain, leading to a reduction in cell and organ size (e.g. the kidney and pancreas) ultimately affecting organ function during the life course (e.g. Barker, 1994; Hales and Barker, 1992). These developmental constraints might be more efficient (or thrifty) where nutritional availability remained low throughout life. It also appeared that restrictions on growth during infancy (age 0–1) would exacerbate the effects of constraints during uterine development—effects that would increase if the child then experienced improved nutrition and rapid catch-up growth (De Boo & Harding, 2006; Eriksson et al., 2000; Gluckman et al., 2005). If the nutritional surfeit, particularly in glucose, continued during later life, then the adult would be at increased risk for T2DM (Hales & Barker, 2001, 2013).

2.2.3 Further development of DOHaD models

Alan Lucas (1991) had already referred to the adaptive significance of a foetus being able to predict its future environment from the quality of early-life nutrition. However, building on the Barker Hypothesis (as it came to be known), Peter Gluckman, Mark Hanson and others took this adaptive approach further. They suggested that the thrifty phenotype essentially represented a 'Predictive Adaptive Response' (PAR), with an individual expecting, through its early life experiences, to meet a constrained environment for which this phenotype was better adapted than if it had not developed that response (Bateson and Gluckman, 2011;

Gluckman et al., 2005, 2007; Gluckman & Hanson, 2006). If, though, the phenotype then encountered an environment richer than expected, the thrifty phenotype might be 'mismatched' to the greater abundance. It is this mismatch, it is argued, that has led to the propensity for adults to develop obesity and other metabolic disorders during adulthood in our contemporary, industrialised environments (Bateson and Gluckman, 2011; Gluckman et al., 2007; Gluckman & Hanson, 2006).

The PAR paradigm has, however, been challenged for its theoretical assumptions by Jonathan Wells (2003, 2006, 2007, 2012), and others (e.g. Kuzawa, 2005; Rickard & Lummaa, 2007; Worthman & Kuzara, 2005), and these challenges have, naturally, been countermanded (Bateson, 2008; Gluckman et al., 2008; Spencer et al., 2006). Likewise, the Foetal Programming Hypothesis has had its critics on the grounds of its adaptationist stance (Jones, 2005), selective bias, sampling confounds, contradictory results among various studies and the way the hypothesis has evolved to encompass different periods of early life (Joseph & Kramer, 1996) (see also Chapter 14).

It is, in fact, possible to use bioarchaeological data to test competing hypotheses within DOHaD. For some time, it has been known that vertebral neural canals (VNCs) complete growth by approximately age 5, and can act as a record of early life conditions, as discussed earlier (Clark et al., 1986; Newman & Gowland, 2015; Watts, 2011). Stunting in relation to early life growth affects the size of these canals as well as thymic and overall immune function (Bourke et al., 2016; Savino, 2002). Clark and colleagues (1986), for example, recorded an association between smaller vertebral neural canals and age at death in a skeletal population from Dickson Mounds located in present-day Illinois (USA) and dating to 950–1300 AD (n = 90). Those individuals with so-called 'stunted' vertebral neural canals were more likely to die before age 35 than those not affected.

Amoroso and Garcia (2018) recently used similar VNC data to evaluate the PAR model. They looked at the size of the VNCs in ninety adult skeletons (forty-four females and forty-six males) derived from the Luís Lopes Collection of Portuguese skeletons, with known ages at death from 20 to 89 years and dating to 1896–1961 at the time of burial; the socio-economic representation of the sample ranged from low to high (Amoroso & Garcia, 2018). Cause of death was known for this sample, split between infectious and non-infectious diseases. The authors found no association between survivorship and VNC size after controlling for covariates such as year of birth and socio-economic status (SES). Without such controls, the associations were significant and positive.

Amoroso and Garcia (2018) interpreted their results as support for the PAR model since it appeared that the physiological disruption evident from earlier in life did not impact later life health negatively. However, it should be noted that their study is distinctly in the minority in its results and in analysing a relatively recent skeletal population where approximately half of individuals died from non-communicable diseases, and where covariates were available to add to the statistical models (cf. Temple, 2014; Watts, 2015). This minority position was acknowledged by the authors. A further consideration is that limited social mobility, typical until relatively recent times, would have meant a consistency of living conditions throughout life. As such, factors such as low SES experienced in early life would have continued to be present and influential at later stages of the life course, including the cause of death (Gage et al., 2016; Sharp et al., 2018). Perhaps unsurprisingly, an earlier study of the same Portuguese sample using DED showed a similar result (Amoroso et al., 2014).

2.3 Life course approaches to health

Another major challenge to the DOHaD model comes from life course studies (e.g. Kuh & Ben-Shlomo, 1997, 2004). Scholars working in this area take issue with the narrowly timed perspective of DOHaD and point, instead, to the extensive period of adult life during which various stressors can impact health and involve significant trade-offs. By taking a life course approach, studies in this genre have also focused on socio-economic differentials between individuals and groups that can have a significant impact on both their ecology and behaviour (Jones et al., 2019; Smith, 2003). Individuals of lower

SES, and/or people subjected to structural violence[2] for a variety of reasons, are more likely to experience conditions that would lead to low birth weight and subsequent metabolic disorders (Alhusen et al., 2016; Earnshaw et al., 2013; Oldroyd, 2005; Wells, 2016). Wells (2016) has referred to such structural inequalities as the 'Metabolic Ghetto'. The socio-economic disparities that make some social groups more vulnerable to low birth weights and later life metabolic diseases lead to the possibility that disparities of this kind could also be visible within the archaeological record.

The Industrial Revolution presents a particularly useful case study marked by an increase in socio-economic disparity, and frequently associated contextual data for burials. DeWitte and colleagues (2016) considered child and adult mortality and survivorship in relation to SES at three industrial era sites from London, England (Chelsea Old Church, St Bride's Fleet Street, St Bride's Farringdon Street). They found reduced survivorship in children from the lower SES site of St Bride's Lower Churchyard, but no significant differences in age at death among adults (compared to the other two high SES samples. These results may reflect selective mortality in the non-adult sample relating to stressors associated with low SES. Any surviving adults would presumably have been more resilient.

Newman and Gowland (2017) also studied sites in London representing one high SES (Chelsea Old Church), two middle SES (St Benet Sherehog and Bow Baptist), and one low SES (Cross Bones). They found no significant differences in growth indicators by SES, although all individuals were below modern growth standards. Children aged 1–5 years at all sites were well represented, especially perinatal individuals at the low-status site of Cross Bones, suggesting either high infant mortality or high fertility rates (or both) (Newman & Gowland, 2017; Sattenspiel & Harpending, 1983). The Cross Bones site and another low SES site (St Bride's Lower Churchyard) also demonstrated particularly high rates of metabolic disease (particularly scurvy)

[2] Systemic conditions, including racism, that differentially impact certain segments of a population, impacting health and well-being.

and/or rickets (Gowland & Newman, 2018), cribra orbitalia and periosteal new bone formation (Newman et al., 2019; Newman & Gowland, 2017). High rates of metabolic disease in perinates as well as high deciduous DED and reduced long bone growth at Cross Bones is consistent with poor maternal health, while peaks of mortality in children aged 1–5 years at other sites may be linked to the timing and type of weaning practices (Hodson & Gowland, 2020; Newman & Gowland, 2017). Industrial-era London thus reveals a complex situation with evidence for SES effects on the development of stress markers from an early age, and effects according to biological sex and environment (DeWitte et al., 2016; Newman et al., 2019; Watts, 2015). The concept of intersectionality (referring to the interconnectedness of different social categories) explored by Yaussy (2019) highlights the need to recognise the complex interaction of a range of variables in shaping growth and development, and ultimately health and disease.

2.3.1 Longitudinal cohort studies and life course analyses in recent populations

Analyses of parameters of the DOHaD model, as well as life course studies, have undoubtedly benefitted from the growing number of longitudinal cohort studies around the world where individuals (especially infants and children) can be followed for years using prospective data collected in a series of 'sweeps' at different time points. In the UK, such cohorts have regularly been mined for appropriate information. Some, such as the early Lothian Birth Cohort Studies (1921 and 1936), were designed to investigate social and genetic factors relating to intelligence, rather than any perceived need to look at physiological development from birth onwards (https://www.lothianbirthcohort.ed.ac.uk/). Other studies that included more broad markers of health useful for DOHaD-related investigations in the UK include the 1946 National Survey of Health and Development (NSHD) (n = 5632, https://www.nshd.mrc.ac.uk/), the 1958 National Child Development Study (NCDS) (n = 17,415, https://cls.ucl.ac.uk/cls-studies/1958-national-child-development-study/), the 1970 British Cohort Study (BCS70) (n = 17,198, https://cls.ucl.ac.uk/

cls-studies/1970-british-cohort-study/), the 1991 Avon Longitudinal Study of Parents and Children (ALSPAC) (n = 15,247, http://www.bristol.ac.uk/alspac/), and the 2000 Millennium Cohort Study (MCS) (n = 18,818, https://cls.ucl.ac.uk/cls-studies/millennium-cohort-study/). All have been mined for DOHaD-related data (e.g. Grijalva-Eternod et al., 2013; Montgomery & Ekbom, 2002; Silverwood et al., 2013). There are also a number of other useful international birth cohorts for DOHaD-related issues (e.g. Lawlor et al., 2009; Pansieri et al., 2020; Richter et al., 2012).

Despite the evident value of longitudinal cohort studies, these are subject to inherent problems. For example, individuals tend to drop out over time or become untraceable, leading to potential subject bias among those who remain. Data sweeps may omit questions that become useful later or lack samples that prove to be valuable as advanced techniques of analyses develop. Over time, samples that were collected may run out or degrade, meaning they can no longer be used to address questions of interest. Changes in methodological approaches may further render earlier results incompatible with later ones, requiring constant critical review of old data.

2.3.2 Life course approaches in bioarchaeology

In the context of bioarchaeology, researchers can target those indicators reflective of early (non-adult) years and consider them in relation to later developing indicators (i.e. those that can be linked to the adult period) (see also Chapter 14). For a cross-sectional approach, a researcher can focus on an age-segmented sample, investigating the pattern of conditions in different age groups in a population. This is obviously not quite the same as a cohort study, since the nature of a cemetery sample is generally attritional rather than capturing a 'snap-shot' of a population at any given time; the exception to this might be catastrophic cemeteries, but even these can be selective and therefore biased in nature (DeWitte & Hughes-Morey, 2012; DeWitte & Wood, 2008). However, the extent to which we can draw broader inferences from evidence on an individual level is limited, and so data are also needed at the population level. A strength of bioarchaeology is to add further context to information drawn from human skeletal remains to better understand any observed patterns. Through this contextualised approach, it is possible to investigate past population patterns using both longitudinal (over the life course of an individual) and cross-sectional data (across multiple individuals).

Examples of such approaches in bioarchaeology using data from adults include a study of skeletal remains from the previously mentioned Dickson Mounds in Illinois, USA, and dating across the Late Woodland period (*c.* 1000–1100 CE) through to the Middle Mississippian period (~1175–1350 CE). During the Late Woodland, inhabitants of the site were semi-sedentary, broad-spectrum foragers but transitioned to a more sedentary agricultural lifestyle during the Mississippian (Goodman et al., 1984; Goodman & Armelagos, 1988). Goodman and Armelagos (1988) identified significantly reduced mean ages at death in individuals exhibiting DED, particularly among those with two or more evident periods of stress, suggesting that recurring episodes related to earlier mortality. This pattern seemed to be most prominent for the later Middle Mississippian period when other skeletal biomarkers (such as porotic hyperostosis, periosteal new bone formation and osteomyelitis), along with overall mortality rates, all appear to indicate a decline in overall health during the agricultural transition (Goodman et al., 1984). A number of other studies have similarly used DEDs and survivorship to demonstrate a relationship between early life stress and risk of death (e.g. Duray, 1996; Gamble et al., 2017; Garland, 2020; Roberts & Steckel, 2019; Rose et al., 1978; Rudney, 1983; Thomas, 2003).

2.4 Exploring explanatory mechanisms between early life development and later life health

Before the twenty-first century, explanatory models for foetal programming pointed primarily to nutritional factors (over- as well as undernutrition), and potential macro-nutrient imbalances during pregnancy (Barker et al., 1993), as well as the impact of increased exposure to glucocorticoids in utero. It was proposed that the latter resulted

from stress activation and perhaps permanently altered responses of the hypothalamic-pituitary-adrenal (HPA) axis (Cottrell & Seckl, 2009; De Boo & Harding, 2006; Edwards et al., 1993; Seckl, 2004). Corticosteroid exposure seemed to be especially linked to risks for hypertension later in life (Seckl, 2004). Both nutritional and glucocorticoid exposure could act synergistically to exacerbate the effects of either alone (Cottrell & Seckl, 2009; De Boo & Harding, 2006) (see also Chapter 14).

More recently, researchers have been focusing on mechanistic factors such as epigenetic changes and even transgenerational epigenetic influences to explain how early life environments can influence the adult phenotype (Cottrell & Seckl, 2009; De Boo & Harding, 2006; Felix & Cecil, 2019; Goyal et al., 2019; Waterland & Michels, 2007). This approach also makes the study of environmental influences (including SES) on past populations particularly pertinent (Thayer & Kuzawa, 2011; Tobi et al., 2009; Waterland & Michels, 2007). Epigenetics refers to modifications of gene expression without alteration of the gene itself. Epigenetic mechanisms include how DNA is packaged (histone modification), the addition or subtraction of specific chemical markers to genes (DNA methylation) and actions of non-coding RNAs (Handy et al., 2011; Vukic et al., 2019; Waterland & Michels, 2007). Most epigenetic studies have focused on DNA methylation of cytosine nucleotides (Jaenisch & Bird, 2003; Weinhold, 2006).

In the field of contemporary human DOHaD studies, analyses have been undertaken of potential epigenetic changes among survivors of the Dutch Hunger Winter. Heijmans and colleagues (2008) showed that the maternally imprinted *IGF2* gene in individuals exposed to the famine around the time of conception was hypomethylated relative to less-exposed siblings, and relative to those exposed during the third trimester. Another set of useful studies focusing on epigenetics originates from work by the Medical Research Council's long-term field station in Keneba, The Gambia, through studying agriculturalists who experience intense seasonal fluctuations in ecology, agricultural labour and nutritional intake (Hennig et al., 2017; Moore, 2020). Here, children born during the summer, a rainy period of intense agricultural work in July and August, have usually been characterised by lower birth weight than children born at other times of the year (Roberts et al., 1982).

In one epigenetic analysis in The Gambia, Waterland & Michels (2007) focused on a suite of genes representing what are called metastable epialleles (i.e. alleles that can differ across related individuals due to epigenetic modifications during development that are highly susceptible to changes in the environment). Since these epialleles represent signatures of the early environment that are subsequently relatively stable and not subject to further change across the life course, they represent a good target for DOHaD-related questions. Waterland and Michels (2007) used peripheral blood leukocytes to analyse DNA methylation patterns of children who had been conceived in the contrasting rainy (hunger) versus dry (harvest) season. The candidate genes examined and believed to represent potential sites of metastable epialleles were all significantly more methylated among children conceived during the rainy hunger season compared to the dry season. There was also significant inter-tissue correlation as well as variation between individuals sampled, whereas the control genes likewise examined for methylation patterns showed inter-individual variation, but not inter-tissue correlation.

In another paper, Dominguez-Salas and colleagues (2014) extended the suite of metastable epialleles under investigation among Gambian children examined, and also assessed the possibility that plasma biomarkers of critical micronutrients in mothers of offspring might affect any observed methylation patterns in the metastable epialleles of observed offspring. Concentrations of three of these micronutrients (vitamin B2, homocysteine and cysteine) that were sampled around the time of conception, as well as the mother's body mass index, predicted methylation patterns in the metastable epialleles in both hair follicles and peripheral blood leukocytes samples from which these patterns were examined.

Studies of potential epigenetic changes across other longitudinal birth cohorts have been mixed in whether they showed evidence of uterine effects on the phenotype and potential links to specific conditions in later life (Felix & Cecil, 2019). Rather than posing specific, targeted questions like the two studies in The Gambia, many of these efforts have used epigenome-wide association studies (EWAS) that target thousands of potential sites of DNA methylation in attempts to discover genes in specific population groups that show differential methylation relative to others (Bianco-Miotto et al., 2017). To date, the most successful studies have found associations between maternal and offspring adiposity but have not yet moved beyond those associations to pinpoint specific disease risks stemming from these associations (e.g. Sharp et al., 2015). This means it is difficult to interpret the cause of these apparent associations. Results from EWAS are therefore hard to associate with potential disease outcomes. At the same time, findings of similar methylation patterns from palaeopopulations, if that were to prove possible, might assist in attempting to reconstruct the phenotype of individuals while they were alive. However, until there is a better understanding of how to interpret methylation patterns in living populations, it will remain difficult to infer much from past ones as well.

2.4.1 Palaeoepigenetics

Intriguing insights are beginning to emerge from a few bioarchaeological studies embracing epigenetics that attempt to forge links between contemporary and past human environments. Given sufficient tissue preservation, it is becoming possible to reconstruct palaeoepigenetic signatures in ancient DNA extracted from preserved soft tissue (Llamas et al., 2012), dental pulp and bone (Smith et al., 2015). These data could illuminate how past environments shaped the phenotype. As of now, the entire methylomes (genome-wide methylation patterns) for both Neanderthals and Denisovans have been mapped from existing genetic data using techniques of inference, whereby it is known that methylated cytosines decay to thymines over time while unmethylated cytosines decay to uracils (Gokhman et al., 2017, 2019). Of course, what is represented is also the methylome from those individuals who unwittingly provided data, and we must assume that greater variation existed at the population level if we were able to sample more broadly.

One problem specific to both epigenetics and palaeoepigenetics, however, is that some epigenetic changes can only be reconstructed using particular tissues (Felix & Cecil, 2019; Gokhman et al., 2017; Waterland et al., 2010; Waterland & Michels, 2007), meaning that skeletal remains consisting mostly of bones and teeth may be inappropriate for certain analyses, and will therefore limit the kinds of epigenetic reconstructions relevant to early life effects on health. Some intriguing examples do exist, however, pointing to the potential of epigenetic applications—including those involving metastable epialleles—that have inferred associations from the living to the past (e.g. Dolinoy et al., 2007). In relation to studies of metastable epialleles in The Gambia, three of the same genes studied in that population (*EXD3*, *RBM46* and *ZNF678*) also appear to be hyper-methylated in Neanderthals and possibly in Denisovans, leading the authors to suggest that these archaic hominins may also have suffered from periodic or chronic nutritional stress that affected the epigenome and potential foetal health (Gokhman et al., 2017, 2019). Studies such as these, using new techniques, hold great promise for bioarchaeology in attempting to search for evidence of early life stressors in relation to indicators of later-life health.

2.5 Future directions

As we have attempted to show here, integrating contemporary studies of early life development and later life health with bioarchaeological research could do much to illuminate how early plasticity affects health across the life course and how the two interact. This is particularly the case, given that the DOHaD model—developed in living populations—can only give us a picture of survivors (presumably the less frail), while the bioarchaeological record gives us the converse—namely, a picture of health (or morbidity) among the frailest (i.e. the non-survivors). By blending our understanding of these two scenarios, particularly in relation to times that were particularly stressful for individuals (e.g. famines or major epidemiological transitions), we can begin to develop a more holistic understanding of the impact of early life development on individual health and well-being.

One area that needs further development is studies in living populations of biomarkers like DEDs

that relate these to other factors useful to bioarchaeologists. Numerous examples already exist relating a higher prevalence of DEDs to low birth weight (e.g. Nelson et al., 2013; Pinho et al., 2012; Prokocimer et al., 2015), stunting and malnutrition (Agarwal et al., 2003; Chaves et al., 2007; Infante et al., 1974) and low SES (Golkari et al., 2016). Some of these studies were explicitly undertaken by anthropologists (e.g. Goodman et al., 1991; Santos & Coimbra, 1999), and even by scholars using longitudinal cohort data from the UK (e.g. Dean et al., 2020; Hassett et al., 2020). If we knew more about how DEDs relate to other specific insults later in life, including disease loads, this might shed light on factors of the past environment that could help with bioarchaeological interpretations.

Some very recent studies among the Tsimane in Bolivia are, in fact, beginning to address this deficit. The Tsimane are forager–horticulturalists living in the Amazon Basin of Bolivia who are gradually acculturating and have been participating in a longitudinal study since 2002 (Leonard et al., 2015). Masterson and colleagues (2017, 2018) examined the dentition for DED on the permanent, central maxillary incisor among 336 Tsimane children and adolescents (aged 10–17), and related these to blood pressure, blood counts, anaemia, helminth infections, stature and markers of metabolic disorders collected prospectively across childhood. The targeted form of DED, however, was described as different from broadly recognised DED forms (the most typical being linear bands across the enamel, and others being broad-spanning plane-form defects and more isolated pit-form defects; see Hillson, 2014 for overview), and was most similar to plane-form defects. This DED differed from the linear enamel hypoplasia usually observed in bioarchaeological specimens, rendering direct analogies to skeletal findings more challenging (Masterson et al., 2017).

Furthermore, these DEDs (located on the central maxillary incisors) related to insults experienced when the individual was aged 1–4, and so were, strictly speaking, outside of the usual temporal focus for DOHaD studies (although analogous to bioarchaeological studies, as noted earlier). Results, nevertheless, countered DOHaD expectations in that, while 16% of participants were overweight, those individuals with enamel defects were significantly less likely to be in this weight category. Similarly, while one-third of participants exhibited at least one marker of metabolic disturbance, individuals with greater enamel defects were significantly less likely to be pre-diabetic, and there was no relationship between DED and blood pressure (Masterson et al., 2018). Instead, the severity of enamel defects (defined by the extent of crown involvement from less than one-third to more than two-thirds) was significantly predictive of both stunting and lower weight among adolescents (Masterson et al., 2018), as well as hookworm infections during childhood (Masterson et al., 2017). In other words, these two studies reinforced associations observed in the bioarchaeological record between DED and stunting.

In addition, studies of the Tsimane have targeted evidence of cribra orbitalia and porotic hyperostosis (Anderson et al., 2019). Preliminary findings from fifty Tsimane men and women aged 45 years and older using CT scans indicate that 4% and 10%, respectively, of the sample showed manifestations of both kinds of lesion. The authors of this study related these findings to helminth infections and high rates of anaemia among the population rather than to direct nutritional factors; this is consistent with an aetiology proposed by Stuart-Macadam (1992a) for bioarchaeological contexts. A more recent publication has suggested an association in particular between both cribra orbitalia/porotic hyperostosis and respiratory infections, and between cribra orbitalia and heart conditions (although the latter may be tied to increased susceptibility of people with heart conditions to respiratory infections) (O'Donnell et al., 2020).

Issues related to stature (and thereby growth) have also been studied in numerous populations in relation to DOHaD, including in developing countries. Such studies can shed light on interpreting stature in bioarchaeological populations. Specific examples that are helpful for bioarchaeologists include work by Emma Pomeroy and colleagues (2012) who, in studying 447 living Peruvian children aged 6–14, used biometrics that parallel those frequently used for skeletal remains (i.e. the humerus, ulna and tibia length) along with limb lengths, stature and head–trunk height.

New genomic and epigenomic studies will further strengthen our ability to elucidate key developmental phases using skeletal data (Franklin, 2010; İşcan, 2005; Procopio et al., 2017). For example, sex estimation in juvenile skeletons may become possible using proteomic approaches using the sexually dimorphic amelogenin protein in tooth enamel (Parker et al., 2019; Stewart et al., 2016, 2017; Gowland et al., 2021). In relation to epigenetics, Gokhman and colleagues (2017) have called for the identification of what they term 'environmentally responsive loci' that could be identified among living populations. These would essentially be signatures of the environment that could be matched to palaeopopulations to reconstruct ancient ecologies. It is only when we gain a better understanding across the life course of how early life stressors leave a permanent mark on those parts of the body visible to bioarchaeologists that we will be able to interpret the palaeopathological record. Nevertheless, the future of bioarchaeology as a sister discipline to EM appears highly promising.

References

Agarwal, K., Narula, S., Faridi, M. & Kalra N. (2003). Deciduous dentition and enamel defects. *Indian Pediatrics*, 40(2), 124–129.

Alfonso, M. P., Thompson, J. L. & Standen, V. G. (2005). Reevaluating Harris lines—a comparison between Harris lines and enamel hypoplasia. *Collegium Antropologicum*, 29(2), 393–408.

Alhusen, J. L., Bower, K. M., Epstein, E. & Sharps, P. (2016). Racial discrimination and adverse birth outcomes: an integrative review. *Journal of Midwifery & Women's Health*, 61(6), 707–720. https://doi.org/10.1111/jmwh.12490

Amoroso, A. & Garcia, S. J. (2018). Can early-life growth disruptions predict longevity? Testing the association between vertebral neural canal (VNC) size and age-at-death. *International Journal of Paleopathology*, 22, 8–17. https://doi.org/10.1016/j.ijpp.2018.03.007

Amoroso, A., Garcia, S. J. & Cardoso, H. F. V. (2014). Age at death and linear enamel hypoplasias: testing the effects of childhood stress and adult socioeconomic circumstances in premature mortality. *American Journal of Human Biology*, 26(4), 461–468. https://doi.org/10.1002/ajhb.22547

Anderson, A. L., Blackwell, A., Trumble, B., Thompson, R., Wann, L., Allam, A., Frohlich, B., Sutherland, M.,

Sutherland, J., Stieglitz, J., Rodriguez, D., Irimia, A., Law, M., Barisano G, Chui, H., Michalik, D., Rowan, C., Lombardi, G., Bedi, R., Garcia, A. R., Cummings, D., Min, J. K., Narula, J., Finch, C. E., Thomas, G. S., Kaplan, A. H. & Gurven, M. (2019). Porotic cranial lesions in living forager-horticulturalists: theoretical pathways and preliminary evidence. *American Journal of Physical Anthropology*, 168, 6–7.

Antonov, A. N. (1947). Children born during The Siege of Leningrad in 1942. *The Journal of Pediatrics*, 30(3), 250–259. https://doi.org/10.1016/S0022-3476(47)80160-X

Armelagos, G. J., Goodman, A. H., Harper, K. N. & Blakey, M. L. (2009). Enamel hypoplasia and early mortality: bioarcheological support for the Barker hypothesis. *Evolutionary Anthropology: Issues, News, and Reviews*, 18(6), 261–271. https://doi.org/10.1002/evan.20239

Barcroft, J. (1947). *Research on prenatal life. Part I.* Blackwell Scientific Publications, Oxford,

Barker D. J. P. (ed). (1992). *Fetal and infant origins of adult disease.* BMJ Books, London.

Barker, D. J. P. (1994). *Mothers, babies and health in later life.* Churchill Livingston, Edinburgh.

Barker, D. J. P. (2002). Fetal programming of coronary heart disease. *Trends in Endocrinology and Metabolism*, 13(9), 364–368. https://doi.org/10.1016/S1043-2760(02)00689-6

Barker, D. J. P., Godfrey, K. M., Gluckman, P. D., Harding, J. E., Owens, J. A. & Robinson, J. S. (1993). Fetal nutrition and cardiovascular disease in adult life. *The Lancet*, 341(8850), 938–941. https://doi.org/10.1016/0140-6736(93)91224-A

Barker, D. J. P. & Osmond, C. (1986). Infant mortality, childhood nutrition, and ischaemic heart disease in England and Wales. *The Lancet*, 327(8489), 1077–1081. https://doi.org/10.1016/S0140-6736(86)91340-1

Bateson, P. (2008). Preparing offspring for future conditions is adaptive. *Trends in Endocrinology & Metabolism*, 19(4), 111. https://doi.org/10.1016/j.tem.2008.02.001

Bateson, P. G. P. & Guckman, P. (2011). *Plasticity, robustness, development and evolution.* Cambridge University Press, Cambridge.

Bianco-Miotto, T., Craig, J. M., Gasser, Y. P., van Dijk, S. J. & Ozanne, S. E. (2017). Epigenetics and DOHaD: from basics to birth and beyond. *Journal of Developmental Origins of Health and Disease*, 8(5), 513–519. https://doi.org/10.1017/S2040174417000733

Blake, K. A. S. (2017). 'The biology of the fetal period: interpreting life from fetal skeletal remains', in S. Han, T. K. Betsinger & A. B. Scott (eds), *The anthropology of the fetus: biology, culture, and society*, pp. 34–58. Berghan Books, New York.

Bogin, B. (1999). Evolutionary perspective on human growth. *Annual Review of Anthropology*, 28(1), 109–153. https://doi.org/10.1146/annurev.anthro.28.1.109

Bogin, B. (2021). *Patterns of human growth*, 3rd edn. Cambridge University Press, Cambridge.

Bogin, B. (1995). 'Plasticity in the growth of Mayan refugee children living in the United States', in C. G. N. Mascie-Taylor & B. Bogin (eds), *Human variability and plasticity*, pp. 46–74. Cambridge University Press, Cambridge.

Bogin, B., Smith, P., Orden, A. B., Varela Silva, M. I. & Loucky, J. (2002). Rapid change in height and body proportions of Maya American children. *American Journal of Human Biology*, 14(6), 753–761. https://doi.org/10.1002/ajhb.10092

Boix, C. & Rosenbluth, F. (2017). 'A bone to pick: using height inequality to test competing hypotheses about political power', in H. G. Klaus, A. R. Harvey & M. N. Cohen (eds), *Bones of complexity: bioarchaeological case studies of social organization and skeletal biology*, pp. 33–51. University of Florida Press, Gainesville.

Boldsen, J. L. (1984). A statistical evaluation of the basis for predicting stature from lengths of long bones in European populations. *American Journal of Physical Anthropology*, 65, 305–311.

Boldsen, J. L. (1990). Body proportions, population structure, and height prediction. *Aldi Tip Dergisi. Journal of Forensic Medicine Istanbul*, 6(1-2), 157–165. https://doi.org/10.1080/03014469800005662

Boldsen, J. L. (1998). Body proportions in a medieval village population: effects of early childhood episodes of ill health. *Annals of Human Biology*, 25(4), 309–317.

Boldsen, J. L. (2007). Early childhood stress and adult age mortality—a study of dental enamel hypoplasia in the medieval Danish village of Tirup. *American Journal of Physical Anthropology*, 132(1), 59–66. https://doi.org/10.1002/ajpa.20467

Bourke, C. D., Berkley, J. A. & Prendergast, A. J. (2016). Immune dysfunction as a cause and consequence of malnutrition. *Trends in Immunology*, 37(6), 386–398. https://doi.org/10.1016/j.it.2016.04.003

Brickley, M. B. (2018). Cribra orbitalia and porotic hyperostosis: a biological approach to diagnosis. *American Journal of Physical Anthropology*, 167(4), 896–902. https://doi.org/10.1002/ajpa.23701

Brickley, M. B. & Ives, R. (2008). *The bioarchaeology of metabolic bone disease.* Academic Press, London.

Brickley, M. B., Ives, R. & Mays, S. A. (2020). *The bioarchaeology of metabolic bone disease*, 2nd edn. Academic Press, London.

Buklijas, T. (2014). Food, growth and time: Elsie Widdowson's and Robert McCance's research into prenatal and early postnatal growth. *Studies in History and Philosophy of Science Part C: Studies in History and Philosophy of Biological and Biomedical Sciences*, 47, 267–277. https://doi.org/10.1016/j.shpsc.2013.12.001

Cares Henriquez, A. & Oxenham, M. F. (2017). An alternative objective microscopic method for the identification of linear enamel hypoplasia (LEH) in the absence of visible perikymata. *Journal of Archaeological Science: Reports*, 14, 76–84. https://doi.org/10.1016/j.jasrep.2017.05.040

Cares Henriquez, A. & Oxenham, M. F. (2019). New distance-based exponential regression method and equations for estimating the chronology of linear enamel hypoplasia (LEH) defects on the anterior dentition. *American Journal of Physical Anthropology*, 168(3), 510–520. https://doi.org/10.1002/ajpa.23764

Chaves, A. M. B., Rosenblatt, A. & Oliveira, O. (2007). Enamel defects and its relation to life course events in primary dentition of Brazilian children: a longitudinal study. *Community Dental Health*, 24(1), 31–36.

Clark, G. A., Hall, N. R., Armelagos, G. J., Borkan, G. A., Panjabi, M. M. & Wetzel, F. T. (1986). Poor growth prior to early childhood: decreased health and life-span in the adult. *American Journal of Physical Anthropology*, 70(2), 145–160.

Clark, G. A., Panjabi, M. M. & Wetzel, F. T. (1985). Can infant malnutrition cause adult vertebral stenosis? *Spine*, 10(2), 165–170.

Cook, D. C. & Buikstra, J. E. (1979). Health and differential survival in prehistoric populations: prenatal dental defects. *American Journal of Physical Anthropology*, 51(4), 649–664.

Cottrell, E. C. & Seckl, J. (2009). Prenatal stress, glucocorticoids and the programming of adult disease. *Frontiers in Behavioral Neuroscience*, 3(19), 1–19. https://doi.org/10.3389/neuro.08.019.2009

D'Ortenzio, L., Ribot, I., Kahlon, B., Bertrand, B., Bocaege, E., Raguin, E., Schattmann, A. & Brickley, M. (2017). The rachitic tooth: the use of radiographs as a screening technique. *International Journal of Paleopathology*, 23, 32–42. https://doi.org/10.1016/J.IJPP.2017.10.001

D'Ortenzio, L., Ribot, I., Raguin, E., Schattmann, A., Bertrand, B., Kahlon, B. & Brickley, M. (2016). The rachitic tooth: a histological examination. *Journal of Archaeological Science*, 74, 152–163. https://doi.org/10.1016/J.JAS.2016.06.006

De Boo, H. A. & Harding, J. E. (2006). The developmental origins of adult disease (Barker) hypothesis. *The Australian and New Zealand Journal of Obstetrics and Gynaecology*, 46(1), 4–14. https://doi.org/10.1111/j.1479-828X.2006.00506.x

Dean, M. C., Humphrey, L., Groom, A. & Hassett, B. (2020). Variation in the timing of enamel formation in modern human deciduous canines. *Archives of Oral Biology*, 114, 104719. https://doi.org/10.1016/j.archoralbio.2020.104719

DeWitte, S. N. (2014). Health in post-Black Death London (1350–1538): age patterns of periosteal new bone formation in a post-epidemic population. *American Journal of Physical Anthropology*, 155(2), 260–267. https://doi.org/10.1002/ajpa.22510

DeWitte, S. N. & Hughes-Morey, G. (2012). Stature and frailty during the Black Death: the effect of stature on risks of epidemic mortality in London, A.D. 1348–1350. *Journal of Archaeological Science*, 39(5), 1412–1419. https://doi.org/10.1016/j.jas.2012.01.019

DeWitte, S. N., Hughes-Morey, G., Bekvalac, J. & Karsten, J. (2016). Wealth, health and frailty in industrial-era London. *Annals of Human Biology*, 43(3), 241–254. https://doi.org/10.3109/03014460.2015.1020873

DeWitte, S. N. & Slavin, P. (2013). Between famine and death: England on the eve of the Black Death—evidence from paleoepidemiology and manorial accounts. *Journal of Interdisciplinary History*, 44(1), 37–60. https://doi.org/10.1162/JINH_a_00500

DeWitte, S. N. & Wood, J. W. (2008). Selectivity of Black Death mortality with respect to preexisting health. *Proceedings of the National Academy of Sciences*, 105(5), 1436–1441. https://doi.org/10.1073/pnas.0705460105

DeWitte, S. N. & Yaussy, S. L. (2017). Femur length and famine mortality in medieval London. *Bioarchaeology International*, 1(3–4), 171–182. https://doi.org/10.5744/bi.2017.1009

Dickinson, H., Moss, T. J., Gatford, K. L., Moritz, K. M., Akison, L., Fullston, T., Hryciw, D. H., Maloney, C. A., Morris, M. J., Wooldridge, A. L., Schjenken, J. E., Robertson, S. A., Waddell, B. J., Mark, P. J., Wyrwoll, C. S., Ellery, S. J., Thornburg, K. L., Muhlhausler, B. S. & Morrison, J. L. (2016). A review of fundamental principles for animal models of DOHaD research: an Australian perspective. *Journal of Developmental Origins of Health and Disease*, 7(5), 449–472. https://doi.org/10.1017/S2040174416000477

DiMeglio, A. (1992). Growth of the spine before age 5 years. *Journal of Pediatric Orthopaedics B*, 1(2), 102–107. https://doi.org/10.1097/01202412-199201020-00003

DiMeglio A. & Bonnel F. (1989). 'Growth of the spine', in A. J. Raimondi, M. Choux & C. Di Rocco (eds), *The pediatric spine I. principles of pediatric neurosurgery*. Springer, New York. https://doi.org/10.1007/978-1-4613-8820-3_3

Dolinoy, D. C., Das, R., Weidman, J. R., & Jirtle, R. L. (2007). Metastable epialleles, imprinting, and the fetal origins of adult diseases. *Pediatric Research*, 61(5 Part 2), 30R–37R. https://doi.org/10.1203/pdr.0b013e31804575f7

Dominguez-Salas, P., Moore, S. E., Baker, M. S., Bergen, A. W., Cox, S. E., Dyer, R. A., Fulford, A. J., Guan, Y., Laritsky, E., Silver, M. J., Swan, G. E., Zeisel, S. H., Innis, S. M., Waterland, R. A., Prentice, A. M. & Hennig, B. J. (2014). Maternal nutrition at conception modulates DNA methylation of human metastable epialleles. *Nature Communications*, 5(1), article 3746. https://doi.org/10.1038/ncomms4746

Dupertuis, C. W. & Hadden, J. A. (1951). On the reconstruction of stature from long bones. *American Journal of Physical Anthropology*, 9(1), 15–54.

Duray, S. M. (1996). Dental indicators of stress and reduced age at death in prehistoric Native Americans. *American Journal of Physical Anthropology*, 99(2), 275–286. https://doi.org/10.1002/(SICI)1096-8644(199602)99:2<275::AID-AJPA5>3.0.CO;2-Y

Earnshaw, V. A., Rosenthal, L., Lewis, J. B., Stasko, E. C., Tobin, J. N., Lewis, T. T., Reid, A. E. & Ickovics, J. R. (2013). Maternal experiences with everyday discrimination and infant birth weight: a test of mediators and moderators among young, urban women of color. *Annals of Behavioral Medicine*, 45(1), 13–23. https://doi.org/10.1007/s12160-012-9404-3

Edwards, C. R., Benediktsson, R., Lindsay, R. & Seckl, J. (1993). Dysfunction of placental glucocorticoid barrier: link between fetal environment and adult hypertension? *The Lancet*, 341(8841), 355–357. https://doi.org/10.1016/0140-6736(93)90148-A

Elliot, M., Souther, S. & Park, E. (1927). Transverse lines in X-ray plates of the long bones of children. *Bulletin of the Johns Hopkins Hospital*, 41, 364–388.

Eriksson, J., Forsén, T., Tuomilehto, J., Osmond, C., & Barker, D. (2000). Fetal and childhood growth and hypertension in adult life. *Hypertension*, 36(5), 790–794. https://doi.org/10.1161/01.HYP.36.5.790

Felix, J. F. & Cecil, C. A. M. (2019). Population DNA methylation studies in the Developmental Origins of Health and Disease (DOHaD) framework. *Journal of Developmental Origins of Health and Disease*, 10(3), 306–313. https://doi.org/10.1017/S2040174418000442

FitzGerald, C., Saunders, S., Bondioli, L. & Macchiarelli, R. (2006). Health of infants in an imperial Roman skeletal sample: perspective from dental microstructure. *American Journal of Physical Anthropology*, 130(2), 179–189. https://doi.org/10.1002/ajpa.20275

Fitzgerald, C. M. & Rose, J. C. (2008). 'Reading between the lines: dental development and subadult age assessment using the microstructural growth markers of teeth', in M. A. Katzenberg & S. R. Saunders (eds), *Biological anthropology of the human skeleton*, 2nd edn, pp. 237–263. John Wiley & Sons, New York.

Fitzgerald, C. M. & Saunders, S. R. (2005). Test of histological methods of determining chronology of accentuated striae in deciduous teeth. *American Journal of Physical Anthropology*, 127(3), 277–290. https://doi.org/10.1002/ajpa.10442

Formicola, V. & Franceschi, M. (1996). Regression equations for estimating stature from long bones of early Holocene European samples. *American Journal of Physical Anthropology*, 100(1), 83–88. https://doi.org/10.1002/(SICI)1096-8644(199605)100:1<83::AID-AJPA8>3.0.CO;2-E.

Forsdahl, A. (1977). Are poor living conditions in childhood and adolescence an important risk factor for arteriosclerotic heart disease? *Journal of Epidemiology & Community Health*, 31(2), 91–95. https://doi.org/10.1136/jech.31.2.91

Franklin, D. (2010). Forensic age estimation in human skeletal remains: current concepts and future directions. *Legal Medicine*, 12(1), 1–7. https://doi.org/10.1016/j.legalmed.2009.09.001

Fully, G. (1956). Une nouvelle méthode de détermination de la taille. *Annales de Médicine Légale*, 35, 266–273.

Gage, S. H., Munafò, M. R. & Davey Smith, G. (2016). Causal inference in Developmental Origins of Health and Disease (DOHaD) research. *Annual Review of Psychology*, 67(1), 567–585. https://doi.org/10.1146/annurev-psych-122414-033352

Gamble, J. A. (2014). *A bioarchaeological approach to stress and health in medieval Denmark: dental enamel defects and adult health in two medieval Danish populations*. PhD Thesis, University of Manitoba, Winnipeg. http://hdl.handle.net/1993/30198

Gamble, J. A. (2020). A life history approach to stature and body proportions in medieval Danes. *Anthropologischer Anzeiger*, 77(1), 27–45. https://doi.org/10.1127/anthranz/2019/0951

Gamble, J. A., Boldsen, J. L. & Hoppa, R. D. (2017). Stressing out in medieval Denmark: an investigation of dental enamel defects and age at death in two medieval Danish cemeteries. *International Journal of Paleopathology*, 17(Suppl C), 52–66. https://doi.org/10.1016/j.ijpp.2017.01.001

Garland, C. J. (2020). Implications of accumulative stress burdens during critical periods of early postnatal life for mortality risk among Guale interred in a colonial era cemetery in Spanish Florida (ca. AD 1605–1680). *American Journal of Physical Anthropology*, 172(4), 621-637. https://doi.org/10.1002/ajpa.24020

Geber, J. (2014). Skeletal manifestations of stress in child victims of the Great Irish Famine (1845–1852): prevalence of enamel hypoplasia, Harris lines, and growth retardation. *American Journal of Physical Anthropology*, 155(1), 149–161. https://doi.org/10.1002/ajpa.22567

Geber, J. (2015). Victims of Ireland's great famine: the bioarchaeology of mass burials at Kilkenny Union Workhouse. *Post-Medieval Archaeology*, 51(3), 525–526. https://doi.org/10.1080/00794236.2017.1383585

Geber, J. (2016). 'Children in a ragged state': seeking a biocultural narrative of a workhouse childhood in Ireland during the Great Famine (1845–1852). *Childhood in the*

Past, 9(2), 120–138. https://doi.org/10.1080/17585716.2016.1205344

Geber, J. & Murphy, E. (2012). Scurvy in the great Irish Famine: evidence of vitamin C deficiency from a mid-19th century skeletal population. *American Journal of Physical Anthropology*, 148(4), 512–524. https://doi.org/10.1002/ajpa.22066

Genovés, S. (1967). Proportionality of the long bones and their relation to stature among Mesoamericans. *American Journal of Physical Anthropology*, 26(1), 67–77. https://doi.org/10.1002/ajpa.1330260109.

Gindhart, P. S. (1969). The frequency of appearance of transverse lines in the tibia in relation to childhood illnesses. *American Journal of Physical Anthropology*, 31(1), 17–22. https://doi.org/10.1002/ajpa.1330310104

Gluckman, P., Beedle, A., Buklijas, T., Low, F. & Hanson, M. (2016). *Principles of evolutionary medicine*, 2nd edn. Oxford University Press, Oxford.

Gluckman, PD, & Hanson, M. (eds). (2006). *Developmental Origins of Health and Disease*. Cambridge University Press, Cambridge.

Gluckman, P. D., Hanson, M. A. & Beedle, A. S. (2007). Early life events and their consequences for later disease: a life history and evolutionary perspective. *American Journal of Human Biology*, 19(1), 1–19. https://doi.org/10.1002/ajhb.20590

Gluckman, P. D., Hanson, M. A., Cooper, C. & Thornburg, K. L. (2008). Effect of in utero and early-life conditions on adult health and disease. *New England Journal of Medicine*, 359(1), 61–73. https://doi.org/10.1056/NEJMra0708473

Gluckman, P. D., Hanson, M. A. & Pinal, C. (2005). The developmental origins of adult disease. *Maternal and Child Nutrition*, 1(3), 130–141. https://doi.org/10.1111/j.1740-8709.2005.00020.x

Gokhman, D., Malul, A. & Carmel, L. (2017). Inferring past environments from ancient epigenomes. *Molecular Biology and Evolution*, 34(10), 2429–2438. https://doi.org/10.1093/molbev/msx211

Gokhman, D., Mishol, N., de Manuel, M., de Juan, D., Shuqrun, J., Meshorer, E., Marques-Bonet, T., Rak, Y. & Carmel, L. (2019). Reconstructing Denisovan anatomy using DNA methylation maps. *Cell*, 179(1), 180–192.e10. https://doi.org/10.1016/j.cell.2019.08.035

Golkari, A., Sabokseir, A., Sheiham, A. & Watt, R. G. (2016). Socioeconomic gradients in general and oral health of primary school children in Shiraz, Iran. *F1000Research*, 5, 767. https://doi.org/10.12688/f1000research.8641.1

Goodman, A. & Armelagos, G. (1988). Childhood stress and decreased longevity in a prehistoric population. *American Anthropologist*, 90, 936–944.

Goodman, A., Lallo, J., Armelagos, G. & Rose, J. (1984). 'Health changes at Dickson Mounds, Illinois (AD 950–1300)', in M. Cohen & G. Armelagos (eds), *Paleopathology at the origins of agriculture*, pp. 271–303. Academic Press, New York.

Goodman, A. H., Thomas, R. B., Swedlund, A. C. & Armelagos, G. J. (1988). Biocultural perspectives on stress in prehistoric, historical, and contemporary population research. *American Journal of Physical Anthropology*, 31(S9), 169–202.

Goodman, A. H, Martinez, C. & Chavez, A. (1991). Nutritional supplementation and the development of linear enamel hypoplasias in children from Tezonteopan, Mexico. *The American Journal of Clinical Nutrition*, 53(3), 773–781. https://doi.org/10.1093/ajcn/53.3.773

Goodman, A. H. & Rose, J. C. (1990). Assessment of systemic physiological perturbations from dental enamel hypoplasias and associated histological structures. *American Journal of Physical Anthropology*, 33(11 S), 59–110. https://doi.org/10.1002/ajpa.1330330506

Gowland, R., Stewart, N. A., Crowder, K. D., Hodson, C., Gron, K. J. & Montgomery, J. (2021). Sex estimation of teeth at different developmental stages using dimorphic enamel peptide analysis. *American Journal of Physical Anthropology* 174(4), 859–869. https://doi.org/10.1002/ajpa.24231

Gowland, R. L. (2015). Entangled lives: implications of the developmental origins of health and disease hypothesis for bioarchaeology and the life course. *American Journal of Physical Anthropology*, 158(4), 530–540. https://doi.org/10.1002/ajpa.22820

Gowland, R. L. & Newman, S. L. (2018). 'Children of the revolution: childhood health inequalities and the life course during industrialization of the 18th and 19th centuries in England', in P. Beauchesne, S. C. Agarwal, & C. S. Larsen (eds), *Children and childhood in bioarchaeology*, pp. 294–329. University Press of Florida, Gainesville.

Goyal, D., Limesand, S. W., & Goyal, R. (2019). Epigenetic responses and the developmental origins of health and disease. *Journal of Endocrinology*, 242(1), T105–T119. https://doi.org/10.1530/JOE-19-0009

Green, M. H. (2014a). Editor's introduction to pandemic disease in the medieval world: rethinking the Black Death. *The Medieval Globe* [online], 1(1), 9–26. https://scholarworks.wmich.edu/tmg/vol1/iss1/3

Green, M. H. (2014b). Taking 'pandemic' seriously: making the Black Death global. *The Medieval Globe* [online], 1(1), article 4. https://scholarworks.wmich.edu/tmg/vol1/iss1/4

Grijalva-Eternod, C. S., Lawlor, D. A. & Wells, J. C. K. (2013). Testing a capacity-load model for hypertension: disentangling early and late growth effects on childhood blood pressure in a prospective birth cohort. *PLoS ONE*, 8(2), e56078. https://doi.org/10.1371/journal.pone.0056078

Grolleau-Raoux, J. L., Crubézy, E., Rougé, D., Brugne, J. F. & Saunders, S. R. (1997). Harris lines: a study of age-associated bias in counting and interpretation. *American Journal of Physical Anthropology*, 103(2), 209–217. https://doi.org/10.1002/(SICI)1096-8644(199706)103:2<209::AID-AJPA6>3.0.CO;2-O

Guatelli-Steinberg, D. (2015). 'Dental stress indicators from micro- to macroscopic', in J. D. Irish & G. R. Scott (eds), *A Companion to Dental Anthropology*, pp. 450–464. John Wiley & Sons, New York.

Gunnell, D., Rogers, J. & Dieppe, P. (2001). Height and health: predicting longevity from bone length in archaeological remains. *Journal of Epidemiology and Community Health*, 55(7), 505–507. http://dx.doi.org/10.1136/jech.55.7.505.

Halcrow, S.E., Tayles, N. & Elliott, G.E. (2017). 'The bioarchaeology of fetuses', in S. Han, T. K. Betsinger & A. B. Scott (eds), *The anthropology of the fetus: biology, culture, and society*, pp. 83–111. Berghan Books, New York.

Hales, C. N. & Barker, D. J. P. (1992). Type 2 (non-insulin-dependent) diabetes mellitus: the thrifty phenotype hypothesis. *Diabetologia*, 35(7), 595–601. https://doi.org/10.1007/BF00400248

Hales, C. N. & Barker, D. J. P. (2001). The thrifty phenotype hypothesis: type 2 diabetes. *British Medical Bulletin*, 60(1), 5–20. https://doi.org/10.1093/bmb/60.1.5

Hales, C. N. & Barker, D. (2013). Type 2 (non-insulin-dependent) diabetes mellitus: the thrifty phenotype hypothesis. *International Journal of Epidemiology*, 42(5), 1215–1222. https://doi.org/10.1093/ije/dyt133

Handy, D. E., Castro, R. & Loscalzo, J. (2011). Epigenetic modifications. *Circulation*, 123(19), 2145–2156. https://doi.org/10.1161/CIRCULATIONAHA.110.956839

Harder, T., Rodekamp, E., Schellong, K., Dudenhausen, J. W. & Plagemann, A. (2007). Birth weight and subsequent risk of type 2 diabetes: a meta-analysis. *American Journal of Epidemiology*, 165(8), 849–857. https://doi.org/10.1093/aje/kwk071

Harris, H. A. (1933). Bone growth in health and disease: the biological principles underlying the clinical, *radiological, and histological diagnosis of perversions of growth and disease in the skeleton*. Oxford University Press, Oxford.

Hassett, B. R., Dean, M. C., Ring, S., Atkinson, C., Ness, A. R., & Humphrey, L. (2020). Effects of maternal, gestational, and perinatal variables on neonatal line width observed in a modern UK birth cohort. *American Journal of Physical Anthropology*, 172(2), 314–332. https://doi.org/10.1002/ajpa.24042

Heijmans, B. T., Tobi, E. W., Stein, A. D., Putter, H., Blauw, G. J., Susser, E. S., Slagboom, P. E. & Lumey, L. H. (2008). Persistent epigenetic differences associated with prenatal exposure to famine in humans. *Proceedings of the National Academy of Sciences*, 105(44), 17046–17049. https://doi.org/10.1073/pnas.0806560105

Hennig, B. J., Unger, S. A., Dondeh, B. L., Hassan, J., Hawkesworth, S., Jarjou, L., Jones, K. S., Moore, S. E., Nabwera, H. M., Ngum, M., Prentice, A., Sonko, B., Prentice, A. M. & Fulford, A. J. (2017). Cohort profile: the Kiang West Longitudinal Population Study (KWLPS)—a platform for integrated research and health care provision in rural Gambia. *International Journal of Epidemiology*, 46(2), e13. https://doi.org/10.1093/ije/dyv206

Hillson, S. (2014). *Tooth development in human evolution and bioarchaeology*. Cambridge University Press, Cambridge.

Hodson, C. M. & Gowland, R. (2020). 'Like mother, like child: Investigating perinatal and maternal health stress in post-medieval London', in R. L. Gowland & S. Halcrow (eds), *The mother-infant nexus in anthropology*, pp. 39–64, Springer International, New York.

Huang, C., Li, Z., Wang, M. & Martorell, R. (2010). Early life exposure to the 1959–1961 Chinese famine has long-term health consequences. *The Journal of Nutrition*, 140(10), 1874–1878. https://doi.org/10.3945/jn.110.121293

Hughes-Morey, G. (2015). Interpreting adult stature in industrial London. *American Journal of Physical Anthropology* 159(1): 126–134. doi:10.1002/ajpa.22840.

Humphrey, L. T. & King, T. (2000). Childhood stress: a lifetime legacy. *Anthropologie (Brno)*, 38(1), 33–49.

Infante, P. F., Latham, M. C., Stephenson, L. S., Simoons, F. J., Clinkscales, E. & Larkin, F. A. (1974). Letters to the editor: enamel hypoplasia in Apache Indian children. *Ecology of Food and Nutrition*, 3(2), 155–164. https://doi.org/10.1080/03670244.1974.9990375

İşcan, M. Y. (2005). Forensic anthropology of sex and body size. *Forensic Science International*, 147(2–3), 107–112. https://doi.org/10.1016/j.forsciint.2004.09.069

Ives, R. & Humphrey, L. (2017). Patterns of long bone growth in a mid-19th century documented sample of the urban poor from Bethnal Green, London, UK. *American Journal of Physical Anthropology*, 163(1), 173–186. https://doi:10.1002/ajpa.23198

Jaenisch, R. & Bird, A. (2003). Epigenetic regulation of gene expression: how the genome integrates intrinsic and environmental signals. *Nature Genetics*, 33(S3), 245–254. https://doi.org/10.1038/ng1089

Jantz, L. M. & Jantz, R. L. (1999). Secular change in long bone length and proportion in the United States, 1800–1970. *American Journal of Physical Anthropology*, 110(1), 57–67. https://doi.org/10.1002/(SICI)1096-8644(199909)110:1<57::AID-AJPA5>3.0.CO;2-1

Jones, J. H. (2005). Fetal programming: adaptive life-history tactics or making the best of a bad start? *American Journal of Human Biology*, 17(1), 22–33. https://doi.org/10.1002/ajhb.20099

Jones, N. L., Gilman, S. E., Cheng, T. L., Drury, S. S., Hill, C. V. & Geronimus, A. T. (2019). Life course approaches to the causes of health disparities. *American Journal of Public Health*, 109(S1), S48–S55. https://doi.org/10.2105/AJPH.2018.304738

Joseph, K. S. & Kramer, M. S. (1996). Review of the evidence on fetal and early childhood antecedents of adult chronic disease. *Epidemiologic Reviews*, 18(2), 158–174. https://doi.org/10.1093/oxfordjournals.epirev.a017923

Kemkes-Grottenthaler, A. (2005). The short die young: the interrelationship between stature and longevity—evidence from skeletal remains. *American Journal of Physical Anthropology*, 128(2), 340–347. https://doi.org/10.1002/ajpa.20146

Kermack, W., McKendrick, A. & McKinlay, P. (2001). Death-rates in Great Britain and Sweden. Some general regularities and their significance. International Journal of Epidemiology, 30(4), 678–683. https://doi.org/10.1093/ije/30.4.678. First printed in 1934, *The Lancet*, 223(5770), 698–703.

Klaus, H. D. (2014). Frontiers in the bioarchaeology of stress and disease: cross-disciplinary perspectives from pathophysiology, human biology, and epidemiology. *American Journal of Physical Anthropology*, 155(2), 294–308. https://doi.org/10.1002/ajpa.22574

Knop, M. R., Geng, T., Gorny, A. W., Ding, R., Li, C., Ley, S. H. & Huang, T. (2018). Birth weight and risk of type 2 diabetes mellitus, cardiovascular disease, and hypertension in adults: a meta-analysis of 7,646,267 participants from 135 studies. *Journal of the American Heart Association*, 7(23), e008870. https://doi.org/10.1161/JAHA.118.008870

Kuh, D., & Ben-Shlomo, Y. (eds). (1997). *A life course approach to chronic disease epidemiology: tracing the origins of ill-health from early to adult life*. Oxford University Press, Oxford.

Kuh D. J. L., & Ben-Shlomo Y. (eds). (2004). *A life course approach to chronic diseases epidemiology*, 2nd edn. Oxford University Press, Oxford.

Kulus, M. J., & Dąbrowski, P. (2019). How to calculate the age at formation of Harris lines? A step-by-step review of current methods and a proposal for modifications to Byers' formulas. *Archaeological and Anthropological Sciences* 11(4), 1169–1185. https://doi.org/10.1007/s12520-018-00773-5

Kuzawa, C. W. (2005). Fetal origins of developmental plasticity: are fetal cues reliable predictors of future nutritional environments? *American Journal of Human Biology*, 17(1), 5–21. https://doi.org/10.1002/ajhb.20091

Lampl, M., Veldhuis, J. D. & Johnson, M. L. (1992). Saltation and stasis: a model of human growth. *Science*, 258(5083), 801–803. https://doi.org/10.1126/science.1439787

Lawlor, D. A., Andersen, A.-M. N. & Batty, G. D. (2009). Birth cohort studies: past, present and future. *International Journal of Epidemiology*, 38(4), 897–902. https://doi.org/10.1093/ije/dyp240

Leonard, W. R., Reyes-García, V., Tanner, S., Rosinger, A., Schultz, A., Vadez, V., Zhang, R. & Godoy, R. (2015). The Tsimané Amazonian Panel Study (TAPS): nine years (2002–2010) of annual data available to the public. *Economics & Human Biology*, 19, 51–61. https://doi.org/10.1016/j.ehb.2015.07.004

Lewis, M. (2017). 'Fetal paleopathology: an impossible discipline?', in S. Han, T. K. Betsinger & A. B. Scott (eds.), *The anthropology of the fetus: biology, culture, and society*, pp. 112–131. Berghan Books, New York.

Lewis, M. (2002). Impact of industrialization: comparative study of child health in four sites from medieval and postmedieval England (A.D. 850 –1859). *American Journal of Physical Anthropology*, 119(3), 211–223. https://doi-org.uml.idm.oclc.org/10.1002/ajpa.10126

Lewis, M., & Roberts, C. (1997). Growing pains: the interpretation of stress indicators. *International Journal of Osteoarchaeology*, 7(6), 581–586. https://doi.org/10.1002/(sici)1099-1212(199711/12)7:6<581::aid-oa325>3.0.co;2-c

Li, C. & Lumey, L. (2017). Exposure to the Chinese famine of 1959–61 in early life and long-term health conditions: a systematic review and meta-analysis. *International Journal of Epidemiology*, 46(4), 1157–1170. https://doi.org/10.1093/ije/dyx013

Llamas, B., Holland, M. L., Chen, K., Cropley, J. E., Cooper, A. & Suter, C. M. (2012). High-resolution analysis of cytosine methylation in ancient DNA. *PLoS One*, 7(1), e30226. https://doi.org/10.1371/journal.pone.0030226

Lorentz, K. O., Lemmers, S. A. M., Chrysostomou, C., Dirks, W., Zaruri, M. R., Foruzanfar, F. & Sajjadi, S. M. S. (2019). Use of dental microstructure to investigate the role of prenatal and early life physiological stress in age at death. *Journal of Archaeological Science*, 104, 85–96. https://doi.org/10.1016/j.jas.2019.01.007

Lucas, A. (1991). Programming by early nutrition in man. *Ciba Foundation Symposium*, 156, 38–55.

Maijanen, H. (2009). Testing anatomical methods for stature estimation on individuals from the W. M. Bass Donated Skeletal Collection. *Journal of Forensic Sciences*, 54(4), 746–752. https://doi.org/10.1111/j.1556-4029.2009.01053.x

Maijanen, H. & Niskanen, M. (2006). Comparing stature-estimation methods on Medieval inhabitants of Westerhus, Sweden. *Fennoscandia Archaeologica*, 23, 37–46.

Mani-Caplazi, G., Schulz, G., Deyhle, H., Bert, M., Hotz, G., Vach, W. & Wittwer-Backofen, U. (2017). 'Imaging of the human tooth cementum ultrastructure of archeological teeth, using hard X-ray microtomography to determine age-at-death and stress periods', in B. Müller and G. Wang (eds), Developments in X-Ray Tomography XI, Proceedings of SPIE Vol. 10391, 103911C (SPIE, Bellingham, WA, 2017). https://doi.org/10.1117/12.2276148.

Masterson, E. E., Fitzpatrick, A. L., Enquobahrie, D. A., Mancl, L. A., Conde, E. & Hujoel, P. P. (2017). Malnutrition-related early childhood exposures and enamel defects in the permanent dentition: a longitudinal study from the Bolivian Amazon. *American Journal of Physical Anthropology*, 164(2), 416–423. https://doi.org/10.1002/ajpa.23283

Masterson, E. E., Fitzpatrick, A. L., Enquobahrie, D. A., Mancl, L. A., Eisenberg, D. T. A., Conde, E. & Hujoel, P. P. (2018). Dental enamel defects predict adolescent health indicators: a cohort study among the Tsimané of Bolivia. *American Journal of Human Biology*, 30(3), e23107. https://doi.org/10.1002/ajhb.23107

Mays, S. (2014). The palaeopathology of scurvy in Europe. *International Journal of Paleopathology*, 5, 55–62. https://doi.org/10.1016/j.ijpp.2013.09.001

Mays, S, Brickley, M. & Ives, R. (2006). Skeletal manifestation of rickets in infants and young children in a historic population from England. *American Journal of Physical Anthropology*, 129(3), 362–374. https://doi.org/10.1002/ajpa.20292

McFadden, C. & Oxenham, M. F. (2020). A paleoepidemiological approach to the osteological paradox: investigating stress, frailty and resilience through cribra orbitalia. *American Journal of Physical Anthropology*, 173(2), 205–217. https://doi.org/10.1002/ajpa.24091

Meiklejohn C. & Babb J. (2011). 'Long bone length, stature and time in the European Late Pleistocene and Early Holocene', in R. Pinhasi & J. T. Stock (eds), *Human bioarchaeology of the transition to agriculture*, pp. 151–175. John Wiley & Sons, Chichester.

Montgomery, S. M. & Ekbom, A. (2002). Smoking during pregnancy and diabetes mellitus in a British longitudinal birth cohort. *British Medical Journal*, 324(7328), 26–27. https://doi.org/10.1136/bmj.324.7328.26

Moore, S. E. (2020). Using longitudinal data to understand nutrition and health interactions in rural Gambia. *Annals of Human Biology*, 47(2), 125–131. https://doi.org/10.1080/03014460.2020.1718207

Nelson, S., Albert, J. M., Geng, C., Curtan, S., Lang, K., Miadich, S., Heima, M., Malik, A., Ferretti, G., Eggertsson, H., Slayton, R. L. & Milgrom, P. (2013). Increased enamel hypoplasia and very low birthweight infants. *Journal of Dental Research*, 92(9), 788–794. https://doi.org/10.1177/0022034513497751

Newman, S. L. & Gowland, R. L. (2017). Dedicated followers of fashion? Bioarchaeological perspectives on socio-economic status, inequality, and health in urban children from the industrial revolution (18th–19th C), England. *International Journal of Osteoarchaeology*, 27(2), 217–229. https://doi.org/10.1002/oa.2531

Newman, S. L. & Gowland, R. L. (2015). The use of non-adult vertebral dimensions as indicators of growth disruption and non-specific health stress in skeletal populations. *American Journal of Physical Anthropology*, 158(1), 155–164. https://doi.org/10.1002/ajpa.22770

Newman, S. L., Gowland, R. L. & Caffell, A. C. (2019). North and south: a comprehensive analysis of non-adult growth and health in the industrial revolution (AD 18th–19th C), England. *American Journal of Physical Anthropology*, 169(1), 104–121. https://doi.org/10.1002/ajpa.23817

Newnham, J. P. & Ross, M.G. (eds). (2009). *Early life origins of human health and disease*. Karger, Basel.

Novak, M., Vyroubal, V., Krnčević, Ž., Petrinec, M., Howcroft, R., Pinhasi, R., & Slaus, M. (2018). Assessing childhood stress in early mediaeval Croatia by using multiple lines of inquiry. *Anthropologischer Anzeiger*, 75(2), 155–167. https://doi.org/10.1127/anthranz/2018/0819

O'Donnell, L., Hill, E. C., Anderson, A. S. & Edgar, H. J. H. (2020). Cribra orbitalia and porotic hyperostosis are associated with respiratory infections in a contemporary mortality sample from New Mexico. *American Journal of Physical Anthropology*, 173(4), 721–733. https://doi.org/10.1002/ajpa.24131

Oldroyd, J. (2005). Low birth weight in South Asian babies in Britain: time to reduce the inequalities. *Fukushima Journal of Medical Science*, 51(1), 1–10. https://doi.org/10.5387/fms.51.1

Ortner, D. J. & Ericksen, M. F. (1997). Bone changes in the human skull probably resulting from scurvy in infancy and childhood. *International Journal of Osteoarchaeology*, 7(3), 212–220. https://doi.org/10.1002/(SICI)1099-1212(199705)7:3<212::AID-OA346>3.0.CO;2-5

Oxenham, M. F. & Cavill, I. (2010). Porotic hyperostosis and cribra orbitalia: the erythropoietic response to iron-deficiency anaemia. *Anthropological Science*, 118(3), 199–200. https://doi.org/10.1537/ase.100302

Palubeckaitė, Z., Jankauskas, R. & Boldsen, J. (2002) Enamel hypoplasia in Danish and Lithuanian Late Medieval/Early Modern samples: a possible reflection of child morbidity and mortality patterns. *International Journal of Osteoarchaeology*, 12(3), 189–201. https://doi.org/10.1002/oa.607

Pansieri, C., Pandolfini, C., Clavenna, A., Choonara, I. & Bonati, M. (2020). An inventory of European birth cohorts. *International Journal of Environmental Research and Public Health*, 17(9), 3071. https://doi.org/10.3390/ijerph17093071

Papageorgopoulou, C., Suter, S. K., Rühli, F. J. & Siegmund, F. (2011). Harris lines revisited: prevalence, comorbidities, and possible etiologies. *American Journal of Human Biology*, 23(3), 381–391. https://doi.org/10.1002/ajhb.21155

Parker, G. J., Yip, J. M., Eerkens, J. W., Salemi, M., Durbin-Johnson, B., Kiesow, C., Haas, R., Buikstra, J. E., Klaus, H., Regan, L. A., Rocke, D. M. & Phinney, B. S. (2019). Sex estimation using sexually dimorphic amelogenin protein fragments in human enamel. *Journal of Archaeological Science*, 101(May), 169–180. https://doi.org/10.1016/j.jas.2018.08.011

Petersen, H. C. (2005). On the accuracy of estimating living stature from skeletal length in the grave and by linear regression. *International Journal of Osteoarchaeology*, 15(2), 106–114. https://doi.org/10.1002/oa.740

Petersen, H. C. (2011). Technical note: a re-evaluation of stature estimation from skeletal length in the grave. *American Journal of Physical Anthropology*, 144(2), 327–330. https://doi.org/10.1002/ajpa.21427

Pinho, J., Filho, F., Thomaz, E., Lamy, Z., Libério, S. & Ferreira, E. (2012). Are low birth weight, intrauterine growth restriction, and preterm birth associated with enamel developmental defects? *Pediatric Dentistry*, 34(3), 244–248.

Pomeroy, E., Stock, J. T., Stanojevic, S., Miranda, J. J., Cole, T. J. & Wells, J. C. K. (2012). Trade-offs in relative limb length among Peruvian children: extending the thrifty phenotype hypothesis to limb proportions. *PLoS ONE*, 7(12), e51795. https://doi.org/10.1371/journal.pone.0051795

Primeau, C., Homøe, P., & Lynnerup, N. (2018). Childhood health as reflected in adult urban and rural samples from medieval Denmark. *Homo*, 69(1–2), 6–16. https://doi.org/10.1016/j.jchb.2018.03.004

Procopio, N., Chamberlain, A. T. & Buckley, M. (2017). Intra- and interskeletal proteome variations in fresh and buried bones. *Journal of Proteome Research*, 16(5), 2016–2029. https://doi.org/10.1021/acs.jproteome.6b01070

Prokocimer, T., Amir, E., Blumer, S. & Peretz, B. (2015). Birth-weight, pregnancy term, pre-natal and natal complications related to child's dental anomalies. *Journal of Clinical Pediatric Dentistry*, 39(4), 371–376. https://doi.org/10.17796/1053-4628-39.4.371

Raxter, M. H., Auerbach, B. M. & Ruff, C. B. (2006). Revision of the Fully technique for estimating statures. *American Journal of Physical Anthropology*, 130(3), 374–384. https://doi.org/10.1002/ajpa.20361

Reitsema, L. J. & McIlvaine, B. K. (2014). Reconciling 'stress' and 'health' in physical anthropology: what can bioarchaeologists learn from the other subdisciplines? *American Journal of Physical Anthropology*, 155(2), 181–185. https://doi.org/10.1002/ajpa.22596

Richter, L. M., Victora, C. G., Hallal, P. C., Adair, L. S., Bhargava, S. K., Fall, C. H., Lee, N., Martorell, R., Norris, S. A., Sachdev, H. S. & Stein, A. D. (2012). Cohort profile: the consortium of health-orientated research in transitioning societies. *International Journal of Epidemiology*, 41(3), 621–626. https://doi.org/10.1093/ije/dyq251

Rickard, I. (2016). 'The developmental origins of health and disease: adaptation reconsidered', in A. Alvergne, C. Jenkinson & C. Faurie (eds), *Evolutionary thinking in medicine: from research to policy and practice*, pp. 75–88. Springer International, Switzerland.

Rickard, I. J., & Lummaa, V. (2007). The predictive adaptive response and metabolic syndrome: challenges for the hypothesis. *Trends in Endocrinology & Metabolism*, 18(3), 94–99. https://doi.org/10.1016/j.tem.2007.02.004

Rinaldo, N., Zedda, N., Bramanti, B., Rosa, I. & Gualdi-Russo, E. (2019). How reliable is the assessment of porotic hyperostosis and cribra orbitalia in skeletal human remains? A methodological approach for quantitative verification by means of a new evaluation form. *Archaeological and Anthropological Sciences* 11, 3549–3559. https://doi.org/10.1007/s12520-019-00780-0

Rivera, F. & Mirazón Lahr, M. (2017). New evidence suggesting a dissociated etiology for *cribra orbitalia* and porotic hyperostosis. *American Journal of Physical Anthropology*, 164(1), 76–96. https://doi.org/10.1002/ajpa.23258

Robb, J., Bigazzi, R., Lazzarini, L., Scarsini, C. and Sonego, F. (2001). Social 'status' and biological 'status': a comparison of grave goods and skeletal indicators from Pontecagnano. *American Journal of Physical Anthropology* 115(3), 213–222. https://doi.org/10.1002/ajpa.1076

Roberts, C. A. & Steckel, R. A. (2019). 'The developmental origins of health and disease: early life health conditions and adult age at death in Europe', in R. Steckel, C. Larsen, C. Roberts & J. Baten (eds), *The backbone of Europe: health, diet, work and violence over two millennia*, pp. 325–351. Cambridge University Press, Cambridge.

Roberts, S. B., Paul, A. A., Cole, T. J. & Whitehead, R. G. (1982). Seasonal changes in activity, birth weight and lactational performance in rural Gambian women. *Transactions of the Royal Society of Tropical Medicine and Hygiene*, 76(5), 668–678. https://doi.org/10.1016/0035-9203(82)90239-5

Rose, J. C., Armelagos, G. J. & Lallo, J. W. (1978). Histological enamel indicator of childhood stress in prehistoric skeletal samples. *American Journal of Physical*

Anthropology, 49(4), 511–516. https://doi.org/10.1002/ajpa.1330490411

Rothschild, B. (2012). Extirpation of the mythology that porotic hyperostosis is caused by iron deficiency secondary to dietary shift to maize. *Advances in Anthropology*, 2(03), 157–160. https://doi.org/10.4236/aa.2012.23018

Rudney, J. D. (1983). Dental indicators of growth disturbance in a series of ancient Lower Nubian populations: changes over time. *American Journal of Physical Anthropology*, 60(4), 463–470. https://doi.org/10.1002/ajpa.1330600408

Santos, R. V. & Coimbra, C. E. A. (1999). Hardships of contact: enamel hypoplasias in Tupí-Mondé Amerindians from the Brazilian Amazonia. *American Journal of Physical Anthropology*, 109(1), 111–127. https://doi.org/10.1002/(SICI)1096-8644(199905)109:1<111::AID-AJPA9>3.0.CO;2-5

Sattenspiel, L. & Harpending, H. C. (1983). Stable populations and skeletal age. *American Antiquity*, 48(3), 489–498.

Savino, W. (2002). The thymus gland is a target in malnutrition. *European Journal of Clinical Nutrition*, 56(S3), S46–49. https://doi.org/10.1038/sj.ejcn.1601485

Schattmann, A., Bertrand, B., Vatteoni, S. & Brickley, M. (2016). Approaches to co-occurrence: scurvy and rickets in infants and young children of 16–18th century Douai, France. *International Journal of Paleopathology*, 12, 63–75. https://doi.org/10.1016/j.ijpp.2015.12.002

Scott, A. B. & Hoppa, R. D. (2015). Brief communication: a re-evaluation of the impact of radiographic orientation on the identification and interpretation of Harris lines. *American Journal of Physical Anthropology*, 156(1), 141–147. https://doi.org/10.1002/ajpa.22635

Seckl, J. R. (2004). Prenatal glucocorticoids and long-term programming. *European Journal of Endocrinology*, 151 (Suppl_3), U49–62. https://doi.org/10.1530/eje.0.151u049

Sharp, G. C., Lawlor, D. A. & Richardson, S. S. (2018). It's the mother!: how assumptions about the causal primacy of maternal effects influence research on the developmental origins of health and disease. *Social Science and Medicine*, 213, 20–27. https://doi.org/10.1016/j.socscimed.2018.07.035

Sharp, G. C., Lawlor, D. A., Richmond, R. C., Fraser, A., Simpkin, A., Suderman, M., Shihab, H. A., Lyttleton, O., McArdle, W., Ring, S. M., Gaunt, T. R., Davey Smith, G. & Relton, C. L. (2015). Maternal pre-pregnancy BMI and gestational weight gain, offspring DNA methylation and later offspring adiposity: findings from the Avon Longitudinal Study of Parents and Children. *International Journal of Epidemiology*, 44(4), 1288–1304. https://doi.org/10.1093/ije/dyv042

Silverwood, R. J., Pierce, M., Hardy, R., Sattar, N., Whincup, P., Ferro, C., Savage, C., Kuh, D. & Nitsch, D. (2013). Low birth weight, later renal function, and the roles of adulthood blood pressure, diabetes, and obesity in a British birth cohort. *Kidney International*, 84(6), 1262–1270. https://doi.org/10.1038/ki.2013.223

Simpson, S. W., Hoppa, R. D. & FitzGerald, C. M. (1999). 'Reconstructing patterns of growth disruption from enamel microstructure', in R. D. Hoppa & C. M. Fitzgerald (eds), *Human growth in the past: studies from bones and teeth*, pp. 241–263. Cambridge University Press, Cambridge.

Slavin, P. (2016). Climate and famines: a historical reassessment. *Wiley Interdisciplinary Reviews: Climate Change* 7(3), 433–447. https://doi.org/10.1002/wcc.395

Smith, C. A. (1947). Effects of maternal undernutrition upon the newborn infant in Holland (1944–1945). *The Journal of Pediatrics*, 30(3), 229–243. https://doi.org/10.1016/S0022-3476(47)80158-1

Smith, G. (ed) (2003). *Health inequalities: lifecourse approaches*. Policy Press, Bristol.

Smith, R. W. A., Monroe, C. & Bolnick, D. A. (2015). Detection of cytosine methylation in ancient DNA from five native American populations using bisulfite sequencing. *PLoS One*, 10(5), e0125344. https://doi.org/10.1371/journal.pone.0125344

Sparen, P., Vågerö, D., Shestov, D. B., Plavinskaja, S., Parfenova, N., Hoptiar, V., Paturot, D. & Galanti, M. R. (2004). Long-term mortality after severe starvation during the siege of Leningrad: prospective cohort study. *British Medical Journal*, 328(11). https://doi.org/10.1136/bmj.37942.603970.9A

Spencer, H., Hanson, M., & Gluckman, P. (2006). Response to Wells: phenotypic responses to early environmental cues can be adaptive in adults. *Trends in Ecology & Evolution*, 21(8), 425–426. https://doi.org/10.1016/j.tree.2006.05.005

Steckel, R. H. (2005). Young adult mortality following severe physiological stress in childhood: skeletal evidence. *Economics and Human Biology*, 3(2), 314–328. https://doi.org/10.1016/j.ehb.2005.05.006

Stewart, N. A., Gerlach, R. F., Gowland, R. L., Gron, K. J. & Montgomery, J. (2017). Sex determination of human remains from peptides in tooth enamel. *Proceedings of the National Academy of Sciences*, 114(52), 13649–13654. https://doi.org/10.1073/pnas.1714926115

Stewart, N. A., Molina, G. F., Mardegan Issa, J. P., Yates, N. A., Sosovicka, M., Vieira, A. R., Line, S. R. P., Montgomery, J. & Gerlach, R. F. (2016). The identification of peptides by nano LC-MS/MS from human surface tooth enamel following a simple acid etch extraction. *RSC Advances*, 6(66), 61673–61679. https://doi.org/10.1039/c6ra05120k

Stinson, S. (2012). Growth variation: biological and cultural factors. In S Stinson, B Bogin & D O'Rourke (eds), *Human biology: an evolutionary and biocultural perspective*, pp. 587–635. John Wiley & Sons, New York.

Stockard, C. R. (1921). Developmental rate and structural expression: an experimental study of twins, 'double monsters' and single deformities, and the interaction among embryonic organs during their origin and development. *American Journal of Anatomy*, 28(2), 115–277. https://doi.org/10.1002/aja.1000280202

Stuart-Macadam, P. (1985). Porotic hyperostosis: representative of a childhood condition. *American Journal of Physical Anthropology*, 66(4), 391–398. https://doi.org/10.1002/ajpa.1330660407

Stuart-Macadam, P. (1989). Porotic hyperostosis: relationship between orbital and vault lesions. *American Journal of Physical Anthropology*, 80(2), 187–193. https://doi.org/10.1002/ajpa.1330800206

Stuart-Macadam, P. (1992a). 'Anemia in past human populations,' in P. Stuart-Macadam & S. Kent (eds), *Diet, demography, and disease: changing perspectives on anemia*, pp. 151–170. De Gruyter, New York.

Stuart-Macadam, P. (1992b). Porotic hyperostosis: a new perspective. *American Journal of Physical Anthropology*, 87(1), 39–47. https://doi.org/10.1002/ajpa.1330870105

Temple, D. H. (2019). Bioarchaeological evidence for adaptive plasticity and constraint: exploring life-history trade-offs in the human past. *Evolutionary Anthropology*, 28(1), 34–46). John Wiley & Sons, Ltd. https://doi.org/10.1002/evan.21754

Temple, D. H. (2014). Plasticity and constraint in response to early-life stressors among late/final Jomon Period foragers from Japan: evidence for life history trade-offs from incremental microstructures of enamel. *American Journal of Physical Anthropology*, 155(4), 537–545. https://doi.org/10.1002/ajpa.22606

Temple, D. H., & Goodman, A. H. (2014). Bioarcheology has a 'health' problem: conceptualizing 'stress' and 'health' in bioarcheological research. *American Journal of Physical Anthropology*, 155(2), 186–191. https://doi.org/10.1002/ajpa.22602

Temple, D. H., Nakatsukasa, M. & McGroarty, J. N. (2012). Reconstructing patterns of systemic stress in a Jomon period subadult using incremental microstructures of enamel. *Journal of Archaeological Science*, 39(5), 1634–1641. https://doi.org/10.1016/j.jas.2011.12.021

Thayer, Z. M. & Kuzawa, C. W. (2011). Biological memories of past environments: epigenetic pathways to health disparities. *Epigenetics*, 6(7), 798–803. https://doi.org/10.4161/epi.6.7.16222

Thomas, R. F. (2003). *Enamel defects, well-being and mortality in a medieval Danish village*. PhD Thesis, Penn State University, University Park, PA. https://etda.libraries.psu.edu/catalog/6069

Tian, G., Guo, C., Li, Q., Liu, Y., Sun, X., Yin, Z., Li, H., Chen, X., Liu, X., Zhang, D., Cheng, C., Liu, L., Liu, F., Zhou, Q., Wang, C., Li, L., Wang, B., Zhao, Y., Liu, D., Zhang, M. & Hu, D. (2019). Birth weight and risk of type 2 diabetes: A dose-response meta-analysis of cohort studies. *Diabetes/Metabolism Research and Reviews*, 35(5), e3144. https://doi.org/10.1002/dmrr.3144

Tobi, E. W., Lumey, L. H., Talens, R. P., Kremer, D., Putter, H., Stein, A. D., Slagboom, P. E. & Heijmans, B. T. (2009). DNA methylation differences after exposure to prenatal famine are common and timing- and sex-specific. *Human Molecular Genetics*, 18(21), 4046–4053. https://doi.org/10.1093/hmg/ddp353

Trotter, M. (1970). 'Estimation of stature from intact long bones', in T. D. Stewart (ed), *Personal identification in mass disasters*, pp. 71–83. Smithsonian Institution Press, Washington, DC.

Trotter, M. & Gleser, G. C. (1958). A re-evaluation of estimation of stature based on measurements of stature taken during life and of long bones after death. *American Journal of Physical Anthropology*, 16(1), 79–123.

Vågerö, D., Koupil, I., Parfenova, N. & Sparen, P. (2013). 'Long-term health consequences following the Siege of Leningrad', in L. Lumey & A. Vaiserman (eds), *Early life nutrition and adult health and development*, pp. 207–225. Nova Science Publishers, New York.

Vukic, M., Wu, H. & Daxinger, L. (2019). Making headway towards understanding how epigenetic mechanisms contribute to early-life effects. *Philosophical Transactions of the Royal Society B: Biological Sciences*, 374(1770), 20180126. https://doi.org/10.1098/rstb.2018.0126

Walker, P. L., Bathurst, R. R., Richman, R., Gjerdrum, T. & Andrushko, V. A. (2009). The causes of porotic hyperostosis and cribra orbitalia: a reappraisal of the iron-deficiency-anemia hypothesis. *American Journal of Physical Anthropology* 139(2), 109–125. https://doi.org/10.1002/ajpa.21031

Wang, P.-X., Wang, J.-J., Lei, Y.-X., Xiao, L., & Luo, Z.-C. (2012). Impact of fetal and infant exposure to the Chinese Great Famine on the risk of hypertension in adulthood. *PLoS ONE*, 7(11), e49720. https://doi.org/10.1371/journal.pone.0049720

Waterland, R. A., Kellermayer, R., Laritsky, E., Rayco-Solon, P., Harris, R. A., Travisano, M., Zhang, W., Torskaya, M. S., Zhang, J., Shen, L., Manary, M. J. & Prentice, A. M. (2010). Season of conception in rural Gambia affects DNA methylation at putative human metastable epialleles. *PLoS Genetics*, 6(12), e1001252. https://doi.org/10.1371/journal.pgen.1001252

Waterland, R. A. & Michels, K. B. (2007). Epigenetic epidemiology of the developmental origins hypothesis. *Annual Review of Nutrition*, 27(1), 363–388. https://doi.org/10.1146/annurev.nutr.27.061406.093705

Watts, R. (2011). Non-specific indicators of stress and their association with age at death in Medieval York: using stature and vertebral neural canal size to examine the effects of stress occurring during different periods of development. *International Journal of Osteoarchaeology*, 21(5), 568–576. https://doi.org/10.1002/oa.1158

Watts, R. (2013). Childhood development and adult longevity in an archaeological population from Barton-upon-Humber, Lincolnshire, England. *International Journal of Paleopathology*, 3(2), 95–104. https://doi.org/10.1016/j.ijpp.2013.05.001

Watts, R. (2015). The long-term impact of developmental stress. Evidence from later medieval and post-medieval London (AD1117–1853). *American Journal of Physical Anthropology*, 158(4), 569–580. https://doi.org/10.1002/ajpa.22810

Weinhold, B. (2006). Epigenetics: the science of change. *Environmental Health Perspectives*, 114(3), A160–A167. https://doi.org/10.1289/ehp.114-a160

Weiss, N. M., Vercollotti, G., Boano, R., Girotti, M., & Stout, S. (2019). Body size and social status in medieval Alba (Cuneo), Italy. *American Journal of Physical Anthropology* 168(3), 595–605. https://doi.org/10.1002/ajpa.23776

Wells, J. C. K. (2012). A critical appraisal of the predictive adaptive response hypothesis. *International Journal of Epidemiology*, 41(1), 229–235. https://doi.org/10.1093/ije/dyr239

Wells, J. C. K. (2007). Flaws in the theory of predictive adaptive responses. *Trends in Endocrinology & Metabolism*, 18(9), 331–337. https://doi.org/10.1016/j.tem.2007.07.006

Wells, J. C. K. (2006). Is early development in humans a predictive adaptive response anticipating the adult environment? *Trends in Ecology & Evolution*, 21(8), 424–425. https://doi.org/10.1016/j.tree.2006.05.006

Wells, J. C. K. (2016). *The metabolic ghetto: an evolutionary perspective on nutrition, power relations and chronic disease*. Cambridge University Press, Cambridge.

Wells, J. C. K. (2003). The thrifty phenotype hypothesis: thrifty offspring or thrifty mother? *Journal of Theoretical Biology*, 221(1), 143–161. https://doi.org/10.1006/jtbi.2003.3183

Weston, D. A. (2012). 'Nonspecific infection in paleopathology: interpreting periosteal reactions', in A. L. Grauer (ed), *A Companion to Paleopathology*, pp. 492–512. Wiley-Blackwell, Oxford.

Whincup, P. H., Kaye, S. J., Owen, C. G., Huxley, R., Cook, D. G., Anazawa, S., Barrett-Connor, E., Bhargava, S. K., Birgisdottir, B. E., Carlsson, S., de Rooij, S. R., Dyck, R. F., Eriksson, J. G., Falkner, B., Fall, C., Forsén, T., Grill, V., Gudnason, V., Hulman, S., Hyppönen, E., Jeffreys, M., Lawlor, D. A., Leon, D. A., Minami, J., Mishra, G., Osmond, C., Power, C., Rich-Edwards, J. W., Roseboom, T. J., Sachdev, H. S., Syddall, H., Thorsdottir, I., Vanhala, M., Wadworth, M. & Yarbrough, D. E.

(2008). Birth weight and risk of type 2 diabetes. *Journal of the American Medical Association*, 300(24), 2886–2897. https://doi.org/10.1001/jama.2008.886

Widdowson, E. M. & McCance, R. A. (1975). A review: new thoughts on growth. *Pediatric Research*, 9(3), 154–156. https://doi.org/10.1203/00006450-197503000-00010

Widdowson, E. M. & Spray, C. M. (1951). Chemical development in utero. *Archives of Disease in Childhood*, 26(127), 205–214. https://doi.org/10.1136/adc.26.127.205

Wood, J. W., Milner, G. R., Harpending, H. C., Weiss, K. M., Cohen, M. N., Eisenberg, L. E., Hutchinson, D. L., Jankauskas, R., Cesnys, G., Katzenberg, M. A., Lukacs, J. R., McGrath, J. W., Roth, E. A., Ubelaker, D. H. & Wilkinson, R. G. (1992). The osteological paradox: Problems of inferring prehistoric health from skeletal samples [and comments and reply]. *Current Anthropology*, 33(4), 343–370. https://doi.org/10.1086/204084

Worthman, C. M. & Kuzara, J. (2005). Life history and the early origins of health differentials. *American Journal of Human Biology*, 17(1), 95–112. https://doi.org/10.1002/ajhb.20096

Yaussy, S. L. (2019). The intersections of industrialization: Variation in skeletal indicators of frailty by age, sex, and socioeconomic status in 18th- and 19th-century England. *American Journal of Physical Anthropology*, 170(1), 116–130. https://doi.org/10.1002/ajpa.23881

Yaussy, S. L. & DeWitte, S. N. (2018). Patterns of frailty in non-adults from medieval London. *International Journal of Paleopathology*, 22, 1–7. https://doi.org/10.1016/j.ijpp.2018.03.008

Yaussy, S. L., DeWitte, S. N. & Redfern, R. C. (2016). Frailty and famine: patterns of mortality and physiological stress among victims of famine in medieval London. *American Journal of Physical Anthropology*, 160(2), 272–283. https://doi.org/10.1002/ajpa.22954

Ządzińska, E., Lorkiewicz, W., Kurek, M. & Borowska-Strugińska, B. (2015). Accentuated lines in the enamel of primary incisors from skeletal remains: a contribution to the explanation of early childhood mortality in a medieval population from Poland. *American Journal of Physical Anthropology*, 157(3), 402–410. https://doi.org/10.1002/ajpa.22731

Zhao, H., Song, A., Zhang, Y., Zhen, Y., Song, G. & Ma, H. (2018). The association between birth weight and the risk of type 2 diabetes mellitus: a systematic review and meta-analysis. *Endocrine Journal*, 65(9), 923–933. https://doi.org/10.1507/endocrj.EJ18-0072

Zhou, L. & Corruccini, R. S. (1998). Enamel hypoplasias related to famine stress in living Chinese. *American Journal of Human Biology*, 10(6), 723–733. https://doi.org/10.1002/(SICI)1520-6300(1998)10:6<723::AID-AJHB4>3.0.CO;2-Q

CHAPTER 3

Acquired spinal conditions in humans: the roles of spinal curvature, the shape of the lumbar vertebrae, and evolutionary history

Kimberly A. Plomp, Ella Been, and Mark Collard

3.1. Introduction

Today, back pain is both common and often serious (Hoy et al., 2014; Muthuri et al., 2018). As many as two-thirds of people in Western countries experience back pain at some point in their lives (Balague et al., 2012; Gore et al., 2012; Webb et al., 2003), and it is thought to be the single greatest contributor to disability worldwide (Buchbiner et al., 2013; Maher et al. 2017; Murray & Lopez, 2013). Because of its prevalence and the fact that it is often debilitating, back pain has substantial economic impacts (Dieleman et al., 2016; Hong et al., 2013; Webb et al., 2003). It has been estimated to cost the UK as much as £12 billion per year in direct and indirect costs (Donaldson, 2008; Maniadakis & Gray, 2000). The equivalent figure in the USA is $90 billion (Davis, 2012). Given the individual and societal impacts of back pain, improving our understanding of its causes is one of the primary foci of back pain research.

Based on the fossil and bioarchaeological records, it is clear that many of the spinal conditions that afflict people today have a long history—in some cases, a very long history. Palaeopathological studies are limited to conditions that leave traces on skeletal remains, but this still leaves a range of acquired conditions that affect the human spine, including arthritis, intervertebral disc herniation, (IDH) and spondylolysis.

Two types of arthritis that affect the spine have been identified in ancient remains. Arthritis is a general term for inflammatory and/or degenerative conditions that affect joints. Arthritis of the vertebral bodies, or spondylosis, exists in human skeletons recovered from archaeological sites dating as far back as 341,000 BP (before present) (e.g. Chapman, 1962; Bourke, 1971; Jankauskas, 1992; Jurmain, 1990; Lovell, 1994; Maat et al., 1995; Rogers et al., 1985; Strouhal & Jungwirth, 1980). It has also been diagnosed in the remains of at least two extinct hominin species, identified on lower lumbar vertebrae of a 2.14 million year old *Australopithecus africanus* specimen, Stw 431 (Odes et al., 2017; Staps, 2002; but see D'Anastasio et al., 2009), and on the cervical vertebrae of the *Homo neanderthalensis* specimen from the site of La Chapelle-aux-Saints, which dates to around 60,000 BP (Trinkaus, 1985).

The other type of spine-affecting arthritis identified in ancient remains is osteoarthritis of the zygapophyseal joints. Osteoarthritis is an arthritic condition that affects only the synovial joints and, in the spine, involves the breakdown of the synovial joints that articulate one vertebra to the next. Like spondylosis, zygapophyseal osteoarthritis is common in human skeletons recovered from archaeological sites (e.g., Bridges, 1994; Gellhorn et al., 2013; Suri et al., 2011; Waldron, 1992; Zhang et al., 2017). It is generally diagnosed through the presence of

Kimberly A. Plomp et al., *Acquired spinal conditions in humans.* In: *Palaeopathology and Evolutionary Medicine.* Edited by Kimberly A. Plomp et al., Oxford University Press. © Oxford University Press (2022). DOI: 10.1093/oso/9780198849711.003.0003

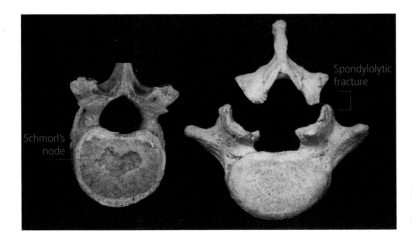

Schmorl's
node

Spondylolytic
fracture

Figure 3.1 Human vertebrae exhibiting a Schmorl's node (left) and spondylolysis (right).

eburnation, or bone polishing, on the joint surface or two or more of the following joint changes: osteophytes, joint contour change and/or porosity (Rogers & Waldron, 1995). So far, the La Chapelle-aux-Saints Neanderthal specimen provides the oldest evidence of zygapophyseal osteoarthritis in the hominin fossil record (Haeusler et al., 2019).

Evidence for IDH is also found in *Homo sapiens* skeletons from archaeological sites (Mays, 2006; Plomp et al., 2012; Šlaus, 2000; Üstündağ, 2009). IDH is a condition where the gel-like substance inside the intervertebral disc, known as the nucleus pulposus, prolapses through the fibrous layers of the disc, called the annulus fibrosis. It can be identified in archaeological skeletons when the disc herniates vertically because this leaves depressions on the vertebral endplate. These depressions are called 'Schmorl's nodes' (Figure 3.1) (Schmorl & Junghans, 1971). In the hominin lineage, the oldest undisputed Schmorl's nodes have been described in the aforementioned La Chapelle-aux-Saints 1 Neanderthal specimen (Haeusler et al., 2019).

Spondylolysis also afflicted people in the past (Mays, 2007; Merbs, 1983; Waldron, 1991). It can be identified in skeletons by a uni- or bilateral separation of the neural arch from the vertebral body at the site of the pars interarticularis; in archaeological human remains the fractured ends need to show evidence of healing to be sure that the fracture is not due to post-mortem damage (Figure 3.1). The oldest evidence of spondylolysis has been identified in a late Upper Palaeolithic skeleton from Italy,

Villabruna-1, which dates to 14,000 BP (Vercellotti et al., 2009, 2014).

A major hurdle in the prevention and treatment of back pain is our limited understanding of why, within a group of ostensibly similar people (i.e., same sex, age, ethnicity), some individuals suffer from back pain while others do not. Clinical studies have looked for patterns in suspected aetiological factors, including genetic predisposition, particular dietary choices, physical activities, and biochemical factors, but few patterns identified have been confirmed by subsequent studies (Hackinger et al., 2017; Nuckley et al., 2008; Nuki, 1999; Riyazi et al., 2005). In fact, to date, the only factor consistently linked to a future episode of back pain is a history of back pain (Stanton et al., 2008). Unfortunately, this means that we are not much closer to understanding the causes of many spinal pathologies than we were thirty years ago.

Back pain is a complex phenomenon: it can occur in any of the five regions of the spine, i.e. the cervical, thoracic, lumbar, sacral and coccygeal regions (Figure 3.2) (Binder, 2007; Katz et al., 2003; Hoy et al., 2014; Manchikanti et al., 2002, 2004; Muthuri et al., 2018; Wild et al., 2006), and it can be chronic or acute (Patrick et al., 2014); have congenital, acquired, or idiopathic causes (Dolan et al., 2013; Giesecke et al., 2004; Modic, 1999; Taskaynatan et al., 2005); and involve soft tissue and/or bone (Dar et al., 2009; Hackinger et al., 2017; Manchikanti et al., 2002, 2004; Martin et al., 2002; Modic, 1999; Nuckley et al., 2008; Nuki, 1999; Riyazi et al.,

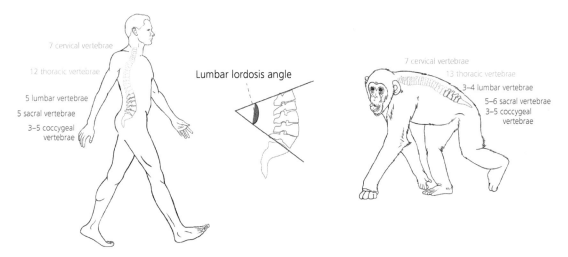

Figure 3.2 An illustration of the five regions (showing modal numbers of vertebrae in each region) and curve of a human spine (left) and chimpanzee spine (right), as well as the lumbar lordosis angle, which is calculated as the angle made between two lines, one running parallel to the superior endplate of L1 and the other running parallel to the inferior endplate of L5 (centre).

2005). This chapter focuses on acquired spinal diseases (ASDs), which are conditions of the spine that develop throughout life either through degeneration or trauma, including arthritis, IDH, and spondylolysis.

We opted to focus on ASDs, to the exclusion of lesions related to infections or development issues, because these conditions have been suggested to afflict humans due to mismatches between our spinal anatomy and our environment, and/or as trade-offs for the ability to walk on two legs (Castillo & Lieberman, 2015; Filler, 2007; Latimer, 2005; Plomp et al., 2015a). Specifically, obligate bipedalism has long been suspected to be an important aetiological factor for ASDs that afflict our species because of the types of stresses it puts on our spines (Been et al., 2019; Castillo & Lieberman, 2015; Filler, 2007; Jurmain, 2000; Keith, 1923; Latimer, 2005; Merbs, 1996; Plomp et al., 2015a). This hypothesis is based partly on the fact that humans experience ASDs far more frequently than non-human hominoids (Filler, 2007; Jurmain, 1989; Lovell, 1990; Lowenstine et al., 2016). For example, spondylosis (please note this is *not* osteoarthritis), also known as vertebral osteophytosis, is reported to affect 48–95% of humans (Cvijetić et al., 2000; Muraki et al., 2009; O'Neill et al., 1999; Prescher, 1998; Sarzi-Puttino et al., 2005). In contrast, spondylosis has

been found to affect only 4% of gorillas, 5% of bonobos, and 2% of chimpanzees in non-human skeletal collections (Jurmain, 2000). Similarly, vertical IDH has been estimated to affect about 50% of modern humans with Western lifestyles but only 2% of chimpanzees and orangutans (Dar et al., 2009; Lovell, 1990). Spondylolysis is unique to humans and does not naturally occur in other animals (Ward et al., 2007).

A number of empirical studies published in the last twenty years have investigated the hypothesised relationship between bipedalism and ASDs (e.g. Been et al., 2019; Masharawi, 2012; Masharawi et al., 2007; Meakin et al., 2008, 2009; Meyer, 2016; Plomp et al., 2015a, 2019a; Scannell & McGill, 2003; Ward & Latimer, 2005; Ward et al., 2007). Collectively, these studies suggest that the relationship is mediated by the nature of the curvature of the spinal column (Been et al., 2019; Meakin et al., 2008). The various studies also suggest that the relationship is influenced by characteristics of the individual vertebrae (Masharawi, 2012; Masharawi et al., 2007; Meakin et al., 2009; Meyer, 2016; Plomp et al., 2015a, 2019a; Scannell & McGill, 2003; Ward & Latimer, 2005; Ward et al., 2007, 2010). The lumbar vertebrae are particularly important in this regard since the incidence of ASD is much higher in the lumbar region of the spine than in the cervical and thoracic regions (Battie et al., 2009; Sparrey et al., 2014), a fact that has led the lumbar region to be called 'the evolutionary weak point' of the human spine (Sparrey et al., 2014, p. 4).

In this chapter, we discuss how the overall shape of the spine (as a column) and features of the lumbar vertebrae may mediate the relationship between bipedalism and some of the most common ASDs suffered by humans. We begin by explaining how spinal shape and the shape of the lumbar vertebrae relate to bipedal posture and locomotion. Next, we outline the findings of clinical studies that have found a relationship between the shapes of the spine and the lumbar vertebrae and the presence of ASDs. Subsequently, we outline palaeopathological and comparative anatomical data that also suggest that spinal curvature and the characteristics of the lumbar vertebrae impact the propensity to develop ASDs. Thereafter, we discuss recent research that suggests that the pathology-linked shapes can be understood in terms of the evolutionary history of our lineage. In the sixth section of the paper, we discuss potential biomechanical explanations for the hypothetical link between the lumbar vertebrae characteristics and IDH and spondylolysis. Finally, we close with a discussion of some potential future research directions.

3.2. Adaptations for bipedalism in the human vertebral column and lumbar vertebrae

When the human vertebral column is considered as an anatomical unit, there are two main features that are thought to be adaptations for bipedalism. One is its distinctive pattern of curvature. While great apes have a C-shaped spine, healthy adult humans have a sinuous spine (Figure 3.2; Box 3.1). The other major feature of the human vertebral column that is thought to be an adaptation for bipedalism is the number of vertebrae in the different regions of the spine (Lovejoy, 2005; Williams, 2012; Williams et al., 2013). Individuals of all the hominoid clade (and most mammals) usually have seven cervical vertebrae, but there is variation in the modal number of thoracic, lumbar, and sacral vertebrae among species (Box 3.2).

Box 3.1   The shape of the sinuous spine

The shape of the human spine is a consequence of the four pre-coccygeal regions (i.e., before the coccyx) of the spinal column having different curves (Abitbol, 1995; Been et al., 2010a, 2017; Keith, 1923; Lovejoy, 2005; Schultz, 1961; Shapiro, 1993a; Ward and Latimer, 2005; Whitcome et al., 2007). The neck or cervical region exhibits lordosis, which is a backward curvature. This results from the intervertebral discs being dorsally wedged (i.e., the discs are craniocaudally shorter at their dorsal border than at their ventral border) (Been et al., 2010a). In contrast, the upper back or thoracic region exhibits kyphosis, which is a forward curvature. This is due to ventral wedging of the vertebral bodies (i.e., the thoracic vertebrae are craniocaudally shorter at their ventral border than at their dorsal border) (Latimer and Ward, 1993). The lower back or lumbar region, like the cervical region, exhibits lordosis. Unlike in the cervical region, however, the lordosis of the lumbar region results from dorsal wedging of both the intervertebral discs and the vertebral bodies (i.e., both the discs and the vertebrae are craniocaudally shorter at their dorsal border than at their ventral border) (Been et al., 2010a). The caudal or inferior-most region of the spinal column, which is formed by the sacrum and coccyx, has a kyphotic curve. This curve results from ventral wedging of the second to fifth sacral vertebrae and all the coccygeal vertebrae and is enhanced by a ventral tilt of the cranial end of the sacrum (Antoniades et al., 2000; Cheng and Song, 2003). The four curves of the human spine are widely accepted to be functionally important (Been et al., 2010a; Latimer and Ward, 1993). They bring the centre of gravity of the body above the hips, unlike it being located ventrally in quadrupeds, and therefore allow the trunk to be balanced above the legs during bipedal walking (Latimer and Ward, 1993; Whitcome et al., 2007). The lumbar curve is particularly important in this regard (Been et al., 2010a, 2019; Latimer and Ward, 1993; Whitcome et al., 2007).

Box 3.2   Numbers of vertebrae in great apes and humans

Humans generally have twelve thoracic, five lumbar, five sacral, and three to five coccygeal vertebrae (Williams et al., 2016). The vertebral formula in great apes varies more between individuals than is typical in humans. Chimpanzees and gorillas typically have thirteen thoracic, three to four lumbar, five to six sacral, and three to five coccygeal vertebrae, while the equivalent figures for bonobos are thirteen to fourteen, three to four, six to seven, and three to five, respectively (Williams et al., 2016). Orangutans usually have twelve thoracic vertebrae, four lumbar vertebrae, five sacral and four to six coccygeal vertebrae (Williams et al., 2016). Thus, humans tend to have a longer lumbar region than the great apes. This has been argued to result in an increased range of motion for flexion and extension (Schultz, 1953; Williams, 2012), which is especially important for maintaining the lumbar lordosis. In addition, it has been proposed that the increased space between the ribcage and the pelvis created by a longer lumbar spine, along with craniocaudally shortened iliac blades, allows for counter-rotation of the trunk relative to the hips, which helps to maintain balance during bipedal walking and running (Bramble and Lieberman, 2004; Williams et al., 2019).

Many of the traits that distinguish the lumbar vertebrae of humans from those of the great apes appear to relate to facilitating and maintaining lumbar lordosis and an upright posture. For example, the orientation of the zygapophyseal facets (Figure 3.3) is thought to be linked to vertebral slippage and rotation in the context of posture and gait (Shapiro, 1993a; Whitcome, 2012). In great apes, the zygapophyseal facets of the lumbar vertebrae are obliquely oriented, while in humans these facets are oriented more towards the sagittal plane, especially in the upper lumbar vertebrae, which has been hypothesised to resist rotation and maintain lumbar lordosis (Ahmed et al., 1990; Been et al., 2010a, Jaumard et al., 2011; Shapiro, 1993a). In addition, in humans, the distance between the zygapophyseal facets gradually increases as one moves down the lumbar spine (Latimer & Ward 1993). This has been suggested to allow for lumbar lordosis without the facets of one vertebra impinging upon the laminae or pars interarticularis of the next vertebra (Latimer & Ward, 1993; Ward & Latimer, 2005; Ward et al., 2007). Also, a larger vertebral foramen would result in the larger inter-facet distances that allow for lumbar lordosis (Latimer & Ward, 1993).

The size and orientation of the transverse processes and the attached muscles of the lumbar vertebrae also seem to play an important role in maintaining

lumbar lordosis (Figure 3.3). In particular, the transverse processes of human lumbar vertebrae are shorter and more dorsally orientated than those of the great apes (Bastir et al., 2017; Jellema et al., 1993; Latimer & Ward, 1993; Plomp et al., 2019a; Ward et al., 2012; Williams & Russo, 2015). This dorsal projection results in greater invagination of the vertebral column (i.e., a ventral curve of the spinal column forward in the thorax) (Jellema et al., 1993; Latimer & Ward, 1993; Ward et al. 2012), which means that the spine is positioned forward in the thorax (Been et al. 2010a; Bogduk et al., 1992; Filler, 2007; Gómez-Olivencia et al., 2017; Sanders, 1998; Shapiro, 1993a, 2007; Whitcome et al., 2007). This increases the length of the lever arms of the erector spinae musculature and increases their ability to extend the spine, resist lateral flexion, and maintain lumbar lordosis during bipedal posture and gait (Argot, 2003; Been et al., 2010a; Benton, 1967; Gómez-Olivencia et al., 2017; Jellema et al., 1993; Latimer & Ward, 1993, 2005; Sanders, 1998; Sanders & Bodenbender, 1994; Shapiro, 1993a, 1995; Ward, 1993; Ward et al., 2012).

Several traits that distinguish the spinous processes of human lumbar vertebrae from those of great apes have likewise been argued to facilitate lumbar lordosis (Figure 3.3). In particular, the spinous processes of human lumbar vertebrae are dorsoventrally shorter (Gómez-Olivencia et al., 2013; Latimer & Ward, 1993; Meyer, 2016; Meyer et al., 2017; Plomp et al., 2019a; Schultz, 1961; Ward, 1991) and have craniocaudally short (or pinched) tips (Plomp et al., 2019a). The shortness of the spinous processes has been hypothesised to decrease the lever arms of the spinal extensor muscles and therefore limit the sagittal mobility of the spine (Argot, 2003; Gómez-Olivencia et al., 2017; Meyer, 2016; Sanders, 1998; Shapiro, 1993a, 2007; Shapiro & Kemp, 2019; Ward, 1991). The craniocaudal pinching of the processes' tips has been suggested to facilitate lumbar lordosis by increasing the spacing between the spinous processes of subjacent vertebrae (Cartmill & Brown, 2017; Erikson, 1963; Gambaryan, 1974; Plomp et al., 2019a; Ritcher, 1970; Shapiro, 1993a).

There are four other traits that differentiate the human lumbar spine from the lumbar spines of the great apes. First, the bodies of the lumbar vertebrae

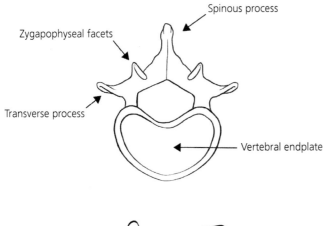

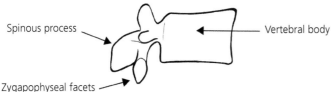

Figure 3.3 Illustration of a typical human lumbar vertebrae showing the terminology and location of vertebral elements.

of humans are dorsoventrally deeper than those of great apes (Hernandez et al., 2009; Latimer & Ward, 1993; Meyer & Williams, 2019; Plomp et al., 2015a, 2019a; Robinson, 1972). Second, the endplates of the lumbar vertebrae of humans are more heart-shaped (i.e., shorter at the midpoint of the sagittal plane compared to the coronal plane) than those of great apes, whose vertebral bodies are more circular in shape (Plomp et al. 2015a, 2019a; Robinson, 1972). Third, the vertebral bodies gradually increase in width from the first to the fifth human lumbar vertebrae (Rose, 1975; Schultz, 1953, 1961). Last, the pedicles of the last two lumbar vertebrae in the human spine are wider than those of the great apes (Figure 3.3) (Been et al., 2010b; Briggs et al., 2004; Davis, 1961; El-Khoury & Whitten, 1993; Panjabi et al., 1993; Sanders & Bodenbender, 1994; Shapiro, 1993a, 1993b; Whyne et al., 1998). All four of these traits have been hypothesised to help the vertebrae withstand the compressive load acting on the lower spine (Been et al., 2010b; Briggs et al., 2004; Davis, 1961; El-Khoury & Whitten, 1993; Hernandez et al., 2009; Latimer & Ward, 1993; Panjabi et al., 1993; Plomp et al., 2012, 2015a,b, 2019b; Rose, 1975; Shapiro, 1991, 1993a; Sanders & Bodenbender, 1994; Whyne et al., 1998).

3.3. Clinical evidence for an impact of spinal and vertebral shape on spinal health

Many of the clinical studies that have investigated the relationship between vertebral shape and spinal health have focused on lumbar lordosis (e.g., Been & Kalichman, 2014; Been et al., 2019; Keller et al., 2005; Scannell & McGill, 2003; Zlolniski et al., 2019). The lordotic angle has been particularly important in these studies. Measured between a line running parallel to the superior endplate of the first lumbar vertebra and a line running parallel to the first sacral endplate (Figure 3.2), this angle is associated with lumbar lordosis such that a large lordotic angle corresponds to a more pronounced lumbar lordosis, whereas a small lordotic angle equals a less pronounced lumbar lordosis. The size of the lordotic angle is highly variable in humans (Been & Kalichman, 2014; Zlolniski et al., 2019) and this variation is associated with the propensity to develop ASDs (Been et al., 2019; Keller et al., 2005; Scannell & McGill, 2003).

One ASD that has been linked with the lordotic angle is osteoarthritis of the zygapophyseal joints. Osteoarthritis is a breakdown of synovial joints, which in the spine are the zygapophyseal and cos-tovertebral joints. Clinically, osteoarthritis preferentially affects individuals with pronounced lumbar lordosis (Roussouly & Pinheiro-Franco, 2011). Its occurrence in the lumbar spine also seems to correlate with zygapophyseal facets that are more sagittally oriented than in healthy individuals (Fujiwara et al., 2001), which may be related to the increased lordosis. Based on these clinical findings, researchers have proposed that a more-pronounced-than-normal lumbar lordosis results in both increased contact between the vertebral facets and a greater amount of shear force acting on the joints, and that this increases the likelihood of the joints breaking down and developing osteoarthritis (Roussouly & Pinheiro-Franco, 2011; Weinberg et al., 2017).

Clinical studies have also suggested that a large lordotic angle may contribute to spondylolysis, which is a cleft in the neural arch caused by a fatigue fracture at the site of the pars interarticularis (Hu et al., 2008). It is particularly common in athletes (Iwamoto et al., 2004), with one study of 100 American adolescent athletes with low back pain finding that 47% had spondylolysis (Micheli & Wood, 1995). Using clinical radiographs, Roussouly and colleagues (2006) found that unusually pronounced lordosis was associated with spondylolysis. In a similar vein to the aforementioned explanations for spinal osteoarthritis, they proposed that a large lordotic angle increases the direct contact between the neural arches of the lumbar vertebrae and ultimately causes the fractures that lead to spondylolysis (Roussouly et al., 2006).

Spondylolysis also has been linked with the shape of the zygapophyseal facets. Specifically, it has been found that the facets of the L4 and L5 vertebrae of individuals with spondylolysis tend to be flatter, more coronally oriented and smaller in the transverse direction than those of individuals without spondylolysis (Grobler et al., 1993; Miyake et al., 1996; Van Roy et al., 2006). As mentioned, the size, shape and orientation of the vertebral facets are associated with the curvature of the spine (Shapiro, 1993a; Whitcome, 2012). In the lumbar spine, the zygapophyseal facets are oriented towards the sagittal plane and become increasingly coronally oriented moving down the lumbar spine, which likely helps to maintain lumbar lordosis (Ahmed et al., 1990; Been et al., 2010a, Jaumard et al., 2011; Latimer & Ward, 1993; Shapiro, 1993a). On this basis, it is thought that the flatness and exaggerated coronal orientation of the facets identified in L4 and L5 vertebrae with spondylolysis may not provide adequate facilitation for, and may instead restrict, the large lordotic angle that is also associated with the condition (Roussouly et al., 2006).

While a number of clinical studies suggest that having an unusually pronounced lumbar lordosis may increase the likelihood of developing zygapophyseal osteoarthritis and spondylolysis, there is also clinical evidence that having a smaller than normal lordosis may negatively impact an individual's spinal health. Several papers have reported that people with evidence of degenerative disc disease and IDH have significantly smaller lordotic angles than those with healthy spines (Barrey et al., 2007; Ergun et al., 2010; Yang et al., 2014). Specifically, these studies have found that individuals with degenerative changes to their discs had an average lordotic angle of 40° while those with disc herniations had an average lumbar lordosis angle of 37° (Endo et al., 2010; Sak et al., 2011; Yang et al., 2014). Both of these angles are considerably smaller than the average lumbar lordosis angle for individuals with healthy lumbar spines. Analyses by Been and colleagues (2010a) and Yang et al. (2014) indicate that the average lordotic angle in healthy humans is 51–53°.

Clinical studies have identified two other traits that appear to be correlated in humans with IDH, a condition where the gel-like substance inside the intervertebral disc prolapses through the fibrous layers of the disc. One of the traits was identified by Harrington and colleagues (2001), who used computed tomography (CT) scans of 97 patients to measure vertebral endplate dimensions and found that individuals with IDH tended to have endplates that are more circular in shape, compared to the more heart-shaped endplate in healthy vertebrae. The other trait was recognised by Pfirrmann and Resnick (2001), who performed an analysis of thoracic and lumbar vertebrae and intervertebral discs from 128 cadavers and discovered that intervertebral disc hernias affected vertebrae with flatter endplates significantly more frequently than vertebrae with more concave endplates.

Based on these clinical studies, it appears that lumbar lordosis plays an important role in spinal health, with a more-pronounced-than-normal lordotic angle possibly leading to spondylolysis and a less-pronounced-than-normal angle potentially increasing the likelihood of IDH. Been and colleagues (2019) proposed what they called the 'Neutral Zone Hypothesis' to explain this pattern. They argued that there is a 'neutral zone' for the lordotic angle in the human spine and that deviations from this zone increase the chances of developing spinal pathologies. They based the neutral zone on a previous study in which members of the same team had calculated the average angle of healthy human spines to be about 51° (Been et al., 2010a). They argued that an individual with a lordosis angle that is at least 10° lower or higher than the average of 51° is at risk of developing spinal pathologies (Been et al., 2019).

3.4. Palaeopathological and comparative anatomical evidence for an impact of spinal and vertebral shape on spinal health

In a similar vein to the clinical studies outlined in Section 3.3, palaeopathological data has been used to investigate whether vertebral shape variation is an aetiological factor in the development of Schmorl's nodes. Plomp and colleagues (2012, 2015b) compared the 2D shape of the superior planar surface of thoracic and lumbar vertebrae with and without Schmorl's nodes from medieval and post-medieval English populations. They found that human vertebrae with Schmorl's nodes differed in shape from those without the lesions. Their analyses indicated that, based on shape traits, vertebrae with Schmorl's nodes could be identified with an accuracy rate of 69–81%. Given these findings, the authors proposed that vertebral shape may be an important factor in the aetiology of IDH and therefore in the development of Schmorl's nodes. They were able to confirm that the superior endplates of vertebrae with Schmorl's nodes tended to be more circular than those of healthy vertebrae. This

aligns with the clinical study performed by Harrington and colleagues (2001). In addition, Plomp and colleagues (2012, 2015b) found that vertebrae with Schmorl's nodes have shorter pedicles and laminae and smaller vertebral foramina than human vertebrae without Schmorl's nodes.

Several palaeopathological and comparative studies investigating the relationship between the condition and vertebral shape have focused on spondylolysis. Similar to the clinical evidence, one study concluded that an unusually high lordotic angle makes individuals more susceptible to developing spondylolysis (Roussouly et al., 2006). Masharawi and colleagues (2007) compared the dorsal and ventral heights of L5 and found that spondylolytic vertebrae tend to have bodies that are more dorsally wedged (i.e., the ventral border of the vertebral body is craniocaudally taller than the dorsal border) than healthy vertebrae. More pronounced wedging should result in a larger lordotic angle, so this result aligns with the findings of Roussouly and colleagues' (2006) clinical study. In addition, Ward and colleagues (2005, 2007, 2010) found a correlation between spondylolysis and a trait that is thought to be related to lumbar lordosis—specifically, that individuals with spondylolysis tend to have reduced transverse spacing between the zygapophyseal facets of adjoining vertebrae compared to those without spondylolysis. Transverse spacing between the facets increases as one moves down the human lumbar spine and this is thought to allow for lumbar lordosis. Given this, Ward and colleagues (2007, 2010) hypothesised that reduced mediolateral spacing leads to the articular processes of one vertebra directly contacting the pars interarticularis of the subjacent one, leading to spondylolysis (Ward & Latimer, 2005; Ward et al., 2007, 2010).

3.5. Evolutionary origins of vertical intervertebral disc herniation and spondylolysis

Based on the results of the clinical, comparative, and palaeopathological studies that have been carried out to date, it appears that variation in vertebral shape influences an individual's propensity to develop a number of ASDs. So, why do some people have vertebral shapes that predispose them to such conditions? Three recent studies have attempted to answer this question by investigating the evolutionary origins of vertebral shape variation in relation to two lesions that have been linked with vertebral shape in clinical, comparative, and palaeopathological studies: Schmorl's nodes and spondylolysis (Plomp et al., 2015b, 2019b, 2020).

Plomp and colleagues (2015a) compared the 2D shape of human final thoracic and first lumbar vertebrae from archaeological populations with those of chimpanzees and orangutans. They used chimpanzee and orangutan vertebrae as comparators to identify evolutionary traits because although it is accepted that humans and chimpanzees shared a common ancestor to the exclusion of other apes, the locomotor behaviour of the common ancestor is debated. The most popular hypotheses are that the ancestor used either knuckle-walking similar to modern chimpanzees, bonobos, and gorillas, or quadrumanous climbing (i.e., using all four feet to grasp branches), such as performed by modern orangutans (Richmond et al., 2001; Thorpe et al., 2007). Plomp et al. (2015a) divided the humans into two groups: one comprising individuals who had Schmorl's nodes, and the other consisting of individuals with no visible spinal pathologies. They found that the vertebrae of humans with Schmorl's nodes were closer in shape to those of chimpanzees than were the healthy human vertebrae. Human vertebrae with Schmorl's nodes and chimpanzee vertebrae were found to have more circular vertebral bodies; shorter, narrower pedicles; and relatively smaller vertebral foramina. The authors argued that because chimpanzees, bonobos and modern humans share an ancestor to the exclusion of all other living species, any vertebral traits that chimpanzees and humans have in common are most likely inherited from their common ancestor. Given this, they asserted, it is reasonable to suppose that humans with Schmorl's nodes experience IDH because their vertebrae were closer to the ancestral shape for the hominin lineage. This ancestral shape, they continued, is not as well adapted to withstand the stresses placed on the spine during bipedalism and thus, increases the likelihood of disc

herniations. Plomp and colleagues (2015a) called this the 'Ancestral Shape Hypothesis'.

Subsequently, Plomp and colleagues (2019b) tested the Ancestral Shape Hypothesis with 3D shape data from the last two thoracic and first lumbar vertebrae of modern humans with and without Schmorl's nodes, chimpanzees and several extinct hominins. As before, they found that modern human vertebrae with Schmorl's nodes shared more similarities in shape with chimpanzee vertebrae than did healthy modern human vertebrae. They also found that the human vertebrae with Schmorl's nodes were closer in shape to the vertebrae of a number of extinct hominins, including Sts 14 (*Australopithecus africanus*), MH1 (*Australopithecus sediba*), SK 853 (*Paranthropus robustus*), UW 101-1733 (*Homo naledi*), Kebara 1 (*H. neanderthalensis*), and Kebara 1 (*H. neanderthalensis*), than were the healthy human vertebrae. They argued that these results provide further support for the Ancestral Shape Hypothesis because they demonstrate that the human vertebrae with Schmorl's nodes do indeed lie towards the ancestral end of the range of shape variation in *H. sapiens*.

The following year, Plomp and colleagues (2020) investigated whether the propensity to develop another spinal condition is affected by individuals' location in the ancestral-to-highly derived spectrum of vertebral shape variation. Building on the putative link between spondylolysis and large lumbar lordosis (Masharawi et al., 2007; Roussouly et al., 2005; Ward et al., 2005, 2007), they hypothesised that spondylolytic vertebrae may have the opposite shape problem to those with Schmorl's nodes. Such vertebrae may, they suggested, exhibit shape traits that are exaggerated adaptations for bipedalism. They called this the 'Overshoot Hypothesis' (Plomp et al., 2020).

To test this, they compared the 3D shape of final lumbar vertebrae of humans, chimpanzees, gorillas, and orangutans. The humans were divided according to whether they had bilateral spondylolysis, Schmorl's nodes on any vertebrae in the spine, or no vertebral lesions. The authors found that, as predicted, the spondylolytic human vertebrae shared fewer similarities in shape with those of great apes than did the healthy human vertebrae. Again, as predicted, they found that vertebrae of humans with Schmorl's nodes sat on the opposite end of the range of variation from the spondylolytic human vertebrae, and that healthy human vertebrae fell between the two groups of pathological vertebrae. Plomp and colleagues (2020) argued that this means that spondylolytic vertebrae show fewer similarities in shape with the vertebrae of great apes than do either healthy human vertebrae or those with Schmorl's nodes. Furthermore, the main ways that spondylolytic vertebrae differ from the other three groups of vertebrae is that they tend to have vertebral bodies that are more dorsally wedged, narrower inter-pedicle distances, more dorsally projecting pedicles, and narrower inter-facet distances. Plomp and colleagues (2020) concluded that their results support the Overshoot Hypothesis.

The findings of these comparative studies suggest that the prevalence of some important ASDs in modern humans is partially explained by the evolution of vertebral shape variation and how well different vertebral shapes withstand the stresses placed upon the bipedal spine. They imply that

we can visualise human vertebral shape variation as a bell-shaped distribution (Figure 3.4). Vertebrae at one end of the distribution display traits that are similar to those of chimpanzee vertebrae and, by extension, the vertebrae of the common ancestor of the hominins. These vertebrae are prone to one type of ASD: IDH. Vertebrae at the other end of the bell-shaped distribution are characterised by traits that are basically exaggerated versions of some of the key vertebral adaptations for bipedalism in humans. These vertebrae are prone to a different type of ASD: spondylolysis. Between the two extremes are vertebrae that are at, or close to, the lineage-specific optimal shape for bipedalism and, therefore, have a lower probability of developing spinal pathologies in response to the stresses of bipedal posture and gait. Importantly, this hypothesis implies that where an individual's vertebral shape sits on this evolutionary spectrum likely influences their spinal health. This hypothesis, which we will refer to hereafter as the 'Evolutionary Shape Hypothesis', clearly overlaps with

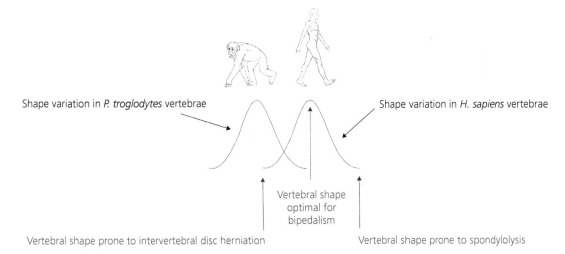

Shape variation in *P. troglodytes* vertebrae

Shape variation in *H. sapiens* vertebrae

Vertebral shape optimal for bipedalism

Vertebral shape prone to intervertebral disc herniation

Vertebral shape prone to spondylolysis

Figure 3.4 The logic of the Evolutionary Shape Hypothesis for back pain. The distribution of vertebral shape variation within humans can be conceptualised as a bell-curve with an ancestral end (left) and a derived end (right). According to the hypothesis, where an individual's vertebral shape sits within this distribution has an important influence on their spinal health. At the centre of the range of variation are vertebrae that have the lineage-specific optimal shape for bipedalism and, therefore, a lower probability of developing spinal disease in response to the stresses of bipedal posture and gait. At the ancestral end, vertebrae differ little from those of chimpanzees and by extension from those of the common ancestor of humans and chimpanzees. People with vertebrae that fall in this part of the distribution have a heightened probability of developing vertical intervertebral disc herniation that can lead to Schmorl's nodes. At the other, highly derived end of the range of variation, vertebrae exhibit exaggerated versions of our species' vertebral adaptations for bipedalism. Individuals with vertebrae that fall in this part of the distribution are more prone to develop the fatigue fractures that cause spondylolysis.

Been and colleagues (2019) Neutral Zone Hypothesis (Section 3.3). Been and colleagues (2019) neutral zone of spinal curvature corresponds to the area under the middle of the bell-curve in Figure 3.4—the area where individuals have vertebrae that are at, or close to, the lineage-specific optimum shape for facilitating bipedal posture and gait.

3.6. Potential biomechanical explanations for links between vertebral traits and acquired spinal diseases

Table 3.1 summarises the vertebral traits found to correlate with the presence of IDH as indicated by Schmorl's nodes, and the vertebral traits that have been found to correlate with the occurrence of spondylolysis. This section outlines some biomechanical hypotheses that attempt to explain how the two sets of traits give rise to the pathologies.

To begin with, vertebrae with Schmorl's nodes tend to have bodies that are less dorsally wedged than the bodies of healthy vertebrae (Plomp et al., 2019a). In principle, this should result in a smaller lordotic angle and therefore a straighter spine (Been et al., 2019). Having a less-pronounced lordotic angle can be expected to influence how the spine absorbs compressive loads during bipedalism (Farfan, 1995; Gracovetsky & Iacono, 1987; Whitcome et al., 2007), as well as increase the load acting on the intervertebral discs (Wei et al., 2013). A straighter lumbar spine should result in a more ventral placement of compressive loads, meaning that most of the loading would occur on the vertebral bodies and intervertebral discs, rather on the bodies, discs, and neural arches (Adams et al., 1994). This may result in an increased propensity to develop disc herniations, because biomechanical tests showed that herniations often occur when the discs are subjected to both compressive and shearing forces (Cholewicki & McGill, 1996). In other words, lumbar vertebrae with bodies that are less dorsally wedged would result in a smaller lordotic angle, which may not be biomechanically well suited to withstand the compressive loading placed on the lower spine during bipedalism (Plomp et al., 2019a).

Another Schmorl's nodes-associated trait of the vertebral body may influence how the vertebrae withstand compressive loads. Lumbar vertebrae with Schmorl's nodes have been found to have more circular vertebral bodies compared to healthy vertebrae, which are more heart-shaped (Plomp et al., 2015a, 2019b). This is significant because more circular endplates can be expected to have a larger diameter compared to more heart-shaped endplates, and increased disc diameter has been argued to foster disc herniation. The explanation for this is LaPlace's Law (Letić, 2012), which states that the ability of a fluid-filled tube to withstand tension decreases with increasing radius. This means that an intervertebral disc/vertebral body that is more circular may be less able to withstand the compressive loads acting

Table 3.1 Summary of the vertebral traits associated with Schmorl's nodes and spondylolysis

	Compared to healthy human vertebrae
Vertebrae with Schmorl's nodes have:	Vertebral bodies that are less dorsally wedged Vertebral bodies that are more circular compared to healthy vertebrae, which are more the heart-shaped Pedicles and laminae tend to be short Transverse processes that are longer and project more laterally Spinous processes that are longer and more cranially oriented Vertebral bodies that are more dorsally wedged resulting in a more pronounced lordotic angle
Vertebrae with spondylolysis have:	Pedicles that project more dorsally Narrower inter-pedicle distances Zygapophyseal facets that are more caudally located Narrower inter-facet distances Inferior endplates with deeper concavities

on the bipedal spine than a heart-shaped vertebral body (Harrington et al., 2001; Plomp et al., 2012, 2015a).

The pedicles and laminae also play an important role in withstanding compressive loads by acting as structural buttresses for the vertebral body (Adams et al., 1994; El-Khoury & Whitten, 1993; Whyne et al., 1998). Plomp and colleagues (2012, 2015a, 2019b) found that the pedicles and laminae of lumbar vertebrae with Schmorl's nodes tended to be relatively short, and suggested that this may make them less able to buttress against loads than the pedicles and laminae of healthy human vertebrae.

The transverse process traits found to be associated with Schmorl's nodes probably relate to the stability of the lumbar spine rather than withstanding loads. For example, comparative analyses have found that mediolaterally longer transverse processes allow for lateral flexion in the lower spine (Argot, 2003; Sanders, 1998; Shapiro, 1993a). Thus, the longer, laterally projecting transverse processes identified in vertebrae with Schmorl's nodes may not provide adequate stability during bipedalism. In addition, transverse processes that project more laterally may be less able to maintain lumbar lordosis than those that project dorsally (Been et al., 2017; Bogduk et al., 1992; Filler, 2007; Sanders, 1998; Sanders & Bodenbender, 1994; Whitcome et al., 2007). Given this, the longer, laterally projecting transverse processes of vertebrae with Schmorl's nodes may increase dorsomobility and result in a lumbar spine that is less stable during bipedalism. The longer, cranially oriented spinous processes of humans with Schmorl's nodes may also cause spinal instability. Specifically, it has been argued that long cranially oriented spinous processes may allow for a greater amount of dorsal mobility in the spine, while short, caudally oriented spinous processes are associated with a less mobile and more stable spine (Argot, 2003; Gómez-Olivencia et al., 2017; Meyer, 2016; Sanders, 1998; Sanders & Bodenbender, 1994; Shapiro, 1993a, 1995, 2007; Shapiro et al., 2005; Ward, 1991). Considering this, Plomp and colleagues (2019b) suggested that the longer, cranially oriented spinous processes of people with Schmorl's nodes may predispose individuals to IDH.

Regarding the traits associated with spondylolysis, Plomp and colleagues (2020) found that L5 vertebrae with spondylolysis have more pronounced dorsal wedging, which can be expected to result in a hyper-lordotic lumbar spine (Masharawi et al., 2007; Plomp et al., 2020). Exaggerated lordosis has been suggested to increase direct contact between the neural arches of the lumbar vertebrae and this has been posited to result in the fatigue fractures that cause spondylolysis (Masharawi et al., 2007). The increased contact between the neural arches is exacerbated by four other traits associated with spondylolysis: narrower inter-pedicle distances, dorsal projection of the pedicles, narrower inter-facet distance, and caudally located zygapophyseal facets (Plomp et al., 2020; Ward et al., 2005, 2007, 2010). The first three of these traits can be expected to result in a mediolaterally narrower neural arch width and therefore, in smaller inter-facet distances (Plomp et al., 2020; Ward et al., 2005, 2007, 2010). Unlike quadrupedal apes, humans have a consecutive increase in inter-facet distances from the top to the bottom of the lumbar spine, which is thought to allow for adequate spacing between the facets in the presence of lumbar lordosis (Latimer & Ward, 1993). Thus, it has been hypothesised that the reduced inter-facet spacing caused by the mediolaterally narrower neural arch width may lead to the articular processes of one vertebra directly contacting the pars interarticularis of the subjacent vertebra, causing the fatigue fracture and, ultimately, spondylolysis (Plomp et al., 2020; Ward et al., 2005, 2007, 2010). The fourth trait—caudally located zygapophyseal facets—can be expected to result in crowding of those joints (Plomp et al., 2020). This has been suggested to increase the contact between the inferior facets of L4 and the pars interarticularis of L5, causing fatigue fractures that can eventually lead to spondylolysis (Plomp et al., 2020).

The last trait associated with spondylolytic vertebrae is a deeper concavity of the inferior endplate (Plomp et al., 2020). Based on clinical studies, this trait may influence the ability of the vertebrae to disperse compressive stress. For example, Liu and colleagues (2007) found that vertebrae with endplates with shallower concavities were better suited to withstand compressive strains and this led to a decrease in the amount of stress placed on the zygapophyseal facets and neural arch. With this in mind, Plomp and colleagues (2020) posited that the increased concavity of vertebral bodies may result in increased loading placed on the posterior vertebral elements, including the pars interarticularis, and ultimately increase the risk of developing spondylolysis.

3.7. Future directions

There are a number of obvious potential avenues for research in the future. One is to evaluate the biomechanical hypotheses outlined in Section 3.6 by analysing human and great ape vertebrae with a combination of dissection, 3D morphometrics, and musculoskeletal modelling. Such a study would help us understand how the traits increase an individual's probability of developing intervertebral disc hernias and spondylolysis. It would also provide insight into the functional anatomy of great ape vertebrae, about which we currently know very little.

A second possible project would be to use medical imaging, geometric morphometrics, and a large sample of healthy and afflicted living humans to develop a predictive model that enables an individual's probability of developing different acquired spine conditions to be calculated based on the shape of their vertebrae. This would allow the formulation of recommendations regarding preventative measures to reduce the likelihood of developing the condition(s).

A third worthwhile endeavour would be to identify the alleles involved in vertebral shape in humans and chimpanzees, and then investigate whether individuals with the vertebral shape associated with IDH share more vertebral shape-related alleles with chimpanzees than do individuals elsewhere in the distribution of vertebral shape variation within humans. This would improve understanding of the genetic basis of specific lumbar diseases and could open up the possibility of large-scale screening for at-risk individuals. The foundations for this project have already been laid by work on other vertebrates (Böhmer, 2017).

Finally, the logic of the Evolutionary Shape Hypothesis may also apply to other acquired diseases affecting the human skeleton—not only in the spine, but also those that affect other parts of the skeleton. The human skeleton differs in many ways from those of the great apes, and some of the differences are in regions commonly afflicted by acquired conditions. As such, it is possible that the link between ancestral and hyper-derived shapes and diseases identified by Plomp and colleagues (2015a, 2019b, 2020) in the vertebrae may hold elsewhere. The knee and hip are good candidates for such a study because they both underwent substantial changes in shape during the shift to bipedalism and are also prone to acquired diseases (Watson et al., 2009). Similarly, the human shoulder differs markedly from that of the great apes and has a different pathology profile (Püschel and Sellers, 2015).

3.8. Conclusions

This chapter has revisited Keith's (1923) classic hypothesis that the stress placed on our spines by our unique mode of locomotion explains why we are so commonly afflicted with back problems. Specifically, we reviewed and synthesised evidence indicating that the relationship between bipedalism and some important ASDs is mediated by the degree of curvature of the lumbar spine and certain shape traits of the lumbar vertebrae. Subsequently, we outlined a revised version of Keith's (1923) Evolutionary Shape Hypothesis, which states that human vertebral shape variation should be viewed as a spectrum, where vertebrae at one end are similar to great ape and fossil hominin vertebrae and vertebrae at the other end exhibit exaggerated adaptations for bipedalism. According to the Evolutionary Shape Hypothesis, where an individual's vertebrae sit on this evolutionary shape spectrum has an important influence on their spinal health. If a person has vertebrae that lie at the ancestral end, they have a higher probability of developing intervertebral disc hernias. Conversely, if their vertebrae lie at the other, highly derived end of the spectrum, they have a greater likelihood of developing spondylolysis.

The Evolutionary Shape Hypothesis not only provides novel insights into what causes back problems, but also has the potential to inform how clinicians manage people with common spinal diseases. As discussed in Section 3.7, the Evolutionary Shape Hypothesis could form the basis of a predictive model for identifying individuals who are at risk of developing different ASDs. In principle, it should be possible to use medical imaging technology, for example, magnetic resonance imaging (MRI) and computerised tomography (CT), alongside geometric morphometrics to analyse a living individual's lumbar vertebrae and assign them to risk categories for IDH and spondylolysis. It should then be possible to devise behavioural strategies that would reduce chances of developing the ASD for which they are at risk, for example, avoiding certain sports. This would represent a substantial step forward in the management of spinal health and therefore of back pain.

More generally, this chapter demonstrated the benefits and applications of palaeopathological and palaeoanthropological data for evolutionary medicine (EM), which is a fast-expanding field that involves the application of Darwinian thinking to medical problems (Nesse & Williams, 1994). Since EM solidified as a field in the early 1990s, a considerable amount of research has illustrated the benefits of looking at health issues with an evolutionary lens. However, so far there have been few attempts to analyse musculoskeletal problems within the framework of EM and consequently very little use of palaeopathological or palaeoanthropological data in EM. This chapter illustrates not only that it is possible to shed new light on musculoskeletal problems with an EM approach, but also that palaeopathological and palaeoanthropological data can be extremely useful for this endeavour.

Furthermore, this chapter demonstrated that palaeopathology can benefit from drawing on evolutionary theory. Despite palaeopathology's ability to provide insight into the health of past peoples, there are a few issues that, historically, have decreased its relevancy to medical professionals. One major criticism levelled at palaeopathology is that it is too focused on individual and unique case studies, rather on hypothesis-driven, population-based research (Rühli et al., 2016). EM offers a robust theoretical framework that can address this criticism. If the palaeopathological and comparative data outlined in this chapter had not been evaluated with an evolutionary perspective, they would have only provided information on how vertebral shape correlates with the presence of ASDs. Instead, interpreting the data within

an evolutionary framework allowed for the development of a novel, productive causal hypothesis for ASDs. In our view, this represents a major advance, and one that can probably be replicated with many of the other skeletal diseases on which palaeopathologists work.

References

Abitbol, M. M. (1995). Lateral view of *Australopithecus afarensis*: primitive aspects of bipedal positional behavior in the earliest hominids. *Journal of Human Evolution*, 28, 211–229.

Adams, M. A., McNally, D. S., Chinn, H. & Dolan P. (1994). Posture and the compressive strength of the lumbar spine. *Clinical Biomechanics*, 9, 5–14

Ahmed, A. D. Duncan, N. A. & Burke, D. L. (1990). The effect of facet geometry on the axial torque-rotation response of lumbar motion segments. *Spine*, 15, 391–401

Antoniades, S. B., Hammerberg, K. W. & DeWald, R. L. (2000). Sagittal plane configuration of the sacrum spondylolisthesis. *Spine*, 35(9), 1085–1091.

Argot, C. (2003). Functional-adaptive anatomy of the axial skeleton of some extant marsupials and the paleobiology of the Paleocene marsupials *Mayulestes ferox* and *Pucadelphys andinus*. *Journal of Morphology*, 255, 279–300

Balague, F., Mannion, A. F., Pellise, F. & Cedraschi, C. (2012). Non-specific low back pain. *The Lancet*, 379, 482–491

Barrey, C., Jund, J., Perrin, G. & Roussouly, P. (2007). Spinopelvic alignment of patients with degenerative spondylolisthesis. *Neurosurgery*, 61(5), 981–986

Bastir, M., Martinez, D., Rios, L., Higuero, A., Barash, A., Martelli, S., Tabernero, A. G., Estalrrich, A., Huguet, R., de la Rasilla, M. & Rosas, A. (2017). Three-dimensional morphometrics of thoracic vertebrae in Neanderthals and the fossil evidence from El Sidron (Asturias, Northern Spain). *Journal of Human Evolution*, 108, 47–61.

Battie, M. C., Videman, T., Kapiro, J., Gibbons, L. E., Gill, K., Manninen, H., Saarela, J. & Peltonen, L. (2009). The twin spine study: contributions to a changing view of disc degermation. *The Spine Journal*, 9(1), 47–59.

Been, E., Barash, A., Marom, A. & Kramer, P. (2010a). Vertebral bodies or discs: which contributes more to human-like lumbar lordosis? *Clinical Orthopaedics and Related Research*, 486, 1822–1829.

Been, E., Gómez-Olivencia, A., Shefi, S., Soudack, M., Bastir, M. & Barash, A. (2017). Evolution of spinopelvic alignment in hominins. *The Anatomical Record*, 300, 900–911.

Been, E. & Kalichman, L. (2014). Lumbar lordosis. *The Spine Journal*, 14(1), 87–97.

Been, E., Peleg, S., Marom, A. & Barash, A. (2010b). Morphology and function of the lumbar spine of the Kebara 2 Neandertal. *American Journal of Physical Anthropology*, 142(4), 549–557

Been, E., Simonovich, A. & Kalichman, L. (2019). 'Spinal posture and pathology in modern humans', in E. Been, A. Gómez-Olivencia & P. A. Kramer (eds), *Spinal evolution*, pp. 310–320. Springer, New York.

Benton, R. S., (1967). 'Morphological evidence for adaptations within the epaxial region of the primates', in H. Vagtborg, (ed), *The baboon in medical research*, pp. 10–20. University of Texas Press, Houston.

Binder, A. I. (2007). Cervical spondylosis and neck pain. *British Medical Journal*, 334, 527–531

Bogduk, N., Macintosh, J. E. & Pearcy, M. J. (1992). A universal model of the lumbar back muscles in the upright position. *Spine*, 17, 897–913.

Böhmer, C. (2017). Correlation between Hox code and vertebral morphology in the mouse: towards a universal model for Synapsida. *Zoology Letters*, 3, 8–19.

Bourke, J. B. (1971). The paleopathology of the vertebral column in Ancient Egypt and Nubia. *Medical History*, 15(4), 363–375

Bramble, D. M. & Lieberman, D. E. (2004). Endurance running and the evolution of Homo. *Nature*, 432, 345–352.

Bridges, P. S. (1994). Vertebral arthritis and physical activities in the prehistoric Southeastern United States. *American Journal of Physical Anthropology*, 93(1), 83.

Briggs, A., Greig, A. M., Wark, J. D., Fazzalari, N. L. & Bennell, K. L. (2004). A review of anatomical and mechanical factors affecting vertebral body integrity. *International Journal of Medical Sciences*, 1, 170

Buchbiner, R., Blyth, F. M., March, L. M., Brooks, P., Woolf, A. D. & Hoy, D. G. (2013). Placing the global burden of low back pain in context. *Best Practice & Research: Clinical Rheumatology*, 27, 575

Cartmill, M. & Brown, K. (2017). 'Posture, locomotion and bipedality: The case of the Gerenuk (*Litocranius walleri*)', in A. Marom & E. Hovers (eds), *Human Paleontology and Prehistory*, pp. 53–70. Springer, New York.

Castillo, E. R. & Lieberman, D. E. (2015). Lower back pain. *Evolution, Medicine, and Public Health* [online], 2015. doi: 10.1093/emph/eou034.

Chapman, F. H. (1962). Incidence of arthritis in a prehistoric middle Mississippian Indian population. *Proceedings of the Indiana Academy of Science*, 72, 59–62

Cheng, J. S. & Song, J. K. (2003). Anatomy of the sacrum. *Journal of Neurosurgery*, 15, 1–4

Cholewick, J. & McGill, S. M. (1996). Mechanical stability of the in vivo lumbar spine: implications for injury and chronic back pain. *Clinical Biomechanics*, 11(1), 1–5

Cvijetić, S., McCloskey, E. & Korsic, M. (2000). Vertebral osteophytosis and vertebral deformities in an elderly population sample. *Wiener Klinische Wochenschrift*, 112(9), 407–412.

D'Anastasio, R, Zipfel, B, Moggi-Cecchi, J, Stanyon, R, Capasso, L, 2009. Possible Brucellosis in an early hominin skeleton from Sterkfontein, South Africa. *PLoS One*, 4(7), e6439.

Dar, G., Peleg, S., Masharawi, Y., Steinberg, N., Hila, M. & Hershovitz, I. (2009). Demographic aspects of Schmorl's nodes: a skeletal study. *Spine*, 34, E312–5

Davis, M. A. (2012). Where the United States spends its spine dollars: expenditures on different ambulatory services for the management of back and neck pain. *Spine*, 37(19), 1693–1701.

Davis, P. R. (1961). Human lower lumbar vertebrae: some mechanical and osteological considerations. *Journal of Anatomy*, 95, 337–344

Dieleman, J. L., Baral, R., Birger, M., Bui, A. L., Bulchis, A., Chapin, A., Hamavid, H., Horst, C., Johnson, E. K., Joseph, J. & Lavado, R. (2016). US Spending on personal health care and public health, 1966–2013. *Journal of the American Medical Association*, 316(24), 2627–2646

Dolan, P., Luo, J., Pollintine, P., Landham, P. R., Stefanakis, M. & Adams, M. A. (2013). Intervertebral disc decompression following endplate damage: implications for disc degermation depend on spinal level and age. *Spine*, 38(17), 1473–1481

Donaldson, L. (2008). *Annual Report of the Chief Medical Officer on the State of Public Health*. Department of Health, London.

El-Khoury, G. Y. & Whitten, C. G. (1993). Trauma to the upper thoracic spine: anatomy, biomechanics, and unique imaging features. *American Journal of Roentgenology*, 160, 95–102

Endo, K., Suzuki, H., Tanaka, H., Kang, Y. & Yamamoto, K. (2010). Sagittal spinal alignment in patients with lumbar disc herniation. *European Spine Journal*, 19, 435–438

Ergun, T., Lakadamyali, H. & Sahin, M. S. (2010). The relation between sagittal morphology of the lumbrosacral spine and the degree of lumbar intervertebral disc degeneration. *Acta Orthopaedic et Traumatologica Turcica*, 44(4), 293–299.

Erikson, G. (1963). Brachiation in New World monkeys and in anthropoid apes. *Symposium of the Zoological Society London*, 10, 135–164.

Farfan, H. F. (1995). Form and function of the musculoskeletal system as revealed by mathematical analysis of the lumbar spine: an essay. *Spine*, 20, 1462–1474.

Filler, A. G. (2007). Emergence and optimization of upright posture among hominiform hominoids and the evolutionary pathophysiology of back pain. *Neurosurgery Focus*, 23, 1–6

Fugiwara, A., Tamai, K., An, H. S., Lim, T., Yoshida, H., Kurihashi, A. & Saotome, K. (2001). Orientation and osteoarthritis of the lumbar facet joint. *Clinical Orthopaedics and Related Research*, 385, 88–94

Gambaryan, P. (1974). *How mammals run.* John Wiley & Sons, New York.

Gellhorn, A. C., Katz, J. N. & Suri, P. (2013). Osteoarthritis of the spine: the facet joints. *National Review of Rheumatology*, 9, 216–224

Giesecke, T., Gracely, R. H., Grant, M., Nachemson, A., Petzke, F., Williams, D. A. & Clauw, D. J. (2004). Evidence of augmented central pain processing in idiopathic chronic low back pain. *Arthritis & Rheumatology*, 50(2), 613–623

Gómez-Olivencia, A., Arlegi, M., Barash, A., Stock, J. T. & Been, E. (2017). The Neanderthal vertebral column 2: the lumbar spine. *Journal of Human Evolution*, 106, 84–101

Gómez-Olivencia, A., Been, E., Arsuaga, J. L. & Stock, J. T. (2013). The Neanderthal vertebral column 1: the cervical spine. *Journal of Human Evolution*, 64, 608–630.

Gore, M., Sadosky, A., Stacey, B. R., Tai, K. S. & Leslie, D. (2012). The burden of chronic low back pain: clinical comorbidities, treatment patterns, and health care costs in usual care settings. *Spine*, 37(11), E668–677.

Gracovetsky, S. A. & Iacono, S. (1987). Energy transfers in the spinal engine. *Medical Engineering & Physics*, 9, 99–114.

Grobler, L. J.., Robertson, P A., Novotny, J. E. & Pope, M. H. (1993). Etiology of spondylolisthesis. Assessment of the role played by lumbar facet joint morphology. *Spine*, 18(1), 80–91

Hackinger, S., Trajanoska, K., Styrkarsdottir, U., Zengini, E., Steinberg, J., Ritchie, G. R. S., Hatzikotoulas, K., Gilly, A., Evangelou, E., Kemp, J. P., arcOGEN Consortium, GEFOS Consortium; Evans, D., Ingvarsson, T., Jonsson, H., Thorsteinsdottir, U., Stefansson, K., McCaskie, A. W., Brooks, R. A., Wilkinson, J. M., Rivadeneira, F. & Zeggini, E. (2017). Evaluation of shared genetic aetiology between osteoarthritis and bone mineral density identifies SMAD3 as a novel osteoarthritis risk locus. *Human Molecular Genetics*, 26, 3850–3858.

Haeusler, M., Trinkaus, E., Fornai, C., Muller, J., Bonneau, N., Boeni, T. & Frater, N. (2019). Morphology, pathology, and the vertebral posture of the La Chapelle-aux-Saints Neandertal. *Proceedings of the National Academy of Sciences*, 116(11), 4923–4927

Harrington, J. F., Sungarian, A., Rogg, J., Makker, V. J. & Epstein, M. H. (2001). The relation between vertebral endplates shape and lumbar disc herniations. *Spine*, 26, 2133–2138.

Hernandez, C. J., Loomis, D. A., Cotter, M. M., Schifle, A. L., Anderson, L. C., Elsmore, L., Kunos, C. & Latimer, B. (2009). Biomechanical allometry in hominoid thoracic vertebrae. *Journal of Human Evolution*, 56, 462–470.

Hong, J., Reed, C., Novick, D. & Happich, M. (2013). Costs associated with treatment of the chronic low back pain: an analysis of the UK general practice research database. *Spine*, 38(1), 75–82

Hoy, D., March, L., Woolf, A., Blyth, F., Brooks, P., Smith, A., Vos, T., Barendregt, J., Blore, J., Murray, C., Burstein, R. & Buchbinder, R. (2014). The global burden of low back pain: estimates the global burden disease 2010 study. *Clinical Epidemiology Research*, 73, 968–974

Hu, S. S., Tribus, C. B., Diab, M. & Ghanayem, A. L. (2008). Spondylolisthesis and spondylolysis. *Journal of Bone and Joint Surgery*, 90, 656–671.

Iwamoto, J., Takeda, T. & Wakano, K. (2004). Returning athletes with severe low back pain and spondylolysis to original sporting activities with conservative treatment. *Scandinavian Journal of Medicine & Science in Sports*, 14, 346–351.

Jankauskas, R. (1992). Degenerative changes of the vertebral column in Lithuanian paleoosteological material. *Anthropologie*, 30(1), 109–119.

Jaumard, N. V., Welch, W. C. & Winkelstein, B. A. (2011). Spinal facet joint biomechanics and mechanotransduction in normal, injury and degenerative conditions. *Journal of Biomechanical Engineering*, 133, 71010.

Jellema, L. M., Latimer, B. & Walker, A. (1993). 'The rib cage', in A. Walker & R. Leakey, (eds), *The Nariokotome* Homo erectus *skeleton*, pp. 294–325. Springer, Berlin.

Jurmain, R. (1990). Paleoepidemiology of a central California prehistoric population from CA-ALA-329: II. Degenerative disease. *American Journal of Physical Anthropology*, 83(1), 83–94.

Jurmain, R. (1989). Skeletal evidence of trauma in African apes, with special reference to the Gombe chimpanzees. *Primates*, 38, 1–4

Jurmain, R. D. (2000). Degenerative joint disease in African great apes: an evolutionary perspective. *Journal of Human Evolution*, 39, 185–203

Katz, V., Schofferman, J. & Reynolds, J. (2003). The sacroiliac joint: a potential cause of pain after lumbar fusion to the sacrum. *Clinical Spine Surgery*, 16(1), 96–99

Keith, A. (1923). Hunterian lectures on man's posture: its evolution and disorders. Lecture II: the evolution of the orthograde spine. *British Medical Journal*, 1, 587

Keller, T. S., Colloca, C. J., Harrison, D. E., Harrison, D. C. & Janik, T. (2005). Influence of spine morphology on intervertebral disc loads and stresses in asymptomatic adults: implications of or the ideal spine. *The Spine Journal*, 5(3), 297–309.

Latimer, B. (2005). The perils of being bipedal. *Annals of Biomedical Engineering*, 33, 3–6.

Latimer, B. & Ward, C. V. (1993). 'The thoracic and lumbar vertebrae', in A. Walker & R. Leakey, (eds), The Nariokotome *Homo erectus* skeleton, pp. 266–293. Springer, Berlin.

Letić, M. (2012). Feeling wall tension in an interactive demonstration of Laplace's law. *Advances in Physiology Education*, 36, 176

Liu, Y. S., Chen, Q. X. & Liu, S. B. (2007). Endplate concavity variation of lumbar motion segments: A finite element analysis. *Journal of Clinical Rehabilitative Tissue Engineering Research*, 12, 8765–8770.

Lovejoy, O. (2005). The natural history of human gait and posture. Part 1: spine and pelvis. *Gait & Posture*, 21, 95–112

Lovell, N. (1990). Patterns of Injury and Illness in the Great Apes: A Skeletal Analysis. Smithsonian Institution, Washington, DC.

Lovell, N. (1994). Spinal arthritis and physical stress at Bronze Age Harappa. *American Journal of Physical Anthropology*, 93, 149–164

Lowenstine, L. J., McManamon, R. & Terio, K. A. (2016). Comparative pathology of aging great apes: bonobos, chimpanzees, gorillas, and orangutans. *Veterinary Pathology*, 53(2), 250–276

Maat, G. J. R., Mastwijk, R. W. & Van Der Velde, E. A. (1995). Skeletal distribution of degenerative changes in vertebral osteophytosis, vertebral osteoarthritis and DISH. *International Journal of Osteoarchaeology*, 5, 289–298

Maher, C., Underwood, M. & Buchbinder, R. (2017). Nonspecific low back pain. *The Lancet*, 389, 736–747.

Manchikanti, L., Boswell, M. V., Singh, V., Pampati, V., Damrom, K. S. & Beyer, C. D. (2004). Prevalence of facet joint pain in chronic spinal pain of cervical, thoracic, and lumbar regions. *BMC Musculoskeletal Disorders*, 5, 1–7.

Manchikanti, L., Singh, V. & Pampati, V. (2002). Evaluation of the prevalence of facet joint pain in chronic thoracic pain. *Pain*, 5(4), 354–359.

Maniadakis, N. & Gray, A. (2000). The economic burden of back pain in the UK. *Pain*, 84, 95–103

Masharawi, Y. (2012). Lumbar shape characterization of the neural arch and vertebral body in spondylolysis: a comparative skeletal study. *Clinical Anatomy*, 25, 224–230.

Masharawi, Y., Alperovitvh-Najenson, D., Steinberg, N, Dar, G., Peleg, S., Rothschild, B., Salame, K. & Hershkovtiz, I. (2007). Lumbar facet orientation in spondylolysis: a skeletal study. *Spine*, 32(6), E176–180.

Mays, S. (2007). Spondylolysis in non-adult skeletons excavated from a medieval rural archaeological site in England. *International Journal of Osteoarchaeology*, 17, 504–513.

Mays, S. (2006). Spondylolysis, spondylolisthesis, and lumbo-sacral morphology in a medieval English skeletal population. *American Journal of Physical Anthropology*, 131, 352–362.

Meakin, J. R., Gregory, J. S.., Aspden, R M., Smith, F.W. & Gilbert, F. J. (2009). The intrinsic shape of the human lumbar spine in supine, sitting, and standing postures: characterization using an active shape model. *Journal of Anatomy*, 215, 206–211.

Meakin, J. R., Smith, F. W., Gilbert, F. J. & Aspden, R. M. (2008). The effect of axial load on the sagittal place curvature of the upright human spine in vivo. *Journal of Biomechanics*, 41(13), 2850–2854.

Merbs, C. (1983). *Patterns of activity-induced pathology in a Canadian Inuit population* (Vol. 119). National Museums of Canada, Ottawa.

Merbs, C. (1996). Spondylolysis and spondylolisthesis: A cost of being an erect biped or clever adaptation? *American Journal of Physical Anthropology*, 101, 201–228.

Meyer, M. R. (2016). 'The cervical vertebrae of KSD-VP-1/1', in Y. Haile-Selassie & D. Su (eds), *The postcranial anatomy of* Australopithecus afarensis, pp. 63–111. Springer, Dordrecht.

Meyer, M. R. & Williams, S. A. (2019). Earliest axial fossils from the genus Australopithecus. *Journal of Human Evolution*, 132, 189–214.

Meyer, M. R., Williams, S. A., Schmid, P., Churchill, S. E. & Berger, L. R. (2017). The cervical spine of *Australopithecus sediba*. *Journal of Human Evolution*, 104, 32–49.

Micheli, L. J. & Wood, R. (1995). Back pain in young athletes: significant differences from adults in causes and patterns. *JAMA Pediatrics*, 149(1), 15–18.

Miyake, R., Ikata, T., Katoh, S. & Morita, T. (1996). Morphologic analysis of the facet joint in the immature lumbosacral spine with special reference to spondylolysis. *Spine*, 21, 783–789.

Modic, M. T. (1999). Degenerative disc disease and back pain. *Magnetic Resonance Imaging Clinic of North America*, 7(3), 481–491.

Muraki, S., Oka, H., Akune, T., Mabuchi, A., En-Yo, Y., Yoshida, M., Saika, A., Suzuki, T., Yoshida, H., Ishibashi, H., Yamamoto, S., Nakamura, K., Kawaguchi, H. & Yoshimura, N. (2009). Prevalence of radiographic lumbar spondylosis and its association with low back pain in elderly subjects of population-based cohorts: the ROAD study. *Annals of Rheumatic Disease*, 68, 1401–1406.

Murray, C. & Lopez, A. D. (2013). Measuring the global burden of disease. *New England Journal of Medicine*, 369, 448–457.

Muthuri, S. G., Kuh, D. & Cooper, R. (2018). Longitudinal profiles of back pain across adulthood and their relationship with childhood factors: evidence from the 1946 British birth cohort. *Pain*, 159(4), 764.

Nesse, R. M. & Williams, G. C. (1994). *Why we get sick: the new science of Darwinian medicine*. Vintage, New York.

Nuckley, D. J., Kramer, P. A.., Rosario, A D., Fabro, N., Baran, S. & Ching, R. P. (2008). Intervertebral disc degeneration in a naturally occurring primate model: radiographic and biomechanical evidence. *Journal of Orthopaedic Research*, 26, 1283–1288.

Nuki, G. (1999). 'Role of mechanical factors in the aetiology, pathogenesis, and progression of osteoarthritis', in J. Y. Reginster, J. P. Pelletier, J. Martel-Pelletier & L. Crasborn (eds), *Osteoarthritis: clinical and experimental aspects*, pp. 110–114. Springer, New York.

O'Neill, T. W., McCloskey, E. V., Kanis, J. A., Bhalla, A. K., Reeve, J., Reid, D. M., Todd, C., Wolff, A. D. & Silman, A. J. (1999). The distribution, determinants, and clinical correlates of vertebral ostephytosis: a population-based survey. *The Journal of Rheumatology*, 26(4), 842–848.

Odes, E. J., Parkinson, A. H., Randolph-Quinney, P. S., Zipfel, B., Jakata, K., Bonney, H. & Berger, L. R. (2017). Osteopathology and insect traces in the *Australopithecus africanus* skeleton StW 431. *South African Journal of Science*, 113(1.2), 1–7.

Panjabi, M., Oxland, T., Takata, K., Goel, V., Duranceau, J. & Krag, M. (1993). Articular facets of the human spine. Quantitative three-dimensional anatomy. *Spine*, 18, 1298–1310.

Patrick, N., Emanski, E. & Knaub, M. A. (2014). Acute and chronic low back pain. *Medical Clinics of North America*, 98(4), 777

Pfirrmann, C. & Resnick, D. (2001). Schmorl's nodes of the thoracic and lumbar spine: Radiographic pathologic study of prevalence, characterization, and correlation with degenerative changes of 1,650 spinal levels in 100 cadavers. *Radiology*, 219, 368–374.

Plomp, K. A., Dobney, K. & Collard, M. (2020). Spondylolysis and spinal adaptations for bipedalism: the Overshoot Hypothesis. *Evolution, Medicine, Public Health*, 1, 35–44.

Plomp, K. A., Dobney, K., Weston, D., Strand Viðarsdóttir, U. & Collard, M. (2019b). 3D shape analyses of extant primate and fossil hominin vertebrae support the Ancestral Shape Hypothesis for intervertebral disc herniation. *BMC Evolutionary Biology*, 19, 1–16.

Plomp, K. A., Roberts, C. A. & Strand Vidarsdottir, U. (2012). Vertebral morphology influences the development of Schmorl's nodes in the lower thoracic vertebra. *American Journal of Physical Anthropology*, 149, 172.

Plomp, K. A., Roberts, C. A. & Strand Viðarsdóttir, U. (2015b). Does the correlation between Schmorl's nodes and vertebral morphology extend into the lumbar spine? *American Journal of Physical Anthropology*, 157, 526.

Plomp, K. A., Strand Viðarsdóttir, U., Weston, D., Dobney, K. M. & Collard, M. (2015a). The ancestral shape hypothesis: an evolutionary explanation for the occurrence of intervertebral disc herniation in humans. *BMC Evolutionary Biology*, 15, 68–78.

Plomp, K. A., Weston, D., Strand Viðarsdóttir, U., Dobney, K. & Collard, M. (2019a). Potential adaptations for bipedalism in modern human thoracic and lumbar vertebrae: a 3D comparative analysis. *Journal of Human Evolution*, 137, 102693.

Prescher, A. (1998). Anatomy and pathology of the aging spine. *European Journal of Radiology*, 27(181), 181–195.

Püschel, T. A. & Sellers, W. I. (2015). Standing on the shoulders of apes: analyzing the form and function of the hominoid scapula using geometric morphometrics and finite element analysis. *American Journal of Physical Anthropology*, 159, 325–341.

Richmond, B. G., Begun, D. R. & Strait, D. S. (2001). Origin of human bipedalism:

the knuckle-walking hypothesis revisited. *American Journal of Physical Anthropology*, 44, 70–105.

Richter, J. (1970). Die fakultative Bipedie der Giraffengazelle *Litocranius walleri sclateri*. Ein Beitrag zur funktionellen Morphologie. *Morphologische Jahrbuch*, 114, 457.

Riyazi, N., Meulenbelt, I., Kroon, H. M., Ronday, K. H., Hellio le Graverand, M. P., Rosendaal, F. R., Breedveld, F. C., Slagboom, P. E. & Kloppenburg, M. (2005). Evidence for familial aggregation of hand, hip, and spine but not knee osteoarthritis in siblings with multiple joint involvement. *Annals of the Rheumatic Diseases*, 64, 438–443.

Robinson, J. T. (1972). *Early hominin posture and locomotion*. University of Chicago Press, Chicago.

Rogers, J. & Waldron, T. (1995). *A field guide to joint disease in archaeology*. John Wiley & Sons, Inc., New York.

Rogers, J., Watt, I. & Dieppe, P. (1985). Paleopathology of spinal osteophytosis, vertebral ankylosis, ankylosing spondylitis, and vertebral hyperostosis. *Annals of the Rheumatic Diseases*, 44, 113–120.

Rose, M. D. (1975). Functional proportions of primate lumbar vertebral bodies. *Journal of Human Evolution*, 4, 21–38.

Roussouly, P., Gollogly, S., Berthonnaud, E., & Dimnet, J. (2005). Classification of the normal variation in the sagittal alignment of the human lumbar spine and pelvis in the standing position. *Spine*, 30, 346–353.

Roussouly, P., Gollogly, S., Berthonnaud, E., Labelle, H. & Weidenbaum, M. (2006). Sagittal alignment of the spine and pelvis in the presence of L5–S1 isthmic lysis and low-grade spondylolisthesis. *Spine*, 31, 2484–2490.

Roussouly, P. & Pinheiro-Franco, J. (2011). Sagittal parameters of the spine biomechanical approach. *European Spine Journal*, 20, 578–585.

Rühli, F. J., Galassi F. M. & Haeusler M. (2016). Palaeopathology: current challenges and medical impact. *Clinical Anatomy*, 29(7), 816–822.

Sak, G. S., Ayoub, C. M., Domloj, N. T., Turbay, M. J., El-Zein, C. & Hourani, M. H. (2011). Effect of age and lordotic angle on the level of lumbar disc herniation. *Advances in Orthopedics*, 2011, 950576.

Sanders, W. J. (1998). Comparative morphometric study of the australopithecine vertebral series Stw-H8/H41. *Journal of Human Evolution*, 34, 249–302.

Sanders, W. J. & Bodenbender, B. E. (1994). Morphometric analysis of lumbar vertebra UMP 67-28: implications for spinal function and phylogeny of the Miocene Moroto hominoid. *Journal of Human Evolution*, 26, 203–237.

Sarzi-Puttini, P., Atzeni, F., Fumagalli, M., Capsoni, F. & Carrabba, M. (2005). Osteoarthritis of the spine. *Seminars in Arthritis and Rheumatism*, 34(6), 38

Scannell, J. P. & McGill, S. M. (2003). Lumbar posture—should it, and can it, be modified? A study of passive tissue stiffness and lumbar position during activities of daily living. *Physical Therapy*, 83, 907–917.

Schmorl, G. & Junghans, H. (1971). *The human spine in health and disease*. Grune and Stratton, New York.

Schultz, A. H. (1953). The relative thickness of the long bones and the vertebrae in primates. *American Journal of Physical Anthropology*, 11, 273–312.

Schultz, A. H. (ed) (1961). 'Vertebral column and thorax', in *Primatologia*, Vol. 4. Karger, Basel.

Shapiro, L, 1991. *Functional morphology of the primate spine with special reference to the orthograde posture and bipedal locomotion*. PhD Thesis, State University of New York at Stony Brook, Stony Brook.

Shapiro, L. (1993a). 'Functional morphology of the vertebral column in primates', in D. L. Gebo (ed), *Postcranial adaptation in non-human primates*, Ch. 5. Northern Illinois University Press, Dekalb.

Shapiro, L. J. (1993b). Evaluation of the "unique" aspects of human vertebral bodies and pedicles with consideration of *Australopithecus africanus*. *Journal of Human Evolution*, 25, 433–470.

Shapiro, L. J. (1995). Functional morphology of indrid lumbar vertebrae. *American Journal of Physical Anthropology*, 98, 323–342.

Shapiro, L. J. (2007). Morphological and functional differentiation in the lumbar spine of lorisids and galagids. *American Journal of Primatology*, 69, 86–102.

Shapiro, L. J. & Kemp, A. D. (2019). Functional and developmental influences on intraspecific variation in catarrhine vertebrae. *American Journal of Physical Anthropology*, 168, 131–144.

Shapiro, L. J., Seiffert, C. V., Godfrey, L. R., Jungers, W. L., Simons, E. L. & Randia, G. F. (2005). Morphometric analysis of lumbar vertebrae in extinct Malagasy strepsirrhines. *American Journal of Physical Anthropology*, 128, 823–839.

Šlaus, M. (2000). Biocultural analysis of sex differences in mortality profiles and stress levels in late Medieval populations from Nova Raca, Croatia. *American Journal of Physical Anthropology*, 111, 193–209.

Sparrey, C. J., Bailey, J. F., Safaee, M., Clark, A. J., Lafage, V., Schwab, F., Smith, J. S. & Ames, C. P. (2014). Etiology of lumbar lordosis and its pathophysiology: a review of the evolution of lumbar lordosis, and the mechanics and biology of lumbar degenerative. *Journal of Neurosurgery*, 36, E1.

Stanton, T. R., Henschke, N., Maher, C. G., Refshauge, K. M., Latimer, J. & McAuley, J. H. (2008). After an episode of acute low back pain, recurrence is unpredictable and not as common as previously thought. *Spine*, 33, 2923–2928.

Staps, D. (2002). The first documented occurrence of spondylosis deformans in an early hominin. *American Journal of Physical Anthropology Suppl*, 34, 146.

Stouhal, E. & Jungwirth, J. (1980). Paleopathology of the Late Roman–Early Byzantine cemeteries at Sayala, Egyptian Nubia. *Journal of Human Evolution*, 9(1), 61–70.

Suri, P., Miyakoshi, A., Hunter, D. J., Jarvik, J. G., Rainville, J., Guermazi, A., Li, L. & Katz, J. N. (2011). Does lumbar spinal degeneration begin with the anterior structures? A study of the observed epidemiology in a community-based population. *BMC Musculoskeletal Disorders*, 12, 1–7.

Taskaynatan, M. A., Izci, Y., Ozgul, A., Hazneci, B., Dursun, H. & Kaylon, T. (2005). Clinical significance of congenital lumbosacral malformations in young male population with prolonged low back pain. *Spine*, 30(8), E210–3.

Thorpe, S. K. S., Holder, R. L. & Crompton, R. H. (2007). Origin of human bipedalism as an adaptation for locomotion on flexible branches. *Science*, 316, 1328–1331.

Trinkaus, E. (1985). Pathology and the posture of the La Chapelle-aux-Saints Neandertal. *American Journal of Physical Anthropology*, 67(1), 19–41.

Üstündağ, H. (2009). Schmorl's nodes in a post-Medieval skeletal sample from Klostermarienberg, Austria. *International Journal of Osteoarchaeology*, 19, 695–710.

Van Roy, P., Barbaix, E., De Maeseneer, M., Pouders, C. & Clarys, J. P. (2006). 'The anatomy of the neural arch of the lumbar spine, with references to spondylolysis, spondylolisthesis, and degenerative spondylolisthesis', in R. Gunzburg & M. Szpalski (eds), *Spondylolysis, spondylolisthesis, and degenerative spondylolisthesis*, pp. 1–10. Wolters Kluwer, Alphen upon Rhine.

Vercellotti, G., Caramella, D., Formicola, V., Fornaciari, G. & Larsen, C. S. (2009). Porotic Hyperostosis in a

Late Upper Palaeolithic Skeleton (Villabruna 1, Italy). *International Journal of Osteoarchaeology*, 20(3), 358–368.

Vercellotti-Monica, G., Simalcsik, A., Bejenaru, L. & Simalcsik, D. (2014). Osteopathies in the population of old Iasi city (Romania): the necropolis of the 'Banu' church, 16th–19th centuries. *Scientific Annals of Alexandru Ioan Cuza University of Iasi, New Series, Section I, Animal Biology*, 60, 91–104.

Waldron, T, 1991. Variations in the rates of spondylolysis in early populations. *International Journal of Osteoarchaeology*, 1, 63–65.

Waldron, T, 1992. Osteoarthritis in a Black Death cemetery in London. *International Journal of Osteoarchaeology*, 2(3), 235–240.

Ward, C, 1993. Torso morphology and locomotion in *Proconsul nyanzae*. *American Journal of Physical Anthropology*, 92, 291–328.

Ward, C, Kimbel, WH, Harmon, EH, Johanson, DC, 2012. New postcranial fossils f Australopithecus afarensis from Hadar, Ethiopia (1999–2007). *Journal of Human Evolution*, 63, 1–51.

Ward, CV, 1991. *The functional anatomy of the lower back and pelvis of the Miocene hominoid* Proconsul nyanzae *from the Miocene of Mfangano Island, Kenya*. PhD Thesis, Johns Hopkins University, Baltimore, MD.

Ward, CV, Latimer, B, 2005. Human evolution and the development of spondylolysis. *Spine*, 30, 1808.

Ward, CV, Latimer, B, Alander, DH, Parker, J, Ronan, JA, Holden, AD, Sanders, C, 2007. Radiographic assessment of lumbar facet distance spacing and spondylolysis. *Spine*, 32, E85–8.

Ward, CV, Mays, SA, Child, S, Latimer, B, 2010. Lumbar vertebral morphology and isthmic spondylolysis in a British medieval population. *American Journal of Physical Anthropology*, 141, 273–280.

Watson, JC, Payne, RC, Chamberlain, AT, Jones, RK, Sellers, WI, 2009. The kinematics of load carrying in humans and great apes: implications for the evolution of human bipedalism. *Folia Primatologica*, 80, 309–328.

Webb, R., Brammah, T., Lunt, M., Urwin, M., Allison, T. & Symmons, D. (2003). Prevalence and predictors of intense, chronic, and disabling neck and back pain in the UK general population. *Spine*, 28, 1195–1202.

Wei, W., Liao, S., Shi, S., Fei, J., Wang, Y. & Chen, C. (2013). Straightened cervical lordosis causes stress concentration: a finite element model study. *Australasian Physical & Engineering Sciences in Medicine*, 36, 27–33.

Weinberg, D. S., Liu, R. W., Xie, K. K., Morris, W. Z., Gebhart, J. J. & Gordon, Z. L. (2017). Increased and decreased pelvic incidence, sagittal facet joint orientations are associated with lumbar spine osteoarthritis in a large cadaveric collection. *International Orthopaedics*, 41, 1593–1600.

Whitcome, K. K. (2012). Functional implications of variation in lumbar vertebral count among hominins. *Journal of Human Evolution*, 62, 486–497.

Whitcome, K. K., Shapiro, L. J. & Lieberman, D. E. (2007). Fetal load and the evolution of lumbar lordosis in bipedal hominins. *Nature*, 450, 1075–1078.

Whyne, C. M., Hu, S. S., Klisch, S. & Lotz, J. (1998). Effect of the pedicle and posterior arch on vertebral strength predictions in finite element modeling. *Spine*, 23, 899–907.

Wild, M. H., Glees, M., Plieschnegger, C. & Wenda, K. (2006). Five-year follow-up examination after purely minimally invasive posterior stabilization of thoracolumbar fractures: a comparison of minimally invasive percutaneously and conventionally open treated patients. *Trauma Surgery*, 127, 335–343.

Williams, S. A. (2012). Placement of the diaphragmatic vertebra in catarrhines: implications for the evolution of the dorsostability in hominoids and bipedalism in hominins. *American Journal of Physical Anthropology*, 148, 111–122.

Williams, S. A., Gómez-Olivencia, A. & Pilbeam, D. R. (2019). 'Numbers of vertebrae in hominoid evolution', in E. Been, A. Gómez-Olivencia & A. Kramer (eds), *Spinal Evolution*, pp. 97–124. Springer, New York.

Williams, S. A., Middleton, E. R., Villamil, C. I. & Shattuck, M. R. (2016) Vertebral numbers and human evolution. *Yearbook of Physical Anthropology*, 159, S19–36.

Williams, S. A., Ostrofsky, K. R., Frater, N., Churchill, S. E., Schmid, P. & Berger, L. R. (2013). The vertebral column of *Australopithecus sediba*. *Science*, 340, 1232996.

Williams, S. A. & Russo, G. A. (2015). Evolution of the hominoid vertebral column: the long and short of it. *Evolutionary Anthropology*, 24, 15–32.

Yang, X., Kong, Q., Song, Y., Liu, L., Zeng, J. & Xing, R. (2014). The characteristics of spinopelvic sagittal alignment in patients with lumbar disc degenerative diseases. *European Spine Journal*, 23, 569–575.

Zhang, H., Merrett, D. C., Jing, Z., Tang, J., He, Y., Yue, H., Yue, Z. & Yang, D. (2017). Osteoarthritis, labour division, and occupational specialization of the Late Shang China—insights from Yinxu (ca. 1250–1046 BC). *PLoS One*, 12(5), e0176329.

Zlolniski, S. L., Torres-Tamayo, N., García-Martínez, D., Blanco-Pérez, E., Mata-Escolano, F., Barash, A., Nalla, S., Martelli, S., Sanchis-Gimeno, J. A. & Bastir, M. (2019). 3D geometric morphometric analysis of variation in the human lumbar spine. *American Journal of Physical Anthropology*, 170, 361–372.

CHAPTER 4

Birthing humans in the past, the present, and future: how birth can be approached holistically through an evolutionary medicine lens

Sarah-Louise Decrausaz and Frances Galloway

4.1 Introduction

Human childbirth is more than a singular biological process—it involves multiple body systems beyond the reproductive system, is supported by extensive and changing cultural practices, and has a complex evolutionary history. Medical interventions in childbirth, such as caesarean sections, save many lives every year (Boerma et al., 2018). However, the value of highly medicalised childbirth practice is continually debated in academic (Dunsworth, 2018) and public health (WHO, 2015) circles. Based on the high prevalence of birth-related deaths with childbirth and associated complications accounting for 9% of maternal mortality globally (WHO, 2014), childbirth remains challenging for both mother and baby. Complications include, but are not limited to, eclampsia, obstetric haemorrhage, and hypertensive disorders (Geller et al., 2018). In a survey of 417 studies across 115 countries, it was found that obstructed labour, or dystocia, accounted for 69,000 deaths between 2003 and 2009 (Say et al., 2014). Every day in 2017, 810 women died from preventable causes related to pregnancy and childbirth (WHO, 2019a).

The risk of maternal and/or infant mortality is highly influenced and further complicated by social and gender inequalities (Wells, 2017). For example, maternal mortality rates in low-income countries in 2017 were 462 per 100,000 live births while the maternal mortality rate was 11 per 100,000 live births in high-income countries (WHO, 2019b). It should also be noted that maternal mortality varies as a result of socio-economic disparities and the history of racial violence in many countries. For instance, Black women are three to four times more likely to die a pregnancy-related death than White women in the USA (Howell, 2018). Black and Asian British women have a higher maternal mortality rate than white British women (Knight et al., 2020). In Canada, pre-term birth occurred in 9.1% of First Nations women from British Columbia compared to 5% of non-First Nations women from the same province (Luo et al., 2004). As a counterpoint to these specific rates of maternal mortality, it is also important to recognise that maternal health has improved overall—maternal mortality rates have dropped by an estimated 44% worldwide from 1990 to 2015, with a decline of 48% for industrialised countries (WHO, 2015).

Mothers may experience long-term health issues as a consequence of childbirth, such as obstetric fistulas (Bomboka et al., 2019) and pelvic floor disorders depending on their delivery style

Sarah-Louise Decrausaz and Frances Galloway, *Birthing humans in the past, the present and future*. In: *Palaeopathology and Evolutionary Medicine*. Edited by Kimberly A. Plomp et al., Oxford University Press. © Oxford University Press (2022). DOI: 10.1093/oso/9780198849711.003.0004

(Blomquist et al., 2018). Furthermore, the human birth experience and the neonatal period has been shown to have key impacts on later-life health. For example, low birth weight infants (babies born after 37 weeks of gestation with a birth weight below 2.5 kg) have been found to be more susceptible to cardiovascular diseases later in life (see Chapter 2 on DOHaD; Barker et al., 1993). In addition, maternal delivery style has been found to influence infant allergies and asthma occurrences (Salam et al., 2006).

Childbirth may seem paradoxical—a key process in the increasing numbers of humans on this planet in the past and today, and yet a risky endeavour for both mother and baby. For decades, anthropologists have cited the apparent evolutionary mismatch between birthing large brain infants through a bipedally adapted pelvis, known as the obstetric dilemma (OD), as the primary factor in making birth such a challenge. However, in recent years, this paradigm is being re-examined to encompass a more biocultural understanding of maternal and infant health (Stone, 2016). This chapter begins by outlining the human birth mechanisms, the OD paradigm, and evolutionary perspectives on maternal and infant health during childbirth. Next, it discusses the fossil and bioarchaeological evidence for maternal and infant health and mortality in the past. Last, it integrates perspectives drawn from the bioarchaeological record with current understanding of human evolution to piece together a holistic, biocultural picture of childbirth and suggests how modern medicine can potentially learn from perspectives derived from evolutionary medicine. It should be noted in this chapter that, while the terms 'mother' and 'woman' are used throughout to refer to a person giving birth to a neonate, the authors of this chapter recognise that gender-diverse people may also become pregnant and be biological parents (Hahn et al., 2019; Hoffkling et al., 2017; Obedin-Maliver & Makadon, 2016). It is beyond the scope of this chapter to discuss the experiences and medical support of gender-diverse pregnant people, but this recognition is made to underline the need both for more inclusive scholarship and pregnancy-related healthcare (Patel & Sweeney, 2021).

4.2 How does research on childbirth in the past contribute to the understanding of the evolution of health and disease?

Birth is the starting point for understanding evolutionary origins for health and disease and is interlaced with health status throughout life, causing both immediate and long-term impacts on health (Wells et al., 2012). Maternal and neonatal health outcomes are influenced by the social and cultural settings in which birth occurs and there is a recurring pattern of the sociality of birth among humans, further reinforcing the impact of political, historical, and structural factors on birth experiences and health outcomes for both mother and infant. Midwifery may act as a means of supporting women through birth in a social manner, but also importantly focuses on the notion that while birth is patterned, the process of birth does not have a standardised rhythm and varies in phases, parity, and individual experiences (Rutherford et al., 2019). By using an evolutionary framework, researchers can examine the impacts of childbirth on health beyond its proximate causes. Examining the impact of connected social, cultural, biological and epidemiological variables on childbirth in past populations can provide a more nuanced understanding of childbirth. Specifically, this understanding recognises the complexity of the human pelvis from an evolutionary perspective and explicitly acknowledges the connections between epidemiological transitions and sociocultural context and their impacts on maternal health.

4.2.1 The human birth mechanism

The human birth mechanism has previously been described as unique due to the incorrect assumption that no other primates experience the same level of biomechanical limitation, with some researchers suggesting that the differences between human and non-human primate birth have been exaggerated (Stoller, 1995). Indeed, a mismatch between the size of a neonate and the maternal birth canal is not even unique to primates: the inter-pubic ligament of free-tailed bats must stretch to fifteen times its original length during birth (Crelin, 1969). Others have suggested that birth in humans differs from many non-human primates due to the birth mechanism, the duration and difficulty of labour, and behaviour before and after labour (Rosenberg, 1992).

Compared to those of non-human primates, the widest breadth of the pelvic inlet in humans is found transversally and the widest breadth of the outlet is on the sagittal plane (Rosenberg & Trevathan, 1995). The head of a human foetus is largest in the sagittal dimension, creating a very tight fit between it and maternal pelvic dimensions. During labour, the foetal head will line up in the transverse plane and then the foetus must rotate to emerge in the sagittal plane to negotiate the zone of tightest fit in the birth canal (Rosenberg & Trevathan, 1995). This is known as the rotational birth mechanism.

The foetus will enter the birth canal facing side-to-side (or slightly oblique) to the posterior and anterior parts of the maternal birth canal and rotates mid-way to emerge facing away from the mother (Trevathan, 2015). This also requires the foetus to have a flexed neck as it passes through the birth canal, as the smallest diameter of the foetal head is the sub-occipito-bregmatic diameter—flexion of the neck allows the foetal head to pass through the pelvic outlet and beneath the subpubic arch of the maternal pelvis (Rosenberg & Trevathan, 2002). The broad and relatively inflexible foetal shoulders must also pass through the birth canal, requiring two stages of foetal rotation (Trevathan & Rosenberg, 2000). Failure to pass the shoulders through the birth canal after the head in living humans may result in shoulder dystocia, and lead to maternal postpartum haemorrhage and brachial plexus injury in the neonate (Ouzounian & Goodwin, 2010). In contrast, Walrath (2003) put forward that birth mechanisms vary now and probably did so in the past. Walrath (2003) suggests instead that the notion of a monotypic birth mechanism has been imported into paleoanthropological discourse from typological thinking in Euro-American biomedical practice and text.

The position of delivery of a neonate differs across primates, including humans. Monkeys have been observed to deliver in a squatting position, reaching down and guiding the foetus out of their birth canal (Rosenberg & Trevathan, 1995). Trevathan (2011) has suggested that if humans were to mimic the same actions of guiding the foetus out of the birth canal

as seen in some non-human primates, the mother risks foetal nerve and muscle injury by pulling the foetus against the normal flexion of its body. The difficulty of labour among primates also varies, as does the duration of labour. The first stage of labour in healthy women today is approximately 7.7 hours for nulliparas and 5.6 hours for multiparas, with the second stage of labour taking approximately fifty-four minutes for nulliparas and eighteen minutes for multiparas (Albers, 1999). Comparatively, labour length in chimpanzees has been reported as one to two hours (Goodall & Athumani, 1980), and as three hours in gorillas (Cole, 2000).

The challenges of the birth mechanism and labour duration in humans suggests that assistance during birth, described as an experience requiring obligate midwifery (Trevathan, 1988), is a key feature of human birth (Rosenberg & Trevathan, 1995, 2002, 2007; Trevathan, 2015). Walrath (2003) notes that the obligate midwifery model championed by Rosenberg and Trevathan highlights that women require human, rather than technological, assistance at the time of childbirth, drawing a distinction between the support of others during birth and the necessity of birth taking place in biomedical setting.

4.2.2 The obstetric dilemma

Much of the scholarly examination of the human birth mechanism from an evolutionary perspective has focused on multiple energetic and morphological factors that must be negotiated. Washburn (1960) suggested that the tight fit between the maternal birth canal and neonatal head is unique to humans because the human pelvis is predominantly adapted for energetic efficiency during bipedal walking. Subsequent to bipedal adaptations, hominins also underwent strong selection for greater brain size (Gruss & Schmitt, 2015), resulting in conflicting pressures rendering childbirth biomechanically challenging and the birth of humans at a more altricial stage due to the mismatch between the size of the infant head and the maternal birth canal (Washburn's OD). Although the OD remained a fixture in characterising human birth as a necessary danger for some time, it has been challenged in recent years.

The main challenges faced by original the OD paradigm relates to its multiple assumptions. First, it assumes that challenging childbirth is unique to humans. However, multiple ODs have been identified in both humans (Rosenberg, 1992) and non-human primates (Trevathan, 2015), including seeking assistance in birth in bonobos and white-headed langurs (Demuruet et al., 2018; Pan et al., 2014). Second, Warrener and colleagues (2015) demonstrated experimentally that there was no locomotor cost imposed by a wider pelvis, suggesting that variation seen in female pelvic shapes (including those limiting successful parturition) is caused by factors other than biomechanical necessity. Fischer and Mitteroecker (2015) examined covariation between human pelvic shape and head circumference among white males and females from the Hamann–Todd skeletal collection. Stature was significantly associated with pelvic shape in both males and females, but head circumference was only significantly associated with pelvic shape in females (Fischer & Mitteroecker, 2015). Results from their work also demonstrated that females with a larger head circumference had birth canals that were shaped to accommodate large-headed infants better (Fischer & Mitteroecker, 2015).

Third, the OD paradigm offers little to no space for a discussion of the plasticity of the human pelvis and the potential for the influence of ecological conditions during growth which might impact pelvic dimensions. Huseynov and colleagues (2016) examined pelvic development in a known-age and known-sex forensic/clinical sample of males and females from a developmental perspective, including individuals ranging from late foetal stages to late adulthood. They found that the pelves of males and females follow largely similar developmental trajectories until puberty when the female pelvis expands in obstetrically relevant dimensions, continuing up to the age of 30 years of age. Thereafter, from about 40 years of age onwards, female pelves resume a pattern of pelvic development similar to males, resulting in a reduction of obstetric dimensions. Huseynov and colleagues (2016) suggested that this developmental plasticity reflects ecological factors during a female's lifetime that mediate obstetric dimensions. Examining pelvic development in living girls, Moerman (1982) collected data on mediolateral inlet breadth, the interspinous diameter, the maximum breadth of the sacral alae

and the inferior breadth of the ischial tuberosities in healthy girls between the ages of eight and 18. They found that, during growth, the pelvic canal expanded at a slower rate than the increase in stature, and that the pelvis was smaller overall among girls who experienced an early menarche. Likewise, Völgyi and colleagues (2010) investigated growth in the breadth of the female pelvis in living Finnish girls between the ages of 10 and 17, finding that the growth velocity of the width of the pelvic canal peaked at 13.5 months prior to menarche and that by the age of 18 years, the girls in their sample had reached their mother's height but not their mothers pelvic canal width. Shirley and colleagues (2020) surveyed the variability of pelvic dimensions in nulliparous South Asian women in the UK from a developmental perspective, finding that size at birth was a poor predictor for pelvic inlet size, but that height-residuals were predictive of bi-acetabular breadth, bi-iliac breadth, and the pelvic inlet, while tibia length significantly predicted all pelvic dimensions measured in the study except interspinous diameter. The results of these studies suggest that the growth pattern of different parts of the pelvis vary and are influenced by age at menarche, which is itself influenced by factors such as body mass index (BMI) (Cooper et al. 1996; Dunger et al., 2005, 2006; Keim et al., 2009) and qualitative dietary factors in mid-childhood (Rogers et al., 2010).

The plasticity of the human pelvis was discussed in greater depth by Wells and colleagues (2012). They proposed that the OD varies in magnitude with different ecological settings and has not been fixed throughout human history, using the advent of agriculture as an illustration of this (Wells et al., 2012). Labour complications are still likely to have occurred prior to agriculture but Wells and colleagues (2012) suggest that the risk created by compromised pelvic capacity was lower prior to the adoption of agriculture. With the advent of agriculture, female growth and development were compromised by poorer dietary quality (Wells et al., 2012). Skeletal evidence shows that the transition to agriculture resulted in a decline in overall health in many populations (Larsen, 2006; Shuler et al., 2012), with an increase in infectious disease (Helle et al., 2014; Larsen, 2006). Wells and colleagues (2012)

suggest that the resulting delay in skeletal maturation led to compromised female pelvic capacity.

There is also great variation in the size and shape
of the human pelvis, including the birth canal.
Evolutionary studies on the variation in pelvic
shape and size in past populations have demonstrated differences in pelvic canal shape with lati-
tude (Kurki, 2013), evidence of protected obstetric
capacity in small-bodied populations (Kurki, 2011),
differences in the level of variation in the canal
compared to the non-canal parts of the pelvis
(Kurki & Decrausaz, 2016), and that neutral evolution through genetic drift and differential migration are largely responsible for the observed pattern
of pelvic morphological diversity (Betti & Manica, 2018). In addition, interactions between nutrition and disease may have impacted pelvic growth
and development in past populations too (Wells
et al., 2012). Baragi and colleagues (2002) identified
pelvic floor shape differences between the skeletons of European-American and African-American
women from the Hamann–Todd Osteological Collection: African-American women were found to
have a shorter bi-spinous breadth and ischial spine-
to-sacrum length. In living women, Rizk and colleagues (2004) found that European/White women
living in the United Arab Emirates had larger pelvic
inlet and outlet breadths than women from non-
European/White ethnicities also living in the United Arab Emirates.

A fourth critique of the OD paradigm is that
it assumes that the solution to a bipedally adapted birth canal and selection for a large neonatal
brain is to truncate gestation, resulting in an altricial
neonate. Dunsworth and colleagues (2012) examined gestational investment among humans and
other primates. Comparing human and non-human
primate neonate body and brain size to maternal body size, Dunsworth and colleagues (2012)
found that neonatal brain and body size increased
in the hominin lineage and that human maternal investment is greater than expected for a primate of our body mass. Dunsworth and colleagues
(2012) used comparative data on maternal metabolic rates from across primate and other mammal
species to show constraints on metabolic rates, and
that pregnant humans approach a metabolic ceiling
duringpregnancy that constrains further energy
output, limiting gestation length. They propose that
energetic constraints of both mother and foetus
are the primary determinants of gestation length
and foetal growth in humans and across mammals,
naming this the energetics of gestation and growth
hypothesis.

4.3 How do bioarchaeology and palaeopathology offer unique perspectives on childbirth?

Childbirth is not merely the event of parturition, but
is also linked to hormonal, biochemical, and musculoskeletal changes throughout the body. Pregnancy and childbirth may also be impacted by
early life factors, such as malnutrition, disease, and
infection. For example, early life ecological and
nutritional factors may influence pelvic capacity
in adulthood (Wells et al., 2012). Examining birth
in deep time, including fossil hominins, requires
researchers to grapple with pelvic morphological
and functional changes, and consider their relevance alongside factors such an energy and growth
requirements. Palaeopathological examinations of
the human skeleton require one to think holistically, considering how nutrition, exposure to disease and infection, and social determinants of
health coalesce, all of which may leave evidence on
skeletal tissue. The holistic perspective brought by
palaeopathology strengthens biomedical research
on childbirth by linking biological and cultural factors that influence childbirth throughout human
history.

4.3.1 Birth in hominins

To date, no fossil remains have been found with
direct evidence of ancestral birth or pregnancy experiences, so we must make indirect inferences based
on the skeletal morphology of hominins. The earliest fossil hominin pelvis comes from the remains
of *Ardipithecus ramidus* from Ethiopia, dated to
4.4 million years ago (Lovejoy et al., 2009; White
et al., 2009). Pelvic structure in this species has
been described as a 'mosaic', displaying features
that resemble both apes and modern humans, and
reflecting a combination of features that enabled
both bipedal locomotion and arboreal living (Lovejoy et al., 2009; White et al., 2009). This includes a
mediolaterally broad ilium (similar to that seen in

mid-late Miocene hominoids) as well as a large iliac fossa, prominent anterior superior and inferior iliac spines, and iliac blades that flare laterally to position the gluteal muscles in a similar manner to later hominins (Lovejoy et al., 2009). DeSilva (2011) estimated the neonatal mass of hominin taxa based on adult brain mass, neonatal brain mass, and neonatal body mass in anthropoids. Based on his findings, *A. ramidus* would have delivered 1.3-kg-sized infants if it followed an ape model of ontogeny, or a 1-kg infant if it followed a human model. Using the body mass estimates, *A. ramidus* would have birthed infants that were 2.1– 3.2% of the mother's body mass, which is within the range of modern African apes (DeSilva, 2011). These data suggest that a low infant-to-maternal mass ratio was the ancestral condition for the African hominins.

The pelvis of *Australopithecus afarensis* is best represented by two specimens from Ethiopia: A.L. 288-1 dated to 3.2 million years ago (Johanson et al., 1982) and KSD-VP-1/1 dated to 3.6 million years ago (Lovejoy et al., 2016; Wall-Scheffler et al., 2020). The iliac blades of the australopith pelvis are wide and laterally rotated compared with extant apes (Lovejoy, 1988), and have a relatively wide interacetabular distance, making the pelvis more broad in the mediolateral plane than the anteroposterior one (Gruss & Schmitt, 2015). The cranial capacity of australopith infants has been estimated to be 180 cc, slightly larger than that of chimpanzee neonates (DeSilva & Lesnik, 2008). There has been debate about the biomechanics of australopith birth. Specifically, Trevathan and Rosenberg (2000) suggest that birth could not be truly non-rotational for australopiths because, similar to humans, the infant neonate would have to have rotated after emergence of the head to allow passage of its wide, rigid shoulders. Conversely, Tague and Lovejoy (1986) explain that australopiths probably had a nonrotational birth mechanism where the neonate's head would have entered the inlet facing laterally and remained in that position until it exited the outlet.

Laudicina and colleagues (2019) reconstructed the birth mechanics of *Australopithecus sediba* using the Malapa Hominin 2 specimen and found that the midplane and outlet diameters of its pelvic canal show anteroposterior expansion, producing a more *Homo*-like canal shape. Similarly, Kibii and colleagues (2011) found the pelvic inlet in the female pelvis of *A. sediba* to be rounder than that of *A. afarensis* and more akin to the pelvis of *Homo* species. However, the relatively small adult brain size in this species suggests that there was not yet a selective pressure for delivering comparatively large-headed infants (Kibii et al., 2011). Nonetheless, in respect to the relationship between the maternal birth canal and neonatal morphology (body and cranium), and even with detailed descriptions of the pelvic canals, it is challenging to make specific statements about ease of birth in australopiths (Wall-Scheffler et al., 2020).

The mechanisms for delivering an infant for early *Homo* species are complicated by continuing taxonomic uncertainty and a lack of comparative cranial remains that can be used to model possible birth processes (Gruss & Schmitt, 2015). In addition, the OD hypothesis has historically driven interpretations of fossil *Homo* pelves (Wall-Scheffler et al., 2020) and its re-evaluation may prove to impact the manner in which the obstetric function of fossil *Homo* pelves are interpreted. The *Homo erectus* partial juvenile skeleton dated to 1.6 million years ago from Nariokotome in Kenya included a pelvic girdle complete enough to be reconstructed (Brown et al., 1985). Walker and Ruff (1993) estimated the sex of this specimen to be male and based on its morphology, suggested that *Homo erectus* had evolved a narrow body breath, similar to the proportions seen in modern humans, as an adaptation for living in the hot and dry eastern African climate. Wall-Scheffler and colleagues (2020) counter that the more vertically aligned iliac blades and narrow pelvis overall are not fully consistent with the long femoral neck seen in the same specimen nor any other early members of the genus *Homo*. Fortunately, a relatively complete female *Homo erectus* pelvis from Gona in the Afar region of Ethiopia dated to between 1.4 to 0.9 million years ago (Simpson et al., 2008) provides an opportunity to examine obstetric capacities of early members of the genus *Homo* in further detail. Simpson and colleagues (2008) describe the pelvic canal of this specimen as particularly large, given its comparatively small acetabular and auricular surface, indicating broad inlet, midplane, and outlet dimensions, a transversely oval inlet, and a nearly round midplane (Simpson et al., 2008). Based on this, the authors estimate that the Gona female could have delivered a neonate with a maximum brain size of 315 cm^3, which is smaller than the average modern human neonate brain size of approximately 367 cm^3 (DeSilva & Lesnik, 2008).

The relationship between neonatal head size and maternal birth canal dimensions among members of *Homo neanderthalensis* differs slightly to *H. erectus* due to the increased size of the pelvis overall, including a wider birth canal (Gruss & Schmidt, 2015). Ponce De León and colleagues (2008) reconstructed the skeleton of a Neanderthal neonate from Mezmaiskaya Cave in Russia that likely died between one and two weeks after birth. Their reconstruction assumed a rotational birth mechanism for Neanderthals and used the female Tabun C1 Neanderthal pelvis (Israel, dated to 60,000–100,000 BP) as a basis for their reconstruction of the Neanderthal birth mechanism. Their results suggest that Neanderthal brain size at birth was similar to that in recent *Homo sapiens* and was most likely subject to similar obstetric constraints. Neanderthal brain growth rates during early infancy were higher, however, and this pattern of growth resulted in larger adult brain sizes, but not in earlier completion of brain growth (Ponce De León et al., 2008). The female Tabun C1 pelvis has a sagittally expanded midplane relative to earlier hominins and a wide pelvic outlet, even when compared to modern humans (Wall-Scheffler et al., 2020). Weaver and Hublin's (2009) reconstruction of the female Neanderthal Tabun C1 pelvis produced a transversely oval inlet and outlet, which differs from the sagittally oval outlet seen in modern humans, and they concluded that Neanderthals could accommodate a modern human neonate, that birth would not be rotational among Neanderthals and that birth was possible in Neanderthal females via pubic bones that are about the same length as those in males but more coronally oriented. In all, Wall-Scheffler and colleagues (2020) suggest that the pelvic evidence from members of the genus *Homo* seems to point to a mediolaterally broad pelvis and pelvic canal as the plesiomorphic condition for the genus, meaning that, while birth may have become more difficult in *Homo*, both the modern human birth canal shape

and the fully rotational birth mechanism are derived features of *Homo sapiens*.

4.3.2 Birth in past populations

It is challenging to assess obstetric pelvic capacity and difficulties related to childbirth in the past. Bones of the pelvic girdle are frequently poorly preserved in archaeological contexts, rendering collecting data on pelvic dimensions difficult, inaccurate, or impossible. In addition, the poor preservation of foetal skeletal remains means that it is equally challenging to examine neonatal size in the context of childbirth in the past. These obstacles make it impossible to examine childbirth in the past, and instead highlight the importance of considering multiple social and biological variables relating to childbirth. Section 4.3.3 is organised broadly by different transitions in health, spotlighting the relevant interlaced factors impacting childbirth using palaeopathological data from different time periods.

4.3.3 The siren song of 'parturition' scarring

For a number of decades, bioarchaeologists have attempted to identify osteological evidence of childbirth on adult female skeletons. In particular, the presence of pitting or rugosity on the dorsal surface of the pelvic bone, as well as alterations to the sacroiliac joint surface, have been argued to indicate evidence of parturition in females (Angel, 1969; Cox, 1989; Holt, 1978; Houghton, 1974; Stewart, 1957). This group of traits, termed 'parturition scarring', has been argued to arise as a result of the strain placed on muscles, ligaments, and bone during childbirth (Ubelaker & De La Paz, 2012). However, a number of studies found that this scarring does not accurately represent parity status in females (Holt, 1978; Snodgrass & Galloway, 2003; Suchey et al., 1979), and has also been identified on male skeletons (Cox, 1989). Instead, parturition scarring has been found to associate positively with age of parous females (Suchey et al., 1979) and with broader bodily dimensions (Cox, 1989).

Maass and Friedling (2014) found that weight-bearing, body size, and pelvic stability seemed to have a greater influence on the presence of 'parturition' scarring among both males and females in a twenty-first-century sample from the Western Cape province, South Africa. Furthermore, Decrausaz (2014) identified parturition scarring on males and females in a modern North American sample and among sixteenth- to eighteenth-century individuals from Spitalfields, London, finding no association between parturition scarring and parity. Furthermore, dorsal pitting correlated weakly with bi-acetabular breadth, mediolateral pelvic inlet, anterior-posterior direction of the pelvic midplane, mediolateral pelvic outlet, and pubic length among females (Decrausaz, 2014). In all, a number of factors other than childbirth have been found to associate with similar scarring on the dorsal pubis, including general age-related changes, urinary tract infections, lumbosacral anomalies, and obesity, as well as repeated minor trauma, surgery, general joint laxity and pelvic instability, variation in sciatic notch angle, and habitual postures, such as squatting (Ubelaker & De La Paz, 2012).

In contrast, Igarashi and colleagues (2020) did not find scarring on the pelves of nulliparous women (i.e., those who have not given birth) and did find some associations between specific types of scarring and numbers of pregnancies in early twentieth-century Japanese women. McFadden and Oxenham's (2017) meta-analysis of studies examining parturition scarring found that dorsal pitting had a significant but weak relationship with parity, though this is likely a product of the moderate association between dorsal pitting and sex in females unrelated to childbirth. More recently, Waltenberger and colleagues (2020) found a significant association between a strong expression of the pre-auricular sulcus and a retroverted position of the acetabulum and suggested that prolonged labour may cause more strongly expressed 'pelvic features' (their proposed new term for parturition scarring). While it is evident that scarring on the surface of the pelvis cannot be used as an unequivocal indication of childbirth, it is possible that its presence may be impacted by childbirth or pregnancy. To this end, features that have been identified as parturition scarring should be more accurately viewed as pelvic scarring.

4.3.4 Estimating birth-related mortality in the past

Since pelvic scarring cannot provide an accurate picture of childbirth in the past, we must instead look to indirect indicators, which is mainly achieved by interpreting burials of young females and infants. Wells (1975) argued that inferences of death in childbirth should be limited to situations where neonatal bones are found in the pelvic cavity of a female skeleton that also shows evidence of obstruction and/or pelvic deformations. However, many scholars relax these protocols and interpret birth-related deaths even without evidence of obstruction (Smith-Oka et al., 2020). For example, Lieverse and colleagues (2015) reported a notable case of dystocic or obstructed childbirth in a young adult female (approximately 20–25 years old) from the middle Holocene hunter-gatherer Cis-Baikal cemetery of Lokomotiv, Siberia. This female skeleton was found with the partial skeletons of two full-term foetuses located in her pelvic area and between her thighs. The foetuses were full term, appear to have been twins, and were not successfully delivered. The relative locations of the foetal remains and their near- or full-term development, strongly suggest an incomplete breech (feet- or bottom-first) birth of the first infant and a normal vertex (head-down) presentation of the second. This conclusion draws on medical evidence of higher rates of complications related to twin births. For instance, twin pregnancies are at a higher risk of foetal growth restriction, miscarriage, and pre-term delivery (Rao et al., 2004).

As anthropologists and archaeologists are moving past the traditional view of the OD to explain birth-related deaths, we are gaining a better understanding of how complex social, economic, and health-related issues influence both maternal and infant health (Stone, 2016). For example, during the Industrial Revolution, the incidence of rickets and osteomalacia increased due to lack of exposure to sunlight. This can result in restrictions of the growth and development of bones, including the pelvis, that can be identified on skeletal remains. Multiple studies have found evidence to suggest that pelvic deformities resulting from rickets contributed to childbirth difficulties for women living in poor urban areas in the UK in the eighteenth and nine-teenth centuries (Brickley et al., 2005, 2007; Brickley et al., 2014). Maternal morbidity was also height-ened by the cultural fashion of wearing corsets and other restrictive clothing that could lead to defor-mations of the ribs, spine, and pelvis (Smith-Oka et al., 2020; Stone, 2020). These deformations ulti-mately led women to experience increased pain and obstruction during birth (Smith-Oka et al., 2020). In fact, these issues have even been cited as the most common reasons for birth-related deaths from the seventeenth to the twentieth centuries (Wells, 1975).

Other indicators of overall health, such as stature, are also considered to be related to maternal and infant mortality rates. Short stature is often iden-tified as evidence of growth impacted by mal-nutrition or chronic health issues, which in turn may impact the development of the female pelvis (Wells, 2015). For instance, Thoms and Godfried (1939) examined the relationship between mater-nal height and pelvic inlet size among ninety-eight women and reported a correlation of 0.64 between maternal height and size of the pelvic inlet. When examining the relationship between pelvic dimen-sions and height among men and women in North-ern Ireland, Holland and colleagues (1982) found height to be significantly positively correlated with year of birth and with four of seven pelvic dimen-sions measured. Illsley (1966) noted that foetal mortality was high among mothers with a short-er stature, pointing to a cluster of influences of low socio-economic status impacting on growth and development. Data from obstetric practice today show that short-statured pregnant women have increased rate of caesarean section delivery (Mah-mood, Campbell & Wilson, 1988), are more likely to deliver macrosomic neonates (Gudmundsson, Hen-ningsson & Lindqvist, 2005), and have a higher rate of cephalopelvic disproportion (Toh-Adam et al., 2012).

Although these examples can provide evidence of obstructed births that led to maternal and infant deaths, many other causes of death during or short-ly after birth (such as post-partum haemorrhage and infections) are unlikely to be identified in the archaeological record. This is due mainly to the lack of preservation in the archaeological record of health complications related to soft tissue, the fragile and porous nature of neonatal skeletons, and potential differences in burial practices of infants between cultural groups (Halcrow, 2018). As such, bioarchaeologists have developed new ways to esti-mate birth-related deaths in the past. Most recently, McFadden and colleagues (2019, 2020) developed an equation for estimating maternal mortality in the past using data from the United Nations database. Specifically, they used age-at-death data and mater-nal mortality rates for seventy-six countries col-lected between 1990 and 1995 to find a correlation between the rates of birth-related deaths and the female–male death ratios of individuals in their childbearing ages (20–24 years). When they applied their equation to bioarchaeological samples from sixteen sites in Southeast Asia, they found that 50% had what they considered to be low maternal mor-tality rates (i.e., fewer than 323 maternal deaths per 100,000), 19% had moderate rates (i.e., 323 to 605 maternal deaths per 100,000) and 31% had high mortality rates (i.e., more than 605 maternal deaths per 100,000). As stated earlier, the maternal mortali-ty rate in low-income countries was 462 per 100,000 live births in 2017 (WHO, 2019a). In addition, Hill and colleagues (2007) collected mortality data from Hiwi hunter–gatherers in Venezuela and Colom-bia between 1985 and 1992 and estimated that their maternal mortality rate was 440 per 100,000 births. These figures indicate that the estimated maternal mortality rate in the past aligns with that of mod-ern populations that lack access to modern medical interventions.

4.4 How can knowledge of the past contribute to modern clinical research on childbirth and obstetric medicine?

Although childbirth is related to many potential health issues, it is not an illness. Instead, childbirth difficulties may be impacted by factors related to poverty, gender inequality, poor nutrition, and poor education (Stone, 2016). Wells (2010) gives exam-ples of the ways in which maternal capital (defined as the phenotypic resources enabling investment in offspring) may be gained or lost across generations, including nutritional perturbations impacting on growth and development, through experiences such as colonialism. For example, Canadian pregnant adolescents of low socio-economic status living rurally have high rates of adverse pregnancy outcomes (Amjad et al., 2019), and poor American women have a greater likelihood of living with chronic health issues that impact negatively on pregnancy and access to prenatal care (Nagahawatte & Goldenberg, 2008). Short maternal stature may result from malnutrition during growth, which is a risk factor for obstructed labour and gestational diabetes, while malnutrition (including rickets, especially in the past) during development can lead to flattened pelves with increased risk for obstructed labour. These risks may be further exacerbated by child marriage, as pregnancy prior to the end of pelvic development makes obstructed labour a greater possibility.

An increase in maternal BMI with respect to obesity has also been identified as an area of concern in maternal health. Between 2013 and 2018, birthing mothers in Spain with an elevated BMI were at higher risk of cephalopelvic disproportion, pre-eclampsia, and induced labour (Ballesta-Castillejos et al., 2020). In Finland, rates of dystocia were higher among severely obese parturients (BMI $\geq 35 \text{ kg/m}^2$) (Hautakangas et al., 2018). There are multiple factors contributing to increasing BMI among women globally. Nagahawatte and Goldenberg (2008) point to relatively high levels of poverty impacting on women's health prior to pregnancy in the US, as women living in lower socio-economic conditions have a greater likelihood of being obese, are less likely to access prenatal care, and are often living with conditions such as hypertension and diabetes. Wells (2017) suggests that the triple burden of malnutrition combined with the global obesity epidemic and gender inequality may play a role in obstructed labour for women around the world today. Alongside these processes, maternal obesity increases the chance of macrosomic offspring (Wells 2017).

Childbirth difficulties may require mothers to deliver their offspring via caesarean section. There has been an increase in the prevalence of caesarean sections globally, with data gathered from 169 countries showing that such deliveries occurred in approximately 16 million births in 2000 and 29.7 million births in 2015 (Boerma et al., 2018). The

World Health Organization has recently released a set of guidelines for non-clinical interventions to reduce the number of unnecessary caesarean deliveries (WHO, 2018). These unnecessary deliveries now account for 18.6% of births in 150 countries, although their frequency can reach 40.5% in Latin America and the Caribbean (Betrán et al., 2016). When the prevalence of caesarean deliveries exceeds 10% of births, they are not associated with decreases in maternal and neonatal mortality rates (Ye et al., 2016), and children born by emergency caesarean section were found to have an increased risk of obesity (Masukume et al., 2018) and developing autoimmune diseases (Kristensen & Henriksen, 2016).

Despite their potential health implications, caesarean sections are still common today for several reasons. Medical indications include length of labour, a lack of continuity-of-care models for childbirth, mismatch between maternal birth canal and neonate size, shoulder dystocia, abnormal foetal heart rates, issues with placental implantation, and abnormal positioning of the foetus (Rosenberg & Trevathan, 2018). Caesarean deliveries are often selected for the convenience of the obstetrician. For instance, 35% of obstetric gynaecologists surveyed across eight hospitals in Mexico City scheduled a caesarean section for convenience (Vallejos Parás et al. 2018). In Brazil, women report opting for caesarean sections because of cultural perspectives on pain management, medical policy, and social perceptions of physical appearance being impacted by pregnancy (Hopkins, 2000). The prevalence of caesarean deliveries in Australia is associated with the notion that it is an easier and more convenient manner of delivering a baby (Walker et al., 2004). In addition, there can be a power imbalance between birthing mothers and doctors, with some doctors possibly using their power to coerce women to deliver via caesarean section (Hopkins, 2000). Hanley and colleagues (2010) examined regional variation in the use of caesarean delivery in the Canadian province of British Columbia and found that its use was explained more by differing approaches in medical decision-making by practitioners rather than by patients.

Zaffarini and Mitteroecker (2019) recently created a model suggesting that secular trends in stature can predict global rates of caesarean deliveries. Their study hypothesises that the variation in secular trends in height in different countries has resulted in an intergenerational change in body size. This stature increase is associated with an increase in mismatch between the size of the neonate and the maternal birth canal, resulting in obstructed labour, and thereby, the need for caesarean deliveries (Zaffarini & Mitteroecker, 2019). Using data from the WHO, UNICEF, and the United Nations (2015), Zaffarini and Mitteroecker (2019) found that between 1971 and 1996, caesarean delivery rate and average stature change had a strong linear positive association. Mitteroecker and colleagues (2016) proposed a theoretical model for the high incidence of obstructed labour, which they called the cliff-edge model of obstetric selection. This model posits there is a positive correlation between neonatal size and maternal birth canal size and once this equilibrium reaches a critical value, natural birth is no longer possible (i.e., the cliff edge). As a result of this highly asymmetric cliff-edge fitness distribution, the symmetrical phenotypic distribution of neonatal size and maternal birth canal dimensions cannot effectively match the fitness distribution. The ideal distribution of neonatal size and maternal birth canal size involves a portion of individuals falling outside of the fitness distribution and experiencing disproportion between foetal head size and maternal birth canal dimensions (Mitteroecker et al., 2016). Using the cliff-edge model, Mitteroecker and colleagues (2016) projected an increase in rates of foetal-maternal birth canal size mismatch due to the regular use of caesarean deliveries acting as a reduction to maternal mortality. However, Grossman (2017) critiqued Mitteroecker and colleagues' (2016) study, outlining that rates of mismatch between the maternal birth canal and the head of the foetus have increased due to several reasons according to the obstetrical literature. He argued that the cliff-edge model did not account for variation in maternal soft tissue that could contribute to obstructed labour, nor the clinical realities of managing different labour stages, and that caesarean rates have also increased due to the loss of surgical training in forceps delivery.

Human birth today is frequently a social event. Human mothers deliver infants with support from other people (Rosenberg & Trevathan, 2002),

including midwives. Midwifery is often considered a necessity in humans due to the biomechanical challenges of childbirth and can also greatly improve childbirth experiences. For example, women in rural Sweden were more likely to have a positive birth experience when they received care from a midwife they knew (Hildingsson et al., 2020). In the UK, midwives are under a professional obligation to support and advocate for women choosing to birth at home (National Institute for Health and Care Excellence 2017; NMC, 2018). Historical sources point to the professionalisation of midwifery in Europe during the late Middle Ages (1100–1400 CE) (Piper, 2010). It should be noted that childbirth occurring in biomedical settings (such as hospitals) is relatively recent (Stone, 2016), meaning that midwifery was more commonly practiced in domestic spaces among past populations. Among industrialised communities during the eighteenth century, childbirth moved to biomedical settings with lying-in hospitals. Maternal mortality increased among women who delivered in these hospitals due to poor hygiene practices, puerperal fever, and haemorrhage (see Stone, 2016 for a detailed review). With the formalisation of biomedicine in the nineteenth century, midwives were often vilified and excluded from assisting women with birth (Stacey, 1988). The formalisation of biomedicine also worked to characterise the female body as a faulty machine, particularly with respect to reproduction, supporting the notion that birth needed to be managed in a biomedical setting (Martin, 2001).

An evolutionary perspective on women's reproductive health is often excluded from medical education (Power et al., 2020), which works to remove a more holistic perspective on the factors that impact key health events such as childbirth. For example, most measurements for the female pelvis used in obstetric education are based on European women, leading to any deviation from these dimensions not being considered normal, and potentially impacting the level of intervention during childbirth for non-European women (Betti, 2021). Understanding the interactions between growth and development and pelvic variability, examined in longitudinal studies (Moerman, 1982; Völgyi et al., 2010) and the impacts of early life on pelvic dimensions in adulthood

(Shirley et al., 2020), may also provide avenues for strengthening ongoing work on how skeletal markers of puberty and development in past populations that could be relevant for reproductive health (see Lewis et al., 2016).

Throughout human evolution, support for the birthing mother was perhaps essential and remains an integral element of maternal health today. Strengthening the capacity of midwives to deliver high-quality maternal and new-born care has been found to improve outcomes for low- and middle-income countries. Nove and colleagues (2021) found that midwife care could prevent 41% of maternal deaths and 39% of neonatal deaths. To ensure that this potential is realised, midwives need to have access to relevant education programmes. The International Confederation of Midwives identifies essential competencies for midwifery practice which highlight the role of the midwife to practice autonomously to support physiological birth, reduce unnecessary interventions, diagnose deviations from normal, refer to doctors as necessary, and provide emergency interventions (ICM, 2018).

4.5 Conclusion

Examining the impact of connected social, cultural, biological, and epidemiological variables on childbirth in past populations can provide a more nuanced understanding of childbirth. Childbirth today is impacted by the same variables that influence health overall, including socio-economic status, generational epidemiological impacts of ongoing processes such as colonialism and gender inequality, and changes to access in medical support. Bioarchaeological approaches bring together skeletal and contextual health evidence, placing the skeleton back into a living body, and acknowledging the multiple factors influencing childbirth. Studies examining growth throughout the reproductive years with a focus on both soft tissues and skeletal changes (e.g., Völgyi et al., 2010) will help to contextualise skeletal changes better. In turn, this could aid bioarchaeological and palaeopathological investigations of pregnancy among mothers and infants who might have died in childbirth. The OD paradigm must be understood as a theoretical concept, no longer as the primary explanation

for the gestation length, childbirth difficulty, and the developmental biology of new-born humans (Dunsworth, 2018). Midwifery practice recognises the importance of compound skeletal, muscular and hormonal actions prior to and during delivery, as well as structural influences on a mother's birth experience. Supporting birthing women with continuity of care is a critical component of the role that midwifery can play in more holistic birth practices (WHO, 2019b). We suggest that future work on childbirth in humans not only acknowledges, but actively incorporates, a holistic research agenda that identifies the multitude of factors that influence childbirth.

References

Albers, L. L. (1999). The duration of labor in healthy women. *Journal of Perinatology*, 19, 114–119.

Amjad, S., Chandra, S., Osornio-Vargas, A., Voaklander, D. & Ospina, M. B. (2019). Maternal area of residence, socioeconomic status, and risk of adverse maternal and birth outcomes in adolescent mothers. *Journal of Obstetrics and Gynaecology of Canada*, 41, 1752–1759.

Angel, J. L. (1969). The bases of paleodemography. *American Journal of Physical Anthropology*, 30, 427–437.

Ballesta-Castillejos A, Gómez-Salgado J, Rodríguez-Almagro J., Ortiz-Esquinas I. & Hernández-Martínez, A. (2020). Relationship between maternal body mass index and obstetric and perinatal complications. *Journal of Clinical Medicine*, 9, 707.

Baragi, R., Delancey, J. O. L., Caspari, R., Howard, D. H. & Ashton-Miller, J. A. (2002). Differences in pelvic floor area between African American and European American women. *American Journal of Obstetrics and Gynecology*, 187, 111–115.

Barker, D. J., Godfrey, K., Gluckman, P., Harding, J. E., Owens, J. A. & Robinson, J. S. (1993). Fetal nutrition and cardiovascular disease in adult life. *The Lancet*, 341, 938–941.

Betrán, A. P., Ye, J., Moller, A. B., Zhang, J., Gülmezoglu, A. M. & Torloni, M. R. (2016). The increasing trend in caesarean section rates: global, regional and national estimates: 1990-2014. *PLoS One*, 11, 1–12.

Betti L. (2021). Shaping birth: variation in the birth canal and the importance of inclusive obstetric care. *Philosophical Transactions of the Royal Society B: Biological Sciences*, 376. doi: 10.1098/rstb.2020.0024

Betti, L. & Manica, A. (2018). Human variation in the shape of the birth canal is significant and geographically structured. *Proceedings of the Royal Society B: Biological Sciences*, 285. doi: 10.1098/rspb.2018.1807

Blomquist, J. L., Muñoz, A., Carroll, M. & Handa, V. L. (2018). Association of delivery mode with pelvic floor disorders after childbirth. *Journal of the American Medical Association*, 320, 2438–2447.

Boerma, T., Ronsmans, C., Melesse, D. Y., et al. (2018). Global epidemiology of use of and disparities in caesarean sections. *The Lancet*, 392, 1341–1348.

Bomboka, J. B., N-Mboowa, M. G. & Nakilembe, J. (2019). Post-effects of obstetric fistula in Uganda: a case study of fistula survivors in KITOVU mission hospital (MASAKA), Uganda. *BMC Public Health*, 19, 1–7.

Brickley, M., Mays, S. & Ives, R. (2007). An investigation of skeletal indicators of vitamin D deficiency in adults: effective markers for interpreting past living conditions and pollution levels in 18th and 19th century Birmingham, England. *American Journal of Physical Anthropology*, 132, 67–79.

Brickley, M., Mays, S. & Ives, R. (2005). Skeletal manifestations of vitamin D deficiency osteomalacia in documented historical collections. *International Journal of Osteoarchaeology*, 15, 389–403.

Brickley, M. B., Moffat, T. & Watamaniuk L. (2014). Biocultural perspectives of vitamin D deficiency in the past. *Journal of Anthropological Archaeology*, 36, 48–59.

Brown, F., Harrist, J., Leakey, R., & Walker, A. (1985). Early *Homo erectus* skeleton from west Lake Turkana, Kenya. *Nature*, 316, 788–792.

Cole, R. E. (2000). Obstetric management of a protracted labor in a captive Western lowland gorilla. *American Journal of Obstetrics and Gynecology*, 182, 1306–1311.

Cooper, C., Kuh, D., Egger, P., Wadsworth, M. & Barker, D. (1996). Childhood growth and age at menarche. *British Journal of Obstetrics*, 103, 814–817.

Cox M. (1989). *An evaluation of the significance of 'scars of parturition' in the Christ Church Spitalfields sample*. PhD Thesis, University of London, London.

Crelin, E. S. (1969). The development of the bony pelvis and its changes during pregnancy and parturition. *Transactions of the New York Academy of Science*, 31, 1049–1058.

Decrausaz, S-L. (2014). *A morphometric analysis of parturition scarring on the human pelvic bone*. Master's thesis, University of Victoria, Victoria.

Demuru, E., Ferrari, P. F. & Palagi, E. (2018). Is birth attendance a uniquely human feature? New evidence suggests that Bonobo females protect and support the parturient. *Evolution and Human Behavior*; 39, 502–510. doi: 10.1016/j.evolhumbehav.2018.05.003

DeSilva, J. M. (2011). A shift toward birthing relatively large infants early in human evolution. Proceedings of the National Academy of Science, 108, 1022–1027.

DeSilva, J. M & Lesnik, J. J. (2008). Brain size at birth throughout human evolution: A new method for estimating neonatal brain size in hominins. *Journal of Human Evolution*, 55, 1064–1074.

Dunger, D. B., Ahmed, M. L. & Ong, K. K. (2006). Early and late weight gain and the timing of puberty. *Molecular & Cellular Endocrinology*, 254–255, 140–145.

Dunger, D. B., Ahmed, M. L. & Ong, K. K. (2005). Effects of obesity on growth and puberty. *Best Practice & Research: Clinical Endocrinology and Metabolism*, 19, 375–390.

Dunsworth, H. M. (2018). There is no 'obstetrical dilemma': towards a braver medicine with fewer childbirth interventions. *Perspectives in Biological Medicine*, 61 249–263.

Dunsworth, H. M., Warrener, A. G., Deacon, T., Ellison, P. T. & Pontzer, H. (2012). Metabolic hypothesis for human altriciality. *Proceedings of the National Academy of Sciences of the United States of America*, 109, 15212–15216.

Fischer, B. & Mitteroecker, P. (2015). Covariation between human pelvis shape, stature, and head size alleviates the obstetric dilemma. *Proceedings of the National Academy of Sciences of the United States of America*, 112, 5655–5660.

Geller, S. E., Koch, A. R., Garland, C. E., MacDonald, E. J., Storey, F. & Lawton, B. (2018). A global view of severe maternal morbidity: moving beyond maternal mortality. *Reproductive Health*, 15. doi: 10.1186/s12978-018-0527-2.

Goodall, J. & Athumani, J. (1980). An observed birth in a free-living chimpanzee (*Pan troglodytes schweinfurthii*) in Gombe National Park, Tanzania. *Primates*, 21, 545–549.

Grossman, R. (2017). Are human heads getting larger? *Proceedings of the National Academy of Sciences of the United States of America*, 114, E1304.

Gruss, L. T. & Schmitt, D. (2015). The evolution of the human pelvis: changing adaptations to bipedalism, obstetrics and thermoregulation. *Philosophical Transactions of the Royal Society B: Biological Sciences*, 370, 1–13.

Gudmundsson, S., Henningsson, A. C. & Lindqvist, P. (2005). Correlation of birth injury with maternal height and birthweight. *British Journal of Obstetrics and Gynaecology*, 112, 764–767.

Hahn, M., Sheran, N., Weber, S., Cohan, D. & Obedin-Maliver, J. (2019). Providing patient-centered perinatal care for transgender men and gender-diverse individuals: a collaborative multidisciplinary team approach. *Obstetrics and Gynecology*, 134, 959–963.

Halcrow, S. E. (2018). Review of *Paleopathology of children: identification of pathological conditions in the human skeletal remains of non-adults*, by Mary Lewis. *Childhood in the Past*, 11, 64–66.

Hanley, G. E., Janssen, P. A. & Greyson, D. (2010). Regional variation in the cesarean delivery and assisted vaginal delivery rates. *Obstetrics and Gynecology*, 115, 1201–1208.

Hautakangas, T., Palomäki, O., Eidstø, K., Huhtala, H. & Uotila, J. (2018). Impact of obesity and other risk factors on labor dystocia in term primiparous women: A case control study. *BMC Pregnancy and Childbirth*, 18, 1–8.

Helle, S., Brommer, J. E., Pettay, J. E., Lummaa, V., Enbuske, M. & Jokela, J. (2014). Evolutionary demography of agricultural expansion in preindustrial northern Finland. *Proceedings of the Royal Society of Medicine B: Biological Sciences*, 281, 20141559–20141569.

Hildingsson, I., Karlström, A. & Larsson, B. (2020). Childbirth experience in women participating in a continuity of midwifery care project. *Women and Birth*, 34(3), e255-e261. doi: 10.1016/j.wombi.2020.04.010

Hill, K., Hurtado, A. M. & Walker, R. S. (2007). High adult mortality among Hiwi hunter-gatherers: Implications for human evolution. *Journal of Human Evolution*, 52, 443–454.

Hoffkling, A., Obedin-Maliver, J. & Sevelius, J. (2017). From erasure to opportunity: A qualitative study of the experiences of transgender men around pregnancy and recommendations for providers. *BMC Pregnancy and Childbirth*, 17. doi: 10.1186/s12884-017-1491-5

Holland, E. L., Cran, G. W., Elwood, J. H. Pinkerton, J. H. & Thompson, W. (1982). Associations between pelvic anatomy, height and year of birth of men and women in Belfast. *Annals of Human Biology*, 9, 113–120.

Holt, C. A. (1978). A re-examination of parturition scars on the human female pelvis. *American Journal of Physical Anthropology*, 49, 91–94.

Hopkins, K. (2000). Are Brazilian women really choosing to deliver by cesarean? *Social Science & Medicine*, 51, 725–740.

Houghton, P. (1974). The relationship of the pre-auricular groove of the ilium to pregnancy. *American Journal of Physical Anthropology*, 41, 381–389.

Howell, E. A. (2018). Reducing disparities in severe maternal morbidity and mortality. *Clinical Obstetrics and Gynecology*, 61, 387–399.

Huseynov, A., Zollikofer, C. P. E., Coudyzer, W., Gascho, D., Kellenberger, C., Hinzpeter, R. & Ponce de León, M. S. (2016). Developmental evidence for obstetric adaptation of the human female pelvis. *Proceedings of the National Academy of Sciences of the United States of America*, 113, 5227–5232.

International Confederation of Midwives. (2018). Essential competencies for midwifery practice 2018 update [online]. https://www.internationalmidwives.org/assets/files/general-files/2018/10/icm-competencies---english-document_final_oct-2018.pdf

Igarashi, Y., Shimizu, K., Mizutaka, S. & Kagawa, K. (2020). Pregnancy parturition scars in the preauricular area and the association with the total number of pregnancies and parturitions. *American Journal of Physical Anthropology*, 171, 260–274.

Illsley, R. (1966). Preventive Medicine in the Perinatal Period: Early Prediction of Perinatal Risk. *Journal of the Royal Society of Medicine*, 59, 181–184.

Johanson, D. C., Lovejoy, C. O., Kimbel, W. H., White, T. D., Ward, S. C., Bush, M. E., Latimer, B. M. & Coppens, Y. (1982). Morphology of the Pliocene partial hominid skeleton (A.L. 288-1) from the Hadar formation, Ethiopia. *American Journal of Physical Anthropology*, 57, 403–451.

Keim, S. A., Branum, A. M., Klebanoff, M. A. & Zemel, B. S. (2009). Maternal body mass index and daughters' age at menarche. *Epidemiology*, 20, 677–681.

Kibii, J., Churchill, S. E., Schmid, P., Carlson, K. J., Reed, N. D., de Ruiter, D. J. & Berger, L. R. (2011). A partial pelvis of *Australopithecus sediba*. *Science*, 333(6048), 1407–1411.

Knight, M., Bunch, K., Kenyon, S., Tuffnell, D. & Kurinczuk, J. J. (2020). A national population-based cohort study to investigate inequalities in maternal mortality in the United Kingdom, 2009-17. *Paediatric and Perinatal Epidemiology*, 34(4), 1–7.

Kristensen, K. & Henriksen, L. (2016). Cesarean section and disease associated with immune function. *Journal of Allergy and Clinical Immunology*, 137, 587–590.

Kurki, H. K. (2011). Compromised skeletal growth? Small body size and clinical contraction thresholds for the female pelvic canal. *International Journal of Paleopathology*, 1, 138–149.

Kurki, H. K. (2013). Bony pelvic canal size and shape in relation to body proportionality in humans. *American Journal of Physical Anthropology*, 151, 88–101.

Kurki, H. K. & Decrausaz, S. L. (2016). Shape variation in the human pelvis and limb skeleton: Implications for obstetric adaptation. *American Journal of Physical Anthropology*, 159, 630–638.

Larsen, C. S. (2006). The agricultural revolution as environmental catastrophe: Implications for health and lifestyle in the Holocene. *Quaternary International*, 150, 12–20.

Laudicina, N. M., Rodriguez, F. & DeSilva, J. M. (2019). Reconstructing birth in *Australopithecus sediba*. *PLoS One*, 14, 1–18.

Lewis, M., Shapland, F. & Watts, R. (2016). On the threshold of adulthood: A new approach for the use of maturation indicators to assess puberty in adolescents from medieval England. *American Journal of Human Biology*, 28, 48–56.

Lieverse, A. R., Bazaliiskii, V. I. & Weber, A. W. (2015). Death by twins: a remarkable case of dystocic childbirth in Early Neolithic Siberia. *Antiquity*, 89, 23–38.

Lovejoy, C. O. (1988). Evolution of human walking. *Scientific American*, 259, 118–125.

Lovejoy, C. O., Latimer, B. M., Spurlock, L. & Haile-Selassie, Y. (2016). 'The pelvic girdle and limb bones of KSD-VP-1/1', in Y. Haile-Selassie & D. Su (eds), *The postcranial anatomy of* Australopithecus Afarensis: *new insights from KSD-VP-1/1,*155–178. Springer Netherlands, Dordrecht.

Lovejoy, C. O., Suwa, G., Spurlock, L., Asfaw, B. & White, T. D. (2009). The pelvis and femur of *Ardipithecus ramidus*: the emergence of upright walking. *Science*, 326, 71–77.

Luo, Z. C., Kierans, W. J., Wilkins, R., Liston, R. M., Uh, S-H. & Kramer, M. S. (2004). Infant mortality among First Nations versus non-First Nations in British Columbia: temporal trends in rural versus urban areas, 1981-2000. *International Journal of Epidemiology*, 33, 1252–1259.

Maass, P. & Friedling, L. J. (2014). Scars of parturition? Influences beyond parity. *International Journal of Osteoarchaeology*, 26, 121–131. doi: 10.1002/oa.2402

Mahmood, T. A., Campbell, D. M. & Wilson, A. W. (1988). Maternal height, shoe size, and outcome of labour in white primigravidas: a prospective anthropometric study. *British Medical Journal*, 297, 515–517.

Martin, E. (2001). *The woman in the body: a cultural analyses of reproduction*. Beacon Press, Boston.

Masukume, G., O'Neill, S. M., Baker, P. N., Kenny, L. C., Morton, S. M. B. & Khashan, A. S. (2018). The impact of caesarean section on the risk of childhood overweight and obesity: new evidence from a contemporary cohort study. *Scientific Reports*, 8 1–9.

McFadden, C. & Oxenham, M. F. (2019). The paleodemographic measure of maternal mortality and a multifaceted approach to maternal health. *Current Anthropology*, 60, 000–000.

McFadden, C. & Oxenham, M. F. (2017). Sex, parity, and scars: a meta-analytic review. *Journal of Forensic Science*, 63, 201–206.

McFadden, C., Van Tiel, B. & Oxenham, M. F. (2020). A stabilized maternal mortality rate estimator for biased skeletal samples. *Anthropological Science*, 128, 113–117.

Mitteroecker, P., Huttegger, S. M., Fischer, B. & Pavlicev, M. (2016). Cliff-edge model of obstetric selection in humans. *Proceedings of the National Academy of Sciences of the United States of America*, 113, 14680–14685.

Moerman, M. (1982). Growth of the birth canal in adolescent girls. *American Journal of Obstetrics and Gynecology*, 143 528–532.

National Institute for Health and Care Excellence. (2017). 'Care of women and their babies during labour care - choosing where to have your baby', in *Intrapartum care for healthy women and babies [CG 190]* [online]. https://www.nice.org.uk/guidance/cg190

Nursing & Midwifery Council (NMC). (2018). Standards for competence for registered midwives [online]. https://www.nmc.org.uk/globalassets/sitedocuments/standards/nmc-standards-for-competence-for-registered-midwives.pdf

Nove, A., Friberg, I. K., De Bernis, L., McConville, F., Moran, A. C., Majjemba, M., ten Hoope-Bender, P., Tracy, S. & Homer, C. S. E. (2021). Potential impact of midwives in preventing and reducing maternal and neonatal mortality and stillbirths: a Lives Saved Tool modelling study. The *Lancet Global Health*, 9, e24–32.

Obedin-Maliver, J. & Makadon, H. J. (2016). Transgender men and pregnancy. *Obstetric Medicine*, 9, 4–8.

Ouzounian, J. & Goodwin, T. M. (2010). 'Shoulder dystocia', in T. M. Goodwin, M. N. Montoro, L. Muderspach, R. Paulson & S. Roy (eds), *Management of common problems in obstetrics and gynecology*, 5th edn, 45–47. Wiley-Blackwell, Hoboken.

Pan, W., Gu, T., Pan, Y., Feng, C., Long, Y., Zhao, Y., Meng, H., Liang, Z. & Yao, M. (2014). Birth intervention and non-maternal infant-handling during parturition in a nonhuman primate. *Primates*, 55, 483–488.

Patel, S. & Sweeney, L. B. (2021). Maternal health in the transgender population. *Journal of Women's Health*, 30, 253–259.

Piper, L. The professionalization of midwifery in the late Middle Ages. (2010). *Canadian Journal of Midwifery Research & Practice*, 9, 21–34.

Ponce de León, M. S., Golovanova, L., Doronichev, V., Romanova, G., Akazawa, T., Kondo, O., Ishida, H. & Zollikofer, C. P. E. (2008). Neanderthal brain size at birth provides insights into the evolution of human life history. *Proceedings of the National Academy of Sciences of the United States of America*, 105, 13764–13768.

Power, M. L., Snead, C., Reed, E. G. & Schulkin, J. (2020). Integrating evolution into medical education for women's health care practitioners. *Evolution, Medicine, & Public Health*, 2020(1), 60–67.

Rao, A., Sairam, S. & Shehata, H. (2004). Obstetric complications of twin pregnancies. *Best Practice & Research Clinical Obstetrics and Gynaecology*, 18, 557–576.

Rizk, D. E. E., Czechowski, J. & Ekelund, L. (2004). Dynamic assessment of pelvic floor and bony pelvis morphologic condition with the use of magnetic resonance imaging in a multiethnic, nulliparous, and healthy female population. *American Journal of Obstetrics and Gynecology*, 191, 83–89.

Rogers, I. S., Northstone, K., Dunger, D. B., Cooper, A. R., Ness, A. R. & Emmett, P. M. (2010). Diet throughout childhood and age at menarche in a contemporary cohort of British girls. *Public Health Nutrition*, 13, 2052–2063.

Rosenberg K. (1992). The evolution of modern human childbirth. *Yearbook of Physical Anthropology*, 35, 89–124.

Rosenberg, K. & Trevathan, W. (1995). Bipedalism and human birth: the obstetrical dilemma revisited. *Evolutionary Anthropology: Issues, News, and Reviews*, 4, 161–168.

Rosenberg, K. & Trevathan, W. (2002). Birth, obstetrics and human evolution. *British Journal of Obstetrics and Gynecology*, 109, 1199–1206.

Rosenberg K. R. & Trevathan, W. R. (2007). An anthropological perspective on the evolutionary context of preeclampsia in humans. *Journal of Reproductive Immunology*, 76, 91–97.

Rosenberg, K. & Trevathan, W. (2018). Evolutionary perspectives on cesarean section. *Evolution, Medicine, & Public Health*, 2018, 67–81.

Rutherford, J. N., Asiodu, I. V. & Liese, K. L. (2019). Reintegrating modern birth practice within ancient birth process: What high cesarean rates ignore about physiologic birth. *American Journal of Human Biology*, 31, 1–11.

Salam, M. T., Margolis, H. G., McConnell, R., McGregor, J. A., Avol, E. L. & Gilliland, F. D. (2006). Mode of delivery is associated with asthma and allergy occurrences in children. *Annals of Epidemiology*, 16, 341–346.

Say, L., Chou, D., Gemmill, A., Tunçalp, Ö., Moller, A-B., Daniels, J., Gülmezoglu, A. M., Temmerman, M. & Alkema, L. (2014). Global causes of maternal death: a WHO systematic analysis. *The Lancet Global Health*, 2, 323–333.

Shirley, M. K., Cole, T. J., Arthurs, O. J., Clark, C. A. & Wells, J. C. K. (2020). Developmental origins of variability in pelvic dimensions: evidence from nulliparous South Asian women in the United Kingdom. *American Journal of Human Biology*, 32, 1–13.

Shuler, K. A., Hodge, S. C., Danforth, M. E., Funkhouser, J. L., Stantis, C., Cook, D. N. & Zeng, P. (2012). In the shadow of Moundville: a bioarchaeological view of the transition to agriculture in the central Tombigbee valley of Alabama and Mississippi. *Journal of Anthropological Archaeology*, 31, 586–603.

Simpson, S., Quade, J., Levin, N., Butler, R., Dupont-Nivet, G., Everett, M. & Semaw, S. (2008). A female *Homo erectus* pelvis from Gona, Ethiopia. Science, 322, 1089–1092.

Smith-Oka, V., Nissen, N. J., Wornhoff, R. & Sheridan, S. (2020). '"I thought I was going to die": examining experiences of childbirth pain through bioarchaeological and ethnographic perspectives', in S. G. Sheridan & L. A. Gregoricka (eds), *Purposeful pain: the bioarchaeology of intentional suffering*, pp. 149–176. Springer, Cham.

Snodgrass, J. J. & Galloway, A. (2003). Utility of dorsal pits and pubic tubercle height in parity assessment. *Journal of Forensic Science*, 48, 1–5.

Stacey, M. (1988). *The sociology of health and healing*. Routledge, London.

Stewart, T. D. (1957). Distortion of the pubic symphyseal surface in females and its effect on age determination. *American Journal of Physical Anthropology*, 15, 9–18.

Stoller, M. (1995). *The obstetric pelvis and mechanism of labor in non-human primates*. PhD Thesis, University of Chicago, Chicago.

Stone, P. K. (2016). Biocultural perspectives on maternal mortality and obstetrical death from the past to the present. *American Journal of Physical Anthropology*, 159, S150–71.

Stone, P. K. (2020). 'Bound to please: the shaping of female beauty, gender theory, structural violence, and bioarchaeological investigations', in S. G. Sheridan & L. A. Gregoricka (eds), *Purposeful pain: the bioarchaeology of intentional suffering*, pp. 39–62. Springer, Cham.

Suchey, J. M., Wiseley, D. V., Green, R. F. & Noguchi, T. T. (1979). Analysis of dorsal pitting in the os publis in an extensive sample of modern American females. *American Journal of Physical Anthropology*, 51, 517–540.

Tague, R. G. & Lovejoy, C. O. (1986). The obstetric pelvis of A.L. 288-1 (Lucy). *Journal of Human Evolution*, 15, 237–255.

Nagahawatte N. T. & Goldenberg, R. L. (2008). Poverty, maternal health, and adverse pregnancy outcomes. *Annals of the New York Academy of Science*, 1136, 80–85.

Thoms, H. & Godfried, M. S. (1939). The interrelationships between fetal weight, size of pelvic inlet, and maternal height. *Yale Journal of Biological Medicine*, 11, 355–362.

Toh-Adam, R., Srisupundit, K. & Tongsong, T. (2012). Short stature as an independent risk factor for cephalopelvic disproportion in a country of relatively small-sized mothers. *Archives of Gynecology and Obstetrics*, 285, 1513–1516.

Trevathan, W. Primate pelvic anatomy and implications for birth. (2015). *Philosophical Transactions of the Royal Society B: Biological Sciences*, 370, 20140065.

Trevathan, W. R. (1988). Fetal emergence patterns in evolutionary perspective. American Anthropologist, 90, 674–681.

Trevathan, W. R. (2011). *The process of parturition. human birth: an evolutionary perspective*, pp. 65–118. Routledge, New York.

Trevathan, W. Rosenberg, K. (2000). The shoulders follow the head: postcranial constraints on human childbirth. *Journal of Human Evolution*, 39, 583–586.

Ubelaker, D. H. & de la Paz, J. S. (2012). Skeletal indicators of pregnancy and parturition: a historical review. *Journal of Forensic Science*, 57, 866–872.

Vallejos Parás, A., Espino y Sosa, S., Jaimes Betancourt, L., Zepeda Tena, C., Cabrera Gaytán, D. A., Arriaga Nieto, L., Valle Alvarado, G., López Cevantes, M. & Durán Arenas, L. (2018). Obstetrician's attitudes about delivery through cesarean section: a study in hospitals at Mexico City. *Perinatología y Reproducción Humana*, 32, 19–26.

Völgyi, E., Tylavsky, F. A., Xu, L., Lu, J., Wang, Q., Alén, M. & Sheng, S. (2010). Bone and body segment lengthening and widening: a 7-year follow-up study in pubertal girls. *Bone*, 47, 773–782.

Walker. A. & Ruff, C. B. (1993). 'The reconstruction of the pelvis', in A. Walker & R. Leakey (eds), *The Nariokotome* Homo erectus *skeleton*, 221–233. Harvard University Press. Cambridge, MA.

Walker, R., Turnbull, D. & Wilkinson, C. (2004). Increasing cesarean section rates: Exploring the role of culture in an Australian community. *Birth*, 31, 117–124.

Wall-Scheffler, C. M., Kurki, H. K. & Auerbach, B. M. (2020). *The evolutionary biology of the human pelvis: an integrative approach*. Cambridge University Press, Cambridge.

Walrath, D. (2003). Rethinking pelvic typologies and the human birth mechanism. *Current Anthropology*, 44, 5–31.

Waltenberger, L., Pany-Kucera, D., Rebay-Salisbury, K. & Mitteroecker, P. (2020). The association of parturition scars and pelvic shape: a geometric morphometric study. *American Journal of Biological Anthropology*, 174, 519–531.

Warrener, A. G., Lewton, K. L, Pontzer, H. & Lieberman, D. E. (1960). A wider pelvis does not increase locomotor cost in humans, with implications for the evolution of childbirth. *PLoS One*, 10, e0118903.

Washburn, S. L. (1960). Tools and human evolution. *Scientific American*, 203, 63–75.

Weaver, T. D. & Hublin, J-J. (2009). Neandertal birth canal shape and the evolution of human childbirth. *Proceedings of the National Academy of Sciences of the United States of America*, 106, 8151–8156.

Wells, C. (1975). Ancient obstetric hazards and female mortality. *Bulletin of the New York Academy of Medicine*, 51, 1235–1249.

Wells, J. C. K. (2015). Between Scylla and Charybdis: renegotiating resolution of the 'obstetric dilemma' in response to ecological change. *Philosophical Transactions of the Royal Society B: Biological Sciences*, 370, 1–12.

Wells, J. C. K. (2010). Maternal capital and the metabolic ghetto: an evolutionary perspective on the transgenerational basis of health inequalities. *American Journal of Human Biology*, 22, 1–17.

Wells, J. C. K. (2017). The new 'obstetrical dilemma': stunting, obesity and the risk of obstructed labour. *Anatomical Records*, 300, 716–731.

Wells, J. C. K, DeSilva, J. M. & Stock, J. T. (2012). The obstetric dilemma: an ancient game of Russian roulette, or a variable dilemma sensitive to ecology? *American Journal of Physical Anthropology*, 149 Suppl, 40–71.

White, T. D., Asfaw, B., Beyene, Y., Haile-Selassie, Y., Lovejoy, C. O., Suwa, G. & Woldegabriel, G. (2009). *Ardipithecus ramidus* and the paleobiology of early hominids. *Science*, 326, 75–86.

World Health Organization (WHO). (2014). *WHO infographic: saving mothers' lives*. World Health Organization, Geneva.

World Health Organization (WHO). (2015). *WHO statement on caesarean section rates (RHR/15.02)*. World Health Organization, Geneva.

World Health Organization (WHO). (2018). *WHO recommendations on non-clinical interventions to reduce unnecessary caesarean sections*. World Health Organization, Geneva.

World Health Organization (WHO) (2019a). *Maternal mortality*. World Health Organization, Geneva.

World Health Organization (WHO) (2019b). *Strengthening quality midwifery education for Universal Health Coverage 2030*.

World Health Organization (WHO), UNICEF, UNFPA, World Bank Group, and the United Nations Population Division (2015). *Trends in Maternal Mortality: 1990 to 2015*. World Health Organization, Geneva.

Ye, J., Zhang, J., Mikolajczyk, R., Torloni, M. R., Gülmezoglu, A. M. & Betran, A. P. (2016). Association between rates of caesarean section and maternal and neonatal mortality in the 21st century: a worldwide population-based ecological study with longitudinal data. *British Journal of Obstetrics and Gynaecology*, 123, 745–753.

Zaffarini, E. & Mitteroecker, P. (2019). Secular changes in body height predict global rates of caesarean section. *Proceedings of the Royal Society B: Biological Sciences*, 286, 20182425.

CHAPTER 5

Isotopic reconstruction of ancient human diet and health: implications for evolutionary medicine

Nicole Burt and Alexandra M. Greenwald

5.1 Introduction

Diet is personally and culturally important, making it a key area of anthropological study. Studying dietary habits and health also helps us understand human physiology, metabolism and the evolution of these systems better. Investigating past diet and health can improve our understanding of our own biology and the ways in which humans have been affecting our own evolution across millennia through the cultural and environmental changes we introduce to food production, processing and consumption.

Understanding the diet and health of our ancestors is directly relevant to modern public health and medicine. Theodosius Dobzhansky (1973) famously proposed in the article of the same title that 'Nothing in biology makes sense except in the light of evolution'. An evolutionary approach should be used to understand current human health and to inform medical interventions that improve public health, while being mindful of evolutionary consequences (Alcock & Schwartz, 2011; Griffith, 2009). This requires an understanding of human diets and associated health outcomes in our evolutionary past. Stable isotope palaeodietary reconstruction and palaeopathological data suggest a diversity of evolved human diets, and that dietary-induced health problems existed in the past. Combining these data can help provide evolutionary medicine (EM) with the evolutionary perspective necessary

for an improved understanding of diet-related conditions like obesity and diabetes.

This chapter focuses on stable isotope analysis as a way to understand the evolution of human diet, the diversity of prehistoric human diets and its potential for application to modern human diet and health. The EM approach to health and diet is a good framework for aligning bioarchaeological research with modern research and intervention. Our goals are to demonstrate the diversity of evolved diets in humans, discuss health outcomes of these diets in the past and to use this understanding to develop an approach that applies these principles to EM and the development of public health interventions in health and nutrition.

5.2 Stable isotope basics

5.2.1 How stable isotope analysis reconstructs diet

Reconstructing the diets of ancient humans on the individual level relies on the analysis of isotopes, for example, carbon and nitrogen.

Biological tissues are generated from the food and water consumed by individuals. Foods necessary for tissue synthesis vary in their stable isotope composition, including in the ratios of ^{13}C to ^{12}C and ^{15}N to ^{14}N. These elements are incorporated into tissues through internal fractionation or enrichment of one isotope relative to the other. Collagen—the

Nicole Burt and Alexandra M. Greenwald, *Isotopic reconstruction of ancient human diet and health*. In: *Palaeopathology and Evolutionary Medicine*. Edited by Kimberly A. Plomp et al., Oxford University Press. © Oxford University Press (2022). DOI: 10.1093/oso/9780198849711.003.0005 primary organic component of bone, dentin and connective tissue—is synthesised primarily from dietary protein, and therefore contains stable isotopic signatures reflective of a large component of individual protein budgets (Ambrose & Norr, 1993; Schoeninger, 1985; Schwarcz, 2000; Tieszen & Fagre, 1993). Apatite, the inorganic component of bone and teeth, reflects the whole diet of an individual via its carbonate component (White & Folkens, 2005).

Bone is formed via a process of resorption and remodelling throughout an individual's life, with tissue turnover occurring every 10–20 years, depending on the skeletal element (Hedges et al., 2007; Manolagas, 2000; White & Folkens, 2005). Consequently, stable isotope measures derived from bone reveal the dietary protein sources consumed during the individual's last decades of life (Eerkens & Bartelink, 2013; Schwarcz & Schoeninger, 1991). Unlike bone, teeth do not experience tissue turnover and replacement; stable isotope ratios derived from either the collagen or apatite components of teeth reflect an individual's diet during the period of development of a specific tooth (Hillson, 1986; White & Folkens, 2005) (Box 5.1).

Box 5.1 Background to stable isotope analysis

Many of the elements incorporated into organisms, and present in the environment, exhibit variation in their mass. All atoms of an element contain the same number of protons and electrons but vary in the number of neutrons in the nuclei. Stable isotopes do not undergo spontaneous radioactive decay, and remain constant in relative abundance across samples of the element (Hoefs, 2009; Schwarcz and Schoeninger, 1991).

Since different isotopes of the same element vary in their mass, differing kinetic and thermodynamic properties cause fractionation during chemical reactions (Schwarcz and Schoeninger, 1991; Urey, 1947). Isotope fractionation causes a detectable difference in the ratio between two stable isotopes of the same element when comparing food and the organism that eats that food. The isotope ratios are measured in fractions of a percent, or 'per-mil' (‰) relative to a standard (Schwarcz and Schoeninger, 1991). Isotope ratio mass spectrometry is used to measure the mass of stable isotopes naturally occurring in organisms in the environment (Katzenberg, 2008). The relative differences in isotope ratios are commonly expressed using the delta symbol (δ). If the difference is less than the international standard—as is the case with carbon-13 values—the δ values will be negative. Stable isotope fractionation occurs during the metabolism of foods, such that the tissues of a consumer have isotopic signatures that vary in predictable ways from the isotopic values of the consumed food. This means that scientists are able to analyse stable isotope ratios in human remains, mostly commonly bones and teeth, and work to determine the diets and associated behaviours of individuals.

Nitrogen isotope ratios, expressed as $\delta^{15}N$, display a trophic-level effect, wherein the collagen of a consumer will be enriched 2–4‰ over the source of dietary protein (Schoeninger, 1985; Schwarcz & Schoeninger, 1991). Different environments will have different isotopic starting points, which is why it is important to reconstruct the local food web or isoscape of the particular region of interest. In a food web within a terrestrial environment, low $\delta^{15}N$ (usually 6–8‰) indicates consumption of primarily plant-based proteins, as is expected of a herbivore like a deer. Higher levels of nitrogen enrichment indicate incorporation of increased levels of animal-derived proteins seen among omnivorous and carnivorous species.

Stable carbon isotopes in collagen weakly track trophic level but show a stronger correlation with the biological or ecological source of dietary protein. Stable carbon isotopes track both marine versus terrestrial sources to the total protein budget and the dietary importance of C_3 versus C_4 plants, permitting archaeologists to trace the importance in the diet of C_4 plants such as maize, sorghum and millet, or animals who foddered on these plants (Schoeninger, 2009).

5.2.2 Stable isotopes, pathology, and health

Stable isotope analysis has also been used to explore pathology in the past by looking for dietary differences that are correlated with disease (D'Ortenzio et al., 2015; Fahy et al., 2017; Katzenberg & Lovell,

1999; Reitsema, 2013). This body of palaeopathology research has two foci: a primary focus identifying correlations between the presence and severity of pathology and an individual's diet to understand how macro—or micro-nutrient deficits may lead to ill health, and a second focus seeking to identify how pathological conditions might change isotopic signatures in the tissues of an affected individual as a result of changes in metabolism (D'Ortenzio et al., 2015, Reitsema, 2013). This chapter includes case studies linking dietary stable isotope data with pathological conditions at the individual and population level in the archaeological record to understand links between diet and health in our evolutionary past. Most links between stable isotopes and health will be indirect, but as methods improve, we are getting closer to having a better understanding of the effects of metabolism (Reitsema, 2013). However, the direct use of stable isotopes for understanding pathology and disease is limited in a modern medical context due to the very invasive nature of testing pathological lesions.

5.3 Diet, pathology, and evolutionary implications

5.3.1 Hominin dietary ecology

The study of the behaviour and lifestyles of our hominin ancestors is an important area of anthropological study and isotope data have been a major contributor to our understanding of their diets (Jaouen, 2018; Sponheimer et al., 2013). Understanding past dietary diversity between and within species is critical to understanding the evolution of human diet and its relationship to our environment and lifestyle (NRC CESCHE, 2010; Potts & Faith, 2015; Power et al., 2018; Zuk, 2013). Isotopic analysis has been applied via both dietary reconstructions, and through the palaeoecological reconstructions of past environments. Direct comparison of diet and behaviour between species is strongest between *Homo neanderthalensis* and *Homo sapiens*. Diet is a cause of health disparities between modern *H. sapiens* populations and has been proposed as one of the key cultural differences between *H. neanderthalensis* and *H. sapiens* and a contributing factor to the better health and survivorship

of *H. sapiens* (Bocquet-Appel & Degioanni, 2013; Hockett, 2012; Melchionna et al., 2018; Shipman, 2008; Villa & Roebroeks, 2014). This idea has persisted in popular consciousness despite clear evidence for dietary diversity and flexibility among Neanderthals, and Neanderthal–*H. sapiens* admixture. This section highlights the dietary diversity in Neanderthals and *H. sapiens* and discusses modern health implications. The research on palaeodiets also sheds light on how contemporary, so-called 'palaeodiets' are generally unscientific and limited (if not erroneous) in their health recommendations, given the wealth of data on the dietary diversity of our evolutionary history.

Neanderthals are thought to have exhibited higher rates of morbidity and infant mortality and lower total fertility rates compared to *H. sapiens* (Hockett, 2012; Villa & Roebroeks, 2014), which could be explained by a static, low-diversity diet. Low dietary diversity could also reduce the ability of Neanderthals to adapt to a changing environment, for example, as that which occurred when the climate warmed and glaciers began receding (Melchionna et al., 2018; Power et al., 2018). If this is an accurate description of Neanderthal diet, what they consumed (or did not consume) could have been a major contributor to extinction. The research on both Neanderthal diet and pathology is key to understanding these relationships. However, research showing variation in both diet and health between Neanderthal communities is growing (Power et al., 2018). With diversity being increasingly part of the *H. neanderthalensis* story it is important to apply this to our theories about the evolutionary relationship of Neanderthals with *H. sapiens*.

Reconstructions of Neanderthal diet using zooarchaeological analysis showed seasonal exploitation of animals with a reliance on terrestrial herbivores (Burke, 2000; Finlayson et al., 2012; Gaudzinski-Windheuser & Kindler, 2012; Hardy & Moncel, 2011; Ready, 2013; Wood et al., 2012). The research was mainly focused on European Neanderthal sites and indicated a large game-centred diet, with little to no evidence of other foods, even the consumption of small mammals. Tool wear analysis and initial residue analysis also indicated a reliance on hunting and butchering of animals

(Beier et al., 2018). Nitrogen stable isotope analysis of Neanderthals confirmed this vision of a meat-centred diet, with individuals having nitrogen signals higher than carnivore bones sampled at the same sites (Bocherens, 2009; Richards & Trinkaus, 2009). The isotope values were very similar in Neanderthals across their European geographical range regardless of ecology (open versus forested sites) and showed less variation within and between groups than was seen in *H. sapiens* from the same time period. If we assume Neanderthals had a metabolism similar to us, it seems impossible that all of their nutritional requirements could be met without significant use of other plant and animal resources (Hardy, 2010; Hockett, 2012; Power et al., 2018). Increasingly, research on Neanderthals has identified greater dietary diversity mapping on to ecological zones than previously identified.

Analysis of Neanderthals in the southern range of the species (the Mediterranean and Middle East) does show evidence of a broadening dietary niche over time. Sites such as Payre in France (125–250 ky BP [thousands of years before present]), Bolomor, Spain (129-175 ky BP), and Fumane Cave, in Italy (44 ky BP) have evidence of small mammal and bird exploitation, rather than a strict focus on large game (Blasco & Peris, 2009; Hardy & Moncel, 2011; Peresani et al., 2011). Modern analysis of plant residues, coprolites and calculus have also provided evidence for a wide spectrum of starches and plants being eaten by later Neanderthals with local/regional diet diversity (Barton, 2000; Madella et al., 2002; Lev et al., 2005; Hardy & Moncel, 2011; Hardy et al., 2012; Henry et al., 2011; Sistiaga et al., 2014). The complex behaviours of fishing and fish processing have also been identified through lithic analysis at Payre (Hardy & Moncel, 2011.) As early as 150,000 BP the remains of processed and burned marine fish and molluscs were found at Neanderthal sites around the Mediterranean (Cortés-Sánchez et al., 2011; Drucker & Bocherens, 2004). The combined chemical, palaeopathological and zooarchaeological evidence seems to indicate that Neanderthals did eat locally available foods and had some diversity in their diets, although not as much as their *H. sapiens* contemporaries.

Stable nitrogen isotope analysis of early human contemporaries shows a wide dietary variation between individuals and groups (Richards & Trinkaus, 2009; Richards et al., 2001). Diverse faunal and flora recovered from these sites clearly indicate a diverse omnivorous diet reflective of local ecology and similar to the patterns we associate with modern hunter-gatherers—although, of course, these latter diets also vary widely based on ecology (Richards & Trinkaus, 2009). These highly localised diets and the ability to utilise new food sources likely played an important part in the global migration of our species. By keeping a versatile and changing diet our ancestors could better supply themselves in any ecological setting or in changing ecological niches. This may indicate that our ancestors were more prepared for the changing environments they faced and outcompeted their Neanderthal contemporaries through better resource exploitation and more rapid cultural transmission. Diet has played a role in the evolutionary success of our species and the diversity of local diets exploited may be a key to a health strategy in adapting to local environments (Chapter 7). Further research that combines both palaeopathological analysis of *H. neanderthalensis* and *H. sapiens* communities with dietary studies is key to understanding this evolutionary story better and applying it to EM.

A flawed lay understanding of palaeodiets with a heavy emphasis on meat protein has led to fad diets that fail to take into account the tremendous diversity in the diets of our ancestors (Zuk, 2013; see also Chapter 7 and references therein). Representing the palaeodiet as a single suite of currently available food resources is not accurate (e.g. Audette et al., 1999; Cordain, 2010). Research on the evolution of human diet has contributed to the understanding in EM that dietary diversity is important for our continued health and highlights the existence of a diversity of healthy human diets, both in our evolutionary past and in the present. Human dietary variation, including our ability to intensify carbohydrate-rich resources, is known to be a key evolutionary strength. Diet and environment are key drivers of our evolutionary past, and a transition to agriculture among many populations worldwide has had far-reaching implications for our foodways and health.

5.3.2 Intensification and agriculture

Following the radiation of *H. sapiens* across the globe and subsequent population increases, many human populations worldwide experienced the need to intensify, or shift, subsistence strategies to extract more calories from a given ecological zone. In some cases that meant expanding diet breadth to include lower-return hunted game and gathered plants, while in other groups intensification involved the transition to intensive plant cultivation and the domestication of animals. Early Indo-European-centric ideas of unilineal cultural evolution that placed agricultural (and subsequently industrial) societies at the pinnacle of human evolution presumed that hunter-gatherer societies were less developed and agricultural societies were necessarily more advanced (Engels, 1884; Morgan, 1877). Increased study of living hunter-gatherers beginning in the mid-twentieth century, along with the application of behavioural ecological models of subsistence behaviour, demonstrated that foraging is an economically rational and less labour-intensive strategy where population density and ecological conditions permit (Lee & DeVore, 1968; Smith & Winterhalder, 1992). Rather than a marker of cultural advancement along a unilineal evolutionary trajectory, adoption of agriculture is an adaptive step in a spectrum of economic intensification associated with population increases, shifts in environmental conditions, or both (Bettinger, 2009, Kelly, 2007; see also Chapter 7).

A greater focus on the benefits of foraging lifeways and the dynamics of subsistence intensification brought new attention to bioarchaeological evidence of the deleterious health impacts of a transition from hunting and gathering wild foods to farming domesticated foods (Larsen, 1995). Declines in dental health (the presence of, or increases in, caries and antemortem tooth loss), childhood growth stunting and skeletal and dental pathologies associated with dietary and physiological stress and degeneration (including cribra orbitalia, porotic hyperostosis, osteoperiostitis, degenerative joint disease and linear enamel hypoplasias) have been noted across numerous populations at the transition from foraging to agricultural lifeways (Cohen & Armelagos, 1984; Cohen & Crane-Kramer, 2007;

Larsen, 1995; Steckel et al., 2002; see also Chapters 2, 6, 14). Increases in infectious and parasitic disease have also been noted, and are linked to increased population densities, which both drive and are supported by agriculture, and the advent of animal husbandry (Larsen, 1995). Close proximity to domesticated animals increased the rates of zoonotic disease transmission (see Chapters 11 and 17). Increases in morbidity coincident with the adoption of agriculture cannot be solely attributed to the consumption of domesticated foods. It is likely that this constellation of inter-related phenomenon, including crowded living conditions in the absence of sanitation systems, contributed to increases in pathological conditions.

The well-publicised findings of the link between agriculture and increases in morbidity have also led directly or indirectly to a host of other fad diets aimed at overcoming the often-exaggerated ill-effects of domesticated grains (e.g. palaeo, ketogenic, Atkins, South Beach, Dukan, Zone) (e.g. Audette et al., 1999; Axe, 2019; Cordain, 2010; Dukan, 2014). This has led to a dramatic swing in popular opinion away from the concept of the agricultural society as superior to an opposite extreme and the rejection of many plant and animal domesticates that do provide important macro- and micronutrients. While the overall trend of general health decline associated with the advent of agricultural lifeways is indisputable, based on bioarchaeological evidence, stable isotope data from archaeological contexts suggest that many agriculturalists included non-domesticated resources in their diets and facultatively switched between economic strategies.

A case study employing stable isotope reconstruction of the diet of an ancient North American population illustrates the diversity of intensification strategies, and the behavioural and dietary flexibility of humans. In prehistoric western North America, the Fremont people occupied the eastern Great Basin and northern Colorado Plateau from AD 400–1350 (Jennings, 1978; Kidder, 1927), and are known for their distinct material remains, including unique basketry and moccasin construction, anthropomorphic clay figurines and a local variety of maize. Fremont subsistence patterns, however, were not well understood and were a source of debate among archaeologists. Some archaeologists argued that they were hunter-gatherers living at the margins of Southwestern agriculturalists, while others argued they were maize agriculturalists in their own right.

To clarify debates about Fremont subsistence behaviour and to reconstruct their diet, Coltrain and Leavitt (2002) conducted stable isotope analysis on Fremont human remains from the Great Salt Lake mortuary assemblage, a burial population excavated near modern day Salt Lake City, Utah in the United States. Variation in carbon enrichment over time indicated that, during some periods, Fremont people were growing maize and consuming it as an important part of their diet. Later, some individuals were pursuing a mixed forager-agriculturalist strategy, while others pursued hunting and gathering. Finally, Fremont peoples transitioned back to foraging wild foods. During periods when maize agriculture was practised, males consumed more maize than females, while both sexes had equal access to non-domesticated animal protein, including terrestrial herbivores and wetland bird and fish species (Coltrain & Leavitt, 2002). Rates of pathology linked to nutritional stress, as well as reduced cross-sectional long-bone robusticity, were higher among individuals consuming maize, although morbidity did not reach levels noted among Near Eastern and European agriculturalists. This may be due to lower density settlements and the absence of domesticated animals and associated zoonotic disease transmission.

While direct causation is nearly impossible to demonstrate in the archaeological record, these shifts in subsistence behaviour are closely correlated with concurrent changes in summer precipitation and lake levels (Coltrain & Leavitt, 2002). Climatic variability and environmental degradation likely shifted economic payoffs such that maize agriculture was no longer sustainable, and foraging became a more efficient strategy. Fremont peoples' repeated and spatially mosaic transitions between foraging wild resources and farming maize demonstrate the adaptive diversity of human subsistence behaviour and diet and refute the notion that subsistence strategies operated on a one-way uni-lineal trajectory. Additionally, isotopic data show

that the Fremont diet during periods of maize cultivation was not devoid of foraged resources. This undoubtedly contributed to a more well-rounded diet than is often presented as a stereotypical agricultural diet by proponents of fad diets who now reject (as far as they are able) domesticated products. The data from Fremont also suggest that inclusion of agricultural resources as part of a balanced diet was not universally deleterious for human health, especially in cases where living conditions did not also contribute to high disease or parasite loads. Contemporary health implications of these findings from the Fremont and other agricultural societies are that domesticated agricultural products are an important part of a diet that includes diverse sources of grains, fruit, vegetables and proteins.

5.3.3 Infant and early childhood diets

Understanding feeding patterns in early life, including the duration of exclusive breastfeeding, timing and type of supplementary food introduction and weaning age can provide insight into the developmental origins of health and disease in ancient and modern populations (see Chapter 2). Additionally, understanding these patterns and their relationship to human life history strategies among ancient populations on evolutionary timescales can aid EM in the development of empirically based recommendations.

Life history strategies encompass a suite of trade-offs all organisms face throughout their lives in the allotment of a limited energy budget to somatic growth, maintenance and reproductive effort (Stearns, 1992) (Box 5.2). Life history strategies vary widely across species, and many individuals within species also exhibit a degree of evolved plasticity in the timing of these trade-offs dependent on local ecological conditions. Examining the relationships between diet, health and life history strategies are important for understanding how cultural and environmental factors influence natural selection. Examining the relationships between life history strategies and breastfeeding offers unique insights into human diet and health in our evolutionary past.

Box 5.2   The quality–quantity trade-off in life history

One of the major life history trade-offs—called the quantity versus quality trade-off—entails choosing a balance between current and future offspring. Unlike many mammals, humans have nested offspring and must allot care to multiple altricial individuals in various states of dependency, while balancing somatic maintenance and potential future reproductive investment. Parents have a limited time and energy budget and must choose a balance of investment that favours either current offspring (quality), or future reproduction (quantity). In this trade-off, individuals are predicted to invest more resources in fewer offspring during periods of environmental and social stability when offspring survival is more sensitive to parental investment (Krebs and Davies, 1978; MacArthur and Wilson, 1967). In stressful environments with high extrinsic mortality, where offspring survival is poorly predicted by parental investment, parents should prioritise number of offspring (Gillespie et al., 2008; Lawson and Mace, 2011; Strassmann and Gillespie, 2002). Reproductive strategies along this spectrum are reflected in parental investment, in the form of breastfeeding and food provisioning to offspring, since resources, especially calories, are a finite resource.

Stable isotopes have been increasingly employed to improve our understanding of parental investment strategies in ancient populations, using weaning age and early childhood diet as proxies. Researchers have focused on reconstructing these behaviours at an individual level using samples from teeth. Recent work has focused on serial micro-sections of single-rooted deciduous teeth and first molars, which develop from birth to nine years, to generate high-resolution reconstructions of the weaning process and early childhood diet, including duration of exclusive breastfeeding, timing and type of supplemental food introduction, age at weaning and the post-weaning diet (Burt & Garvie-Lok, 2013; Burt, 2015; Eerkens et al., 2011; Greenwald, 2017; Greenwald et al., 2016a). This method relies on the trophic-level effect exhibited by $\delta^{15}N$ and minimises the degree of time-averaging across individuals' childhoods by taking ten to twenty serial samples across the tooth.

An additional advantage of this method is that it examines infant and childhood diet in individuals who survived to adulthood, thus removing the bias of restricting analysis to individuals who died as subadults.

Mortality and diet are so closely linked that in skeletal populations of juveniles, age of peak mortality in the sample has often been assumed to be the age at weaning for that population. Breastmilk is a unique source of pathogen-free and nutritionally complete food—a combination not available in weaning foods that may lead to increases in morbidity and mortality during a child's transition from breastmilk to household foods. Previous research by Burt (2015) applied stable isotope analysis to bone to reconstruct individual dietary profiles of sixty-two children from the fourteenth–sixteenth-century site of Fishergate House, York, UK. Initial analysis of the site suggested a weaning age of around 4–6 years—later than contemporary populations—and coincident with high rates of sinusitis indicative of respiratory infections. However, stable isotope analysis of dentinal serial samples demonstrated that weaning actually occurred before the age of 2 years, on average. Post-weaning diets exhibited surprisingly high nitrogen values, indicating there may have been more meat or fish in the diet than expected for children. These findings suggest that high rates of sinusitis cannot be directly tied to weaning age, but that earlier weaning ages may have made individuals vulnerable to illnesses later in childhood.

Studies of hunter-gatherers living in what is now the San Francisco Bay Area, California, USA, across evolutionary timescales also find a strong relationship between breastfeeding behaviour and health. Findings across ten archaeological sites spanning 7000 years of occupation indicate a high degree of inter-individual variation in the duration of exclusive breastfeeding, from approximately three to eight months, with an average of around six months (Greenwald, 2017, Greenwald & Hinde, 2018). Supplemental foods provided to infants still consuming breastmilk closely track adult diets in their respective sites and ecological zones (Eerkens et al., 2011; Greenwald et al., 2016a; 2016b). In Central California, these supplemental foods were primarily C$_3$ plants (acorns and small seeds), terrestrial game and marine invertebrates.

Weaning age in ancient central California was variable at the inter-individual level and across time. Average weaning age in the region was approximately 3 years of age, which is comparable to ethnographic observations of breastfeeding behaviour in historically recorded and extant living hunter-gatherer populations (Eerkens et al., 2011, Greenwald et al., 2016a, 2016b, Greenwald, 2017). However, weaning age drops to an average of 2 years of age during a high-stress drought period in the region that led to food shortages, increases in morbidity, interpersonal violence and, counter-intuitively, population growth (Greenwald, 2017). This finding is supported by evidence that during the Medieval Climate Anomaly (MCA), offspring were less sensitive to investment (i.e. increased duration of breastfeeding did not increase survival), which is consistent with the hypothesised quantity–quality trade-off, wherein parents invested less in each individual offspring and produced greater numbers of them during periods of environmental and social stress (Greenwald, 2017). Rates of pathologies indicative of nutritional stress or illness in early childhood, such as enamel hypoplasia, cribra orbitalia and porotic hyperostosis, are more prevalent during the high-stress drought period, and occur in higher rates among individuals weaned in the 12–30-month age range (Greenwald, 2017; Schwitalla, 2013). These results suggest that breastmilk consumption in infancy and early childhood is vital to children's health, and that public health interventions to address high rates of infant and child mortality attributable to preventable disease and malnutrition have the potential to both reduce death rates as well as birth rates, and foster higher levels of parental investment in children (Kramer & Kakuma, 2002). These data from ancient hunter-gatherers also indicate that variability in the timing and type of supplemental food introduction, and the timing of breastfeeding cessation, is a normal, evolved aspect of human dietary ecology.

5.4 Isotope applications in evolutionary medicine

5.4.1 Stable isotope analysis for studying health in a modern context

As the science of stable isotope analysis has matured, the application of the methods and techniques to new questions and areas of inquiry has grown. We have established how stable isotope analysis is traditionally used to reconstruct diets and health interactions in the past, but it is also being used increasingly in the present. Health researchers and community members want better ways of tracking nutritional quality and quantity. Without understanding what food is being eaten by individuals, we cannot isolate how metabolism and illness may be affecting the utilisation of that food. The obvious approach is to use surveys or interviews, but research indicates that both children and adults may misrepresent the calories they eat, what kinds of food they eat or simply cannot recall what they have been eating (Archer et al., 2013; Cook et al., 2000; Jahren et al., 2006). Bias in survey data is hard to correct, as individual values and perceptions of the participants cause them to misrepresent their diet, consciously or unconsciously. For instance, men may overestimate the amount of meat they eat as this is seen as a positive masculine quality, while teen girls often significantly under report their calorie intake. This is why non-invasive analysis of biological samples, collected with the consent of patients, are prized by scientists and medical doctors alike for providing a reconstruction of both nutrient levels and diet. To understand the complex relationships between genetics, development, behaviour and metabolism better, we need to create non-invasive and accurate measures of what people are eating. Carbon and nitrogen stable isotope analysis, extensively used for dietary reconstruction in past populations, could provide a cost effect solution to this problem.

5.4.2 Carbon isotopes and obesogenic diets

Common sources of sugar in the US (high fructose corn syrup (HFCS), cane sugar and processed corn sugar) currently come from plants with distinct carbon stable isotope signals. Research has shown that added sugar in diets can be detected using stable isotope analysis (Jahren et al., 2006; Yeung et al., 2010; Davy & Jahren, 2016; Jahren et al., 2014; Kuhnle, 2018). These sugars are linked to chronic disease and obesity, making carbon stable isotope analysis a potential tool, with appropriate consent, for early detection of high-risk diets or for tracking the success of dietary interventions focused on reducing sugar intake. Stable isotope analysis is inexpensive and requires very small samples (hair, fingernails, dried blood spots from finger pricks or small amounts of drawn blood). By using stable carbon analysis, a non-biased assessment of current and changing diets can be analysed. Collaboration with a medical practitioner would result in individual consultation as part of standard medical practice and covered by Health Insurance Portability and Accountability Act (HIPAA) medical data sharing protocols in the US. Research data is collected, de-identified and/or anonymised in compliance with research ethics standards and regulated by the Internal Review Board (IRB) system. This means individuals would know they contributed to data but could not be specifically identified or targeted by participations. Rules and regulation for research and medical participation vary by country and will have different levels of privacy based on local ethical standards.

Many Native American communities, for example, the Yup'ik in Alaska, are experiencing health issues associated with shifts away from their traditional diets. Nash and colleagues (2014) found that hair and blood could both be used to predict sugar intake using a sample of sixty-eight Yup'ik participants aged 14–79 years. Hedrick and colleagues (2015) found that as Health Eating Index 2010 scores increased, carbon ratios decreased and, as dietary nutrition improved, the carbon biomarker also decreased, indicating less added sugar in the healthier diets. These kind of applications shows the potential of carbon to track dietary changes in real time and relate them to changes in health. These tools could be powerful in helping communities take control of their own health with the consent of individuals involved and community-run efforts.

Results such as these clearly indicate the need for more work on stable carbon as a biomarker that focuses on both the potential and limitations of the method (Davy & Jahren, 2016; O'Brien, 2015; Kuhnle, 2018). Categorising diverse diets across time and space is both interesting and important but understanding how to identify actual consumed diets that are obesogenic is critical to developing timely intervention into diet that can result in the prevention of chronic disease.

5.4.3 Nitrogen isotopes and early infant health

We suggest the potential to leverage the well-documented ability of nitrogen isotope ratios to track breastmilk consumption among contemporary mother–infant dyads struggling with feeding (Fuller, 2006). Many mothers, especially primiparous ones, encounter early difficulties in breastfeeding, and the quantity of breastmilk transmitted to a struggling infant is nearly impossible to measure with cases of direct milk transfer (Cohen et al., 2018; Hackman et al., 2015; Kitano et al., 2016; Riordan et al., 2005). Currently, milk transfer is measured by weighing an infant pre-and post-feed (Riordan et al., 2005). This coarse-grained method cannot account for digestive difficulties post-feeding and cannot measure actual macro-nutrient metabolism.

Using rapid, inexpensive and non-invasive isotopic sampling of paired serial samples of fingernails and/or hair from mothers and infants can track baseline maternal diet and infant dietary signatures in utero and postpartum (Fuller, 2006). Using stable isotope analysis for tracking breastfeeding and early childhood diet is dependent on non-invasive sampling methods that span the time period of interest. Tissues collected at birth and immediately postpartum will reflect dietary signatures in utero, whereas hair growth commencing immediately postpartum will reflect diet and/or metabolic issues after birth. In addition to detecting nitrogen enrichment associated with breastmilk consumption, this method would also permit the detection of tissue catabolism associated with insufficient caloric intake and indicate the need for medical intervention and/or supplementation with specially processed donor breastmilk or formula developed for specific metabolic disorders. This method has potential in infants who may be struggling to feed to differentiate between failure to thrive due to insufficient nutrition and modest growth trajectories that track to the lower percentiles on the World Health Organization growth charts.

5.5 Culturally sensitive and inclusive applications of evolutionary medicine

There is a lack of consensus on what constitutes a healthy, nutritious human diet, which is almost certainly because there is a diversity of evolved diets, and individual metabolism plays a key role in health outcomes (Carrol et al., 2018; Katz & Meller, 2014). As mentioned, humans are omnivores and our evolutionary history reflects that of a flexible omnivore adapted to many ecological conditions. As we move forward in understanding variation in human diet and nutrition, we hope to do so in a way that decolonises and broadens what is seen as normal or healthy. An examination of healthy eating cookbooks and dietary recommendations reveals a heavy bias towards a Western, Educated, Industrialised, Rich and Democratic (WEIRD) perspective (Henrich et al., 2010). Limited access to nutritious food by historically marginalised people in the Western world is also often the direct result of harmful historic interventions and policies. As we push forward in understanding diet and health, it is important that a broader, more inclusive and decolonising approach be used to develop culturally appropriate interventions informed by EM to improve health outcomes.

5.5.1 Decolonising foodways: native foodways renaissance

Native peoples indigenous to North America living in both urban and reservation areas disproportionately experience the ill effects of food deserts—areas with little or no access to affordable fresh produce, lean proteins and complex carbohydrates. This leads to higher rates of obesity, diabetes and overall higher morbidity and mortality rates (Indian Health Service, 2018; Mundel & Chapman, 2010,

Tso, 2014). Understanding the cause of these trends is of the utmost importance to the biomedical community and we propose taking an EM approach to better understand the evolutionary causes that shaped the mechanism that link diet to health (but see Chapter 7 on the importance of considering proximate factors, such as socio-economic status, and Chapter 2 on structural violence).

Higher rates of diabetes noted in Native populations may be explained, in part, by the thrifty phenotype phenomenon, wherein a disproportionate number of low birthweight individuals develop non-insulin-dependent diabetes (Hales & Barker, 1992; Savage et al., 1979). The causal mechanism is hypothesised to be nutritional deprivation in utero leading to impaired development of the pancreas (McCance & Widdowson, 1974). An alternative evolutionary explanation for the prevalence of diabetes among Indigenous subjects with low birthweights may reflect selective survival of low birthweight infants who are genetically predisposed to developing diabetes; the mechanisms that permit rapid catch-up growth and survival also lead to development of insulin resistance later in life (Lillioja et al., 1993; McCance et al., 1994; see also Chapter 2). This explains the evolutionary mechanism that may cause these patterns, but not the systems that led to their presence in Native peoples.

A paucity of nutritious food and the prevalence of associated health problems are the direct result of the legacy of colonialism. Removing people from their traditional territories and establishing Euro-American agricultural or extractive industries effectively erased opportunities for Indigenous people to pursue their traditional subsistence, whether farming domesticated crops or foraging for wild resources. Once placed on reservations that either dramatically reduced traditional territories, or completely removed a people from their ancestral territory, food subsidies were necessary, and most often consisted of allotments of wheat flour, sugar, and lard or cooking oil. Erasure of traditional food gathering, farming and ecological knowledge through forced assimilation often prevented/prevents the supplementation of these macronutrient rich, but micro-nutrient poor, resources with fresh produce and complete protein (Mihesuah & Hoover, 2019).

While Indigenous cultures have adapted and thrived in the face of historical trauma, nutrition-related diseases continue to plague many communities. In recent years, a number of Indigenous activists and organisations have called for the decolonising of native foodways. Organisations such as Nātifs (North American Traditional Indigenous Food Systems, 2019), founded by Native chef Sean Sherman, work to create systems that return knowledge of traditional foodways to communities through education, practise and economic development so that Indigenous people have access to, and make a living from, the traditional foods of their ancestors and do not have to rely on the calorie-rich, but nutritionally poor, foods so commonly available in food deserts (https://www.natifs.org). Archaeological evidence of diet and ancestral health states can offer additional lines of evidence in support of community-held traditional cultural knowledge regarding traditional foods. In addition to addressing nutrition-related disease, decolonising foodways and food sovereignty movements acknowledge the mental and physical health impacts on native people resulting from a legacy of historical trauma, and work to rectify the systems that continue to contribute to ill health (Mihesuah & Hoover, 2019; https://www.natifs.org). This movement, and the organisations supporting it, demonstrate the importance of understanding the diversity of evolved human diets across cultures and translating that understanding into culturally appropriate interventions to improve public health within contemporary populations.

5.5.2 Decolonising foodways: translating USDA recommendations

Most nutrition research and recommendations produced by WEIRD countries are viewed as representative of the 'normal' healthy diet and are used as a basis for dietary guidelines globally (FAO UN, 2019). Even within WEIRD countries, a lack of diversity and cultural sensitivity in food recommendations is a major barrier to healthy eating in many non-white middle-class communities (Byrne, 2018). The examples of healthy food portrayed in the media, and often by medical practitioners, have two main flaws that can result in health loss rather

than gain. The first, as mentioned, is the promotion of healthy diet options that lack diversity and are specifically dominated by a white middle-class ideal. The second is a focus on weight loss, rather than on developing a healthy diet and lifestyle habits. Medical professionals are increasingly involved in prescribing diets and foods, but often they lack the background in evolution, anthropology or nutrition science to ensure recommendations are based on strong science and target individual patient needs (Schuetz, 2017; Trinquart et al., 2016). If food is seen as a medicine, it is important that recommendations link to scientific reality and are socially responsible.

The trend in recommendations is likely related to the identities of researchers and practitioners. The Academy of Nutrition and Dietetics reports that 77.8% of all registered dieticians are White (CDR, 2019). This almost certainly has an effect on the kinds of foods and recipes being developed for healthy eating. However, as shown through this review of diet and human evolution, there is no perfect diet for humans, but there exist many options. It is possible to have healthy and culturally diverse dietary recommendations. It is clear that more scientifically rigorous and longitudinal nutrition research is needed to understand the mechanisms linking diet and health, not just correlations between food behaviours and poor health outcomes such as obesity. One struggle in dietary research is the unreliable information people give about their diet, which has led to a shift to look for biomarkers to provide more objective data (Combs et al., 2013, Trepanowski & Ioannidis, 2018). Stable isotope analysis could be a potential method for more accurately reconstructing diets in living people. To advance diet and nutrition research in an EM framework, we must leverage advances in stable isotope analysis, and an improved understanding of the cultural importance of food, including the creation of culturally inclusive dietary recommendations offered by the medical community.

5.6 Conclusions

This chapter presented case studies from the archaeological record pairing isotopic reconstruction of diet and palaeopathology that provide valuable insights into human dietary adaptations and health, with important implications and novel applications for EM and public health. A key adaptation of *H. sapiens* is our flexible and diverse diet. The successful radiation of *H. sapiens* across nearly every ecological zone on Earth is tied to our successful exploitation of dietary resources within those environments. Intensification of resources, including domestication of carbohydrate-rich plant foods, is an adaptive response to population increase and depletion of higher-ranking resources, and does not inevitably lead to poor health outcomes. Modern fad diets rejecting carbohydrate-rich foods suffer from a reductive misunderstanding of our evolutionary past. Isotopic reconstruction of breastfeeding behaviour and health outcomes also demonstrates human adaptive flexibility in the timing and type of supplemental food introduction and the cessation of breastfeeding and highlights the critical importance of breastfeeding for infant and child health. The archaeological case studies highlighted in this chapter point to a critical need in EM and public health to support initiatives like those presented, which build from an understanding that diverse groups have adapted to diverse diets, and that culturally sensitive approaches to diet-based health interventions are necessary in the battle against conditions such as obesity and diabetes.

References

Alcock, J. & Schwartz, M. D. (2011). A clinical perspective in evolutionary medicine: what we wish we had learned in medical school. *Evolution Education Outreach*, 4, 574–579.

Ambrose, S. H. & Norr, L. (1993). 'Experimental evidence for the relationship of the carbon isotope ratios of whole diet and dietary protein to those of bone collagen and carbonate', in J. B. Lambert & G. Grupe (eds), *Prehistoric human bone: archaeology at the molecular level*, pp. 1–37. Springer-Verlag, New York.

Archer, E., Hand, G. A. & Blair, S. N. (2013). Validity of U.S. nutritional surveillance: National Health and Nutrition Examination Survey caloric energy intake data, 1971–2010. *PLoS One*, 8(10), 1–12.

Audette, R., Gilchrist, T. & Eades, M. E. (1999). *NeanderThin: eat like a caveman to achieve a lean, strong, healthy body*. St. Martin's Press, New York.

Axe, J. (2019). *Keto diet: your 30-day plan to lose weight, balance hormones, boost brain health, and reverse disease.* Orion Spring, New York.

Barton, R. N. E. (2000). 'Mousterian hearths and shellfish: late Neanderthal activities on Gibraltar', in C. B. Stringer, R. N. E. Barton & J. C. Finlayson (eds), Neanderthals on the Edge: Papers from a Conference Marking the 150th Anniversary of the Forbes' Quarry Discovery, Gibraltar, pp. 211–220. Oxbow Books, Oxford.

Beier, J., Anthes N., Wahl, J. & Harvati, K. (2018). Similar cranial trauma prevalence among Neanderthals and Upper Palaeolithic modern humans. *Nature*, 563, 686–690.

Bettinger, R. L. (2009). *Hunter-gatherer foraging: five simple models*. Eliot Werner Publications, New York.

Blasco, R. & Fernández Peris, J. (2009). Middle Pleistocene bird consumption at Level XI of Bolomor Cave (Valencia, Spain). *Journal of Archaeological Science*, 36, 2213–2223.

Bocherens, H. (2009). 'Neanderthal dietary habits: review of the isotopic evidence', in J. J. Hublin & M. P. Richard (eds), *The evolution of Hominin diets*, pp. 241–250. Springer, Dordrecht.

Bocquet-Appel, J. P., & Degioanni, A. (2013). Neanderthal demographic estimates. *Current Anthropology*, 54(8), s202–s213.

Burke, A. (2000). Hunting in the middle Palaeolithic. *Journal of Osteoarchaeology*, 10(5), 281–285.

Burt, N. M. (2015). Individual dietary patterns during childhood: an archaeological application of a stable isotope microsampling method for tooth dentin. *Journal of Archaeological Science*, 53, 277–290.

Burt, N. M. & Garvie-Lok, S. (2013). A new method of dentine microsampling of deciduous teeth for stable isotope ratio analysis. *Journal of Archaeological Science*, 40(11), 3854–3864.

Byrne, C. (2018). Nutrition advice won't help if it's not culturally sensitive. *Self* [online], https://www.self.com/story/culturally-sensitive-nutrition-advice

Carrol, G. M. A., Inskip, S. A. & Waters-Rist, A. (2018). Pathophysiological stable isotope fractionation: assessing the impact of anemia on enamel apatite $\delta^{18}O$ and $\delta^{13}C$ values and bone collagen $\delta^{15}N$ and $\delta^{13}C$ values. *Bioarchaeology International*, 2(2), 117–146.

Cohen, M.N. & G.J. Armelagos. (1984). *Palaeopathology at the origins of agriculture*. Academic Press, New York.

Cohen, M.N. & Crane-Kramer, G. M. M. (2007). *Ancient health: skeletal indicators of agricultural and economic intensification*. University Press of Florida, Gainesville.

Cohen, S.S., Alexander, D.D. Krebs, N.F. Young, B. E., Cabana, M. D., Erdmann, P., Hays, N. P., Bezold, C. P., Levin-Sparenberg, E., Turnin. M. & Saavedra, J. M.

(2018). Factors associated with breastfeeding initiation and continuation: a meta-analysis. *The Journal of Pediatrics*, 203, 190–196.

Coltrain, J. B. & Leavitt, S. W. (2002). Climate and diet in Fremont prehistory: economic variability and abandonment of maize agriculture in the Great Salt Lake Basin. *American Antiquity*, 67(3), 453–485.

Combs Jr., G. F., Trumbo, P. R., McKinley, M. C., Milner, J., Studenski, S., Kimura, T., Watkins, S. M. & Raitin, D. J. (2013). Biomarkers in nutrition: new frontiers in research and application. *Annals of the New York Academy of Sciences*, 1278, 1–10.

Commission on Dietetic Registration (CDR). (2019). Registry statistics: Registered dietitian (RD) and registered dietitian nutritionist (RDN) by Demographics [online]. https://www.cdrnet.org/registry-statistics?id=2579&actionxm=ByDemographics

Cook, A., Pryer, J. & Shetty, P. (2000). The problem of accuracy in dietary surveys: analysis of the over 65 UK National Diet and Nutrition Survey. *Journal Epidemiology Community Health*, 54, 611–616.

Cordain, L. (2010). *The paleo diet: lose weight and get healthy by eating the foods you were designed to eat.* John Wiley & Sons, Hoboken.

Cortés-Sánchez, M., Morales-Muñiz, A., Simón-Vallejo, M. D. Lozano-Francisco, M. C., Vera-Peláez, J. L., Finlayson, C., Rodríguez-Vidal, J., Delgado-Huertas, A., Jiménez-Espejo, F. J., Martínez-Ruiz, F., Aranzazu Martínez-Aguirre, M., Pascual-Granged, A. J., Bergadà-Zapata, M., Gibaja-Bao, J. F., Riquelme-Cantal, J. A., López-Sáez, J. A., Rodrigo-Gámiz, M., Sakai, S., Sugisaki, S., Finlayson, G., Fa, D. A. & Bicho, N. F. (2011). Earliest known use of marine resources by Neanderthals. *PLoS One*, 6(9): e54026.

D'Ortenzio, L., Brickley, M., Schwarcz, H. & Prowse, T. (2015). You are not what you eat during physiological stress: isotopic evaluation of human health. *American Journal of Physical Anthropology*, 157(3), 374–388.

Davy, B. & Jahren, H. (2016). New markers of dietary added sugar intake. *Current Opinion Clinical Nutrition Metabolic Care*, 19, 282–288.

Dobzhansky, T. (1973). Nothing in biology makes sense except in the light of evolution. *The American Biology Teacher*, 35(3), 125–129.

Drucker, D. & Bocherens, H. (2004). Carbon and nitrogen stable isotopes as tracers of change in diet breadth during Middle and Upper Palaeolithic in Europe. *International Journal of osteoarchaeoloy*, 14, 162–177.

Dukan, P. (2014). The Dukan Diet Made Easy: Cruise Through Permanent Weight Loss—and Keep It Off for Life! Harmony/Random House, Toronto.

Eerkens, J. W. & Bartelink, E. J. (2013). Sex-biased weaning and early childhood diet among Middle Holocene hunter-gatherers in central California. *American Journal of Physical Anthropology*, 152, 471–483.

Eerkens, J. W., Berget, A. G. & Bartelink, E. J. (2011). Estimating weaning and early childhood diet from serial micro-samples of dentin collagen. *Journal of Archaeological Science*, 38, 3101–3111.

Engels, F. (1884). *Der Ursprung der Familie, des Privateigenthums und des Staats* (The origin of the family, private property and the state). Verlag der Schweizerischen Volksbuchhandlung, Zurich.

Fahy, G. E., Deter, C., Pitfield, R., Miszkiewics, J. J. & Mahoney, P. (2017). Bone deep: variation in stable isotope ratios and histomorphometric measurements of bone remodelling within adult humans. *Journal of Archaeological Science*, 87, 10–16.

Finlayson, C., Brown, K., Blasco, R., Rosell, J., Negro, J. J., Bortolotti, G. R., Finlayson, G., Sánchez Marco, A., Pacheco, F. G., Rodríguez Vidal, J., Carrión, J. S. & Fa, D. A. (2012). Birds of a feather: Neanderthal exploitation of raptors and corvids. *PLoS One*, 7(9), 1–9.

Food and Agriculture Organization of the United Nations (FAO UN). (2019). Food-based dietary guidelines. http://www.fao.org/nutrition/nutrition-education/food-dietary-guidelines/en/.

Fuller, B.T., Fuller, J. L., Harris, D. A. & Hedges, R. E. M. (2006). Detection of breastfeeding and weaning in modern human infants with carbon and nitrogen stable isotope ratios. *American Journal of Physical Anthropology*, 129, 279–293.

Gaudzinski-Windheuser, S. & Kindler, L. (2012). Research perspectives for the study of Neandertal subsistence strategies based on the analysis of archaeozoological assemblages. *Quaternary International*, 247, 59–68.

Gillespie, D. O. S., Russell, A. F. & Lummaa, V. (2008). When fecundity does not equal fitness: evidence of an offspring quantity versus quality trade-off in pre-industrial humans. *Proceedings of the Royal Society B: Biological Sciences*, 275, 713–722.

Greenwald, A. M. (2017). *Isotopic reconstruction of weaning age and childhood diet among ancient California foragers: life history strategies and implications for demographics, resource intensification, and social organization.* PhD Thesis, University of California, Davis, Davis, CA.

Greenwald, A. M., DeGeorgey, A., Martinez, M. C., Eerkens, J. W., Bartelink, E. J., Simons, D., Alonzo, C. & Garibay, R. (2016a). Maternal time allocation and parental investment in an intensive hunter-gatherer subsistence economy. In A. M. Greenwald & G. R. Burns (eds), *Reconstructing lifeways in ancient California: stable isotope evidence of foraging behavior, life history strategies, and kinship patterns*, pp. 11–30. Center for Archaeological Research at Davis Publications, Davis, CA.

Greenwald, A. M., Eerkens, J. W. & Bartelink, E. J. (2016b). Stable isotope evidence of juvenile foraging in prehistoric Central California. *Journal of Archaeological Science: Reports*, 7, 146–154.

Greenwald, A. M. & Hinde, K. (2018). 'Evolutionary insights into breastfeeding from an archaeological hunter-gatherer population.' *4th Annual meeting of the International Society for Evolution, Medicine, & Public Health.* Park City, UT, 1–4 August.

Griffith, P. D. (2009). In what sense does 'Nothing make sense except in the light of evolution'? *Acta Biotheoretica*, 57(1-2), 11–32.

Indian Health Service. (2018). Fact Sheets: Disparities [online]. https://www.ihs.gov/newsroom/factsheets/disparities/.

Hackman, N. M., Schaefer, E. W. Beiler, J. S. Rose, C. M. Paul, I. M. (2015). Breastfeeding outcome comparison by parity. *Breastfeeding Medicine*, 10(3), 156–162.

Hales, C. N. & Barker, D. J. P. (1992) Type 2 (non-insulin-dependent) diabetes mellitus: the thrifty phenotype hypothesis. *Diabetologia*, 35, 595–601.

Hardy, B. L. (2010). Climatic variability and plant food distribution in Pleistocene Europe: Implications for Neanderthal diet and subsistence. *Quaternary Science Reviews*, 29, 662–679.

Hardy, B.L. & Moncel, M. H. (2011). Neanderthal use of fish, mammals, birds, starchy plants, and wood 125-250,000 years ago. *PLoS One*, 6(8), 1–10.

Hardy, K., Buckley, S., Collins, M. J., Estalrrich, A., Brothwell, D., Copeland, L., García-Tabernero, A., García-Vargas, S., de la Rasilla, M., Lalueza-Fox, C., Huguet, R., Bastir, M., Santamaría, D., Madella, M., Wilson, J., Fernández Cortés, Á. & Rosas, A. (2012). Neanderthal medics? Evidence of food, cooking, and medicinal plants entrapped in dental calculus. *The Science of Nature*, 99(8), 617–626

Hedges, R. E. M., Clement, J. G., Thomas, C. D. L. & O'Connell, T. C. (2007). Collagen turnover in the adult femoral mid-shaft: modeled from anthropogenic radio-carbon tracer measurements. *American Journal of Physical Anthropology*, 133, 808–816.

Hedrick, V. E., Zoellner, J. M., Jahren, A. H., Woodford, N. A., Bostic, J. N. & Davy, B. M. (2015). A dual-carbon-and-nitrogen stable isotope ratio model is not superior to a single-carbon stable isotope ratio model for predicting added sugar intake in Southwest Virginian adults. *The Journal of Nutrition*, 145(6), 1362–1369.

Henrich, J., Heine, S. J. & Norenzayan, A. (2010). The weirdest people in the world? *Behavioral and Brain Sciences*, 33(6), 61–135.

Henry, A. G., Brooks, A. S. & Piperno, D. R. (2011). Microfossils in calculus demonstrate consumption of plants and cooked foods in Neanderthal diets (Shanidar III,

Iraq; Spy I and II, Belgium). *Proceedings of the National Academy of Sciences of the United States of America*, 108(2), 486–491.

Hillson, S. (1986). *Teeth*. Cambridge University Press, Cambridge.

Hockett, B. (2012). The consequences of Middle Palaeolithic diets on pregnant Neanderthal women. *Quaternary International*, 264, 78–82.

Hoefs, J. (2009). *Stable isotope geochemistry*. Springer Verlag, Berlin.

Jahren, A. H., Bostic, J. N. & Davy, B. M. (2014). The potential for a carbon stable isotope biomarker of dietary sugar intake. *Journal of Analytical Atomic Spectrometry*, 29(5), 795–227.

Jahren, A.H., Saudek, C., Yeung, E. H., Kao, W. H. L., Kraft, R. A. & Caballero, B. (2006). An isotopic method for quantifying sweeteners derived from corn and sugar cane. *American Journal Clinical Nutrition*, 84, 1380–1384.

Jaouen, K. (2018). What is our toolbox of analytical chemistry for exploring ancient hominin diets in the absence of organic preservation? *Quaternary Science Reviews*, 197, 307–318.

Jennings, J. D. (1978). *Prehistory of Utah and the Eastern Great Basin. University of Utah Anthropological Papers No. 98*. University of Utah Press, Salt Lake City.

Katz, D. L. & Meller, S. (2014). Can we say what diet is best for health? *Public Health*, 35, 83–103.

Katzenberg, M. A. (2008). 'Stable isotope analysis: a tool for studying past diet, demography, and life history', in M. A. Katzenberg & S. R. Saunders (eds), *Biological anthropology of the human skeleton*, 2nd edn, pp. 413–442. John Wiley & Sons, Inc., New York.

Katzenberg, M.A., and Lovell, N. (1999) Stable isotope variation in pathological bone. *International Journal of Osteoarchaeology*. 9(5), 316–324.

Kelly, R. L. (2007). *The foraging spectrum: diversity in hunter-gatherer lifeways*, 2nd edn. Persheron Press, New York.

Kidder, A. V. (1927). Southwestern archaeological conference. *Science*, 66, 489–491.

Kitano, N., Nomura, K., Kido, M., Murakami, K., Ohkubo, T., Ueno, M. & Sugimoto, M. (2016). Combined effects of maternal age and parity on successful initiation of exclusive breastfeeding. *Preventative Medicine Reports*, 3, 121–126.

Kramer, M. S. & Kakuma, R. (2002). *The optimal duration of exclusive breastfeeding: a systematic review*. World Health Organization, Geneva.

Krebs, J. R. & Davies, N. B. (1978). *Behavioral Ecology: An Evolutionary Approach*. Blackwell Publishing, Oxford.

Kuhnle, G. G. C. (2018). Stable isotope ratios: nutritional biomarkers and more. *The Journal of Nutrition*, 148, 1883–1885

Larsen, C. S. (1995). Biological changes in human populations with agriculture. *Annual Review of Anthropology*, 24, 185–213.

Lawson, D. W. & Mace R. (2011). Parental investment and the optimization of human family size. *Philosophical Transactions of the Royal Society B* 366:333–343.

Lee, R. B. & DeVore, I. (1968). *Man the hunter*. Aldine de Gruyter, New York.

Lev, E., Kislev, M. E. & Bar-Yosef, O. (2005), Mousterian vegetal food in Kebara cave, Mt. Camel. *Journal Archaeological Science*, 32, 475–484.

Lillioja, S., Mott, D. M., Spraul, M., Ferraro, R., Foley, J. E., Ravussin, E., Knowler, W. C. & Bogardus, C. (1993) Insulin resistance and insulin secretory dysfunction as precursors of non-insulin dependent diabetes mellitus: prospective studies in Pima Indians. *New England Journal of Medicine*, 329, 1988–1992.

MacArthur, R. H. & Wilson, E. O. (1967) *The theory of island biogeography*. Princeton University Press, Princeton, NJ.

Madella, M., Jones, M. K., Goldberg, P., Goren, Y. & Hovers, E. (2002). The exploitation of plant resources by Neanderthals in Amud Cave (Israel): the evidence from phytolith studies. *Journal of Archaeological Science*, 29(7), 703–719.

Manolagas, S. (2000). Birth and death of bone cells: basic regulatory mechanisms and implications for the pathogenesis and treatment of osteoporosis. *Endocrine Reviews*, 21, 115–137.

McCance, D. R., Pettitt, D. J., Hanson, R. L., Jacobsson, L. T. H., Knowler, W. C. & Bennett, P. H. (1994) Birth weight and non-insulin dependent diabetes: thrifty genotype, thrifty phenotype, or surviving small baby genotype? *British Medical Journal*, 308, 942

McCance, R. A. & Widdowson, E. M. (1974) The determinants of fetal growth and form. *Proceedings of the Royal Society* London, 185, 1–17.

Melchionna, M., Di Febbraro, M., Carotenuto, F., Rook, L., Mondanaro, A., Castiglione, S., Serio, C., Vero, V. A., Tesone, G., Piccolo, M., Felizola Diniz-Filho, J. A. & Raia, P. (2018). Fragmentation of Neanderthals' pre-extinction distribution by climate change. *Palaeogeography, Palaeoclimatology, Palaeoecology*, 496, 146–154.

Mihesuah, D. A. & Hoover, E. (2019) *Indigenous food sovereignty in the United States: restoring cultural knowledge, protecting environments, and regaining health*. University of Oklahoma Press, Norman.

Morgan, L. H. (1877). Ancient society. MacMillan & Company, London.

Mundel, E. & Chapman, G. E. (2010). A decolonizing approach to health promotion in Canada: the case of the Urban Aboriginal Community Kitchen Garden Project. *Health Promotion International*, 25(2), 166–173

Nash, S. H., Kristal, A. R., Hopkins, S. E., Boyer, B. B. & O'Brien, D. M. (2014). Stable isotope models of sugar intake using hair, red blood cells, and plasma, but not fasting plasma glucose, predict sugar intake in a Yup'ik study population. *The Journal of Nutrition*, 68(1), 75–80.

National Research Council of the National Academies; Committee on the Earth System Context for Hominin Evolution (NRC CESCHE). (2010). *Understanding climate's influence on human evolution*. National Academies Press (US), Washington, DC. https://www.ncbi.nlm.nih.gov/books/NBK208099/.

North American Traditional Indigenous Food Systems (NāTIFS) [online]. (2019). https://www.natifs.org.

O'Brien, D. M. (2015). Stable isotope ratios as biomarkers of diet for health research. *Annual Review of Nutrition*, 35, 565–594.

Peresani, M., Fiore, I., Gala, M., Romandini, M. & Tagliocozzo, A. (2011). Late Neandertals and the intentional removal of feathers as evidenced from bird bone taphonomy at Fumane Cave 44 ky B.P., Italy. *Proceedings of the National Academy of Sciences of the United States of America*, 108, 3888–3893.

Potts, R. & Faith, J. T. (2015). Alternating high and low climate variability: the context of natural selection and speciation in Plio-Pleistocene hominin evolution. *Journal of Human Evolution*, 87, 5–20.

Power, R. C., Salazar-García, D. C., Rubini, M., Darlas, A., Harvati, K., Walker, M., Hublin, J-J. & Henry, A. G. (2018). Dental calculus indicates widespread plant use within the stable Neanderthal dietary niche. *Journal of Human Evolution*, 119, 27–41.

Ready, E. (2013). Neandertal foraging during the late Mousterian in the Pyrenees: new insights based on faunal remains from Gatzarria Cave. *Journal of Archaeological Science*, 40, 1568–1578.

Reitsema, L. J. (2013). Beyond diet reconstruction: stable isotope applications to human physiology, health, and nutrition. *American Journal of Human Biology*, 25, 445–456.

Richards, M. P., Pettitt, P. B., Stiner, M.C. & Trinkaus, E. (2001). Stable isotope evidence for increasing dietary breadth in the European mid-Upper Palaeolithic. *Proceedings of the National Academy of Sciences of the United States of America*, 98(11), 6528–6532.

Richards, M. P. & Trinkaus, E. (2009). Isotopic evidence for the diets of European Neanderthals and early modern humans. *Proceedings of the National Academy of Sciences of the United States of America*, 106(38), 16034–16039.

Riordan, J., Gill-Hopple, K. & Angeron, J. (2005). Indicators of effective breastfeeding and estimates of breast milk intake. *Journal of Human Lactation*, 21(4), 406–412.

Savage, P. J., Bennett, P. H., Senter, R. G. & Miller, M. (1979). High prevalence of diabetes in young Pima Indians: evidence of phenotypic variation in a genetically isolated population. *Diabetes*, 28, 939–942.

Schoeninger, M. J. (2009). Stable isotope evidence for the adoption of maize agriculture. *Current Anthropology*, 50, 633–640.

Schoeninger, M. J. (1985). Trophic effects on $^{15}N/^{14}N$ and $^{13}C/^{12}C$ ratios in human bone collagen and strontium levels in bone mineral. *Journal of Human Evolution*, 14, 515–525.

Schuetz, P. (2017). Food for thought: why does the medical community struggle with research about nutritional therapy in the acute care setting? *BMC Medicine*, 15(1), 38.

Schwarcz, H.P. (2000). Some biochemical aspects of carbon isotopic palaeodiet studies', in S. H. Ambrose & M. A. Katzenberg (eds), *Biogeochemical approaches to palaeodietary analysis*, pp. 189–209. Kluwer Academic, New York.

Schwarcz, H.P. and Schoeninger, M.J. (1991). Stable isotope analyses in human nutritional ecology. *Yearbook of Physical Anthropology*, 34, 283–321.

Schwitalla, A. W. (2013) *Global warming in California: a lesson from the medieval climatic anomaly*, 17, Center for Archaeological Research at Davis, University of California, Davis.

Shipman, P. (2008). Separating 'us' from 'them': Neanderthal and modern human behavior. *Proceedings of the National Academy of Sciences of the United States of America*, 105(38), 14241–14242.

Sistiaga, A., Mallol, C. Galván, B. & Summons, R. E. (2014). The Neanderthal meal: a new perspective using faecal biomarkers. *PLoS One*, 9(6), e101045.

Smith, E. A. and Winterhalder, B. (1992). 'Natural selection and decision making: some fundamental principles', in E. A. Smith & B. Winterhalder (eds), *Evolutionary ecology and human behavior*, pp. 25–60. Aldine de Gruyter, New York.

Sponheimer, M., Alemseged, Z., Cerling, T. E., Grine, F. E., Kimbel, W. H., Leakey, M. G., Lee-Thorpe, J. A., Manthi, F. K., Reed, K. E., Wood, B. A. & Wynn, J. (2013). Isotopic evidence of early hominin diets. *Proceedings of the National Academy of Sciences of the United States of America*, 110(26), 10513–10518.

Stearns, S. C. (1992). *The evolution of life histories*. Oxford University Press, Oxford.

Steckel, R. H., Rose, J. C., Larsen, C. S. & Walker, P. L. (2002). Skeletal health in the western hemisphere from 4000 B.C. to the present. *Evolutionary Anthropology*, 11, 142–155.

Strassmann, B. I. & Gillespie, B. (2002) Life-history theory, fertility and reproductive success in humans. *Proceedings of the Royal Society B: Biological Sciences*, 269, 553–562.

Tieszen, L.L., & Fagre, T. (1993). 'Effect of diet quality and composition on the isotopic composition of respiratory CO_2, bone collagen, bioapatite, and soft tissues', in J. B. Lambert & G. Grupe (eds), *Prehistoric human bone: archaeology at the molecular level*, pp. 121–155. Springer-Verlag: New York.

Trepanowski, J. F. & Ioannidis, J. P. A. (2018). Perspective: limiting dependence on nonrandomized studies and improving randomized trials in human nutrition research: why and how. *Advanced Nutrition*, 9, 367–377.

Trinquart, L., Johns, D. M. & Galea, S. (2016). Why do we think we know what we know? A metaknowledge analysis of the salt controversy. *International Journal of Epidemiology*, 45(1), 251–260.

Tso, M. (2014). Dine food sovereignty: decolonization through the lens of food. *Scripps Senior Theses*, 348 [online]. https://scholarship.claremont.edu/scripps_theses/348.

Urey, H. C. (1947). The thermodynamic properties of isotopic substances. *Journal of the Chemical Society*, 1947, 562.

Villa, P. & W. Roebroeks. (2014). Neandertal demise: an archaeological analysis of the modern human superiority complex. *PLoS One*, 9(4), e96424.

White, T. D. & Folkens, P. A. (2005). *The human bone manual*. Academic Press, Boston.

Wood, R. E., Higham, T. F. G., de Torres, T., et al. (2012). A new date for the Neanderthals from El Sidrón Cave (Asturias, Northern Spain). *Archaeolmetry*, 55(1), 148–158.

Yeung, E. H., Saudek, C. D., Jahren, A. H., Kao, W. H., Islas, M., Kraft, R., Coresh, J., Anderson, C. A. M. (2010). Evaluation of a novel isotope biomarker for dietary consumption of sweets. *American Journal of Epidemiology*, 172(9), 1045–1052.

Zuk, M. (2013). *Palaeofantasy: what evolution really tells us about sex, diet and how we live*, Norton & Company, New York.

CHAPTER 6

Developmental, evolutionary and behavioural perspectives on oral health

Tanya M. Smith and Christina Warinner*

6.1 Teeth as childhood record keepers

The milestones of birth, tooth emergence and the loss of deciduous teeth (also known as baby, milk or primary teeth) encompass an incredibly vulnerable time for human children. Born toothless, during the first few months of life babies rely exclusively on their mothers for nutrition. In small-scale traditional societies, soft nursing foods are introduced around six months, close to the age that infants' first teeth emerge (Sellen & Smay, 2001). For the next few years it seems that everything goes into the mouth for a taste or a nibble. Babies aren't just learning about food during this time—they are also encountering new immunological challenges. Their digestive and immune systems are developing rapidly, and bouts of gastric distress and illness are unavoidable, although mitigated by protective chemicals and antibodies in the mother's milk (Andreas et al., 2015; Caballero-Flores et al., 2019). During the transition from breast milk to

* We appreciate the careful comments and suggestions of the editors and two reviewers, as well as Don Reid, Michael Foley, Janet Fink and Lulu Cook who aided with an earlier version. Research assistance was provided by Barbara Teßman, Hans-Ulrich Luder, Christophe Boesch, Robin Feeney, Leslea Hlusko and the Phoebe A. Hearst Museum of Anthropology at the University of California, Berkeley. Financial assistance was provided by the Werner Siemens Foundation, the German Research Foundation, the US National Science Foundation (BCS-1643318, BCS-1516633), Griffith University and Harvard University.

an adult diet, new foods test these systems further, particularly in countries where clean water and sewage disposal are limited (Black et al., 2003; Humphrey, 2008). Here we discuss how these formative experiences can leave behind an indelible set of microscopic impressions.

6.1.1 Developmental disruptions begin at birth

Deciduous teeth begin calcifying in utero, as does the first permanent molar (Massler et al., 1941). These teeth preserve lines that indicate the exact position of enamel- and dentine-forming cells at birth (Figure 6.1). Scientists discovered these 'neonatal lines' by examining hundreds of baby teeth from human children, finding similar dark lines in a comparable position in nearly all teeth (reviewed in Hurnanen et al., 2017; Smith, 2018). Figure 6.1 shows that postnatal disturbances may also imprint the position of cells that were actively secreting at that time, which histologists generically call 'accentuated lines' (e.g. Schwartz et al., 2006; Smith & Boesch, 2015). Such alterations in enamel and dentine structure can be used to age a disruptor since teeth also record daily cellular secretions—timestamps that can be measured and counted for accurate age estimates (reviewed in Smith, 2013; Birch & Dean, 2014).

One remarkable use of the neonatal line is to determine whether an infant was born prematurely.

Tanya M. Smith and Christina Warinner, *Developmental, evolutionary and behavioural perspectives on oral health.* In: *Palaeopathology and Evolutionary Medicine.* Edited by Kimberly A. Plomp et al., Oxford University Press. © Oxford University Press (2022). DOI: 10.1093/oso/9780198849711.003.0006

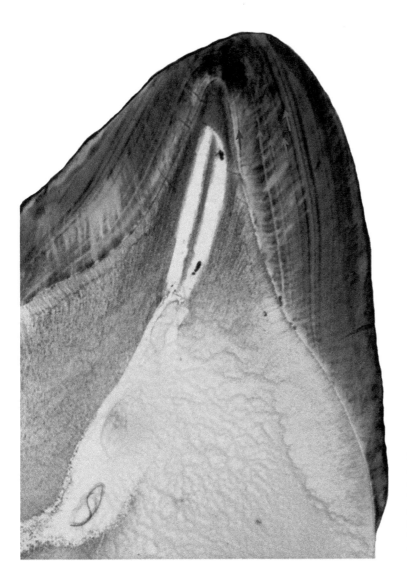

Figure 6.1 Early-life developmental disruptions in the enamel of a first molar cusp. The neonatal line (blue arrow) is followed by several dark, contrasted, accentuated lines (red arrows) formed during the first year of life.
Human tooth courtesy of Robin Feeney.
Reproduced with permission from Smith (2018).

For example, Birch and Dean (2014) looked inside the deciduous teeth of a colleague's twins through histological sectioning and light microscopy. They located the neonatal line of each twin and counted daily growth lines to determine the number of days these teeth had been forming prior to birth. The authors came up with counts of seventy-five and seventy-four days for the two twins, which was sixty-five and sixty-six days shorter than the average prenatal formation time for third premolars. They inferred that the twins were born approximately two months prematurely. After completing their study, they reviewed medical records for the children, and learned that the twins were born fifty-eight days before their predicted due date.

6.1.2 What can these disruptions tell us about childhood?

Because most permanent tooth crowns form between birth and 6 years of age, they are a powerful window into the health and well-being of young children. Major disruptions during or after birth can produce 'hypoplasias'—areas of depressed or missing enamel where a group of enamel-forming cells stop secreting prematurely, causing a pit,

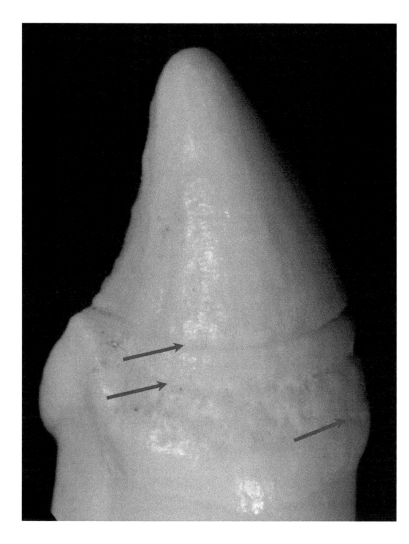

Figure 6.2 Developmental defects known as hypoplasias on the surface of a canine crown. Note the circular ring (upper arrow) as well as the pitted areas (lower arrows) that also mostly encircle the tooth crown.
Chimpanzee specimen courtesy of Christophe Boesch. Reproduced with permission from Smith (2018).

plane, or ring around the tooth surface (Figure 6.2) (e.g. Hillson, 1996, 2014; Towel & Irish, 2020). Glancing through a microscope at the teeth of prehistoric humans reveals that childhood has always been a vulnerable experience, and probably even more so in the past (reviewed in Goodman & Rose, 1991; Hillson, 2014). Numerous studies have documented accentuated lines and hypoplasias in pre-industrial modern humans, fossil hominins and even great apes (e.g. Hillson, 2014; Smith & Boesch, 2015; McGrath et al., 2018, 2021).

Scholars emphasise that the presence and/or frequency of developmental defects should be interpreted with care prior to concluding that an individual or group of individuals was 'stressed' (Schwartz et al., 2006; Smith & Boesch, 2015; Edinborough & Rando, 2020; Towel & Irish, 2020). Disruptors can be organised into several broad categories: specific illnesses or symptoms, nutritional deficiencies and physical or psychological stresses (Table 6.1). As we discuss later, many of these causes have been deduced from experiments on laboratory animals or observations of captive primates with associated medical records (Box 6.1). Careful studies of people with documented early histories are quite rare, but they make for illuminating reading.

In the twin study of Birch and Dean (2014), the authors also counted daily increments after the twins' births to assign ages to their accentuated lines. Once they estimated those ages, they

Table 6.1 Suggested causes of developmental defects (hypoplasias and accentuated lines).

Type	Specific Cause
Nutritional	Malnutrition
	Vitamin A deficiency
	Vitamin D deficiency
Disease	Chickenpox
	Diabetes
	Diphtheria
	Measles
	Neonatal haemolytic anaemia (immune blood disease)
	Parasite infection
	Pneumonia
	Rubella
	Scarlet fever
	Scurvy
	Smallpox
	Syphilis
	Whooping cough
Symptom	Allergies
	Bone inflammation
	Convulsions
	Diarrhoea
	Fever
	Vomiting
Event	Antibiotic administration
	Birth
	Birth injury
	Electrical burns
	Enclosure (room) transfer (e.g. in captive animals)
	Eye injury
	Fluorochrome (biomarker) administration
	Hospitalisation (admittance or discharge)
	Inoculation (with pathogenic virus or bacteria)
	Ionising radiation
	Parturition (giving birth)
	Trauma
	Weaning (cessation of breastfeeding)

Note: There may be at least 500 studies of enamel defects and nearly 100 identified causes, which have not been comprehensively reviewed here (see Goodman and Rose, 1991).

compared them to medical records and found that approximately one-third of the lines could be related to a known event. For example, a near perfect match was obtained for lines on the day that three routine vaccinations were administered to both twins. A similar correspondence was found when the twins were transferred out of intensive care, when one twin was discharged from the hospital, as well as when the other was readmitted for

a surgical procedure. Gastrointestinal upsets, eye problems and incidences of vomiting all seemed to match up with accentuated lines. These symptoms occurred within a week of the estimated ages of the lines, and in several cases, they happened on the exact day a line was formed—a striking example of the intimate recording system in teeth.

Box 6.1   Formation of hypoplastic defects.

The best evidence for why hypoplastic defects on tooth surfaces form comes from experiments of laboratory animals (reviewed in Goodman and Rose, 1991; Hillson, 2014). At the time there was considerable interest in understanding rickets, a disease that impairs normal skeletal formation in children, particularly at latitudes where exposure to sunlight is limited (Holick, 2006). Exposure of the skin to ultraviolet light is necessary for the formation of vitamin D, which is required for the absorption of calcium and phosphorus to calcify bone (reviewed in Prentiss, 2013). Most vitamin D is accrued by the body as a result of UV light exposure, with some of the vitamin ingested in the diet. This deficiency was especially problematic during the height of the nineteenth-century Industrial Revolution in Europe, when extreme air pollution reduced UV light exposure. Researchers used dogs as an experimental model, decreasing vitamin D to mimic rickets and inadvertently discovering that this deficiency produced hypoplasias in their teeth. Similar nutritional and hormonal manipulations of mice led to dental defects, as did excessive fluoride supplementation and intentional parasitic infections in sheep. Another troubling example is the case of human children born with syphilis. This potentially fatal disease, which is caused by the movement of bacteria from an infected mother into the developing foetus, produces extreme hypoplasias and dental deformities, including Hutchinson teeth (reviewed in Hillson, 2014; Ioannou et al., 2016).

There appears to be a similar sensitivity in the tooth development of non-human primates. For example, a captive gorilla showed a precise age match between accentuated lines and an eye injury and surgery (Schwartz et al., 2006). Subsequent hospitalisations and moves to new enclosures matched the ages of accentuated lines formed prior to this individual's accidental death.

Smith (2013) documented developmental disruptions in a young monkey, finding accentuated lines that corresponded to dates of an enclosure transfer, leg injury and tail amputation. Moreover, other lines marked bouts of dehydration and diarrhoea that led to multiple hospitalisations and eventual euthanasia. It may seem surprising that transfers to new rooms or enclosures appear in teeth alongside indications of illnesses. These experiences appear to have measurable effects on youngsters' development, pointing to a degree of environmental sensitivity of which their caregivers may be unaware. Importantly, accentuated lines that form due to these different causes are identical in appearance; additional biochemical and elemental analyses are required to gain further insight into possible explanations (e.g. Austin et al., 2016; Smith et al., 2021).

6.1.3 Early life dental disruptions and disease

Scholars have pointed to evidence from developmental defects to support the hypothesis that stressful events during early life have negative consequences for adult health (reviewed in Armelagos et al., 2009; Temple, 2019). Unfortunately, incorrect assumptions about tooth formation, limited understanding of the complex aetiology of dental defects and biases in bioarchaeological data undermine most of the initial evidence provided for this intriguing idea (Hillson, 1996; Reid & Dean, 2000; DeWitte & Stojanowski, 2015; Towle & Irish, 2020). An exception is a recent analysis of twelve Iranian subadults' deciduous canines, which showed a weakly positive association between prenatal developmental defects and earlier age at death (Lorentz et al., 2019)—but this type of analysis cannot be extended to adult individuals. Early research on accentuated lines failed to control for thin section obliquity and heterogeneity (e.g. Smith, 2004, their Figure 3.1), which may obscure identification of accentuated lines (including the neonatal line). Influential studies of enamel hypoplasias in deciduous and permanent teeth relied on models of tooth growth that have been shown to be in error (Reid & Dean, 2000; AlQahtani et al., 2014). The good news is that bioarchaeologists are now well positioned to revisit ideas about the developmental origins

of health, particularly in recent British children—whose tooth development has been extensively documented with modern approaches over the past two decades.

6.1.4 Dental disruptions and early-life diets

Biological anthropologists and archaeologists have been especially interested in using developmental defects to study breastfeeding practices in ancient hominins and modern humans (e.g. Katzenberg et al., 1996; Skinner, 1996). Knowing when mothers wean their babies helps us to understand how energy is used for growth and reproduction over the life course. From the weanlings' perspective, the end of breastfeeding is thought to be stressful since the flow of nutritional benefits and maternal antibodies ceases, along with the psychological comfort of nursing. Weaning too soon without sufficient resources can lead to malnutrition, and consuming non-maternal foods and liquids carries a risk of infection from food-borne pathogens (Black et al., 2003).

Although weaning is frequently inferred as a cause of hypoplasias, evidence from humans is equivocal. For example, 27 children of eighteenth- and nineteenth-century enslaved African-Americans were found to have developed hypoplasias between about 1.5 and 4.5 years of age, yet historic reports indicate they would have been weaned between the ages of 9 and 12 months (Blakey et al., 1994). It is not known what might have caused these dental disruptions over the following few years, although one can imagine that they faced innumerable hardships. Part of the difficulty in making sense of accentuated lines and hypoplasias is the fact that there are numerous examples where an apparent link or association between observations has led to a mistaken belief that one thing caused another; in other words, correlation does not necessarily indicate causation. Children often show hypoplasias occurring between about 1 to 4 years of age—when mothers in many societies cease breastfeeding—yet this association does not prove a direct causal relationship. Studies of non-human primates with documented histories have yet to produce any direct evidence that hypoplasias are a result of the weaning process (e.g. Smith & Boesch, 2015), nor has this become apparent from elemental proxies of nursing behaviour in teeth (reviewed in Smith et al., 2021).

Anthropologists have also investigated whether malnutrition causes developmental defects (Goodman & Rose, 1991; May et al., 1993). Here we find more evidence that deserves further study. For example, rural Mexican and Guatemalan children given food supplements showed fewer hypoplasias than children who did not participate in the nutrition study. Scientists have also documented increases in hypoplasias during major cultural transitions, and point to potential deficiencies in the nutritional status or health of these populations (Larsen, 1995; Hillson, 2014). The development of agriculture beginning 10,000–12,000 years ago is the most extreme shift in nutrition and health over the history of our species (reviewed in Page et al., 2016; see also Chapters 5 and 7). Prior to this time, humans obtained food exclusively through hunting and collecting wild foods. During the initial agricultural 'revolution', people began to cultivate cereals, rice and other plants. They settled into permanent dwellings to tend crops and led more sedentary lifestyles. Some early agriculturalists decreased the breadth of their diet, incorporated more carbohydrates and lived in larger communities, where diseases could spread more easily. With this adoption of agriculture, the prevalence of hypoplasias generally increased, particularly in Native North American communities that relied on maize-intensive farming—although exceptions have been documented (reviewed in Larsen, 1995; Lukacs, 2012; Smith et al., 2016).

Other transitions suspected to affect teeth include periods of social stratification or contact between different populations. Littleton (2005) documented hypoplasias on the teeth of Australian Aboriginal children born between 1890 and 1960. She found that dental defects increased as social contact with Europeans became more frequent. This trend was particularly noticeable during the Second World War when certain Aboriginal communities were relocated to government settlements and forced to cease hunting and gathering. Although rations were provided, limited nutritional quality and cramped living conditions led to poor health, and hypoplasias appeared at earlier ages and more frequently than in pre-contact populations. It would be worth exploring whether other communities forced to give up their traditional livelihoods show similar dental records as a result of systematic violence and disenfranchisement.

Importantly though, these studies do not establish that poor nutrition causes developmental defects in teeth. Malnourished children have weakened immune systems and are more prone to infections (reviewed in Larsen, 1995). Given the apparent sensitivity of developing teeth, additional research is needed to untangle the potential interplay between subsistence methods, dietary quality, and immune function.

6.2 Those pesky wisdom teeth

The final readily observable phase of human dental development is the eruption of our third molars (M3s) at approximately 18–20 years of age—inspiring the moniker 'wisdom teeth' as a signifier of adulthood. For centuries humans have suffered from two M3 disorders: failure to form, known as 'agenesis', and failure to erupt into the mouth, known as 'impaction'. Agenesis is more straightforward to diagnose than impaction—it is simply the complete lack of a tooth. When impaction is suspected from a simple visual inspection, we must rule out the possibilities that a tooth may not be in place because it has not finished forming, or because it never formed in the first place. For teeth that do not erupt within the expected age ranges, assessment of X-ray images is essential to decipher the underlying problem (Figure 6.3).

Agenesis and impaction of M3s each affect one out of every four people living today (see meta-analyses in Carter & Worthington, 2015, 2016), including the authors. In the case of CW, all four of her M3s failed to form, as did three of her second molars, an extreme form of molar agenesis, while TS had all four M3s surgically removed at age 20 after developing a rapid and painful lower jaw infection due to M3 impaction. Agenesis and impaction appear in certain groups more often than in others (Carter & Worthington, 2015, 2016). For example, it is very common for Asian and Native

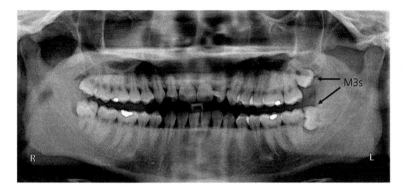

Figure 6.3 Panoramic X-ray revealing impacted M3s in a human patient. Bright regions in other teeth are fillings. Image Credit: Ka-ho Chu (copyright free). Modified from https://commons.wikimedia.org/wiki/File:Impacted_wisdom_teeth.jpg, downloaded 3 September 2016.

American individuals to be missing their M3s, while Africans and Aboriginal Australians show lower incidences of agenesis. Similarly, impaction affects Middle Eastern and Asian individuals more often than African populations. Missing and impacted wisdom teeth are thought to be recent developmental disorders (reviewed in Sergi, 1914; Lieberman, 2011; Carter, 2016). Aside from a few potential cases in the distant past, these conditions rarely troubled any hominin species other than our own.

6.2.1 Why are M3 problems so common in some of us?

One evolutionary reason commonly given for missing or impacted wisdom teeth is the belief that our jaws have reduced in size faster than our teeth have, leading to arrested or stunted M3 development. This theory points to our recent adoption of a soft-textured diet, which requires lower chewing forces than harder and coarser ancestral diets, as a primary cause of jaw shrinkage (Lieberman, 2011). Bones respond to mechanical pressure, so consuming a harder diet during tooth development leads to larger, stronger jaws (Makaremi et al., 2015). Moreover, alveolar bone, which supports the teeth, is known to be particularly responsive to pressure and tension (Maleeh et al., 2016). This is why dentistry has long used steady mechanical pressure (in the form of braces and retainers) to move and realign teeth within the jaw. Third molars are impacted more often in small-jawed individuals than in large-jawed individuals (Carter, 2016). Additionally, impaction disproportionately affects women, who tend to have smaller jaws than men. In short,

wisdom teeth do not erupt properly when there isn't room for them.

Another idea is that large-scale changes in our diet decreased tooth wear, and that this phenomenon led to more variable dental development (Brace, 1964). A final hypothesis is that natural selection led to the loss of M3s due to the health risks of impaction (Calcagno & Gibson, 1988). Those of us who have experienced the unpleasant consequences of wisdom tooth impaction first hand understand why some dentists and oral surgeons often recommend removing them altogether. Before modern medical intervention, adults who developed wisdom teeth likely experienced higher mortality from infections than those who never formed them. Infections of the gums or jaws can spread to the head, neck or bloodstream, which can be fatal when left untreated (Calcagno & Gibson, 1988; Kunkel et al., 2007; Otte, 2017). Even surgical removal of M3s once entailed significant risk. In a 1936 review of 622 people with impacted wisdom teeth, it was reported that 42 deaths occurred due to complications after surgery, a mortality rate of nearly 7% (Henry & Morant, 1936).

It appears that there is a cultural explanation for smaller jaws and impacted teeth. Carter (2016) found fivefold more M3 impaction in living humans compared to early agriculturalists, and others have estimated that it has become ten times more likely since industrial food production began (reviewed in Lieberman, 2011). Impaction of any tooth type also occurs with some regularity in captive and domesticated animals, which typically eat a softer diet than their wild counterparts (O'Regan & Kitchener, 2005). Thus wisdom tooth impaction appears to be a true disease of civilisation, or a mismatch between our modern culture and biology—and M3s are not the only part of our body to become maladaptive recently (Lieberman, 2013).

Making sense of M3 agenesis is more difficult than figuring out why impaction has become so common. Carter (2016) observed that larger-jawed individuals who lived during the agricultural transition were more likely to be missing their M3s than smaller-jawed individuals—the opposite pattern of what she found for tooth impaction and jaw size. This suggests that it was not simply the lack of space that was inhibiting tooth formation. Yet during the industrial transition, smaller-jawed individuals were more likely to be missing their M3s. Carter (2016) concluded that this second result is consistent with the theory that natural selection might have been acting to favour individuals who were missing M3s, particularly as people with small jaws who retained them were at greater risk of impaction, infection and possibly death.

Broader studies of the oral health of additional individuals who lived during the development and intensification of agriculture and industrialisation might help to confirm these conclusions. For example, gum and tooth infections often leave characteristic signatures in the alveolar bone around teeth—clues that we explore later. In addition, studies of genes involved in tooth agenesis may also help to clarify its evolutionary history (De Coster et al., 2009; Haga et al., 2013).

6.3 Our teeth are never alone: the oral cavity as a microbial culture dish

Although we may think of our oral cavity as simply a part of our body, it is also home to billions of bacteria—the oral microbiome. From the moment we are born, our mouths become colonised with saliva-loving bacteria, and with the eruption of the first deciduous tooth, many more species of bacteria form biofilms known as dental plaque. You may have had the experience of running a fingernail or toothpick over your front teeth first thing in the morning and noticed the sticky plaque that formed overnight. They are not random assortments of bacteria, but rather complex communities of hundreds of species that grow in a coordinated and ordered

manner (Kolenbrander et al., 2006; Welch et al., 2019). We acquire these bacteria during early life primarily from our mothers and caregivers (Ferretti et al., 2018), but also through contact with innumerable objects babies put into their mouths, despite the best efforts of their salivary glands to flush them away. Once established, the oral microbiome is remarkably stable throughout adulthood (Zaura et al., 2014), although lifestyle and dietary factors can lead to health-associated microbial perturbations, which we discuss later in this chapter.

6.3.1 The oral microbiome in health

While many people spend a good deal of time and money trying to remove their dental plaque through toothbrushing and other interventions, it is actually thought that these bacteria co-evolved with us in a primarily beneficial, and even protective, role (Zaura et al., 2014; Marsh et al., 2016). The oral cavity is an inviting place—warm, moist and full of nutrients—just the sort of place in which bacteria colonise and thrive. Saliva and gingival crevicular fluid are the most important defenders of our oral cavity, and they contain numerous antimicrobial compounds that collectively kill and aggregate foreign bacteria for removal, with an estimated success rate of more than 80 billion bacteria swallowed per day (Amerongen & Veerman, 2002; Warinner et al., 2015). In fact, the antimicrobial properties of saliva are why it is thought that animals have a natural instinct to lick wounds (Brand & Veerman, 2013). Although foreign bacteria are largely susceptible to such defences, our own bacteria have coevolved to circumvent and evade such attacks, and in fact many oral bacteria consume salivary components as their primary nutrient source (Marsh et al., 2016). By densely occupying our oral tissues, oral bacteria then provide a second line of defence from foreign bacteria by either outcompeting them for resources or killing them directly through the secretion of toxins, activities collectively known as 'colonisation resistance' (He et al., 2013; Zaura et al., 2014).

The importance of saliva in maintaining a stable, health-associated microbiome is most obvious in cases where it is absent, such as in individuals with

Sjögren's syndrome—an autoimmune disorder that leads to underproduction of saliva and tears. Individuals with Sjogren's syndrome experience massive, unregulated overgrowth of opportunistic and pathogenic bacteria and fungi throughout the oral cavity, regardless of oral hygiene measures. This results in greatly elevated rates of dental caries, thrush (candidiasis) and oral inflammation, leading to widespread tooth loss early in life (Baker et al., 2014; Matthews et al., 2008). By contrast, people with normal salivary production maintain stable communities of oral bacteria for years, retaining their teeth long into adulthood and even senescence.

6.3.2 The microbial biogeography of the oral cavity

Within the oral cavity, there are several distinct habitats or ecological niches, including the tongue, gingiva (gums), buccal mucosa (cheeks), saliva and teeth, each home to a slightly different microbial community (Human Microbiome Consortium Project, 2012; Welch et al., 2020). Teeth support the most complex assortment of microbes, including more than 700 species (Dewhirst et al., 2010), and yet teeth are also the most difficult tissue for bacteria to colonise. The enamel crowns of teeth are made of nearly pure hydroxyapatite, the calcium phosphate mineral that also makes up about 70% of the mass of your bones (Ten Cate, 2003). Bacteria cannot permanently attach themselves to such minerals and simply slide off with the slightest pressure. Instead, saliva plays a major role in helping bacteria to colonise the teeth (Box 6.2) (Ruhl et al., 2004; Siquiera et al., 2012).

Recently it was shown that the high proportion of *Streptococcus* among early colonisers in humans is correlated with the acquisition of amylase-binding protein genes, which enable them to bind to the salivary alpha-amylase present in the acquired enamel pellicle (AEP) and also to use salivary amylase to digest dietary starch for themselves (Fellows Yates et al., 2021). Such findings suggest that the human oral microbiome may have evolved to become starch-adapted early in human evolution, providing clues from an unexpected source about the lives and habits of our early ancestors.

> **Box 6.2   Salivary proteins**
>
> Saliva is rich in proteins, and many of these proteins have multiple functions. Salivary proline-rich proteins, for example, bind up and neutralise dietary tannins, plant defensive compounds that interfere with digestion (Butler, 1989; Carlson, 1993). They also spontaneously bind to enamel, coating the teeth in a thin layer known as the acquired enamel pellicle (AEP) (Siquiera et al., 2012). Other proteins also contribute to the AEP, including salivary alpha-amylase (an enzyme that breaks down starch), agglutinins and mucins (defensive proteins that aggregate bacteria), and statherin (a protein that supports tooth mineralisation). Over time, a few oral bacteria have evolved the ability to bind to the proteins of the AEP, allowing them to colonise this otherwise inaccessible surface. In humans, two groups of bacteria have achieved this feat: *Actinomyces* and *Streptococcus*. Known as early colonisers, they form the basal layer of dental plaque, and all of the remaining bacterial species found in plaque are anchored in one way or another to this initial scaffold (Kolenbrander et al., 2006; Siquiera et al., 2012).
>
> In many ways, dental plaque is an evolutionary marvel, with each species having coevolved specific adhesins and receptors that enable it to attach to other unrelated species within the dental plaque biofilm. This ultimately allows a multi-species plaque to build up like a three-dimensional jigsaw puzzle (Kolenbrander, 2006; Welch et al., 2019). Some bacteria, like the giant cells of *Fusobacterium nucleatum*, can attach to more than a dozen other species in the plaque biofilm, making them a kind of superstar among oral bacteria (Kolenbrander et al., 2006). Other taxa that play outsized roles in structuring the plaque biofilm include *Leptotrichia*, *Capnocytophaga*, and especially *Corynebacterium* (Welch et al., 2020). Additional lesser known taxa, including members of Anaerolinaceae, *Campylobacter*, *Ottowia*, *Prevotella*, *Pseudopropionibacerium*, and *Selenomonas*, were recently shown to be highly conserved among both apes and New World monkeys, and thus are presumably functionally important in the primate oral microbiome (Fellows Yates et al., 2021).

Although serving as coevolutionary partners in our mouths, our oral bacteria can take on more sinister roles if translocated elsewhere. This is in part because they are fairly good at evading our immune system and, if uncaught, will readily colonise new tissues and build biofilms in undesired locations. Bloodborne oral bacteria such as *Streptococcus* are known to contribute to the development of arterial plaques, and they are a leading cause of heart valve infection (Duval & Leport, 2008). *Fusobacterium nucleatum* is particularly dangerous outside of the oral cavity, and tends to form purulent infections and abscesses if transferred through bites (Talan et al., 2003; Abrahamian & Goldstein, 2011). It is also a leading cause of stillbirth and pre-term birth (Han, 2015). Changing hormones during pregnancy can cause gum inflammation, which may contribute to this dangerous transfer to the placenta (Han, 2011; Balan et al., 2018).

The oral cavity is also the natural habitat of numerous bacteria implicated in acute respiratory, neurological and systemic infections, including *Streptococcus pyogenes* (strep throat, scarlet fever, toxic shock syndrome), *Streptococcus pneumoniae* (ear infections, pink eye, bronchitis, pneumonia, bacterial meningitis), *Haemophilus influenzae* (ear infections, pink eye, sinusitis, pneumonia) and *Neisseria meningitidis* (epidemic bacterial meningitis) (reviewed in Warinner, 2016). These bacteria are known as pathobionts—commensal bacteria that are normally harmless, but which can cause disease in individuals who are weakened or otherwise immunocompromised, with children and the elderly being especially vulnerable. Most people carry one or more of these 'frenemy' species in their mouths at all times, and (although it is not a pleasant thought) it is very likely that your dental plaque currently contains all of them.

6.3.3 The formation of dental calculus

Dental plaque has a boom-bust lifecycle, and, left alone over time, the innermost layers of plaque will begin to calcify as hydroxyapatite and other minerals precipitate from calcium phosphates dissolved in saliva (White, 1991, 1997). This process kills the once-thriving bacterial communities and establishes a new mineral surface upon which early colonisers initiate a new biofilm. This process repeats over and over, building up layers of calcified plaque known as tartar or 'calculus' (Figure 6.4a, b). Being made of hydroxyapatite, calculus is just as hard as bone, and it preserves as well as the rest of the skeleton in the archaeological record. Dental calculus is found nearly ubiquitously among living and ancient human populations, and the oldest calculus described to date has been found on the teeth of Miocene apes dating as far back as 12 million years ago (Hershkovitz et al., 1997; Fuss et al., 2018).

Excess plaque and calculus challenge our oral health, yet in many ways they are a boon for anthropologists (Velsko & Warinner, 2017). Calculus research is deepening our understanding of

(a)   (b)   (c)

Figure 6.4 Dental calculus. (a) Buildup of calcified dental plaque (calculus) on the teeth of a lower jaw. Calculus typically forms along the gumline, indicating where the gingiva would have been during life. This individual also shows indications of several dental anomalies, including a prominent horizontal hypoplasia on the canine and malocclusion of the M3 from crowding. Historic human mandible from Mogador (Morocco) from the Rudolf Virchow Collection (BGAEU-RV 3480). (b) Magnified view of dental calculus in cross-section using backscattered scanning electron microscopy. Note the sequential buildup of calculus, which begins on the tooth surface (upper left, out of frame) and continues outward (bottom right), layer by layer. Burial soil (dark gray with inclusions) adheres to the outermost layer of the dental calculus. Medieval dental calculus from Dalheim, Germany (individual B78) from the collections of the University of Zürich's Institute for Evolutionary Medicine. (c) Further magnified view of calcified dental plaque bacteria within the dental calculus shown in (b).

ancient diets as we learn that food proteins and plant microfossils can be excavated from teeth thousands or even millions of years after death (e.g. phytoliths, starch granules and seed fragments: Henry et al., 2011, 2012; Warinner et al., 2014b; Hendy et al., 2018; Jeong et al., 2018; Scott et al., 2021). Furthermore, airborne microcharcoal and pollen, and even craft-related materials like paint pigments and textile fibres, have been recovered from dental calculus—revealing novel aspects of the daily activities of past people (Radini et al., 2017, 2019). Most importantly, calculus traps and preserves bacteria (Figure 6.4c), allowing scientists to study oral bacterial evolution and track diseases prior to recorded history, including diseases that do not visibly affect the skeleton (e.g. Adler et al., 2013; Cornejo et al., 2013; Warinner et al., 2014b; Weyrich et al., 2017).

6.4 Caries, gum disease and tooth loss: additional diseases of civilisation?

Dental plaque is the main culprit in caries formation. The bacteria in plaque feed off saliva, fluid from our gums and ingested sugars—producing acids that contribute to the destruction of enamel, dentine and cementum (the hard 'glue' that attaches tooth roots to the ligament and bone of the tooth socket) (Hillson, 2008; Lukacs, 2012). Cariogenic (caries-causing) species of bacteria like *Lactobacillus* and *Streptococcus mutans* are particularly efficient at breaking down dietary sugars and forming lactic acid that lowers the pH of plaque. Caries begin when this increasing acidity causes the loss of nearby minerals, forming a weakened area called a lesion. As the lesion undergoes further exposure to acid, the underlying enamel and dentine begin to dissolve, eventually forming a hollowed-out region, or cavity (Figure 6.5). When left untreated, entire crowns may be destroyed, and roots may become so degraded that their bony attachments loosen, leading to tooth loss.

6.4.1 Caries prevalence in the past

Caries are often described as a disease of our modern lifestyle, although whether they are of modern origin or the remnant of an ancient menace is still

up for debate. Evidence in favour of the idea that caries originated recently comes from a study of the caries-causing bacteria *S. mutans* in thirty-four Europeans who lived during the last few thousand years (Adler et al., 2013). The dental calculus of twelve individuals from 4,000–7,600 years ago did not contain *S. mutans*, and it was only identified in a few of the more recent individuals. Given the fact that *S. mutans* is very common in humans living today, the authors hypothesised that this bacterium became widespread after the Industrial Revolution. A genetic study of modern strains of *Streptococcus* reached a different conclusion, estimating that *S. mutans* became common sometime between 3,000 and 14,000 years ago (Cornejo et al., 2013). This period includes the origins of agriculture, a dietary transition that may have created a favourable environment for *S. mutans* in the mouths of early farmers.

These two studies are consistent with tying the prevalence of *S. mutans* to dietary shifts that may enhance its role in caries formation, although they implicate different cultural transitions. One of the challenges of studying *S. mutans* and other cariogenic bacteria in archaeological calculus is that their DNA is rarely found in more than trace amounts. This is likely because cariogenic bacteria like *S. mutans* produce high amounts of acid as a by-product of their metabolism, and in addition to dissolving the tooth enamel during caries formation, this acid also interferes with calculus formation. However, as methodological sensitivity improves and full genomes of ancient *S. mutans* strains become available, it will hopefully be possible to use ancient DNA to resolve its evolutionary history.

Examinations of tooth pathology suggest that caries is not unique to agricultural or industrial practices. Caries occur with some frequency in recent and ancient hunter-gatherers, hominins and chimpanzees—our fruit-loving closest living relatives (Figure 6.5b) (Lovell, 1990). Hominins living several million years ago had caries from time to time (reviewed in Lukacs, 2012), although less frequently than people living today. A striking example comes from the first hominin fossil to be discovered in Africa, known as the Broken Hill skull and which is dated to 300,000 years ago (Grün et al., 2020). This unlucky member of the genus *Homo*

(a)

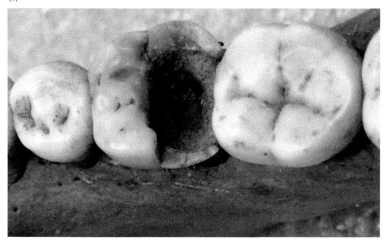

(b)

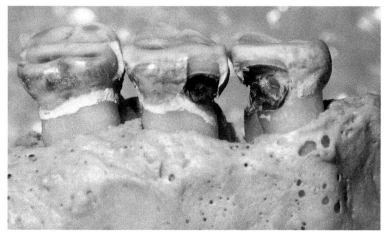

Figure 6.5 Caries of the crown and root in molar teeth. (a) Modern human jaw showing caries in the crown.
(b) Chimpanzee jaw showing interproximal caries that have impacted the roots. Irregular receded bone beneath the crowns shows indications of periodontal disease.
(a) Courtesy of the Phoebe A. Hearst Museum of Anthropology and the Regents of the University of California, (Cat. no. 12-6971B.2).
(b) Chimpanzee specimen courtesy of Christophe Boesch; reproduced with permission from Smith (2018).

shows caries in ten of its remaining eleven upper teeth (Puech et al., 1980; Lacy, 2014). Moreover, a study of ancient hunter-gatherers from Morocco documented an exceptionally high incidence of caries—51% of the teeth of fifty-two adults showed them—similar to the levels of modern industrialised populations (Humphrey et al., 2014). In a twist that might disappoint fans of the Paleo Diet®, these 13,700–15,000-year-old humans likely relied on wild starchy foods that caused their caries, including acorns and pine nuts. A similarly-aged male hunter-gatherer recovered in Italy shows the earliest evidence of caries treatment—a remarkable, partially drilled molar lesion (Oxilia et al., 2015). It would be helpful to learn whether *S. mutans* is present in

the calculus of the Broken Hill individual and these ancient hunter-gatherers, since they all lived before the invention of agriculture.

Scholars who favour a recent origin for caries often point to an influential survey of human archaeological populations that practiced different subsistence styles (Turner, 1979). Turner reported that the frequency of caries was very low in hunter-gatherers (2% of 47,672 teeth examined), increasing slightly in a mixed-subsistence group (4% of 58,137 teeth) and more in agricultural groups (9% of 504,095 teeth). These numbers represent conservative estimates, since it is often necessary to use radiography to locate early caries formation that is not externally visible, which was not possible in Turner's study. Importantly, he omitted modern people consuming processed diets, as he felt that this would lead to artificially inflated rates. A more recent meta-analysis of groups practicing different subsistence methods also reported a greater caries prevalence in agriculturalists than in hunter-gatherers (Marklein et al., 2019), but found considerable variation within subsistence groups, as did Turner (1979)—meaning that caries prevalence is not a particularly effective predictor of subsistence practices. Numerous factors have been posited to account for this variation, including worldwide differences in cultigens, gathered foods, and processing methods, as well as variation in fertility and the genetics of saliva and tooth enamel (Lukacs, 2012; Marklein et al., 2019).

6.4.2 Modern perspectives on caries

Regardless of when caries originated, rates in children and adults have skyrocketed over the past several thousand years, particularly with the accessibility of refined flour and sugar during the past two centuries (Hillson, 2008; Warinner, 2016). Epidemic proportions followed the advent of commercial sugar production. Britain was one of the first major importers of sugar, and once the first 'sugar tax' was reduced in the nineteenth century, it became more affordable for the general public. Dental remains of British individuals show fivefold increases of caries in children and threefold increases in adults over the last few hundred years (see Figure 6 in Hillson, 2008).

Another illustrative example comes from a comparison of two contemporary Mayan communities in Mexico (Vega Lizama & Cucina, 2014). One village relies on a traditional agricultural economy dominated by the cultivation of maize, while the other village has access to global food markets, which include processed foods and artificially sweetened drinks. When the teeth of individuals of the same sex and age group in each village were compared, those in the village with processed foods had more caries in nine of eleven comparisons. To be clear, caries affected more than half the adults in both villages, but the authors attributed the nearly ubiquitous incidence in the global-market village to the popularity of refined sugars and sodas. Sadly,

this phenomenon is playing out across the world as many traditional rural communities join the global economy.

Although research has demonstrated a causal relationship between diet and caries, other behavioural, environmental and physiological factors may also play a role (Lukacs & Largaespada, 2006; but see Temple, 2016). For example, women tend to develop caries more often than men. Behavioural differences in traditional food gathering and preparation roles may lead to more snacking by women (reviewed in Lukacs & Largaespada, 2006; Oziegbe & Schepartz, 2021). Women also produce higher amounts of oestrogen, a hormone believed to affect saliva production. Because saliva is a key part of the body's defence against the bacteria that cause caries, such differences in women's saliva can reduce its protective effects, and pregnancy can further exacerbate caries risk. One folkloric adage is that 'for every child a tooth is lost', which points to the belief that childbearing exerts a special toll on the oral health of women—an observation borne out by studies of increasing rates of gum disease during pregnancy (Silk et al., 2008; Balan et al., 2018), as well as comparisons of high and low parity woman in northwest Nigeria (Oziegbe and Schepartz, 2021). Another important factor is the presence of *S. mutans*, since pregnant women often show elevated bacteria levels (reviewed in Lukacs & Largaespada, 2006). Changes in diet and oral health care may be a perfect storm for caries formation—particularly in populations where woman bear large numbers of children. In this case, traditional folklore appears to have been validated by modern science.

6.4.3 Gum disease and tooth loss

The good news is that caries rarely lead to fatalities for those with access to modern medical care. At their worst they create painful infections of the pulp and supporting tissues, as well as tooth loss (Nainar, 2015). Yet caries are not the only reason we lose our teeth. Prolonged gum inflammation—driven by plaque and calculus formation, sugary diets, stress, smoking, genetics and certain bacteria—is a major cause of tooth loss (Pihlstrom et al., 2005; Nelson, 2016). The invading bacteria trigger an immune response, causing inflammation that leads to the loss of the surrounding gum, bone and ligament (periodontal tissues) over time, particularly when these other factors are present. This inflammation-driven destruction of oral tissues, known as periodontal disease, comes in two forms.

When inflammation is limited to the gums and other soft tissues, the condition is called gingivitis, a mild form of the disease that is reversible with improved oral hygiene. However, the condition becomes more serious when inflammation extends to the surrounding bone. When this occurs, the condition is called periodontitis, and much of the destruction that occurs is permanent (Kinane, 2001; Tonetti et al., 2018). Interestingly, the loss of bone is not caused directly by the bacteria, but rather by the increasingly uncoordinated action of the immune system, which progressively destroys the body's own tissues as inflammation increases and cell communication becomes chaotic (Van Dyke & Serhan, 2003; Scott & Krauss, 2012). Of greater concern is the fact that periodontal disease is linked to several life-threatening diseases, including diabetes, heart disease and cancer—complex diseases that are further exacerbated by dysregulation of the immune system (Beck et al., 2019; Genco et al., 2020). While these medical conditions do not typically leave traces in fossils, the presence of diseased alveolar bone is an important indicator of poor periodontal heath.

Scholars have also pondered whether periodontal disease has ancient origins, or is a result of our modern lifestyle (Clarke et al., 1986; Adler et al., 2013). Possible evidence from the hominin fossil record can be found at least 1.8 million years ago in Eurasia, and includes several individuals recovered in Spain from over a span of one million years (Martinón-Torres et al., 2011; Gracia-Téllez et al., 2013; Margvelashvili et al., 2016). In these individuals, the bone around the tooth roots shows resorption, or loss, thought to be a consequence of prolonged gum inflammation (see, for example, the teeth in Figure 6.5b). Studies of tooth calculus suggest that the bacterial species associated with periodontal disease have been relatively stable since the introduction of agriculture (Adler et al., 2013; Warinner et al., 2014a). Yet dental remains recovered from archaeological excavations may show less

periodontal disease prior to the Industrial Revolution than is found in modern populations (Clarke et al., 1986). Diagnoses of periodontal disease in adults living today vary greatly depending on how it is defined. One estimate suggests that the mild inflammatory form, gingivitis, affects 90% of adults worldwide (Pihlstrom et al., 2005), while the more severe inflammatory form, periodontitis, is reported to affect more than 10% of adults over 30 years of age (Albandar et al., 1999).

Although recent trends are suggestive, additional research could help clarify how current disease levels are related to bacterial prevalence or to specific dietary changes, as we have seen for impacted M3s and caries. In recent years, an international research team was invited to supervise a creative experiment that bears upon the issue of oral health in the past (Baumgartner et al., 2009). Ten people from Switzerland engaged in four weeks of 'Stone Age' living, which took place in an archeologically informed environment, including simple huts, clothing and mostly wild foods. Their health was evaluated at the beginning of the study and again at the end. The intrepid group was given a basic supply of whole grains, as well as salt, herbs, honey, milk and fresh raw meat. The participants supplemented their provisioned food with locally foraged items, and they had to prepare meals themselves without modern implements. Barring access to toothbrushes, dental floss, or toothpicks, some people attempted to clean their teeth with twigs. Interestingly, the group members' gums were not more inflamed than when they began four weeks earlier, although their plaque levels were elevated. While the design of the study would leave most academics sceptical—particularly as it formed the basis of a Swiss reality television show—the authors suggested that the elimination of refined sugars may counterbalance the deleterious effects of a more basic home-care routine (Baumgartner et al., 2009). Over four weeks the bacterial communities of their mouths changed, and certain strains associated with gingivitis became less prevalent while other micro-organisms flourished. The team concluded that a lack of dental care does not necessarily lead to gingival inflammation, particularly when processed foods are avoided.

A final aspect of oral health that can be easily identified in prehistoric humans and ancient hominins is tooth loss. Prior to the advent of agriculture, tooth loss due to generalised wear and extensive abrasion (rubbing and scraping) was particularly common (Anderson, 1965; Nelson, 2016). Once agriculture intensified, tooth wear lessened but caries increased and became the main cause of tooth loss. When wear or decay reaches the pulp space inside the crown, the soft tissues become susceptible to bacterial invasion and infection. As the pulp becomes inflamed, its blood vessels, nerves and dentine-forming cells die. The infection may then spread through the root canal and out into the surrounding periodontal ligament, loosening the connection between the tooth roots and their bony sockets. Infections may also destroy the nearby bone, creating sizable holes when left untreated (Figure 6.6).

Regardless of cause, once teeth are lost, the body resorbs the alveolar bone that surrounded the roots, leading to reduction in the height of the jaws (Figure 6.7). A jaw that has lost its teeth is known as 'edentulous'. Over time, it becomes impossible to determine if teeth were lost because of heavy wear, caries and pulp infection, periodontal disease or intentional removal (reviewed in Smith,

2018). Partially edentulous jaws are found in several ancient hominins, including a 1.8-million-year-old *Homo erectus* individual who only had a single tooth left at death (Lordkipanidze et al., 2005). Another example is the French Neanderthal from La Chapelle-aux-Saints, who retained only a few front teeth (but see Tappen, 1985). The discoveries of these individuals has inspired a lively discussion about whether ancient hominins took care of their elderly members (Lebel et al., 2001; DeGusta, 2002), since toothless individuals would have had difficulty eating wild foods that hadn't been softened or cooked. While our oral health has been negatively impacted by the recent adoption of a carbohydrate-rich diet, these hunter-gatherers did not have it easy either. Heavy wear, pulp infections, trauma, toothaches and lost teeth appear to have been a fact of life for hominins (Margvelashvili et al., 2016), particularly our own species—as we've been surviving to reach 'old age' more often than any species that preceded us (Caspari & Lee, 2004).

A recent economic study of lost productivity ranked tooth loss as the costliest dental problem across the globe (Listl et al., 2015). Global incidences

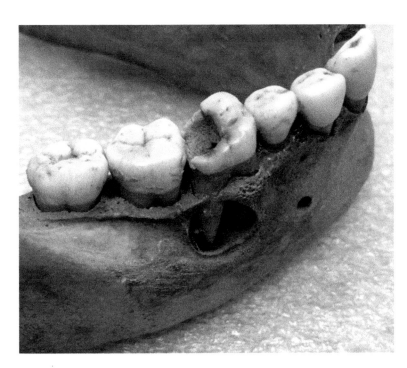

Figure 6.6 Bone infection (abscess) likely due to the destruction of the enamel in a modern human lower jaw. This molar shows a deep cavity (illustrated Figure 6.4) and infection of the pulp leading to a lesion at the base of the tooth root with extensive bone loss (circular abscess). The resorption of bone beneath the crown is also indicative of periodontal disease or gum inflammation due to the infection.
Reproduced with permission from Smith (2018). Courtesy of the Phoebe A. Hearst Museum of Anthropology and the Regents of the University of California, (Cat. no. 12-6971B.2).

(a)

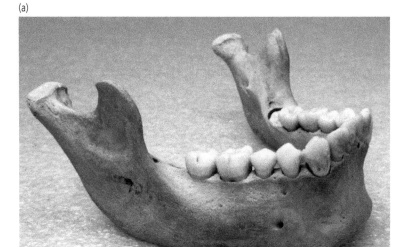

(b)

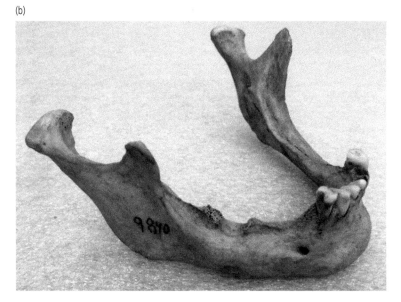

Figure 6.7 Comparison of an adolescent human jaw (top; M3s are unerupted) with a partially edentulous jaw (bottom) from an aged adult.
Courtesy of the Phoebe A. Hearst Museum of Anthropology and the Regents of the University of California, (Cat. nos. 12-9042(0), 12-9840(0)).

in 2010 were estimated to have led to the loss of US$63 billion, edging out periodontal disease at US$54 billion and adult caries at US$25 billion, although obviously the three conditions are inter-related. Clinicians and dental researchers have long endeavoured to create affordable replacements for lost teeth. Although artificial teeth—including modern bridges and dentures—have improved the quality of life for millions of people, tooth replacement has had a rather gruesome history. In the eighteenth century it was common for destitute individuals to sell their healthy teeth to dentists, who would pull them out and implant them in the mouths of wealthy individuals (Garfield, 1969). Victor Hugo's novel *Les Misérables* dramatized this unsavoury practice, which wasn't particularly successful, since teeth require intact blood vessels, nerves, and ligaments to function properly. Moreover, transplanted teeth often carried syphilis, leading to dangerous infections in their new owners.

6.5 Use it or lose it: evolutionary and behavioural perspectives on orthodontics

Despite the occasional caries or lost tooth, fossil hominins and great apes tend to have neatly fitting teeth. In contrast, nearly two-thirds of all Americans show 'malocclusion', or misaligned teeth, yet another disease of modern civilisation (Corrucini, 1991; Rose & Roblee, 2009). These positional anomalies, which some prefer to term 'occlusal variation' (Lukacs, 2012), do not always result in a loss of chewing efficiency (Türp et al., 2008). Many of us have fretted through our preteen years with awkward metal braces, while younger generations perhaps decided to straighten teeth with removable Invisalign® trays. It might surprise you to know that malocclusion treatments over the last century have been heavily influenced by evolutionary theories about dental crowding, and recently these have been called into question.

In the 1920s, the up-and-coming Australian orthodontist Percy Raymond Begg conducted an influential study of hundreds of indigenous Australians, who were traditionally hunter-gatherers and known to have large and heavily worn teeth (Begg, 1954). He noticed that malocclusion was less common in their teeth than in other contemporary human populations. Begg surmised that excessive tooth wear helped to reduce dental crowding. This is because teeth do not just wear down on their chewing surfaces. Wear also occurs between adjacent teeth as they rub against one another, which eventually shortens the length of the whole tooth row. This creates room for teeth to drift slightly forward within the jaw as they continue their slow process of eruption. Begg concluded that humans consuming soft diets suffered from having overly large chewing surfaces, crowding jaws during adulthood, since their diets did not lead to sufficient tooth wear. He regarded the 'textbook model' of unworn human teeth in perfect occlusion as unnatural—a bold and controversial position to take as an aspiring clinician. This conviction led him to begin removing premolar and molar teeth from his patients in order to create space for other teeth to properly fit together in the jaw, which remains a common orthodontic practice today.

An alternative evolutionary hypothesis for malocclusion has come from observations of face, jaw and tooth size during the development of agriculture. A study of ancient Egyptians found the facial skeleton decreased in size and shifted back towards the spinal column as their reliance on agriculture increased (Carlson & Van Gerven, 1977). As discussed, the diets of many agriculturists were less diverse than hunter-gatherers, with a heavy consumption of cereals and starchy plants that were cooked or processed into soft digestible forms. Scholars have extrapolated from the patterns in Egyptians and our fossil ancestors that the development of smaller faces was caused by a reduction in chewing stresses. Thus, the problem for modern humans is not a lack of tooth wear; rather, it is that we do not grow our jaws large enough to hold our teeth.

This idea is further strengthened by comparisons of teeth and diets across generations of recent humans (Corrucini, 1991). When older individuals who grew up eating a harder, more traditional diet are compared with younger generations who ate softer processed diets, the younger generations show greater malocclusion. Similar results are found when comparing rural Indian children on a traditional diet with those that grew up in an urban environment with more processed foods. These comparisons provide important evidence that misaligned teeth are caused by the lack of sufficient chewing forces during development. They are also consistent with links between jaw size, tooth crowding and chewing forces; recall that recent small-jawed individuals are more likely to have M3 impaction (Carter, 2016). Moreover, animals fed soft diets show reduced facial sizes and crowded teeth when compared to their wild counterparts (Rose & Roblee, 2009), which may also help explain why captive animals have long been noted to have poorer dental health (Colyer, 1947).

Interestingly, a recent study has linked malocclusion in modern humans with novel speech production (Blasi et al., 2019). It turns out that labiodentals, sounds produced by positioning the lower lip against the upper teeth, as in 'f' or 'v', are most easily formed by misaligned anterior teeth. These consonants occur in almost half of all modern languages. The notion that the jaws of modern children are not growing large enough to hold their teeth is starting to gain traction in the anthropological, clinical and evolutionary biology research communities (Corruccini, 1999; Lieberman, 2011; Makaremi et al., 2015). Some children are now given jaw-strengthening exercises as orthodontic experiments, which was not an option for braces-baring teenagers of the twentieth century. We will be fascinated to learn if this new interpretation of the past can help us to overcome a disease of modern civilisation—sparing adolescents the physical and social discomfort of having their teeth corrected.

6.6 Conclusions and future directions

Numerous studies have elucidated how teeth form, increase in size and emerge into the mouth, only to begin wearing down and changing with age (reviewed in Hillson, 1996; Ten Cate et al., 2003; Smith, 2018). Over the course of crown and root development during childhood, bodily disturbances may lead to microscopic defects or—less commonly—teeth may fail to form or erupt properly. Once our teeth do emerge, they are colonised by bacteria, which form thick layers of plaque that calcify regularly, resulting in dental calculus accretions that grow like the rings of a tree (Warinner et al., 2015). Over time, oral bacteria and dietary habits contribute to caries and gum inflammation, which may ultimately lead to tooth loss. Some of these experiences have plagued us since ancient times, while others are believed to be diseases of modern civilisation.

Understanding the evolution of oral health and behavioural roots of dental pathology provides new perspectives on human health and disease, as the development of agriculture and subsequent production of food on an industrial scale have profoundly affected our physiology and anatomy, including our teeth. Comparisons with our non-human primate relatives reveal that, while features like caries are at an epidemic level, they are not unique to humans. Palaeopathology offers a unique perspective on this subject matter, as intimate knowledge of past human populations and our evolutionary lineage provide critical context for understanding the causes and origins of modern diseases. For example, microscopic records of daily cellular activity during tooth development permit an unparalleled glimpse into the childhoods of ancient humans, and grappling with the fine details of tooth growth provides insights into much more than our past. Such palaeopathological studies also point to the power of tooth microstructure for forensic investigations (e.g. Katzenberg et al., 2005; Birch & Dean, 2014).

Consideration of such trends in oral health also contribute to modern clinical research and medicine. Recovery of ancient proteins and DNA from diverse bacteria in our mouths hold great promise to explain the delicate balance achieved between bacterial colonisation and the immune system that shapes our oral microbiome, as well as the impact of major events and transitions in human history and evolution that have influenced, changed and (at times) strained this host-microbial relationship. Oral biologists are also busy uncovering the environmental influences that lead to tooth loss, dental decay, and misalignment. This is particularly timely, since dental development in human populations may well be accelerating (Cardoso et al., 2010; Sasso et al., 2013; Vucic et al., 2014), which could lead to greater incidences of these problems.

Efforts to understand our dental health by studying the cellular dance that begins before we are born may inform future attempts to prevent and repair the dental abnormalities and diseases that afflict us today. One growing area of research focuses on using the cells from a patient's own body to grow replacement tissues or organs, easing the unmet demand for donated organs and lessening the body's rejection of foreign cells. This may solve the nearly universal problem of tooth loss, which is a serious public health problem. The field of restorative dentistry has recently benefitted from recent breakthroughs in stem cell research and tissue engineering (reviewed in Smith, 2018; Shao et al., 2019). Innovative mechanisms for creating replacement teeth may eventually supersede dentures, implants or bridges—expensive appliances that lack the vitality or sensory capabilities of healthy teeth.

There is no question that humans continue to evolve and adapt, in part through rapid cultural innovations that have allowed us to spread around the globe and swell rapidly in number, but a great deal still remains to be understood about the long-term impact of the agricultural and Industrial Revolutions on human health. Our dentitions mark modern behaviours in myriad ways, serving as a time capsule for future studies of our curious species.

References

Abrahamian, F. M. & Goldstein, E. J. (2011). Microbiology of animal bite wound infections. *Clinical Microbiology Reviews*, 24, 231–246.

Adler, C. J., Dobney, K., Weyrich, L. S., Kaidonis, J., Walker, A. W., Haak, W., Bradshaw, C. J. A., Townsend, G., Sołtysiak, A., Alt, K. W., Parkhill, J. & Cooper, A. (2013). Sequencing ancient calcified dental plaque shows changes in oral microbiota with dietary shifts of the Neolithic and industrial revolutions. *Nature Genetics*, 45, 450–455.

Albandar, J. M., Brunelle, J. A. & Kingman, A. (1999). Destructive periodontal disease in adults 30 years of age and older in the United States, 1988–1994. *Journal of Periodontology*, 70, 13–29.

AlQahtani, S. J., Hector, M. P. & Liversidge, H. M. (2014). Accuracy of dental age estimation charts: Schour and Massler, Ubelaker, and the London atlas. *American Journal of Physical Anthropology* 154, 70–78.

Amerongen, A. N. & Veerman, E. C. I. (2002). Saliva–the defender of the oral cavity. *Oral Diseases*, 8, 12–22.

Anderson, J. E. (1965). Human skeletons of Tehuacán. *Science*, 148, 496–497.

Andreas, N. J., Kampmann, B. & Mehring Le-Doare, K. (2015). Human breast milk: a review on its composition and bioactivity. *Early Human Development*, 91, 629–635.

Armelagos, G. J., Goodman, A. H., Harper, K. N. & Blakey, M. L. (2009). Enamel hypoplasia and early mortality: bioarchaeological support for the Barker hypothesis. *Evolutionary Anthropology* 18, 261–271.

Austin, C., Smith, T. M., Farahani, R. M. Z., Hinde, K., Carter, E. A., Lee, J., Lay, P. A., Kennedy, B. J., Sarrafpour, B., Wright, R. J., Wright, R. O. & Arora, M. (2016). Uncovering system-specific stress signatures in primate teeth with multimodal imaging. *Science Reports*, 6, 18802.

Baker, O. J., Edgerton, M., Kramer, J. M. & Ruhl, S. (2014). Saliva-microbe interactions and salivary gland dysfunction. *Advances in Dental Research*, 26, 7–14.

Balan, P., Chong, Y. S., Umashankar, S., Swarup, S., Loke, W. M., Lopez, V., He, H. G. & Seneviratne, C. J. (2018). Keystone species in pregnancy gingivitis: a snapshot of oral microbiome during pregnancy and postpartum period. *Frontiers in Microbiology*, 9, 2360.

Baumgartner S., Imfeld, T., Schicht, O., Rath, C., Persson, R. E. & Persson, G.R. (2009). The impact of the Stone Age diet on gingival conditions in the absence of oral hygiene. *Journal of Periodontology*, 8, 759–767.

Beck, J. D., Papapanou, P. N., Philips, K. H. & Offenbacher, S. (2019). Periodontal medicine: 100 years of progress. *Journal of Dental Research*, 98, 1053–1062.

Begg, P. R. (1954). Stone Age man's dentition. *American Journal of Orthodontics*, 40, 373–383.

Birch, W. & Dean, M. C. (2014). A method of calculating human deciduous crown formation times and of estimating the chronological ages of stressful events occurring during deciduous enamel formation. *Journal of Forensic and Legal Medicine*, 22, 127–144.

Black, R. E., Morris S. S. & Bryce, J. (2003). Where and why are 10 million children dying every year? *Lancet*, 361, 2226–2234.

Blakey, M. L., Leslie, T. E. & Reidy, J. P. (1994). Frequency and chronological distribution of dental enamel hypoplasia. *American Journal of Physical Anthropology*, 95, 371–383.

Blasi, D. E., Moran, S., Moisik, S. R., Widmer, P., Dediu, D. & Bickel, B. (2019). Human sound systems are shaped by post-Neolithic changes in bite configuration. *Science*, 363, eaav3218.

Brace, C. L. (1964). The probable mutation effect. *American Naturalist*, 98, 453–455.

Brand, H. S. & Veerman, E. C. (2013). Saliva and wound healing. *Chinese Journal of Dental Research*, 16, 7–12.

Butler, L. G. (1989). 'Effects of condensed tannin on animal nutrition', in R. W. Hemingway & J. J. Karchesy (eds), *Chemistry and significance of condensed tannins*, pp. 391–402. Springer, Boston.

Caballero-Flores, G., Sakamoto, K., Zeng, M. Y., Wang, W., Hakim, J., Matus-Acuña, V., Inohara, N. & Núñez, G. (2019). Maternal immunization confers protection to the offspring against an attaching and effacing pathogen through delivery of IgG in breast milk. *Cell Host & Microbe*, 25, 313–323.

Calcagno, J. M. & Gibson, K. R. (1988). Human dental reduction: natural selection or the probable mutation effect. *American Journal of Physical Anthropology*, 77, 505–517.

Cardoso, H. F. V., Heuzé, Y. & Júlio, P. (2010). Secular change in the timing of dental root maturation in Portuguese boys and girls. *American Journal of Human Biology*, 22, 791–800.

Carlson, D. M. (1993). Salivary proline-rich proteins: biochemistry, molecular biology, and regulation of expression. *Critical Reviews in Oral Biology & Medicine*, 4, 495–502.

Carlson, D. S. & van Gerven, D. P. (1977). Masticatory function and post-Pleistocene evolution in Nubia. *American Journal of Physical Anthropology*, 46, 495–506.

Carter, K. (2016). *The evolution of third molar agenesis and impaction.* PhD Thesis, Harvard University, Cambridge, MA.

Carter, K. & Worthington, S. (2015). Morphologic and demographic predictors of third molar agenesis: a systematic review and meta-analysis. *Journal of Dental Research*, 94, 886–894.

Carter, K. & Worthington, S. (2016). Predictors of third molar impaction: a systematic review and meta-analysis. *Journal of Dental Research*, 95, 267–276.

Caspari, R. & Lee, S.-H. (2004). Older age becomes common late in human evolution. *Proceedings of the National Academy of Sciences of the United States of America*, 101, 10895–10900.

Clarke, N. G., Carey, S. E., Srikandi, W., Hirsch, R. S. & Leppard, P. I. (1986). Periodontal disease in ancient populations. *American Journal of Physical Anthropology*, 71, 173–183.

Colyer, F. (1947). Dental disease in animals. *British Dental Journal*, 82, 2–35.

Cornejo O. E., Lefebure, T., Bitar, P. D. P., Lang, P., Richards, V. P., Eilertson, K., Do, T., Beighton, D., Zeng, L., Ahn, S-J., Burne, R. A., Siepel, A., Bustamente, C. D. & Stanhope, M. J. (2013). Evolutionary and population genomics of the cavity causing bacteria Streptococcus mutans. *Molecular Biology and Evolution*, 30, 881–893.

Corrucini, R. S. (1991). 'Anthropological aspects of orofacial and occlusal variations and anomalies', in M. A. Kelley & C. S. Larson (eds), *Advances in dental anthropology*, pp. 143–68. Wiley-Liss, New York.

Corrucini, R. S. (1999). *How anthropology informs the orthodontic diagnosis of malocclusion's causes.* Edwin Mellen Press, Lewiston.

De Coster, P. J., Marks, L. A., Martens, L. C. & Huysseune, A. (2009). Dental agenesis: genetic and clinical perspectives. *Journal of Oral Pathology & Medicine*, 38, 1–17.

DeGusta, D. (2002). Comparative skeletal pathology and the case for conspecific care in Middle Pleistocene hominids. *Journal of Archaeological Science* 29, 1435–1438.

Dewhirst, F. E., Chen, T., Izard, J., Paster, B. J., Tanner, A. C. R., Yu, W-H., Lakshmanan, A. & Wade, W. G. (2010). The human oral microbiome. *Journal of Bacteriology*, 192, 5002–5017.

DeWitte, S. N. & Stojanowski, C. (2015). The osteological paradox 20 years later: past perspectives, future directions. *Journal of Archaeological Research*, 23, 397–450.

Duval, X. & Leport, C. (2008). Prophylaxis of infective endocarditis: current tendencies, continuing controversies. *The Lancet Infectious Diseases*, 8, 225–232.

Edinborough, M. & Rando, C. (2020). Stressed out: reconsidering stress in the study of archaeological human remains. *Journal of Archaeological Science*, 121, 105197.

Fellows Yates, J., Velsko, I., Aron, F., Hofman, C., Austin, R., Arthur, J., Crevecoeur, I., Dalén, L., Gonzalez Morales, M., Guschanski, K., Henry, A.G., Humphrey, L.T., Mann, A.E., Nägele, K., Parker, C. E., Posth, C., Rougier, H., Semal, P., Stock, J., Guy Strauss, L., Weedman, A. K., Wrangham, R., Curtis, M., Diez, J. C., Gibbon, V., Menedez, M., Peresani, M., Roksandic, M., Walker, M. J., Power, R. C., Lewis, C. M., Sankaranarayan, K., Salazar-Garcia, D. C., Krause, J., Herbig, A. & Warinner, C. (2021). The evolution and changing ecology of the African hominid oral microbiome. *Proceedings of the National Academy of Sciences of the United States of America*, 118 (20), e2021655118.

Ferretti, P., Pasolli, E., Tett, A., Asnicar, F., Gorfer, V., Fedi, S., Armanini, F., Truong, D. T., Manara, S., Zolfo, M., Beghini, F., Bertorelli, R., De Sanctis, V., Bariletti, I., Canto, R., Clementi, R., Cologna, M., Crifò, T., Cusumano, G., Gottardi, S., Innamorati, C., Masè, C., Postai, D., Savoi, D., Duranti, S., Lugli, G. A., Mancabelli, L., Turroni, F., Ferrario, C., Milani, C., Mangifesta, M., Anzalone, R., Viappiani, A., Yassour, M., Vlamakis, H., Xavier, R., Collado, C. M., Koren, O., Tateo, S., Soffiati, M., Pedrotti, A., Ventura, M., Huttenhower, C., Bork, P. & Segata, N. (2018). Mother-to-infant microbial transmission from different body sites shapes the developing infant gut microbiome. *Cell Host & Microbe*, 24, 133–145.

Fuss, J., Uhlig, G. & Böhme, M. (2018). Earliest evidence of caries lesion in hominids reveal sugar-rich diet for a Middle Miocene dryopithecine from Europe. *PloS One*, 13 (8), e0203307.

Garfield, S. (1969). *Teeth teeth teeth.* Simon and Schuster, New York.

Genco, R. J., Graziani, F. & Hasturk, H. (2020). Effects of periodontal disease on glycemic control, complications, and incidence of diabetes mellitus. *Periodontology 2000*, 83, 59–65.

Goodman, A. H. & Rose, J. C. (1991). 'Dental enamel hypoplasias as indicators of nutritional stress', in M. A. Kelley & C. S. Larson (eds), *Advances in dental anthropology*, pp. 279–93. Wiley-Liss, New York.

Gracia-Téllez, A., Arsuaga, J.-L., Martínez, I., Martín-Francés, L., Martinón-Torres, M., Bermúdez de Castro, J-M., Bonmatí, A. & Lira, J. (2013). Orofacial pathology in *Homo heidelbergensis*: the case of Skull 5 from the Sima de los Huesos site (Atapuerca, Spain). *Quaternary International*, 295, 83–93.

Grün, R., Pike, A., McDermott, F., Eggins, S., Mortimer, G., Aubert, M., Kinsley, L., Joannes-Boyau, R., Rumsey, M., Denys, C., Brink, J., Clark, T. & Stringer, C. (2020). Dating the skull from Broken Hill, Zambia, and its position in human evolution. *Nature*, 580, 372–375.

Haga, S., Nakaoka, H., Yamaguchi, T., Yamamoto, K., Kim, Y-I., Samoto, H., Ohno, T., Katayama, K., Ishida, H., Park, S-B., Kimura, R., Maki, K. & Inoue, I. (2013). A genome-wide association study of third molar agenesis in Japanese and Korean populations. *Journal of Human Genetics*, 58, 799–803.

Han, Y. W. (2011). Oral health and adverse pregnancy outcomes–what's next? *Journal of Dental Research*, 90, 289–293.

Han, Y. W. (2015). *Fusobacterium nucleatum*: a commensal-turned pathogen. *Current Opinion in Microbiology*, 23, 141–147.

He, X., McLean, J. S., Guo, L., Lux, R. & Shi, W. (2013). The social structure of microbial community involved in colonization resistance. *The ISME Journal* 8, 564–574.

Hendy, J., Warinner, C., Barnes, I., Collins, M. J., Fiddyment, S., Fischer, R., Haga, R., Hofman, C. A., Holst, M., Chaves, E., Klaus, L., Larson, G., Mackie, M., McGrath, K., Mundorff, A. Z., Radini, A., Rao, H., Traschel, C., Velsko, I. M. & Speller, C. M. (2018). Proteomic evidence of dietary sources in ancient dental calculus. *Proceedings of the Royal Society B: Biological Sciences*, 285, 20180977.

Henry, A. G., Brooks, A. S. & Piperno, D. R. (2011). Microfossils in calculus demonstrate consumption of plants and cooked foods in Neanderthal diets (Shanidar III, Iraq; Spy I and II, Belgium). *Proceedings of the National Academy of Sciences of the United States of America*, 108, 486–491.

Henry, A. G., Ungar, P. S., Passey, B. H., Sponheimer, M., Rossouw, L., Bamford, M., Sandberg, P., de Ruiter, D. J. & Berger, L. (2012). The diet of *Australopithecus sediba*. *Nature*, 487, 90–93.

Henry, C. B. & Morant, G. M. (1936). A preliminary study of the eruption of the mandibular third molar tooth in man based on measurements obtained from radiographs, with special reference to the problem of predicting cases of ultimate impaction of the tooth. *Biometrika*, 28, 378–427.

Hershkovitz, I., Kelly, J., Latimer, B., Simpson, S., Polak, J. & Rosenberg, M. (1997). Oral bacteria in Miocene Sivapithecus. *Journal of Human Evolution*, 33, 507–512.

Hillson, S. (1996). *Dental anthropology*. Cambridge University Press, Cambridge.

Hillson, S. (2008). 'The current state of dental decay', in J. D. Irish and G. C. Nelson (eds), *Technique and application in dental anthropology*, pp. 111–135. Cambridge University Press, Cambridge.

Hillson, S. (2014). *Tooth development in human evolution and bioarchaeology*. Cambridge University Press, Cambridge.

Holick, M. (2006). Resurrection of vitamin D deficiency and rickets. *The Journal of Clinical Investigation*, 116, 2062–2072.

Human Microbiome Consortium Project. (2012). Structure, function and diversity of the healthy human microbiome. *Nature*, 486, 207–214.

Humphrey, L. T. (2008). 'Enamel traces of early lifetime events', in H. Schutkowski (ed), *Between biology and culture*, pp. 186–206. Cambridge University Press, Cambridge.

Humphrey, L. T., de Groote, I., Morales, J., Barton, N., Collcutt, S., Ramsey, C. B. & Bouzouggar, A. (2014). Earliest evidence for caries and exploitation of starchy plant foods in Pleistocene hunter-gatherers from Morocco. *Proceedings of the National Academy of Sciences of the United States of America*, 111, 954–959.

Hurnanen, J., Visnapuu, V., Sillanpää, M., Löyttyniemi, E. & Rautava, J. (2017). Deciduous neonatal line: width is associated with duration of delivery. *Forensic Science International*, 271, 87–91.

Ioannou, S., Sassani, S., Henneberg, M. & Henneberg, R. J. (2016). Diagnosing congenital syphilis using Hutchinson's method: differentiating between syphilitic, mercurial, and syphilitic-mercurial dental defects. *American Journal of Physical Anthropology*, 159, 617–629.

Jeong, C., Wilkin, S., Amgalantugs, T., Bouwman, A. S., Taylor, W. T. T., Hagan, R. W., Bromage, S., Tsolmon, S., Trachsel, C., Grossmann, J., Littleton, J., Makarewicz, C. A., Krigbaum, J., Burri, M., Scott, A., Davaasambuu, G., Wright, J., Irmer, F., Myagmar, E., Boivin, N., Robbeets, M., Rühli, F. J., Krause, J., Frohlich, B., Hendy, J. & Warinner, C. (2018). Bronze Age population dynamics and the rise of dairy pastoralism on the eastern Eurasian steppe. *Proceedings of the National Academy of Sciences of the United States of America*, 115, E11248–55.

Katzenberg, M., Herring, D. A. & Saunders, S. R. (1996). Weaning and infant mortality: evaluating the skeletal evidence. *Yearbook of Physical Anthropology*, 39, 177–199.

Katzenberg, M. A., Oetelaar, G., Oetelaar, J., FitzGerald, C., Yang, D. & Saunders, S. R. (2005). Identification of historical human skeletal remains: a case study using skeletal and dental age, history and DNA. *International Journal of Osteoarchaeology*, 15, 61–72.

Kinane, D. F. (2001). Causation and pathogenesis of periodontal disease. *Periodontology 2000*, 25, 8–20.

Kolenbrander, P. E., Palmer Jr., R. J., Rickard, A. H., Jakubovics, N. S., Chalmers, N. I. & Diaz, P. I. (2006). Bacterial interactions and successions during plaque development. *Periodontology 2000*, 42, 47–79.

Kunkel, M., Kleis, W., Morbach, T. & Wagner, W. (2007). Several third molar complications including death—lessons from 100 cases requiring hospitalization. *Journal of Maxillofacial Surgery*, 65, 1700–1706.

Lacy, S. A. (2014). The oral pathological conditions of the Broken Hill (Kabwe) 1 cranium. *International Journal of Paleopathology* 7, 57–63.

Larsen, C. S. (1995). Biological changes in human populations with agriculture. *Annual Review of Anthropology*, 24, 185–213.

Lebel, S., Trinkaus, E., Faure, M., Fernandez, P., Guérin, C., Richter, D., Mercier, N., Valladas, H. & Wagner, G. A. (2001). Comparative morphology and paleobiology of Middle Pleistocene human remains from the Bau de l'Aubesier, Vaucluse, France. *Proceedings of the National Academy of Sciences of the United States of America*, 98, 11097–11102.

Lieberman, D. E. (2011). *The evolution of the human head*. Harvard University Press, Cambridge, MA.

Lieberman, D. E. (2013). *The story of the human body*. Vintage Books, New York.

Listl, S., Galloway, J., Mossey, P. A. & Marcenes, W. (2015). Global economic impact of dental diseases. *Journal of Dental Research*, 94, 1355–1361.

Littleton, J. (2005). Invisible impacts but long-term consequences: hypoplasia and contact in Central Australia. *American Journal of Physical Anthropology*, 126, 295–304.

Lordkipanidze, D., Vekua, A., Ferring, R., Rightmire, G. P., Agusti, J., Kiladze, G., Mouskhelishvili, A., Nioradze, M., Ponce de León, M. S., Tappen, M. & Zollikofer, C. P. E. (2005). Anthropology: the earliest toothless hominin skull. *Nature*, 434, 717–718.

Lorentz, K. O., Lemmers, S. A. M., Chrysostomou, C., Dirks, W., Zaruri, M. R., Foruzanfar, F. & Sajjadi, S. M. S. (2019). Use of dental microstructure to investigate the role of prenatal and early life physiological stress in age at death. *Journal of Archaeological Science*, 104, 85–96.

Lovell, N. C. (1990). *Patterns of injury and illness in great apes*. Smithsonian Institution Press, Washington, DC.

Lukacs, J. R. (2012). 'Oral health in past populations: Context, concepts and controversy', in A. L. Grauer (ed), *A companion to paleopathology*, pp. 553–581. Wiley-Blackwell, Chichester.

Lukacs, J. R. & Largaespada, L. L. (2006). Explaining sex differences in dental caries prevalence: saliva, hormones, and 'life-history' etiologies. *American Journal of Human Biology*, 18, 540–555.

Makaremi, M., Zink, K. & de Brondeau, F. (2015). Apport des contraintes masticatrices fortes dans la stabilisation de l'expansion maxillaire/The importance of elevated masticatory forces on the stability of maxillary expansion. *Revue d'Orthopédie Dento Faciale*, 49, 11–20.

Maleeh, I., Robinson, J. & Wadhwa, S. (2016). 'Role of alveolar bone in mediating orthodontic tooth movement and relapse', in B. Shroff (ed), *Biology of orthodontic tooth movement*, pp. 1–12. Springer International Publishing, Geneva.

Margvelashvili, A., Zollikofer, C. P. E., Lordkipanidze, D., Tafforeau, P. & Ponce de Leon, M. S. (2016). Comparative analysis of dentognathic pathologies in the Dmanisi mandibles. *American Journal of Physical Anthropology*, 160, 229–253.

Marklein, K. E., Torres-Rouff, C., King, L. M. & Hubbe, M. (2019). The precarious state of subsistence. *Current Anthropology*, 60, 341–368.

Marsh, P. D., Do, T., Beighton, D. & Devine, D. A. (2016). Influence of saliva on the oral microbiota. *Periodontology 2000*, 70, 80–92.

Martinón-Torres, M., Martín-Francés, L., Gracia, A., Olejniczak, A., Prado-Simón, L., Gómez-Robles, A., Lapresa, M., Carbonell, E., Arsuaga, J. L. & Bermúdez de Castro, J. M. (2011). Early Pleistocene human mandible from Sima del Elefante (TE) cave site in Sierra de Atapuerca (Spain): a palaeopathological study. *Journal of Human Evolution*, 61, 1–11.

Massler, M., Schour, I. & Poncher, H. G. (1941). Developmental pattern of the child as reflected in the calcification pattern of the teeth. *American Journal of Diseases of Children*, 629, 33–67.

Mathews, S. A., Kurien, B. T. & Scofield, R. H. (2008). Oral manifestations of Sjögren's syndrome. *Journal of Dental Research*, 87, 308–318.

May, R. L., Goodman, A. H. & Meindl, R. S. (1993). Response of bone and enamel formation to nutritional supplementation and morbidity among malnourished Guatemalan children. *American Journal of Physical Anthropology*, 92, 37–51.

McGrath, K., El-Zaatari, S., Guatelli-Steinberg, D., Stanton, M. A., Reid, D. J., Stoinski, T. S., Cranfield, M. R., Mudakikwa, A. & McFarlin, S. C. (2018). Quantifying linear enamel hypoplasia in Virunga Mountain gorillas and other great apes. *American Journal of Physical Anthropology*, 166, 337–352.

McGrath, K., Limmer, L. S., Lockey, A.-L., Guatelli-Steinberg, D., Reid, D. J., Witzel, C., Bocaege, E., McFarlin, S. C. & El Zaatari, S. (2021). 3D enamel profilometry reveals faster growth but similar stress severity in Neanderthal versus *Homo sapiens* teeth. *Scientific Reports*, 11, 522.

Nainar, S. M. H. (2015). Is it ethical to withhold restorative dental care from a child with occlusoproximal caries lesions into dentin of primary molars? *Pediatric Dentistry*, 37, 329–331.

Nelson, G. C. (2016). 'A host of other dental diseases and disorders', in J. D. Irish & G. R. Scott (eds), *A companion to dental anthropology*, pp. 465–83. John Wiley and Sons, Chichester.

O'Regan, H. J. & Kitchener, A. C. (2005). The effects of captivity on the morphology of captive, domesticated and feral mammals, *Mammal Review*, 35, 215–230.

Otte, M. (2017). *Teeth: the story of beauty, inequality and the struggle for oral health in America*. The New Press, New York.

Oxilia G., Peresani, M., Romandini, M., Matteucci, C., Debono Spiteri, C., Henry, A. G., Schulz, D., Archer, W., Crezzini, J., Boschin, F., Boscato, P., Jaouen, K., Dogandzic, T., Broglio, A., Moggi-Cecchi, J., Fiorenza, L., Hublin, J-J., Kullmer, O. & Benazzi, S. (2015). Earliest evidence of dental caries manipulation in the Late Upper Palaeolithic. *Scientific Reports*, 5, 12150.

Oziegbe, E. O. & Schepartz, L. A. (2021). Association between parity and tooth loss among northern Nigerian Hausa women. *American Journal of Physical Anthropology*, 174, 451–462.

Page, A. E., Viguier, S., Dyble, M., Smith, D., Chaudhary, N., Salali, G. D., Thompson, J., Vinicius, L., Mace, R. & Bamberg Migliano, A. (2016) Reproductive trade-offs in extant hunter-gatherers suggest adaptive mechanism for the Neolithic expansion. *Proceedings of the National Academy of Sciences*, 113, 4694–4699.

Pihlstrom, B. L., Michalowicz, B. S. & Johnson, N. W. (2005). Periodontal diseases. *Lancet*, 366, 1809–1820.

Prentiss, A. (2013). Nutritional rickets around the world. *Journal of Steroid Biochemistry & Molecular Biology*, 136, 201–206.

Puech, P.-F., Albertini, H. & Mills, N. T. W. (1980). Dental destruction in Broken-Hill man. *Journal of Human Evolution*, 9, 33–39.

Radini, A., Nikita, E., Buckley, S., Copeland, L. & Hardy, K. (2017). Beyond food: the multiple pathways for inclusion of materials into ancient dental calculus. *American Journal of Physical Anthropology*, 162, 71–83.

Radini, A., Tromp, M., Beach, A., Tong, E., Speller, C., McCormick, M., Dudgeon, J. V., Collins, M. J., Rühli, F., Kröger, R. & Warinner, C. (2019). Medieval women's early involvement in manuscript production suggested by lapis lazuli identification in dental calculus. *Science Advances*, 5, eaau7126.

Reid, D. J. & Dean, M. C. (2000) Brief communication: the timing of linear hypoplasias on human anterior teeth, *American Journal of Physical Anthropology*, 113, 135–139.

Rose, J. C. & Roblee, R. D. (2009). Origins of dental crowding and malocclusions: an anthropological perspective. *Compendium of Continuing Education in Dentistry*, 30, 292–300.

Ruhl, S., Sandberg, A. L. & Cisar, J. O. (2004). Salivary receptors for the proline-rich protein-binding and lectin-like adhesins of oral actinomyces and streptococci. *Journal of Dental Research*, 83, 505–510.

Sasso, A., Špalj, S., Maričić, B. M., Sasso, A., Cabov, T. & Legović, M. (2013). Secular trend in the development of permanent teeth in a population of Istria and the Littoral region of Croatia. *Journal of Forensic Science* 58, 673–677.

Schwartz, G. T., Reid, D. J., Dean, M. C. & Zihlman, A. L. (2006). A faithful record of stressful life events preserved in the dental developmental record of a juvenile gorilla. *International Journal of Primatology*, 27, 1201–1219.

Scott, A., Power, R. C., Altmann-Wendling, V., Artzy, M., Martin, M. A. S., Eisenmann, S., Hagan, R., Salazar-García, D. C., Salmon, Y., Yegorov, D., Milevski, I., Finkelstein, I., Stockhammer, P. W. & Warinner, C. (2021). Exotic foods reveal contact between South Asia and the Near East during the second millennium BCE. *Proceedings of the National Academy of Sciences*, 118, e2014956117.

Scott, D. A. & Krauss, J. (2012). Neutrophils in periodontal inflammation. *Frontiers in Oral Biology*, 15, 56–83.

Sellen, D. W. & Smay, D. B. (2001). Relationship between subsistence and age at weaning in 'preindustrial' societies. *Human Nature*, 12, 47–87.

Sergi, S. (1914). Missing teeth inherited. *The Journal of Heredity*, 5, 559–560.

Shao, C., Jin, B., Mu, Z., Lu, H., Zhao, Y., Wu. Z., Yan, L., Zhang, Z., Zhou, Y., Pan, H., Liu, Z. & Tang, R. (2019). Repair of tooth enamel by a biomimetic mineralization frontier ensuring epitaxial growth. *Science Advances*, 5, eaaw9569.

Silk, H., Douglass, A. B., Douglass, J. M. & Silk, L. (2008). Oral health during pregnancy. *American Family Physician*, 77, 1139–1144.

Siqueira, W. L., Custodio, W. & McDonald, E. E. (2012). New insights into the composition and functions of the acquired enamel pellicle. *Journal of Dental Research*, 91, 1110–1118.

Skinner, M. (1996). Developmental stress in immature hominines from late Pleistocene Eurasia: evidence from enamel hypoplasia. *Journal of Archaeological Science*, 23, 833–852.

Smith, M. O., Kurtenbach, K. J. & Vermaat, J. C. (2016). Linear enamel hypoplasia in Schroeder Mounds (11HE177): a Late Woodland period site in Illinois. *International Journal of Paleopathology*, 14, 10–23.

Smith, T. M. (2004). *Incremental Development of Primate Tooth Enamel*. PhD Thesis, Stony Brook University, New York. https://paleoanthro.org/static/dissertations/tanya%20smith.pdf

Smith, T. M. (2013). Teeth and human life-history evolution. *Annual Review of Anthropology*, 42, 191–208.

Smith, T. M. (2018). *The tales teeth tell: development, evolution, behavior*. MIT Press, Cambridge.

Smith, T. M. & Boesch, C. (2015). Developmental defects in the teeth of three wild chimpanzees from the Taï forest, *American Journal of Physical Anthropology*, 157, 556–570.

Smith, T. M., Cook, L., Dirks, W., Green, D. & Austin, C. (2021). Teeth reveal juvenile diet, health and neurotoxicant exposure retrospectively: What biological rhythms and chemical records tell us. *BioEssays*, 43, e2000298.

Talan, D. A., Abrahamian, F. M., Moran, G. J., Citron, D. M., Tan J. O., Goldstein, E. J. C.; Emergency Medicine Human Bite Infection Study Group. (2003). Clinical presentation and bacteriologic analysis of infected human bites in patients presenting to emergency departments. *Clinical Infectious Diseases*, 37, 1481–1489.

Tappen, N. C. (1985). The dentition of the 'old man' of La Chapelle-aux-Saints and inferences concerning Neanderthal behavior. *American Journal of Physical Anthropology*, 67, 43–50.

Temple, D. H. (2019). Bioarchaeological evidence for adaptive plasticity and constraint: exploring life-history trade-offs in the human past. *Evolutionary Anthropology*, 28, 34–46.

Temple, D. H. (2016). 'Caries: the ancient scourge', in J. D. Irish & G. R. Scott (eds), *A companion to dental anthropology*, pp. 433–49. John Wiley and Sons, Chichester.

Ten Cate, A. R., Sharpe, P. T., Roy, S. & Nancy, A. (2003). 'Development of the teeth and its supporting structures', in A. Nanci (ed), *Ten Cate's oral histology*, pp. 79–110. Mosby, St. Louis.

Tonetti, M. S., Greenwell, H. & Kornman, K. S. (2018). Staging and grading of periodontitis: framework and proposal of a new classification and case definition. *Journal of Periodontology*, 89, S159–72.

Towel, I. & Irish, J. D. (2020). Recording and interpreting enamel hypoplasia in samples from archaeological and palaeoanthropological contexts. *Journal of Archaeological Science*, 114, 105077.

Turner, C. G. (1979). Dental anthropological indications of agriculture among the Jomon people of central Japan. *American Journal of Physical Anthropology*, 51, 619–636.

Türp, J. C., Greene, C. S. & Strub, J. R. (2008). Dental occlusion: a critical reflection on past, present and future concepts. *Journal of Oral Rehabilitation*, 35, 446–453.

Van Dyke, T. E. & Serhan, C. N. (2003). Resolution of inflammation: a new paradigm for the pathogenesis of periodontal diseases. *Journal of Dental Research*, 82, 82–90.

Vega Lizama, E. M. & Cucina, A. (2014). Maize dependence or market integration? Caries prevalence among indigenous Maya communities with maize-based versus globalized economies. *American Journal of Physical Anthropology*, 153, 190–202.

Velsko, I. M. & Warinner, C. G. (2017). Bioarchaeology of the human microbiome. *Bioarchaeology International*, 1, 86–99.

Vucic, S., de Vries, E., Eilers P. H. C., Willemsen, S. P., Kuijpers, M. A. R., Prahl-Andersen, B., Jaddoe, V. W. V., Hofman, A., Wolvius, E. B. & Ongkosuwito, E. M. (2014). Secular trend of dental development in Dutch children. *American Journal of Physical Anthropology*, 155, 91–98.

Warinner, C. (2016). Dental calculus and the evolution of the human oral microbiome. *Journal of the California Dental Association*, 44, 411–420.

Warinner, C., Hendy, J., Speller, C., Cappellini, E., Fischer, R., Trachsel, C., Arneborg, J., Lynnerup, N., Craig, O. E., Swallow, D. M., Fotakis, A., Christensen, R. J., Olsen, J. V., Liebert, A., Montalva, N., Fiddyment, S., Charlton, S., Mackie, M., Canci, A., Bouwman, A., Rühli, F., Gilbert, M. T. P. & Collins, M. J. (2014b). Direct evidence of milk consumption from ancient human dental calculus. *Scientific Reports*, 4, 7104.

Warinner, C., Rodrigues, J. F. M., Vyas, R., Trachsel, C., Shved, N., Grossmann, J., Radini, A., Hancock, Y., Tito, R. Y., Fiddyment, S., Speller, C., Hendy, J., Charlton, S., Luder, H. U., Salazar-García, D. C., Eppler, E., Seiler, R., Hansen, L. H., Samaniego Castruita, J. A., Barkow-Oesterreicher, S., Teoh, K. Y., Kelstrup, C. D., Olsen, J. V., Nanni, P., Kawai, T., Willerslev, E., von Mering, C., Lewis Jr., C. M., Collins, M. J., Gilbert, M. T. P., Rühli, F. & Cappellini, E. (2014a). Pathogens and host immunity in the ancient human oral cavity. *Nature Genetics*, 46, 336–346.

Warinner, C., Speller, C. & Collins, M. J. (2015). A new era in palaeomicrobiology: prospects for ancient dental calculus as a long-term record of the human oral microbiome. *Philosophical Transactions of the Royal Society B: Biological Sciences*, 370, 20130376.

Welch, J. L. M., Dewhirst, F. E., Borisy, G. G. (2019). Biogeography of the oral microbiome: the site-specialist hypothesis. *Annual Review of Microbiology*, 73, 335–358.

Welch, J. L. M., Ramírez-Puebla, S. T. & Borisy, G. G. (2020). Oral microbiome geography: micron-scale habitat and niche. *Cell Host & Microbe*, 28, 160–168.

Weyrich, L. S., Duchene, S., Soubrier, J., Arriola, L., Llamas, B., Breen, J., Morris, A. G., Alt, K. W., Caramelli, D., Dresely, V., Farrell, M., Farrer, A. G., Francken, M., Gully, N., Haak, W., Hardy, K., Harvati, K., Held, P., Holmes, E. C., Kaidonis, J., Lalueza-Fox, C., de la Rasilla, M., Rosas, A., Semal, P., Soltysiak, A., Townsend, G., Usai, D., Wahl, J., Huson, D. H., Dobney, K. & Cooper, A. (2017). Neanderthal behaviour, diet, and disease inferred from ancient DNA in dental calculus. *Nature*, 544, 357–361.

White, D. J. (1991). Processes contributing to the formation of dental calculus. *Biofouling*, 4, 209–218.

White, D. J. (1997). Dental calculus: recent insights into occurrence, formation, prevention, removal and oral health effects of supragingival and subgingival deposits. *European Journal of Oral Sciences*, 105, 508–522.

Zaura, E., Nicu, E. A., Krom, B. P. & Keijser, B. J. (2014). Acquiring and maintaining a normal oral microbiome: current perspective. *Frontiers in Cellular and Infection Microbiology*, 4, Article 85.

CHAPTER 7

Palaeoecology: considering proximate and ultimate influences on human diets and environmental responses in the early Holocene Dnieper River region of Ukraine

Malcolm C. Lillie and Sarah Elton*

7.1 Introduction

Much has been written about the importance of palaeoecology and palaeoenvironments in shaping human evolutionary history. 'Palaeoenvironment' is often used as a catch-all term that variously includes abiotic environmental elements such as those related to climate (e.g. rainfall and temperature), elevation, and the landscape (e.g. the presence of lakes and rivers, often reconstructed from sediment studies), and biotic components such as vegetation, invertebrates and vertebrates. Climate, seasonality and particular landscapes have all been invoked as selection pressures in early hominin evolution (e.g. DeMenocal, 1995; Foley, 1993; Winder et al., 2015). For modern humans, most often the focus of evolutionary medicine (EM), there are clear examples of adaptations to their environments, such as genotypes that facilitate survival at high altitude (Beall, 2007). Palaeoecology is defined as the interaction of

living organisms, including humans, with their environments, and is commonly considered alongside palaeoenvironmental reconstruction. Palaeoecology provides a wealth of examples of selective pressures within all periods of human evolution, including the roles of parasites, bacteria, viruses and other pathogens in disease causation and subsequent adaptation (see Chapters 8, 9, 10, 11 and 17). Consideration of diet is also central to understanding palaeoecology. A great deal of attention has been paid to the role of diet in human evolution (Chapter 5), from its influences on brain evolution and digestive tract morphology (Aiello & Wheeler, 1995), to discussions of the role that 'Stone Age diets' play in contemporary human health (Eaton & Konner, 1985; Eaton & Eaton, 1999; Eaton et al., 1999; Cordain et al., 2000).

Palaeoecology and palaeoenvironments provide the context for evolution and the selective pressures that shape it. In formulating relevant hypotheses, it can be challenging to accommodate the immense temporal and spatial variation that humans have experienced over the course of our evolution (Elton, 2008). Indeed, some prominent and well-established narratives of human adaptation within EM (such as the 'Stone Age diet') tend to underplay the variation and diversity inherent in human populations in favour of perspectives that

* We thank Dr Chelsea Budd (Archaeology Section, Department of Historical, Philosophical and Religious Studies, Umeå University, Sweden), Dr Inna Potekhina (Institute of Archaeology, NAS, Kiev, Ukraine) and Professor Alex Nikitin (Department of Biology, GVSU, Michigan, USA) for discussions linked to the context of the Ukrainian materials considered in this paper. We also thank the reviewers and editors for their constructive comments.

Malcolm C. Lillie and Sarah Elton, *Palaeoecology*. In: *Palaeopathology and Evolutionary Medicine*. Edited by Kimberly A. Plomp et al., Oxford University Press. © Oxford University Press (2022). DOI: 10.1093/oso/9780198849711.003.0007 stress 'mismatch' between the modern condition and a largely homogeneous past (Eaton & Konner, 1985; Eaton & Eaton, 1999; Eaton et al., 1999; Cordain et al., 2000). We argue that human evolutionary history is characterised by the exploitation of diverse habitats, consumption of diverse diets and adoption of varied cultural and behavioural practices, a view that has been discussed elsewhere in the EM literature (Elton, 2008; Foley, 1995; Gowlett, 2003; Turner & Thompson, 2013, 2014; Ulijaszek et al., 2012). It is within this framework of palaeoenvironmental and palaeoecological diversity that we must consider the evolution of health and disease.

In this chapter, we bring together bioarchaeological data and evolutionary theory to give an overview of early Holocene human palaeoecology in changing environments. Given the immense temporal and spatial variation in environments across the globe, and across time, we necessarily focus on archaeological sites from one region and a restricted time period, Mesolithic and early Neolithic assemblages from contemporary Ukraine. To date, central and eastern Europe has not been studied extensively in the context of evolutionary ecology and medicine, yet its archaeological record provides an ideal chance to examine detailed trends in diet and skeletal pathology in ecological and environmental context. Globally, it is clear that environmental diversity resulted in an equally diverse range of cultural adaptations and behavioural responses as human groups adapted their food procurement strategies. In turn, this may be reflected in biological variation in both space and time. Variability and fluctuations in the environment, ecology, prey species, plant biomass and aquatic resources combine to ensure that multidisciplinary approaches, open to the concept of variation, are required if we are to develop realistic and reliable interpretations of human responses in these changing and changeable palaeoenvironments. Study of the human skeleton offers the perfect opportunity to investigate biological responses to that variability and fluctuation in past populations, with stable isotope and palaeopathological analysis giving a window into the diets and health statuses of individuals and populations. Thus, using palaeopathological and isotopic studies of Ukrainian prehistoric populations, we examine dietary variation and

health status in the context of Holocene warming and cooling, and discuss the impact that these might have had on cultural developments across the Mesolithic and Neolithic in Ukraine. We further use our case study of Ukraine to consider the relative importance of ultimate (evolutionary) and proximate (immediate) effects in determining the health of individuals and populations during the early Holocene 'transition' period.

7.1.1 Background

The Holocene (the current geological epoch, which began ~ 12,000 years ago) is seen as a time of transition in many parts of the world from hunter-gathering mobile ways of life to a more sedentary and agriculture-based existence. Human populations in the earliest Holocene (referred to by archaeologists as the Mesolithic period) are, like those earlier in the Palaeolithic, generally characterised by mobile hunting and gathering (or combinations of hunting, fishing and foraging strategies, e.g. Zilhão et al., 2020). Somewhat later, in the Neolithic, it is commonly understood that many groups became more sedentary, with the adoption of agriculture becoming widespread. In the earlier EM literature, considerable emphasis was placed on the transition to farming and its implications for human diet and health (Eaton & Konner, 1985; Eaton & Eaton, 1999; Eaton et al., 1999; Cordain et al., 2000), although some authors drew attention to the lack of a 'revolutionary' transition or argued that the relatively recent Industrial Revolution may have had more profound impacts (Elton, 2008, 2016; Strassman & Dunbar, 1999). The bioarchaeological literature documents a decline in health with the transition to farming, also known as the first epidemiological transition (e.g. Cohen & Armelagos, 1984; Cohen & Crane-Kramer, 2007; Roberts, 2015).

Significantly, though, numerous studies have shown considerable overlap between foraging and farming at the Mesolithic–Neolithic boundary throughout central and eastern Europe. Despite the generalisation that Mesolithic equals hunting and gathering and Neolithic equals farming, Dennell (1983,15) stated that '. . . much confusion has arisen over how to distinguish food-extraction from food-production', and in some areas this remains a valid observation. There is increasing awareness that the key features of what we now recognise as farming—sedentary life, decreased mobility, food storage, domestication of plants and animals—have a prehistory that extends much further back than was originally assumed, into the very earliest Holocene and even the Palaeolithic (e.g. Boaretto et al., 2009; Brown et al., 2009; Dobney & Larson, 2006; Revedin et al., 2010; Wu et al., 2012), and that there was no single point at which these behaviours were adopted in combination across the world. To this day, there exists in some regions a lack of a clear dichotomy between modern hunting and farming groups, which is not simply due to biases in research strategies or a lack of targeted study. Thus, reducing twenty-first-century insults to human health simply to a 'mismatch' between the transition to agriculture and our biology is a potentially uninformative generalisation. Further, it may downplay the serious proximate effects of variations in the environment on individuals (Elton, 2008), including differential socio-economic status and the inequality that results. Careful study of the archaeological, palaeoecological and palaeoenvironmental records will help to understand the proximate and ultimate factors that contributed to health in the past, which in turn will facilitate greater awareness of how to influence health status positively in contemporary populations.

The Holocene epoch is an interglacial (warm period) that succeeded the significant glacial period that reached a peak at ~ 20,000 BP (before present), after which northern hemisphere ice sheets started to retreat and sea and moisture levels rose. A short and abrupt cold period, the Younger Dryas, which started at ~ 12,900 BP and ended at 11,700 BP, marked the end of the late Pleistocene and the beginning of the Holocene. Within the Holocene, temperatures fluctuated, and include a warming period (the Holocene Climatic Optimum [HCO] between ~ 9000 and 5000 BP), and several colder pulses known as Bond Events (Bond et al., 1997). The first cold Bond Event at 8200 BP is cited as having a significant impact on farming groups in the Near East and Europe (e.g. Roffet-Salque et al., 2018). The significance of the HCO and the cold Bond Events has recently been correlated both to the collapse and development of different cultures in Ukraine (Nikitin 2020). Other studies have suggested that far

from the 8200 BP event causing collapse, site abandonment and migration, the early farming communities of southwestern Asia showed resilience to abrupt and severe climate change (Flohr et al., 2016). For the Near East, the impact of the 8200 BP event can be seen in changes in husbandry and consumption practices at Çatalhöyük East, with visible reductions in cattle herd sizes accompanied by an increase in caprine (sheep/goat) herd sizes (Roffet-Salque et al., 2018). Such shifts are wholly consistent with adaptability and resilience.

In general, the shift from the Late Glacial into the Holocene saw a marked improvement in the environments of Europe, with expansion of warmth-loving species (humans included) from glacial refugia, and the replacement of open grassland habitats with closed canopy deciduous woodlands alongside expansion or contraction of specific ecoregions or zones, such as the forest/forest steppe in Ukraine. Khotinsky (1984) notes that in the former USSR, the Late Glacial is characterised by a continental climate, with evidence for hyperzonality, that is, a mixture of tundra, forest and steppe environments (with some taxa that are found in present day steppe and desert zones), determined by water drainage and whether the land was north or south facing. In the Holocene there was a relatively rapid shift to more homogeneous zonal environments, with bands of tundra, forest and steppe, in the pattern that is more familiar today (Khotinsky, 1984). During the Mesolithic period, the zones were characterised by tundra in the northern parts of the former USSR, steppe in the southern regions, and a belt of birch, pine, spruce and larch forest in between.

There was an expansion of temperate forests during the Mesolithic, but Ukraine lacks direct evidence for plant-based subsistence strategies (e.g. as outlined by Zvelebil [1994] for some other Mesolithic sites in Europe). Exploitation of grasses in Ukraine at this time has been inferred from indirect evidence, such as sickle gloss or cereal imprints in pottery (Zvelebil and Dolukhanov, 1991). Nonetheless, the veracity of the identification of cereal imprints in relation to domesticated species and the adoption of agriculture in Ukraine has been questioned by the recent work of Motuzaite Matuzeviciute (2020), with the earliest evidence for domesticated cereals being linked to the appearance of the Linearbandkeramik culture in western Ukraine in the second half of the sixth millennium BCE. However, Motuzaite Matuzeviciute (2020) notes that the extent of these influences remains to be determined and that only limited evidence of cereal use, attributed to the Sredny Stog culture, exists for eastern Ukraine prior to the fourth millennium BCE.

As might be anticipated, alongside the shifts in vegetation at the Pleistocene–Holocene transition, we see concomitant changes in the nature of the biomass, seasonality in resource availability and a shift away from the large herds of animals, such as reindeer, mammoth, bison and giant deer, which characterised the more open landscapes of the Late Glacial to less easily exploited solitary animals or small herds (e.g. aurochsen, red deer, wild horse and wild donkey, gazelle, mountain goat, wild boar, elk and brown bear) that occupied the steppe, forest and tundra zones (Zaliznyak, 2020a,b). In addition, it has been noted that during the Mesolithic period, fishing and plant gathering increased in importance compared to hunting (Dolukhanov & Khotinskiy, 1984). In essence, despite a broad range of species being available for exploitation, the nature of the landscape changes meant that the larger game animals were more dispersed in the environment and less predictable in terms of their availability than in the Late Glacial period. This necessitated a concomitant shift in the nature of food procurement strategies on the part of human groups and changes in the nature of the toolkits that were used in subsistence and other tasks (e.g. Lillie, 2015). Such a switch in subsistence activity may have had important implications for human adaptation and health.

7.2 Environments and health status along the Dnieper River

We focus on the Dnieper Rapids region of Ukraine, which preserves large cemetery populations, many belonging to the Neolithic Dnieper-Donets culture that spans the period 10,400–5470 cal BP (~ 10,000–3500 cal BCE). These cemeteries provide the opportunity for in-depth study of palaeopathological evidence alongside stable isotope data to provide dietary context, within a region for which palaeoenvironments are generally well understood.

Our case study concentrates on four cemeteries (the Epipalaeolithic cemetery of Vasilyevka III, the Mesolithic cemetery of Vasilyevka II and the Neolithic (Mariupol-type) cemeteries of Yasinovatka and Dereivka I; see Figure 7.1) where the dating, pathological and stable isotope data provide detailed insights into individual life histories and population dynamics.

Vasilyevka III is located on the left bank of the Dnieper, to the south of the town of Dniepropetrovsk, and is dated to 10,400–9920 cal BP (10,000 to ~ 9300 cal BCE) (Jacobs, 1993, Lillie et al., 2003), at the Late Glacial—Holocene transition, with evidence that it was in use into the Mesolithic period at ~ 8850 cal BP (8280–7967 cal BCE) (Mathieson et al., 2018). During the periods that Vasilyevka III was being used for interment (probably the final stages of the Younger Dryas, with temperatures rising into the Holocene, plus the early Holocene itself), a cool/dry to warm/dry climate persisted. The environment is initially characterised by poorly watered steppe away from the river valleys in the southern part of Ukraine (the Black Sea lowlands), with a central belt of forest-steppe, and the Polissya forest lowlands in the North (Zaliznyak, 2020a). During the period ~ 9000–5000 BP temperatures increased across the northern hemisphere, reaching a peak at ~ 8000 BP. Significant warming marks the onset of the earliest Holocene in Ukraine, and the cool climate and moist soils promoted the expansion of birch-pine in the north and the expansion of pine in the south, with birch-pine woodlands extending ~ 150 km along the left bank of the Dnieper (Zaliznyak, 2020b).

Zaliznyak (2020b) notes the extent of the changes across the Pleistocene–Holocene transition were not as marked in the southern part of Ukraine as the north, although the loss of the cold-tundra forests resulted in the disappearance of herd animals such as bison, saiga and donkey in the Black Sea lowland zone. In general, the early Holocene in Ukraine at 10,300 BP is characterised by decreased continentality and a reduction in hyperzonality (Velichko, 1973, cited by Dolukanov & Khotinskiy, 1984). By ~ 8,000 BP there was a transition to a more humid, increasingly warm and differentiated landscape. At this time, broadleaved forests expanded into the lower Dnieper valley and steppe cover became more mesophilous (avoiding extremes of moist or dry

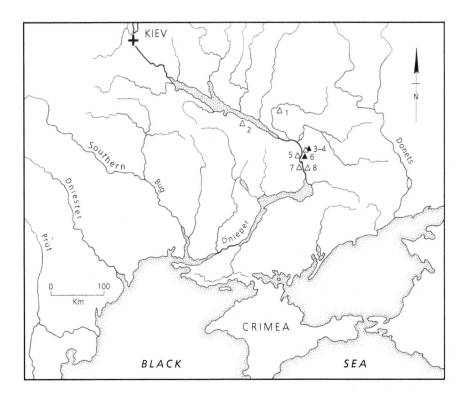

Figure 7.1 The Dnieper Rapids region, showing locations of cemeteries studied by Lillie (1998a). 1—Osipovka, 2—Dereivka I and II, 3—Vasilyevka V, 4—Vasilyevka III and II, 5—Nikolskoye, 6—Marievka, 7—Vovnigi II, 8—Yasinovatka. ▲Mesolithic, △—Neolithic.

environments) (Kremenetski, 1995), with vegetation that preferred more moderate climates. At ~ 8200 BP (~ 8000 BP in northern Eurasia), a Bond Event (Bond et al., 1997) ushered in a period of dry and cold conditions for a period of ~ 60 years, with a milder interval of a few decades followed by progressive warming that took ~ 70 years (Kobashi et al., 2007; Thomas et al., 2007). However, while it has been suggested that this may also have occurred in the lower mid-latitudes—much farther south than previously thought—the actual impact of this remains to be established, and the climatic fluctuation itself may have simply been an element of longer-term climatic instability (Wainwright & Ayala, 2019).

Vasilyevka II is located on the left bank of the Dnieper, in the Rapids region near the village of Vasilyevka, and is dated to the later Mesolithic at 9190–8020 cal BP (~ 7000–6200 cal BCE) (Telegin & Potekhina, 1987; Lillie & Jacobs, 2006). Based on the available dating, the Vasilyevka II assemblage

at the Dnieper Rapids occurred at a time when the climate was nearing its optimum, which was reached between 6000 and 4600 BP. At this point, the steppe vegetation became more mesophytic when compared to that of the early Neolithic periods (cf. Dolukhanov & Khotinskiy, 1984). The period leading up to the climatic optimum appears to be reflected in the increased concentration of cemetery activity in and around the Dnieper Rapids region across the later Mesolithic and Neolithic periods. However, currently there is very little evidence to suggest that the 8200 BP Bond Event exerted any tangible impacts on the populations exploiting the Dnieper region, beyond the fact that the cemetery evidence is limited across the period of inferred climatic instability.

The Neolithic Mariupol-type cemetery of Yasinovatka is located on a high terrace ~ 25–30 m above the left bank of the river, near the Dnieper Rapids. It is dated to 7476–6990 cal BP (~ 5800–4780 cal

BCE) (Lillie, 1998a; Lillie et al., 2009). In contrast, the Neolithic Mariupol-type cemetery of Dereivka I is located on the right bank of the Omelnik river at the point where it joins the Dnieper near the village of Dereivka, ~ 120 km upstream of the other sites included in our case study. It is dated to 7253–6700 cal BP (~ 5270–4750 cal BCE), although this site also contains Mesolithic burials, with a date of 8340–7875 cal BP (6360–5879 cal BCE) on burial 84 (OxA-6161) (Lillie 1998a, Lillie et al., 2009). This site is located closer to the forest-steppe boundary at this time, but it is feasible to suggest that the conditions adjacent to the Dnieper may well have been similar throughout its middle and lower reaches due to the existence (and persistence) of forests along the valley throughout the periods discussed. However, we might anticipate some differing isotope values for biota at the forest-steppe boundary when contrasted with those encountered in the steppe zone near the Dnieper Rapids, irrespective of the forested nature of the river valley itself.

Both Neolithic cemeteries in our case study were established in the period of climatic stability between the 8200 BP and the 6300 BP Bond Events, and indeed there are probably seven cemeteries in total that were established at the Rapids, or at the edge of the forest-steppe zone, during this period. This may well reflect the fact that optimum conditions existed for the expansion of populations into the region across this period, as the available dating indicates that activity and settlement along the Dnieper continued past the 6300 BP Bond Event.

In the following sections, which discuss the individual cemetaries, we focus on stable isotope data for diet and palaeopathological evidence for compromised health for both males and females. Most of the palaeopathological evidence is documented on dentitions, with other conditions less frequent or altogether absent. Further information about skeletal preservation and other aspects of the sites and material can be found in the relevant cited literature. It should be noted that the excavations generally revealed excellent preservation of the cranial region, and the cranial material was preferentially curated in the collections available for study. There was differential postcranial preservation, and (for dating purposes) only certain long bones were retained.

Due to the free-draining nature of the loess soils, the post-cranial material often could not be lifted, and age and sex estimation was undertaken *in situ*.

7.2.1 Vasilyevka III

Forty-four individuals were originally excavated from this cemetery, with twenty-one individuals available for study. These twenty-one individuals comprised ten males, nine females, one 'non-adult' individual and an individual of indeterminate age and sex (Lillie et al., 2003). The palaeopathological evidence suggests that the individuals at Vasilyevka III consumed protein-rich diets (attested to by calculus deposition on their teeth; Hillson, 1979), with no evidence for the consumption of cariogenic foodstuffs such as carbohydrates or soft sticky foods (given the complete absence of caries on the teeth). Low levels of sub-optimal juvenile health are indicated by infrequent occurrences of enamel hypoplasia (Goodman et al., 1984a, b; Goodman & Armelagos, 1985a,b). The rates of expression of these non-specific indicators of juvenile physiological stress suggest that, in general, the human groups in the Dnieper region were probably not overly stressed in their developmental years, at least in dietary terms. The age distribution of the incidence of the enamel hypoplasias suggests that prolonged weaning, through to the ages of 4.0–5.5 years, may have been in place for this population, with occasional instances of sub-optimal health prior to this age range (e.g. a female and a male individual who exhibit hypoplasia correlated to ~ 3–3.5 years of age), perhaps reflecting causes (e.g. sub-optimal living conditions) unrelated to the post-weaning diet (Lillie, 1998a). There was no indication of porotic hyperostosis/cribra orbitalia (bone changes on the cranial bones and in the orbits, respectively) on any of the crania examined from Vasilyevka III (Lillie, 1998a).

Stable isotope analysis of twenty-one individuals from Vasilyevka III (Lillie et al., 2003) produced results for fourteen individuals. The average $\delta^{13}C$ value was −22.2±0.2‰ and average $\delta^{15}N$ value was 12.7±0.6‰ (Lillie et al., 2003), both of which indicate a diet with a significant input from freshwater resources, which links straightforwardly back to the evidence for calculus deposition on the detentions studied. This is perhaps unsurprising given the position of these cemeteries close to the Dnieper Rapids, giving direct access to freshwater resources. Interestingly, two of the three individuals with the highest nitrogen isotope values at Vasilyevka III were from males aged 50+, perhaps suggesting that these older people were consuming diets that comprised significant inputs of fish, a soft foodstuff. The fact that a female aged 18–22 exhibits the second highest nitrogen value of 13.10‰ is less easy to interpret in this context (i.e. she is not an old individual), but the value is not too much higher than other females in the skeletal sample, which range from 11.66 to 13.10‰ (six individuals). This may indicate that while males may have hunted, perhaps expending a lot of energy for unpredictable returns, females were exploiting a more reliable and easily accessible food resource through fishing. In addition, the older male individuals in the society, who were potentially no longer able to hunt, were also involved in these activities in some way.

At Vasilyevka III, Nuzhnyi (1989, 1990) identified evidence for interpersonal violence on five of the individuals from the cemetery population, with arrowheads and slotted bone points in association with the remains. In two instances, the deceased had microliths embedded in their lumbar vertebrae, indicating that they had been shot in the back, and another individual had arrow and spear points embedded in their skeleton. Furthermore, at the cemetery sites of Vasilyevka I and Voloshkoe, evidence for impact damage from arrows has also been recovered. While these latter cemeteries remain undated in absolute terms, the stratigraphic context and type of artefacts led Nuzhnyi (1989, 1990) to suggest that they are broadly contemporary in chronological age with Vasilyevka III. The evidence for interpersonal violence may well indicate increased competition for resources at the Late Glacial–Holocene transition in the region, and the positioning of the cemeteries at the Rapids could be interpreted as an assertion of ancestral rights of access to the reliable freshwater resources available at this location (Lillie, 1998a).

Overall, despite the chronological position of Vasilyevka III near the Late Glacial–Holocene boundary, the data suggest that there is minimal evidence for dietary stressors in the skeletal remains studied and that, while there is good evidence for the consumption of protein-rich diets, the complete absence of evidence for dental caries indicates that cariogenic foodstuffs were not integral to the diet of these individuals. Perhaps the most important indicator of some form of 'stress' in terms of access to resources is the prevalence of bone damage related to interpersonal violence and the fact that the cemeteries themselves are located at the Dnieper Rapids, in a riparian (riverbank) environment where a broad range of resources were available for exploitation.

7.2.2 Vasilyevka II

Demographic and palaeopathological analyses of the Vasilyevka II cemetery population (fourteen individuals–five females and nine males) demonstrate that they lived well past the mean average life expectancy for the Mesolithic period, with three individuals (two males and a female) reaching the 50–60-year age bracket at death (assessed via molar wear seriation and cranial suture closure), and a further three males and one female surviving into the 40–50-year age category. Interestingly, one of the older males also has evidence for healed cranial surgery (trepanation) and linear enamel hypoplasia on the left upper first molar (Lillie, 1998b). This hypoplastic event correlates with a non-specific metabolic disruption occurring at 2.0–2.5 years of age (Goodman et al., 1987). Despite this childhood stressor, this individual clearly lived to a mature age, and even subsequently survived the surgical intervention associated with a trepanning procedure (Lillie, 1998b).

At Vasilyevka II, there is no evidence for caries on the dentitions of this group, nor porotic hyperostosis, which is a non-specific indicator of anaemia (Lillie, 1998a; Stuart-Macadam, 1992). The diseases that may cause this response include hereditary haemolytic anaemias, thalassemia, sickle-cell anaemia, iron deficiency anaemia, and diarrhoeal disease (Mensforth et al. 1978; Goodman et al. 1984b; Hengen, 1971; Stuart-Macadam 1992; Walker 1986). To assess protein consumption, the preserved teeth (n = 324) were studied for calculus deposition, although 147 teeth could not be considered due to factors such as exhibiting post-mortem damage, advanced functional-related wear affecting the crowns (where calculus mainly accumulates) or a low functional age (i.e. the individual was below ~ 24 years of age and calculus deposition was not evidenced). Despite these limitations, all male teeth that could be studied exhibited calculus, while females had more teeth with no evidence for calculus deposition. There are limitations to the data in that there are no females represented in the 25–40-year age categories and, while the population analysed is small, the lack of females in these age groups is unusual (Lillie, 1998a).

There is a clear lack of evidence on the dentitions for the consumption of dietary foodstuffs that would have initiated dental caries and, as only one individual exhibited a single hypoplasia, there is also a lack of evidence for childhood stress at Vasilyevka II. The presence of post-cranial skeletal remains at Vasilyevka II offers a rare opportunity for the study of pathology in this region of the skeleton, as skulls usually dominate in the curated Dnieper Rapids human remains, due to preservation issues in the loess soils of the steppe region. The only other pathological evidence on these remains was osteoarthritis, which was visible on the joints of the left femur, humerus and ulna of one individual. Meiklejohn and Zvelebil (1991) previously noted that osteoarthritis is the most common pathology found on Mesolithic post-cranial skeletal remains in Europe. The evidence of disease from Vasilyevka II is limited due to the small number of individuals, but the data do conform to the generally low levels of pathology that were in evidence at Vasilyevka III.

Stable isotope analysis carried out on fourteen individuals from the cemetery of Vasilyevka II (Lillie & Jacobs, 2006) revealed isotope ratios with a mean of $-20.94\pm0.49\permil$ for $\delta^{13}C$ and $13.39\pm0.62\permil$ for $\delta^{15}N$, with the individual with the highest $\delta^{15}N$ value being a male aged 20–30 years. The Vasilyevka II individuals all have $\delta^{15}N$ values above $12.35\permil$, and the data indicate relatively uniform diets among males and females, with freshwater fish and animal proteins being important dietary elements (Lillie & Jacobs, 2006). The $\delta^{15}N$ values from Vasilyevka II differ slightly from the Epipalaeolithic–Earlier Mesolithic values from Vasilyevka III, with the later Mesolithic values at $13.39\pm0.62\permil$ being higher when compared to the chronologically earlier values of $12.47\pm0.61\permil$. The elevated nitrogen from Vasilyevka II may indicate a higher input from freshwater fish in the later Mesolithic when compared to the Epipalaeolithic period. The $\delta^{13}C$ value, at $-20.94\pm0.49\permil$, is more positive than at Vasilyevka III, where values of $-22.2\pm0.2\permil$ are recorded. This shift may well reflect variations in the resources that were available for exploitation by the populations in this region of Ukraine, linked to the changing environments that occurred during the time that Vasilyevka II was in use (Velichko et al., 2009; Kotova & Makhortykh, 2010; Budd & Lillie, 2020).

7.2.3 Yasinovatka

There were sixty-eight individuals excavated from the Neolithic cemetery of Yasinovatka across its use. The cemetery area was identified as a sub-rectangular pit highlighted by red staining associated with the ochre applied to Neolithic burials in the region (Telegin and Potekhina, 1987). Two distinct burial contexts were distinguished, with a number of discrete grave pits of oval form overlain by a large collective burial pit. For the current study, thirty-one individuals with preserved teeth were assessed (comprising twenty males, six females and five individuals of indeterminate biological sex, although two individuals had teeth that were too worn to allow detailed study) (Lillie 1998a, Budd et al., 2020). No carious lesions were identified in the preserved dentitions, which again suggests that their diet was high in proteins and/or low in carbohydrates. At this cemetery, only one of the studied individuals lacked calculus (the presence of which in archaeological remains is generally common) and six of the thirty-one individuals studied had evidence for enamel hypoplasia. Of the individuals with calculus, all males and females exhibited deposits across the grades, from light to heavy, although a substantial number of teeth (130/441) had no evidence for any calculus despite some high functional ages (i.e. older than ~ 45 years of age; Lillie 1998a). The timing for enamel hypoplasia formation appeared variable, occurring as early as 2.5 years of age in some individuals, and later (up to 5 years of age) in others, suggesting that low levels of juvenile stress were occurring in early childhood. However, this cannot be reliably linked to weaning stress, as opposed to any of the other child-hood stressors that might cause hypoplasia, such as hypoparathyroidism, vitamin A and D deficiencies, fever, maternal diabetes, neonatal asphyxia and jaundice, nephrotic syndrome and gastroenteritis (Goodman et al., 1984a, b; Goodman et al., 1987; Goodman, 1993; Seow, 1997, 2014).

Interestingly, statistical analysis of the observed versus expected frequencies of calculus for males versus females at Yasinovatka (Lillie, 1998a) high-lights a reversal in the general trends observed for the remainder of the skeletal series from the Dnieper region. Namely, it is the females who exhibit the highest prevalence of heavy calculus. The fact that an 18–25-year-old female exhibits heavy calculus deposition suggests preferential access to dietary proteins for females from an early age. Such heavy deposits would suggest high protein intakes, as would the overall wider distribution of calculus seen in individuals from Yasinovatka. As males out-number females at a ratio of 4:1 in the assemblage studied, this observation suggests that the male–female expression of calculus at Yasinovatka reflects the consumption of (at the very least) equivalent levels of protein-rich foodstuffs between the sexes in this cemetery population. However, it is tempting to suggest that at Yasinovatka, females are central to securing dietary protein, potentially through fish-ing, and/or that the males at this site are not actively engaged in the procurement of dietary proteins to the same degree as females. Obviously, caution is needed, given the low numbers of individuals avail-able for study, but some intriguing trends appear at Yasinovatka that are not seen elsewhere in the skeletal series studied to date.

Where crania were preserved (twenty-four indi-viduals), no evidence for cribra orbitalia or porot-ic hyperostosis was found and, despite the fact that fish and deer tooth pendants are ubiquitous in the burial inventory at Yasinovatka, levels of fish consumption do not appear to have resulted in increased parasitic infestation (e.g. Seow 1997, 2014). As this observation runs contra to the evi-dence from Dereivka I (considered in Section 7.2.4), we might hypothesise that the group using Yasino-vatka as a burial ground were perhaps more peri-patetic than the people who were burying their dead at Dereivka I.

At Yasinovatka, samples from thirty-eight indi-viduals produced reliable isotope measurements. These results, with $\delta^{13}C$ at $-22.3\pm0.9‰$ and $\delta^{15}N$ at $13.7\pm1.5‰$, are commensurate with data from the site of Nikolskoye (another key Mariupol-type cemetery that fully overlaps with the dating evi-dence from Yasinovatka and Dereivka), and are more positive than those from Dereivka I. The data indicate that freshwater aquatic proteins formed the mainstay of the diet at Yasinovatka (Budd et al., 2020). The material culture inventory from Yasinovatka includes bivalve shells and teeth of freshwater fish (Telegin & Potekhina, 1987), and as many Dnieper-Donets Mariupol-type cemeter-ies have evidence for the exploitation of these resources, these data reinforce the importance of, and dependence on, the freshwater environment (Lillie, 2020; Budd & Lillie, 2020). Nonetheless, as recorded at other locations, there is variation among the population. Two individuals produced lower values than most of the population analysed. A young adult male (aged 18–25 years), who stands out in having decorative plates of boar tusk that appear to have been sewn on to the arm of his cloth-ing, produced a $\delta^{13}C$ value of $-23.6‰$ and a $\delta^{15}N$ value of $11.4‰$, and a female aged 20–25 years, had a $\delta^{13}C$ value of $-22.4‰$ and a $\delta^{15}N$ value of $7.4‰$. There is clear variability in diet at Yasinovat-ka and, in the latter two individuals, C_3 terrestrial proteins formed the basis of their diets, although the lower $\delta^{15}N$ value of the female suggests she ate proportionately less animal protein than the male. Recent assessment of this cemetery (Budd et al., 2020) has also suggested that these individuals may have been consuming terrestrial plants and animals from areas away from the Rapids, i.e. that poten-tially they were recent immigrants into the region. Intriguingly, these individuals date as the earliest burials at Yasinovatka (Lillie et al., 2020).

7.2.4 Dereivka I

For the Neolithic period, the site of Dereivka I offers insights into dietary pathways away from the concentration of cemeteries that are located ~ 120 km downstream at the Dnieper Rapids. This is the largest Mariupol-type cemetery in the Dnieper system. This site also stands out from the other cemeteries studied by Lillie (1998a) in that it is the only location where the palaeopathological evidence highlights a combination of a degree of sedentism associated with what appears to be an increased risk of parasite loading. Of a total of 104 adult individuals excavated, Lillie (1998a) analysed sixty-two individuals comprising forty-two males, thirteen females and seven individuals of indeterminate biological sex. As with the other cemeteries in the Dnieper region, there is a complete absence of caries on the dentitions of the individuals studied. However, calculus is again in evidence, although at this location all the individuals with teeth exhibited some degree of calculus deposition. In relation to the rates of calculus, an interesting aspect of the dentitions from Dereivka I is the fact that five individuals in the younger age range (12–22 years of age) have lower levels of calculus deposition than the five people in the 18–25 years of age category. This distribution suggests heavier protein consumption patterns for some than have generally been recorded elsewhere in the cemetery samples studied. Importantly, statistical analysis indicates that male–female expression of calculus is equivalent (Lillie, 1998a), suggesting that equality in access to dietary protein is occurring at Dereivka I. This dietary equivalence may also help to account for the single instance of enamel hypoplasia, indicative of a non-specific childhood stress event.

At Dereivka I, a total of eighty individuals were represented by preserved crania suitable for the study of pathology and, of this number, four (5% of the cemetery population) had cribra orbitalia (Hengen, 1971; Stuart-Macadam, 1991, 1992). When coupled with the absence of corresponding osseous involvement in the facial region, the levels of cribra orbitalia recorded at Dereivka I indicate that hereditary causes, haematological and circulatory disorders can be discounted as the cause for lesion development; this is because the population-based expression exceeds 1% (cf. Klepinger, 1992). While several factors may have influenced cribra orbitalia development (Stuart-Macadam, 1989; Walker et al., 2009), the dependence on fish in the diet of these people at Dereivka, coupled with increased sedentism, could have led to increased pathogen loads (fish-borne parasites) and the initiation of an adaptive response

in the immune system (cf. Kent, 1986; Macchiarelli, 1989; Stuart-Macadam 1992; Walker 1986). Evidence from elsewhere in Europe, such as the Danish Ertebølle and Danubian Iron Gates, has demonstrated that a heavy reliance on aquatic/marine resources can lead to reductions in the quality of fisher-forager diets due to increased parasite load (Meiklejohn & Zvelebil, 1991), and the evidence from Dereivka I lends additional support to these observations.

At Dereivka I, which lies at the boundary between the forest-steppe and steppe zones, the atypical nature of the pathologies discussed are not reflected in the isotope ratios that have been recorded to date (Lillie et al., 2011; Budd & Lillie, 2020). Of the 104 individuals available for study, so far only thirteen have been analysed for diet using stable isotope analysis. The results for $\delta^{13}C$ and $\delta^{15}N$ are –23.2±0.6‰ and 12±1‰, respectively, indicating that aquatic proteins did contribute a significant proportion to the diet at Dereivka, but these values are lower than those for the earlier Mesolithic site of Vasilyevka II—in fact, they are lower than those found at Yasinovatka and Nikolskoye, at 13.4±0.6‰ (Lillie et al., 2011; Budd et al., 2020). Budd and Lillie (2020) have argued, based on the diet isotope data, that the location of Dereivka, ~120 km to the northwest of the Rapids sites, near the forest-steppe boundary, and on a tributary of the Dnieper, may potentially have facilitated access to different fish species, and also to a broader spectrum of terrestrial plant and animal species.

The nitrogen isotope values from two individuals at Dereivka I, at 9.9‰ and 10.5‰ respectively, suggest that some of the individuals consumed diets that were focused on terrestrial resources (Lillie et al., 2011). As one individual from this site has been shown to group with Anatolian and European Neolithic farmers at the whole genome level (Mathieson et al., 2018), it is entirely feasible that at this relatively early date in the Neolithic of the Dnieper region we are seeing the first immigration of individuals with southwestern origins into this steppe region. The fact that Dereivka I is at the forest-steppe/steppe boundary, at a location that mirrors the subsequent spread and distribution of the Trypillia farming culture in Ukraine, may well be more than mere coincidence. As such, we might hypothesise that variability in diet at Dereivka I could indicate a shift towards a more terrestrial resource-focused subsistence strategy, at least for some individuals, occurring at a time (7253–6700 cal BP (4949–4799 cal BCE)) when the initial stages of contact with western farmers were happening in this region.

7.3 Discussion

Using the evidence from the Dnieper River region in Ukraine, we attempt to disentangle the climate/environmental and archaeological data with a view to identifying significant impacts on the sociocultural and economic trajectories of the groups in this region across the period 10,400–5700 cal BP (~ 10,000–3800 cal BCE), and in turn assess how these may have influenced evolutionary fitness. The data underline that even in a limited geographic area and relatively short period of time (~ 6000 years or so), human diets varied within and between groups, even if the basic spectrum of available resources was similar. In the Dnieper region, red and roe deer, elk, aurochs and wild boar were all hunted across the Mesolithic period, as seen in faunal evidence, and archaeological evidence further attests to the exploitation of freshwater resources such as carp, pearl roach and freshwater molluscs (Telegin & Potekhina, 1987). This is consistent with other Mesolithic sites in central and eastern Europe. At Zvejnieki in Latvia, Larsson and colleagues (2017, 60) report that elk and beaver dominate the faunal remains at this Mesolithic settlement site, while pike is the most common fish species. The Danubian Iron Gates sites (Serbia and Romania) have evidence for the consumption of sturgeon, carp, catfish and freshwater mussels (Cook et al., 2009). While C_3 terrestrial resources such as red deer, aurochs and wild boar were important, Bonsall and colleagues (2015, 696) argue that substantial amounts of fish were consumed by the Iron Gates populations at Vlasac, Lepenski Vir, Hajdučka Vodenica and Padina across the Mesolithic, with the possibility that marine resources (including sturgeon) increased in importance during the later Mesolithic period (Cook et al., 2009). In Karelia and north-western Russia, Mesolithic economies were oriented towards the hunting of forest animals (such as reindeer, red deer, boar and beaver) and waterfowl, with fishing also being of some importance (Dolukhanov, 2017). Indeed, it has been argued that 'fishing and gathering were of great importance everywhere' in the forest zone of the former USSR (Dolukhanov & Khotinskiy, 1984, 321).

Fish certainly comprised an important part of the diet in the Dnieper region across the Epipalaeolithic (the transitional period at the very earliest part of the Holocene) through to the Eneolithic (the period at the junction between the latest Neolithic and Copper Age) periods (Lillie & Richards, 2000; Lillie et al., 2003; Lillie & Jacobs, 2006; Lillie et al., 2011; Budd & Lillie, 2020; Budd et al., 2020). However, the stable isotope data reviewed here indicate that these resources were not of equal importance for all people and all groups. Older males at Vasilyevka III seemingly ate more fish than did other members of that population (possibly because of ease of mastication) (Lillie, 1998a, 2020). A couple of individuals at both Dereivka I and Yasinovatka had diets dominated by terrestrial rather than aquatic foodstuffs. All people at Vasilyevka II had a greater emphasis on fish in the diet than those at the earlier site of Vasilyevka III and the later site of Dereivka I. This may reflect the preferences of people in those different groups. There is a wealth of ethnographic data indicating that humans do not simply eat the foods available to them, but make cultural choices about their diet, governed, among other things, by symbolic value, prestige and enculturated taste (see review in Ulijaszek et al., 2012). However, it may also reflect short-term shifts in resource availability that could have been caused by the climatic fluctuations, or Bond Events, occurring between ~ 8200 and 6300 BP.

Across the early- to mid-Holocene, the climate, environment and landscape of central and eastern Europe are characterised by relative stability punctuated by a series of global and localised climatic perturbations that vary in their duration, severity and potential impact on human populations, which are identified in part by differences in stable isotope signatures in the humans living in the region. The difference in diet at Vasilyevka II (8200–8020 cal BP) might be a response to the developing Bond Cooling Event that occurred around 8200 BP. This could be because patterns of terrestrial mammal availability shifted, forcing a greater reliance on fish. Although the geographic distribution of freshwater fish will shift with long-term thermal changes, fish initially respond to temperature changes through shifts in behaviour and micro-environment within a given water source (Pletterbauer et al., 2018). Therefore, whereas the marked environmental shift that characterised the Late Glacial to the Holocene period led to a significant restructuring of the available biomass (and by extension human subsistence) (Dolukhanov, 1997), the opportunism and flexibility that has such a long evolutionary history in hominins (Elton, 2008; Potts, 1998; Ulijaszek et al., 2012) potentially facilitated a shorter-term dietary response to the more modest climatic perturbations of the Holocene.

One important question for those interested in EM, and indeed human health and disease more generally, is how that short-term variation may have influenced health outcomes, proximately and ultimately. The available palaeopathological data from the Dnieper series indicate generally low rates of pathology, although caveats obviously exist in relation to the absence of curation of post-cranial remains due to factors such as preservation (Lillie, 1998a). The pathologies in evidence accord well with known rates of pathology identified in hunter-gatherer skeletons excavated from sites in central and northern Europe. In those groups, outside of the Mediterranean region, caries rates are generally low (Meiklejohn & Zvelebil, 1991). The age at which stress episodes occur in the Dnieper series (evidenced by enamel hypoplasias) is commensurate with Mesolithic populations elsewhere, such as those at Skateholm I and II (Meiklejohn & Zvelebil, 1991; Lillie, 1998a). This indicates that, overall, there was no health (and, by implication, fitness) cost or benefit to small variations in diet over time and space. However, a combination of the diet consumed and human behaviour may have resulted in adverse health outcomes. The identification of cribra orbitalia at Dereivka I is not unusual for a population exploiting fish resources wherein increased pathogen (parasite) loads could have been present. This situation is especially significant given the fact that the Dereivka population appears to

have been more sedentary than the groups downstream in the Dnieper system at the Rapids, and the temporally overlapping group at Yasinovatka. Although early exposure to parasites may help to prevent autoimmune conditions (the 'Hygiene Hypothesis'; see Chapter 11), long-term research on contemporary hunter-gatherer populations suggests that parasite load increases interbirth intervals and reduces overall health and well-being, potentially impacting on evolutionary fitness (Hurtado et al., 2008).

The concept of 'transition' is influential in EM. It is closely aligned to 'mismatch', with the former often argued to lead to the latter. When thinking about palaeodiets, the transition from mobile hunting and gathering to a sedentary agrarian existence is emphasised. The Neolithic is often viewed as a transition point, with hunting and gathering giving way to a more sedentary agrarian subsistence pattern. While influences upon cultural evolution are exerted by a range of mechanisms, including social, economic and ecological variables, there is little doubt that the relative significance of these factors varied across the transition from the late Pleistocene to Holocene periods throughout Europe (Dolukhanov, 1984; Dolukhanov & Khotinskiy, 1984). Being located between western and eastern Europe, Ukraine experienced influences from incipient farming cultures in the west from ~ 6500 cal BCE onwards. However, the rate of spread and adoption of farming is protracted and variable in both a west-east direction and a south-north direction such that, while contacts with farming groups are increasing across the latter part of the seventh millennium and into the sixth millennium BCE in the south and west of Ukraine, populations to the east and north retain a fisher-hunter-gatherer subsistence lifestyle.

Lillie (1998a) has previously suggested that, if not for the differences in material culture, there is little to distinguish the skeletal remains from Mesolithic sites from those of the early Neolithic period in the Dnieper region of Ukraine. These populations were clearly following very similar lifeways, with a reliance on freshwater resources continuing from the Epipalaeolithic at Vasilyevka III through to the end of the Neolithic period and into the Eneolithic period in this region. The Dnieper-Donets populations who buried their dead in the Mariupol-type cemeteries align closely with Mesolithic groups in the Danube and Baltic regions. However, at the Neolithic transition, while some aspects of material culture suggest that changes are occurring, unlike in areas to the west, these populations retain the essentially Mesolithic subsistence strategies that are attested at sites like Olenii Ostrov in Karelia and Zvejnieki in Latvia. At the former, recent stable isotope analysis (Budd & Lillie, in prep) has shown that the Mesolithic individuals interred at this location have elevated nitrogen ratios that are fully commensurate with those in evidence at Mesolithic sites in the Danubian Iron Gates sites (e.g. Vlasac and Schela Cladovei; see Cook et al., 2009; Lillie et al., 2011), the Mesolithic period burials at Zvejnieki in northern Latvia (Eriksson et al., 2003) and the Dnieper Rapids series (Lillie et al., 2011; Budd & Lillie, 2020).

In the context of transition, it is the change in terrestrial biomass between the Pleistocene and Holocene that is more important for diet and behaviour in the Dnieper Rapids region than the shift between the Mesolithic and Neolithic periods. The increasing importance of aquatic resources in the Mesolithic came about because of the increasingly dispersed nature of large game animals and the need to exploit a 'broader spectrum' of resources (Binford, 1968; Flannery, 1969). Of course, exploitation of small game and fish was not a new phenomenon in the Mesolithic and had its roots in the Pleistocene (see review in Ulijaszek et al., 2012), but its relative importance probably increased (sensu Stiner, 2001). This may have helped shape behaviour, with groups no longer travelling large distances following game herds but instead staying closer to home and extracting resources from a more settled and secure base (as seen in the Danube region; e.g. Bonsall et al., 1997; Cook et al., 2001, 2002, 2009; Bonsall et al., 2015).

What the Mesolithic–Neolithic data from Ukraine indicate is that potentially adverse health outcomes may arise not from dramatic shifts in subsistence (the premise on which 'Stone Age diet' arguments are built), but from small, potentially transient behavioural changes, such as a decrease in mobility in one group compared to another, that may interact with diet (or its indirect effects). All the groups represented in the Mariupol-type cemeteries

in the Dnieper River region have the same fundamental diet, albeit with variations in the relative importance of terrestrial and aquatic resources. The palaeopathological data described here are largely homogeneous among these groups (and align with findings in other European hunter-fisher-forager groups), except for the evidence for cribra orbitalia at Dereivka. This health stress, and a condition that develops in the growing years of childhood, was not obvious from the dietary data, and it is only with the combination of archaeological, isotopic and palaeopathological data that inferences about its cause can be made. In this case, a greater exposure to fish-borne parasites because of a probable increase in settled living is evidenced. What is interesting from an evolutionary perspective is that this was not a consistent, long-lasting change, or indeed one that was mirrored at the contemporary sites of Yasinovatka and Nikolskoye (the latter site continues in use until the Eneolithic transition in this region, 5700–5470 cal BP) (Lillie, 2020). Thus, although the reproductive success of the Dereivka population may have been adversely affected by increased pathogen load, there is no evidence that this was a consistent health risk in the Neolithic of the region that led to a long-term selective pressure. This in turn lends weight to arguments that, in many cases, we should not view changes of diet and human behaviour exclusively in terms of 'revolutions' that promoted mismatch, and that we should also think about people's health through a proximate as well as an ultimate lens when considering how evolutionary approaches may inform our understanding of health, both in the past and today.

7.4 Conclusion

Undeniably, pathogens and diets have exerted selective pressures on humans, but variation and variability in environments and behaviours may have also applied profound pressures on humans, leading to selection for plasticity in morphology, physiology and behaviour (reviewed in Elton, 2008). Plasticity has fitness advantages in rapidly changing environments and the potential range of responses within and among populations to varied and varying habitats and situations is so great that we must critique the notion of adaptation to a monotypic ancestral environment. However, this is not to say that plastic responses in all combinations have similarly beneficial (or even neutral) outcomes. The data we present here, based on a detailed examination of health and dietary indicators from populations on a relatively small geographic and temporal scale, at a key juncture in human history, indicate that modest changes in behaviour may synergise to create adverse health outcomes. In this way, understanding the ecology of human populations may be as important as evolution in appreciating the trajectory of human health. Using the palaeopathological record to identify small scale differences in health among populations sheds light on these ecological differences. The take-home message from the work described here is that, although an evolutionary perspective over a long time-span provides undeniably important context, we must not overlook the impact of changes on a small, proximate scale that affects the health of individuals. For the most part, it is these proximate effects on individuals that are the subject of treatment by clinicians, and it is hard to see how this could be altered substantially without detriment to health and well-being. One key challenge for the future is thus to link proximate and evolutionary perspectives to improve our understanding of health in the past and to influence health status positively in contemporary populations.

References

Aiello, L. C. & Wheeler, P. (1995). The expensive tissue hypothesis: the brain and the digestive system in human and primate evolution. *Current Anthropology*, 36, 199–221.

Beall, C. M. (2007). Two routes to functional adaptation: Tibetan and Andean high-altitude natives. *Proceedings of the National Academy of Sciences of the United States of America*, 104(Suppl. 1), 8655–8660. doi: 10.1073/pnas.0701985104

Binford L. R. (1968). 'Post-Pleistocene adaptations', in S. R. Binford & L. R. Binford (eds), *New perspectives in archaeology*, pp 313–341. Aldine, Chicago.

Boaretto, E. Wu, X., Yuan, J., Bar-Yosef, O., Chu, V., Pan, Y., Liu, K., Cohen, D., Jiao, T., Li, S., Gu, H., Goldberg, P. & Weiner, S. (2009). Radiocarbon dating of charcoal and bone collagen associated with early pottery at Yuchanyan Cave, Hunan Province, China. *Proceedings of the National Academy of Sciences of the United States of America*, 106, 9595–9600.

Bond, G., Showers, W., Cheseby, M., Lotti, R., Almasi, P., De Menocal, P., Priore, P., Cullen, H., Hajdas, I. & Bonani, G. (1997). A pervasive millennial scale cycle in the North Atlantic Holocene and glacial climates. *Science*, 294, 2130–2136.

Bonsall, C., Lennon, R. J., McSweeney, K., Stewart, C., Harkness, D. D., Boroneant, V., Payton, R. W., Bartosiewicz, L. & Chapman, J. C. (1997). Mesolithic and Early Neolithic in the Iron Gates: a palaeodietary perspective. *Journal of European Archaeology*, 5(1), 50–92.

Bonsall, C., Macklin, M. G., Boroneant, A., Pickard, C., Bartosiewicz, L., Cook, G. T. & Higham, T. F. G. (2015). Holocene climate change and prehistoric settlement in the lower Danube valley. *Quaternary International*, 378, 14–21.

Brown, T. A., Jones, M. K., Powell, W. & Allaby, R. G. (2009). The complex origins of domesticated crops in the Fertile Crescent. *Trends in Ecology and Evolution*, 24, 103–109.

Budd, C. & Lillie, M. C. (2020). 'The prehistoric populations of Ukraine: stable isotope studies of fisher-hunter-forager and pastoralist—incipient farmer dietary pathways', in M. C. Lillie & I. D. Potekhina (eds), *Prehistoric Ukraine: from the first hunters to the first farmers*, pp. 283–307. Oxbow Books, Oxford. ISBN: 9781789254587

Budd, C. & Lillie, M. C. in prep. Yuzhnyi Olenii Ostrov, Lake Onega, Karelia: first diet isotope analysis of the Mesolithic fisher-hunter-forager population. Antiquity.

Budd, C. E., Potekhina, I. D. & Lillie, M. C. (2020). Continuation of fishing subsistence in the Ukrainian Neolithic: diet isotope studies at Yasinovatka, Dnieper Rapids. *Archaeological and Anthropological Sciences*, 12(2), article id. 64. https://doi.org/10.1007/s12520-020-01014-4

Cohen, M. & Armelagos, G. (eds). (1984). *Palaeopathology at the origins of agriculture*. Academic Press, London.

Cohen, M. N., & Crane-Kramer, G. M. M. (2007). *Ancient health: skeletal indicators of agricultural and economic intensification*. University Press of Florida, Gainesville.

Cook, G., Bonsall, C., Pickard, C., McSweeney, K., Bartosiewicz, L. & Boroneant, A. (2009). 'The Mesolithic–Neolithic transition in the Iron Gates, Southeast Europe: calibration and dietary issues', in P. Crombé, M. Van Strydonck, J. Sergant, M. Bats & M. Boudin (eds), *Chronology and evolution within the Mesolithic of northwest Europe*, pp. 497–515. Cambridge Scholars Publishing, Newcastle Upon Tyne.

Cook, G. T., Bonsall, C., Hedges, R. E. M., McSweeney, K., Boroneant, V., Bartosiewicz, L. & Pettitt, P. (2002). Problems of dating human bones from the Iron Gates. *Antiquity*, 76, 77–85.

Cook, G. T., Bonsall, C., Hedges, R. E. M., McSweeney, K., Boroneant, V. & Pettitt, P. B. (2001). A freshwater diet-derived ^{14}C reservoir effect at the Stone Age Sites in the Iron Gates gorge. *Radiocarbon*, 43, 453–460.

Cordain, L., Brand Miller, J., Eaton, S. B., Mann, N., Holt, S. H. A. & Speth, J. D. (2000). Plant-animal subsistence ratios and macronutrient energy estimations in world-wide hunter-gatherer diets. *American Journal of Clinical Nutrition*, 71, 682–692.

DeMenocal, P. B. (1995). Plio-Pleistocene African climate. *Science*, 270, 53–59.

Dennell, R. (1983). *European economic prehistory: A new approach*. Academic Press, London.

Dobney, K. & Larson, G. (2006). Genetics and animal domestication: new windows on an elusive process. *Journal of Zoology*, 269, 261–271.

Dolukhanov, P. M. (1997). The Pleistocene–Holocene transition in Northern Eurasia: environmental changes and human adaptations. *Quaternary International*, 41/41, 181–191.

Dolukhanov, P. M. (2017). 'Postglacial human colonisation of Fennoscandia', in V. M. Kotlyakov, A. A. Velichko & S.A. Vasil'ev (eds), *Human colonisation of the Arctic: the interaction between early migration and the palaeoenvironment*, pp. 7–14. Academic Press, London.

Dolukhanov, P. M. (1984). 'Upper Pleistocene and Holocene cultures of the Russian Plain and Caucasus: ecology, economy and settlement pattern', in F. Wendorf & A. E. Close (eds), *Advances in world archaeology: volume 1*, pp. 323–58. Academic Press, New York.

Dolukhanov, P. M. & Khotinsky, N. A. (1984). 'Human cultures and the natural environment in the USSR during the Mesolithic and Neolithic', in H. E. Wright & C. W. Barnosky (eds), *Late Quaternary environments of the Soviet Union*. pp. 319–27. University of Minnesota Press, Minneapolis.

Eaton, S. B. & Eaton, S. B. I. (1999). 'The evolutionary context of chronic degenerative diseases', in S. C. Stearns (ed), *Evolution in health and disease*, pp. 251–259. Oxford University Press, Oxford.

Eaton, S. B., Eaton, S. B. I. & M. J. Konner. (1999). 'Palaeolithic nutrition revisited', in W. R. Trevathan, E. O. Smith & McKenna, J. J. (eds), *Evolutionary medicine*, pp. 313–332. Oxford University Press, Oxford.

Eaton, S. B. & Konner, M. (1985). Paleolithic nutrition. A consideration of its nature and current implications. *New England Journal of Medicine*, 312, 283–289.

Elton, S. (2008). 'Environments, adaptation, and evolutionary medicine: should we be eating a Stone Age diet?', in S. Elton & P. O'Higgins (eds), *Medicine and evolution: current applications, future prospects*, pp. 9–33. CRC Press, Boca Raton.

Elton, S. (2016). Mismatch or mass marketing? Stone Age diets, industrialisation and ultra-processed foods.

Toplum ve Hekim (Community and Physician), 31(5), 323–333.

Eriksson, G., Lõugas, L. & Zagorska, I. (2003). Stone Age hunter-fisher-gatherers at Zvejnieki, northern Latvia: radiocarbon, stable isotope and archaeozoology data. *Before Farming*, 2003/1(2), 1–25.

Flannery, K. V. (1969). Origins and ecological effects of early domestication in Iran and the Near East. in P. J. Ucko & G. W. Dimbleby (eds), *The domestication and exploitation of plants and animals*, pp. 73–100. Aldine, Chicago.

Flohr, P., Fleitmann, D., Matthews, R., Matthews, W. & S. Black. (2016). Evidence of resilience to past climate change in Southwest Asia: early farming communities and the 9.2 and 8.2 ka events. *Quaternary Science Reviews*, 136, 23–39.

Foley, R. A. (1995). The adaptive legacy of human evolution: a search for the environment of evolutionary adaptedness. *Evolutionary Anthropology*, 4, 194–203.

Foley, R. A. (1993). 'The influence of seasonality on hominid evolution', in S. J. Ulijaszek & S. Strickland (eds), *Seasonality and human ecology*, pp. 17–37. Cambridge University Press, Cambridge.

Goodman, A. H. (1993). On the interpretation of health from skeletal remains. *Current Anthropology*, 34(3), 281–288.

Goodman, A. H., Allen, L. H., Hernandez, G. P., Amador, A., Arriola, L. V., Chavez, A. & Pelto, G. H. (1987). Prevalence and age at development of enamel hypoplasias in Mexican children. *American Journal of Physical Anthropology*, 72, 7–19.

Goodman, A. H. & Armelagos, G. J. (1985a). The chronological distribution of enamel hypoplasias: a comparison of permanent incisor and canine patterns. *Archives of Oral Biology*, 30, 503–507.

Goodman, A. H. & Armelagos, G. J. (1985b). Factors affecting the distribution of enamel hypoplasias within the human permanent dentition. *American Journal of Physical Anthropology*, 68, 479–493.

Goodman, A. H., Armelagos, G. J. & Rose, J. C. (1984a). The chronological distribution of enamel hypoplasias from prehistoric Dickson Mounds populations. *American Journal of Physical Anthropology*, 65, 259–266.

Goodman, A. H., Martin, D. L. & Armelagos, G. J. (1984b). 'Indications of stress from bones and teeth', in M. N. Cohen, & G. J. Armelagos (eds.), *Palaeopathology at the origins of agriculture*. pp. 13–49. Academic Press, Inc., Orlando.

Gowlett, J. A. J. (2003). What actually was the Stone Age Diet? *Journal of Nutritional & Environmental Medicine*, 13(3), 143–147. doi: 10.1080/13590840310001619338

Hengen, O. P. (1971). Cribra orbitalia: pathogenesis and probable etiology. *Homo* 22:57–75.

Hillson, S. W. (1979). Diet and dental disease. *World Archaeology* 11(2):147–162.

Hurtado, A. M., Frey, M., Hurtado, I., & J. Baker. (2008). 'The role of helminths in human evolution', in S. Elton & P. O'Higgins (eds), *Medicine and evolution: current applications, future prospects*, pp. 83–97. CRC, Boca Raton.

Jacobs, K. (1993). Human postcranial variation in the Ukrainian Mesolithic-Neolithic. *Current Anthropology*, 34, 311–324.

Kent, S. (1986). The influence of sedentism and aggregation on porotic hyperostosis and anaemia: a case study. *Man (N.S.)*, 21, 605–636.

Khotinsky, N. A. (1984). 'Holocene vegetation history', in H. E. Wright & C. W. Barnosky (eds), *Late Quaternary environments of the Soviet Union*. pp. 179–200. University of Minnesota Press, Minneapolis.

Klepinger, L. L. (1992). 'Innovative approaches to the study of past human health and subsistence strategies', In S. R. Saunders & M. A. Katzenberg (eds), *Skeletal biology of past peoples: research methods*, pp. 121–30. Wiley Liss, New York.

Kobashi, T., Severinghaus, J. P., Brook, E. J., Barnola, J-M. & Grachev, A. M. (2007). Precise timing and characterization of abrupt climate change 8200 years ago from air trapped in polar ice. *Quaternary Science Reviews*, 26, 1212–1222.

Kotova, N. & Makhortykh, S. (2010). Human adaptation to past climate changes in the northern Pontic steppe. *Quaternary International*, 220(1-2), 88–94.

Kremenettski, C. V. (1995). Holocene vegetation and climate history of southwestern Ukraine. *Review of Palaeobotany and Palynology*, 85, 289–301.

Larsson, L., Nilsson Stutz, L., Zagorska, I., Bērziņš, V. & Cerina, A. (2017). New aspects of the Mesolithic—Neolithic cemeteries and settlement at Zvejnieki, Northern Latvia. *Acta Archaeologica*, 88(1), 57–93.

Lillie, M. C. (1998b). Cranial surgery dates back to Mesolithic. *Nature*, 391, 854.

Lillie, M. C. (1998a). The Dnieper Rapids region of Ukraine: a consideration of chronology, dental pathology and diet at the Mesolithic-Neolithic transition. Unpublished PhD thesis, Sheffield University, Sheffield.

Lillie, M. C. (2015). *Hunters, fishers and foragers in Wales: towards a social narrative of Mesolithic lifeways*. Oxbow Books, Oxford.

Lillie, M. C. (2020). 'Palaeopathology of the prehistoric populations of Ukraine', in M. C. Lillie and I. D. Potekhina (eds), *Prehistoric Ukraine: from the first hunters to the first farmers*. pp. 235–282. Oxbow Books, Oxford.

Lillie, M.C., Budd, C. & Potekhina, I. D. 2020. 'Radiocarbon dating of sites in the Dnieper Region and western Ukraine', in M. C. Lillie & I. D. Potekhina (eds), *Prehistoric Ukraine: from the first hunters to the first farmers*, pp. 187–233. Oxbow Books, Oxford.

Lillie, M. C, Budd, C. & Potekhina, I. D. (2011). Stable isotope analysis of prehistoric populations from the cemeteries of the Middle and Lower Dnieper Basin, Ukraine. *Journal of Archaeological Science*, 38(1), 57–68.

Lillie, M. C., Budd, C. E., Potekhina, I. D. & Hedges, R. E. M. (2009). The radiocarbon reservoir effect: new evidence from the cemeteries of the Middle and Lower Dnieper Basin, Ukraine. *Journal of Archaeological Science*, 36, 256–264.

Lillie, M. C. & Jacobs, K. (2006). Stable isotope analysis of fourteen individuals from the Mesolithic cemetery of Vasilyevka II, Dnieper Rapids region, Ukraine. *Journal of Archaeological Science*, 33, 880–886.

Lillie, M. C. & Richards, M. P. (2000). Stable isotope analysis and dental evidence of diet at the Mesolithic-Neolithic transition in Ukraine. *Journal of Archaeological Science*, 27, 965–972.

Lillie, M. C., Richards, M. P. & Jacobs, K. (2003). Stable isotope analysis of twenty-one individuals from the epipalaeolithic cemetery of Vasilyevka III, Dnieper Rapids region, Ukraine. *Journal of Archaeological Science*, 30, 743–752.

Macchiarelli, R. (1989). Prehistoric 'fish-eaters' along the eastern Arabian coasts: dental variation, morphology, and oral health in the Ra's al-Hamra community (Qurum, Sultanate of Oman, 5th-4th millennia BCE). *American Journal of Physical Anthropology*, 78, 575–594.

Mathieson, I., Alpaslan-Roodenberg, S., Posth, C., Szécsényi-Nagy, A., Rohland, N., Mallick, S., Olalde, I., Broomandkhoshbacht, N., Candilio, F., Cheronet, O., Fernandes, D., Ferry, M., Gamarra, B., González Fortes, G., Haak, W., Harney, E., Jones, E., Keating, D., Krause-Kyora, B., Kucukkalipci, I., Michel, M., Mittnik, A., Nägele, K., Novak, M., Oppenheimer, J., Patterson, N., Pfrengle, S., Sirak, K., Stewardson, K., Vai, S., Alexandrov, S., Alt, K. W., Andreescu, R., Antonović, D., Ash, A., Atanassova, N., Bacvarov, K., Balázs Gusztáv, M., Bocherens, H., Bolus, M., Boroneant, A., Boyadzhiev, Y., Budnik, A., Burmaz, J., Chohadzhiev, S., Conard, N. J., Cottiaux, R., Čuka, M., Cupillard, C., Drucker, D. G., Elenski, N., Francken, M., Galabova, B., Ganetsovski, G., Gély, B., Hajdu, T., Handzhyiska, V., Harvati, K., Higham, T., Iliev, S., Janković, I., Karavanić, I., Kennett, D. J., Komšo, D., Kozak, A., Labuda, D., Lari, M., Lazar, C., Leppek, M., Leshtakov, K., Lo Vetro, D., Los, Ivaylo Lozanov, D., Malina, M., Martini, F., McSweeney, K., Meller, H., Mendušić, M., Mirea, P., Moiseyev, V., Petrova, V., Price, T. D., Simalcsik, A., Sineo, L., Šlaus, M., Slavchev, V., Stanev, P., Starović, A., Szeniczey, T., Talamo, S., Teschler-Nicola, M., Thevenet, C., Valchev,

I., Valentin, F., Vasilyev, S., Veljanovska, F., Venelinova, S., Veselovskaya, E., Viola, B., Virag, C., Zaninović, J., Zäuner, S., Stockhammer, P. W., Catalano, G., Krauß, R., Caramelli, D., Zariṇa, G., Gaydarska, B., Lillie, M. C., Nikitin, A. G., Potekhina, I. D., Papathanasiou, A., Borić, D., Bonsall, C., Krause, J., Pinhasi, R. & Reich, D. (2018). The genomic history of Southeastern Europe. *Nature*, 555, 197–203. https://doi.org/10.1038/nature25778

Meiklejohn, C. & Zvelebil, M. (1991). 'Health status of European populations at the agricultural transition and the implications for the adoption of farming', in H. Bush & M. Zvelebil (eds), *Health in past societies: biocultural interpretations of human skeletal remains in archaeological contexts*, pp. 129–144. Tempus Reparatum, Oxford.

Mensforth, R. C., Lovejoy, C. O., Lallo, J. & Armelagos, G. J. (1978). The role of constitutional factors, diet, and infectious disease in the etiology of porotic hyperostosis and periosteal reactions in prehistoric infants and children. *Medical Anthropolology* 2: 1–59.

Motuzaite Matuzeviciute, G. (2020). 'The adoption of agriculture: archaeobotanical studies and the earliest evidence for domesticated plants', in M. C. Lillie & I. D. Potekhina (eds), *Prehistoric Ukraine: from the first hunters to the first farmers*, pp. 309–325. Oxbow Books, Oxford.

Nikitin, A. (2020). 'The genetic landscape of Ukraine from the Early Holocene to the Early Metal ages', in M. C. Lillie & I. D. Potekhina (eds), *Prehistoric Ukraine: from the first hunters to the first farmers*, pp. 327–339. Oxbow Books, Oxford.

Nuzhnyi, D. 1990. 'Projectile damage on Upper Palaeolithic microliths and the use of the bow and arrow among Pleistocene hunters in the Ukraine', in B. Gräslund (ed), *The Interpretative Possibilities of Microwear Studies. Proceedings of the International conference of lithic use-wear analysis*. Uppsala, Sweden, 15-17th February. pp. 113–24.

Nuzhnyi, D. (1989). L'utilisation des microlithes géométriques et non géométriques comme armatures de projectiles. *Bulletin de la Société Préhistorique Française*, 86, 88–96.

Pletterbauer F., Melcher A. & Graf, W. (2018). 'Climate change impacts in riverine ecosystems', in S. Schmutz & J. Sendzimir (eds), *Riverine ecosystem management*, pp. 203–223. Aquatic Ecology Series, vol 8. Springer Open, Cham.

Potts, R. (1998). Variability selection in hominid evolution. *Evolutionary Anthropology*, 7, 81–96.

Revedin, A., Aranguren, B., Becattini, R., Longo, L., Marconi, E., Lippi, M. M. & Svoboda, J. (2010). Thirty-thousand-year-old evidence of plant food processing. *Proceedings of the National Academy of Sciences of the United States of America*, 107, 18815–18819. http://doi.org/10.1073/pnas.1006993107

Roberts, C. A. (2015) 'What did agriculture do for us? The bioarchaeology of health and diet', in G. Barker & C. Goucher (eds), *The Cambridge world history. volume 2: A world with agriculture*, 12, 000 bce-500 CE, pp. 93–123. Cambridge University Press, Cambridge.

Roffet-alque, M., Marciniak, A., Valdes, P. J., Pawłowska, K., Pyzel, J., Czerniak, L., Krüger, M., Roberts, C. N., Pitter, S. & Evershed, R. P. (2018). Evidence for the impact of the 8.2-kyBP climate event on Near Eastern early farmers. *Proceedings of the National Academy of Sciences of the United States of America*, 115(35), 8705–8709.

Seow, W. K. (1997). Clinical diagnosis of enamel defects: Pitfalls and practical guidelines. *International Dental Journal*, 47, 173–182.

Seow, W. K. (2014). Developmental defects of enamel and dentine: challenges for basic science research and clinical management. *Australian Dental Journal*, 59(1 Suppl),143–154.

Stiner, M. C. (2001). Thirty years on the 'Broad Spectrum Revolution' and Paleolithic demography. *Proceedings of the National Academy of Sciences of the United States of America*, 98, 6993–6996. doi: 10.1073/pnas.121176198.

Strassmann, B. I. & Dunbar, R. I. M. (1999). Human evolution and disease: putting the Stone Age in perspective', in S. C. Stearns (ed), *Evolution in Health and Disease*, pp. 91–101. Oxford University Press, Oxford.

Stuart-Macadam, P. (1991). Anaemia in Roman Britain: Poundbury Camp', in H. Bush & M. Zvelebil (eds), *Health in past societies: biocultural interpretations of human skeletal remains in archaeological contexts*, pp. 101–113. British Archaeological Reports, International Series 567. Tempus Reparatum: Oxford.

Stuart-Macadam, P. (1989). 'Nutritional deficiency diseases: a survey of scurvy, rickets and iron deficiency anemia', in M. Y. Iscan, & K. A. R. Kennedy (eds), *Reconstruction of life from the skeleton*, pp. 201–22. Alan Liss, New York.

Stuart-Macadam, P. (1992). Porotic hyperostosis: a new perspective. *American Journal of Physical Anthropology*, 87, 39–47.

Telegin, D. Ya. & Potekhina, I. D. (1987). *Neolithic cemeteries and populations in the Dnieper Basin*. BAR: Oxford.

Thomas, E. R., Wolff, E. W., Mulvaney, R., Steffensen, J. P., Johnsen, S. J., Arrowsmith, C., White, J. W. C., Vaughn, B. & Popp, T. (2007). The 8.2 ka event from Greenland ice cores. *Quaternary Science Reviews*, 26, 70–81.

Turner, B. L. & Thompson, A. L. (2014). Beyond the Paleolithic prescription: authors' reply to Commentary. *Nutrition Reviews*, 72, 287–288. http://doi.org/10.1111/nure.12113

Turner, B. L. & Thompson, A.L. (2013). Beyond the Paleolithic prescription: incorporating diversity and flexibility in the study of human diet evolution. *Nutrition Reviews*, 71, 501–510. doi: 10.1111/nure.12039.

Ulijaszek, S. J., Mann, N. & Elton, S. (2012). *Evolving human nutrition: implications for public health.* Cambridge University Press, Cambridge.

Velichko, A. A. (1973). *Natural process during the Pleistocene.* Nauka, Moscow.

Velichko, A. A., Kurenkova, E. I. & Dolukhanov, P.M. (2009). Human socio-economic adaptation to environment in Late Palaeolithic, Mesolithic and Neolithic Eastern Europe. *Quaternary International*, 203(1-2), 1–9.

Wainwright, J. & Ayala, G. (2019). Teleconnections and environmental determinism: Was there really a climate-driven collapse at Late Neolithic Çatalhöyük? *Proceedings of the National Academy of Sciences of the United States of America*, 116(9), 3343–3344.

Walker, P. L. (1986). Porotic hyperostosis in a marine-dependent Californian Indian population. *American Journal of Physical Anthropology*, 69, 345–354.

Walker, P. L., Barthurst, R. R., Richman, R., Gjedrum, T. & Andrushko, V. A. (2009). The causes of porotic hyperostosis and cribra orbitalia: a reappraisal of the iron-deficiency-anemia hypothesis. *American Journal of Physical Anthropology*, 139, 109–125.

Winder, I. C., Devès, M. H., King, G. C. P., Bailey, G. N., Inglis, R. H. & Meredith-Williams, M. G. (2015). Evolution and dispersal of the genus Homo: a landscape approach. *Journal of Human Evolution*, 87, 48–65. doi: https://doi.org/10.1016/j.jhevol.2015.07.002

Wu, X., Zhang, C., Goldberg, P., Cohen, D., Pan, Y., Arpin, T. & Bar-Yosef, O. (2012). Early pottery at 20,000 years ago in Xianrendong Cave, China. *Science*, 29, 1696–1700.

Zaliznyak, L. (2020b). 'Landscape change, human-landsca-pe interactions and societal developments in the Mesolithic period', in M. C. Lillie & I. D. Potekhina (eds), *Prehistoric Ukraine: from the first hunters to the first farmers*, pp. 87–110. Oxbow Books, Oxford.

Zaliznyak, L. (2020a). 'The Upper Palaeolithic', in M. C. Lillie & I. D. Potekhina (eds), *Prehistoric Ukraine: from the first hunters to the first farmers*, pp. 63–85. Oxbow Books, Oxford.

Zilhão, J., Angelucci, D. E., Araújo Igreja, M., Arnold, L. J., Badal, E., Callapez, P., Cardoso, J.L., d'Errico, F., Daura, J., Demuro, M., Deschamps, M., Dupont, C., Gabriel, S., Hoffmann, D. L., Legoinha, P., Matias, H., Monge Soares, H. M., Nabais, M., Portela, P., Queffelec, A., Rodrigues, F. & Souto, P. (2020). Last interglacial Iberian Neandertals as fisher-hunter-gatherers. *Science*, 367, eaaz7943.

Zvelebil, M. (1994). Plant use in the Mesolithic and its role in the transition to farming. *Proceedings of the Prehistoric Society*, 60(1), 35–74.

Zvelebil, M. & Dolukhanov, P. M. (1991). The transition to farming in eastern and northern Europe. *Journal of World Prehistory*, 5(3), 233–278.

CHAPTER 8

Human resistance and the evolution of plague in Medieval Europe

Kirsten Bos and Sharon N. DeWitte

8.1 Plague in the past and today

Epidemics of plague, including the well-known fourteenth-century 'Black Death' and its persistence in human populations today, present excellent opportunities to examine several phenomena of interest to the field of evolutionary medicine (EM). Some examples include examination of both proximate and ultimate causes of vulnerability to disease, discerning microevolutionary processes related to human disease and coevolution between humans and pathogens, assessing the interaction of disease with developmental plasticity and life history trade-offs, analysis of pathogen phylogeny, and determination of the effects of cultural context on disease experiences and outcomes (Gluckman et al., 2011; Grunspan et al., 2018; Kuzawa, 2008; Muehlenbein, 2010; Trevathan, 2007; Trevathan et al., 2008).

Many historical plague outbreaks were of such an extraordinary scale (with respect to mortality, societal disruption, and other factors) that they are described (often meticulously) in contemporary records and were severe enough to have potentially produced measurable changes in human biology, demography, health, and evolution. The timing of historical plague epidemics is, for many locations, firmly established, and they were relatively brief events. These factors, combined with the growing number of excavated plague burial grounds (see Kacki, 2019) and expanding historiography of plague, allow anthropologists to examine conditions (at the genetic, individual, and population levels) before and after plague epidemics, providing

a degree of temporal resolution to the contexts of their emergence and consequences, which is not possible for many other diseases of interest. Leveraging these multiple lines of evidence in palaeopathological research on plague aligns with what Trevathan (2007, 140) describes as the unique anthropological contribution to EM, that is, 'a focus on the body as the result of evolutionary and developmental processes in specific sociocultural and socio-political contexts'.

Plague, the same 'Black Death' that once devastated Europe, is not a thing of the past: despite vaccination attempts and available antibiotic treatments, it persists on every continent inhabited by humans except Europe and Australia, and it remains responsible for extremely rapid mortality from sporadic individual infections to large-scale epidemics that require immediate measures of control. Plague is an acute infection of the Gram-negative bacterium *Yersinia pestis*, named after Alexandre Yersin, the scientist popularly credited with its first scientific description. This pathogen was the cause of many historical plague pandemics. Much of what we know regarding plague ecology and pathophysiology comes from work carried out at the height of an intense nineteenth- to early-twentieth-century outbreak of the disease that emerged in China and then spread worldwide, becoming a pandemic. We know that:

1. The disease is an acquired zoonotic infection from rodents where bacteria are transmitted

Kirsten Bos and Sharon N. DeWitte, *Human resistance and the evolution of plague in Medieval Europe*. In: *Palaeopathology and Evolutionary Medicine*. Edited by Kimberly A. Plomp et al., Oxford University Press. © Oxford University Press (2022). DOI: 10.1093/oso/9780198849711.003.0008 between individuals via flea and other ectoparasite bites.

2. Certain species of wild rodents are resistant to the bacterium and hence operate as disease 'reservoirs'.

3. The rodent species most closely associated with infection in humans, namely rats, are also fully susceptible to infection.

4. The bacterium has multiple clinical forms:
   - primarily pulmonary—an infection of the lungs;
   - septicaemic—an infection of the circulatory system; and
   - bubonic—an infection of the lymphatic system.
   - pharyngeal and gastrointestinal—infection of the pharynx and GI tract, respectively.

5. Each form differs in symptom presentation and prognosis.

Plague has become one of the most studied infectious diseases in both modern and ancient contexts, and important contributions have come from both the physical and molecular study of its victims. Its connections to world history have played heavily into its notoriety and into our understanding of how great a threat it can pose if left uncontrolled.

There has been a burgeoning production of genomic data from past and currently circulating strains of *Y. pestis*. Ancient DNA studies of medieval people themselves are growing in number, and though the data are currently limited, continuation of these studies may provide insights into genetic vulnerabilities to historical plague epidemics, the genetic changes produced by differential survival during pandemics (some of which may be relevant to understanding susceptibility to other diseases today), or the dynamics of coevolution between humans and *Y. pestis* (see also Bouwman & Rühli, 2016, regarding the application of palaeogenetics to EM).

8.1.1 History of the Black Death

The Black Death stands as one of the few past pandemics that continues to occupy the modern imagination. Though synonymous today with plague,

the term the 'Black Death' commonly refers not to the disease itself, but rather to a distinct event in Eurasia ~ 1346–1353 that was part of a larger Afro-Eurasian pandemic (Benedictow, 2004) and caused exceptionally high mortality, conventionally attributed to the onslaught of a single highly virulent pathogen. This term was not used by contemporaries, but rather was coined more recently to refer to the medieval pandemic (Varlik, 2021). According to Varlik (2021), while references to the blackness or bleakness of plague date to the medieval period, the first use of the German term 'der schwarze Tod', which translates to the Black Death, appears to date to the eighteenth century. Though some have argued that the term references dark marks that appeared on the skin of plague victims, Varlik (2021) argues the term might have a more symbolic meaning, i.e. 'black' as a reference to death and dying in general. Molecular evidence indicates that the strain that caused the Black Death emerged in the twelfth or thirteenth century (Bos et al., 2011; Spyrou et al., 2019), with East or Central Asia as its likely focus (Green, 2020). Under this scenario, the Black Death was the culmination of several decades of preceding outbreaks and westward spread of the disease in susceptible populations (Green, 2020; Syprou et al., 2019). For decades it was widely argued that the disease was initially introduced to the Mediterranean in 1347 by Italian merchants following their exposure during the siege of Caffa on the Crimean peninsula, during which Mongols of the Golden Horde (Ulus of Jochi) catapulted the bodies of plague victims over the city walls (Wheelis, 2002). However, recent evidence supports contaminated grain from the port of Tana on the Sea of Azov as the most likely source of the pandemic (Barker, 2021; Green, 2020). Having reached new fertile ground, the disease dispersed throughout Europe with incredible speed, killing an estimated 30–60% of affected populations before the outbreak subsided in late 1353 (Benedictow, 2004; DeWitte & Kowaleski, 2017; Green, 2015). Human experience with the disease was, however, far from over. The pestilence soon became a regular feature of the medieval and post-medieval worlds, returning as local outbreaks time and time again throughout Europe and parts of Western Eurasia until as late as the nineteenth century (referred to, in combination, as the Second Pandemic), after which it seemingly disappeared from these areas (Varlik, 2020).

8.1.2 Consequences of the Black Death

The extraordinarily high mortality caused by the Black Death produced or accelerated major social, economic and demographic changes. Rather than providing an exhaustive summary of all the work that has been done to date on the effects of the Black Death, we provide a few illustrative examples. In England, the Black Death produced a labour shortage, and the high demand for labour following the epidemic increased real wages relative to pre-Black Death conditions, while prices for food, goods and housing fell. This led to improvements in diet and housing conditions (Bailey, 1996; Dyer, 2002), and possibly improved general levels of health (see Section 8.3.3; similar post-pandemic trends have also been suggested for European populations affected by the Plague of Justinian; see Steckel et al., 2019). Improved economic conditions also likely contributed to increases in migration after the Black Death, particularly from rural to urban environments as young people sought varied employment opportunities and better wages (Dyer, 2002, 2005). There is also evidence that expanded employment opportunities and higher wages (particularly for women) in the aftermath of the Black Death motivated the adoption of the European Marriage Pattern in England and other parts of north-western Europe. This is characterised by, among other things, a relatively late average age at marriage (and thus late average ages at first reproduction) and high female celibacy, both of which might have contributed to reduced fertility rates following the Black Death (Bailey, 1996; de Moor & Van Zanden, 2010).

With respect to the economic repercussions of the epidemic, Egypt provides a contrasting situation. The primary driver of its economy was agricultural production dependent upon a labour-intensive irrigation system. Depopulation caused by the Black Death and subsequent outbreaks of plague severely disrupted and led to the decay of the irrigation system and ultimately caused a collapse of the economic infrastructure (Borsch, 2005, 2014). Borsch (2005) emphasises the importance of context. In England, the pre-Black Death condition of overpopulation and diminishing returns on agricultural production paved the way for improvements in standards of living following depopulation. This was not the case in Egypt, where the pre-Black Death conditions were essentially 'the reverse of the agrarian labor function for England' (Borsch, 2005, 53), leading to collapse in the face of depopulation. These two examples highlight the variable effects produced by the Black Death and demonstrate how highly those effects depended upon pre-existing conditions (see also the variation between England, the Netherlands, Italy and Spain detailed by Campbell, 2016). More work needs to be done to clarify how experiences of and outcomes of the Black Death varied by historical and cultural context.

8.1.3 Differences between the Second and Third Pandemics

Although genetic analyses confirm that *Y. pestis* impacted human health for millennia, sufficiently large, well-dated assemblages of the skeletal remains of known plague victims and/or historical records of the resolution needed for in-depth analysis are currently available for only the medieval/post-medieval outbreaks in the fourteenth to eighteenth centuries, inclusive of the Black Death (known as the Second Pandemic) and the nineteenth to twenty-first-century epidemics (including the Third Pandemic, which originated in China in the late nineteenth century and lasted until the ~ 1950s). The First Pandemic, the Plague of Justinian, started in ~ 500 AD.

The characteristics of the Second and Third Pandemics are far from identical, with existing documentary and clinical evidence indicating important epidemiological differences between them. One major difference is the much higher estimated mortality during the Black Death and several other outbreaks during the Second Pandemic compared to any outbreak that has occurred since the nineteenth century (the mortality rates we discuss here are the proportions of the entire at-risk population that died from plague, not case-fatality rates, which are proportions of people diagnosed with a disease who die from it). Many of the Second Pandemic outbreaks predate the regular, formal collection of vital rate data via birth, marriage and death registries. For example, the London Bills of Mortality only date from the late 1500s and were intermittent until 1603 (Ogle, 1892), that is, not long before the Great Plague of London in 1665, which is considered to be the end of the Second Pandemic in England (Appleby, 1980). However, other sources of data allow for assessment of plague mortality rates during the first centuries of the Second Pandemic. Much of the work done to date by historians to estimate these rates has focused on English populations because of the relatively high quality and variety of documentary sources that survive from the medieval and post-medieval periods in England (DeWitte & Kowaleski, 2017). These sources include records of heriots or death duties that were paid by peasants on a Manorial estate; mortality rates can be calculated by comparing the number of payments of heriot at the time of the Black Death to the total number of manorial tenants liable to heriot during the same period. Other sources of data include taxes (head-tax or tithing penny) paid by male tenants or serfs (agricultural labourers) above 12 years of age on some estates; mortality rates can be calculated by comparing the sums paid during plague years to those paid during pre-plague years (see details in DeWitte & Kowaleski, 2017).

Documentary evidence is not without flaws. For example, sources of evidence are not equally available from all regions in England, nor do they represent all individuals at risk of death in the affected populations. Heriot (and other) payments may produce underestimates of mortality rates given the efforts made by some people to avoid detection and payments (Hatcher, 1986). Nonetheless, analyses of these data have produced an average estimate of about 50% mortality for males during the Black Death in England; upward corrections have been made based on the assumption that mortality rates for women (who are disproportionately affected by plague in some (but not all) populations today), children or poor people— who are underrepresented or totally missing from existing documentary sources—would have been even higher (DeWitte & Kowaleski, 2017). Evidence

from heriot payments, wills and appointments of clergy to vacated benefices have produced estimates of mortality rates during the plague of 1361 in England in the range of 10–30% (DeWitte & Kowaleski, 2017).

It is possible to examine plague mortality rates directly for early twentieth-century plague outbreaks, rather than relying (as in the case of medieval plague) on proxies thereof. The relevant data for some of the earliest outbreaks come from volumes of the *Statistical Abstract Relating to British India* that list deaths by cause and province. In India in 1904, plague caused mortality rates of about 2–3% (Her Majesty's Stationary Office, 1905), which amounted to a very large number of deaths, but represents mortality in a much lower proportion of the at-risk population compared to the Black Death. Even the lowest estimates of mortality caused by the Black Death and other medieval and post-medieval plagues are many times higher than those of any outbreak during the Third Pandemic, which begs the question of whether such differences reflect something inherent to circulating strains of medieval *Y. pestis* that enhanced virulence. While examination of early 20th Century outbreaks allows us to control for the beneficial effects of antibiotic treatment on mortality levels during more recent plague outbreaks, this does not control for any changes since the post-medieval period in understandings of disease causation or transmission, hygiene, nutrition, or other factors that might have led to reductions in plague mortality. That is, there are myriad factors exogenous to the pathogen that can affect heterogeneity in risk of exposure to the pathogen, susceptibility to infection, and risk of death, and thus affect overall mortality rates. Thus, the question of why the Second Pandemic caused higher rates of mortality requires further study.

Another major epidemiological difference between the Second and Third Pandemic is the much faster rate of spread of the disease during the former. For example, the Black Death spread across Europe, from the Mediterranean to the British Isles and Scandinavia, in about six years, with it spreading throughout England in eighteen months at an average velocity of 1 km/day (see references in DeWitte & Kowaleski, 2017). In contrast, in the eighteenth and nineteenth centuries, it took over 100 years for plague to spread from western China to Hong Kong (Benedict, 1996). The highest estimated median spread velocity during the Third Pandemic is 41.04 km/year (Xu et al., 2019), or 0.11 km/day, a fraction of the velocity estimated for the Black Death in England. The rate of geographic spread of modern plague is determined in part by its primary mode of transmission via insect vectors.

Though it is often assumed that the Black Death and later medieval plagues were spread by rats and fleas, much still remains unknown about the transmission dynamics during this pandemic. Importantly, there is yet no conclusive evidence regarding the exact animal reservoir(s) and insect vector(s) that were involved. Many animals are natural reservoirs of plague and many ectoparasites can transmit the disease to humans, including the human louse. Determining which were involved in the spread of the Black Death would further our understanding of why the epidemic spread so much more quickly than is true of more recent plague outbreaks. Model-based research is yielding insights that might resolve debates about transmission of the Black Death (Dean et al., 2018). Genetic analyses of faunal assemblages contemporaneous to plague outbreaks might also yield insight into past models of transmission and ecology that go beyond human reservoirs, which dominate the narratives of current archaeological and historical research.

8.1.4 Why study plague in the past?

In addition to being fascinating in and of itself, the Black Death is worthy of study because of the continued persistence of plague in human populations today. A recent outbreak of plague in 2017 in Madagascar killed more than 200 people (of over 2000 cases) and was characterised by an unprecedentedly high proportion (77%) of pneumonic cases (the most fatal form of the disease) (Nguyen et al., 2018). The threat that this disease still poses motivates us to learn as much as possible about how plague has behaved in all of its historical and modern iterations and what factors have shaped plague epidemiology and its outcomes. Given the density of historical literature on this disease, inclusive of descriptions regarding early attempts at disease management, and the availability of skeletal remains of

the victims of historical plague epidemics, much can be learned about human vulnerability to and responses to epidemic threats, and the short- and long-term biological effects of such selective forces at the population level.

Although documentary sources provide a wealth of information about the Black Death, there are aspects of plague in the past that are inaccessible from historical evidence alone. Studies using biological evidence from the Black Death—human skeletal remains of plague victims and molecules of the pathogen contained therein—have the potential to yield information missing from historical documents that could shape planning and response strategies in the face of plague, and perhaps other infectious disease outbreaks. For example, research using the remains of people who died during the Black Death yields unique perspectives on the effects that health and nutritional status, including the long-term outcomes of developmental stressors (as predicted by the developmental origins of health and disease hypothesis, discussed in Chapter 2), had on risks of death during the epidemic, as described in Section 8.3.1. Understanding these phenomena can improve our understanding of the mechanisms driving individual- and population-level effects of the Black Death, and in turn might be leveraged in health promotion and disease prevention campaigns in ways that benefit people today. Accuracy in the use of such demographic information requires confirmation that plague was responsible for a given large-scale mortality event. For this, ancient DNA offers the highest resolution in detection of ancient plague infections.

8.2 Ancient DNA analyses of medieval human populations

8.2.1 Detecting plague in archaeological contexts

Recovery of molecular remnants of bacterial pathogens in archaeological human remains was pursued immediately following early indications that DNA can survive for prolonged periods in certain biological samples and be accessed via PCR (polymerase chain reaction) (Pääbo, 1989; Salo et al., 1994; Spigelman & Lemma, 1993). As the precise cause of the Black Death was a point of contention among many scholars (Cohn, 2002; Scott & Duncan 2001; Twigg, 1984), ancient DNA (aDNA) evidence was enthusiastically sought to provide some additional resolution on the pathogens that could be identified in victims of this mortality event. Early work reported on the successful recovery of *Y. pestis* DNA from the dental pulp chamber of victims of both the Black Death and the Great Plague of Marseille (1720–1722) (Drancourt et al., 1998; Raoult et al., 2000) via targeted exploration of known variable positions in the genome. Together these results were seen as support for the involvement of *Y. pestis* in the Black Death; however, as the pathogen still has a wide geographic distribution in sylvatic (wild rodent) reservoirs today and is studied in several laboratories around the world, the results are difficult to authenticate since the methods used could not unequivocally demonstrate that the recovered molecules were ancient.

Advancements in DNA sequencing introduced a dramatic change in the scale of recoverable data and the analyses that were possible with archaeological and archival (preserved soft tissue) biological samples. The feasibility and value of genomic-level investigations were demonstrated by work on early hominins (Green et al., 2015; Reich et al., 2010), and similar efforts directed toward ancient pathogens soon followed, with plague at the forefront. *Yersinia pestis* was the first ancient pathogen isolated from archaeological material to be investigated at the genomic level, which enabled a robust demonstration that the recovered DNA was ancient (Schuenemann, Bos et al., 2011) and permitted a confident evaluation of its relationship to the modern globally distributed genotypes of this pathogen (Bos, Schuenemann et al., 2011). Subsequent investigations soon followed that explored similar approaches in genome-level retrieval of other ancient human pathogens, which include bacteria (Bos et al., 2014; DeVault et al., 2014; Schuenemann et al., 2013, 2018), viruses (Duggan et al., 2016; Muhlemann et al., 2018), and now also eukaryotes such as the plasmodium responsible for certain types of malaria (Gelabert et al., 2018), all from a variety of time periods extending as far back as the Neolithic period (5000 years BP [before present]; Rascovan et al., 2019). Of these, plague remains the most

intensively studied infectious disease, likely due to the study of known plague-related burials based on documentary contexts in historical periods, its exceedingly high concentrations in the blood during acute fatal infection (Sebbane et al., 2005) that increase the chances of DNA survival in archaeological tissues and the abundance of knowledge made available through the prolific study of well over 100 modern strains (Cui et al., 2013).

8.2.2 Where has ancient plague DNA been found?

For decades it was assumed that the Plague of Justinian (sixth to eighth centuries AD) was humankind's first introduction to the scourge of *Y. pestis* infection, and aDNA work has since confirmed its involvement in infections from this time period (Feldman et al., 2016; Keller & Spyrou et al., 2019; Wagner et al., 2014; Weichmann & Grupe, 2004). Unexpected findings of plague in Neolithic and Bronze Age human remains from across Eurasia, however, have prompted revisions to long-standing established models of plague evolution and demonstrated the crucial role bioarchaeology can play in addressing questions of pathogen emergence and evolution that are central to studies of EM (Rasmussen et al., 2015; Spyrou et al., 2018; Valtueña et al., 2017). Identification of genes that seem to be uniquely missing in these deep temporal lineages led to the proposal that these early varieties had a decreased potential for transmission via a flea vector, the main way that plague is transmitted today (Rasmussen et al., 2015). The establishment of a transmission model in these time periods is further complicated by the observation that the evolution of plague pathogens followed the pattern of broad human population movements that characterise the time period (Valtueña et al., 2017). Taken at face value, this would imply that plague in the pre-historic periods was not the result of sporadic infections acquired via interactions with dispersed environmental reservoirs, but rather an indication that plague somehow travelled alongside their susceptible human (or associated mammalian) hosts in some capacity. These observations are in contrast to the current established models of plague ecology, where plague infections in humans and other mammals are contingent upon chance exposure to infected fleas from wild rodent reservoir species, and where epidemics are the result of spillover events related to climatically driven population booms and busts in natural reservoirs (Keeling & Gilligan, 2000). As some of the plague-associated Neolithic human populations followed the tradition of nomadic pastoralism, regular exposure to rodents to the extent of permitting pathogen co-evolution among the two hosts is unlikely, leaving open hypothetical scenarios of disease movement that differ from modern models of plague ecology.

The greatest density of ancient plague genomes in both time and space derives from contexts in medieval and post-medieval Europe. Characterisation of the *Y. pestis* strains associated with these outbreaks, which persisted until the mid-eighteenth century, has revealed a prolific lineage descended from the Black Death that accounts for all epidemics in Western Europe that postdate the 'pestis secunda' of 1358–1363 (Bos et al., 2016; Giffin et al., 2020; Guellil et al., 2021; Haensch et al., 2010; Morosova et al., 2020; Namouchi et al., 2018; Siefert et al., 2016; Spryou, Keller et al., 2019; Spyrou et al., 2016; Susat et al., 2020). While the predominance of Western European Second Pandemic plague lineages in current datasets is unquestionably the result of a sampling bias, their close genetic relationships are interpreted as evidence for a local reservoir of plague in Europe that became established following the Black Death and persisted until the dawn of the Industrial Era (Spyrou et al., 2016). A competing theory proposes repeated introductions of plague from reservoirs in East Asia, where movement of plague facilitated by climate-related expansions and collapses of wild rodent populations across Eurasia is thought to have fuelled the Second Pandemic mortality events of Western Europe (Schmid et al., 2015). Such multiple introductions of plague from a non-European source have also been argued based on genomic evidence, though in contexts that post-date the Black Death (Giffin, Lankapalli et al., 2020; Guellil et al., 2021). Eurasia houses several natural plague reservoirs today, each hosting a rather diverse collection of plague lineages (Eroshenko et al., 2017; Kutyrev et al., 2018). Further support for consideration of these areas as

source populations for the Second Pandemic of Western Europe could come from the identification of a lineage in skeletal remains closely related to one of the many lineages currently resident in Eurasia, most notably the Caucasus. Such a finding remains elusive, but this possibility nonetheless highlights the potential for bioarchaeology to contribute to one of the core principles of EM—tracing phylogenetic relationships among pathogens (Grunspan et al., 2018).

A comparison of the Medieval and modern outbreaks of plague presents several important inconsistencies and open questions. In the light of the focus of this volume, we will consider the three questions most relevant to EM:

1. Why was the scale of mortality from the Black Death so much higher than typically observed in modern outbreaks?
2. How did the plague spread so quickly during these periods?
3. Why did the plague disappear from Europe?

One potential explanation is that all of these phenomena are the result of genetic changes related to the bacterium's pathogenicity. Analysts have inspected ancient molecular data for genetic features unique to the medieval lineages that might explain such trends, but as yet few clues have surfaced. Medieval plague is nearly identical to modern plague, and the few genetic changes that exist seem not to affect portions of the genome that could influence the shape or function of a protein (Bos, Schuenemann et al., 2011).

Despite these findings, genetic changes that are potentially adaptive have been observed from genomes that derive from outbreaks toward the end of the Second Pandemic. Plague genotypes from both mid-seventeenth-century London and early-eighteenth-century France have revealed a common sizeable deletion of 50 kb in a genomic region that houses several ion transport channels and genes involved in motility (Spyrou et al., 2019). On its own, this observation could be evidence of a sporadic lineage-specific genomic deletion, which is a common feature of *Y. pestis* (Zhou & Yang, 2009). However, the observation of a parallel deletion in material from the waning period of the First Pandemic (Keller et al., 2019) encourages a more thorough evaluation of the deletion's biological effect, its adaptive significance, and in turn its potential role in the disappearance of plague from Europe in the terminal periods of both the First and the Second Pandemics.

8.2.3 What effect did the Black Death have on human genetic variation?

During the Black Death, millions of people died in Europe in less than ten years. For example, England alone experienced an estimated two million plague-related deaths during the epidemic (Campbell, 2016). Such prolific mortality events are rare in our history, and mortality of this scale from disease in a narrow temporal window leaves open the possibility of a selection event, making the epidemic an excellent potential case study for examining phenomena of interest to EM. Evidence for the event could be detected as a shift in allele frequency in immune-related genes across the population if certain genotypes have a differential influence over survivability. The human leukocyte antigen (HLA) locus has regulation of host immunity as its primary function and its status as the most polymorphic section of our genome is surely reflective of adaptive immune responses and selection events in our history. Malaria, for example, is suspected to have afflicted humans for millennia, as evidenced by adaptive signals at the HLA and other immune loci in populations resident in areas where the disease is relatively common (Garamzegi et al., 2014). Studies on the HLA that draw from archaeological samples are beginning to surface (Krause-Kyora et al., 2018), and this relatively new angle for understanding immune variation in the past may reveal important clues to adaptive changes in immune function in response to different selective pressures.

The immune system has a complex basis and functions via coordination of many systems outside of the HLA. It has been observed that the T-cell receptor mutation CCR5-Δ32, which confers resistance to cell invasion by the human immunodeficiency virus (HIV), has a frequency of 5–14% in areas hit by the Black Death. Regardless of this

association, selection for the genotype occurred far earlier than the emergence of HIV in humans. It has been proposed that another infectious disease, possibly plague, produced the selective pressure that led to the high frequency of the deletion in certain human groups (Stephens et al., 1998), although Galvani and Slatkin (2003) have proposed high mortality from other diseases of the medieval and post-medieval era, such as smallpox. A subsequent investigation, however, demonstrated no increased signal of selection at this locus when considered in the context of neutral variation across the human genome, and posited a first appearance of the allele earlier than 5000 years BP (Sabeti et al., 2005). Further PCR-based analyses of human skeletal samples have not yielded evidence of substantial changes in frequencies of the mutation before, during and after the Black Death (Bouwman et al., 2017; Hummel et al., 2005; Kremeyer et al., 2005), as would be expected if the Black Death played a primary role in driving patterns of currently observed frequencies. Detection of plague in skeletal remains as old as 5000 years indicates multiple opportunities for pandemics and host genetic selection in our history, where a dynamic relationship may have existed deeper in our evolutionary past.

In addition to plague as a selective pressure, the massive mortality caused by the Black Death might have created detectable declines in genetic diversity and is known to have led to increases in migration in some contexts (Kowaleski, 2014), both of which can produce changes in human genetic variation at the local level. Researchers have begun to examine genetic signals of these demographic events using human skeletal remains that pre- and post-dates the Black Death in Northern Europe, though thus far genetic continuity has been observed throughout these periods (Klunk et al., 2019).

8.3 Bioarchaeological analyses of plague

8.3.1 Why was the Black Death so deadly?

As mentioned (Section 8.1.4), it is not yet entirely clear why the Black Death and subsequent outbreaks during the Second Pandemic produced mortality rates that were much higher (i.e. ten or more times higher) than have been observed during the Third Pandemic. Previous bioarchaeological research has attempted to address this, specifically by determining whether medieval populations were vulnerable to the Black Death because of declining health in the face of repeated famines and increasing social inequality before the epidemic. Analyses of human skeletal remains from London, dated to the eleventh to thirteenth centuries, that are representative of all ages, sexes and high and low socio-economic status has revealed that survivorship (the proportion of people who survived to each age, typically measured in years) decreased in the first half of the thirteenth century compared to the eleventh and twelfth centuries (DeWitte, 2015, 2018). At the same time, frequencies of skeletal markers of developmental stress (linear enamel hypoplasia and short tibiae, as a proxy for short achieved adult stature), which reflect exposure to factors like malnutrition or infection during childhood or adolescence, increased significantly. In combination, these results suggest that health in general declined in the first half of the thirteenth century. Although these skeletal samples pre-date the Black Death by several decades, numerous severe subsistence crises occurred in England in the interim, such as a prolonged famine from 1256 to 1260 and the Great Famine in 1315–1317, each of which killed an estimated 10–15% of the population of England and Wales (Campbell, 2016). The decades leading up to the Black Death were also characterised by large increases in population size (relative to agricultural production as medieval famines would likely not have been frequent nor severe enough to curtail population growth, see Watkins & Menkin, 1985), reductions in the size of land holdings and attendant reductions in the ability of families to produce sufficient food to feed themselves, increased proportions of families living at or below the poverty line and periodic taxation necessitated by military operations that was particularly burdensome for poorer inhabitants of England (Campbell, 2016; Rigby, 2006). These trends suggest that the declines in health apparent in the London skeletal remains studied likely did not reverse prior to the Black Death. The cumulative effects of famine, increasing population pressure and increasing social inequality and poverty might have heightened vulnerability to infectious disease and led to high mortality levels during the Black Death. These findings, which are relevant to our understanding of proximate mechanisms shaping vulnerability to plague in the medieval period, might eventually be integrated with palaeogenetic data that are informative about ultimate (evolutionary) mechanisms.

8.3.2 Who died during the Black Death?

Since it killed so many people quickly, it is often assumed that the Black Death killed its victims irrespective of age, social status, gender and other biosocial categories. However, bioarchaeological and historical analyses of mortality patterns during the epidemic have challenged this assumption. Analyses of the skeletal remains of people who died in London during the Black Death, ~ 1349–1350, suggest that previous exposure to physiological stress (e.g. malnutrition and infection) increased risks of mortality, and that risks of death varied across age, at least among adults, during the epidemic (DeWitte, 2010; DeWitte & Hughes-Morey, 2012; DeWitte & Wood, 2008; Godde et al., 2020). Castex and Kacki (2016) describe consistent, very low proportions of infants in Black Death burial assemblages, but whether this reflects levels of infant mortality during the epidemic remains unclear. Thus far, bioarchaeological analyses have not discerned any difference in risk of Black Death mortality between the sexes (Castex & Kacki, 2016; DeWitte, 2009; Godde et al., 2020). However, Curtis and Roosen (2017) found evidence, via analysis of historical data from the Netherlands, that more females died during the Black Death and later outbreaks in the medieval period compared to periods of normal mortality, which suggests a sex-selective effect during the epidemics.

Importantly, from the perspective of EM, some of the stress markers used to detect selective effects of the Black Death represent interruptions in development and/or are reflective of aetiologies that affect individuals only early in life (e.g. cribra orbitalia, linear enamel hypoplasia, and short adult stature). Their association with elevated risks of mortality during the Black Death provides evidence of the later-life outcomes of adverse conditions (malnutrition or disease) during critical developmental periods, as predicted by the developmental origins of health and disease (DOHaD) hypothesis (Chapter 2). Humans are characterised by developmental plasticity, an adaptation that allows for survival in geographically or temporally variable ecological conditions, but survival at one age may come at a cost to health at later ages (Kuzawa, 2008; McDade et al., 2016). The very fact that adults who died during the Black Death had skeletal evidence of early life stressors indicates they were resilient enough to survive those stressors and thereby reach adulthood, but this survival might have necessitated trade-offs with respect to growth and development that produced long-term negative effects on their immune systems, ultimately increasing their vulnerability to plague (regarding immunity as an evolved reaction norm, see Muehlenbein, 2010).

Together, many bioarchaeological and palaeopathological studies suggest that the Black Death was selective, at least in some contexts, with respect to age, sex and/or pre-existing health condition. However, we should be cautious about overgeneralising these findings. A recent meta-analysis by Bramanti and colleagues (2018) highlights regional variation in medieval plague mortality patterns generated by bioarchaeological studies of medieval plague mortality patterns. Thus, more work needs to be done to assess general versus location-specific mortality patterns during the Black Death.

8.3.3 What effects did the Black Death have on health and demography?

Given its very high mortality levels and evidence that, at least in some contexts, the Black Death disproportionately killed frail people of all ages, the epidemic might have exerted either a short-term harvesting effect ('weeding out' the frailest people) or a longer-term selective effect (assuming some heritability of any of the underlying mechanisms or conditions producing frailty). In either case, it is possible that the Black Death led to improved average health in the surviving population, perhaps for multiple generations afterwards, by powerfully shaping population-level patterns of human biology or genetic variation. Moreover, as described earlier, there is historical evidence that standards of living improved in England as a result of the depopulation caused by the Black Death, and this might also have led to improved health in general.

Bioarchaeological analysis of human skeletal remains from London has revealed evidence of increased survivorship after the Black Death (~ 1350–1540) compared to conditions beforehand (~ 1200–1250) (DeWitte, 2014, 2018). Additionally, following the Black Death, frequencies of developmental stress markers declined significantly, at least for males. These patterns suggest that there were improvements in health, at least temporarily, after the Black Death in London. The exact mechanisms that might have produced these changes remain to be determined. Recent analysis of temporal trends in skeletal indicators of pubertal timing suggest that following the Black Death in London, there was a decrease in the average age at menarche and a decrease in achieved adult stature for females (DeWitte & Lewis, 2020). Evidence from living populations indicates that improvements in standards of living and decreases in disease burdens are associated with decreases in age at menarche (e.g. Cho et al., 2010; Ellis, 2004; Sohn, 2016) and in some contexts early age at menarche is associated with relatively short stature (e.g. Kang et al., 2019; Schooling et al., 2010). These findings might thus reflect substantial changes in nutritional status, disease burden and physiological (and perhaps even psychological) health after the Black Death that influenced pubertal timing (and by inference, reproduction), perhaps at the expense of female growth. More broadly, these findings demonstrate the potential for bioarchaeological research to contribute to our understanding of human life history trade-offs and of the adaptive nature of a reproductive system that is sensitive to environmental conditions (Gluckman et al., 2011; Trevathan, 2007).

Skeletal samples can also yield data that are informative about patterns of migration at the time of the Black Death. For example, via analysis of strontium and oxygen isotopes, Kendall and colleagues (2013) found that 16.7% of the individuals they analysed from a London Black Death cemetery were immigrants, with potential source populations ranging from the surrounding hinterlands to more distant regions of Britain. Analyses of samples from medieval London and various cities in Denmark have revealed that mitochondrial DNA diversity in these contexts was high before, during and after the Black Death, which might reflect consistent, high levels of female migration into cities before and after the epidemic (Klunk et al., 2019). These studies are at the forefront of discoveries about temporal trends in, the environmental and socioeconomic contexts of, and the consequences of migration at the time of major medieval demographic crises.

8.4 Conclusion

This chapter highlighted how bioarchaeological research, including aDNA analyses, has yielded important insights on medieval plague, but there is still much to learn about this disease that powerfully shaped human history and continues to affect people today. Improvements in molecular and bioarchaeological analytical methods and the integration of data from these conventionally disparate fields, allows us to address questions about the evolution of plague and about the mechanisms (e.g. genetic, epigenetic, pathophysiological, immunological, nutritional) that drove changes in health and demography in the medieval period in the context of plague epidemics.

The contributions of aDNA research to our understanding of plague has provided important details on its genomic changes over time and has revealed a temporal and geographic distribution that was inscrutable from historical documents alone. Techniques in ancient DNA acquisitions continue to improve, and DNA recovery from time periods (Meyer et al., 2016), locations (Gallego Llorente et al., 2015) and unusual sources (Jensen et al., 2019) continue to challenge our concept of temporal and climatic boundaries in relation to molecular preservation. Advances in laboratory processing have increased the number of archaeological samples that can be analysed in parallel, and sample sizes in the hundreds are becoming more common in ancient DNA publications (Mathieson et al., 2015; Valtueña et al., 2017). Such efforts open the possibility of broad pathogen screening of full cemeteries, which will yield a better understanding of plague epidemiology in the past, as well as provide quantitation of human immune variation over

time. Non-targeted methods of pathogen screening show great promise for the detection of unsuspected diseases in archaeological samples (Huebler et al., 2019; Wood & Salzberg, 2014), and these may one day reveal a spectrum of infections that circulated or co-circulated (Giffin, Lankapalli et al., 2020) in past human populations. Such investigations of disease ecology and the potential identification of co-morbidities in archaeological contexts become central in understanding the dynamics of human health in the past, especially with reference to the Black Death, the mortality and speed of travel of which differ radically from what is observed in modern pandemics of the same pathogen. The absence of a genetic signal to account for obvious differences in virulence between ancient and modern infections invites a more in-depth investigation of factors other than genetics that can account for such differences.

Bioarchaeological studies of plague can leverage data informative about diet (e.g. via carbon and nitrogen stable isotope analyses) and migration (e.g. via strontium and lead isotopes) to reveal the synergistic effects of malnutrition, host immune responses and exposure to biosocial stressors in multiple contexts across the life course to provide greater resolution on vulnerability or resilience to plague at the individual and population levels. Contextualising the results of aDNA studies within the larger framework of human population dynamics and social and economic contexts—information about which is accessible via bioarchaeological and historical analyses—not only improves our understanding of how and why medieval plague affected people in the past, but also will allow us to draw parallels between conditions that existed several hundred years ago and today. Ultimately, and ideally, a richer understanding of plague in the past has the potential to enhance interventions in living populations.

References

Appleby, A. B. (1980). The disappearance of plague: a continuing puzzle. *The Economic History Review*, 33(2), 161–173. doi: 10.1111/j.1468-0289.1980.tb01821.x

Bailey, M. (1996). T. S. Ashton Prize: joint winning essay. Demographic decline in late Medieval England:

some thoughts on recent research. *The Economic History Review*, 49(1), 1–19.

Barker, H. (2021). Laying the corpses to rest: grain, embargoes, and Yersinia pestis in the Black Sea, 1346–1348. *Speculum*, 96, 97–126. https://doi.org/10.1086/711596

Benedict, C. (1996). *Bubonic plague in nineteenth-century China*. Stanford University Press, Stanford.

Benedictow, O. J. (2004). *The Black Death, 1346–1353: the complete history*. Boydell & Brewer, London.

Borsch, S. (2005). *The Black Death in England and Egypt: A Comparative Study*. American University Press, Cairo.

Borsch, S. (2014). Plague depopulation and irrigation decay in Medieval Egypt. *The Medieval Globe*, 1, 125–156.

Bos, K. I., Harkins, K. M., Herbig, A., Coscolla, M., Weber, N., Comas, I., Forrest, S. A., Bryant, J. M., Harris, S. R., Schuenemann, V. J., Campbell, T. J., Majander, K., Wilbur, A. K., Guichon, R. A., Wolfe Steadman, D. L., Collins Cook, D., Niemann, S., Behr, M. A., Zumarraga, M., Bastida, R., Huson, D., Nieselt, K., Young, D., Parkhill, J., Buikstra, J. E., Gagneux, S., Stone, A. C. & Krause, J. (2014). Pre-Columbian mycobacterial genomes reveal seals as a source of New World human tuberculosis. *Nature*, 514, 494.

Bos, K. I., Herbig, A., Sahl, J., Waglechner, N., Fourment, M., Forrest, S. A., Klunk, J., Schuenemann, V. J., Poinar, D., Kuch, M., Golding, G. B., Dutour, O., Keim, P., Wagner, D. M., Holmes, E. C., Krause, J. & Poinar, H. N. (2016). Eighteenth-century genomes reveal the long-term persistence of an historical plague focus. *Elife [online]*, 5, e12994

Bos, K. I., Schuenemann, V. J., Golding, G. B., Burbano, H. A., Waglechner, N., Coombes, B. K., McPhee, J. B., DeWitte, S. N., Meyer, M., Schmedes, S., Wood, J., Earn, D. J. D., Herring, D. A., Bauer, P., Poinar, H. N. & Krause, J. (2011). A draft genome of Yersinia pestis from victims of the Black Death. *Nature*, 478, 506–510.

Bouwman, A. & Rühli, F. (2016). Archaeogenetics in evolutionary medicine. *Journal of Molecular Medicine*, 94(9), 971–977. https://doi.org/10.1007/s00109-016-1438-8

Bouwman, A., Shved, N., Akgül, G., Rühli, F. & Warinner, C. (2017). Ancient DNA investigation of a Medieval German cemetery confirms long-term stability of CCR5-Δ32 allele frequencies in Central Europe. *Human Biology*, 89(2), 119–124.

Bramanti, B., Zedda, N., Rinaldo, N. & Gualdi-Russo, E. (2018). A critical review of anthropological studies on skeletons from European plague pits of different epochs. *Scientific Reports*, 8(1), 1–12. doi: 10.1038/s41598-018-36201-w

Campbell, B. M. S. (2016). *The great transition: climate, disease and society in the late-medieval world*. Cambridge University Press, Cambridge.

Castex, D. & Kacki, S. (2016). Demographic patterns distinctive of epidemic cemeteries in archaeological samples', *Microbiology Spectrum [online]*, 4(4). doi: 10.1128/microbiolspec.PoH-0015-2015

Cho, G. J., Park, H. T., Shin, J. H., Hur, J. Y., Kim, Y. T., Kim, S. H., Lee, K. W. & Kim, T. (2010). Age at menarche in a Korean population: Secular trends and influencing factors. *European Journal of Pediatrics*, 169(1), 89–94. http://dx.doi.org.pallas2.tcl.sc.edu/10.1007/s00431-009-0993-1

Cohn, S. K., Jr. (2002). *The Black Death transformed: disease and culture in early Renaissance Europe*. Hodder Education, London.

Cui, Y., Yu, C., Yan, Y., Li, D., Li, Y., Jombart, T., Weinert, L. A., Wang, Z., Guo, Z., Xu, L., Zhang, Y., Zheng, H., Qin, N., Xiao, X., Wu, M., Wang, X., Zhou, D., Qi, Z., Du, Z., Wu, H., Yang, X., Cao, H., Wang, H., Wang, J., Yao, S., Rakin, A., Li, Y., Falush, D., Balloux, F., Achtman, M., Song, Y., Wang, J. & Yang, R. (2013). Historical variations in mutation rate in an epidemic pathogen, *Yersinia pestis*. *Proceedings of the National Academy of Sciences of the United States of America*, 110, 577–582.

Curtis, D. R. & Roosen, J. (2017). The sex-selective impact of the Black Death and recurring plagues in the Southern Netherlands, 1349–1450. *American Journal of Physical Anthropology*, 164(2), 246–259. doi: 10.1002/ajpa.23266.

de Moor, T. & Van Zanden, J. L. (2010). Girl power: the European marriage pattern and labour markets in the North Sea region in the late medieval and early modern period. *The Economic History Review*, 63(1), 1–33. doi: 10.1111/j.1468-0289.2009.00483.x

Dean, K. R., Krauer, F., Walløe, L., Lingjærde, O. C., Bramanti, B., Stenseth, N. C. & Schmid, B. V. (2018). Human ectoparasites and the spread of plague in Europe during the Second Pandemic. *Proceedings of the National Academy of Sciences of the United States of America*, 115(6), 1304–1309. doi: 10.1073/pnas.1715640115

DeVault, A. M., Golding, G. B., Waglechner, N., Enk, J. M., Kuch, M., Tien, J. H., Shi, M., Fisman, D. N., Dhody, A. N., Forrest, S., Bos, K. I., Earn, D. J. D., Holmes, E. C. & Poinar, H. N. (2014). Second-pandemic strain of *Virbrio cholerae* from the Philadelphia cholera outbreak of 1849. *New England Journal of Medicine*, doi: 10.1056/NEJMoa1308663

DeWitte, S. N. (2009). The effect of sex on risk of mortality during the Black Death in London, A.D. 1349–1350. *American Journal of Physical Anthropology*, 139(2), 222–234. doi: 10.1002/ajpa.20974

DeWitte, S. N. (2010). Sex differentials in frailty in medieval England. *American Journal of Physical Anthropology*, 143(2), 285–297. doi: 10.1002/ajpa.21316

DeWitte, S. N. (2014). Mortality risk and survival in the aftermath of the Medieval Black Death. *PLoS One*, 9(5), e96513. doi: 10.1371/journal.pone.0096513

DeWitte, S. N. (2015). Setting the stage for medieval plague: pre-Black Death trends in survival and mortality. *American Journal of Physical Anthropology*, 158(3), 441–451. doi: 10.1002/ajpa.22806.

DeWitte, S. N. (2010). Sex differentials in frailty in medieval England. *American Journal of Physical Anthropology*, 143(2), 285–297. doi: 10.1002/ajpa.21316.

DeWitte, S. N. (2018). Stress, sex, and plague: patterns of developmental stress and survival in pre- and post-Black Death London. *American Journal of Human Biology*, 30(1), e23073. doi: 10.1002/ajhb.23073

DeWitte, S. N. & Hughes-Morey, G. (2012). Stature and frailty during the Black Death: the effect of stature on risks of epidemic mortality in London, A.D. 1348–1350. *Journal of Archaeological Science*, 39(5), 1412–1419.

DeWitte, S. N. & Kowaleski, M. (2017). Black Death bodies. *Fragments: Interdisciplinary Approaches to the Study of Ancient and Medieval Pasts*, 6, 1–37.

DeWitte, S. N. & Lewis, M. (2020). Medieval menarche: changes in pubertal timing before and after the Black Death. *American Journal of Human Biology*, 33, e23439. https://doi.org/10.1002/ajhb.23439

DeWitte, S. N. & Wood, J. W. (2008). Selectivity of Black Death mortality with respect to preexisting health. *Proceedings of the National Academy of Sciences of the United States of America*, 105(5), 1436–1441. doi: 10.1073/pnas.0705460105.

Drancourt, M., Aboudharam, G., Signoli, M., Dutour, O. & Raoult, D. (1998). Detection of 400-year-old *Yersinia pestis* DNA in human dental pulp: an approach to the diagnosis of ancient septicemia. *Proceedings of the National Academy of Sciences of the United States of America*, 95, 12637–12640.

Duggan, A. T., Perdomo, M. F., Piombino-Mascali, D., Marciniak, S., Poinar, D., Emery, M. V., Buchmann, J. P., Duchêne, S., Jankauskas, R., Humphreys, M., Golding, G. B., Southon, J., Devault, A., Rouillard, J-M., Sahl, J. W., Dutour, O., Hedman, K., Sajantila, A., Smith, G. L., Holmes, E. C. & Poinar, H. N. (2016). 17th century variola virus reveals the recent history of smallpox. *Current Biology*, 26, 3407–3412

Dyer, C. (2005). *An age of transition? Economy and society in England in the Later Middle Ages*. Oxford University Press, Oxford.

Dyer, C. (2002). *Making a living in the Middle Ages: the people of Britain 850–1520*. Yale University Press, New Haven.

Ellis, B. J. (2004). Timing of pubertal maturation in girls: an integrated life history approach. *Psychological Bulletin*, 130(6), 920–958. https://doi.org/10.1037/0033-2909.130.6.920

Eroshenko G. A., Nosov, N. Y., Krasnov, Y. M., Oglodin, Y. G., Kukleva, L. M., Guseva, N. P., Kuznetsov, A. A., Abdikarimov, S. T., Dzhaparova, A. K. & Kutyrev, V. (2017). *Yersinia pestis* strains of ancient phylogenetic branch 0.ANT are widely spread in the high-mountain plague foci of Kyrgyzstan. *PLoS One*, 12(10): e0187230. doi: 10.1371/journal.pone.0187230

Feldman, M., Harbeck, M., Keller, M., Spyrou, M. A., Rott, A., Trautmann, B., Scholz, H. C., Päffgen, B., Peters, J., McCormick, M., Bos, K., Herbig, A. & Krause, J. (2016). A high-coverage *Yersinia pestis* genome from a sixth-century Justinian Plague victim. *Molecular Biology and Evolution*, 33, 2911–2923. doi.org/10.1093/molbev/msw170

Gallego Llorente, M., Jones, E. R., Eriksson, A., Siska, V., Arthur, K. W., Arthur, J. W., Curtis, M. C., Stock, J. T., Coltorti, M., Pieruccini, P., Stretton, S., Brock, F., Higham, T., Park, Y., Hofreiter, M., Bradley, D. G., Bhak, J., Pinhasi, R. & Manica, A. (2015). Ancient Ethiopian genome reveals extensive Eurasian admixture in Eastern Africa. *Science*, 350(6262), 820–822. doi:10.1126/science.aad2879

Galvani, A. P. & Slatin, M. (2003). Evaluating plague and smallpox as historical selective pressures for the CCR5-Δ32 HIV-resistance allele. *Proceedings of the National Academy of Sciences of the United States of America*, 100(25), 15276–15279.

Garamszegi, L. Z. (2014). Global distribution of malaria-resistant MHC-HLA alleles: the number and frequencies of alleles and malaria risk. *Malaria Journal*, 13, 349

Gelabert, P., Sandoval-Velasco, M., Olalde, I., Fregel, R., Rieux, A., Escosa, R., Aranda, C., Paaijmans, K., Mueller, I., Gilbert, M. T. P. & Lalueza-Fox, C. (2016). Mitochondrial DNA from the eradicated European *Plasmodium vivax* and *P. falciparum* from 70-year-old slides from the Ebro Delta in Spain. *Proceedings of the National Academy of Sciences of the United States of America*, 113(41), 11495–11500.

Giffin K., Lankapalli A. K., Sabin, S., Spyrou, M. A., Posth, C., Kozakaitė, J., Friedrich, R., Miliauskienė, Ž., Jankauskas, R., Herbig, A. & Bos, K. I. (2020). A treponemal genome from an historic plague victim supports a recent emergence of yaws and its presence in 15th century Europe. *Scientific Reports*, 10(1), 9499. doi: 10.1038/s41598-020-66012-x

Gluckman, P. D., Low, F. M., Buklijas, T., Hanson, M. A. & Beedle, A. S. (2011). How evolutionary principles improve the understanding of human health and disease. *Evolutionary Applications*, 4(2), 249–263. https://doi.org/10.1111/j.1752-4571.2010.00164.x

Godde, K., Pasillas, V. & Sanchez, A. (2020). Survival analysis of the Black Death: social inequality of women and the perils of life and death in Medieval London. *American Journal of Physical Anthropology*, 173(1), 168–178. https://doi.org/10.1002/ajpa.24081

Green, M. H. (2015). 'Editor's introduction to 'Pandemic disease in the medieval world: Rethinking the Black Death', in M. H. Green (ed), *Pandemic disease in the medieval world: Rethinking the Black Death*, pp. 9–26. Arc Medieval Press, Kalamazoo.

Green, M. H. (2020). The four Black Deaths. *The American Historical Review*, 125, 1601–1631. https://doi.org/10.1093/ahr/rhaa511

Grunspan, D. Z., Nesse, R. M., Barnes, M. E. & Brownell, S. E. (2018). 'Core principles of evolutionary medicine: a Delphi study'. *Evolution, Medicine, and Public Health*, 2018(1), 13–23. https://doi.org/10.1093/emph/eox025

Guellil, M., Kersten, O., Namouchi, A., Luciani, S., Marota, I., Arcini, C. A., Iregren, E., Lindemann, R. A., Warfvinge, G., Bakanidze, L., Bitadze, L., Rubini, M., Zaio, P., Zaio, M., Neri, D., Stenseth, N. C. & Bramanti, B. (2021). A genomic and historical synthesis of plague in 18th-century Eurasia. *Proceedings of the National Academy of Sciences of the United States of America*, 117, 28328–28335. doi.org/10.1073/pnas.2009677117

Haensch, S., Bianucci, R., Signoli, M., Rajerison, M., Schultz, M., Kacki, S., Vermunt, M., Weston, D. A., Hurst, D., Achtman, M., Carniel, E. & Bramanti, B. (2010). Distinct clones of *Yersinia pestis* caused the Black Death. *PLoS Pathogens [online]*. doi.org/10.1371/journal.ppat.1001134

Hatcher, J. (1986). Mortality in the fifteenth century: some new evidence. *The Economic History Review*, 39(1), 19–38. doi: 10.1111/j.1468-0289.1986.tb00395.x

Hübler, R., Key, F. M., Warriner, C., Bos, K. I., Krause, J. & Herbig, A. (2019). HOPS: automated detection and authentication of pathogen DNA in archaeological remains. *Genome Biology*, 20, article no. 280. doi: https://doi.org/10.1186/s13059-019-1903-0

Hummel, S., Schmidt, D., Kremeyer, B., Herrmann, B. & Oppermann, M. (2005). Detection of the CCR5-Delta 32 HIV resistance gene in Bronze Age skeletons. *Genes and Immunity*, 6, 371–374.

Jensen, T. Z. T., Niemann, J., Højholt Iversen, K., Fotakis, A. K., Gopalakrishnan, S., Vågene, Å., Winther Pedersen, M., Sinding, M-H. S., Ellegaard, M. R., Allentoft, M. E., Lanigan, L. T., Taurozzi, A. J., Holtsmark Nielsen, S., Dee, M. W., Mortensen, M. N., Christensen, M. C., Sørensen, S. A., Collins, M. J., Gilbert, M. T. P., Sikora, M., Rasmussen, S. & Schroeder, H. (2019). A 5700-year-old human genome and oral microbiome from chewed birch pitch. *Nature Communications*, 5520, article no. 5520. doi:10.1038/s41467-019-13549-9

Kacki, S. (2019). 'Black Death: cultures in crisis', in C. Smith (ed), Encyclopedia of *global archaeology*, pp. 1–12. Springer, Cham. https://doi.org/10.1007/978-3-319-51726-1_2858-1

Kang, S., Kim, Y. M., Lee, J. A., Kim, D. H., & Lim, J. S. (2019). Early menarche is a risk factor of short stature in young Korean female: epidemiologic study. *Journal of Clinical Research in Pediatric Endocrinology*, 11(3), 234–239. https://doi.org/10.4274/jcrpe.0274

Keeling, M. J. & Gilligan, C.A. (2000). Metapopulation dynamics of bubonic plague. *Nature*, 407, 903–906.

Keller, M., Spyrou, M. A., Scheib, C. L., Kröpelin, A., Haas-Gebhard, B., Päffgen, B., Haberstroh, J., Ribera i Lacomba, A., Raynaud, C., Cessford, C., Durand, R., Stadler, P., Nägele, K., Bates, J. S., Trautmann, B., Inskip, S. A., Peters, J., Robb, J. E., Kivisild, T., Castex, D., McCormick, M., Bos, K. I., Harbeck, M., Herbig, A. & Krause, J. (2019). Ancient *Yersinia pestis* genomes from across Western Europe reveal early diversification during the First Pandemic (541–750). *Proceedings of the National Academy of Sciences of the United States of America*, 116, 12363–12372. doi: 10.1073/pnas.1820447116

Kendall, E. J., Montgomery, J., Evans, J. A., Stantis, C. & Mueller, V. (2013). Mobility, mortality, and the Middle Ages: identification of migrant individuals in a 14th-century Black Death cemetery population. *American Journal of Physical Anthropology*, 150(2), 210–222. doi: 10.1002/ajpa.22194

Klunk, J., Duggan, A. T., Redfern, R., Gamble, J., Boldsen, J. L., Golding, G. B., Walter, B. S., Eaton, K., Stangroom, J., Rouillard, J-M., Devault, A., DeWitte, S. N. & Poinar, H. N. (2019). Genetic resiliency and the Black Death: no apparent loss of mitogenomic diversity due to the Black Death in medieval London and Denmark. *American Journal of Physical Anthropology*, 169(2), 240–252.

Kowaleski, M. (2014). Medieval people in town and country: new perspectives from demography and bioarchaeology. *Speculum*, 89(3), 573–600. doi: 10.1017/S0038713414000815

Krause-Kyora, B., Nutsua, M., Boehme, L., Pierini, F., Pedersen, D. D., Kornell, S-C., Drichel, D., Bonazzi, M., Möbus, L., Tarp, P., Susat, J., Bosse, E., Willburger, B., Schmidt, A. H., Sauter, J., Franke, A., Wittig, M., Caliebe, A., Nothnagel, M., Schreiber, S., Boldsen, J. L., Lenz, T. L. & Nebel, A. (2018). Ancient DNA study reveals HLA susceptibility locus for leprosy in medieval Europeans. *Nature Communications*, 9, 1569.

Kremeyer, B., Hummel, S. & Herrmann, B. (2005). Frequency analysis of the Δ32ccr5 HIV resistance allele in a medieval plague mass grave. *Anthropologischer Anzeiger*, 63(1), 13–22.

Kutyrev, V. V., Eroshenko, G. A., Motin, V. L., Nosov, N. Y., Krasnov, J. M., Kukleva, L. M., Nikiforov, K. A., Al'khova, Z. V., Oglodin, E. G. & Guseva, N. P. (2018). Phylogeny and classification of *Yersinia pestis* through the lens of strains from the plague foci of Commonwealth of Independent States. *Frontiers in Microbiology*, 9, 1106.

Kuzawa, C. W. (2008). The developmental origins of adult health: intergenerational inertia in adaptation and disease', in W. R. Trevathan, E. O. Smith & J. J. McKenna (eds), *Evolutionary medicine: new perspectives*, pp. 1–54. Oxford University Press, Oxford.

Mathieson, I., Lazaridis, I., Rohland, N., Mallick, S., Patterson, N., Roodenberg, S. A., Harney, E., Stewardson, K., Fernandes, D., Novak, M., Sirak, K., Gamba, C., Jones, E. R., Llamas, B., Dryomov, S., Pickrell, J., Arsuaga, J. L., Bermúdez de Castro, J. M., Carbonell, E., Gerritsen, F., Khokhlov, A., Kuznetsov, P., Lozano, M., Meller, H., Mochalov, O., Moiseyev, V., Rojo Guerra, M. A., Roodenberg, J., Vergès, J. M., Krause, J., Cooper, A., Alt, K. W., Brown, D., Anthony, D., Lalueza-Fox, C., Haak, W., Pinhasi, R. & Reich, D. (2015). Genome-wide patterns of selection in 230 ancient Eurasians. *Nature*, 528, 499

McDade, T. W., Georgiev, A. V. & Kuzawa, C. W. (2016). Trade-offs between acquired and innate immune defenses in humans. *Evolution, Medicine, and Public Health*, 2016(1), 1–16. https://doi.org/10.1093/emph/eov033

Meyer, M., Kircher, M., Gansauge, M-T., Li, H., Racimo, F., Mallick, S., Schraiber, J. G., Jay, F., Prüfer, K., de Filippo, C., Sudmant, P. H., Alkan, C., Fu, Q., Do, R., Rohland, N., Tandon, A., Siebauer, M., Green, R. E., Bryc, K., Briggs, A. W., Stenzel, U., Dabney, J., Shendure, J., Kitzman, J., Hammer, M. F., Shunkov, M. V., Derevianko, A. P., Patterson, N., Andrés, A. M., Eichler, E. E., Slatkin, M., Reich, D., Kelso, J. & Pääbo, S. (2012). A high-coverage genome sequence from an archaic Denisovan individual. *Science*, 338, 222–226.

Meyer, M., Arsuaga, J-L., de Filippo, C., Nagel, S., Aximu-Petri, A., Nickel, B., Martínez, I., Gracia, A., Bermúdez de Castro, J. M., Carbonell, E., Viola, B., Kelso, J., Prüfer, K. & Pääbo, S. (2016). Nuclear DNA sequences from the Middle Pleistocene Sima de los Huesos hominins. *Nature*, 531(7595), 504–507

Morosova, I., Kasianov, A., Bruskin, S., Neukamm, J., Molak, M., Batieva, C., Pudło, A., Rühli, F. J. & Schuenemann, V. J. (2020). New ancient Eastern European *Yersinia pestis* genomes illuminate the dispersal of plague in Europe. *Philosophical Transactions of the Royal Society B: Biological Sciences*, 375, 20190569. doi.org/10.1098/rstb.2019.0569

Muehlenbein, M. P. (2010). 'Evolutionary medicine, immunity, and infectious disease', in M. P. Muehlenbein (ed), *Human evolutionary biology*, pp. 459–490. Cambridge University Press, Cambridge.

Mühlemann, B., Jones, T. C., de Barros Damgaard, P., Allentoft, M. E., Shevnina, I., Logvin, A., Usmanova, E., Panyushkina, I. P., Boldgiv, B., Bazartseren, T., Tashbaeva, K., Merz, V., Lau, N., Smrčka, V., Voyakin, D., Kitov, E., Epimakhov, A., Pokutta, D., Vicze, M., Price, T. D., Moiseyev, V., Hansen, A. J., Orlando, L., Rasmussen, S., Sikora, M., Vinner, L., Osterhaus, A. D. M.

E., Smith, D. J., Glebe, D., Fouchier, R. A. M., Drosten, C., Sjögren, K-G., Kristiansen, K. & Willerslev, E. (2018) Ancient hepatitis B viruses from the Bronze Age to the Medieval period. *Nature*, 557, 418–423.

Namouchi A., Guellil, M., Kersten, O., Hänsch, S., Ottoni, C., Schmid, B. V., Pacciani, E., Quaglia, L., Vermunt, M., Bauer, E. L., Derrick, M., Jensen, A. Ø., Kacki, S., Cohn, Jr., S. K., Stenseth, N. C. & Bramanti, B. (2018). Integrative approach using *Yersinia pestis* genomes to revisit the historical landscape of plague during the Medieval Period. *Proceedings of the National Academy of Sciences of the United States of America*, 115(50), E11790–E11797. doi.org/10.1073/pnas.1812865115

Nguyen, V. K., Parra-Rojas, C. & Hernandez-Vargas, E. A. (2018). The 2017 plague outbreak in Madagascar: data descriptions and epidemic modelling. *Epidemics*, 25, 20–25. doi: 10.1016/j.epidem.2018.05.001.

Her Majesty's Stationary Office. (1905). *Statical abstract relating to British India from 1894-95 to 1903-04*. Her Majesty's Stationary Office, London.

Ogle, W. (1892). An inquiry into the trustworthiness of the old bills of mortality. *Journal of the Royal Statistical Society*, 55(3), 437. doi: 10.2307/2979466.

Pääbo, S. (1989). Ancient DNA: extraction, characterization, molecular cloning, and enzymatic amplification. *Proceedings of the National Academy of Sciences of the United States of America*, 86, 1939–1943.

Raoult, D., Aboudharam, G., Crubézy, E., Larrouy, G., Ludes, B. & Drancourt, M. (2000). Molecular identification by 'suicide PCR' of *Yersinia pestis* as the agent of medieval Black Death. *Proceedings of the National Academy of Sciences of the United States of America*, 97, 12800–12803. doi: 10.1073/pnas.220225197

Rascovan N, Sjögren, K.-G., Kristiansen, K., Nielsen, R., Willerslev, E., Desnues, C. & Rasmussen, S. (2019). Emergence and spread of basal lineages of *Yersinia pestis* during the Neolithic decline. *Cell*, 176, 295–305. https://doi.org/10.1016/j.cell.2018.11.005

Rasmussen, S., Allentoft, M. E., Nielsen, K., Orlando, L., Sikora, M., Sjögren, K-G., Pedersen, A. G., Schubert, M., Van Dam, A., Kapel, C. M. O., Nielsen, H. B., Brunak, S., Avetisyan, P., Epimakhov, A., Khalyapin, M. V., Gnuni, A., Kriiska, A., Lasak, I., Metspalu, M., Moiseyev, V., Gromov, A., Pokutta, D., Saag, L., Varul, L., Yepisko-posyan, L., Sicheritz-Pontén, T., Foley, R. A., Mirazón Lahr, M., Nielsen, R., Kristiansen, K. & Willerslev, E. (2015). Early divergent strains of Yersinia pestis in Eurasia 5,000 years ago. *Cell*, 163, 571–582.

Reich D., Green, R. E., Kircher, M., Krause, J., Patterson, N., Durand, E. Y., Viola, B., Briggs, A. W., Stenzel, U., Johnson, P. L. F., Maricic, T., Good, J. M., Marques-Bonet, T., Alkan, C., Fu, Q., Mallick, S., Li, H., Meyer, M., Eichler, E. E., Stoneking, M., Richards, M., Talamo, S., Shunkov, M. V., Derevianko, A. P., Hublin, J-J., Kelso,

J., Slatkin, M. & Pääbo, S. (2010). Genetic history of an archaic hominin group from Denisova Cave in Siberia. *Nature*, 468, 1053–1060. doi.org/10.1038/nature09710

Rigby, S. H. (2006). 'Introduction: social structure and economic change in late medieval England', in R. Horrox & W. M. Ormond (eds), *A social history of England 1200–1500*, pp. 1–30. Cambridge University Press, Cambridge.

Sabeti, P. C., Walsh, E., Schaffner, S. F., Varilly, P., Fry, B., Hutcheson, H. B., Cullen, M., Mikkelsen, T. S., Roy, J., Patterson, N., Cooper, R., Reich, D., Altshuler, D., O'Brien, S. & Lander, E. S. (2005). The case for selection at CCR5-Δ32. *PLoS Biology*, 3, e378

Salo, W. L., Aufderheide, A. C., Buikstra, J. & Holcomb, T. A. (1994). Identification of *Mycobacterium tuberculosis* DNA in a pre-Columbian Peruvian mummy. *Proceedings of the National Academy of Sciences of the United States of America*, 91, 2091–2094

Schmid, B. V., Buntgen, U., Easterday, W. R., Ginzler, C., Walløe, L., Bramanti, B. & Stenseth, N. C. (2015). Climate-driven introduction of the Black Death and successive plague reintroductions into Europe. *Proceedings of the National Academy of Sciences of the United States of America*, 112, 3020–3025.

Schooling, C. M., Jiang, C. Q., Lam, T. H., Zhang, W. S., Adab, P., Cheng, K. K. & Leung, G. M. (2010). Leg length and age of puberty among men and women from a developing population: The Guangzhou Biobank Cohort study. *American Journal of Human Biology*, 22(5), 683–687. https://doi.org/10.1002/ajhb.21067

Schuenemann, V. J., Avanzi, C., Krause-Kyora, B., Seitz, A., Herbig, A., Inskip, S., Bonazzi, M., Reiter, E., Urban, C., Dangvard Pedersen, D., Taylor, G. M., Singh, P., Stewart, G. R., Velemínský, P., Likovsky, J., Marcsik, A., Molnár, E., Pálfi, G., Mariotti, V., Riga, A., Belcastro, M. G., Boldsen, J. L., Nebel, A., Mays, S., Donoghue, H. D., Zakrzewski, S., Benjak, A., Nieselt, K., Cole, S. T. & Krause, J. (2018). Ancient genomes reveal a high diversity of *Mycobacterium leprae* in medieval Europe. *American Journal of Physical Anthropology*, 165, 245–45.

Schuenemann, V. J., Bos, K., DeWitte, S., Schmedes, S., Jamieson, J., Mittnik, A., Forrest, S., Coombes, B. K., Wood, J. W., Earn, D. J. D., White, W., Krause, J. & Poinar, H. N. (2011). Targeted enrichment of ancient pathogens yielding the pPCP1 plasmid of *Yersinia pestis* from victims of the Black Death. *Proceedings of the National Academy of Sciences of the United States of America*, 201105107

Schuenemann, V. J., Singh, P., Mendum, T. A., Krause-Kyora, B., Jäger, G., Bos, K. I., Herbig, A., Economou, C., Benjak, A., Busso, P., Nebel, A., Boldsen, J. L., Kjellström, A., Wu, H., Stewart, G. R., Taylor, G. M., Bauer, P., Lee, O. Y-C., Wu, H. H. T., Minnikin, D. E., Besra, G.

S., Tucker, K., Roffey, S., Sow, S. O., Cole, S. T., Nieselt, K. & Krause, J. (2013). Genome-wide comparison of medieval and modern *Mycobacterium leprae*. *Science*, 341, 179–183.

Scott S. & Duncan, C. J. (2001). *Biology of plagues: evidence from historical populations*. Cambridge University Press, Cambridge.

Sebbane F., Gardner, D., Long, D., Gowen, B. B. & Hinnebusch, B. J. (2005). Kinetics of disease progression and host response in a rate model of bubonic plague. *American Journal of Pathology*, 166(5), 1427–1439. https://doi.org/10.1016/S0002-9440(10)62360-7

Siefert L., Wiechmann, I., Harbeck, M., Thomas, A., Grupe, G., Projahn, M., Scholz, H. C. & Riehm, J. M. (2016). Genotyping *Yersinia pestis* in historical plague: evidence for long-term persistence of *Y. pestis* in Europe from the 14th to the 17th century. *PLoS One*, 11(1), e0145194. https://doi.org/10.1371/journal.pone.0145194

Sohn, K. (2016). Sexual stature dimorphism as an indicator of living standards? *Annals of Human Biology*, 43(6), 537–541. https://doi.org/10.3109/03014460.2015.1115125

Spigelman, M. & Lemma, E. (1993). The use of the polymerase chain reactions (PCR) to detect *Mycobacterium tuberculosis* in ancient skeletons. *International Journal of Osteoarchaeology*, 3, 137–143. doi.org/10.1002/oa.1390030211

Spyrou, M. A., Keller, M., Tukhbatova, R. I., Scheib, C. L., Nelson, E. A., Andrades Valtueña, A., Neumann, G. U., Walker, D., Alterauge, A., Carty, N., Cessford, C., Fetz, H., Gourvennec, M., Hartle, R., Henderson, M., von Heyking, K., Inskip, S. A., Kacki, S., Key, F. M., Knox, E. L., Later, C., Maheshwari-Aplin, P., Peters, J., Robb, J. E., Schreiber, J., Kivisild, T., Castex, D., Lösch, S., Harbeck, M., Herbig, A., Bos, K. I. & Krause, J. (2019). Phylogeography of the second plague pandemic revealed through analysis of historical *Yersinia pestis* genomes. *Nature Communications*, 10, article no. 4470. doi: 10.1038/s41467-019-12154-0

Spyrou, M. A., Tukhbatova, R. I., Feldman, M., Drath, J., Kacki, S., Beltrán de Heredia, J., Arnold, S., Sitdikov, A. G., Castex, D., Wahl, J., Gazimzyanov, I. R., Nurgaliev, D. K., Herbig, A., Bos, K. I. & Krause, J. (2016). Historical *Y. pestis* genomes reveal the European Black Death as the source of ancient and modern plague pandemics. *Cell Host & Microbe*, 19, 874–881

Spyrou, M. A., Tukhbatova, R. I., Wang, C-C., Valtueña, A. A., Lankapalli, A. K., Kondrashin, V. V., Tsybin, V. A., Khokhlov, A., Kühnert, D., Herbig, A., Bos, K. I. & Johannes Krause (2018). Analysis of 3800-year-old *Yersinia pestis* genomes suggests Bronze Age origin for bubonic plague. *Nature Communications*, 9, 2234

Steckel, R. H., Larsen, C. S., Roberts, C. A. & Baten, J. (eds). (2019). *The backbone of Europe: health, diet, work and violence over two millennia*. Cambridge University Press, Cambridge.

Stephens, J. C., Reich, D. E., Goldstein, D. B., Shin, H. D., Smith, M. W., Carrington, M., Winkler, C., Huttley, G. A., Allikmets, R., Schriml, L., Gerrard, B., Malasky, M., Ramos, M. D., Morlot, S., Tzetis, M., Oddoux, C., di Giovine, F. S., Nasioulas, G., Chandler, D., Aseev, M., Hanson, M., Kalaydjieva, L., Glavac, D., Gasparini, P., Kanavakis, E., Claustres, M., Kambouris, M., Ostrer, H., Duff, G., Baranov, V., Sibul, H., Metspalu, A., Goldman, D., Martin, N., Duffy, D., Schmidtke, J., Estivill, X., O'Brien, S. J. & Dean, M. (1998). Dating the origin of the CCR5-Δ32 AIDS-resistance allele by the coalescence of haplotypes. *The American Journal of Human Genetics*, 62, 1507–1515

Susat, J., Bonczarowska, J. H., Pētersone-Gordina, E., Immel, A., Nebel, A., Gerhards, G. & Krause-Kyora, B. (2020). *Yersinia pestis* strains from Latvia show depletion of the *pla* virulence gene at the end of the second pandemic. *Scientific Reports*, 10, article no. 14628.

Trevathan, W. R. (2007). Evolutionary medicine. *Annual Review of Anthropology*, 36(1), 139–154. https://doi.org/10.1146/annurev.anthro.36.081406.094321

Trevathan, W. R., Smith, E. O. & McKenna, J. J. (2008). 'Introduction and overview of evolutionary medicine', in W. R. Trevathan, E. O. Smith & J. J. McKenna (eds), *Evolutionary medicine: new perspectives*, pp. 1–54. Oxford University Press, Oxford.

Twigg, G. (1984). *The Black Death: a biological reappraisal*. Batsford Academic and Educational, London.

Valtueña, A. A., Mittnik, A., Key, F. M., Haak, W., Allmńe, R., Belinskij, A., Daubaras, M., Feldman, M., Jankauskas, R., Janković, I., Massy, K., Novak, M., Pfrengle, S., Reinhold, S., Šlaus, M., Spyrou, M. A., Szécsényi-Nagy, A., Tõrv, M., Hansen, S., Bos, K. I., Stockhammer, P. W. & Herbig, A. (2017). The Stone Age plague and its persistence in Eurasia. *Current Biology*, 27, 3683–3691.

van der Valk, T., Pečnerová, P., Díez-del-Molino, D., Bergström, A., Oppenheimer, J., Hartmann, S., Xenikoudakis, G., Thomas, J. A., Dehasque, M., Sağlıcan, E., Fidan, F. R., Barnes, I., Liu, S., Somel, M., Heintzman, P. D., Nikolskiy, P., Shapiro, B., Skoglund, P., Hofreiter, M., Lister, A. M., Götherström, A. & Dalén, L. (2021). Million-year-old DNA sheds light on the genomic history of mammoths. *Nature*, 591, 265–269. doi.org/10.1038/s41586-021-03224-9

Varlik, N. (2020). Rethinking the history of plague in the time of COVID-19. *Centaurus*, 62(2), 285–293. https://doi.org/10.1111/1600-0498.12302

Varlik N. 2021. 'Why Is Black Death black? European gothic imaginaries of "Oriental" plague', in C. Lynteris (ed), *Plague image and imagination from medieval to modern times*, pp. 11–36. Springer Nature, Cham.

Wagner, D. M., Klunk, J., Harbeck, M., Devault, A., Waglechner, N., Sahl, J. W., Enk, J., Birdsell, D. N., Kuch, M., Lumibao, C., Poinar, D., Pearson, T., Fourment, M., Golding, B., Riehm, J. M., Earn, D. J. D., Dewitte, S., Rouillard, J-M., Grupe, G., Wiechmann, I., Bliska, J. B., Keim, P. S., Scholz, H. C., Holmes, E. C. & Poinar, H. (2014). *Yersinia pestis* and the plague of Justinian 541–543 AD: a genomic analysis. *The Lancet Infectious Diseases*, 14(4), 319–326. doi: 10.1016/S1473-3099(13)70323-2.

Watkins, S. C. & Menken, J. (1985). Famines in historical perspective. *Population and Development Review*, 11(4), 647–675. https://doi.org/10.2307/197345

Weichmann I. & Grupe G. (2004). Detection of *Yersinia pestis* in two early medieval skeletal finds from Achheim (Upper Bavaria, 6th century A.D.). *American Journal of Physical Anthropology*, 126, 48–55. doi.org/10.1002/ajpa.10276

Wheelis, M. (2002). Biological warfare at the 1346 Siege of Caffa. *Emerging Infectious Diseases*, 8(9), 971–975.

Wood D. E. & Salzburg S. L. (2014). Kraken: ultrafast metagenomic sequence classification using exact alignments. *Genome Biology*, 15, article no. R46.

Xu, L., Stige, L. C., Leirs, H., Neerinckx, S., Gage, K. L., Yang, R., Liu, Q., Bramanti, B., Dean, K. R., Tang, H., Sun, Z., Stenseth, N. C. & Zhang, Z. (2019). Historical and genomic data reveal the influencing factors on global transmission velocity of plague during the Third Pandemic. *Proceedings of the National Academy of Sciences of the United States of America*, 116, 11833–11838. doi: 10.1073/pnas.1901366116

Zhou, D. & Yang, R. (2009). Molecular Darwinian evolution of virulence in *Yersinia pestis*. *Infection and Immunity*, 77, 2242–2250. doi: 10.1128/IAI.01477-08

Leprosy is down but not yet out: new insights shed light on its origin and evolution

Charlotte Roberts, David M. Scollard, and Vinicius M. Fava

Today leprosy is no more a dreaded disease as it is curable ... and the disease is capable of being eliminated as a public health problem through organized case detection and treatment.

Yawalkar, 2009, 11.

9.1 Introduction

Leprosy is a mycobacterial infection caused by *Mycobacterium leprae* and *Mycobacterium lepromatosis*. It affects the skeleton and therefore can be detected in the archaeological record. In exploring its origin and evolution, it has been the subject of modern and ancient DNA analyses that have established that it originated in Africa or the Near East 100,000 years ago. In human skeletal remains it has a several thousand-year history. It is a declining disease today and yet it lingers in some parts of the world, often because of poverty, a lack of education about the infection, inadequate access to the antibiotics used to treat it and stigma. While the evolution of leprosy has seen attention, particularly from DNA research, there remain gaps in knowledge in both the palaeopathological and DNA records.

Studying any infection's origin and evolution has much to offer us in terms of insights to the impact of it on the people affected, both past and present. *Why We Get Sick* by Nesse and Williams (1994) was ground-breaking in so many ways and was predicted to revolutionise attitudes to illness and prompt new research. In their opening chapter, they ask:

Why, in a body of such exquisite design, are there a thousand flaws and frailties that make us vulnerable to disease? If evolution by natural selection can shape sophisticated mechanisms such as the eye, heart, and brain, why hasn't it shaped ways to prevent nearsightedness, heart attacks, and Alzheimer's disease (Nesse and Williams, 1994, 3)?

The authors suggested that taking an evolutionary approach to understanding disease might help the world better manage unhealthy states (Nesse, 2008) and could provide the answers to contemporary questions about health that so far have been evasive. This same approach is now being taken towards leprosy.

In their closing remarks, Nesse and Williams (1994, 237) make recommendations for future textbook descriptions of disease, suggesting that historical legacies contributing to susceptibility to disease should be included. However, they do not specifically mention how palaeopathology might contribute knowledge to the evolution of our health. Since palaeopathology is the study of diseases in past human remains from prehistory to the more recent past, surely it has a contribution to make, alongside evolutionary biology, to

Charlotte Roberts, David M. Scollard and Vinicius M. Fava, *Leprosy is down but not yet out*. In: *Palaeopathology and Evolutionary Medicine*. Edited by Kimberly A. Plomp et al., Oxford University Press. © Oxford University Press (2022). DOI: 10.1093/oso/9780198849711.003.0009 understanding disease today. For example, in their chapter on Norwegian dietary guidelines, Mysterud and colleagues note that '... knowledge from academic disciplines like evolutionary biology and paleopathology has not been considered relevant by those designated as experts in formulating official guidelines for healthy living for the Nordic populations' (Mysterud et al., 2008, 106).

Recently, palaeopathologists have started to explicitly address questions generated from an evolutionary medicine framework (e.g., Holloway et al., 2011; Hunt et al., 2018; Jablonski & Chaplin, 2018; Roberts & Steckel, 2019; Wander et al., 2009; Zuckerman et al., 2012). Increasingly, we are also exploring the long view of disease in human populations using biomolecular approaches to understand pathogens and host susceptibility (Bos et al., 2014; Müller et al., 2014; Perrin, 2015; Schuenemann et al., 2013), but also developing understanding of disease from a genetic perspective (e.g. Brosch et al., 2002; Monot et al., 2005, 2009), including leprosy. As Nesse (2008) says, perspectives on infections that explore host–pathogen co-evolution, including those from palaeopathology, can help to address antibiotic resistance.

With this backdrop, this chapter looks at the present and past of leprosy and attempts to merge the evidence together to understand both better. It aims to:

1. Introduce what we know about leprosy, including current genetic research.
2. Introduce what is known about leprosy in the past through the archaeological record.
3. Discuss the genetic evidence, both past and present, and what it tells us about the origin, evolution and history of leprosy.
4. Discuss how current understanding of leprosy, including data from palaeopathology, can contribute to modern clinical research and medicine.

9.2 Leprosy: present and past

Leprosy can be considered as two conjoined diseases. The first is a bacterial infection of the skin and nerves caused by *M. leprae* and *M. lepromatosis*; the second is a peripheral neuropathy (nerve damage) that results from this infection and sometimes leads to functional impairment and permanent disabilities (Scollard et al., 2006). Unless otherwise specified, both mycobacteria will be referred to in this chapter as *M. leprae* (Brennan & Spencer, 2019). The neuropathy caused by leprosy may persist long after the infection has resolved and is responsible for most of the disability and disfigurement associated with leprosy. Nerve damage and other lesions caused by a leprosy infection can be identified and diagnosed in human remains, although this can be challenging for several reasons. First, while leprosy can affect the skeleton, only 3–5% of untreated people may develop bone damage. Thus, as palaeopathologists, we really only see the 'tip of the iceberg' of leprosy in the past. Second, the issue of differential diagnoses (e.g., mistaking leprosy for venereal syphilis) can prevent confident identification of leprosy lesions in human skeletons (the potential for misdiagnoses is discussed further in Chapter 1). Last, there remain some gaps in our knowledge of how the two mycobacteria affect the skeleton, and in particular, whether leprosy caused by *M. lepromatosis* leads to the same bone damage as *M. leprae*.

Clinically, leprosy presents with various cutaneous lesions that can resemble a wide range of other illnesses, with symptoms ranging from diffuse thickening of the skin to nodules (i.e., lumps on the skin), and eventually to macules (i.e. flat areas of colour changes). Physicians often fail to include leprosy in the differential diagnosis process, along with other rare diseases that are outside their realm of experience. One reason for this is that leprosy resembles many other more common diseases, such as ringworm. The distinguishing feature of the skin lesions is a reduction or total loss of sensation within them due to the neurological involvement of *M. leprae*. Notably, unless complications called 'reactions' develop, leprosy is a very indolent infection and typically does not cause pain or fever. The absence of these symptoms likely results in delays in seeking medical care, which usually occurs only when skin lesions are of cosmetic concern, or when anaesthesia or paralysis due to nerve injury affects the patient's ability to work, e.g., inability to use a pen or other tool. For obvious reasons, skin lesions are not detectable in archaeological skeletal remains but could be seen in present on the skin of a preserved body. However, skin lesions accepted to be related

Table 9.1 Traits of *M. leprae* and human hosts that may have resulted from evolutionary adaptation

Trait	Benefit to	Disadvantage to	Comment
Pathogen requires intracellular environment		*M. leprae*	limits survival and transmission; mainly infects macrophages and Schwann cells
Pathogen has limited range of species infected		*M. leprae*	limits survival and transmission
Most humans are genetically resistant to leprosy		*M. leprae*	limits *M. leprae* to a niche of susceptible hosts
Pathogen has extremely slow growth	*M. leprae/host*		for *M. leprae*, it delays detection and promotes transmission for the host there is no sepsis
Infects endothelial cells	*M. leprae*	Host	facilitates infection of nerves
Infects Schwann cells	*M. leprae*	Host	Injury to nerves, disability of host
No fever in host	*M. leprae*	Host	delays detection and treatment
No pain in host	*M. leprae*	Host	delays detection and treatment
Not fatal for host	*M. leprae/host*	Host	prolongs pathogen survival without being lethal to the host as a host disadvantage, long-term survival may come with long-term injuries
Limited or no host resistance (lepromatous)	*M. leprae*		allows unlimited proliferation of *M. leprae*
Host resistance (tuberculoid) is CMI, granuloma formation		Host	granulomas injure tissue—in leprosy, nerve tissue

to leprosy have been recorded in art (see Boeckl, 2021) and in descriptions in documents (Dharmendra, 1947).

In the late 1940s, dapsone became the first drug recognised to kill *M. leprae* and to cure infection. A multiple drug (antibiotic) treatment (MDT), or oral administration of dapsone, rifampicin and clofazimine, has been widely used to treat leprosy since the 1980s (Lockwood, 2019). MDT produces the most rapid response and when used properly; it prevents the development of drug resistance, although alternative agents are also available if needed. Recommendations for treatment duration range from six months to two years, depending on the patient's disease type (classification) and on national recommendations. Global implementation of MDT has been successful, and in 2019 only 202,185 new 'cases' were reported by the World Health Organization (WHO, 2020). The impact of MDT on leprosy frequency since its introduction is clear. Even with dapsone treatment, despite leprosy 'cases' rising by 93% from 2.8 million to 5.4 million between 1976 and 1985, by 1989 there were 3.9 million cases, a drop of 28% (Noordeen, 1990). However, today many people with leprosy remain unreported, and the total number is certainly higher than the 2019 figures.

Although the global prevalence of leprosy has declined, it has not been eliminated, and it remains a challenge for some parts of the world, especially Brazil, India and Indonesia (WHO, 2019). Leprosy has a long history, with evidence in archaeological human remains several thousands of years old (Roberts, 2020), thus providing an opportunity for a deep evolutionary perspective to be taken on this infection's rise and fall.

9.2.1 The bacterium

M. leprae is a well-adapted pathogen that occupies a unique niche in its selection of species (humans, armadillos and red squirrels) and its cellular niche, as an obligate intracellular organism that primarily infect macrophages and Schwann cells (Table 9.1, Box 9.1).

Box 9.1 The leprosy bacterium and its transmission

The leprosy bacterium, *M. leprae*, appears microscopically as a short, acid-fast rod, and it is morphologically very similar to its microbial relative, *M. tuberculosis*. *M. leprae* prefers a cooler growth temperature than most human

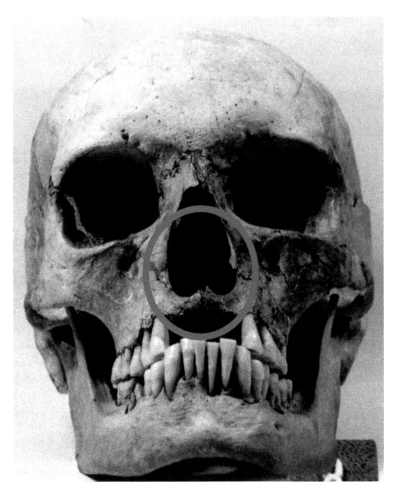

Figure 9.1 Leprosy related facial bone damage (loss of depth of the maxillary alveolar bone anteriorly and antemortem loss of anterior teeth, marked with a blue circle) in a medieval skeleton buried in the cemetery associated with the Naestved leprosy hospital, Denmark. Part of the rhinomaxillary syndrome.

pathogens (approximately 33°C), partially explaining the localisation of lesions to cooler parts of the skin, such as peripheral nerves (*M. leprae* is the only human bacterial pathogen that can infect these nerves). The growth rate of *M. leprae* is the slowest of any human pathogen, with a doubling time of approximately thirteen days. Understanding the bacterium's origin and evolution has been enhanced over the last twenty years, through the sequencing of the genomes from modern samples of *M. leprae* and *M. lepromatosis* (Cole et al., 2001; Han et al., 2015). Fragments of the DNA of *M. leprae* have also been detected in human remains (skeletons and preserved bodies) using molecular methods such as the polymerase chain reaction (PCR) and next-generation sequencing (Singh et al., 2018; and see Economou et al., 2013a,b;

Schuenemann et al., 2013; Taylor et al., 2006). This work is discussed further in Section 9.3. Current evidence suggests that infection enters a human host via the nose and upper respiratory tract. Early lesions of soft tissue and bone are seen in the naso- and oropharynx, including the maxillary dentition (Scollard et al., 1999). The bacilli are understood to spread via the bloodstream and establish foci of infection in cool sites of the skin, peripheral nerves and the skeleton. This process can be detected in palaeopathological evidence of damage to the bones of the upper jaw (maxilla); inflammatory changes to the oral and nasal surfaces of the palate; widening of the nasal aperture; loss of the anterior nasal spine and, rarely, stunting of the development of the anterior tooth roots, known as leprogenic odontodysplasia (Figure 9.1).

9.2.2 Transmission

The exact means of transmission of *M. leprae* remains undetermined but, contrary to popular myth, leprosy is not highly contagious (Box 9.1). In the Western hemisphere (southern United States, Mexico and Central and South America), some nine-banded armadillos are natural hosts for *M. leprae* infections, that is, where the pathogen can be commonly found and where it can complete its development (Truman et al., 2011). Current evidence suggests that a certain strain of leprosy (3I) 'arrived' in America with people from the Old World, thus explaining its presence in armadillos. However, the infection was only first observed in free-living armadillos of the *Dasypus novemcinctus* species in the USA in 1975 (Walsh et al., 1975). Furthermore, through DNA analysis that compared *M. leprae* in wild armadillos and patients with leprosy, a unique *M. leprae* genotype was found in most of the twenty-eight armadillos studied and two-thirds of twenty-five humans. The patients lived in the same areas where exposure to armadillos was possible (Truman et al., 2011). It is thought that armadillos transmit leprosy to humans in the Americas today (Da Silva et al., 2018; Kerr et al., 2015; Truman et al., 2011). Interestingly, the source of leprosy in armadillos in Texas and Louisiana is unknown but it could have come from untreated patients with leprosy before treatment was implemented. Meyers and colleagues (1984) suggested that indirect contact with leprosy via clothing and bandages from people with leprosy was a possibility. Identification of more than one unique 'armadillo strain' in the United States indicates more than one instance of transmission of *M. leprae* from humans to armadillos (Sharma et al., 2015).

Along with armadillos, there are other animal hosts of leprosy. For example, in Africa, *M. leprae* has been recently found in two geographically distant populations of wild chimpanzees with no prolonged direct contact with humans (Hockings et al., 2020). It has also been identified in sooty mangabey monkeys (*Cercocebus atys*) and other primates (Walsh, 2020). *M. leprae* has also been found in red squirrels in the UK (Avanzi et al., 2016), but there is as yet no known transmission from squirrels to humans. Thus, the American nine-banded armadillo and British/Irish red squirrels are natural hosts for *M. leprae*, and do not need humans for the pathogen to survive. The affected skin of British red squirrels appears similar in appearance to the skin of people with lepromatous leprosy, and genetic analysis of squirrels has revealed a similar DNA sequence to *M. lepromatosis* (Meredith et al., 2014), including the 3I strain that is also found in wild armadillos and humans (Kerudin et al., 2019; Taylor et al., 2018; Avanzi et al., 2016; Inskip et al., 2015, 2017; Schuenemann et al., 2013; Taylor & Donoghue, 2011; Truman et al., 2011; Monot et al., 2009). How long *M. leprae* has been harboured by red squirrels and from where they contracted leprosy, or whether it is possible for them to transmit leprosy to humans, is still unknown, making it a relevant area of enquiry in archaeological research. In terms of the vectors transmitting leprosy from animals to humans, and vice versa, cayenne ticks (*Amblyomma mixtum*) and reduviid bugs have been shown to sustain viable *M. leprae* after infection in laboratory settings (Ferreira et al., 2018; da Silva Neumann et al., 2016). Interestingly, no animal reservoirs have so far been identified in Asia and continental Europe. This geographical range supports the recent suggestion that environmental sources may be an important factor in *M. leprae* infections (e.g., Turankar et al., 2019), something not usually considered when looking at the archaeological evidence for leprosy.

9.2.3 The immune system and leprosy

In humans, the ability to develop immune resistance to *M. leprae* is determined by genetic factors. Approximately 95% of contemporary living adults have innate immunity to leprosy and will not develop it even after heavy, prolonged exposure (Moraes et al., 2017). Among people who are susceptible, the incubation time averages between five and seven years but may be considerably longer. In these susceptible patients, cell-mediated immunity (CMI) provides varying degrees of resistance (Weiss et al., 2020). This can be thought of as a spectrum, with one end of that spectrum representing patients with high CMI, which results in tuberculoid leprosy (TT), where bacilli are rare and difficult to find.

The other, opposite end of the spectrum represents patients with little or no CMI, which can result in lepromatous leprosy (LL) and an astonishing number of bacilli. In between these two extremes on the spectrum is the intermediate form of infection known as borderline leprosy (Scollard et al., 2006) (Figures 9.2 and 9.3).

The lower panel of Figure 9.3 shows the range of *M. leprae* across the spectrum in the same biopsies stained for acid-fast bacilli, which stain red. In TT patients, bacilli are rare and difficult to find. In LL patients they are exceedingly abundant throughout the tissue. The bacterial load increases incrementally across the spectrum from TT to LL (left to right), corresponding to the decline in organisation of the inflammation and the decline in the strength of the cellular immune response (Fite stain, original magnification = 1000×).

The wide range of CMI in *M. leprae*, and the regulation of these processes, are highly complex and not completely understood. While patients may have

expanding skin lesions and gradual, non-painful loss of peripheral nerve function, the bacilli are often well tolerated, and thus, the patient may not feel ill and not display any signs. In about half of patients, *M. leprae* infection and the CMI response are also complicated by sudden episodes of immunologically mediated 'reactions', whereby people with leprosy can become quite ill with fever, malaise, inflammation and ulceration of skin, nerves, joints and other organs. When brought to the emergency room, physicians unaware of their underlying diagnosis often suspect that such patients are suffering from bloodstream infection (sepsis). During these reactions, nerve injury may develop rapidly and result in permanent loss of sensation or motor nerve function within days or weeks (Walker, 2020).

In palaeopathology, methods have been developed to identify people at both ends of the immune spectrum based on skeletal lesions (e.g. see Matos, 2009; Møller-Christensen, 1961), but individuals

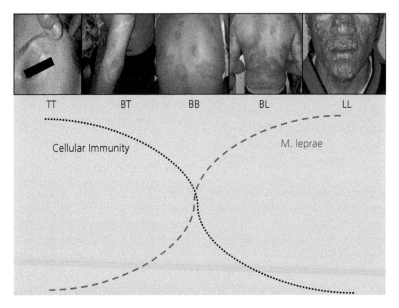

Figure 9.2 The Clinical, Immunological, and Bacteriological Spectrum of Leprosy.
Leprosy is classified according to a five-part system described by Ridley and Jopling (1966). At one extreme, patients with few lesions, strong cellular immunity (CMI) and rare bacilli are classified as polar tuberculoid (TT). At the other extreme, patients have numerous papules or nodules, abundant bacilli and little or no CMI, and these are classified as polar lepromatous (LL). Between these two poles is a wide borderline group, further divided into borderline tuberculoid (BT), mid-borderline (BB) and borderline lepromatous (BL). A patient's position on this spectrum is genetically determined according to their ability to mount a cellular immune response to *M. leprae*; those with strong immunity have rare bacilli, and those with little or no immunity have a very high number of *M. leprae* in the skin and peripheral nerves.

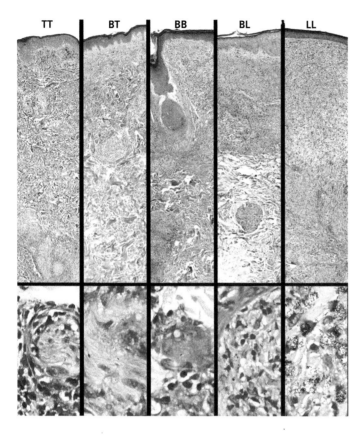

Figure 9.3 The Histopathological Spectrum of Leprosy.

The histological appearance of the inflammatory responses across the leprosy spectrum are illustrated in the upper panel by hematoxylin and eosin-stained sections of skin biopsies. Well-organised granulomas are seen in the skin of TT patients, while disorganised collections of macrophages and lymphocytes are seen in LL patients. A gradual decline in organisation of the infiltrate is seen across the spectrum from TT to LL. The degree or organisation of the inflammation corresponds to the strength of the cellular immune response in each category, as shown in Figure 9.2. (H/E stain, original magnification = 63×). The lower panel shows the range of *M. leprae* across the spectrum in the same biopsies stained for acid-fast bacilli, which stain red. In TT patients, bacilli are rare and difficult to find. In LL patients they are exceedingly abundant throughout the tissue. The bacterial load increases incrementally across the spectrum from TT to LL, corresponding to the decline in organization of the inflammation and the decline in the strength of the cellular immune response.

Acronyms as specified in Figure 9.2. From Scollard et al., *Clinical Microbiology Reviews*, 19:338-381, 2006; author's right to republish under https://journals.asm.org/author-self-archiving-permissions.

with lepromatous leprosy currently contribute most of the evidence in archaeological human remains. As people in the past became more resistant to leprosy through exposure, tuberculoid leprosy may have become more common. Currently there is only one report of leprosy in a preserved body (i.e., with soft tissue)—an Egyptian mummy (Smith, 1908) (the majority is in skeletons), which is surprising considering the amount of research undertaken on mummies.

9.2.4 Nerve injury

Nerve injury in leprosy is caused by infection within and around peripheral sensory, motor and autonomic nerves, and is the hallmark sign that clinically distinguishes leprosy from other skin diseases (Job, 1989). Despite this, however, the mechanism of this affinity for peripheral nerves is limited and not understood fully (Scollard et al., 2015). Understanding has been limited by the minimal

opportunities to study affected nerves in humans. The nine-banded armadillo provides the only animal model of the pathogenesis of *M. leprae* infection. However, new tools are enabling the study and correlation of events occurring in epidermal nerve fibres, dermal nerves and nerve trunks, including neurophysiologic parameters, bacterial load, and changes in gene transcription in both neural and inflammatory cells. The armadillo model is thus likely to enhance future understanding of the mechanisms of nerve injury in leprosy and offers a means of testing proposed interventions.

Nerve involvement develops slowly and insidiously at first, with loss of sensation in the hands and feet (Yawalkar, 2009). At this point, the patient will not feel any pain in affected extremities and may not be aware of ulcers, cuts, burns, scalding or even partial amputation of fingers or toes. Fingers and toes do not 'drop off', but the bones may be affected and can shorten as a response to ulcers and/or repeated trauma of the soft tissues leading to bone absorption

(Paterson & Job, 1964). As the disease progresses, a secondary bacterial infection may result from open wounds on the hands and feet and this could eventually affect the bones and joints, thus paving the way for further damage and, consequently, structural instability of the extremities. In archaeological human remains, these effects can be identified in the skeleton (Møller-Christensen, 1961). If sensory loss involves the eye, irritants will not be felt, and any subsequent chronic injury could lead to corneal scarring and blindness.

Based on comparison of palaeopathological bone lesions with clinical radiographs of people with motor nerve damage-linked fixed flexion deformities of the fingers and toes, it was also inferred that destructive lesions in the phalanges of the feet and hands of archaeological skeletons may indicate these flexion deformities in life (Andersen & Manchester, 1987). This motor nerve damage leads to weakness and paralysis of muscles, muscle groups and tendons; a loss of synergistic action and balance; subluxation (partial dislocation) and dislocation of the interphalangeal joints (hyperflexion); and hyperextension of the metacarpophalangeal (in hands), metatarsophalangeal (in feet) and interphalangeal joints (in hands and feet), all potentially causing this recognisable bone damage. Similarly, foot drop can occur in leprosy (Andersen, 1964). The cause is a paralysis of the lateral popliteal nerve at the popliteal fossa (behind the knee). The condition is a muscular weakness or paralysis that makes it difficult to lift the front part of the foot and toes. It can also lead to an affected person dragging their foot on the ground during walking. However, it can have many causes and is a sign of an underlying problem rather than a condition itself. Leprosy-related drop foot may cause bone formation on the dorsal tarsal surfaces (Andersen & Manchester, 1988).

Leprosy is rarely fatal and when death does occur, it is typically due to immune and inflammatory responses to the bacilli or secondary infections, rather than to the leprosy infection itself. One reason why *M. leprae* is generally non-fatal is because it grows exceedingly slowly and avoids infecting vital organs such as the heart, lungs or brain, most likely because these organs are too warm to provide suitable growth conditions.

9.2.5 Stigma and leprosy

People with leprosy have faced stigma in the past, as can be seen in historic descriptions and illustrations (e.g., Grön 1973) (Figure 9.4), and this stigma continues even today. The stigma associated with leprosy in many communities can impact the early diagnosis of the infection and subsequent effective implementation or treatment, because people are reluctant to access relevant healthcare facilities (Rao, 2015). Travel distances and domestic and other work responsibilities may also be barriers. Furthermore, it has been identified that there is a low chance of leprosy eradication without 'some radical changes in the implementation of WHO operational guidelines using community-based participatory techniques' (Rao, 2015, 55). Today, early diagnosis, treatment and medical management of complications make leprosy far less damaging than it was historically, although it is still strongly stigmatised in many parts of the world (van Brakel, 2019).

There are also many ancient documentary representations of stigma, and a common reference used is the apparent description of leprosy (the Hebrew word *tzaraat*) in the Bible, now believed to refer to people with many illnesses that could have included leprosy. However, today it is leprosy that has continued to carry the stigma. In the more distant past it was often the case that stigma resulted in ostracism of patients, who were sent to isolated places such as islands (Figure 9.5) and leprosaria. In the most highly affected people, the combination of neurological injuries could leave a patient blind, with deformed and anaesthetised limbs, and unable to speak. It has been argued that this combination of progressive, severe disability, unrelieved by death, made many people consider leprosy a 'fate worse than death'. This view contributed significantly to the stigma and ostracism characteristically associated with this disease and seen in the Bible and other historical records that borrowed information from that source throughout history (Kellersberger, 1951). However, Grzybowski and Nita (2016) have shown that *tzaraat* does not relate to modern-day leprosy or any other contemporary disease. *Tzaraat* was mistranslated as leprosy and was later associated with uncleanliness (Lie, 1938). It is a word that encompassed disfigurative conditions of

Figure 9.4 Lithograph by Talbaux after Bevalet, of an early-nineteenth-century woman with leprosy from Iceland. Illustration from the medical atlas section in Robert, E., (1851). *Voyage en Islande et au Groenland publie par ordre du Roi sous la direction de M Paul Gaimard*. Paris.
© Science Museum / Science & Society Picture Library

the skin, the hair of the beard and head, and clothing made of linen or wool.

We should also be mindful that there is evidence from the archaeological record that the signs of leprosy may have been mistaken for other diseases such as syphilis or tuberculosis (TB); skeletons with these latter diseases have been recovered from medieval leprosaria cemeteries (e.g., St Mary Magdalen, Chichester, England; see Magilton et al., 2008). Of particular importance is the large proportion of palaeopathological evidence for leprosy deriving mostly from non-leprosaria cemeteries where people were buried 'normally' for the place and time; 'special' (not the norm for the period and place) burials of individuals with leprosy,

for example of high status, are noted too (e.g., England; see Duhig, 1998). This suggests that, contrary to much historical evidence, communities could be accepting of people with leprosy, thus raising the question about how stigmatised the infection was in the past.

9.3 The origin and evolution of leprosy

Genetic research using both modern and ancient DNA (aDNA) has provided evidence to suggest that leprosy originated in East Africa or the Near East 100,000 years ago, with the Silk Road playing an important role in its spread via human migration (Monot et al., 2005, 2009) (Figure 9.6). Whether the

Figure 9.5 Robben Island where people with leprosy were incarcerated in the nineteenth century.
National Archives and Records of South Africa, Arcadia.

original host was a human or other animal is not currently known (see discussion in Monot et al., 2009). After its emergence, it is thought that the disease spread to North Africa, west to Europe and east to Asia, with no evidence of leprosy in the Americas before the late eighteenth century. In fact, it is hypothesised that leprosy was introduced into the Americas with the colonisation of the continents by Europeans through the movement of people in the last 500 years. This suggestion is supported by recent aDNA analyses, which identified the same *M. leprae* strains that are in modern America in archaeological skeletons from Denmark, England, and Sweden (see a summary of the aDNA research, with references, in Roberts, 2020). Research on lineages/strains of *M. leprae* identified in skeletons using aDNA analysis is particularly important for exploring the history of leprosy. Still, the highly conserved aspect of the *M. leprae* genome points to host genetic variability and environmental factors as crucial for defining susceptibility to leprosy. Therefore, more work on identifying such variability in people affected and pinpointing environmental risk factors will to progress this research (e.g., see Singh et al., 2015).

Based on modern and aDNA work, *M. leprae* is estimated to have emerged as a human pathogen more than 5000 years ago (summary in Spyrou et al., 2019), and Köhler et al., (2017) have reported the earliest European evidence in Hungary dated to between 5780 and 5650 years ago (radiocarbon date). Moving from Europe along the Silk Road, other areas, including China, Egypt, India, Iran, Thailand and Uzbekistan, all have examples of skeletal evidence that may/does indicate leprosy. Nevertheless, the evidence consists of only a few skeletons and one Egyptian mummy, with the 'probable' Iranian evidence being the oldest yet documented globally (6200–5700 cal BC). The suggestion that leprosy originated in Africa or the Near East needs corroboration with early evidence from the palaeopathological record, and yet only Iran, Sudan (possibly) and Turkey have very early evidence. Most modern-day countries in this region either have no or very little evidence, which may reflect the relative lack of palaeopathological work in those places.

As mentioned, leprosy as an outcome of exposure to *M. leprae* is strongly dependent on the genetic background of the host, plus environmental factors, which can be very diverse (Fava et al., 2019a). Research on aDNA also highlights that there was a diversity of strains of *M. leprae* present in Europe. In particular, a low diversity of strains at non-leprosarium sites and high diversity at the local level in some leprosaria have been noted (Prengle

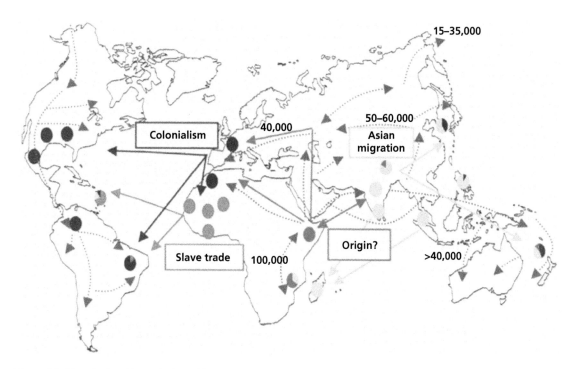

Figure 9.6 Dissemination of leprosy in the world.
The circles indicate the country of origin of the samples examined and their distribution into the four single nucleotide polymorphism (SNP) types, which are colour coded. The coloured arrows indicate the direction of human migrations predicted by, or inferred from, SNP analysis: grey arrows correspond to the migration routes of humans derived from genetic, archaeological and anthropological studies, with the estimated time of migration in years.
From Monot et al. (2005). On the origin of leprosy. *Science*, 308,1040–1042. Reprinted with permission from the American Association for the Advancement of Science/Rightslink.

et al., 2021; Schuenemann et al., 2013, 2018). For example, the recent study of the genetic variation of *M. leprae* in medieval Europe, focusing on bacterial strains from previously unstudied regions and based on newly sequenced ancient genomes, found high diversity of *M. leprae* strains at leprosarium sites (Prengle et al., 2021). It appears that some people with leprosy were coming to leprosaria from far distant places, thus contributing to the strain diversity seen in Europe (see Roffey et al., 2017). Travel and contact between people necessarily are—and were—important for transmission, and there are some correlations between skeletons with leprosy and the movement of people. For example, expansion of the Roman Empire from Italy led to people moving into other areas of Europe, perhaps taking diseases like leprosy with them.

The 3I *M. leprae* strain, also seen in armadillos and red squirrels today (Avanzi et al., 2016; Monot et al., 2005; see also Truman et al., 2011), is the most common in medieval Europe and has been around for approximately 1200 years (Prengle et al., 2021). Based on documentary and archaeological evidence, leprosy apparently reached its peak incidence in Europe in the late medieval period (twelfth to sixteenth centuries AD), and then declined (Roberts, 2020). Unfortunately, little research has so far been undertaken on strain diversity outside Europe, leaving us with a largely incomplete picture of leprosy's evolution from a global perspective. However, more work is needed to identify how widespread local strain diversity was in the past beyond Europe. The identification of a strain of *M. leprae* in modern red squirrels from southern England that is closely related to a strain found in the skeleton of an early medieval male, also from southern England, indicates the possibility of a historic animal reservoir (Avanzi et al., 2016). Thus, we should not totally dismiss the possibility that (environmental) animal reservoirs could contribute to complexity in strain diversity, as suggested by Singh and colleagues (2015). Contrary to this, the existence of an animal reservoir in past millennia in the 'Old World' is not yet established, including in the suggested leprosy origin regions of the Near East and Africa. However, as yet there is no evidence that one of the two successful known animal reservoirs of *M. leprae*, the armadillo from the 'New World' (see Section 9.2.2), has been influential in the evolution of the genetic variation of the bacterium anywhere in the world. The strains seen in the Americas in both armadillos and humans are European, and leprosy evolved for millennia in the 'Old World' before the armadillo reservoir existed. However, there may be successful animal reservoirs yet to be found, and other environmental sources that might have been relevant where leprosy originated.

In relation to the decline of leprosy, thousands of years of host–pathogen co-evolution led to *M. leprae* losing a significant part of its genome and strain diversity as it specialised as a human pathogen (Cole et al., 2001). A consequence was that the majority of the human population became resistant, leading to its eventual decline. Segregation of patients due to the stigma associated with the disease, coupled with lower reproduction rates of infected individuals, may have led to the selection of individuals genetically resistant to leprosy. Another 'decline hypothesis' still debated is that other epidemics (e.g. TB), may have favoured the survival of people resistant to mycobacterial infections (e.g. Chaussinand, 1948; Donoghue et al., 2005; Leitman et al., 1997; Manchester, 1984; see also Crespo et al., 2019 and Wilbur et al., 2002 showing challenges with data). Exposure to environmental mycobacteria could have acted as a 'natural' vaccine, training our innate immune system to be more resistant to leprosy, as shown for BCG (Bacillus Calmette–Guérin) in TB (Crespo et al., 2019; Kaufmann et al., 2018). However, we know that people in the past had both TB and leprosy infections, as illustrated by palaeopathological and aDNA analyses that detected leprosy and TB co-infection in second-century BC skeletons from Dakhleh Oasis, Egypt (Donoghue et al., 2005; Molto, 2002). This ancient population also revealed cribra

Figure 9.7 Cribra orbitalia (holes) in an eye socket.

orbitalia (Figure 9.7), an indicator of 'stress' due to anaemia of many potential causes, including infection (Stuart-Macadam, 1992). This suggests the population could have had poor nutrition, as well as the mycobacterial infections. Iron is an essential nutrient for mycobacterial metabolism, and mutations in the *SLC11A1* gene (formerly *NRAMP₁*) encoding an essential iron transporter are an established susceptibility factor for both leprosy and TB (Abel et al., 1998; Bellamy et al., 1998). Pathogens causing infections need iron to survive and be successful infective organisms. Deficiency of iron in the blood illustrates an adaptation to microorganism invasion (Stuart-Macadam, 1992). When faced with heavy pathogen loads people become hypoferraemic and more susceptible to iron deficiency anaemia. It is tempting to speculate a role for *SLC11A1* as a major factor for mycobacterial susceptibility at Dakhleh Oasis. However, these links make an interesting story combining the genetics of leprosy, TB, and iron transport (also see Wilbur et al., 2008) that would benefit from further analyses, especially focusing on aDNA.

Due to the marked conservation of the *M. leprae* genome, all people with leprosy are virtually affected by the same bacteria. Even so, it is possible to separate *M. leprae* strains into different ancestral lineages using a few selected SNPs. In the past decade with the development of molecular techniques in sequencing entire human genomes, the scientific community has identified multiple genomic variants controlling leprosy susceptibility (Fava et al. 2019a). Among the strongest leprosy susceptibility associations are variants located in human leukocyte antigen (HLA) genes. HLA genes encode proteins that are responsible for presenting antigens to the T (white blood) cells of the immune system and are central in the adaptive immune response to pathogens (Weiss et al., 2020). HLA proteins are highly polymorphic, making it challenging to identify a single protein mutation associated with leprosy. Therefore, combinations of amino acid changes in HLA genes (e.g., HLA alleles) are frequently used as markers. The HLA class II allele HLA-DRB1*15:01 is a strong risk factor for leprosy in humans (Wang et al., 2016) and has been shown to be common in European people dating to 1000–1600 AD (Krause-Kyora et al., 2018). Hence, HLA risk factors for leprosy appear to have been conserved since this period. Modern HLA sequencing techniques have recently refined the HLA association with leprosy in Southeast Asians to four amino acids (Dallmann-Sauer et al., 2020). This study provided a unifying view of the complex relationship of leprosy susceptibility and HLA alleles, with the four amino acids being shown to be sufficient to explain the majority of leprosy–HLA associations. It would be interesting to evaluate the presence of this particular set of amino acids in the medieval period using aDNA analysis.

The selective pressure caused by constant host–pathogen interactions has shaped our immune system to mount stronger inflammatory responses against pathogens (Barreiro & Quintana-Murci, 2010). The fitness of our immune system relies on the delicate balance between too much versus too little inflammation. The co-existence between

humans and *M. leprae* has likely influenced both organisms to adapt in an evolutionary arms-race paradigm, although this can be difficult to support with data. One line of evidence supporting the suggestion that we adapted in response to *M. leprae* infections is the apparent positive selection of the *TLR1* (I602S) mutation. This mutation has been associated with protection against leprosy (Barreiro & Quintana-Murci, 2010; Wong et al., 2010), as it is one of the receptors recognising *M. leprae* antigens. However, *TLR1* also recognises antigens from other bacteria, making it difficult to pinpoint leprosy as the main selective pressure since both leprosy and TB have been epidemic in past human populations humans (Barreiro & Quintana-Murci, 2010; Matheson et al., 2015). The potential co-morbidity of leprosy and TB prompts the question of whether leprosy eventually declined because our immune system had already somewhat adapted to fight TB over the past 2000 years. This would mean that humans may have adapted in response to mycobacteria, but perhaps not specifically *M. leprae*.

Past human adaptation to pathogens is hypothesised to have partly contributed to the increased incidence of immune-mediated and neurodegenerative diseases today. This correlation occurred via the selection of genetic variants enhancing host immune response (Fava et al., 2019a). An interesting observation in aDNA studies of HLA genes was that HLA-DRB1*15:01 co-occurs with HLA-DQB1*06:02. These two HLA alleles are strong risk factors for immune-mediated and Parkinson's diseases but are also associated with resistance to leprosy, making it, in effect, a trade-off (Wang et al., 2016; Tsuchiya et al., 2001; Kaushansky & Ben-Nun, 2014; Sulzer et al., 2017). In addition, there are other examples of mirrored genetic associations between infectious diseases and immune-mediated diseases, and gene–environment interactions in humans. For example, variants in the *PRKN*, *NOD2*, *LACC1* and *TLR1* genes have been consistently found in association with leprosy resistance (Alter et al., 2013; Zhang et al., 2009; Sales-Marques et al., 2014; Wong et al., 2010). Mutations in *PRKN* that confer resistance to leprosy are the best-known cause of early-onset Parkinson's disease (Nédélec et al., 2016; Fava, 2019a, 2019b), while variants in *NOD2* and *LACC1* are strong risk factors for inflammatory bowel disease (Jostins et al., 2012). Further, regulatory variants for the *PRKN* gene expression were shown as major risk factors for leprosy, while rare mutations were associated with leprosy reactions (Alter et al., 2013; Fava et al., 2019b). Essentially, neuroinflammation in Parkinson's Disease (a mutation in the *PRKN* gene) and peripheral nerve damage due to inflammation in Type 1 leprosy reactions share overlapping genetic control of their pathogenicity. In fact, Mi and colleagues (2020) have suggested in their study that defect in the regulation of innate pro-inflammation genes contributed to risk of Type 1 reaction onset. Relevant, too, is a population study on current inhabitants of two Croatian islands that were former leprosaria; here an increased frequency of *PRKN* promoter variants were associated with leprosy (Bakija-Konsuo et al., 2011). This observation reinforces the role of *PRKN* on the occurrence of leprosy with possible positive selection during the medieval period in Europe. Another interesting example of selection due to host adaptation to mycobacteria is a mutation in the *TLR1* gene. The *TLR1* I602S mutation impairs the translocation of the protein to the cell surface and is associated with resistance to leprosy (Wong et al., 2010). The mutation is rare in African populations but most frequent among Europeans. Its higher frequency in Europeans, coupled with the strong association with leprosy, suggests positive selection due to a mycobacterial disease in the past. Studies using aDNA could offer new perspectives in evaluating the hypothesis of resistance to ancient infections versus modern immune mediated diseases. For instance, by contrasting the frequency of mutations in leprosy aDNA versus non-leprosy aDNA and modern human DNA, one could evaluate the selective pressure caused by *M. leprae* or another infection such as TB. An exciting venue to explore would be to evaluate aDNA of leprosy versus modern DNA of Crohn's disease in Nordic countries where leprosy was prevalent in the medieval period and immune-mediated conditions such as inflammatory bowel disease are a public health concern today (Ng et al., 2017). Taken together, genetic studies of aDNA and mapping of positive selection are starting to unravel the history of human adaptation to leprosy. However, while aDNA studies

of pathogens causing infectious diseases have an almost thirty-year history, starting in the early 1990s with TB (Spigelman & Lemma, 1993) and leprosy (Rafi et al., 1994), we are now seeing much more nuanced studies, with more to come in the future.

9.4 Leprosy in the past

There have been considerable advances in knowledge of the clinical and genetic aspects of leprosy, but what do we know about the past? On the basis of extant evidence from historical documents and skeletal remains showing signs of leprosy, it is not possible to surmise how frequent leprosy actually was, nor is it possible to do so based on the number of skeletons with leprosy. However, there are many caveats to the data, including that only a small proportion of untreated people who contract leprosy will develop bone changes, that the high resistance form of leprosy may not cause bone changes and that people may have died before bone changes could become manifest (see Roberts, 2020,186–9 for a full outline of the limitations). Furthermore, some parts of the world see relatively little palaeopathological work compared to places like Europe. On the other hand, historical writings suggest leprosy was endemic in Europe. If leprosaria could be used as an indicator of leprosy frequency—very debatable because there are challenges to using such data as objective scientific data—Lendrum's (1952) estimate of 20,000 leprosaria in Europe suggests a high frequency (but see Tabuteau, 2010). On the basis of the skeletal evidence of leprosy, we might also think about demographic patterns of leprosy in the past when compared to those patterns today (more adult men and children). However, on the basis of skeletal evidence of leprosy, which is selective and incomplete, and depending on which parts of the world have detected evidence, it seems inappropriate to speculate on how many men, women and children were affected in the past.

9.4.1 The evidence

Leprosy has had a long history, which is evidenced through a diverse range of sources: primary evidence (human remains, genetic data, archaeological sites, i.e., leprosaria), and secondary information (written documents, artistic representations). All these sources have their advantages and limitations, and we have assumed that the disease we recognise today has not changed over its long history with humans. For example, literary or artistic descriptions and depictions of non-specific skin rashes, or deformities of the face, hands and feet, may not necessarily be representations of a leprosy infection (see Mitchell, 2017 on criteria for validation). According to Dharmendra (1947), some reliable historical evidence does exist that describes sensory nerve damage and deformities (e.g., Indian Ayurvedic medical sources, such as the Suśruta-saṃhitā and Charak-saṃhitā dated to the fourth and first centuries AD, respectively). However, here we focus on the skeletal evidence.

Most archaeological skeletons with evidence of leprosy infections are concentrated in the cemeteries of leprosaria, while relatively small numbers have been identified outside of these contexts (e.g., Møller-Christensen, 1969; Magilton et al., 2008; Roffey & Tucker, 2012). When synthesising the evidence of leprosy in archaeological skeletons, it should be noted that this can only provide a snapshot of the infection's impact on our ancestors (see global overview in Roberts, 2020). Skeletal evidence for leprosy has so far been identified on three continents, Africa, Asia, and Europe, although it is unfortunate that little palaeopathological research has been undertaken in some regions; its full geographical range is thus unknown (see Buikstra & Roberts, 2012 on the global history of palaeopathology). As of yet, there has been no confirmed skeletal evidence of a leprosy infection from the Americas. Most evidence is found in Northern Europe, a region today that has a low prevalence of the disease with no cases within living memory (e.g., see WHO, 2019).

Much of the European evidence dates to the late medieval period, although early medieval evidence is also quite common (e.g., in Britain between the sixth and eleventh centuries AD) compared to the post-sixteenth century, when the prevalence declined dramatically. In relative terms, some countries have revealed more evidence than others, probably reflecting the number of palaeopathologists working 'on the ground' (e.g., Denmark, Hungary, Sweden, UK). That said, there is evidence several thousand years old in Hungary, India, and Turkey, and possibly Iran and Sudan but, apart from Hungary and Sudan, these parts of the world do not see much palaeopathological work. This is unfortunate when we start to think about the origin, evolution and transmission of leprosy over deep time. Palaeopathology clearly provides us with the primary evidence for leprosy in the past, notwithstanding that skeletons with no bone lesions may represent people with leprosy in life who had not developed bone damage at death (Wood et al., 1992). However, and importantly, initial studies of skeletons with leprosy can be combined with the collection of bone and tooth samples (aDNA and isotopes) and, more recently, dental calculus (Fotakis et al., 2020). These all can potentially provide information about the bacterial strain that infected the person, and their origin and movement during life. In turn, the resulting combined data can be used alongside modern DNA evidence to explore the bacteria's origin, evolution and spread.

9.5 Conclusions

Integrating clinical, genetic and palaeopathological evidence has allowed researchers to explore the origin, evolution and history of leprosy. In addition to the great advances in our knowledge of leprosy, Fine's (1982, 161) statement remains valid: 'there are few diseases that have a richer cultural heritage than leprosy, or are steeped in as many myths'. However, there is much that remains unanswered or unresolved, for example, the contribution of environmental sources of *M. leprae*, past and present, and disease transmission, including from other animals. It is perhaps biomolecular research on DNA samples from living people with leprosy and from the bones of the ancient dead that is beginning to strengthen our understanding of the evolution of this infection. There are a number of take-home messages that provide a narrative for thinking about how this current overview contributes to the understanding of the evolution of health and disease—in particular leprosy—how palaeopathology offers a unique perspective or new understanding and how this new understanding contributes to modern medicine.

First, there are various clinical manifestations of leprosy in people today, from having the infection in their bodies but no recognisable disease (subclinical infection) to developing plainly apparent leprosy. In addition, individuals have different immunological responses across a wide immunological spectrum, with correspondingly diverse manifestations of the disease, depending on their ability to generate resistance. Genetic factors affect the response of a person to *M. leprae* (or *M. lepromatosis*), as seen through biomolecular work, and must have done so in the past. However, *M. lepromatosis* aDNA has not yet been identified from archaeological human remains, although molecular data suggest *M. leprae*, *M. lepromatosis* and their last common ancestor probably both evolved alongside humans (Han & Silva, 2014).

Second, using modern and aDNA data, phylogenetic analyses have indicated that some people harbouring leprosy were moving across the globe for some considerable distance. For example, in her study of skeletons showing leprosy from Swedish archaeological sites, Caroline Arcini (2018) shows that none were born and raised in the local area where they were buried. Economou and colleagues (2013a,b) further note a medieval strain of leprosy that may have derived from Asia, thus suggesting that the person with that strain may have moved a long distance from their origin during their lives to be eventually buried in what is now modern day Sweden. Inskip and colleagues (2015) also describe the skeleton of a person with leprosy whose origin was outside England. Lastly, Rofffey and colleagues (2017) report a 'pilgrim burial', with a scallop shell in his grave. He had bone changes of leprosy and was buried in a cemetery associated with a leprosy hospital. Using aDNA and stable isotope data, it was suggested that the infected individual was not local to the area where he was buried.

Clearly, people were on the move, and they took infections with them. Finding *M. leprae* strains originating far from where a person was buried shows that leprosy must have spread with people who travelled (e.g. see Donoghue et al., 2015; Donoghue, 2019). Stable isotope analysis in archaeology is making us realise how mobile people were in the past, and that mobility contributed to the transmission of infectious disease. While modern leprosy DNA research revealed different *M. leprae* strains

in different parts of the globe (Monot et al., 2005, 2009), it has not been until relatively recently that these scientific data have been available to address the ancient origins of leprosy. The DNA and isotopic data from the remains of ancient people with leprosy have started to corroborate the modern DNA data and, in so doing, are providing a deep time perspective on the bacteria's evolution and spread. More studies combining these methods are needed to further explore the mobility of people with leprosy, particularly in areas of the world suggested by Monot and colleagues (2005, 2009) to be leprosy's origin, as well as along migratory and trade routes of human populations. While we note that researchers are using biomolecular markers for different bacterial strains of leprosy, past and present, on a more local scale these markers are being used today to assess exposure to leprosy, and tracking and tracing people with leprosy to document transmission patterns (e.g., Salipante & Hall, 2011).

Third, in understanding the socio-political, economic and environmental context in which people with leprosy lived in the past, it has been possible to show that most skeletons discovered so far with evidence of leprosy are associated with relatively recent time periods in urban Europe. These were places where population densities were often high, and inequality and poverty were likely common (e.g., Rawcliffe, 2013; Rosen, 1993), compromising people's immune systems. Leprosy would not have been solely a disease of the poor, however, as skeletons of what appear to be high-status people in medieval England have also been diagnosed with leprosy, indicating that this did not prevent them from having the infection and that the infection did not diminish their social status to the point that it was not recognised at death.

Few leprosy hospital cemeteries have been excavated, and the majority of skeletons have been found buried normally for their community, place and time period (see Roberts, 2020). This is a relatively recent discovery, as palaeopathologists increasingly seek to link evidence for disease in skeletons with funerary context. People with leprosy were not necessarily stigmatised as many historical sources have intimated. This evidence sits well alongside re-evaluations of historical data by medical historians (e.g., Demaitre, 2007; Rawcliffe, 2006).

Last, some discussion here should address the question of whether taking an evolutionary approach to leprosy, past and present, using DNA analysis might help understand antibiotic resistance. While drug resistance can and does develop in leprosy, particularly when the bacilli are exposed to only one anti-microbial drug (i.e. monotherapy), drug resistance occurs at a very low level. Partly this is undoubtedly due to the use of multi-drug regimens since the 1980s (approximately thirty years after the introduction of the first effective agent, dapsone), but the exceedingly slow growth rate of the bacilli is probably also a factor. For these reasons, leprosy is probably a very poor model for the effects of antibiotic resistance, unlike TB.

9.6 Future directions

In relation to a palaeopathological perspective for leprosy, there needs to be more effort directed at identifying human remains with leprosy indicators in their bones in 'palaeopathologically empty' parts of the world, and especially in regions relevant to hypotheses of the spread of leprosy. If the historical and genetic data are to be believed, we would not expect to see leprosy in human remains prior to very recent times in the Americas. It is also important to consider skeletal evidence for leprosy in concert with burial context to continue to build our understanding of how infected people were viewed by their communities in the past. This area of research has seen much more attention in recent years and has contributed to dispelling ideas derived from historical representations of leprosy, including that affected people were all stigmatised.

Further aDNA studies also need to be undertaken to test hypotheses for the origin and evolution of leprosy and fill the gaps in genetic knowledge along proposed transmission routes across the world. In addition, it would be beneficial to find aDNA evidence for *M. lepromatosis* in human remains. Sequencing of the genome of this organism has shown it to have evolved more recently than *M. leprae*, and Han and Silva (2014) suggest that both causative organisms for leprosy (*M. leprae* and *M. lepromatosis*) and their last common ancestor probably evolved with humans. Work with aDNA could further explore the *SLC11A1* gene in relation

to the presence/absence of cribra orbitalia and HLA genes in relation to leprosy susceptibility in past populations. Again, in concert with stable isotope analyses, more aDNA work will help to determine the bacterium's mobility patterns.

Finally, focusing on a 'One Health' approach (Mackenzie & Jeggo, 2019) and appreciating the relationship between leprosy in humans and other animals better would help us understand its transmission. For example, archaeozoologists might be persuaded to look for evidence of leprosy in armadillo and squirrel bones. Were these animals important for leprosy transmission in the past? Indeed, are red squirrels a 'leprosy hazard' for humans today?

References

Abel, L., Sánchez, F. O., Oberti, J., Thuc, N. V., Hoa, L. V., Lap, V. D., Skameme, E., Lagrange, P. H. & Schurr, E. (1998). Susceptibility to leprosy is linked to the human *NRAMP₁* gene. *Journal of Infectious Diseases*, 177, 133–145.

Alter, A., Fava, V.M., Huong, N. T., Singh, M., Orlova, M., Thuc, N. V., Katoch, K., Thai, V. H., Ba, N. N., Abel, L., Mehra, N., Alcaïs, A. & Schurr, E. (2013). Linkage disequilibrium pattern and age-at-diagnosis are critical for replicating genetic associations across ethnic groups in leprosy. *Human Genetics*, 132, 107–116.

Andersen, J. G. (1964). Foot drop in leprosy. *Leprosy Review*, 35, 41–46.

Andersen, J. G. & Manchester, K. (1987). Grooving of the proximal phalanx in leprosy: a palaeopathological and radiological study. *Journal of Archaeological Science*,14, 77–82.

Andersen, J. G. & Manchester, K. (1988). Dorsal tarsal exostoses in leprosy: a palaeopathological and radiological study. *Journal of Archaeological Science*,15, 51–56.

Arcini, C. (2018). *The Viking age: a time of many faces*. Oxbow, Oxford.

Avanzi, C., Del-Pozo, J., Benjak, A., Stevenson, K., Simpson, V. R., Busso, P., McLuckie, J., Loiseau, C., Lawton, C., Schoening, J., Shaw, D. J., Piton, J., Vera-Cabrera, L., Velarde-Felix, J. S., McDermott, F., Gordon, S. V., Cole, S. T. & Meredith, A. L. (2016). Red squirrels in the British Isles are infected with leprosy bacilli. *Science*, 354(6313), 744–747.

Bakija-Konsuo, A., Mulić, R., Boraska, V., Pehlic, M., Huffman, J. E., Hayward, C., Marlais, M., Zemunik, T. & Rudan, I. (2011). Leprosy epidemics during history increased protective allele frequency of PARK2/PACRG genes in the population of the Mljet Island, Croatia. *European Journal of Medical Genetics*, 54, e548–552. doi:10.1016/j.ejmg.2011.06.010.

Barreiro, L. B. & Quintana-Murci, L. (2010). From evolutionary genetics to human immunology: how selection shapes host defence genes. *Nature Reviews Genetics*, 11(1),17–30. doi: 10.1038/nrg2698.

Bellamy, R., Ruwende, C., Corrah, T., McAdam, K. P., Whittle, H.C. & Hill, A. V. (1998). Variations in the *NRAMP₁* gene and susceptibility to tuberculosis in West Africans. *New England Journal of Medicine*, 338, 640–644.

Boeckl, C. (2021). *Images of leprosy: disease, religion & politics in European art*. Truman State University Press, Kirksville.

Bos, K. I., Harkins, K. M., Herbig, A., Coscolla, M., Weber, N., Comas, I., Forrest, S. A., Bryant, J. M., Harris, S. R., Schuenemann, V. J., Campbell, T. J., Majander, K., Wilbur, A. K., Guichon, R. A., Wolfe Steadman, D. L., Collins Cook, D., Niemann, S., Behr, M. A., Zumarraga, M., Bastida, R., Huson, D., Nieselt, K., Young, D., Parkhill, J., Buikstra, J. E., Gagneux, S., Stone, A. C. & Krause, J. (2014). Pre-Columbian mycobacterial genomes reveal seals as a source of New World human tuberculosis. *Nature*, 514(7523), 494–497. doi:10.1038/nature13591.

Brennan, P. J. & Spencer, J. S. (2019). 'The physiology of Mycobacterium leprae', in D. M. Scollard & T. P. Gillis (eds), *International textbook of leprosy* [online], Chapter 5.1. https://internationaltextbookofleprosy.org.

Brosch, R., Gordon, S. V., Marmiesse, M., Brodin, P., Buchrieser, C., Eiglmeier, K., Garnier, T., Gutierrez, C., Hewinson, G., Kremer, K., Parsons, L. M., Pym, A. S., Samper, S., van Soolingen, D. & Cole, S. T. (2002). A new evolutionary scenario for the *Mycobacterium tuberculosis* complex. *Proceedings of the National Academy of Sciences of the United States of America*, 99(6), 3684–3689.

Buikstra, J. E. & Roberts, C.A. (eds). (2012). *A global history of paleopathology: pioneers and prospects*. Oxford University Press, New York.

Chaussinand, R. (1948). Tuberculose et lèpre, maladies antagoniques: éviction de la lèpre par la tuberculose. *International Journal of Leprosy and Other Mycobacterial Diseases*, 16, 431–438.

Cole, S. T., Eiglmeier, K., Parkhill, J., James, K. D., Thomson, N. R., Wheeler, P. R., Honoré, N., Garnier, T., Churcher, C., Harris, D., Mungall, K., Basham, D., Brown, D., Chillingworth, T., Connor, R., Davies, R. M., Devlin, K., Duthoy, S., Feltwell, T., Fraser, A., Hamlin, N., Holroyd, S., Hornsby, T., Jagels, K., Lacroix, C., Maclean, J., Moule, S., Murphy, L., Oliver, K., Quail, M. A., Rajandream, M.-A., Rutherford, K. M., Rutter, S., Seeger, K., Simon, S., Simmonds, M., Skelton, J., Squares, R., Squares, S., Stevens, K., Taylor, K.,

Whitehead, S., Woodward, J. R. & Barrell, B. G. (2001). Massive gene decay in the leprosy bacillus. *Nature*, 409, 1007–1011. doi:10.1038/35059006.

Crespo, F., White, J. & Roberts, C. A. (2019). Revisiting the tuberculosis and leprosy cross-immunity hypothesis: expanding the dialogue between immunology and paleopathology. *International Journal of Paleopathology*, 26, 37–47.

Da Silva, M. B., Portela, J. M., Li, W., Jackson, M., Gonzalez, M., Hidalgo, A. S., Belisle, J. T., Bouth, R. C., Gobbo, A. R., Barreto, J. G., Minervino, A. H. H., Cole, S. T., Avanzi, C., Busso, P., Frade, M. A. C., Geluk, A., Salgado, C. G. & Spencer, J. S. (2018). Evidence of zoonotic leprosy in Pará, Brazilian Amazon, and risks associated with human contact or consumption of armadillos. *PLoS Neglected Tropical Diseases*, 12(6), e0006532. https://journals.plos.org/plosntds/article?id=10.1371/journal.pntd.0006532.

Dallmann-Sauer, M., Fava, V. M., Gzara, C., Orlova, M., Thuc, N. V., Thai. V. H., Alcaïs, L. A., Cobat, A. & Schurr, E. (2020). The complex pattern of genetic associations of leprosy with HLA class I and class II alleles can be reduced to four amino acid positions. *PLoS Pathogens* [online]. https://doi.org/10.1371/journal.ppat.1008818.

Da Silva Neumann, A., de Almeida Dias, F., da Silva Ferreira, J., Nogueira Brum Fontes, A., Sammarco Rosa, P., Macedo, R. E., Oliveira, J. H., Lima de Figueiredo Teixeira, R., Vidal Pessolani, M. C., Moraes, M. O., Suffys, P. N., Oliveira, P. L., Ferreira Sorgine, M. H. & Lara, F. A. (2016). Experimental infection of *Rhodnius prolixus* (Hemiptera, Triatominae) with *Mycobacterium leprae* indicates potential for leprosy transmission. *PLoS One*, 11(5), e0156037.

Demaitre, L. (2007). *Leprosy in premodern medicine: A malady of the whole body*. Johns Hopkins University Press, Baltimore.

Dharmendra. (1947). Leprosy in ancient Indian medicine. *International Journal of Leprosy*, 15, 424–430.

Donoghue, H. D. (2019). Tuberculosis and leprosy associated with historical human population movements in Europe and beyond: an overview based on Mycobacterial ancient DNA. *Annals of Human Biology*, 46, 120–128. doi.org/10.1080/03014460.2019.1624822.

Donoghue, H. D., Marcsik, A., Matheson, C., Vernon, K., Nuorala, E., Molto, J. E., Greenblatt, C. L. & Spigelman, M. (2005). Co-infection of *Mycobacterium tuberculosis* and *Mycobacterium leprae* in human archaeological samples: a possible explanation for the historical decline of leprosy. *Proceedings of the Royal Society B: Biological Sciences*, 272, 389–394. doi.org/10.1098/rspb.2004.2966.

Donoghue, H. D., Taylor, G. M., Marcsik, A., Molnár, E., Pálfi, G., Pap, I., Teschler-Nicola, M., Pinhasi, R., Erdal,

Y. S., Veleminský, P., Likovský, J., Belcastro, M. G., Mariotti, V., Riga, A., Rubini, M., Zaio, P., Besra, G., Lee, O., Wu, H. H., Minnikin, D., Bull, I., O'Grady, J. & Spigelman, M. (2015). A migration-driven model for the historical spread of leprosy in medieval Eastern and Central Europe. *Infection, Genetics and Evolution*, 31, 250–256.

Duhig, C. (1998). 'The human skeletal material', in T. Malin & J. Hines (eds), *The Anglo-Saxon cemetery at Edix Hill (Barrington A), Cambridgeshire*, pp. 154–199. Council for British Archaeology Research Reports, vol. 112. Council for British Archaeology, York.

Economou, C., Kjellstrom, A., Lidén, K. & Panagopoulos, I. (2013a). Ancient-DNA reveals an Asian type of *Mycobacterium leprae* in medieval Scandinavia. *Journal of Archaeological Science*, 40, 465–470.

Economou, C., Kjellstrom, A., Lidén, K. & Panagopoulos, I. (2013b). Corrigendum to 'Ancient-DNA reveals an Asian type of *Mycobacterium leprae* in medieval Scandinavia' [J. Archaeol. Sci. 40(1): 465–470]. *Journal of Archaeological Science*, 40, 2867.

Fava, V. M., Dallmann-Sauer, M. & Schurr, E. (2019a). Genetics of leprosy: today and beyond. *Human Genetics*, 139(6–7), 835–846. doi:10.1007/s00439-019-02087-5.

Fava, V. M., Xu, Y. Z., Lettre, G., Thuc, N. V., Orlova, M., Thai, V. H., Tao, S., Croteau, N., Eldeeb, M. A., MacDougall, E. J., Cambri, G., Lahiri, R., Adams, L., Fon, E. A., Trempe, J-F, Cobat, A., Alcaïs, A., Abel, L. & Schurr, E. (2019b). Pleiotropic effects for Parkin and LRRK2 in leprosy type-1 reactions and Parkinson's disease. *Proceedings of the National Academy of Sciences of the United States of America*, 116, 15616–15624. doi:10.1073/pnas.1901805116.

Ferreira, J. D. S., Souza Oliveira, D. A., Santos, J. P., Uzedo Ribeiro, C. C. D., Baêta, B. A., Teixeira, R. C., da Silva Neumann, A., Sammarco Rosa, P., Vidal Pessolani, M. C., Moraes, M. O., Bechara, G. H., de Oliveira, P. L., Ferreira Sorgine, M. H., Suffys, P. N., Brum Fontes, A. N., Bell-Sakyi, L., Fonseca, A. H. & Lara, F. A. (2018). Ticks as potential vectors of *Mycobacterium leprae*: use of tick cell lines to culture the bacilli and generate transgenic strains. *PLoS Neglected Tropical Diseases* [online]. doi: 10.1371/journal.pntd.0007001.

Fine, P. E. M. (1982). Leprosy: the epidemiology of a slow bacterium. *Epidemiologic Reviews*, 4, 161–188.

Fotakis, A. K., Denham, S. D., Mackie, M., Orbegozo, M. I., Mylopotamitaki, D., Gopalakrishnan, S., Sicheritz-Pontén, T., Olsen, J. V., Cappellini, E., Zhang, G., Christophersen, A., Gilbert, M. T. P. & Vågene, Å. J. (2020). Multi-omic detection of *Mycobacterium leprae* in archaeological human dental calculus. *Philosophical Transactions of the Royal Society B: Biological Sciences*, 375, 20190584. http://dx.doi.org/10.1098/rstb.2019.0584.

Grön, K. (1973). Leprosy in literature and art. *International Journal of Leprosy*, 41, 249–283.

Grzybowski, A. & Nita, N. (2016). Leprosy in the Bible. *Clinics in Dermatology*, 34, 3–7.

Han, X. Y., Mistry, N. A., Thompson, E. J., Tang, H. L., Khanna, K. & Zharng, L. (2015). Draft genome of new leprosy agent *Mycobacterium lepromatosis*. *Genome Announcements*, 3(3), e00513–e00515.

Han, X. Y. & Silva, F. J. (2014). On the age of leprosy. *PLoS Neglected Tropical Diseases*, 8(2), e2544.

Hockings, K. J., Mubemba, B., Avanzi, C., Pleh, K., Düx, A., Bersacola, E., Bessa, J., Ramon, M., Metzger, S., Patrono, L. V., Jaffe, J. E., Benjak, A., Bonneaud, C., Busso, P., Couacy-Hymann, E., Gado, M., Gagneux, S., Johnson, R. C., Kodio, M., Lynton-Jenkins, J., Morozova, I., Kätz-Rensing, K., Regalla, A., Said, A. R., Schuenemann, A. J., Sow, S. O., Spencer, J. S., Ulrich, M., Zoubi, H., Cole, S. T., Wittig, R. M., Calvignac-Spencer, S. & Leendertz, F. H. (2020). Leprosy in wild chimpanzees. *bioRxiv* [preprint online]. https://doi.org/10.1101/2020.11.10.374371.

Holloway, K. L., Henneberg, R. J., de Barros Lopez, M. & Henneberg, M. (2011). Evolution of human tuberculosis: a systematic review and meta-analysis of paleopathological evidence. *Homo*, 62, 402–458.

Hunt, K., Roberts, C. A. & Kirkpatrick, C. (2018). Taking stock: a systematic review of archaeological evidence of cancers in human and early hominin remains. *International Journal of Paleopathology*, 21, 12–26.

Inskip, S., Taylor, G. M. & Stewart, G. (2017). Leprosy in pre-Norman Suffolk, UK: biomolecular and geochemical analysis of the woman from Hoxne. *Journal of Medical Microbiology*, 66(11), 1640–1649.

Inskip, S. A., Taylor, G. M., Zakrzewski, S. R., Mays, S. A., Pike, A. W. G., Llewellyn, G., Williams, C. M., Lee, O. Y-C., Wu, H. H. T., Minnikin, D. E., Besra, G. S. & Stewart, G. R. (2015). Osteological, biomolecular and geochemical examination of an Early Anglo-Saxon case of lepromatous leprosy. *PLoS One*, 10(5), e0124282. https://journals.plos.org/plosone/article?id=10.1371/journal.pone.0124282.

Jablonski, N. G. & Chaplin, G. (2018). The roles of vitamin D and cutaneous vitamin D production in human evolution and health. *International Journal of Paleopathology*, 23, 54–59.

Job, C. K. (1989). Nerve damage in leprosy. *International Journal of Leprosy and Other Mycobacterial Diseases*, 57, 532–539.

Jostins, L., Ripke, S., Weersma, R. K., Duerr, R. H., McGovern, D. P., Hui, K. Y., Lee, J. C., Schumm, L. P., Sharma, Y., Anderson, C. A., Essers, J., Mitrovic, M., Ning, K., Cleynen, I., Theatre, E., Spain, S. L., Raychaudhuri, S., Goyette, P., Wei, Z., Abraham, C., Achkar, J-P., Ahmad, T., Amininejad, L., Ananthakrishnan, A. N., Andersen,

V., Andrews, J. M., Baidoo, L., Balschun, T., Bampton, P. A., Bitton, A., Boucher, G., Brand, S., Büning, C., Cohain, A., Cichon, S., D'Amato, M., De Jong, D., Devaney, K. L., Dubinsky, M., Edwards, C., Ellinghaus, D., Ferguson, L. R., Franchimont, D., Fransen, K., Gearry, R., Georges, M., Gieger, C., Glas, J., Haritunians, T., Hart, A., Hawkey, C., Hedl, M., Hu, X., Karlsen, T. H., Kupcinskas, L., Kugathasan, S., Latiano, A., Laukens, D., Lawrance, I. C., Lees, C. W., Louis, E., Mahy, G., Mansfield, J., Morgan, A. R., Mowat, C., Newman, W., Palmieri, O., Ponsioen, C. Y., Potocnik, U., Prescott, N. J., Regueiro, M., Rotter, J. I., Russell, R. K., Sanderson, J. D., Sans, M., Satsangi, J., Schreiber, S., Simms, L. A., Sventoraityte, J., Targan, S. R., Taylor, K. D., Tremelling, M., Verspaget, H. W., De Vos, M., Wijmenga, C., Wilson, D. C., Winkelmann, J., Xavier, R. J., Zeissig, S., Zhang, B., Zhang, C. K., Zhao, H., International IBD Genetics Consortium (IIBDGC); Silverberg, M. S., Annese, V., Hakonarson, H., Brant, S. R., Radford-Smith, G., Mathew, C. G., Rioux, J. D., Schadt, E. E., Daly, M. J., Franke, A., Parkes, M., Vermeire, S., Barrett, J. C. & Cho, J. H. (2012). Host-microbe interactions have shaped the genetic architecture of inflammatory bowel disease. *Nature*, 491, 119–124.

Kaufmann, E., Sanz, J., Dunn, J. L., Khan, N., Mendonça, L. E., Pacis, A., Tzelepis, F., Pernet, E., Dumaine, A., Grenier, J-C., Mailhot-Léonard, F., Ahmed, E., Belle, J., Besla, R., Mazer, B., King, I. L., Nijnik, A., Robbins, C. S., Barreiro, L. B. & Divangahi, M. (2018). BCG educates hematopoietic stem cells to generate protective innate immunity against tuberculosis. *Cell*, 72, 176–190.

Kaushansky, N. & Ben-Nun, A. (2014). DQB1*06:02-associated pathogenic anti-myelin autoimmunity in multiple sclerosis-like disease: potential function of DQB1*06:02 as a disease-predisposing allele. *Frontiers of Oncology*, 4, 280: doi:10.3389/fonc.2014.00280.

Kellersberger, E. R. (1951). The social stigma of leprosy. *Annals of the New York Academy of Sciences*, 54, 126–133. doi:10.1111/j.1749-6632.

Kerr, L., Kendall, C., Sousa, C.A., Frota, C. C., Graham, J., Rodrigues, L., Lima Fernandes, R. & Lima Barreto, M. (2015). Human-armadillo interaction in Ceara, Brazil: potential for transmission of *Mycobacterium leprae*. *Acta Tropica*, 152, 74–79.

Kerudin, A., Müller, R., Buckberry, J., Knüsel, C. & Brown, T. (2019). Ancient *Mycobacterium leprae* genomes from mediaeval England. *Journal of Archaeological Science*, 112, 105035. doi.org/10.1016/j.jas.2019.105035.

Köhler, K., Marcsik, A., Zádori, P., Biro, G., Szeniczey, T., Fábián, S., Serlegi, G., Marton, T., Donoghue, H. D. & Hajdu, T. (2017). Possible cases of leprosy from the Late Copper Age (3780–3650 cal bc) in Hungary. *PLoS One*, 12, e0185966.

Krause-Kyora, B., Nutsoa, M., Boehme, L., Kornell, S-C., Drichel, D., Bonazzi, M., Möbus, L., Tarp, P., Susat, J., Bosse, E., Willburger, B., Schmidt, A. H., Sauter, J., Franke, A., Wittig, M., Caliebe, A., Nothnagel, M., Schreiber, S., Boldsen, J. L., Lenz, T. L. & Nebel, A. (2018). Ancient DNA study reveals HLA susceptibility locus for leprosy in medieval Europeans. *Nature Communications*, 9, 1569. doi:10.1038/s41467-018-03857-x.

Leitman, T., Porco, T. & Blower, S. (1997). Leprosy and tuberculosis: the epidemiological consequences of cross-immunity. *American Journal of Public Health*, 87, 1923–1927.

Lendrum, F. C. (1952). The name 'leprosy'. *American Journal of Tropical Medicine and Hygiene*, 1, 999–1008.

Lie, H. P. (1938). On leprosy in the Bible. *Leprosy Review*, 9, 25–67.

Lockwood, D. N. J. (2019). 'Treatment of leprosy', in D. M. Scollard & T. P. Gillis (eds), *International textbook of leprosy* [online], Chapter 2.6. https://internationaltextbookofleprosy.org.

Mackenzie, J. S. & Jeggo, M. (2019). The One Health approach—why is it so important? *Tropical Medicine and Infectious Disease*, 4(2). 10.3390/tropicalmed4020088.

Magilton, J. R., Lee, F. & Boylston, A. (eds). (2008). *'Lepers outside the gate': excavations at the cemetery of the hospital of St James and St Mary Magdalene, Chichester, 1986–87 and 1993*. Council for British Archaeology Research Reports, vol. 158. Council for British Archaeology, York.

Manchester, K. (1984). Tuberculosis and leprosy in antiquity: an interpretation. *Medical History*, 28, 162–173.

Mathieson, I., Lazaridis, I., Rohland, N., Mallick, S., Patterson, N., Roodenberg, S. A., Harney, E., Stewardson, K., Fernandes, D., Novak, M., Sirak, K., Gamba, C., Jones, E. R., Llamas, B., Dryomov, S., Pickrell, J., Arsuaga, J. L., Bermúdez de Castro, J. M., Carbonell, E., Gerritsen, F., Khokhlov, A., Kuznetsov, P., Lozano, M., Meller, H., Mochalov, O., Moiseyev, V., Rojo Guerra, M. A., Roodenberg, J., Vergès, J. M., Krause, J., Cooper, A., Alt, K. W., Brown, D., Anthony, D., Lalueza-Fox, C., Haak, W., Pinhasi, R. & Reich, D. (2015). Genome-wide patterns of selection in 230 ancient Eurasians. *Nature*, 528, 499–503. https://doi.org/10.1038/nature16152.

Matos, V. M. J. (2009). *Odiagnóstico retrospectivo da lepra: complimentaridade clínica e paleopatológica no arquivo médico do Hospital-Colónia Rovisco Pais (Século X. X., Tocha, Portugal) e na colecção de esqueletos da leprosaria medieval de St Jørgen's (Odense, Dinamarca) [The retrospective diagnosis of leprosy: clinical and paleopathological complementarities in the medical files from the Rovisco Pais Hospital-Colony (20th century, Tocha, Portugal) and in the skeletal collection from the Medieval leprosarium of St. Jorgen's (Odense, Denmark)]*. PhD Thesis, Universidade de Coimbra, Faculdade de Ciências e Technologia, Coimbra.

Meredith, A., Del Pozo, J., Smith, S., Milne, E., Stevenson, K. & McLuckie, J. (2014). Leprosy in red squirrels in Scotland. *Veterinary Record*, 175, 285–286.

Meyers, W. M., Binford, C. H., Walsh, G. P., Wolf, R. H., Gormus, B. J., Martin, L. N. & Gerone, P. J. (1984). 'Animal models in leprosy', in L. Leive & D. Schlessinger (eds), *Microbiology*, pp. 307–311. American Society for Microbiology, Washington DC.

Mi, Z., Liu, H. & Zhang, F. (2020). Advances in the immunology and genetics of leprosy. *Frontiers in Immunology* [online]. https://doi.org/10.3389/fimmu.2020.00567.

Mitchell, P. D. (2017). Improving the use of historical written sources in paleopathology. *International Journal of Paleopathology*, 19, 88–95.

Møller-Christensen, V. (1961). *Bone changes in leprosy*. Munksgaard, Copenhagen.

Møller-Christensen, V. (1969). 'Provisional results of the examination of the whole Naestved leprosy hospital churchyard—ab. 1250–1550 A. D.', in W. Kock (ed), Nordisk Medicinhistorik Årsbok, Part 4, pp. 29–36. Museum of Natural History: Stockholm.

Molto, J. E. (2002). 'Leprosy in Roman period skeletons from Kellis 2, Dakhleh, Egypt', in C. A. Roberts, M. E. Lewis. & K. Manchester (eds), *The past and present of leprosy: archaeological, historical, palaeopathological and clinical approaches*, pp. 179–192. British Archaeological Reports International Series, vol. 1054. Archaeopress, Oxford.

Monot, M., Honoré, N., Garnier, M. T., Araoz, R., Coppée, J-Y., Lacroix, C., Sow, S., Spencer, J. S., Truman, R. W., Williams, D. L., Gelber, R., Virmond, M., Flageul, B., Cho, S.-N., Ji, B., Paniz-Mondolfi, A., Convit, J., Young, S., Fine, P. E., Rasolofo, V., Brennan, P. J., & Cole, S. T. (2005). On the origin of leprosy. *Science*, 308, 1040–1042.

Monot, M., Honoré, N., Garnier, T., Zidane, N., Sherafi, D., Paniz-Mondolfi, A., Matsuoka, M., Taylor, G. M., Donoghue, H. D., Bouwman, A., Mays, S., Watson, C., Lockwood, D., Khamesipour, A., Dowlati, Y., Jianping, S., Rea, T. H., Vera-Cabrera, L., Stefani, M. M., Banu, S., Macdonald, M., Sapkota, B. R., Spencer, J. S., Thomas, J., Harshman, K., Singh, P., Busso, P., Gattiker, A., Rougemont, J., Brennan, P. J. & Cole, S. T. (2009). Comparative genomic and phylogeographic analysis of *Mycobacterium leprae*. *Nature Genetics*, 41, 1282–1289.

Moraes, M. O., Silva, L. R. B. & Pinheiro, R. O. (2017). 'Innate immunology', in D. M. Scollard & T. P. Gillis (eds), *International textbook of leprosy* [online], Chapter 6.1. https://internationaltextbookofleprosy.org.

Müller, R., Roberts, C. A., & Brown T. A. (2014). Genotyping of ancient *Mycobacterium tuberculosis* strains reveals historic genetic diversity. *Proceedings of the Royal Society B: Biological Sciences*, 281(1781), 20133236. doi: 10.1098/rspb.2013.3236.

Mysterud, I., Viljen Poleszynski, D., Lindberg, F. A. & Bruset, S.A. (2008). 'To eat or what not to eat, that's the question. A critique of the official Norwegian dietary guidelines', in W. R. Trevathan, E. O Smith & J. J. McKenna (eds), *Evolutionary medicine and health: new perspectives*, pp. 96–115. Oxford University Press, Oxford.

Nédélec, Y., Sanz, J., Baharian, G., Szpiech, Z. A., Pacis, A., Dumaine, A., Grenier, J-C., Freiman, A., Sams, A. J., Hebert, S., Sabourin, A. P., Luca, F., Blekhman, R., Hernandez, R. D., Pique-Regi, R., Tung, J., Yotova, V. & Barreiro, L. B. (2016). Genetic ancestry and natural selection drive population differences in immune responses to pathogens. *Cell*, 167(3), 657–669. doi: 10.1016/j.cell.2016.09.025.

Nesse, R. M. (2008). 'The importance of evolution for medicine', In W. R. Trevathan, E. O. Smith & J. J. McKenna (eds), *Evolutionary medicine and health: new perspectives*, pp. 416–431. Oxford University Press, Oxford.

Nesse, R. M. & Williams, G. C. (1994). *Why we get sick. The new science of Darwinian medicine*. Vintage Books, New York.

Ng, S. C., Shi, H. Y, Hamidi, N., Underwood, F. E., Tang, W., Benchimol, E. I., Panaccione, R., Ghosh, R., Wu, J. C. Y., Chan, F. K. L., Sung, J. J. Y. & Kaplan, G. G. (2017). Worldwide incidence and prevalence of inflammatory bowel disease in the 21st century: a systematic review of population-based studies. *The Lancet*, 390, 2769–2778.

Noordeen, S. K. (1990). MDT and leprosy control. *Indian Journal of Leprosy*, 62, 448–458.

Paterson, D. E. & Job, C. K. (1964). 'Bone changes and absorption in leprosy: a radiological, pathological and clinical study', in R. G. Cochrane & T. F. Davey (eds), *Leprosy in theory and practice*, pp. 425–446. John Wright and Sons, Bristol.

Perrin, P. (2015). Human and tuberculosis co-evolution: an integrative view. *Tuberculosis*, 95(Suppl 1), S112–S116.

Prengle, S., Neukamm, J., Guellil, M., Keller, M., Molak, M., Avanzi, C., Kushniarevich, A., Montes, N., Neumann, G. U., Reiter, E., Tukhbatova, R. I., Berezina, N. Y., Buzhilova, A. P., Korobov, D. S., Suppersberger Hamre, S., Matos, V. M. J., Ferreira, M. T., González-Garrido, L., Wasterlain, S. N., Lopes, C., Santos, A. L., Antunes-Ferreira, N., Duarte, V., Silva, A. M., Melo, L., Sarkic, N., Saag, L., Tambets, K., Busso, P., Cole, S. T., Avlasovich, A., Roberts, C. A., Sheridan, A., Cessford, C., Robb, J., Krause, J., Scheib, C. L., Inskip, S. A. & Schuenemann, V. J. (2021). *Mycobacterium leprae* diversity and population dynamics in medieval Europe from novel ancient genomes. *BMC Biology* [online], 19, article no. 220.

Rafi, A., Spigelman, M., Stanford, J., Lemma, E., Donoghue, H. & Zias, J. (1994). DNA of *Mycobacterium leprae* detected by PCR in ancient bone. *International Journal of Osteoarchaeology*, 4, 287–290.

Rao, P. S. S. (2015). Perspectives on the impact of stigma in leprosy: strategies to improve access to health care. *Research and Reports in Tropical Medicine*, 6, 49–57.

Rawcliffe, C. (2006). *Leprosy in medieval England*. Boydell Press, Woodbridge.

Rawcliffe, C. (2013). *Urban bodies: communal health in late medieval towns and cities*. Boydell Press, Woodbridge.

Ridley, D. S. & Jopling, W. H. (1966). Classification of leprosy according to immunity. A five-group system. *International Journal Leprosy and Other Mycobacterial Diseases*, 34(3), 255–273.

Roberts, C. A. (2020). *The past and present of leprosy*. University Press of Florida, Gainesville.

Roberts, C. A. & Steckel, R. H. (2019). 'The developmental origins hypothesis and the history of health project', in R. H. Steckel, C. S. Larsen, C. A. Roberts & J. Baten (eds). *The backbone of Europe. Health, diet, work and violence over two millennia*, pp. 325–351. Cambridge University Press, Cambridge.

Roffey, S. & Tucker, K. (2012). A contextual study of the medieval hospital and cemetery of St Mary Magdalen, Winchester, England. *International Journal of Paleopathology*, 2, 170–180.

Roffey, S., Tucker, K., Filipek-Ogden, K., Montgomery, J., Cameron, J, O'Connell, T., Evans, J., Marter, P. & Taylor, G. M. (2017). Investigation of a medieval pilgrim burial excavated from the leprosarium of St Mary Magdalen, Winchester, UK. *PLoS Neglected Tropical Diseases*, 11(1), e0005186. https://journals.plos.org/plosntds/article?id=10.1371/journal.pntd.0005186.

Rosen, G. (1993). *A history of public health*. Johns Hopkins University Press, London.

Sales-Marques, C., Salomão, H., Fava, V. M., Alvarado-Arnez, L. E., Pinheiro Amaral, E., Chester Cardoso, C., Foschiani Dias-Batista, I. M., da Silva, W. L., Medeiros, P., da Cunha Lopes Virmond, M., Lana, F. C. F., Pacheco, A. G., Moraes, M. O., Távora Mira, M. & Pereira Latini, A. C. (2014). NOD2 and CCDC122-LACC1 genes are associated with leprosy susceptibility in Brazilians. *Human Genetics*, 133, 1525–1532. doi:10.1007/s00439-014-1502-9.

Salipante, S. J. & Hall, B. G. (2011). Towards an epidemiology of *Mycobacterium leprae*: strategies, successes, and shortcomings. *Infection, Genetics and Evolution*, 11, 1505–1513.

Schuenemann, V. J., Avanzi, C., Krause-Kyora, B., Seitz, A., Herbig, A., Inskip, S., Bonazzi, M., Reiter, E., Urban, C., Dangvard Pedersen, D., Taylor, G. M., Singh, P., Stewart, G. R., Velemínský, P., Likovsky, J., Marcsik, A., Molnár, E., Pálfi, G., Mariotti, V., Riga, A., Belcastro, M. G., Boldsen, J. L., Nebel, A., Mays, S., Donoghue, H. D., Zakrzewski, S., Benjak, A., Nieselt, K., Cole, S. T. &

Krause, J. (2018). Ancient genomes reveal a high diversity of *Mycobacterium leprae* in medieval Europe. *American Journal of Physical Anthropology*, 165, 245–45.

Schuenemann, V. J., Singh, P., Mendum, T. A., Krause-Kyora, B., Jäger, G., Bos, K. I., Herbig, A., Economou, C., Benjak, A., Busso, P., Nebel, A., Boldsen, J. L., Kjellström, A., Wu, H., Stewart, G. R., Taylor, G. M., Bauer, P., Lee, O. Y-C., Wu, H. H. T., Minnikin, D. E., Besra, G. S., Tucker, K., Roffey, S., Sow, S. O., Cole, S. T., Nieselt, K. & Krause, J. (2013). Genome-wide comparison of medieval and modern *Mycobacterium leprae*. *Science*, 341, 179–183.

Scollard, D. M. & Skinsnes, O. K. (1999). Oropharyngeal leprosy in art, history, and medicine. *Oral Surgery, Oral Medicine, Oral Pathology and Oral Radiology*, 87(4), 463–470.

Scollard, D. M., Adams, L. B., Gillis, T. P., Krahenbuhl, J. L., Truman, R. W. & Williams, D. L. (2006). The continuing challenges of leprosy. *Clinical Microbiology Reviews*, 19(2), 338–381.

Scollard, D. M., McCormick, G. & Allen, J. L. (2015). Localization of *Mycobacterium leprae* to endothelial cells of epineurial and perineurial blood vessels and lymphatics. *The American Journal of Pathology*, 154(5), 1611–1620.

Scollard, D. M., Truman, R. W. & Ebenezer, G. J. (2015). Mechanisms of nerve injury in leprosy. *Clinical Dermatology*, 33(1), 46–54.

Sharma, R., Singh, P., Loughry, W. J., Lockhart, J. M., Inman, W. B., Duthie, M. S., Pena, M. T., Marcos, L. A., Scollard, D. M., Cole, S. T. & Truman, R. W. (2015). Zoonotic leprosy in the Southeastern United States. *Emerging Infectious Diseases*, 21, 2127–2134.

Singh, P., Benjaka, A., Schuenemann, V. J., Herbig, A., Avanzi, C., Busso, P. Niesel, K., Krause, J., Vera-Cabrera, L. & Cole, S. T. (2015). Insight into the evolution and origin of leprosy bacilli from the genome sequence of *Mycobacterium lepromatosis*. *Proceedings of the National Academy of Science of the United States of America*, 112, 4459–4464.

Singh, P., Tufariello, J. & Wattam, A. R. (2018). 'Genomics insights into the biology and evolution of leprosy bacilli', in D. M. Scollard & T. P. Gillis (eds), *International textbook of leprosy* [online], Chapter 8.2. https://internationaltextbookofleprosy.org.

Smith, G. E. (1908). 'Report on the human remains', in G. A. Reisner (ed), *The archæological survey of Nubia: report for 1907-1908*, pp. 125–35. National Print Department, Cairo.

Spigelman, M. & Lemma, E. (1993). The use of the polymerase chain reaction (PCR) to detect *Mycobacterium tuberculosis* in ancient skeletons. *International Journal of Osteoarchaeology*, 3, 137–143.

Spyrou, M. A., Bos, K. I., Herbig, A. & Krause, J. (2019). Ancient pathogen genomics as an emerging tool for infectious disease research. *Nature Reviews Genetics*, 20, 323–340. doi:10.1038/s41576-019-0119-1.

Stuart-Macadam, P. (1992). Porotic hyperostosis: a new perspective. *American Journal of Physical Anthropology*, 87, 39–47.

Sulzer, D., Alcalay, R.N., Garretti, F., Sulzer, D., Alcalay, R. N., Garretti, F., Cote, L., Kanter, E., Agin-Liebes, J. P., Liong, C., McMurtrey, C., Hildebrand, W. H., Mao, X., Dawson, V. L., Dawson, T. M., Oseroff, C., Pham, J., Sidney, J., Dillon, M. B., Carpenter, C., Weiskopf, D., Phillips, E., Mallal, S., Peters, B., Frazier, A., Lindestam Arlehamn, C. S. & Sette, A. (2017). T cells from patients with Parkinson's disease recognize α-synuclein peptides [published correction appears in *Nature*, 549, (7671), 292]. *Nature*, 546(7660), 656–661. doi:10.1038/nature22815.

Tabuteau, B. (2010). Vingt mille léproseries au Moyen Âge? Tradition française d'un poncif historiographique. *Revue belge de Philologie et d'Histoire Année*, 88(4), 1293–1300.

Taylor, G. M. & Donoghue, H. D. (2011). Multiple loci variable number tandem repeat (VNTR) analysis (MLVA) of *Mycobacterium leprae* isolates amplified from European archaeological human remains with lepromatous leprosy. *Microbes and Infection*, 13, 923–929.

Taylor, G. M., Murphy, E. M., Mendum, T. A., Pike, A. W. G., Linscott, B., Wu, H., O'Grady, J., Richardson, H., O'Donovan, E., Troy, C. & Stewart, G. R. (2018). Leprosy at the edge of Europe: biomolecular, isotopic and osteoarchaeological findings from medieval Ireland. *PLoS One*, 13, e.0209495.

Taylor, G. M., Watson, C. L., Bouwman, A. S., Lockwood, D. N. J. & Mays, S. A. (2006). Variable nucleotide tandem repeat (VNTR) typing of two palaeopathological cases of lepromatous leprosy from mediaeval England. *Journal of Archaeological Science*, 33, 1569–1579.

Truman, R. W., Singh, P., Sharma, R., Busso, P., Rougemont, J., Paniz-Mondolfi, A., Kapopoulou, A., Brisse, S., Scollard, D. M., Gillis, T. P. & Cole, S. T. (2011). Probable zoonotic leprosy in the southern United States. *New England Journal of Medicine*, 364, 1626–1633.

Tsuchiya, N., Kawasaki, A., Tsao, B. P., Komata, T., Grossman, J. M. & Tokunaga, K. (2001). Analysis of the association of HLA-DRB1, TNFalpha promoter and TNFR2 (TNFRSF1B) polymorphisms with SLE using transmission disequilibrium test. *Genes and Immunity*, 2, 317–322. doi:10.1038/sj.gene.6363783.

Turankar, R. P., Lavania, M., Darlong, J., Siva Sai, K. S. R., Sengupta, U. & Jadhav, R. S. (2019). Survival of *Mycobacterium leprae* and association with Acanthamoeba from environmental samples in the inhabitant areas of active leprosy cases: a cross-sectional study from endemic pockets of Purulia, West Bengal. *Infection, Genetics and Evolution*, 72, 199–204.

van Brakel, W. H., Peters, R. M. H. & da Silva Pereira, Z. B. (2019). 'Stigma related to leprosy – a scientific view', in D. M. Scollard & T. P. Gillis (eds), *International textbook of leprosy* [online], Chapter 4.4. https://internationaltextbookofleprosy.org.

Walker, S. L. (2020). 'Leprosy reactions', in D. M. Scollard & T. P. Gillis (eds), *International textbook of leprosy* [online], Chapter 6.1. https://internationaltextbookofleprosy.org.

Walsh, D. W. (2020). 'Overview of animal models', in D. M. Scollard & T. P. Gillis (eds), *International textbook of leprosy* [online], Chapter 10.1. https://internationaltextbookofleprosy.org.

Walsh, G. P., Storrs, E. E., Burchfield, H. P., Cotrell, E. H., Vidrine, M. F. & Binford, C. H. (1975). Leprosy-like disease occurring naturally in armadillos. *Journal of the Reticuloendothelial Society*, 18, 347–351.

Wander, K., Shell-Duncan, B. & McDade, T. W. (2009). Evaluation of iron deficiency as a nutritional adaptation to infectious disease. An evolutionary medicine perspective. *American Journal of Human Biology*, 21, 172–179.

Wang, Z., Sun, Y., Fu, X., Yu, G., Wang, C., Bao, F., Yue, Z., Li, J., Sun, L., Irwanto, A., Yu, Y., Chen, M., Mi, Z., Wang, H., Huai, P., Li, Y., Du, T., Yu, W., Xia, Y., Xiao, H., You, J., Li, J., Yang, Q., Wang, N., Shang, P., Niu, G., Chi, X., Wang, X., Cao, J., Cheng, X., Liu, H., Liu, J. & Zhang, F. (2016). A large-scale genome-wide association and meta-analysis identified four novel susceptibility loci for leprosy. *Nature Communications*, 7, 13760. https://doi.org/10.1038/ncomms13760.

Weiss, D. I., Do, T. H., de Andrade Silva, B. J., Teles, R. M. B., Andrade, P. R., Ochoa, M. T. & Modlin, R. L. (2020). 'Adaptive immune response in leprosy', in D. M. Scollard & T. P. Gillis (eds), *International textbook of leprosy* [online], Chapter 6.2. https://internationaltextbookofleprosy.org.

Wilbur, A. K., Buikstra, J. E. & Stojanowski, C. (2002). 'Mycobacterial disease in North America: an epidemiological test of Chaussinand's cross-immunity hypothesis', in C. A. Roberts, M. Lewis & K. Manchester (eds), *Proceedings of the 3rd international congress on the evolution and palaeoepidemiology of infectious diseases*, pp. 251–262. Archaeopress, Oxford.

Wilbur, A. K., Farnbach, A. W., Knudson, K. J. & Buikstra, J. E. (2008). Diet, tuberculosis, and the paleopathological record. *Current Anthropology*, 49, 963–991.

Wong, S. H., Gochhait, S., Malhotra, D., Pettersson, F. H., Teo, Y. Y., Khor, C. C., Rautanen, A., Chapman, S. J., Mills, T. C., Srivastava, A., Rudko, A., Freidin, M. B., Puzyrev, V. P., Ali, S., Aggarwal, S., Chopra, R., Reddy, B. S. N., Garg, V. K., Roy, S., Meisner, S.,

Hazra, S. K., Saha, B., Floyd, S., Keating, B. J., Kim, C., Fairfax, B. P., Knight, J. C., Hill, P. C., Adegbola, R. A., Hakonarson, H., Fine, P. E. M., Pitchappan, R. M., Bamezai, R. N. K., Hill, A. V. S. & Vannberg, F. O. (2010). Leprosy and the adaptation of human toll-like receptor 1. *PLoS Pathogens*, 6, e1000979. doi:10.1371/journal.ppat.1000979.

Wood, J. W., Milner, G. R., Harpending, H. C. & Weiss, K. M. (1992). The osteological paradox: Problems of inferring prehistoric health from skeletal samples. *Current Anthropology*, 33, 343–370.

World Health Organization. (2019). Global leprosy update, 2018: moving toward a leprosy-free world. *Weekly Epidemiological Record*, 94 (35–36), 389–412.

World Health Organization. (2020). Global leprosy (Hansen disease) update, 2019: time to step-up prevention initiatives. *Weekly Epidemiological Record*, 95, 417–440.

Yawalkar, S. J. (2009). *Leprosy for medical practitioners and paramedical workers*, 8th rev. edn. Novartis Foundation for Sustainable Development, Basel.

Zhang, F. R., Huang, W., Chen, S. M., Sun, L-D., Liu, H., Li, Y., Cui, Y., Yan, X-X., Yang, H-T., Yang, R-D., Chu, T-S., Zhang, C., Zhang, L., Han, J-W., Yu, G-Q., Quan, C., Yu, Y-X., Zhang, Z., Shi, B-Q., Zhang, L-H., Cheng, H., Wang, C-Y., Lin, Y., Zheng, H-F., Fu, X-A., Zuo, X-B., Wang, Q., Long, H., Sun, Y-P., Cheng, Y-L., Tian, H-Q., Zhou, F-S., Liu, H-X., Lu, W-S., He, S-M., Du, W-L., Shen, M., Jin, Q-Y., Wang, Y., Low, H-Q., Erwin, T., Yang, N-H., Li, J-Y., Zhao, X., Jiao, Y-L., Mao, L-G., Yin, G., Jiang, Z-X., Wang, X-D., Yu, J-P., Hu, Z-H., Gong, C-H., Liu, Y-Q., Liu, R-Y., Wang, D-M., Wei, D., Liu, J-X., Cao, W-K., Cao, H-Z., Li, Y-P., Yan, W-G., Wei, S-Y., Wang, K-J., Hibberd, M. L., Yang, S., Zhang, X-J. & Liu, J-J. (2009). Genomewide association study of leprosy. *New England Journal of Medicine*, 361, 2609–2618. doi:10.1056/NEJMoa0903753.

Zuckerman, M. K., Turner, B. L. & Armelagos, G. J. (2012). 'Evolutionary thought in paleopathology and the rise of the biocultural approach', in A. L. Grauer (ed), *A companion to paleopathology*, pp. 34–57. Wiley-Blackwell, Chichester.

CHAPTER 10

Preventable and curable, but still a global problem: tuberculosis from an evolutionary perspective

Charlotte A. Roberts, Peter D.O. Davies, Kelly E. Blevins, and Anne C. Stone

10.1 Introduction

This chapter focuses on the bacterial infection tuberculosis (TB). In 1988, Smith declared that TB had been 'conquered' in the industrialised world (1988), following a long onslaught on the world's population over several thousands of years. However, by 1993 the World Health Organization (WHO) had declared a global emergency on this disease (WHO, 1994). It remains present in developed and developing parts of the world, being a top cause of death; TB also tops the list of causes of death from one organism (WHO, 2019, 1). TB is a re-emerging infectious disease in high-income countries and a neglected one in low—and middle-income countries (Macedo Couto et al., 2019).

TB also affects other animals globally and is a major concern for veterinary science because of the number of domestic animals affected (WHO, 2020). It is an infection that is easily encompassed under the concept of One Health (Cassidy, 2018; Zinsstag et al., 2011; see also the One Health website, OIE, 2021), a movement that started in 2004 (Destoumieux-Garzón et al., 2018) and recognises the interdependence of humans, animals and ecosystem health. One Health 'seeks to improve health and well-being through the integrated management of disease risks at the interface between humans, animals and the natural environment' (Iossa and White, 2018, 858).

This chapter discusses how palaeopathological, clinical and veterinary research, alongside that

coming from modern and ancient pathogenetic analysis, contributes to an evolutionary understanding of TB. In particular, palaeopathology offers a unique perspective, since skeletal markers for TB in human remains provide direct evidence of the infection, while ancient DNA (aDNA) analysis provides insight into the evolution of the pathogen. This provides a long view, or deep time perspective, of the evolution of this ancient disease—one that is not used/discussed in normal clinical practice. Alongside archaeological and historical evidence, palaeopathology also highlights the risk factors for TB inherent in environments ranging in time and space. Having knowledge of TB's origin and evolution and how, when and why it spread into communities around the world is important and can help us understand why the disease remains with us today. This chapter looks at the present and past of TB and attempts to merge the different lines of evidence together to address salient questions of evolutionary medicine (EM). In doing so, we address three questions:

1. How does our research on TB contribute to an understanding of the evolution of health and disease?
2. How does palaeopathology offer a unique perspective and new understanding of TB?
3. How does this new understanding contribute to modern clinical research and medicine?

Charlotte A. Roberts et al., *Preventable and curable, but still a global problem.* In: *Palaeopathology and Evolutionary Medicine.*
Edited by Kimberly A. Plomp, et al., Oxford University Press. © Oxford University Press (2022). DOI: 10.1093/oso/9780198849711.003.0010

10.2 What is known about tuberculosis (TB) today

TB is an infectious disease caused by an organism of the *Mycobacterium tuberculosis* complex (MTBC). The human predilected members of the MTBC (*M. tuberculosis* lineages 1–8 and *Mycobacterium canettii*) are basal. This means that they branch closer to the root of the MTBC phylogeny—the family tree of evolutionary relationships among different groups of organisms—when compared to animal adapted strains, such as *Mycobacterium microti*, *Mycobacterium pinnipedii* and *Mycobacterium bovis*, which are more derived (Brosch et al., 2002). Although TB can infect any organ of the body, it usually affects the lungs, where it can result in infectious TB.

10.2.1 How TB affects the body

TB is contracted when the bacteria are inhaled into the lungs or ingested (Box 10.1) through infected animal products (De la Cruz et al., 2021; Wilson, 2021). For example, in countries where milk and meat are not TB checked, the infection can be acquired through the gastro-intestinal track (GIT) when tainted products are consumed. TB infection may spread via the blood and lymphatic systems to any part of the body beyond the lungs/GIT (resulting in extra-pulmonary TB). For example, TB meningitis, the form of TB that attacks the central nervous system (brain and spinal cord) is by far the most serious (Thwaites, 2014). Unfortunately, identification of meningitis in archaeological human remains is challenging. Bone lesions on the internal or endocranial surface of the skull may provide clues, although it has been suggested that other respiratory disease may also cause these lesions (e.g. Hershkovitz et al., 2002; Schultz, 1999). Similarly to the rib lesions described in Box 10.1, endocranial bone change has not been widely accepted as necessarily pathognomonic/specific for TB (e.g. Lewis, 2004). Nevertheless, in advanced chronic secondary TB infection, the bacteria can affect the bones and joints.

Box 10.1 TB aetiology, symptoms, and identification in the palaeopathological record

Inhaling or ingesting TB bacteria results in the primary stage of TB, which has no detectable symptoms in

patients. In 90% of people infected, healing occurs with no development into active disease, and there is no further disease (Lalvani et al., 2021). Today, post-primary TB in adults (10% of those infected) is key for transmission to take place between humans (De la Cruz et al., 2021). In clinical terms, this form of TB is defined by 'the presence of an antigen-specific immune response at the start of the disease process' (De la Cruz et al., 2021, 239). People with secondary TB have either experienced reactivation of primary TB, or they are experiencing re-infection. Symptoms for a person in the secondary stage may include cough, fever, night sweats, weight loss, pain and breathlessness. It was this post-primary TB that killed some of England's brightest minds, such as the nineteenth-century poet John Keats and two of the three literary Brontë sisters (e.g. see Daniel, 2017). For TB to be detectable in archaeological human remains, the infection has to progress to its post-primary (or secondary) stage, although aDNA analysis could potentially detect primary TB at the time of death. Of interest here is that new bone formation can be present on the internal (visceral) surfaces of ribs, assumed to be a reaction to inflammation of the lung tissue in post-primary TB (e.g., see Kelley and Micozzi, 1984; Roberts et al., 1994; Santos and Roberts, 2006). In these studies focusing on documented late-nineteenth/early-twentieth-century skeletal collections with assigned causes of death, it was concluded that TB was the most likely cause of the lesions. However, very few clinical studies have shown an association between rib enlargement (assumed to be the result of the new bone formation described here) and a TB diagnosis. It is notable that in one study, some patients with rib enlargement had chronic pulmonary infections apart from TB (Eyler et al., 1996), as noted by Roberts et al., (1994). The challenge for clinicians in detecting this new bone (if they so wished to) is that the bone formation can be very subtle and will not always be manifest in a radiograph.

In palaeopathological analysis, it is the spine (particularly the bodies of the lower thoracic and lumbar vertebrae) that is the most commonly affected part of the skeleton (also called Pott's disease). However, while TB can leave identifiable lesions in archaeological human remains, it will only be manifest in a small proportion of (untreated) people (Roberts and Buikstra, 2019; see also Lewis, 2018, 155–164, on TB in children's skeletons), although higher frequencies of TB affecting the skeleton in children have been recorded (e.g. Bernard, 2003; Roberts and Bernard, 2015 based on twentieth-century medical records from a children's sanatorium). If TB is not treated promptly today (not an option in antiquity), bone damage and deformity can result, as they did in the past. Of course, TB-related pathological changes to the bodily organs and other soft tissues are usually not detected in an archaeological context, except in the cases where soft tissue is preserved, such as in mummies, and/or if aDNA analysis is used (Salo et al., 1994; Zink et al., 2007). For example, Salo and colleagues (1994) found DNA of *M. tuberculosis* in a lymph node sample from a 1000-year-old Peruvian female. Since there was no specific evidence of skeletal change related to TB found on this individual, such information about a specific person with the infection would not have been reported if not for the use of microbial DNA analysis.

10.2.2 Epidemiology, prevention, and treatment

Globally, 10 million people fell ill with TB in 2019, but numbers have been declining very slowly over more recent years (WHO, 2020). Of these 10 million people, men were more affected than women, children younger than 15 years of age accounted for 12% of the people who developed post-primary stage TB and people between 15 to 35 years of age were the most commonly affected. As TB incidence declines in a country, it becomes more commonly diagnosed in older individuals because earlier birth cohorts had much higher rates of TB than those born more recently. Conversely, in modern day developing countries where the infection can be endemic, it is more commonly seen in children and younger adults (Snow et al., 2018). There is also a critical issue of co-morbidity, with 8.2% of people with TB in 2019 also living with human immunodeficiency virus (HIV) infection (WHO, 2020). The WHO regions of South-East Asia, Africa and the Western Pacific contributed most to that global TB total, with a lesser contribution from the Eastern Mediterranean, the Americas and Europe. Seven countries—India, Indonesia, China, the Philippines, Pakistan, Nigeria, Bangladesh, and South Africa—contributed two-thirds of that global total. While the number of people affected is declining slowly each year, COVID-19 is now affecting recent progress in that decline (WHO, 2020, xvi). The economic impact of the pandemic is predicted to worsen at least two of the key determinants of TB incidence: GDP

(Gross Domestic Product) per capita and undernutrition. Globally, between 2015 and 2019, there has been a decline of 9% in TB infections, but the decline varies between geographical, political and economic boundaries. For example, in the WHO European region, the reduction rate for TB infection has reached 19%, which is nearing the 2020 milestone of 20% set in 2014 by the WHO's *The End TB Strategy*.

Today, the relative importance of risk factors for TB vary according to geography and the state of economic development of a country. For example, today in some parts of sub-Saharan Africa, HIV infection is by far the most important factor increasing the chance of TB infection; HIV infection increases the risk of active TB by approximately a hundredfold. In other countries, risk factors are, in decreasing order of importance: an origin in a high-TB incidence country; history of homelessness; history of illicit drug use; history of imprisonment; HIV infection; other causes of immunosuppression; and an advanced age (Abubakar and Aldridge, 2014). An increased risk of contracting TB is clear is also for healthcare workers, along with being associated with tobacco smoking and malnutrition, and the consumption of unpasteurised milk and cheese (Theron et al., 2021). For the infection to then progress to TB disease, HIV co-morbidity remains important, along with diabetes mellitus, cigarette smoking, exposure to biomass fuel stoves, and vitamin D deficiency.

Often classed as a disease of the poor, the most important risk factors for TB in the past were likely to have been similar to today: poverty, crowded living conditions, malnutrition, vitamin D deficiency, people being on the move (migration) and specific occupations such as working with infected animals (Geddes, 2019; Huang et al., 2016; Roberts and Buikstra, 2019; Roberts and Brickley, 2018). Air pollution has also been linked to TB today (e.g., see Blount et al., 2017; Fuller, 2018; Popovic et al., 2019). For example, Fuller (2018) reported that fine particulate matter (< 2.5 μm; also known as $PM_{2.5}$) that results from natural and human generated pollution (e.g. from burning fuels, including by road vehicles) was most frequently associated with statistically significant effects on TB outcomes. This could have been significant for people in the past, too, considering that there is palaeopathological evidence of air pollution affecting the respiratory tract (sinuses, e.g. Panhuysen et al., 1997 and Roberts, 2007; ribs, e.g. Lambert, 2002; both, e.g. Davies-Barrett, 2018), and the soft tissues of the lungs in preserved archaeological bodies (Aufderheide, 2000; Monsalve et al., 2018).

The TB vaccination, BCG (Bacillus Calmette-Guérin), was developed from *M. bovis* nearly 100 years ago and is still the only option for prevention. Most countries of the world have a national vaccination policy (see Zwerling et al., 2011), but efficacy can vary (Tanner and McShane, 2021). Antibiotic treatment for TB started in the 1950s, with several antibiotics in combination being the current norm for treatment, or multi-drug therapy (MDT) being the current norm for treatment. However, within the last thirty years, antibiotic-resistant TB has become an increasing problem. Multi-drug-resistant TB (MDRTB) is resistant to isoniazid and rifampicin, and sometimes this can lead to extreme drug-resistant TB (XDRTB). While there are more than a dozen older and less effective drugs still available to treat TB, in the last few years two new drugs, bedaquiline and pretomanid, have been developed and brought into the treatment of patients with XDRTB (Pym et al., 2021). Drug resistance poses huge problems for the control of TB. The three countries with the largest share of the global challenge of drug resistance (MDR/RR-TB multidrug and rifampicin-resistant TB) during 2019 were India, China and the Russian Federation. In people newly diagnosed with TB across the world in 2019 there were 3.3% with drug resistance, with 17.7% of previously treated people also affected (WHO, 2020). The most important risk factors for MDR/RR-TB are a history of previous treatment, being in hospital or prison and HIV infection. In addition, living in poverty and in highly populated urban environments are also risk factors. This is a critical consideration, because globally today, many more (often poor) people are living in densely populated urban environments that have a high risk for TB (WHO, no date a). By 2050 it is estimated that 68% of the world's population will live in urban areas, which will pose a considerable challenge for controlling TB in the future. Considering that both children and adults are equally at risk (Dean et al., 2017), getting a handle on MDR/RR-TB is vital for our future.

10.2.3 The genetics of TB

For over twenty years, rapid DNA techniques have been available that allow confirmation of a diagnosis of TB, and DNA or culture of the pathogen enables the sensitivity of the organism to be determined so that specific treatments can be planned (Chin et al., 2018). Inferences about the evolutionary history of TB from modern genetic data were surprisingly limited prior to the late 1990s, with most analyses focused on understanding drug resistance or designing clinical diagnostic methods for differentiating *M. tuberculosis* from other mycobacteria. Until relatively recently, it was assumed that humans originally contracted TB from their domesticated animals at the advent of farming in the Neolithic—which led to the first epidemiological transition several thousand years ago, specific dating depending on place (e.g., Cockburn, 1963). However, Gordon and colleagues (1999) and Brosch and colleagues (2002) showed that human strains of TB did not evolve from *M. bovis* (the TB species found commonly in cattle), but it was in fact, the other way around. Since then, genetic, and particularly genomic, analyses of *M. tuberculosis* strains have shown they cluster into eight lineages that primarily affect humans, with evolution of animal-adapted lineages (often defined as ecotypes or species such as *M. bovis*) occurring within the radiation of human-specific strains (Brites et al., 2018; Hershberg et al., 2008; Nebenzahl-Guimaraes et al., 2016; Ngabonziza et al., 2020) (Figure 10.1). This indicates that the current diversity of strains in animals was seeded by a human strain rather than the reverse.

Thousands of modern *M. tuberculosis* genomes have now been sequenced, and analyses of these have also informed us about mutation rates, transmission biases, adaptive pressures and epidemiological clusters (e.g. Manson et al., 2017; Pepperell et al., 2013; Roetzer et al., 2013; Trauner et al., 2017). For example, although past genetic analyses have suggested that there is limited genetic diversity in *M. tuberculosis* as a result of a low mutation rate, more recent studies that directly sequenced patient sputum have revealed that individual patients can have comparatively higher diversity of strains (Ley et al., 2019; Nimmo et al., 2019; Séraphin et al., 2019). Simulations of

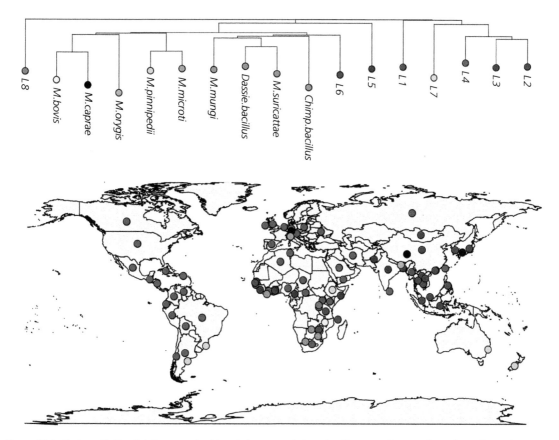

Figure 10.1 Maximum likelihood phylogeny made with a whole genome single nucleotide variant alignment of 23,452 sites from eighteen genomes at 500 bootstrap replicates and rooted to L8. For human MTBC lineages (L1–L8), the circles are coloured by the most prevalent strain in that country (as reported in Gagneux et al., 2006, and Reed et al., 2009). *Mycobacterium bovis* has a global distribution in domestic cattle and wildlife but in this figure has only been plotted in Malawi, where the most basal lineage has been isolated. All other animal lineages are plotted where they have been isolated from naturally infected animals. Plot generated in R version 4.1.2 using the ape, phytools, and phylotools packages (Paradis and Schliep, 2019; Revell, 2012; Zhang et al., 2017).
Image by Kelly E. Blevins.

population genetic parameters corroborate this observation of greater-than-expected diversity within hosts and identify progeny skews (the genetic bottleneck that occurs when one bacterium contributes the majority of offspring in the next generation or next infection), rather than a low mutation rate, as a mechanism causing low diversity between patients (Morales-Arce et al., 2020). Genomic analyses have also been employed to understand susceptibility to TB and other severe disease in the human host (Abel et al., 2018; Boisson-Dupuis et al., 2018; Moller et al., 2018; McHenry et al., 2020; Kerner et al., 2021). For example, homozygotes of a variant of the tyrosine

kinase 2 (*TYK2*) gene most common in Europeans today have impairment of the IL-23 responsive pathway, increasing the susceptibility to severe TB (Boisson-Dupuis et al., 2018). Analyses of ancient genomes show that this allele has declined in frequency in Europeans since the Bronze Age, possibly because of increasing morbidity and mortality from TB (Boisson-Dupuis et al., 2018, Kerner et al., 2021).

Investigations into the pressures acting on *M. tuberculosis* reveal the importance of purifying or negative selection (the selective removal of deleterious alleles, see Pepperell et al., 2013), and analyses of positive selection acting on TB have focused on drug resistance because of its clinical importance. Unlike bacteria (including *M. canettii*) that exchange resistance loci via horizontal gene transfer and recombination, resistance arises via *de novo* mutation in the MTBC since it reproduces clonally (Boritsch et al., 2016; Eldholm et al., 2014; Manson et al., 2017; Mortimer and Pepperell, 2014; Mortimer et al., 2018). In particular, resistance to the antibiotic isoniazid almost always arises first with mutations in the *katG* gene (Manson et al., 2017). This gene is the predominant mechanism of isoniazid resistance in our circulating strains. People with isoniazid-resistant TB with non-*katG* mutations have faster responses to treatment (Jaber et al., 1996). Mutations at many other genes are involved in resistance to this and other drugs used for TB treatment (Bardill et al., 2018; Miotto et al., 2017). Compensatory mutations that reduce the fitness cost of these changes have also been identified (Comas et al., 2012). Signals of positive selection were also found at other loci such as the *lldD2* gene (encoding the enzyme lactate dehydrogenase, which may be involved in metabolic adaptation to the environment) with changes in the promotor (where gene transcription begins) in multiple sublineages of L4 and other lineages (Brynildsrud et al., 2018; Osorio et al., 2013). In addition, proline-glutamate (PE) and proline-proline-glutamate (PPE) genes may be affected by positive and diversifying selection. The PE and PPE genes encode proteins that are thought to play a role in virulence and interactions with the host immune system (Brennan 2017). For example, experimental evidence indicates that deletion of the *ppe38* gene results in the loss of PE_PGRS and PPE-MPTR secretion and increases the virulence of some strains, particularly some L2 (Beijing) strains (Ates et al., 2018). Evaluating these changes in ancient individuals may help elucidate the selection pressures on *M. tuberculosis* as it changes hosts over time, and it may enable the use of ancient genomic epidemiology to trace the spread of an outbreak through time.

In summary, TB is a complex but usually treatable disease. It remains common in the poorest countries and in the most vulnerable citizens of developed countries. Diagnosis of TB depends on the awareness of attending medical staff and supporting infrastructure. If better finances and support at national and international level were available and poverty could be eliminated, it is a disease that could be virtually banished from human populations.

10.3 What do we know about TB in the past?

TB in the past can be recognised through bone damage in excavated remains of humans from archaeological sites and through the extraction and analysis of ancient biomolecules, such as DNA. This evidence is complimented by secondary evidence from historical writing and images (e.g. see Anderson et al., 2012; Pálfi et al., 1999; Roberts and Buikstra, 2003, 2021). The advantages and limitations of all these sources of information are wide ranging.

10.3.1 The evidence from palaeopathology

Visual/macroscopic and/or biomolecular detection of TB in bones provides primary evidence for the infection in past societies. Prior to describing the evidence of TB in human remains (with a focus on the skeleton), it should be remembered that absence of evidence is not necessarily evidence of absence. Explanations for this might include: some TB infected individuals may have died before skeletal lesions manifested; some parts of the world do not see much palaeopathological analysis, if any; methods of laying the dead to rest may make remains relatively invisible to the archaeologist (e.g. cremation; burial in 'watery' environments); and the climate/environment may not be conducive to preservation of human remains over hundreds or thousands of years since burial. Thus, for all these reasons relatively little evidence of TB has been found in preserved bodies compared to skeletons (but see, for example, Kay et al., 2015, Hungary; Salo et al., 1994, Chile; and Zink et al., 1999, Egypt).

Clinical data suggest that active TB infection results in skeletal changes (Box 10.2) in 1–21% of affected individuals, depending on the study population. This varies widely based on geographic region, population differences and availability of antibiotic treatment (Davidson & Horowitz, 1970; Davies et al., 1984; Jaffe, 1972; McLain & Isada, 2004; Resnick & Niwayama, 1995). TB may only affect the skeleton of a relatively small number of untreated infected people (3–5% estimate for data collected in the 1940s and 1950s, see Jaffe, 1972), and therefore in some parts of the ancient world we may see the 'tip of the iceberg' of past TB. Therefore, TB-related bone changes will under-represent the real frequency of ancient TB in the past. For example, in their TB (biomolecular) mycolic acid analyses of rib and soil samples from 21 burials from the eighteenth–nineteenth-century Newcastle Infirmary burial ground in north-east England, Gernaey and colleagues (1999) found more evidence for TB (5/21; 23.8%) than in macroscopic analysis of the total skeletons excavated (2/210; 0.95%). The 23.8% correlated much better with the 27% of individuals buried there who were documented to have died from TB in the contemporary infirmary records. The relationship between MTBC DNA recovery and other variables, including lesion presence, type, location and severity, remains unclear. Some studies have recovered microbial DNA from skeletons without any TB bone lesions (e.g., Faerman et al 1999; Nelson et al., 2020), thus illustrating how the 'osteological paradox' is important for understanding health and well-being in the past, including the impact of TB (Wood et al., 1992). For example, what does a positive aDNA result for TB in a skeleton without any obvious TB-related lesions mean for the experience of that once living person?

Box 10.2 Changes to bone as a result of TB

Notably in disease processes, bones can only be affected in three basic ways: abnormal bone formation, bone destruction or a mixture of the two, making diagnosis challenging, but TB generally causes more bone destruction than formation. For example, Roberts and Buikstra (2003, their Table 3.3) list the potential diagnoses for the spinal damage seen in TB. The lower thoracic and lumbar spine is the most affected part of the skeleton (Figures 10.2a,b), followed by the major weight-bearing joints (hip and knee). TB may spread from the spine into the adjacent soft tissues, and specifically the psoas muscle, forming an abscess (Resnick, 1995). In this case TB can cause bone formation and/or destruction of the relevant part of the spine, and also of the pelvic bones and/or the proximal end of the femur, reflecting where the psoas muscle sits anatomically in the body. Beyond

the spine and major joints, any bone of the body can be involved, including the flat bones (e.g. skull destruction), long bones (e.g. as seen in hypertrophic pulmonary osteoarthropathy: bone formation; see, for example, Assis et al., 2011) and short bones (spina ventosa: bone formation and destruction; Resnick, 1995). The ribs may be targeted in TB, too, but in clinical texts, destruction of bone is described for ribs rather than the bone formation mentioned earlier because the destruction is easier to see radiologically (Fitzgerald and Hutchinson, 1992; see also Gooderham et al., 2020 on an archaeological example of a twentieth-century Portuguese person with a documented cause of death of pulmonary TB). The prevalence of new bone formation (Figure 10.2c) recorded on ribs in documented collections (e.g. Kelley and Micozzi, 1984; Mariotti et al., 2015; Matos and Santos, 2006; Roberts et al., 1994; Santos and Roberts, 2006) and the archaeological record seemingly contradicts the clinical observation that it is rare for ribs to be affected in TB (e.g. Johnston and Rothstein, 1952). This may be because clinical literature tends to record its presence on the basis of bone destruction rather than bone formation, the latter often displayed as subtle bone formation that is not radiologically identified. A high prevalence of rib TB, however, may be expected if TB was as common in the past as historical sources indicate. Palaeopathology should always be cognisant of the potential differential diagnoses that may apply to all TB-related lesions, even though more current research is concluding that TB could be responsible for bone changes that have been generally considered non-specific for TB (e.g. new bone formation on ribs and alterations to the endocranial surface; see Pedersen et al., 2019, and Spekker et al., 2018, 2020a, 2020b).

TB seems to have affected the bones more frequently in people today than in the past. This is because they are living longer as a result of treatment and therefore have more chance to develop skeletal lesions (Steyn et al., 2013). Using data collected from the Pretoria Bone (University of Pretoria; started in 1942) and the Raymond A. Dart (University of the Witwatersrand; started early twentieth century) Collections in South Africa, a both with skeletons representing poor people known to have died from TB, it was found that rib lesions (bone formation) were more common in recent times, while spinal lesions were decreasing (Steyn et al., 2013). This may suggest that if people are treated, they will not develop the extra-pulmonary (spinal) bone damage as commonly as

(a)   (b)   (c)

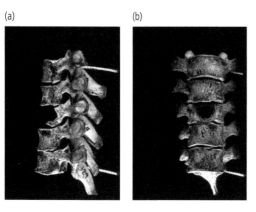

Figure 10.2 (a) Lateral view of spine affected by TB in individual 129; (b) Anterior view of spine affected by TB in individual 129; (c) Ribs of individual 129 showing enlargement of their vertebral ends and new bone formation, possibly related to TB of the lungs.
All images late nineteenth/early twentieth century, from the Robert J. Terry Collection. Photos courtesy of Charlotte Roberts.

was the case before modern medical treatment was available. In a complimentary study by Holloway and colleagues (2013) based on the Galler Collection of skeletons (created 1925–1977; Zürich, Switzerland) spanning three time periods (pre-1946 with no treatment available; 1946–1950; and post–1950), it was concluded that with the introduction of drug therapy for TB, the number of individuals with multiple TB skeletal lesions in the spine decreased from 80% pre-treatment to 25% post-treatment. In concluding this section, we note that the identification of TB in skeletons or preserved bodies is the precursor to aDNA analysis. Scholars focusing on both methods have to work together, and increasingly are, to achieve the best possible outcomes for the data produced. This naturally goes alongside the archaeological and historical contextualisation of those data.

10.3.2 The evidence from aDNA analyses of the MTBC

Recent advances in laboratory methods and computational power have enabled the recovery of whole ancient MTBC genomes (Orlando et al., 2021; Stone & Ozga, 2019). These biomolecular methods can be used to detect signatures of MTBC in archaeological human remains. In the mid-2000s, (quantitative) polymerase chain reaction ((q)PCR) assays and mycolic acid profiles were used to identify MTBC in archaeological remains including in skeletal tissue, calcified pleura and mummified lung tissue (Donoghue et al., 2005; Gernaey et al., 1999; Harkins et al., 2015; Müller et al., 2014a,b; Salo et al., 1994;

Spigelman & Lemma, 1993). Additionally, alongside the growing skeletal evidence for TB, these early studies refuted the notion that TB was not present in the Americas prior to European contact (Salo et al., 1994). The increasing popularity of using destructive biomolecular analyses has been met with alarm from some quarters, leading to the publication of guidelines emphasising authentication of data and the need for clear research designs and methods (Roberts & Ingham, 2008; Stone et al., 2009; Wilbur et al., 2009). There have been many publications outlining the benefits and harms of DNA analysis, including the ethics of destructive sampling per se and, in particular, doing so without community consultation (e.g. Austin et al., 2019; Bardill et al., 2018; Prendergast & Sawchuk, 2018; Squires et al., 2019; Tsosie et al., 2020; Wagner et al., 2020). Additionally, environmental contamination still poses a problem for authenticating MTBC aDNA and is a concern for any method of ancient TB detection. In particular, MTBC shares genomic regions with non-pathogenic or opportunistically pathogenic environmental mycobacteria (Müller et al., 2015, 2016; Nelson et al., 2020), many of which are not characterised.

10.3 Synthesis of the skeletal and biomolecular evidence of TB from a global perspective

10.3.1 The origin of TB: palaeopathological and ancient and modern molecular data

Genetic evidence to date suggests that TB's origin lies in Africa (Figure 10.1), because it is there that we find the most basal lineage (L8) and greatest diversity of strains today (Comas et al., 2015; Ngabonziza et al., 2020), However, we should remember that this may or may not have been the case in the past. However, a combination of molecular and palaeopathological data can be used to estimate when TB first infected humans. Prior to the analysis of ancient genomes, estimating the rate of nucleotide substitution (i.e. changes in the DNA) for evolutionary analyses of *M. tuberculosis* and the time from the most recent common ancestor (TMRCA) proved difficult, with estimates ranging from about 15,000 to several million years ago (Comas et al., 2013; Gutierrez et al., 2005; Hughes et al., 2002; Kapur et al., 1994; Wirth et al., 2008). For example, Gutierrez and colleagues (2005) inferred that *M. tuberculosis* from eastern Africa represent extant bacteria of a more diverse ancestral species, as old as three million years, from which the MTBC evolved. Such analyses estimated evolutionary rates using methods that employed divergence estimates for *Escherichia coli* and *Salmonella typhimurium* or assumed a correlation between the expansion of humans out of Africa and the expansion of *M. tuberculosis* strains. Better calibration points for the *M. tuberculosis* molecular clock are provided by ancient *M. tuberculosis* genomes that are accurately dated using historical and archaeological methods. In particular, Bos and colleagues (2014) published three roughly contemporaneous genomes from pre-contact Peru and the radiocarbon dates for these function as calibration points, as do the genome data from an infected Hungarian woman with a known date of death in the eighteenth century AD (Chan et al., 2013; Kay et al., 2015). These data, along with recent genome data from a calcified lung nodule from a seventeenth-century bishop in Lund (Sabin et al., 2020), support a TMRCA for the MTBC that is less than 6000 years old (Bos et al., 2014). These divergence time estimates are in accordance with substitution rates estimated by other researchers (Duchene et al., 2016; Menardo et al., 2019; Morales-Arce et al., 2020).

The earliest evidence of TB found in human remains is from the Mediterranean area, and particularly Israel (9250–8160 years ago, see Hershkovitz et al., 2008), Italy (around 6000 years ago, see Canci et al., 1996; 6500 years ago, see Sparacello

et al., 2017), Jordan (5150–4200 years ago, see Ortner, 1979) and Egypt (2000 years ago, see Morse et al., 1964). Northern Europe also has early evidence from Hungary (7000 years ago, see Masson et al., 2013), Germany (6450–6775 years ago, see Nicklisch et al., 2012) and Poland (~ 7000 years ago, see Gladykowska-Rzeczycka, 1999). In fact, virtually all modern-day countries in Europe (Mediterranean and Northern European areas) have some evidence for TB in archaeological human remains, but most of that evidence comes from the late and post-medieval periods (twelfth to nineteenth centuries—see Roberts and Buikstra, 2021). Asia and the Pacific Islands have evidence of TB in human remains relatively late compared to Europe, including the Mediterranean area. However, recently published information provides the earliest evidence to date from China (Shanghai) dated to 5900–5200 years ago (Okazaki et al., 2019). This adds to the sparse evidence from China that is published (see Table 2 in Li et al., 2019, who also add new Chinese evidence). Modern day Hawaii, Japan, Korea, Papua New Guinea and Thailand have also revealed, albeit younger, evidence of TB in human remains.

If DNA data suggest a TMRCA of less than 6000 years for *M. tuberculosis*, could TB have affected humans for more than 6000 years? Reports identifying TB in human remains prior to this time (e.g. Hershkovitz et al., 2008) have been debated due to questionable palaeopathological and molecular evidence (Stone et al., 2009; Wilbur et al., 2009). However, convincing palaeopathological evidence (i.e. diagnostic lesions) and TB DNA data from skeletons that have radiocarbon dates older than 6000 years do exist, albeit they are, to date, rather limited (e.g. Masson et al., 2013; Nicklisch et al., 2012; Sparacello et al., 2017). More skeletal evidence and new ancient genome data could potentially resolve this debate. In particular, there could have been other, more diverse MTBC lineages in the past that have not yet been sampled that would push the TMRCA back. However, the current TMRCA may reflect the time of a population bottleneck or selective sweep event (when alleles become fixed and variation reduces because of a beneficial mutation that spreads in the population), during which older, unknown MTBC lineages were lost, rather than the time when *M.*

tuberculosis was first found in human populations. Estimating TMRCA and other divergence dates are important for understanding the potential timing of pathogen emergence in human populations in the past (i.e. zoonotic jumps) and inferring significant range expansions or the divergence of specific lineages.

10.3.2 TB and the epidemiological transitions

It is generally accepted that with the move to agriculture and sedentism in the Neolithic (the first epidemiological transition), population numbers and density increased (Roberts, 2015), and this perhaps provided the opportunity for *M. tuberculosis* to 'jump' into humans from an animal, to be transmitted more readily between people and for *M. tuberculosis* to further 'jump' from humans to other species, including our domesticates (but see Brosch et al, 2002 discussed above). Cows and other domesticates could then transmit animal adapted strains such as *M. bovis* back to humans. After the first epidemiological transition, we observe TB in skeletons of people who lived in these settled permanent communities, as opposed to groups that were on the move (hunter-gatherers). However, TB, as documented in skeletons at least, did not become much more common until people started to live in urban environments where population density and poverty increased (Roberts & Buikstra, 2003, 2021, and references within).

Once industrialisation occurred (the second epidemiological transition; in Europe during the mid-eighteenth through to the early twentieth centuries), non-communicable (chronic) diseases (NCDs; e.g., cancers, cardiovascular disease, diabetes, cardiovascular, and non-infectious respiratory diseases) started to rise and replace infections (Barrett et al., 1998). Tobacco use, physical inactivity, unhealthy diets and the harmful use of alcohol are key risk factors for NCDs. Nevertheless, TB even became, for some, a fashionable disease to have, especially for women. Becoming pale and thin as the infection took hold was apparently more attractive to men (see Day, 2017; Roberts, 2020). The decline in TB in the early twentieth century came before the availability of antibiotics for treatment, this fact sensibly and logically argued to have been related to better living conditions and diet (Davies et al., 1999; Marmot, 2004). Even so, TB was still killing a considerable number of people between the mid-nineteenth century and well into the twentieth century, as a result of consuming infected milk and meat from cattle (Atkins, 2016, and see Figure 10.3), but also via droplet infection from human to human. There is skeletal evidence of TB in this period (e.g. Santos & Roberts, 2006; Waldron, 1993), and historical documents, such as the London Bills of Mortality (Boyce, 2020) record TB ('consumption') as a common cause of death (also see Roberts and Cox, 2003, and Woods & Shelton, 1997). We now live in the third epidemiological transition (Harper & Armelagos, 2010), and have seen TB return along with new infections emerging, for example, the recent outbreaks of viral infections (coronaviruses) causing severe acute respiratory syndrome (SARS) in 2002, Middle East respiratory syndrome (MERS) in 2012 and SARS-CoV-2 in late 2019/early 2020 (National Institutes of Allergies and Infectious Diseases, 2021).

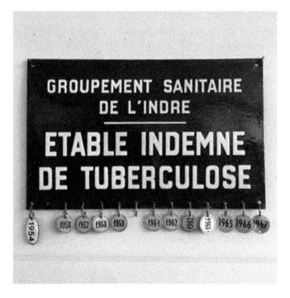

Figure 10.3 'Health certificate' from a stable located in the Indre department of France, declared free of bovine tuberculosis; the small discs correspond to the years when there was no TB recorded. Photo taken by Charlotte Roberts.

Relevant to all these transitions, we should ask what the study of diseases like TB in archaeological animal remains can offer to discussions and debates about TB's origin and history. Clearly, we have already seen that animals other than humans are key to our understanding of such diseases, and this is discussed in greater depth in Chapter 17. However, as noted by Thomas (2019), there is a dearth of evidence for TB in archaeological animal remains, compared to that in human remains, although lesions on ribs of cows have been noted that suggest pulmonary disease (bone formation) and possibly TB caused *by M. bovis*. Diagnosing disease in archaeological animal bones is challenging. For example, *M. bovis* rarely leads to bone damage (Thomas, 2019). There is also a lack of a standardised methodology for recording ribs and vertebrae (unlike for human remains), and animal bones found on archaeological sites are usually fragmentary and disarticulated due to butchery, but palaeopathology within archaeozoology is fortunately developing (see Thomas, 2012, 2019). Since we know that many human pathogens originate(d) in other animals (Dong et al., 2020) or can be transmitted back and forth between humans and other animals, such as MTBC, investigating archaeological animal bones for infectious disease evidence is important. Indeed, aDNA research has identified *M. bovis* in humans in Siberia (Taylor et al., 2007) and *M. pinnipedii*, which affects seals and sea lions, in humans in Peru (Bos et al., 2014), again emphasising the need to take the 'One Health' approach to understanding infectious diseases like TB (Macedo Couto et al., 2019).

10.3.3 Evidence for the distribution and spread of TB around the world

Exploring the phylogeny and distribution of MTBC strains has been the focus of aDNA analyses, particularly in Europe. For example, Müller and colleagues (2014a, their Table 1) extracted DNA from nine skeletal remains with lesions that are specific or non-specific for TB across archaeological sites in England and one from France. The skeletons dated from between the second and the nineteenth centuries CE. The authors used PCR assays and molecular cloning to target single nucleotide polymorphisms and large sequence polymorphisms to genotype all ten strains as Lineage 4 (see Gagneux & Small, 2007 for a summary of MTBC molecular

markers and lineage designations). Furthermore, Kay and colleagues (2015) and Sabin and colleagues (2020) recovered whole MTBC genomes from mummified eighteenth-century Hungarian and seventeenth-century Swedish remains with preserved soft tissue and skeletal evidence of TB, respectively. These whole genomes belong to 'generalist' sub-lineages of Lineage 4, which today are concentrated in Europe and areas that were colonised by European nations (Brynildsrud et al., 2018; Stucki et al., 2016).

Interestingly, all genotyped or reconstructed ancient MTBC strains from Europe belong to a single lineage: Lineage 4 (Bouwman et al., 2012; Kay et al., 2015; Müller et al., 2014a; Sabin et al., 2020), which demonstrates temporal continuity in this region as well as the overall long-term success of Lineage 4 strains affecting humans. The homogeneity of ancient MTBC strains in Europe is also interesting, given the high genetic diversity of contemporaneous circulating *M. leprae* genomes. Strains representing four of the six major branches of the *M. leprae* phylogeny, including the most basal branch, have been recovered from across medieval Europe (Mendum et al., 2014; Schuenemann et al., 2013, 2018).

In the Americas, the evidence for TB comes quite a bit later (Roberts & Buikstra, 2003, 2021, and references within). The earliest information comes from South America (Chile, Colombia, Peru and Venezuela), with Chile (the Caserones site in the Atacama Desert) providing the earliest data from these modern-day countries (700 CE). There is little evidence in Mesoamerica, but skeletons from two areas of North America have produced quite a lot of evidence for TB: the eastern region (particularly the mid-continent region that includes modern day northern Texas and portions of Nebraska, Kansas and Oklahoma), and the Southwest region (Arizona and New Mexico and parts of adjacent states such as California). In the past, as early as 1100 CE, they both had large population centres, which would have promoted the maintenance and spread of TB.

Today, the majority of MTBC strains circulating in the Americas are closely related to European strains, indicating an introduction during contact and colonisation (Brynildsrud et al., 2018; Gagneux & Small, 2007). The skeletal evidence alongside early PCR data have clearly established that MTBC was present in the pre-conquest Americas, leading to questions about the origin and phylogenetic placement of these early strains. To address this question, Bos and colleagues (2014) reconstructed three MTBC genomes from skeletal remains from the pre-conquest sites in the Osmore River Valley, southern coastal Peru (1028–1280 CE), and found that these ancient lineages are most closely related to *M. pinnipedii*, a member of the MTBC known to infect seals and sea lions today. The recovery of a pinniped-associated strain from human skeletal remains suggests that a zoonotic transmission occurred between seals or sea lions and humans on South American coasts. Three additional MTBC genomes recovered from further inland in the Osmore Valley (1000–1476 CE) and two highland sites in Colombia (1256–1640 CE; Vågene et al., in press), suggest that this zoonotic transmission successfully spread inland with humans, and potentially into North America (Blevins et al., 2018, 2019, 2021). Additional pre-conquest MTBC genomes are now needed to assess the number of zoonotic 'jumps', and address questions about the movement of strains among humans in the Americas, along with the process of adaptation of strains to human hosts. In addition, future research should investigate whether there were cycles of outbreaks between humans and other inland mammalian hosts and reservoirs.

Studies performed on both ancient and modern genomes indicate that migrations, population interactions and social upheavals have affected the distribution of MTBC strains. This is particularly true in the post-contact Americas. For example, Pepperell and colleagues (2011) found that the spread of TB into Canadian First Nations communities was mediated through the fur trade. One specific strain within the Lineage 4 group was likely introduced by French fur traders, but epidemic outbreaks occurred later, in concert with social upheavals. Furthermore, the patterns found were consistent with genetic drift and population structure rather than positive selection. Thus, the Lineage 4 groups thought to have arisen in Europe spread around the globe via European colonialism (Brynildsrud et al., 2018; Hershberg et al., 2008; O'Neill et al., 2019; Stucki et al., 2016).

10.4. Discussion

Who could ever have imagined that Smith's (1988) declaration that TB was not an infectious disease of significance anymore could be so quickly superseded by the WHO declaring a TB global emergency in 1993? Even now (in 2021), TB is present in every part of the world (WHO, no date b) and multi-drug resistance remains a real challenge. TB prevention, surveillance and treatment programmes have also felt the negative effect from the onslaught of the current COVID-19 pandemic (Jain et al., 2020). In this discussion, we focus on a number of key findings from palaeopathology and biomolecular analysis, consider the data alongside evolutionary theory and suggest future research avenues.

The palaeopathology of TB shows a long history of the disease in humans. However, recent modern and ancient TB genetic evidence has overtaken ideas of a much older disease. It has been possible to establish that TB became a challenge for our ancestors after the transition to agriculture, and during urbanisation and eventually industrialisation. New discoveries in the future may alter the current picture and may fill gaps in knowledge of TB in some parts of the world. For example, more focus on exploring human and non-human archaeological remains of pre-agricultural periods would be warranted because there are some reports, albeit scarce, of TB in ancient hunter gatherers (Guichón et al., 2015). In addition, the lack of data from Africa is somewhat surprising, especially considering the number of palaeopathological analyses that have been performed in Egypt since the first reports in the seventeenth century (Baker & Judd, 2012).

The paucity of data from the remains of animals affected by TB would further benefit from more zooarchaeological work to ultimately contribute to a One Health approach to TB and its co-evolution. For example, it is well known that wild animals can contract TB today (Malone & Gordon, 2017), and one might thus expect to see TB in humans and other animals in pre-farming cultures. There is a paucity of extant data from macroscopic and biomolecular (aDNA) analyses of non-human remains (but see Bathurst & Barta, 2004, who report on evidence for TB in a sixteenth-century dog from Ontario,

Canada) plus little evidence for animal strains in archaeological humans (but see Taylor et al.. 2007, who record *M. bovis* in Iron Age human remains from Aymyrlyg, South Siberia, dated to ~ 2000 years BP). Nonetheless, there is much potential in trying to move archaeozoology forward in addressing the impact of animal TB on humans. Early writers have described TB in animals (e.g., Columnella in the first century AD; see Swabe, 1999), and products of animals such as milk, meat, their hides and dung, all of which could have harboured MTBC species, including *M. bovis*. Thus, the presence of animal TB must be much higher than currently recognised in the human and non-human palaeopathological records.

Many of the varied TB risk factors today can also be identified in the past by interrogating the context in which people with TB were living. For example, high population density, poor living conditions, poverty, malnutrition, risky occupations, contact with animals, migration, vitamin D deficiency, parasite and other pathogen co-infections and air pollution were all potential risk factors in medieval Europe for contracting TB. It is important to be cognisant that these extrinsic hazards can intersect, as they can with intrinsic factors (e.g., age, sex, genetic make-up, immune system strength). This is an area in the palaeopathology of TB that could be better considered, especially due to intersectional approaches emphasised in modern settings (Rice et al., 2019). For example, a number of extrinsic risk factors may be prevalent within a person's lived environment (e.g., air pollution and their place of work) and they may be malnourished and affected by a lack of vitamin D. It has been proposed that the early use of fire, the smoke of which could have irritated the lungs of humans, made them more susceptible to contracting TB (Chisholm et al., 2016). While lesions on ribs are not pathognomonic for TB in skeletons, indicating a pulmonary disease *per se*, this proposal would benefit from more palaeopathological work. This could include analysing the dental calculus of skeletons with rib lesions to try to detect particulate matter (see Gerrard et al., 2018).

What have we learnt about TB from aDNA studies? First, this is a team approach representing different disciplines, and one where the initial step is to identify a question that we can answer using this type of analysis. Second, identifying

skeletons/mummies appropriate to the question is then necessary (time period and location), as is analysis that occurs preferably where there is some likelihood that the DNA of MTBC organisms will be preserved. While aDNA analysis attracts headline news, this does not mean 'traditional' palaeopathology should be excluded from methodological repertoires, especially for MTBC research. Notably, for aDNA analysis, whereas many pathogen aDNA discoveries have been made using samples of teeth (e.g., Margaryan et al., 2018; Schuenemann et al., 2018; Spyrou et al., 2019), MTBC aDNA has not been recovered reliably from the dental pulp chamber. This suggests that MTBC does not result in high-bacterial-load septicaemia, unlike plague or *Salmonella enterica* (Bos et al., 2011; Key et al., 2020; Spyrou et al., 2016; Vågene et al., 2018). Therefore, MTBC aDNA recovery will continue to rely on:

1. palaeopathological expertise in TB lesion identification and sampling methodology, through dry bone analysis or imaging/endoscopic sampling of preserved bodies;
2. a clear project design and familiarity with (bio)archaeological evidence of TB risk factors for the site in question (e.g., dense population, animal domestication, trade networks, occupational hazards); and
3. a close regard for ethical concerns when conducting DNA research.

The latter especially relates to sampling and/or exporting skeletal remains across the globe to laboratories for destructive analyses (e.g. see Claw et al., 2017; Squires et al., 2019; Tsosie et al., 2020; Wagner et al., 2020).

The recovery and analysis of ancient MTBC genomes have enabled us to begin tracking the movement of TB with humans and other animals around the world, alongside the use of data from modern genomic analyses. Across Europe, MTBC recovery suggests that Lineage 4 strains have predominated since at least the second century CE, highlighting the long-term success of this lineage (Müller et al., 2014a). In the Americas, MTBC aDNA analysis has yielded the unexpected discovery that TB was likely introduced via zoonotic transmission from pinnipeds to humans. The many examples of skeletal TB from inland archaeological sites across

North and South America suggest that this spill-over, or others, resulted in sustained human-to-human transmission. Young dating estimates generated using ancient genomes calibrated by radiocarbon dates and mutation rate modelling indicate that MTBC spread throughout the Americas relatively quickly (Bos et al., 2014; Kay et al., 2015; Morales-Arce et al., 2020). Despite this rapid introduction and widespread transmission, the predominance of L4 in the Americas today means that these earlier forms were likely entirely replaced in humans. However, many questions remain about the dynamics of the spill-over(s) and adaptation of the pinniped-associated clade to human hosts and the widespread strain introduction and replacement during colonisation of the Americas. Geneticists, palaeopathologists and archaeologists are aptly positioned to address these questions and expand our knowledge of the origin, evolution and history of this disease. In addition, it would be instructive to combine the pathogen aDNA data with stable isotope data in skeletal remains to reflect on and address the mobility of people, alongside the hard evidence of TB in their skeletons. This would allow a more nuanced analysis of the impact of migration on the spread of TB (e.g. see Richards & Montgomery, 2012 for a review of stable isotope and palaeopathological analyses; and Goude et al., 2020 and Quinn, 2017 for examples of studies of mobility and TB in the past).

In view of this discussion synthesising clinical, genetic and archaeological/historical data to explore the origin, evolution and history of the MTBC, we can clearly learn from different disciplines to nuance understandings of the present and the past. In closing this discussion, we return to the original three questions pertinent to this volume: how our research on TB contributes to the understanding of the evolution of health and disease; how palaeopathology offers a unique perspective or new understanding of TB; and how this new understanding contributes to modern clinical research and medicine.

Using palaeopathological, clinical and veterinary research alongside that coming from modern and ancient pathogenetic analysis, we have been able to contribute to an evolutionary understanding of TB in relation to humans and other animals. Increased surveillance in animal populations has helped document the geographic range of animal adapted MTBC lineages, which have been identified on all continents except Antarctica. The promiscuity of MTBC is undeniable. For example, *Mycobacterium caprae* has been detected in naturally infected deer, boar, goats, wolves and cattle (Ciaravino et al., 2020; Dorn-In et al., 2020; Magnani et al., 2020; Orłowska et al., 2017). Concern over sustained transmission in animal populations resulting from spill-overs of MTBC infections from humans to animals or animals to animals will undoubtedly increase after the COVID-19 pandemic that started in 2020. Critical questions remain about host-adapted and more generalised MTBC lineages, such as how the latter become the former. This is important because animal lineages could diversify and establish themselves in human populations.

Palaeopathological markers for TB in human remains can provide direct evidence of the infection, as can aDNA analysis, but it is important to be clear when recording the palaeopathological evidence as to whether any bone changes are active or healed/chronic at the time of death. Data from aDNA will provide an indication of whether a person had the disease in their body, again at the time of death. The latter helps with adding to our knowledge of who had the disease (even without bone changes), whether they might have had a latent infection, and what the DNA information tells us about phylogeography. This provides a long view, or deep time perspective, of the evolution of this ancient disease, one that is not normally considered in clinical practice. Alongside archaeological and historical evidence, it has also highlighted the risk factors inherent in environments in different times and places in which these people lived. Studying MTBC aDNA allows us to see how these processes have played out in the past, thereby giving us direct insight into the rate and speed at which zoonotic spill-overs occurred and how successful they were. MTBC in the pre-conquest Americas is an exciting model for tracking the introduction and spread of TB across the continents. In addition, while the colonial-period evidence of MTBC is scarce, it will be key for understanding the eventual replacement of pinniped-like strains by Lineage 4 in the Americas today.

Having the knowledge of TB's origin and evolution and how, when and why it spread around the world and into communities is important, including for modern clinical research and medicine. It provides empirical evidence that helps us to understand why it remains with us today. From an evolutionary medicine view, infections such as TB have often been the focus of narratives about strains of bacteria becoming resistant to antibiotics used to treat humans. Nesse and Williams (1994; see also De Martino et al., 2019) state that 'Bacteria can evolve as much in a day as we can in a thousand years and this gives us a grossly unfair handicap in the arms race.' They also note that (at their time of writing) TB was more difficult to control with antibiotics, and that economically advanced societies had already had 'a golden age of relief from bacterial disease' (Nesse & Williams, 1994; see also Hall et al., 2018). In relation to antibiotic resistance, Trevathan and colleagues (2008) note potential obstacles to having successful treatments for infections, including people not taking their full antibiotic courses, and bacteria mutating quickly to become antibiotic-resistant strains. Whether the data, and particularly those deriving from modern and aDNA, can help us to better appreciate antibiotic resistance seen in people today and help develop more effective therapy is a question yet to be answered. As a result of antibiotic resistance, this situation can lead to resistant strains that result in an infection that is difficult to manage at the patient level. Evolutionary processes that underly the need to complete a full course of antibiotics are relevant here for both patients and practitioners.

10.5 Conclusions

TB has been with human and animal populations for millennia and does not seem to be leaving anytime soon. No one discipline can tell us all we wish to know about its origin, evolution and history. Teamwork is essential for nuanced interpretations. While palaeopathology's long history will be extended into the future, and clinical and veterinary medicine will benefit from new innovations in the management of TB in both humans and other animals (e.g., new therapeutics), the more recent development of aDNA analysis over the last thirty or so years will continue to make significant

impacts on our understanding of health and wellbeing, both past and present. Indeed, the COVID-19 pandemic has shown how science has come to the fore in addressing the challenges of this viral infection.

References

Abel, L., Fellay, J., Haas, D. W., Schurr, E., Srikrishna, G., Urbanowski, M., Chaturvedi, N., Srinivasan, S., Johnson, D. H. & Bishai, W. R. (2018). Genetics of human susceptibility to active and latent tuberculosis: present knowledge and future perspectives. *The Lancet Infectious Disease*, 18(3), e64–e75.

Abubakar, I. & Aldridge, R. (2014). 'Control of tuberculosis in low incidence countries', in P. D. O. Davies, S. B. Gordon & G. Davies (eds), *Clinical tuberculosis*, 5th edn, pp. 361–374. CRC Press, London.

Anderson, J., Barnes, E. & Shackleton, E. (2012). *The art of medicine: over 2,000 years of images and imagination*. University of Chicago Press, Chicago.

Assis, S., Santos, A. L. & Roberts, C. A. (2011). Evidence of hypertrophic osteoarthropathy in individuals from the Coimbra Skeletal Identified Collection (Portugal). *International Journal of Paleopathology*, 1, 155–163.

Ates, L. S., Dippenaar, A., Ummels, R., Piersma, S. R., van der Woude, A. D., van der Kuij, K., Le Chevalier, F., Mata-Espinosa, D., Barrios-Payán, J., Marquina-Castillo, B., Guapillo, C., Jiménez, C. R., Pain, A., Houben, E. N. G., Warren, R. M., Brosch, R., Hernández-Pando, R. & Bitter, W. (2018). Mutations in ppe38 11block PE_PGRS secretion and increase virulence of *Mycobacterium tuberculosis*. *Nature Microbiology*, 3, 181–188.

Atkins, P. J. (2016). *A history of uncertainty: bovine tuberculosis in Britain, 1950 to the present*. Winchester University Press, Winchester.

Aufderheide, A. C. (2000). *The scientific study of mummies*. Cambridge University Press, Cambridge.

Austin, R. M., Sholts, S. B., Williams, L., Kistler, L. & Hofman, C. A. (2019). Opinion: to curate the molecular past, museums need a carefully considered set of best practices. *Proceedings of the National Academy of Sciences of the United States of America*, 116(5), 1471–1474.

Baker, B. J. & Judd, M. A. (2012). 'Development of paleopathology in the Nile Valley', in J. E. Buikstra & C. A. Roberts (eds), *A global history of paleopathology: pioneers and prospects*, pp. 209–234. Oxford University Press, Oxford.

Bardill, J., Bader, A. C., Garrison, N. A., Bolnick, D. A., Raff, J. A., Walker, A., Malhi, R. S., & the Summer internship for Indigenous peoples in Genomics (SING) Consortium. (2018). Advancing the ethics of paleogenomics. *Science*, 360 (6387), 384–385.

Barrett, R., Kuzawa, C. W., McDade, T. & Armelagos, G. (1998). Emerging and re-emerging infectious diseases: the third epidemiologic transition. *Annual Review of Anthropology*, 27, 247–271.

Bathurst, R. R. & Barta, J. L. (2004). Molecular evidence of tuberculosis induced hypertrophic osteopathy in a 16th-century Iroquoian dog. *Journal of Archaeological Science*, 31, 917–925.

Bernard, M-C. (2003). *Tuberculosis: a demographic analysis and social study of admissions to a children's sanatorium (1936-1954) in Stannington, Northumberland.* Unpublished PhD Thesis, Durham University, Durham.

Blevins, K. E., Buikstra, J., Stone, A. C. & Mansilla Lory, J. (2018). Searching for tuberculosis at a Mesoamerican postclassic urban center. *American Journal of Physical Anthropology*, 165(S66), 31.

Blevins, K. E., Nelson, E. A., Herbig, A., Krause, J., Buikstra, J. E., Mansilla Lory, J., Bos, K. I. & Stone, A. C. (2019). 'Mycobacterium tuberculosis complex genomes from the Postclassic Basin of Mexico (1300–1521 CE) in Mexico City'. *4th Mexico Population Genomics Meeting.* Mexico City, Mexico, 11 January.

Blevins, K. E., Nelson, E. A., Herbig, A., Krause, J., Buikstra, J. E., Mansilla Lory, J., Bos, K. I. & Stone, A. C. (2021). 'Skeletal and molecular evidence of the Mycobacterium tuberculosis complex from Tenochtitlan-Tlatelolco, a late postclassic Mesoamerican urban center (1350–1521 CE)'. *48th annual Meeting of the Paleopathology Association*, Los Angeles, CA, 14–15 April. Virtual, 6, 7, 12, 16, 22 and 23 April.

Blount, R. J., Pascopella, L., Catanzaro, D. G., Barry, P. M., English, P. B., Segal, M. R., Flood, J., Meltzer, D., Jones, B., Balmes, J. & Nahid, P. (2017). Traffic-related air pollution and all-cause mortality during tuberculosis treatment in California. *Environmental Health Perspectives* [online], 125. doi.org/10.1289/EHP1699

Boisson-Dupuis, S., Ramirez-Alejo, R., Li, Z., Patin, E. & Rao, G. (2018). Tuberculosis and impaired IL-23–dependent IFN-γ immunity in humans homozygous for a common *TYK2* missense variant. *Science Immunology*, 3(30), eaau8714. doi: 10.1126/sciimmunol.aau8714

Boritsch, E. C., Khanna, V., Pawlik, A., Honore, N., Navas, V. H., Ma, L., Bouchier, C., Seemann, T., Supply, P., Stinear, T. P. & Brosch, R. (2016). Key experimental evidence of chromosomal DNA transfer among selected tuberculosis-causing mycobacteria. *Proceedings of the National Academy of Sciences of the United States of America*, 113, 9876–9881.

Bos, K. I., Harkins, K. M., Herbig, A., Coscolla, M., Weber, N., Coscolla, M., Weber, N., Comas, I., Forrest, S. A., Bryant, J. M., Harris, S. R., Schuenemann, V. J., Campbell, T. J., Majander, K., Wilbur, A. K., Guichon, R. A., Wolfe Steadman, D. L., Collins Cook, D., Niemann,

S., Behr, M. A., Zumarraga, M., Bastida, R., Huson, D., Nieselt, K., Young, D., Parkhill, J., Buikstra, J. E., Gagneux, S., Stone, A. C. & Krause, J. (2014). Pre-Columbian mycobacterial genomes reveal seals as a source of New World human tuberculosis. *Nature*, 514, 494–497.

Bos, K. I., Schuenemann, V. J., Golding, G. B., Burbano, H. A., Waglechner, N., Coombes, B. K., McPhee, J. B., DeWitte, S. N., Meyer, M., Schmedes, S., Wood, J., Earn, D. J. D., Herring, D. A., Bauer, P., Poinar, H. N. & Krause, J. (2011). A draft genome of *Yersinia pestis* from victims of the Black Death. *Nature*, 478, 506–510.

Bouwman, A. S., Bunning, S. L., Müller R., Holst, M., Caffell, A. C., Roberts, C. A. & Brown, T. A. (2012). The genotype of a historic strain of Mycobacterium tuberculosis . *Proceedings of the National Academy of Sciences of the United States of America*, 109, 18511–6.

B oyce, N. (2020). Bills of mortality: tracking disease in early modern London. *The Lancet*, 395, 1186–1187.

Brennan, M. J. (2017). The enigmatic PE/PPE multigene family of mycobacteria and tuberculosis vaccination. *Infection and Immunity*, 85(6), e00969–16. doi: 10.1128/IAI.00969-16

Brites, D., Loiseau, C., Menardo, F., Borrell, S., Boniotti, M. B., Warren, R., Dippenaar, A., Parsons, S. D. C., Beisel, C., Behr, M. A., Fyfe, J. A., Coscolla, M. & Gagneux, S. (2018). A new phylogenetic framework for the animal-adapted *Mycobacterium tuberculosis* complex. *Frontiers of Microbiology*, 9, 2820. https://doi.org/10.3389/fmicb.2018.02820

Brosch, R., Gordon, S.V., Marmiesse, M., Brodin, P., Buchrieser, C., Eiglmeier, K., Garnier, T., Gutierrez, C., Hewinson, G., Kremer, K., Parsons, L. M., Pym, A. S., Samper, S., van Soolingen, D. & Cole, S. T. (2002). A new evolutionary sequence for the *Mycobacterium tuberculosis* complex. *Proceedings of the National Academy of Sciences of the United States of America*, 99(6), 3684–3689.

Brynildsrud, O. B., Pepperell, C. S., Suffys, P., Grandjean, L., Monteserin, J., Debech, N., Bohlin, J., Alfsnes, K., Pettersson, J. O.-H., Kirkeleite, I., Fandinho, F., Aparecida da Silva, M., Perdigao, J., Portugal, I., Viveiros, M., Clark, T., Caws, M., Dunstan, S., Thai, P. V. K., Lopez, B., Ritacco, V., Kitchen, A., Brown, T. S., van Soolingen, D., O'Neill, M. B., Holt, K. E., Feil, E. J., Mathema, B., Balloux, F. & Eldholm, V. (2018). Global expansion of *Mycobacterium tuberculosis* lineage 4 shaped by colonial migration and local adaptation. *Science Advances*, 4(10), eaat5869.

Canci, A., Minozzi, S. & Borgognini Tarli, S. (1996). New evidence of tuberculous spondylitis from Neolithic Liguria (Italy). *International Journal of Osteoarchaeology*, 6, 497–501.

Cassidy, A. (2018). 'Humans, other animals and 'One Health' in the early 21st century', in A. Woods, M. Bresalier, A. Cassidy & R. Mason Dentinger (eds), *Animals and the shaping of modern medicine: one health and its histories*, pp.193–236. Palgrave Macmillan, Cham.

Chan, J. Z., Sergeant, M. J., Lee, O. Y., Minnikin, D. E., Besra, G. S., Pap, I., Spigelman, M., Donoghue, H. D. & Pallen, M. J. (2013). Metagenomic analysis of tuberculosis in a mummy. *New England Journal of Medicine*, 369(3), 289–290.

Chin, K. L., Sarmkineto, M. E., Norazmi, M. N. & Acosta, A. (2018). DNA markers for tuberculosis diagnosis. *Tuberculosis*, 113, 139–152.

Chisholm, R. H., Trauer, J. M., Curnoe, D. & Tanaka, M. M. (2016). Controlled fire use in early humans might have triggered the evolutionary emergence of tuberculosis. *Proceedings of the National Academy of Sciences of the United States of America*, 113, 9051–9056.

Ciaravino, G., Vidal, E., Cortey, M., Martín, M., Sanz, A., Mercader, I., Perea, C., Robbe-Austerman, S., Allepuz, A. & Pérez de Val, B. (2020). Phylogenetic relationships investigation of *Mycobacterium caprae* strains from sympatric wild boar and goats based on whole genome sequencing. *Transboundary and Emerging Diseases*: doi: 10.1111/tbed.13816

Claw, K. G., Lippert, D., Bardill, J., Cordova, A., Fox, K., Yracheta, J. M., Bader, A. C., Bolnick, D. A., Malhi, R. S., TallBear, K. & Garrison, N. A. (2017). Chaco Canyon dig unearths ethical concerns. *Human Biology*, 89, 177–180.

Cockburn, A. (1963). *The evolution and eradication of infectious disease*. Johns Hopkins University Press, Baltimore.

Comas, I., Borrell, S., Roetzer, A., Rose, G., Malla, B., Kato-Maeda, M., Galagan, J., Niemann, S. & Gagneux, S. (2012). Whole-genome sequencing of rifampicin-resistant *Mycobacterium tuberculosis* strains identifies compensatory mutations in RNA polymerase genes. *Nature Genetics*, 44(1), 106–110.

Comas, I., Coscolla, M., Luo, T., Borrell, S., Holt, K. E., Kato-Maeda, M., Parkhill, J., Malla, B., Berg, S., Thwaites, G., Yeboah-Manu, D., Bothamley, G., Mei, J., Wei, L., Bentley, S., Harris, S. R., Niemann, S., Diel, R., Aseffa, A., Gao, Q., Young, D. & Gagneux, S. (2013). Out-of-Africa migration and Neolithic coexpansion of *Mycobacterium tuberculosis* with modern humans. *Nature Genetics*, 45(10), 1176–1182.

Comas, I., Hailu, E., Kiros, T., Bekele, S., Mekonnen, W., Gumi, B., Tschopp, R., Ameni, G., Hewinson, R. G., Robertson, B. D., Goig, G. A., Stucki, D., Gagneux, S., Aseffa, A., Young, D. & Berg, S. (2015). Population genomics of *Mycobacterium tuberculosis* in Ethiopia contradicts the virgin soil hypothesis for human tuberculosis in Sub-Saharan Africa. *Current Biology*, 25(24), 3260–3266.

Daniel, T. (2017). 'Tuberculosis in history: did it change the way we live?', in D. Schlossberg (ed), *Tuberculosis and nontuberculous mycobacterial infections*, 7th edn, pp. 3–10. ASM Press, Washington, DC.

Davidson, P. T. & Horowitz, I. (1970). Skeletal tuberculosis: a review with patient presentations and discussion. *The American Journal of Medicine*, 48, 77–84.

Davies, P. D. O., Humphries, M. J., Byfield, S. P., Nunn, A. J., Darbyshire, J. H., Citron, K. M. & Fox, W. (1984). Bone and joint tuberculosis: a survey of notifications in England and Wales. *The Journal of Bone and Joint Surgery*, 66, 326–330.

Davies, R. P. O., Tocque, K., Bellis, M. A., Rimmington, T. & Davies, P. D. O. (1999). 'Historical declines in tuberculosis: improving social conditions or natural selection?', in G. Pálfi, O. Dutour, J. Deák & I. Hutás (eds), *Tuberculosis: past and present*, pp. 89–92. Golden Book Publishers and Tuberculosis Foundation, Budapest/Szeged.

Davies-Barrett, A. M. (2018). *Respiratory disease in the Middle Nile Valley: a bioarchaeological analysis of the impact of environmental and sociocultural change from the Neolithic to medieval periods*. Unpublished PhD Thesis, Durham University, Durham.

Day, C. A. (2017). *Consumptive chic: a history of beauty, fashion, and disease*. Bloomsbury, London.

De La Cruz, C. S., Seaworth, B. & Bothamley, G. (2021). 'Pulmonary tuberculosis', in L. Friedman, M. Dedicoat & P. D. O. Davies (eds), *Clinical tuberculosis*, 6th edn, pp. 237–248. CRC Press, Abingdon.

De Martino, M., Lodi, L., Galli, L. & Chiappini, E. (2019). Immune response to *Mycobacterium tuberculosis*: A narrative review. *Frontiers of Pediatrics*, 7, 350. doi:10.3389/fped.2019.00350

Dean, A. S., Cox, H. & Zignol, M. (2017). Epidemiology of drug-resistant tuberculosis. *Advances in Experimental and Medical Biology*, 1019, 209–220. doi: 10.1007/978-3-319-64371-7_11

Destoumieux-Garzón, D., Mavingui, P., Boetsch, G., Boissier, J., Darriet, F., Duboz, P., Fritsch, C., Giradoux, P., Le Roux, F., Morand, S., Paillard, C., Pontier, D., Sueur, C. & Voituron, A. (2018). The One Health concept: 10 years old and a long road ahead. *Frontiers in Veterinary Science* [online]. doi: 10.3389/fvets.2018.00014

Dong, R., Pei, S., Yin, C., He, R. L. & Yau, S. S. (2020). Analysis of the hosts and transmission paths of SARS-CoV-2 in the COVID-19 outbreak. *Genes (Basel)*, 11 (6), 637. doi:10.3390/genes11060637

Donoghue, H. D., Marcsik, A., Matheson, C., Vernon, K., Nuorala, E., Molto, J. E., Greenblatt, C. L. & Spigelman, M. (2005). Co-infection of *Mycobacterium tuberculosis* and *Mycobacterium leprae* in human archaeological samples: a possible explanation for the historical decline of leprosy. *Proceedings of the*

Royal Society B: Biological Sciences, 272(1561), 389–394. doi.org/10.1098/rspb.2004.2966

Dorn-In, S., Körner, T., Büttner, M., Hafner-Marx, A., Müller, M., Heurich, M., Varadharajan, A., Blum, H., Gareis, M. & Schwaiger, K. (2020). Shedding of *Mycobacterium caprae* by wild red deer (*Cervus elaphus*) in the Bavarian alpine regions, Germany. *Transboundary and Emerging Diseases*, 67(1), 308–317.

Duchene, S., Holt, K. E., Weill, F. X., Le Hello, S., Hawkey, J., Edwards, D. J., Fourment, M. & Holmes, E. C. (2016). Genome-scale rates of evolutionary change in bacteria. *Microbial Genomics*, 2 (11), e000094.

Eldholm, V., Norheim, G., von der Lippe, B., Kinander, W., Dahle, U. R., Caugant, D. A., Mannsåker, T., Mengshoel, A. T., Dyrhol-Riise, A. M. & Balloux, F. (2014). Evolution of extensively drug-resistant *Mycobacterium tuberculosis* from a susceptible ancestor in a single patient. *Genome Biology*, 15(11), 490.

Eyler, W. R., Monsein, L. H., Beute, G. H., Tilley, B., Schultz, L. R. & Schmitt, W. G. (1996). Rib enlargement in patients with chronic pleural disease. *American Journal of Roentgenology*, 167, 921–926.

Faerman, M., Jankauskas, R., Gorski, A., Bercovier, H. & Greenblatt, C. (1999). 'Detecting *Mycobacterium tuberculosis* in medieval skeletal remains from Lithuania', in G. Pálfi, O. Dutour, J. Deák & I. Hutás (eds), *Tuberculosis: past and present*, pp. 371–376. Golden Book Publishers and Tuberculosis Foundation, Budapest/Szeged.

Fitzgerald, R. & Hutchinson, C. E. (1992). Tuberculosis of the ribs: computed tomographic findings. *British Journal of Radiology*, 65, 82–84.

Fuller, G. (2018). *The invisible killer: the rising global threat of air pollution—and how we can fight back*. Melville House UK, London.

Gagneux, S., DeRiemer, K., Van, T., Kato-Maeda, M., de Jong, B. C., Narayanan, S., Nicol, M., Niemann, S., Kremer, K., Gutierrez, M. C., Hilty, M., Hopewell, P. C. & Small, P. M. (2006). Variable host-pathogen compatibility in *Mycobacterium tuberculosis*. *Proceedings of the National Academy of Sciences of the United States of America*, 103(8), 2869–2873.

Gagneux, S. & Small, P. M. (2007). Global phylogeography of *Mycobacterium tuberculosis* and implications for tuberculosis product development. *The Lancet Infectious Diseases*, 7(5), 328–337. http://linkinghub.elsevier.com/retrieve/pii/S1473309907701081.

Geddes, L. (2019). *Chasing the sun: the new science of sunlight and how it shapes our bodies and minds*. Profile Books/Wellcome Collection, London.

Gernaey, A. M., Minnikin, D. E., Copley, M. S., Ahmed, A. M. S., Robertson, D. J., Nolan, J. & Chamberlain, A. T. (1999). 'Correlation of the occurrence of mycolic acids with tuberculosis in an archaeological population', in G. Pálfi, O. Dutour, J. Deák & I. Hutás (eds), *Tuberculosis: past and present*, pp. 275–282. Golden Book Publishers and Tuberculosis Foundation, Budapest/Szeged.

Gerrard, C. M., Graves, P., Millard, A., Annis, R. & Caffell, A. (2018). *Lost lives, new voices. Unlocking the secrets of the Scottish soldiers from the Battle of Dunbar 1650*. Oxbow Books, Oxford.

Gladykowska-Rzeczycka, J. J. (1999). 'Tuberculosis in the past and present in Poland', in G. Pálfi, O. Dutour, J. Deák & I. Hutás (eds), *Tuberculosis: past and present*, pp. 561–573. Golden Book Publishers and Tuberculosis Foundation, Budapest/Szeged.

Gooderham, E, Marinho, L., Spake, L., Fisk, S., Prates, C., Sousa, S., Oliviera, C., Santos, A. L. & Cardoso, H. V. F. (2020). Severe skeletal lesions, osteopenia and growth deficit in a child with pulmonary tuberculosis (mid-20th century, Portugal). *International Journal of Paleopathology*, 30, 47–56.

Gordon, S. V., Brosch, R., Billault, A., Garnier, T., Eiglmeier, K. & Cole, S. T. (1999). Identification of variable regions in the genomes of tubercle bacilli using bacterial artificial chromosome arrays. *Molecular Microbiology*, 32, 643–655.

Goude, G., Dori, I., Sparacello, V. S., Starnini, E. & Varalli, A. (2020). Multi-proxy stable isotope analyses of dentine microsections reveal diachronic changes in life history adaptations, mobility, and tuberculosis-induced wasting in prehistoric Liguria (Finale Ligure, Italy, northwestern Mediterranean). *International Journal of Paleopathology*, 28, 99–111.

Guichón, R. A., Buikstra, J. E., Stone, A. C., Harkins, K. M., Suby, J. A., Massone, M., Prieto Lglesias, A., Wilbur, A., Constantinescu, F. & Rodríguez Martín, C. (2015). Pre-Columbian tuberculosis in Tierra del Fuego? Discussion of the paleopathological and molecular evidence. *International Journal of Paleopathology*, 11, 92–101.

Gutierrez, M. C., Brisse, S., Brosch, R., Fabre, M., Omaïs, B., Marmiesse, M., Supply, P. & Vincent, V. (2005). Ancient origin and gene mosaicism of the progenitor of *Mycobacterium tuberculosis*. *PLoS Pathogens* 1(1), e5. doi:10.1371/journal.ppat.0010005

Hall, W., McDonnell, A. & O'Neill, J. (2018). *Superbugs: an arms race against bacteria*. Harvard University Press, Cambridge, MA.

Harkins, K.M., Buikstra, J.E., Campbell, T., Bos, K. I., Johnson, E. D., Krause, J. & Stone, A. C. (2015). Screening ancient tuberculosis with qPCR: challenges and opportunities. *Philosophical Transactions of the Royal Society B: Biological Sciences*, 370(1660), 20130622. doi.org/10.1098/rstb.2013.0622

Harper, K. & Armelagos, G. J. (2010). The changing disease-scape in the third epidemiological transition.

International Journal of Environmental Research and Public Health, 7, 675–697.

Hershberg, R., Lipatov, M., Small, P. M., Sheffer, H., Niemann, S., Homolka, S., Roach, J. C., Kremer, K., Petrov, D. A., Feldman, M. W. & Gagneux, S. (2008). High functional diversity in *Mycobacterium tuberculosis* driven by genetic drift and human demography. *PLoS Biology*, 6(12), e311.

Hershkovitz, I., Donoghue, H. D., Minnikin, D. E., Besra, G. S., Lee, O. Y., Gernaey, A. M., Galili, E., Eshed, V., Greenblatt, C. L., Lemma, E., Bar-Gal, G. K. & Spigelman, M. (2008). Detection and molecular characterization of 9,000-year-old *Mycobacterium tuberculosis* from a Neolithic settlement in the Eastern Mediterranean. *PLoS One*: doi:10.1371/journal.pone.0003426

Hershkovitz, I., Greenwald, C. M., Latimer, B., Jellema, L. M., Wish-Baratz, S., Eshed, V., Dutour, O. & Rothschild, B. M. (2002). Serpens endocrania symmetrica (SES): a new term and a possible clue for identifying intrathoracic disease in skeletal populations. *American Journal of Physical Anthropology*, 118, 201–216.

Holloway, K. L., Link, K., Rühli, F. & Henneberg, M. (2013). Skeletal lesions in human tuberculosis may sometimes heal: an aid to palaeopathological diagnoses. *PLoS One*, 8, e62798.

Huang, S-J., Wang, X-H., Liu, Z-D., Cao, W-L., Han, Y, Ma, A-G. & Xu, S-F. (2016). Vitamin D deficiency and the risk of tuberculosis: a meta-analysis. *Drug Design Development and Therapy*, 11, 91–102.

Hughes, A. L., Friedman, R. & Murray, M. (2002). Genomewide pattern of synonymous nucleotide substitution in two complete genomes of *Mycobacterium tuberculosis*. *Emerging Infectious Diseases*, 8(11), 1342–1346.

Iossa, G. & White, P. C. L. (2018). The natural environment: a critical missing link in national action plans on antimicrobial resistance. *Bulletin of the World Health Organization*, 96, 858–860.

Jaber, M., Rattan, A. & Kumar, R. (1996). Presence of *katG* gene in resistant *Mycobacterium tuberculosis*. *Journal of Clinical Pathology*, 49, 945–947.

Jaffe, H. L. (1972). *Metabolic, degenerative and inflammatory diseases of bones and joints*. Lea and Febiger, Philadelphia.

Jain, V. K., Iyengar, K. P., Samy, D. A. & Vaishya, R. (2020). Tuberculosis in the era of COVID-19 in India. *Diabetes and Metabolic Syndrome*, 14, 1439–1443.

Johnston, M. P. & Rothstein, E. (1952). Tuberculosis of the rib. *Journal of Bone and Joint Surgery*, 34A, 878–882.

Kapur, V., Whittam, T. S. & Musser, J. M. (1994). Is *Mycobacterium tuberculosis* 15,000 years old? *The Journal of Infectious Diseases*, 170(5), 1348–1349.

Kay, G., Sergeant, M., Zhou, Z., Chan, J. Z., Millard, A., Quick, J., Szikossy, I., Pap, I., Spigelman, M., Loman, N. J., Achtman, M., Donoghue, H. D. & Pallen, M. J. (2015). Eighteenth-century genomes show that mixed infections were common at time of peak tuberculosis in Europe. *Nature Communications*, 6: 6717, doi.org/10.1038/ncomms7717

Kelley, M. A. & Micozzi, M. S. (1984). Rib lesions in chronic pulmonary tuberculosis. *American Journal of Physical Anthropology*, 65, 381–386.

Kerner, G., Laval, G., Patin, E., Boisson-Dupuis, S., Abel, L., Casanova, J. L. & Quintana-Murci, L. (2021). Human ancient DNA analyses reveal the high burden of tuberculosis in Europeans over the last 2,000 years. *American Journal of Human Genetics*, 108(3), 517–524. doi: 10.1016/j.ajhg.2021.02.009.

Key, F. M., Posth, C., Esquivel-Gomez, L. R., Hübler, R., Spyrou, M. A., Neumann, G. U., Furtwängler, A., Sabin, S., Burri, M., Wissgott, A., Lankapalli, A. K., Vågene, Å. J., Meyer, M., Nagel, S., Tukhbatova, R., Khokhlov, A., Chizhevsky, A., Hansen, S., Belinsky, A. B., Kalmykov, A., Kantorovich, A. R., Maslov, V. E., Stockhammer, P. W., Vai, S., Zavattaro, M., Riga, A., Caramelli, D., Skeates, R., Beckett, J., Gradoli, M. G., Steuri, N., Hafner, A., Ramstein, M., Siebke, I., Lösch, S., Erdal, Y. S., Alikhan, N. F., Zhou, Z., Achtman, M., Bos, K., Reinhold, S., Haak, W., Kühnert, D., Herbig, A. & Krause, J. (2020). Emergence of human-adapted *Salmonella enterica* is linked to the Neolithization process. *Nature Ecology and Evolution*, 4(3), 324–333.

Lalvani, A., Fraser, C. & Pareek, M. (2021). 'Diagnosis of latent TB infection', in L. Friedman, M. Dedicoat & P. D. O Davies, eds. *Clinical tuberculosis*, 6th edn, pp. 153–171. CRC Press, Abingdon.

Lambert, P. (2002). Rib lesions in a prehistoric Puebloan sample from Southwestern Colorado. *American Journal of Physical Anthropology*, 117, 281–292.

Lewis, M. E. (2004). Endocranial lesions in non-adult skeletons: understanding their aetiology. *International Journal of Osteoarchaeology*, 14, 82–97.

Lewis, M. E. (2018). *Paleopathology of children: identification of pathological conditions in the human skeletal remains of non-adults*. Academic Press, London.

Ley, S. D., de Vos, M., Van Rie, A. & Warren, R. M. (2019). Deciphering within-host microevolution of *Mycobacterium tuberculosis* through whole-genome sequencing: the phenotypic impact and way forward. *Microbiology and Molecular Biology Reviews*, 83(2), e00062-18. doi: 10.1128/MMBR.00062-18

Li, M., Chen, L., Zhao, D. & Roberts, C. A. (2019). A male adult skeleton from the Han Dynasty in Shaanxi, China (202 BC–220 AD) with bone changes that possibly represent spinal tuberculosis. *International Journal of Paleopathology*, 27, 9–16.

Macedo Couto, R., Ranzani, O. T. & Waldman, E. A. (2019). Zoonotic tuberculosis in humans: control, surveillance, and the One Health approach. *Epidemiological Review*, 41, 130–144.

Magnani, R., Cavalca, M., Pierantoni, M., Luppi, A., Cantoni, A. M., Prosperi, A., Pacciarini, M., Zanoni, M., Tamba, M., Santi, A. & Bonardi, S. (2020). Infection by *Mycobacterium caprae* in three cattle herds in Emilia-Romagna Region, Northern Italy. *Italian Journal of Food Safety*, 9(1), 9–11.

Malone, K. M. & Gordon, S. V. (2017). *Mycobacterium tuberculosis* complex members adapted to wild and domestic animals. *Advances in Experimental Medicine and Biology*, 1019, 135–154.

Manson, A. L., Cohen, K. A., Abeel, T., Desjardins, C. A., Armstrong, D. T., Barry, C. E. 3rd, Brand, J.; TBResist Global Genome Consortium, Chapman, S. B., Cho, S. N., Gabrielian, A., Gomez, J., Jodals, A. M., Joloba, M., Jureen, P., Lee, J. S., Malinga, L., Maiga, M., Nordenberg, D., Noroc, E., Romancenco, E., Salazar, A., Ssengooba, W., Velayati, A. A., Winglee, K., Zalutskaya, A., Via, L. E., Cassell, G. H., Dorman, S. E., Ellner, J., Farnia, P., Galagan, J. E., Rosenthal, A., Crudu, V., Homorodean, D., Hsueh, P. R., Narayanan, S., Pym, A. S., Skrahina, A., Swaminathan, S., Van der Walt, M., Alland, D., Bishai, W. R., Cohen, T., Hoffner, S., Birren, B. W. & Earl, A. M. (2017). Genomic analysis of globally diverse *Mycobacterium tuberculosis* strains provides insights into the emergence and spread of multidrug resistance. *Nature Genetics*, 49(3), 395–402.

Margaryan, A., Hansen, H. B., Rasmussen, S., Sikora, M., Moiseyev, V., Khoklov, A., Epimakhov, A., Yepiskoposyan, L., Kriiska, A., Varul, L., Saag, L., Lynnerup, N., Willerslev, E. & Allentoft, M. E. (2018). Ancient pathogen DNA in human teeth and petrous bones. *Ecology and Evolution*, 8, 3534–3542.

Mariotti, V., Zuppello, M., Pedrosi, M. E., Bettuzzi, M., Brancaccio, R., Peccenini, E., Morigi, M. P. & Belcastro, M. G. (2015). Skeletal evidence of tuberculosis in a modern identified human skeletal collection (Certosa Cemetery, Bologna, Italy). *American Journal of Physical Anthropology*, 157, 389-401. https://doi.org/10.1002/ajpa.22727

Marmot, M. (2004). *The status syndrome: how social standing affects our health and longevity*. Times Books, New York.

Masson, M., Molnár, E., Donoghue, H. D., Besra, G. S., Minnikin, D. E., Wu, H. H., Lee, O. Y., Bull, I. D. & Pálfi, G. (2013). Osteological and biomolecular evidence of a 7000-year-old case of hypertrophic pulmonary osteopathy secondary to tuberculosis from Neolithic Hungary. *PLoS One*, https://doi.org/10.1371/journal.pone.0078252

Matos, V. & Santos, A. L. (2006). On the trail of pulmonary tuberculosis based on rib lesions: results from the human identified skeletal collection from the Museu Bocage (Lisbon, Portugal). *American Journal of Physical Anthropology*, 130, 190–200. https://doi.org/10.1002/ajpa.20309.

McHenry, M. L., Bartlett, J., Igo Jr., R. P., Wampande, E. M., Benchek, P., Mayanja-Kizza, H., Fluegge, K., Hall, N. B., Gagneux, S., Tishkoff, S. A., Wejse, C., Sirugo, G., Boom, W. H., Joloba, M., Williams, S. M. & Stein, C. M. 2020. Interaction between host genes and *Mycobacterium tuberculosis* lineage can affect tuberculosis severity: evidence for coevolution? *PLoS Genetics*, 16(4), e1008728.

McLain, R. F. & Isada, C. (2004). Spinal tuberculosis deserves a place on the radar screen. *Cleveland Clinic Journal of Medicine*, 71(7), 537–539.

Menardo, F., Duchene, S., Brites, D. & Gagneux, S. (2019). The molecular clock of *Mycobacterium tuberculosis*. *PLoS Pathogens*, 15(9), e1008067.

Mendum, T. A., Schuenemann, V. J., Roffey, S., Taylor, G. M., Wu, H., Singh, P., Tucker, K., Hinds, J., Cole, S. T., Kierzek, A. M., Nieselt, K., Krause, J. & Stewart, G. R. (2014). *Mycobacterium leprae* genomes from a British medieval leprosy hospital: towards understanding an ancient epidemic. *BMC Genomics*, 15, 270. https://doi.org/10.1186/1471-2164-15-270

Miotto, P., Tessema, B., Tagliani, E., Chindelevitch, L., Starks, A. M., Emerson, C., Hanna, D., Kim, P. S., Liwski, R., Zignol, M., Gilpin, C., Niemann, S., Denkinger, C. M., Fleming, J., Warren, R. M., Crook, D., Posey, J., Gagneux, S., Hoffner, S., Rodrigues, C., Comas, I., Engelthaler, D. M., Murray, M., Alland, D., Rigouts, L., Lange, C., Dheda, K., Hasan, R., Ranganathan, U. D. K., McNerney, R., Ezewudo, M., Cirillo, D. M., Schito, M., Köser, C. U. & Rodwell, T. C. (2017). A standardised method for interpreting the association between mutations and phenotypic drug resistance in *Mycobacterium tuberculosis*. *European Respiratory Journal*, 50(6), 1701354. doi: 10.1183/13993003.01354-2017

Moller, M., Kinnear, C. J., Orlova, M., Kroon, E. E., van Helden, P. D., Schurr, E. & Hoal, E. G. (2018). Genetic resistance to *Mycobacterium tuberculosis* infection and disease. *Frontiers of Immunology*, 9, 2219. https://doi.org/10.3389/fimmu.2018.02219

Monsalve, M. V., Humphrey, E., Yueyeng, S., Schulze, H. G., Atkins, C. G., Blades, M. W., Konorov, S. O., Turner, R. F. B. & Walker, D. C. (2018). Deposits in the lungs of Kwäday Dän Ts'inchį man: characterization by a combination of analytical microscopical methods. *American Journal of Physical Anthropology*, 167, 337–347.

Morales-Arce, A. Y., Harris, R. B., Stone, A. C. & Jensen, J. D. (2020). Evaluating the contributions of purifying selection and progeny-skew in dictating within-host *Mycobacterium tuberculosis* evolution. *Evolution*, 74(5), 992–1001.

Morse, D., Brothwell, D. R. & Ucko, P. J. (1964). Tuberculosis in ancient Egypt. *American Review of Respiratory Diseases*, 90(4), 526–541.

Mortimer, T. D. & Pepperell, C. S. (2014). Genomic signatures of distributive conjugal transfer among mycobacteria. *Genome Biology and Evolution*, 6(9), 2489–2500.

Mortimer, T. D., Weber, A. M. & Pepperell, C. S. (2018). Signatures of selection at drug resistance loci in *Mycobacterium tuberculosis*. *mSystems*, 3(1), e00108–17. doi: 10.1128/mSystems.00108-17

Müller, R., Roberts, C. A. & Brown, T. A. (2014a). Biomolecular identification of ancient *Mycobacterium tuberculosis* complex DNA in human remains from Britain and continental Europe. *American Journal of Physical Anthropology*, 153(2), 178–189. https://doi.org/10.1002/ajpa.22417

Müller, R., Roberts, C. A. & Brown, T. A. (2014b). Genotyping of ancient *Mycobacterium tuberculosis* strains reveals historic genetic diversity. *Proceedings of the Royal Society B: Biological Sciences*, 281(1781), article no. 20133236. doi.org/10.1098/rspb.2013.3236

Müller, R., Roberts, C. A. & Brown, T. A. (2015). Complications in the study of ancient tuberculosis: non-specificity of IS6110 PCRs. *Science and Technology of Archaeological Research* 1(1), 1–8.

Müller, R., Roberts, C. A. & Brown, T. A. (2016). Complications in the study of ancient tuberculosis: presence of environmental bacteria in human archaeological remains. *Journal of Archaeological Science*, 68, 5–11.

National Institutes of Allergies and Infectious Diseases. (2021). Coronaviruses. https://www.niaid.nih.gov/diseases-conditions/coronaviruses

Nebenzahl-Guimaraes, H., Yimer, S. A., Holm-Hansen, C., de Beer, J., Brosch, R. & van Soolingen, D. (2016). Genomic characterization of *Mycobacterium tuberculosis* lineage 7 and a proposed name: 'Aethiops vetus'. *Microbial Genomics*, 2(6), e000063.

Nelson, E. A., Buikstra, J. E., Herbig, A., Tung, T. A. & Bos, K. I. (2020). Advances in the molecular detection of tuberculosis in pre-contact Andean South America. *International Journal of Paleopathology*, 29, 128–140.

Nesse, R. M. & Williams, G. C. (1994). *Why we get sick: the new science of Darwinian medicine*. Vintage Books, New York.

Ngabonziza, J. C. S., Loiseau, C., Marceau, M., Jouet, A., Menardo, F., Tzfadia, O., Antoine, R., Niyigena, E. B., Mulders, W., Fissette, K., Diels, M., Gaudin, C., Duthoy, S., Ssengooba, W., André, E., Kaswa, M. K., Habimana, Y. M., Brites, D., Affolabi, D., Mazarati, J. B., de Jong, B. C., Rigouts, L., Gagneux, S., Meehan, C. J. & Supply, P. (2020). A sister lineage of the *Mycobacterium tuberculosis* complex discovered in the African Great Lakes region.

Nature Communications, 11(1), 2917. doi.10.1038/s41467-020-16626-6

Nicklisch, N., Maixner, F., Ganslmeier, R., Friederich, S., Dresely, V., Meller, H., Zink, A. & Alt, K. W. (2012). Rib lesions in skeletons from early Neolithic sites in Central Germany: on the trail of tuberculosis at the onset of agriculture. *American Journal of Physical Anthropology*, 149, 391–404.

Nimmo, C., Shaw, L. P., Doyle, R., Williams, R., Brien, K., Burgess, C., Breuer, J., Balloux, F. & Pym, A. S. (2019). Whole genome sequencing *Mycobacterium tuberculosis* directly from sputum identifies more genetic diversity than sequencing from culture. *BMC Genomics*, 20(1), 389. https://doi.org/10.1186/s12864-019-5782-2

O'Neill, M. B., Shockey, A., Zarley, A., Aylward, W., Eldholm, V., Kitchen, A. & Pepperell, C. S. (2019). Lineage specific histories of *Mycobacterium tuberculosis* dispersal in Africa and Eurasia. *Molecular Ecology*, 28(13), 3241–3256.

Okazaki, K., Takamuku, S., Yonemoto, S., Itahashi, Y., Gakuhari, T., Yoneda, M. & Chen, J. (2019). A paleopathological approach to early human adaptation for wet-rice agriculture: the first case of Neolithic spinal tuberculosis at the Yangtze River Delta of China. *International Journal of Paleopathology*, 24, 236–244.

Orlando, L., Allaby, R., Skoglund, P., Der Sarkissian, C., Stockhammer, P. W., Ávila-Arcos, M. C., Fu, Q., Krause, J., Willerslev, E., Stone, A. C. & Warinner, C. (2021). Ancient DNA analyses. *Nature Reviews Methods Primers*, 1(1), 1–26.

Orłowska, B., Augustynowicz-Kopeć, E., Krajewska, M., Zabost, A., Welz, M., Kaczor, S. & Anusz, K. (2017). *Mycobacterium caprae* transmission to free-living grey wolves (*Canis lupus*) in the Bieszczady Mountains in Southern Poland. *European Journal of Wildlife Research*, 63(1), 2–6.

Ortner, D. J. (1979). Disease and mortality in the Early Bronze Age people of Bab edh-Dhra, Jordan. *American Journal of Physical Anthropology*, 51, 589–598.

Osorio, N. S., Rodrigues, F., Gagneux, S., Pedrosa, J., Pinto-Carbo, M., Castro, A. G., Young, D., Comas, I. & Saraiva, M. (2013). Evidence for diversifying selection in a set of *Mycobacterium tuberculosis* genes in response to antibiotic- and nonantibiotic-related pressure. *Molecular Biology and Evolution*, 30(6), 1326–1336.

Pálfi, G., Dutour, O., Deák, J. & Hutás, I. (eds). (1999). *Tuberculosis: past and present*. Golden Book Publishers and Tuberculosis Foundation, Budapest/Szeged.

Panhuysen, R., Conenen, V., Bruintjes, T. (1997). Chronic maxillary sinusitis in medieval Maastricht, The Netherlands. *International Journal of Osteoarchaeology*, 7, 610–614.

Paradis, E. & Schliep, K. (2019). Ape 5.0: An environment for modern phylogenetics and evolutionary analyses in R. *Bioinformatics*, 35(3), 526–528.

Pedersen, D. D., Milner, G. R., Kolmos, H. J. & Boldsen, J. L. (2019). The association between skeletal lesions and tuberculosis diagnosis using a probabilistic approach. *International Journal of Paleopathology*, 27, 88–100.

Pepperell, C. S., Casto, A. M., Kitchen, A., Granka, J. M., Cornejo, O. E., Holmes, E. C., Birren, B., Galagan, J. & Feldman, M. W. (2013). The role of selection in shaping diversity of natural *M. tuberculosis* populations. *PLoS Pathogens*, 9(8), e1003543.

Pepperell, C. S., Granka, J. M., Alexander, D. C., Behr, M. A., Chui, L., Gordon, J., Guthrie, J. L., Jamieson, F. B., Langlois-Klassen, D., Long, R., Nguyen, D., Wobeser, W. & Feldman, M. W. (2011). Dispersal of *Mycobacterium tuberculosis* via the Canadian fur trade. *Proceedings of the National Academy of Sciences of the United States of America*, 108(16), 6526–6531.

Popovic, I., Soares Magalhaes, R. J., Ge, E., Marks, G. B., Dong, G. H., Wei, X. & Knibbs, L. D. (2019). A systematic literature review and critical appraisal of epidemiological studies on outdoor air pollution and tuberculosis outcomes. *Environmental Research*, 170, 33–45.

Prendergast, M. E. & Sawchuk, E. (2018). Boots on the ground in Africa's ancient DNA 'revolution': archaeological perspectives on ethics and best practices. *Antiquity*, 92(363), 803–815.

Pym, A. S., Nimmo, C. & Millard, J. (2021). 'New developments in drug treatment', in L. Friedman, M. Dedicoat & P. D. O Davies (eds), *Clinical tuberculosis*, 6th edn, pp. 203–216. CRC Press, Abingdon.

Quinn, K. (2017). *A bioarchaeological study of the impact of mobility on the transmission of tuberculosis in Roman Britain*. Unpublished PhD thesis, Durham University, Durham.

Reed, M. B., Pichler, V. K., Mcintosh, F., Mattia, A., Fallow, A., Masala, S., Domenech, P., Zwerling, A., Thibert, L., Menzies, D., Schwartzman, K. & Behr, M. A. (2009). Major *Mycobacterium tuberculosis* lineages associate with patient country of origin. *Journal of Clinical Microbiology*, 47(4), 1119–1128.

Resnick, D. (1995). 'Hemaglobinopathies and other anemias', in D. Resnick (ed), *Diagnosis of bone and joint disorders*, 3rd edn, pp. 2107–2146. Saunders, Philadelphia.

Resnick, D. & Niwayama, G., (1995). 'Osteomyelitis, septic arthritis, and soft tissue infection: Organisms', in D. Resnick (ed), *Diagnosis of bone and joint disorders*, 3rd edn, pp. 2448–2558. Saunders, Philadelphia.

Revell, L. J. (2012). phytools: An R package for phylogenetic comparative biology (and other things). *Methods in Ecology and Evolution*, 3(2), 217–223.

Rice, C., Harrison, E. & Friedman, M. (2019). Doing justice to intersectionality in research. *Cultural Studies ↔ Critical Methodologies*, 19, 409–420.

Richards, M. & Montgomery, J. (2012). 'Isotope analysis and palaeopathology: a short review and future developments', in J. E. Buikstra & C. A. Roberts (eds), *A global history of paleopathology: pioneers and prospects*, pp. 718–731. Oxford University Press, Oxford.

Roberts, C. & Ingham, S. (2008). Using ancient DNA analysis in palaeopathology: a critical analysis of published papers, with recommendations for future work. *International Journal of Osteoarchaeology*, 18, 600–613.

Roberts, C. A. (2007). A bioarchaeological study of maxillary sinusitis. *American Journal of Physical Anthropology*, 133, 792–807.

Roberts, C. A. (2015). 'What did agriculture do for us? The bioarchaeology of health and diet', in G. Barker & C. Goucher (eds), *The Cambridge world history. volume 2: a world with agriculture, 12,000 BCE–500 CE*, pp. 93–123. Cambridge University Press, Cambridge.

Roberts, C. A. (2020). 'No pain, no gain: ideals of beauty and success: fashionable but debilitating diseases: tuberculosis past and present', in S. Sheridan & L. Gregoricka (eds), *Purposeful pain: the bioarchaeology of intentional suffering*, pp. 21–38. Springer, Cham.

Roberts, C. A. & Bernard, M-C. (2015). Tuberculosis: a demographic analysis and social study of admissions to a children's sanatorium (1936–1954) in Stannington, Northumberland, England. *Tuberculosis Special Issue*, 95, S105–108.

Roberts, C. A. & Brickley, M. (2018). 'Infectious and metabolic diseases: a synergistic bioarchaeology', in A. Katzenberg & A. L. Grauer (eds), *Biological anthropology of the human skeleton*, 3rd edn, pp. 415–446. Wiley-Blackwell, Chichester.

Roberts, C. A. & Buikstra, J. E. (2003). *The bioarchaeology of tuberculosis: a global view on a re-emerging disease*. University Press of Florida, Gainesville.

Roberts, C. A. & Buikstra, J. E. (2019). 'Bacterial infections', in J. E. Buikstra (ed), *Ortner's identification of pathological conditions in human skeletal remains*, pp. 321–439. Academic Press, London.

Roberts, C. A. & Buikstra, J. E. (2021). 'History of tuberculosis from the earliest times to the development of drugs', in L. Friedman, M. Dedicoat & P. D. O. Davies (eds), *Clinical tuberculosis*, 6th edn, pp. 3–15. CRC Press, Abingdon.

Roberts, C. A. & Cox, M. (2003). *Health and disease in Britain: from prehistory to the present day*. Sutton Publishing, Stroud.

Roberts, C. A., Lucy, D. & Manchester, K. (1994). Inflammatory lesions of ribs: an analysis of the Terry

Collection. *American Journal of Physical Anthropology*, 95(2), 169–182.

Roetzer, A., Diel, R., Kohl, T. A., Rückert, C., Nübel, U., Blom, J., Wirth, T., Jaenicke, S., Schuback, S., Rüsch-Gerdes, S., Supply, P., Kalinowski, J. & Niemann, S. (2013). Whole genome sequencing versus traditional genotyping for investigation of a *Mycobacterium tuberculosis* outbreak: a longitudinal molecular epidemiological study. *PLoS Medicine*, 10, e1001387.

Sabin, S., Herbig, A., Vågene, A. J., Ahlstrom, T., Bozovic, G., Arcini, C., Kühnert, D. & Bos, K. I. (2020). A seventeenth-century *Mycobacterium tuberculosis* genome supports a Neolithic emergence of the *Mycobacterium tuberculosis* complex. *Genome Biology*, 21(1), 201.

Salo, W. L., Aufderheide, A. C., Buikstra, J. E. & Holcomb, T. A. (1994). Identification of *Mycobacterium tuberculosis* DNA in a pre-Columbian mummy. *Proceedings of the National Academy of Sciences of the United States of America*, 91, 2091–2094.

Santos, A. L. & Roberts, C. A. (2006). Anatomy of a serial killer: differential diagnosis of tuberculosis based on rib lesions of adult individuals from the Coimbra Identified Skeletal Collection, Portugal. *American Journal of Physical Anthropology*, 130, 38–49.

Schuenemann, V. J., Avanzi, C., Krause-Kyora, B., Seitz, A., Herbig, A., Inskip, S., Bonazzi, M., Reiter, E., Urban, C., Dangvard Pedersen, D., Taylor, G. M., Singh, P., Stewart, G. R., Velemínský, P., Likovsky, J., Marcsik, A., Molnár, E., Pálfi, G., Mariotti, V., Riga, A., Belcastro, M. G., Boldsen, J. L., Nebel, A., Mays, S., Donoghue, H. D., Zakrzewski, S., Benjak, A., Nieselt, K., Cole, S. T. & Krause, J. (2018). Ancient genomes reveal a high diversity of *Mycobacterium leprae* in medieval Europe. *PLoS Pathogens*, 14(5), e1006997–17.

Schuenemann, V. J., Singh, P., Mendum, T. A., Krause-Kyora, B., Jäger, G., Bos, K. I., Herbig, A., Economou, C., Benjak, A., Busso, P., Nebel, A., Boldsen, J. L., Kjellström, A., Wu, H., Stewart, G. R., Taylor, G. M., Bauer, P., Lee, O. Y-C., Wu, H. H. T., Minnikin, D. E., Besra, G. S., Tucker, K., Roffey, S., Sow, S. O., Cole, S. T., Nieselt, K. & Krause, J. (2013). Genome-wide comparison of medieval and modern Mycobacterium leprae. *Science*, 341(6142), 179–183.

Schultz, M. (1999). 'The role of tuberculosis in infancy and childhood in prehistoric and historic populations', in G. Pálfi, O. Dutour, J. Deák & I. Hutás (eds), *Tuberculosis: past and present*, pp. 503–507. Golden Book Publishers and Tuberculosis Foundation, Budapest/Szeged.

Séraphin, M. N., Norman, A., Rasmussen, E. M., Gerace, A. M., Chiribau, C. B., Rowlinson, M-C., Lillebaek, T. & Lauzardo, M. (2019). Direct transmission of within-host *Mycobacterium tuberculosis* diversity to secondary cases can lead to variable between-host heterogeneity

without de novo mutation: a genomic investigation. *EBioMedicine*, 47, 293–300.

Smith, F. B. (1988). *The retreat of tuberculosis 1850–1950*. Croom Helm, London.

Snow, K. J., Sismanidis, C., Denhol, J., Sawyer, S. M. & Graham, S. M. (2018). The incidence of tuberculosis among adolescents and young adults: a global estimate. *European Respiratory Journal*, 51, 1702352. doi.10.1183/13993003.02352-2017

Sparacello, V. S., Roberts, C. A., Kerudin, A. & Müller, R. (2017). A 6,500-year-old Middle Neolithic child from Pollera Cave (Liguria, Italy) with probable multifocal osteoarticular tuberculosis. *International Journal of Paleopathology*, 17, 67–74.

Spekker, O., Hunt, D. R., Paja, L., Molnár, E., Pálfi, G. & Schultz, M. (2020a). Tracking down the White Plague. The skeletal evidence of tuberculous meningitis in the Robert J. Terry anatomical skeletal collection. *PLoS One*, 15(3), e0230418. doi:10.1371/journal.pone.0230418

Spekker, O., Hunt, D. R., Varadi, O. A., Berthon, W., Molnár, E. & Pálfi, G. (2018). Rare manifestations of spinal tuberculosis in the Robert J. Terry skeletal collection (National Museum of Natural History, Smithsonian Institution, Washington DC, USA). *International Journal of Osteoarchaeology*, 28, 343–353.

Spekker, O., Schultz, M., Paja, L., Váradi, O. A., Molnár, E., Pálfi, G. & Hunt, D. R. (2020b). Tracking down the White Plague. Chapter two: the role of endocranial abnormal blood vessel impressions and periosteal appositions in the palaeopathological diagnosis of tuberculous meningitis. *PLoS One*, 15(9), e0238444, doi:10.1371/journal.pone.0238444

Spigelman, M. & Lemma, E. (1993). The use of the polymerase chain reaction (PCR) to detect *Mycobacterium tuberculosis* in ancient skeletons. *International Journal of Osteoarchaeology*, 3(2), 137–143.

Spyrou, M. A., Bos, K. I., Herbig, A. & Krause, J. (2019). Ancient pathogen genomics as an emerging tool for infectious disease research. *Nature Reviews Genetics*, 20, 323–340.

Spyrou, M. A., Tukhbatova, R. I., Feldman, M., Drath, J., Kacki, S., Beltrán de Heredia, J., Arnold, S., Sitdikov, A. G., Castex, D., Wahl, J., Gazimzyanov, I. R., Nurgaliev, D. K., Herbig, A., Bos, K. I. & Krause, J. (2016). Historical *Y. pestis* genomes reveal the European Black Death as the source of ancient and modern plague pandemics. *Cell Host and Microbe*, 19(6), 874–881.

Squires, K., Booth, T. & Roberts, C. A. (2019). 'The ethics of sampling human skeletal remains for destructive analyses: a UK perspective', in K. Squires, D. Errickson & N. Márquez-Grant (eds), *Ethical challenges in *the analysis of human remains*, pp. 265–297. Elsevier, Amsterdam.

Steyn, M., Scholtz, Y., Botha, D. & Pretorius, S. (2013). The changing face of tuberculosis: trends in tuberculosis-associated skeletal changes. *Tuberculosis*, 93, 467–474.

Stone, A. C. & Ozga, A. T. (2019). 'Ancient DNA in the study of ancient disease', in J. E. Buikstra & C. A. Roberts (eds), *A global history of paleopathology: pioneers and prospects*, pp. 183–210. Oxford University Press, Oxford.

Stone, A. C., Wilbur, A. K., Buikstra, J. E. & Roberts, C. A. (2009). Tuberculosis and leprosy in perspective. *Yearbook of Physical Anthropology*, 52, 66–94.

Stucki, D., Brites, D., Jeljeli, L., et al. (2016). *Mycobacterium tuberculosis* lineage 4 comprises globally distributed and geographically restricted sublineages. *Nature Genetics*, 48(12), 1535–1543. doi.10.1038/ng.3704

Swabe, J. (1999). *Animals, disease and human society*. Routledge, London.

Tanner, R. & McShane, H. (2021). 'BCG and other vaccines', in L. Friedman, M. Dedicoat & P.D.O. Davies (eds), *Clinical tuberculosis*, 6th edn, pp. 217–233. CRC Press, Abingdon.

Taylor, G. M., Murphy, E., Hopkins, E., Rutland, P. & Christov, Y. (2007). First report of *Mycobacterium bovis* DNA in human remains from the Iron Age. *Microbiology*, 153, 1243–1249.

Theron, G., Cohen, T. & Dye, C. (2021). 'Epidemiology', in L. Friedman, M. Dedicoat & P.D.O. Davies (eds), *Clinical tuberculosis*, 6th edn, pp. 17–37. CRC Press, Abingdon.

Thomas, R. (2012). 'Non-human paleopathology', in J. E. Buikstra & C. A. Roberts (eds), *A global history of paleopathology: pioneers and prospects*, pp. 652–664. Oxford University Press, Oxford.

Thomas, T. (2019). 'Nonhuman animal paleopathology—are we so different?', in J. E. Buikstra (ed), *Ortner's identification of pathological conditions in human skeletal remains*, pp. 652–664. Academic Press, London.

Thwaites, G. (2014). 'Tuberculosis of the central nervous system', in P. D. O. Davies, S. B. Gordon & G. Davies (eds), *Clinical tuberculosis*, 5th edn, pp. 151–165. CRC Press, Abingdon.

Trauner, A., Liu, Q. & Via, L. E. (2017). The within-host population dynamics of Mycobacterium tuberculosis vary with treatment efficacy. *Genome Biology*, 18(1), 71. doi: 10.1186/s13059-017-1196-0

Trevathan, W. R., Smith, E. O. & McKenna, J. J. (2008). 'Introduction and overview of evolutionary medicine', in W. R. Trevathan, E. O. Smith, J. J. McKenna (eds), *Evolutionary medicine and health. New perspectives*, pp. 1–54. Oxford University Press, Oxford.

Tsosie, K. S., Begay, R. L., Fox, F. & Garrison, N. A. (2020). Generations of genomes: advances in paleogenomics technology and engagement for Indigenous people of the Americas. *Current Opinion in Genetics and Development*, 62, 91–96. doi:10.1016/j.gde.2020.06.010

Vågene, A. J., Herbig, A., Campana, M. G., Robles García, N. M., Warinner, C., Sabin, S., Spyrou, M. A., Andrades Valtueña, A., Huson, D., Tuross, N., Bos, K. I. & Krause, J. (2018). *Salmonella enterica* genomes from victims of a major sixteenth-century epidemic in Mexico. *Nature Ecology and Evolution*, 2, 520–528.

Vågene, Å. J., Honap, T. P., Harkins, K. M., et al. In press. Geographically dispersed zoonotic tuberculosis in pre-contact New World human populations. *Nature Communications*.

Wagner, J. K., Colwell, C., Claw, K. G., Stone, A. C., Bolnick, D. A., Hawks, J., Brothers, K. B. & Garrison, N. A. (2020). Fostering responsible research on ancient DNA. *American Journal of Human Genetics*, 107, 183–195.

Waldron, T. (1993). 'The health of adults', in T. Molleson & M. Cox (eds), *The Spitalfields Project: volume 2: the anthropology: the middling sort*. Council for British Archaeology Research Report 86, pp. 67–79. Archaeopress, York.

Wilbur, A. K., Bouwman, A. S., Stone, A. C., Roberts, C. A., Pfister, L. A. & Buikstra, J. E. (2009). Deficiencies and challenges in the study of ancient tuberculosis DNA. *Journal of Archaeological Science*, 36(9), 1990–1997.

Wilson, C. (2021). 'Animal tuberculosis', in L. Friedman, M. Dedicoat & P. D. O. Davies (eds), *Clinical tuberculosis*, 6th edn, pp. 415–435. CRC Press, Abingdon.

Wirth, T., Hildebrand, F., Allix-Béguec, C., Wölbeling, F., Kubica, T., Kremer, K., van Soolingen, D., Rüsch-Gerdes, S., Locht, C., Brisse, S., Meyer, A., Supply, P. & Niemann, S. (2008). Origin, spread and demography of the *Mycobacterium tuberculosis* complex. *PLoS Pathogens*, 4(9), e1000160. doi: 10.1371/journal.ppat.1000160

Wood, J. W., Milner, G. R., Harpending, H. C. & Weiss, K. M. (1992). The osteological paradox: problems of inferring prehistoric health from skeletal samples. *Current Anthropology*, 33, 343–370.

Woods, R. & Shelton, N. (1997). *An atlas of Victorian mortality*. Liverpool University Press, Liverpool.

World Health Organization. (2014). *The end TB strategy*. World Health Organization, Geneva.

World Health Organization. (2019). *Global tuberculosis report 2019*. World Health Organization, Geneva.

World Health Organization. (2020). *Global tuberculosis report 2020*. World Health Organization, Geneva.

World Health Organization. (no date a). Urban health. https://www.who.int/health-topics/urban-health (accessed 21.7.2021).

World Health Organization. (no date b). Tuberculosis key facts. https://www.who.int/news-room/fact-sheets/detail/tuberculosis (accessed 21.7.2021).

World Health Organization Global Tuberculosis Programme. (1994). *TB: a global emergency, WHO report on the TB epidemic.* World Health Organization, Geneva.

World Organisation for Animal Health (OIE). (2021). Global health risks and tomorrow's challenges. https://www.oie.int/en/what-we-do/global-initiatives/one-health/.

Zhang, J., Pei, N. & Mi, X. (2017). *phylotools: phylogenetic tools for eco-phylogenetics* (version 0.2.2).

Zink, A., Haas, C. J., Hagedorn, H. G., Szeimies, U. & Nerlich, A. G. (1999). 'Morphological and molecular evidence for pulmonary and osseous tuberculosis in a male Egyptian mummy', in G. Pálfi, O. Dutour, J. Deák & I. Hutás (eds), *Tuberculosis: past and present*, pp. 379–391.

Golden Book Publishers and Tuberculosis Foundation, Budapest/Szeged.

Zink, A. R., Molnár, E., Motamedi, N., Pálfy, G., Marcsik, A. & Nerlich, A. G. (2007). Molecular history of tuberculosis from ancient mummies and skeletons. *International Journal of Osteoarchaeology*, 17, 380–391.

Zinsstag, J., Schelling, E., Waltner-Toews, D. & Tanner, M. (2011). From 'One Medicine' to 'One Health' and systemic approaches to health and well-being. *Preventive Veterinary Medicine*, 101(3-4), 148–156.

Zwerling, A., Behr, M.A., Verma, A., Brewer, T. F., Menzies, D. & Pai, M. (2011). The BCG world atlas: a database of global BCG vaccination policies and practices. *PLoS Medicine* [online]. doi:10.1371/journal.pmed.1001012

Evolutionary perspectives on human parasitic infection: from ancient parasites to modern medicine

Marissa L. Ledger and Piers D. Mitchell

11.1 Introduction

Human-infecting pathogens are important to human evolutionary studies due to the reliance of these organisms on humans, the pathological effects as a result of infection and transmissibility between people, all potentially resulting in significant pressures on evolutionary trajectories. These intimate relationships with humans, sometimes lasting for decades in the same host, allow them to be useful proxies for past human migrations, the evolution of immunity-related genes and behaviours allowing for disease propagation. Parasites, here defined as organisms that live in or on a host and that acquire nutrients from that host, include ectoparasites, helminths and protozoa. Key examples of parasites infecting humans discussed in this chapter are given in Table 11.1. Parasites are particularly well suited as proxies for human evolutionary and anthropological questions due to the depth of our common evolution, the host specificity of some species and their complex life cycles (Figure 11.1).

This chapter focuses on evidence for human parasite infections in the past and how archaeological evidence can be used to better understand parasitic disease today. The chapter concentrates on intestinal helminths, but also discusses protozoa such as malaria and trypanosomes. It first presents evidence supporting the long co-evolutionary history and ubiquity of parasites in past populations that has been gleaned through fossil, archaeological and genetic evidence. It then considers the evidence for the co-evolutionary outcomes of human parasitic infection and the impact parasites have had on immune function and health. Finally, it discusses and makes recommendations for how palaeoparasitology can be used to inform public health initiatives, clinical medicine and future translational research.

11.2 Evidence for parasites throughout human history

Palaeoparasitology differs from most work in palaeopathology in that skeletal remains cannot be used to reliably retrieve evidence for most parasitic infections. Rather, recovery of preserved endoparasite eggs, cysts, protozoan antigens or ectoparasites provide evidence for their presence in the past (Figure 11.2). The predominant samples used in palaeoparasitology to find preserved parasites are palaeofaeces or coprolites (preserved faeces), sediment from the pelves of skeletons where the intestines were located and later decomposed, sediment from sewers and latrines and the intestinal contents or soft tissues of mummies (Bouchet et al., 2003; Camacho et al., 2018; Reinhard et al., 1986).

Marissa L. Ledger and Piers D. Mitchell, *Evolutionary perspectives on human parasitic infection*. In: *Palaeopathology and Evolutionary Medicine*. Edited by Kimberly A. Plomp et al., Oxford University Press. © Oxford University Press (2022). DOI: 10.1093/oso/9780198849711.003.0011

Table 11.1 Key examples of protozoa and helminths (worms) infecting humans. Helminths are subdivided into common taxonomic groups including nematodes (roundworms), cestodes (tapeworms) and trematodes (flukes). The most common definitive hosts and route of transmission are listed.

HELMINTHS			
Parasite	Common Name/Disease	Definitive Host(s)	Transmission
Nematodes			
Ascaris lumbricoides	Roundworm	Humans	Faecal–oral
Ancylostoma duodenale/ Necator sp.	Hookworm	Humans	Larvae penetrate skin
Enterobius vermicularis	Pinworm	Humans	Faecal–oral
Trichuris trichiura	Whipworm	Humans	Faecal–oral
Cestodes			
Dibothriocephalus sp.	Fish tapeworm	Mammals	Undercooked fish
Taenia saginata/Taenia asiatica	Beef tapeworm	Humans	Undercooked beef
Taenia solium	Pork tapeworm	Humans	Undercooked pork
Hymenolepis nana	Dwarf tapeworm	Humans and rodents	Faecal–oral or insect ingestion
Trematodes			
Fasciola spp.	Liver fluke	Sheep, cattle, humans	Aquatic plants
Dicrocoelium dendriticum	Lancet liver fluke	Sheep, deer, humans	Ants
Schistosoma spp.	Schistosomiasis	Species dependent (humans, primates, monkeys, cattle)	Larval form penetrates skin in water
PROTOZOA			
Parasite	Common Name/Disease	Definitive/Reservoir Host(s)	Transmission
Cryptosporidium spp.	-	Vertebrates	Faecal–oral
Entamoeba histolytica	Amoebic dysentery	Primates	Faecal–oral
Giardia duodenalis	Giardiasis	Mammals	Faecal–oral
Plasmodium falciparum/ ovale/malariae/vivax	Malaria	Humans	*Anopheles* mosquito bite
Trypanosoma cruzi	Chagas disease	Humans, dogs, cats, rodents	Bite of triatomine bug

So far, the earliest evidence for helminths infecting humans comes from sediment dated to 30,000 years ago from in a cave in France (Bouchet et al., 1996). However, parasites were certainly infecting humans and their hominin ancestors long before that. In the absence of preserved parasites in-situ with hominin remains, multiple approaches have been used to understand parasites that likely infected hominins. First, if we consider the parasites found in non-human primates in Africa, those that are similar to species found in humans today are likely to have affected the ancestors of the entire hominid lineage, including humans and non-human primates. These are called heirloom parasites (Pritchard et al., 2020), and examples include dwarf tapeworm (*Hymenolepis nana*), hookworm (*Ancylostoma duodenale* and *Necator* sp.), pinworm (*Enterobius* sp.), roundworm (*Ascaris* sp.),

schistosomes (*Schistosoma mansoni*), threadworm (*Strongyloides* sp.), whipworm (*Trichuris* sp.) and protozoa causing dysentery (*Entamoeba* sp.) (see Box 11.1 and Mitchell, 2013). Many of these parasites later evolved to have species that are specific to humans. A second approach to identify parasites that infected hominin ancestors is modern phylogenetics. For example, genetic analysis of *Taenia* tapeworms has shown that the species infecting humans, pork tapeworm (*Taenia solium*) and beef tapeworm (*Taenia saginata* and *Taenia asiatica*), are most closely related to the species that infect hyenas and lions, respectively (Terefe et al., 2014). This analysis suggests that human *Taenia* species evolved in Africa about 1–2.5 million years ago as humans were scavenging or hunting the same animals as these predators (Hoberg et al., 2001).

(a) Humans as necessary or sole definitive host

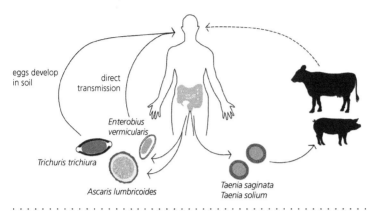

eggs develop
in soil

direct
transmission

Enterobius
vermicularis

Trichuris trichiura

Ascaris lumbricoides

Taenia saginata
Taenia solium

(b) Humans as one possible host

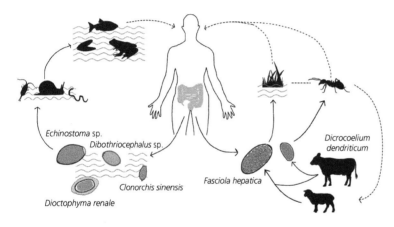

Echinostoma sp.

Dibothriocephalus sp.

Clonorchis sinensis

Dioctophyma renale

Fasciola hepatica

Dicrocoelium
dendriticum

Figure 11.1 Simplified overview of common parasite life cycles.
(a) The left shows parasites that can be transmitted directly from human to human without an animal intermediate host, while the right shows parasites that only have humans as their definitive host but require an animal as intermediate host. (b) Includes parasites that can infect humans along with many other animal hosts; on the left are those acquired from raw or undercooked fish and other aquatic animals and on the right those that are shared with common domesticated animals. Solid lines with arrows indicate routes of transmission. Dashed lines indicate that the parasite is acquired from ingestion of animals or plants carrying the parasite.

Box 11.1   Heirloom parasites

Roundworm (*Ascaris lumbricoides*) and whipworm (*Trichuris trichiura*) are generally considered to be heirloom parasites because of early evidence for infection in the human past alongside the worldwide distribution of both of the parasites today and their archaeological evidence (Araújo et al., 2013; Mitchell, 2013). These parasites are transmitted by the faecal–oral route and need a period of time to develop in the soil before they become infective. This has led to the suggestion that though they were present, their prevalence would have remained low in hunter-gatherer groups that were constantly moving and thus not exposed to high levels of faecal waste in

their living spaces (Perry, 2014). Other species such as the human pinworm (*Enterobius vermicularis*) can be passed directly between humans without a necessary period of development outside the host, and therefore they could more easily be carried by early hunter-gatherer groups (Araújo et al., 2013; Mitchell, 2013). Zoonotic parasites are those that are transmitted between animals and humans. Examples of common zoonotic parasites which are also likely to have infected hominins are *Taenia* spp. (beef and pork tapeworms), *Fasciola* liver fluke and certain species of *Schistosoma* (e.g. *Schistosoma haematobium* and *Schistosoma mansoni*) (Ledger & Mitchell, 2019; Mitchell, 2013). Some of these parasites, such as *Schistosoma*, are more geographically restricted due to their reliance on specific animal intermediate hosts.

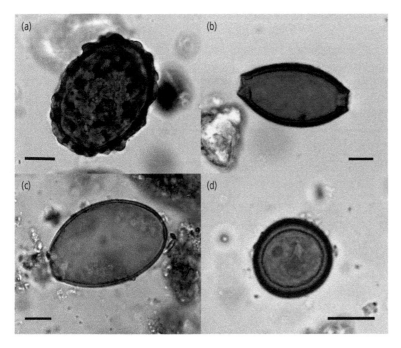

Figure 11.2 Examples of helminth eggs found in archaeological samples.
(a) roundworm; (b) whipworm; (c) fish tapeworm; (d) beef/pork tapeworm. Black scale bars indicate 20 μm.

Parasites that infected hominins after they migrated to regions outside of Africa are called souvenir parasites. Most of these are species that are geographically restricted due to their reliance on specific intermediate hosts, such as Chagas disease (*Trypanosoma cruzi*) in South America, *Paragonimus westermani* in East Asia and *Schistosoma japonicum* in East Asia. For example, Chagas disease uses triatomine insects (*Triatoma infestans*, *Triatoma dimidiata*, *Rhodnius prolixus*) as its intermediate host, the most important species of which are restricted to South and Central America (Rassi et al., 2010). *Schistosoma japonicum* relies on a specific species of snail, *Oncomelania hupensis*, as its first intermediate host, which is found in limited geographical regions in East Asia (Zhu et al., 2015). Thus, humans could only be exposed to such parasites once they migrated to these regions.

For most of the human past, our ancestors were living in small groups relying on hunting and gathering for subsistence. Unfortunately, we have little direct archaeological evidence for parasitic infections in these early hunter-gatherer groups. After the *Ascaris* eggs found in Palaeolithic cave sediment in France (Bouchet et al., 1996), we have evidence for whipworm (*Trichuris* sp.), another soil-transmitted helminth, found in Mesolithic samples from Europe (Bergman, 2018; Dark, 2004). For zoonotic parasites, Palaeolithic hunter-gatherers were likely commonly infected due to a diet reliant on wild animals and insects. It has been suggested that helminths that are rare today, such as *Echinostoma* sp. and giant kidney worm (*Dioctophyma renale*), were more common causes of infection in the past due to this reliance on wild animals and insects as food sources (Fugassa et al., 2010; Reinhard et al., 2013; Sianto et al., 2012). In Mesolithic sediment samples from a hunter-gatherer lake island site in Ireland (7420–6360 cal BP), fish tapeworm (*Dibothriocephalus* sp.) eggs were found in multiple samples indicating the human population's reliance on local fish as a food source (Perri et al., 2018).

Changes in parasitic infection that coincide with major changes in human lifestyle, such as the agricultural transition (that started ~ 12,000 years ago; also see Chapter 7) when humans adopted agriculture, are being investigated. With the adoption of agriculture, sedentism and increasing population sizes, transmission of pathogens is generally expected to have increased as new pathogens were supported in human communities (Armelagos et al., 1991; Larsen, 2018). However, many parasites would not follow the same patterns as those epidemic bacterial and viral pathogens that require large populations sizes for transmission (e.g. influenza, measles, smallpox). Parasites can be maintained in relatively small population sizes, in part because they have relatively long reproductive lifespans, use intermediate hosts and have long transmission periods (Anderson & May, 1992). Heirloom parasites, especially those that are transmitted by the faecal–oral route, likely infected humans before and after the agricultural transition. Even though these parasites were infecting humans long before the adoption of agriculture and shift to sedentary lifestyles, growing population sizes and densities likely made it more common for people to come into contact with faecal material and increase transmission of species spread by the faecal–oral route. In one of the earliest farming communities studied thus far in the Near East, where there is evidence of dense living and disposal of faecal waste beside dwellings, whipworm (*Trichuris trichiura*) eggs have been found in human coprolites (Ledger et al., 2019a). Roundworm and whipworm have been identified in many Neolithic sites in Europe (Anastasiou, 2015; Maicher et al., 2019). Zoonotic parasites also continued to infect humans in the Neolithic period (Anastasiou, 2015; Harter-Lailheugue et al., 2005; Le Bailly et al., 2003), possibly due to a diet consisting of a mix of wild and domesticated animals (Ledger & Mitchell, 2019; Reinhard et al., 2013).

Diet-related or food-borne parasites have infected humans throughout our evolution. Parasites contracted from eating raw or undercooked fish and shellfish, such as fish tapeworm (*Dibothriocephalus* sp.), *Echinostoma* sp., Chinese liver fluke (*Clonorchis sinensis*) and *Paragonimus westermani* (lung fluke), have archaeological evidence dating from the Neolithic period onwards and would have infected hunter-gatherers prior to that (Ledger & Mitchell, 2019). For example, *Taenia* tapeworms were acquired by hominins in Africa and continue to infect people across the world today (Hoberg et al., 2001; Michelet & Dauga, 2012). In general, a more diverse parasite burden is found in Neolithic communities compared to later time periods due to consumption of undercooked food, wild animals and insects. These include parasite species

that are rare today such as *Capillaria*, *Echinostoma* and the giant kidney worm (Fugassa et al., 2010; Le Bailly et al., 2003; Ledger et al., 2019b; Maicher et al., 2019). With increased animal husbandry and reliance on farmed animals and fish, it has been suggested that parasites acquired from domesticates, such as *Taenia* tapeworms, *Fasciola* liver fluke and fish tapeworm (*Dibothriocephalus*), also increased (Sianto et al., 2009).

As populations continued to grow after the agricultural transition and people settled into villages and towns that became increasingly urbanised and connected, there were continual changes in parasitic infections. In the Roman and medieval periods in Europe, analyses have shown the ubiquity of helminths transmitted by the faecal–oral route, namely roundworm and whipworm (Mitchell, 2015; 2017). It appears that towns in the Roman and medieval periods did not have adequate systems to remove faecal material from densely populated living spaces. Certain practices such as the use of human faeces as crop fertiliser, dumping of excrement in streets and contamination of drinking water by sewage allowed these parasites to flourish (Graff et al., 2020; Ledger et al., 2018; Taylor, 2015; Williams et al., 2017). In the Americas, hygiene-related parasites are also commonly found in past populations. In contrast to Europe, where roundworm and whipworm are the most frequently identified, pinworm is one of the most frequently identified parasites in the Americas (Araújo et al., 2015). Though fewer time periods have been studied in East Asia, roundworm and whipworm have been found in archaeological soil samples and mummies, further attesting to their ubiquity in the human past (Kim et al., 2016; Seo & Shin, 2015). Whipworm and roundworm remained common in China over the last 2000 years, only decreasing in prevalence since the 1990s with the introduction of effective sanitation. Chinese liver fluke (*Clonorchis sinensis*) was just as common in ancient samples, but dramatically decreased in prevalence prior to the introduction of effective sanitation. This is thought to have been due to changes in cultural preferences to eat cooked fish rather than raw fish and draining of marshland areas, which are required for the fluke to complete its life cycle (Yeh & Mitchell, 2016; Zhan et al., 2019). Whipworm and roundworm are heirloom parasites with a long co-evolutionary history with hominins, possibly owing to their success in infecting humans through a diverse range of ecologies and sociocultural practices.

As territories were conquered and empires established, another major impact on infectious disease spread was long-distance trade allowing for the movement of people and foodstuffs at an increasing rate. Evidence for the Chinese liver fluke (*Clonorchis sinensis*) at a Silk Road relay station in north-west China suggests movement of this parasite along the Silk Road 2000 years ago (Yeh et al., 2016). The Chinese liver fluke could not have been endemic in this semi-desert region of China because of the arid climate, which cannot support its life cycle, nor the snail that acts as the first intermediate host. Therefore, it was likely brought by infected merchants or government officials from eastern or southern China where the parasite was endemic (Yeh et al., 2016). In the medieval period, the discovery of fish tapeworm eggs in thirteenth-century crusader latrines in Acre, Israel led to the proposal that it was brought there in the intestines of crusaders from northern Europe, where the parasite was common at that time (Mitchell et al., 2011). Similarly, the presence of *Schistosoma mansoni* eggs (causing schistosomiasis) in a latrine in France dated to 1450–1550 CE suggests movement of an infected individual from Africa. This is because *S. mansoni* is endemic in Africa and the parasite is not typically found in mainland Europe because its snail intermediate host is not often found there (Bouchet et al., 2002). Parasite species such as these have remained geographically restricted due to their need for very specific ecological niches that contain their intermediate and definitive hosts. However, their geographical distribution may potentially increase in the future with the effects of climate change and increasing travel (Box 11.2).

Box 11.2 Parasites today

Today parasite infections are amongst the largest contributors to morbidity worldwide. In particular, soil-transmitted helminths such as roundworm (*Ascaris lumbricoides*), whipworm (*Trichuris trichiura*) and hookworm (*Necator americanus/Ancylostoma duodenale*), together with schistosomiasis, are major contributors to disease burden in developing countries. The Global Burden of Diseases Study in 2019 reported intestinal nematode infections to account for 1.97 million DALYs (disability-adjusted life years), a decrease from 2.84 million in 2017 (GBD 2019 Diseases and Injuries Collaborators, 2020). In 2019, they estimated that 1 billion people were infected with intestinal nematodes, 140 million with schistosomiasis and 33.5 million with food-borne trematodes, including *Fasciola* and *Clonorchis* among others (GBD 2019 Diseases and Injuries Collaborators, 2020). Today, infections with soil-transmitted helminths are largely concentrated in sub-Saharan Africa, Asia and Latin America in impoverished communities where sanitation infrastructure and access to clean water are lacking (Pullan & Brooker, 2012).

11.3 Human-parasite co-evolution

The negative health effects, growth and developmental impacts and economic burden of parasite infections are well recognised. Helminth infections and protozoan infections vary in their symptomatology. The most common infections are caused by the soil-transmitted helminths roundworm (*Ascaris lumbricoides*), whipworm (*Trichuris trichiura*) and hookworm (*Necator americanus/Ancylostoma duodenale*), and these disproportionately affect school-aged children (Hotez et al., 2006; Pabalan et al., 2018). Infection with these helminths result in nutrient deficiencies, stunting of growth, anaemia and cognitive impairments, and have been linked to decreased school attendance (Hotez et al., 2008; Jourdan et al., 2018; Owada et al., 2017; Stephenson et al., 2000; Taylor-Robinson et al., 2019). The health impacts and high prevalence in endemic regions have resulted in the World Health Organization (WHO) implementing mass drug administration aimed at treating children in endemic regions for soil-transmitted helminth infections (WHO, 2012). Helminth infection during pregnancy can have negative impacts on the mother and foetus by perpetuating nutritional stress causing anaemia in the mother and neonate, as well as increasing the risk for other infections, premature birth weight and low birth weight (Aderoba et al., 2015; Brooker et al., 2008; Gyorkos et al., 2012; Taghipour et al., 2020).

Given the high rates of infection in pre-reproductive years and impacts on pregnancy and foetal outcomes, we would expect that parasites can, and have, exerted a significant selective pressure (see Fumagalli et al., 2010 for further discussion).

11.3.1 Human evolution in response to parasite infection

Infectious diseases have consistently been a major selective pressure on humans (Karlsson et al., 2014). One signature of the selective forces exerted by pathogens is the fixing of variants that result in resistance or tolerance to these pathogens (Pittman et al., 2016). Numerous gene variants that offer protection against malaria have been identified, including the Duffy antigen receptor for chemokines (DARC), variations in haemoglobin such as the sickle cell allele and thalassemia, enzyme dysfunction such as glucose-6-phosphate dehydrogenase (G6PD deficiency) and others (see review by Kwiatkowski, 2005). These have become classic examples of pathogens driving human evolution. However, less work has been done on common soil-transmitted helminths (e.g. roundworm, whipworm, hookworm) compared to protozoa such as malaria. Candidate gene-association studies and genome-wide linkage studies have produced variants of interest that appear to offer protection against roundworm and whipworm and further work in this area may reveal more (Dold & Holland, 2011; Mangano & Modiano, 2014).

Of particular interest in modern medicine and immunology is the link that some of these resistance genes have with autoimmune diseases. For this reason, we know the most about genes that predispose us to autoimmune diseases but have also been found to increase resistance to parasitic infection. For example, specific variants on the *APOL1* gene protect humans from infection with African trypanosomiasis (caused by the protozoan *Trypanosoma brucei*) and are under positive selection (Genovese et al., 2010; Nadkarni et al., 2018). These variants allow for lysing (i.e. breakdown) of the invading *Trypanosoma* parasites, but also increase the risk of renal disease (Genovese et al., 2010). Genotypes encoding a deficiency in inhibitory Fcγ receptors have been shown to be protective against the malaria parasite (*Plasmodium* spp.) but also predispose to autoimmune diseases, particularly systemic lupus erythematosus (Clatworthy et al., 2007; Willcocks et al., 2010). Fcγ receptors are present on various immune cells and bind the Fc domain of antibodies, in particular IgG, resulting in intracellular signalling. The polymorphism for non-functional receptors is found more commonly in African and Asian populations where malaria is endemic, indicating that reducing malaria susceptibility may have resulted in increased allele frequencies. Further to this, Machado and colleagues (2012) have shown a link between helminth diversity and polymorphisms in genes encoding the Fcγ receptors, suggesting that helminth infection resulted in selection for variants that impact immune response to helminths but also predispose to autoimmune disease.

Epigenetic changes to immune genes have also been discovered, and evidence exists that helminth infection during pregnancy induces epigenetic changes (see Chapter 2) that can be passed to the foetus, possibly contributing to immunomodulation early in life. Epigenetic changes inhibiting the Th1 immune response have been noted in children with roundworm (*Ascaris lumbricoides*) and schistosome infection, and in the case of schistosomiasis these persist for at least six months after treatment (DiNardo et al., 2018). Helminth infection during pregnancy can alter neonatal immune response and has been shown to decrease allergic reactions in neonates (Elliott et al., 2005; Smits et al., 2010; Mpairwe et al., 2011). However, this decrease in allergic disease has not consistently been shown to persist beyond the early childhood period. For example, one cohort study done in Uganda showed that treatment of helminth infection in pregnant mothers resulted in increased atopy in infants (Mpairwe et al., 2011), but decreased allergy-related disease in their children at 9 years of age (Namara et al., 2017). Further studies are definitely necessary to fully understand these relationships and the impact of factors including timing of infection, intensity of infection, recurrence of infections and parasite species involved (Smits et al., 2010).

11.3.2 Parasite evolution in response to the human host

Human activities have had major impacts on parasite evolution. Long-distance human migrations serve to move parasite species outside of their endemic regions and can result in increased genetic diversity. Comparison of mitochondrial genomes of *Plasmodium vivax* (a causative agent of malaria) from different endemic regions today suggests that the genetically diverse *P. vivax* malaria lineages found in the Americas resulted from admixture between lineages carried by people migrating from Africa and Australasia in the past (Rodrigues et al., 2018). More recently, co-infection of individuals with two different parasite species has resulted in documented occurrences of hybridisation between parasite species (Detwiler & Criscione, 2010), and it has been suggested that these events drive the emergence and diversification of parasites (King et al., 2015). A recent outbreak of schistosomiasis in Corsica, France, was shown to be due to a hybrid form of *Schistosoma haematobium* (a human-infecting species typically restricted to tropical and sub-tropical regions) and *Schistosoma bovis* (a species that infects livestock) (Oleaga et al., 2019). This hybrid form of the parasite was introduced to Corsica by an infected individual from Senegal and was then responsible for multiple outbreaks of the hybrid species (Boissier et al., 2016; Oleaga et al., 2019).

The wide-scale use of anti-helminthic medications to decrease the prevalence of parasite infections is a clear example of human activities driving parasite evolution. Increased efforts to treat and prevent malaria infections in endemic regions have successfully decreased rates of infection (Alonso & Noor, 2017). However, decades of use of various anti-malarials have resulted in numerous genetic changes to *Plasmodium falciparum* and *Plasmodium vivax* (two of the species causing malaria) that confer resistance to these drugs (Hyde, 2005; Menard & Dondorp, 2017). Mass drug administration (i.e. praziquantel for schistosomiasis and albendazole or mebendazole for soil-transmitted helminths) have surprisingly not resulted in meaningful resistance in these helminths so far (Webster et al., 2014). However, the selective pressure exerted by these relatively recent wide-scale treatment programmes have been noted along with clinical reports of treatment failures in some patients (Alonso et al., 2006) and increasing resistance of whipworm (*Trichuris trichiura*) to albendazole and mebendazole (Moser et al., 2017). Humans can have particularly marked effects on these parasites as they are the sole definitive host for these species (see Table 11.1).

Helminths have evolved sophisticated mechanisms to enable invasion, reproduction and evasion of the human host immune system, thereby maintaining infection, sometimes for decades. Experimental evidence has also shown that helminths use the gut microbiota as cues that they have reached the appropriate area of the intestines to establish infection (Hayes et al., 2010). Similarly, once this appropriate environment is reached, helminths also promote certain bacteria that aid their survival to flourish in the intestines (Harnett & Harnett, 2017). Excretory/secretory products from helminths can successfully induce T_{reg} cell populations in the human host and necessary cytokines to modulate unnecessary inflammation, maintain immune homeostasis and promote tissue repair (Jackson et al., 2009; Harnett & Harnett, 2017). Numerous excretory and secretory proteins of helminths have been identified that result in this immunomodulation (Harnett, 2014; Harnett & Harnett, 2017; Maizels & McSorley, 2016). Currently, many of these experimental studies are based on mouse models. For example, using mice infected with *Echinostoma caproni*, it was discovered that host-derived antibodies bound to the surface of the worm could be trapped under a layer of excretory/secretory products and then degraded (Cortés et al., 2017). Responses such as these could effectively allow for establishment of chronic infections. Excreted and secreted proteins have also been shown to block immune cell signalling, degrade antibodies and stop the migration of antigen-presenting cells (King & Li, 2018). These excreted and secreted proteins can be species-specific (Cuesta-Astroz et al., 2017). Therefore, we may expect that in heirloom parasites, some of these products may have evolved directly in response to human immune defences.

11.4. Applications of palaeoparasitology and human–parasite evolution

11.4.1. The hygiene hypothesis

One of the major co-evolutionary outcomes of parasitic infection is immune tolerance in the human host. Immune tolerance allows for long-term infection leading to decreased energy expenditure and potential immune damage by the host. This has likely co-evolved between humans and other mammals and their parasites and has been linked to potential beneficial effects that parasites can have on other disease processes, especially autoimmune diseases (Smallwood et al., 2017; Sutherland et al., 2011). To some extent immune tolerance would be energetically favourable for both the parasite and host, allowing parasites to establish chronic infections and preventing humans from mounting unsuccessful, energetically costly, long term immune attacks.

Human immune responses can be categorised as Type 1 (induction of Th1 cells), Type 2 (induction of Th2 cells) or Type 3 (induction of Th17 cells), depending on cytokine signals given to helper T cells (Annunziato et al., 2015). In general, when humans are infected with a helminth, the worm causes natural killer cells to release cytokines that induce a Th2 response from helper T cells (Finlay et al., 2014). In contrast, many invading bacteria and viruses cause secretion of cytokines that induce a Th1 response from helper T cells. Extracellular bacteria and fungi elicit a Th17 response (Annunziato et al., 2015). The activation of one arm will inhibit the other creating a sort of balance or polarisation of the immune response based on the invading pathogen (Romagnani, 2004). An important additional component to this is regulatory T cells (T_{reg}) that can keep the immune response in check and prevent unnecessary activation or attack of one's own cells (Murray et al., 2009; Zhou et al., 2009). Various helminth species induce T_{reg} cells and IL-10. IL-10 can cause an antibody isotype switch from IgE proinflammatory response to IgG4, thus avoiding harmful inflammatory damage to host tissues, and this has been suggested to occur in chronic helminth infections (Maizels & Yazdanbakhsh, 2003; McSorley & Maizels, 2012).

The long co-evolutionary history between humans and certain micro-organisms has formed the basis of major hypotheses such as the *hygiene hypothesis*, originally proposed by Strachan (1989), and the *old friends hypothesis* (Maizels, 2020; Rook, 2009; Sironi & Clerici, 2010). These hypotheses are (in part) based on the knowledge that our immune system has evolved in the presence of certain pathogens and environmental antigens. In an era of increasing rates of autoimmune and allergic diseases and decreasing infectious diseases, epidemiological studies resulted in the original

hygiene hypothesis that suggested that current hygienic environments and lack of exposure to antigens were resulting in increased prevalence of allergic disease (Strachan, 1989). Later, the hygiene hypothesis was adapted with an evolutionary focus to suggest that organisms that have infected or colonised humans for most of our evolutionary history and are now lacking in modern populations may be responsible for increased autoimmune and atopic disease (Maizels, 2020; Rook, 2010). These so-called old friends include helminths, mycobacteria, gut microbiota, hepatitis A and *Salmonella* (Rook, 2010). Work in palaeoparasitology has confirmed that there are numerous helminths and protozoa that have infected humans throughout our evolution and infected our ancestors before the origins of *Homo sapiens*. The hygiene or old friends hypotheses have fuelled research into the effects of parasitic infection on the human immune system and modulation of other diseases.

Epidemiological studies have supported the inclusion of helminths in the hygiene or old friends hypotheses, showing the link between helminth infection and decreased allergic response and autoimmunity (Yazdanbakhsh et al., 2002). Children infected with schistosomes and living in communities with high risk for parasite infection have been shown to have decreased allergic reactivity to common allergens, such as cockroach and dust mites, on skin prick tests (Araújo et al., 2000; Cooper et al., 2003; Nyan et al., 2001; van den Biggelaar et al., 2000). In a systematic review, Feary and colleagues (2011) noted that the species of infecting parasite may be important in decreased rates of atopy. Smits and colleagues (2010) have also noted that the species of helminth plays a role in the type of atopic disease from which they confer protection, but also that some can *increase* rates of atopic disease. In particular, in their review of meta-analyses, they noted that whipworm, hookworm and schistosomes protected individuals from skin reactivity to allergens, while hookworm was associated with reduction in asthma. On the other hand, infection with *Toxocara*, a nematode that does not use humans as a definitive host, resulted in increased skin reactivity to allergens (Gonzalez-Quintela et al., 2006). Palaeoparasitological data has shown the temporal and geographic ubiquity of soil-transmitted helminths in the human past and these human-specific heirloom parasites have had the longest periods of time to co-evolve with the human host, thus possibly explaining their impacts on human immune function and their ability to dampen allergic responses (Mitchell, 2013; Pritchard et al., 2020).

Similarly, in the case of autoimmune diseases, helminth infection has been shown to be beneficial for prevention or reduction in severity of diseases such as multiple sclerosis and Type 1 diabetes. First, in mice, infection with *Trichinella spiralis* and *Heligmosomoides polygyrus* can protect from Type 1 diabetes and decrease its severity (Saunders et al., 2007; Liu et al., 2009). A similar pattern was seen for people in India, where a reduced prevalence of lymphatic filariasis was found in individuals with Type 1 diabetes (Aravindhan et al., 2010). Finally, Correale and Farez (2011) found that patients with multiple sclerosis who became concurrently infected with helminths (including dwarf tapeworm, whipworm, roundworm, *Strongyloides stercoralis* and pinworm) had decreased relapses over a period of seven years compared to uninfected individuals with multiple sclerosis. Again, many of these species, especially roundworm and whipworm, are heirloom parasites of humans commonly found in archaeological samples. Helminth infection has also been linked to better outcomes in inflammatory modulated diseases such as cardiovascular disease, Type 2 diabetes and inflammatory bowel disease (Gurven et al., 2016).

Studies like these have led to an intense interest in using live helminths or helminth-derived products as treatments for a range of autoimmune and inflammatory diseases. Clinical trials have been undertaken to study the efficacy of treating inflammatory bowel disease (ulcerative colitis and Crohn's disease), coeliac disease, multiple sclerosis, psoriasis and rheumatoid arthritis with helminths (see reviews in Fleming & Weinstock, 2015; Smallwood et al., 2017). These trials have used experimental infections of primarily two helminths: pig whipworm (*Trichuris suis*) and human hookworm (*Necator americanus*). Early trials, with few study patients with inflammatory bowel disease, were promising. In one randomised controlled trial, patients with ulcerative colitis were given pig whipworm ova every two weeks for twelve weeks and showed improvement in disease index over a placebo group (Summers et al., 2005). This was a follow-up to a smaller trial including patients with Crohn's and ulcerative colitis, which showed improvement in disease index with repeated dosing of pig whipworm ova (Summers et al., 2003). However, larger trials have had mixed results with a general trend of minor improvements, but not for all patients (Huang et al., 2018). In two randomised controlled trials where patients with Crohn's disease were given increasing doses of pig whipworm ova (up to 7500 ova), infection resulted in no obvious benefits based on gastrointestinal signs and symptoms (Sandborn et al., 2013; Schölmerich et al., 2017). Similarly, effects have been mixed in trials using helminth infection to treat coeliac disease, multiple sclerosis and allergic rhinitis (see Smallwood et al., 2017). From an evolutionary perspective, it may be the case that pig whipworm as a therapeutic agent (rather than human whipworm) has lower efficacy because human whipworm has had more time to develop intricate interactions with its human host. However, pig whipworm has been the ideal choice for wide-scale treatments with live helminths based on its safety (Weinstock & Elliot, 2013). Animal models have shown promise for the use of helminth excretory and secretory products as autoimmune disease treatments, and these are becoming the focus of future clinical trials on helminth therapy and remove the necessity to infect individuals with viable helminths (Kahl et al., 2018).

11.4.2 A role for palaeoparasitology

As demonstrated, palaeoparasitology research provides direct evidence for parasite infections in the past. It clarifies which parasites have been long-term pathogens of humans, which have been more recently acquired through migrations and changing ecosystems and how prevalence and distribution has broadly changed through time. At the same time, research in medical sciences and immunology informed by evolutionary studies shows how parasites, especially helminths, have co-evolved with humans to have very intricate (and even some beneficial) interactions with our immune systems. This evolutionary approach to parasitic infection can guide future research and contribute to a better understanding of the impacts of parasitic infection on modern populations.

Palaeoparasitological research has largely focused on how past environments and cultural practices impact disease dynamics. Answering such questions can guide the focus of public health initiatives. For example, what environments have aided parasite transmission? And what social or cultural practices have consistently contributed to infections? Certain environments and sociocultural practices are important in establishing or allowing for parasite species to flourish throughout human history. For example, increasing urbanisation, poor disposal of faecal material and the use of faeces as fertiliser in Europe during the Roman and medieval periods seems to be a major contributor to the ubiquity of roundworm and whipworm eggs recovered from archaeological samples in these time periods (Ledger et al., 2020; Mitchell, 2015). Many researchers note that control of similar sanitation factors and access to clean water are vitally important measures alongside administration of anthelminthic medications for reducing the burden of soil-transmitted helminths in endemic areas today (Strunz et al., 2014; Vaz Nery et al., 2019).

Early evidence for schistosomiasis comes from pelvic soil studied from burials in Syria dated between 6500 and 6000 years BP. The finding of schistosomiasis at the same time as the construction of crop irrigation channels in this area led to the proposal that construction of these channels with intensification of agriculture created standing water that allowed for establishment of schistosomiasis in the region (Anastasiou et al., 2014). Similarly, it has been proposed that rock shelters inhabited by humans in prehistoric North and South America created an ideal ecological niche for triatomine bugs, which could then transmit the endoparasite *Trypanosoma cruzi* which causes Chagas disease (Reinhard and Araújo, 2015). Later, Chagas disease increased after European migrations that brought the introduction of mud and daub buildings, within which triatomine bugs found an ideal habitat (Araújo et al., 2009). These examples provide key evidence for major changes we have made to our landscapes and living environments that allowed for increased transmission of certain parasite species. Not only can they direct us to key

areas of current environments that can be targeted to decrease transmission, but also they can be used as predictive factors to determine which populations are (or could be) at risk for infection as a result of urbanisation, economic developments or climate change. While some parasite species have infected humans in nearly all of their environments, such as the soil-transmitted helminths, others are very much related to specific environmental changes that we have created.

Palaeoparasitology can also help inform further research into the immune effects of parasite infection, which in turn may aid in drug development for a range of autoimmune and inflammatory diseases. The underlying concepts driving the reasoning to focus on parasites as potential treatments have been based in evolutionary theory, and the helminth species used to develop these into therapies can also be informed by evolutionary studies. The focus on helminths as regulators of the human immune system, the absence of which may be contributing to rising rates of allergy and autoimmunity was first based on the known extensive length of time and ubiquity of infections in the human past. Palaeoparasitology is uniquely positioned to provide evidence for which specific parasites were infecting humans in different areas of the world throughout time, and this evidence can be used to direct focus on parasites that have the longest history with our species. Many clinical trials testing intentional infection with helminths to treat autoimmune and allergic conditions have used pig whipworm (*T. suis*) and hookworm (Smallwood et al., 2017). While the use of pig whipworm is justified due to its close relationship with human whipworm and more importantly, its safety, as it cannot establish long-term infection in humans, we hypothesise that heirloom species specific to humans would have the largest beneficial effects. Humans were no doubt exposed to pig whipworm in the past, especially after the domestication of pigs, but this species never evolved to infect humans. Thus, its effects on the immune system are not likely to be as complex as those caused by human whipworm. Other authors note that infection by human-specific helminth species have been shown to reduce rates of atopic disease while infection by those that do not use humans as a primary host (e.g. *Toxocara canis*, a roundworm that infects dogs) actually increase atopic disease (Smits et al., 2010). The next steps in this area of research will be to determine if excreted/secreted products of our heirloom parasites can create new treatment options for autoimmune diseases.

Most of the data that have been used to gain an understanding of parasite infection in past populations come from recovery of preserved eggs or cysts in archaeological material. Palaeogenetics is also making major contributions to our knowledge of parasite evolution and human–parasite co-evolution. The uptake of ancient DNA (aDNA) studies on parasites has lagged behind most other pathogens, although the use of aDNA to characterise ancient parasites is increasing (Anastasiou & Mitchell, 2013; Côté & Le Bailly, 2018; Flammer et al., 2020; Sabin et al., 2020; Søe et al., 2018). Studies using aDNA that focus on parasites will make important contributions to future work in palaeoparasitology, as they will allow for detection of species for which eggs and cysts do not preserve well, increased resolution of phylogenetic reconstructions, spread and distribution of parasites with past human groups and calibration of rates of evolution and major genetic changes through time for different parasite species. This type of work can provide important complementary data to that being learned from modern parasite genomics, as has been done for other pathogens (Spyrou et al., 2019).

11.5 Conclusions

It has been shown that parasites, specifically intestinal helminths and protozoa, are long-standing human companions, and many species have been passed down along phylogenetic lines from our primate ancestors. Palaeoparasitology confirms that parasite infections were common in the human past, and they remain significant causes of disease in many parts of the world today. This means they have had millions of years to co-evolve with humans. Evolutionary medicine aims to use knowledge gained from evolutionary biology to inform and contribute to medical and epidemiological research in order to better understand human health and disease (Stearns, 2012).

Palaeopathology, specifically palaeoparasitology, can contribute to these goals by providing direct evidence for human–parasite interactions throughout human evolution. It can provide insights into the health impacts of parasite infection, human activities that have modified parasite transmission and long-term epidemiological patterns.

From the perspective of the medical sciences, there is increased focus on using the immune-modulating effects of helminths to design treatments for autoimmune and allergic diseases in line with the hygiene hypothesis. This is an obvious place where an evolutionary perspective has had an impact on medical research and drug development. Views of parasites as solely harmful pathogens are changing, and the role that they have to play in our immune system development and continual functioning are becoming better understood. Applying an evolutionary approach to increasing rates of atopic and autoimmune disease has ultimately allowed for the generation of hypotheses that have driven clinical trials, such as using helminths to treat inflammatory bowel disease. Palaeoparasitology can play a key role in determining which species of helminths may be the most promising focus for further investigation of immune effects and potential molecules for drug development.

From the perspective of public health, palaeoparasitology can contribute to identification of important disease prevention and public health measures. Research in palaeoparasitology has provided valuable insight into the sociocultural factors involved in parasite transmission over a long time scale, which can point us towards areas where we can focus public health initiatives. Identification of environments throughout human history that have aided parasite transmission, as well as social or cultural practices that have contributed to infections, can help inform approaches that could be taken to mitigate infections today.

References

Aderoba, A. K., Iribhogbe, O. I., Olagbuji, B. N., Olokor, O. E., Ojide, C. K. & Ande, A. B. (2015). Prevalence of helminth infestation during pregnancy and its association with maternal anemia and low birth weight.

International Journal of Gynaecology and Obstetrics, 129(3), 199–202.

Alonso, D., Muñoz, J., Gascón, J., Valls, M. E. & Corachan, M. (2006). Failure of standard treatment with praziquantel in two returned travelers with *Schistosoma haematobium* infection. *The American Journal of Tropical Medicine and Hygiene*, 74(2), 342–344.

Alonso, P. & Noor, A. M. (2017). The global fight against malaria is at crossroads. *The Lancet*, 390 (10112), 2532–2534.

Anastasiou, E., Lorentz, K. O., Stein, G. J. & Mitchell, P. D. (2014). Prehistoric schistosomiasis parasite found in the Middle East. *The Lancet Infectious Diseases*, 14, 553–554.

Anastasiou, E. (2015). 'Parasites in European populations from prehistory to the Industrial Revolution', in P. D. Mitchell (ed), *Sanitation, latrines and intestinal parasites in past populations*, pp. 203–217. Ashgate, Farnham.

Anastasiou, E. & Mitchell, P. D. (2013). Paleopathology and genes: investigating the genetics of infectious diseases in excavated human skeletal remains and mummies from past populations. *Gene*, 528, 33–40.

Anderson, R. M. & May, R. M. (1992). *Infectious diseases of humans: dynamics and control.* Oxford Science Publications, Oxford.

Annunziato, F., Romagnani, C., & Romagnani, S. (2015). The 3 major types of innate and adaptive cell-mediated effector immunity. *The Journal of Allergy and Clinical Immunology*, 135 (3), 626–635.

Araújo, A., Jansen, A. M., Reinhard, K. & Ferreira, L. F. (2009). Paleoparasitology of Chagas disease – a review. *Memorias do Instituto Oswaldo Cruz*, 104(Suppl 1), 9–16.

Araújo, A., Reinhard, K., Ferreira, L. F., Pucu, E. & Chieffi, P. P. (2013). Paleoparasitology: the origin of human parasites. *Arquivos de Neuro-Psiquiatria*, 71, 722–726.

Araújo, A., Ferreira, L. F., Fugassa, M., Leles, D., Sianto, L., de Souza, S. M. M., Dutra, J., Iñiguez, A. & Reinhard, K. (2015). 'New World paleoparasitology', in P. D. Mitchell (ed), *Sanitation, latrines and intestinal parasites in past populations*, pp. 165–202. Ashgate, Farnham.

Araújo, M. I., Lopes, A. A., Medeiros, M., Cruz, A. A., Sousa-Atta, L., Solé, D. & Carvalho, E.M . (2000). Inverse association between skin response to aeroallergens and *Schistosoma mansoni* infection. *International Archives of Allergy and Immunology*, 123, 145–148.

Aravindhan, V., Mohan, V., Surendar, J., Rao, M. M., Ranjani, H., Kumaraswami, V., Nutman, T. B. & Babu, S. (2010). Decreased prevalence of lymphatic filariasis among subjects with type-1 diabetes. *American Journal of Hygiene and Tropical Medicine*, 83, 1336–1339.

Armelagos, G. J., Goodman, A. H. & Jacobs, K. H. (1991). The origins of agriculture: population growth during a period of declining health. *Population and Environment*, 13, 9–22.

Bergman, J. (2018). Stone age disease in the north – human intestinal parasites from a Mesolithic burial in Motala, Sweden. *Journal of Archaeological Science*, 96, 26–32.

Boissier, J., Grech-Angelini, S., Webster, B. L., Allienne, J. F., Huyse, T., Mas-Coma, S., Toulza, E., Barré-Cardi, H., Rollinson, D., Kincaid-Smith, J., Oleaga, A., Galinier, R., Foata, J., Rognon, A., Berry, A., Mouahid, G., Henneron, R., Moné, H., Noel, H. & Mitta, G. (2016). Outbreak of urogenital schistosomiasis in Corsica (France): an epidemiological case study. *The Lancet Infectious Diseases*, 16(8), 971–979.

Bouchet, F., Baffier, D. & Girard, M. (1996). Paléoparasitologie en contexte pléistocène: premières observations à la Grande Grotte d'Arcy-sur-Cure (Yonne), France. *Comptes Rendus de l'Academie de Sciences*, 319, 147–151.

Bouchet, F., Harter, S., Paicheler, J. C., Aráujo, A. & Ferreira, L. F. (2002). First recovery of *Schistosoma mansoni* eggs from a latrine in Europe (15–16th centuries). *The Journal of Parasitology*, 88, 404–405.

Bouchet, F., Guidon, N., Dittmar, K., Harter, S., Ferreira, L. F., Chaves, S. M., Reinhard, K. & Aráujo, A. (2003). Parasite remains in archaeological sites. *Memorias do Instituto Oswaldo Cruz*, 98(Suppl 1), 47–52.

Brooker, S., Hotez, P. J. & Bundy, D. A. P. (2008). Hookworm-related anaemia among pregnant women: a systematic review. *PLoS Neglected Tropical Diseases*, 2(9), e291.

Camacho, M., Aráujo, A., Morrow, J., Buikstra, J. & Reinhard, K. (2018). Recovering parasites from mummies and coprolites: an epidemiological approach. *Parasites & Vectors*, 11, 248.

Clatworthy, M. R., Willcocks, L., Urban, B., Langhorne, J., Williams, T. N., Peshu, N., Watkins, N. A., Floto, R. A. & Smith, K. G. (2007). Systemic lupus erythematosus-associated defects in the inhibitory receptor FcγRIIb reduce susceptibility to malaria. *Proceedings of the National Academy of Sciences of the United States of America*, 104, 7169–7174.

Cooper, P. J., Chico, M. E., Rodrigues, L. C., Ordonez, M., Strachan, D., Griffin, G. E. & Nutman, T. B. (2003). Reduced risk of atopy among school-age children infected with geohelminth parasites in a rural area of the tropics. *The Journal of Allergy and Clinical Immunology*, 111, 995–1000.

Correale, J. & Farez, M. F. (2011). The impact of parasite infections on the course of multiple sclerosis. *Journal of Neuroimmunology*, 233, 6–11.

Cortés, A., Muñoz-Antolí, C., Molina-Durán, J., Esteban, J. G. & Toledo, R. (2017). Antibody trapping: a novel mechanism of parasite immune evasion by the trematode Echinostoma caproni. *PLoS Neglected Tropical Diseases*, 11, e0005773.

Côté, N. M.-L. & Le Bailly, M. (2018). Palaeoparasitology and palaeogenetics: review and perspectives for the study of ancient human parasites. *Parasitology*, 145, 656–664.

Cuesta-Astroz, Y., Oliveira, F. S., Nahum, L. A. & Oliveira, G. (2017). Helminth secretomes reflect different lifestyles and parasitized hosts. *International Journal for Parasitology*, 47, 529–544.

Dark, P. (2004). New evidence for the antiquity of the intestinal parasite Trichuris (whipworm) in Europe. *Antiquity*, 78, 676–681.

Detwiler, J. T. & Criscione, C. D. (2010). An infectious topic in reticulate evolution: introgression and hybridization in animal parasites. *Genes*, 1 (1), 102–123.

DiNardo, A. R., Nishiguchi, T., Mace, E. M., Rajapakshe, K., Mtetwa, G., Kay, A., Maphalala, G., Secor, W. E., Mejia, R., Orange, J. S., Coarfa, C., Bhalla, K. N., Graviss, E. A., Mandalakas, A. M. & Makedonas, G. (2018). Schistosomiasis induces persistent DNA methylation and tuberculosis-specific immune changes. *Journal of Immunology*, 201, 124–133.

Dold, C. & Holland, C. V. (2011). Investigating the underlying mechanism of resistance to *Ascaris* infection. *Microbes and Infection*, 13, 624–631.

Elliott, A. M., Mpairwe, H., Quigley, M. A., Nampijja, M., Muhangi, L., Oweka-Onyee, J., Muwanga, M., Ndibazza, J. & Whitworth, J. A. (2005). Helminth infection during pregnancy and development of infantile eczema. *The Journal of the American Medical Association*, 294, 2028–2034.

Feary, J., Britton, J. & Leonardi-Bee, J. (2011). Atopy and current intestinal parasite infection: a systematic review and meta-analysis. *Allergy*, 66, 569–578.

Finlay, C. M., Walsh, K. P. & Mills, K. H. G. (2014). Induction of regulatory cells by helminth parasites: exploitation for the treatment of inflammatory diseases. *Immunological Reviews*, 259, 206–230.

Flammer, P. G., Ryan, H., Preston, S. G., Warren, S., Přichystalová, R., Weiss, R., Palmowski, V., Boschert, S., Fellgiebel, K., Jasch-Boley, I., Kairies, M. S., Rümmele, E., Rieger, D., Schmid, B., Reeves, B., Nicholson, R., Loe, L., Guy, C., Waldron, T., Macháček, J., Wahl, J., Pollard, M., Larson, G. & Smith, A. L. (2020). Epidemiological insights from a large-scale investigation of intestinal helminths in medieval Europe. *PLoS Neglected Tropical Diseases*, 14(8), e0008600.

Fleming, J. O. & Weinstock, J. V. (2015). Clinical trials of helminth therapy in autoimmune diseases: rationale and findings. *Parasite Immunology*, 37, 277–292.

Fugassa, M. H., Beltrame, M., Sardella, N., Civalero, M. & Aschero, C. (2010). Paleoparasitological results from coprolites dated at the Pleistocene–Holocene transition as source of paleoecological evidence in Patagonia. *Journal of Archaeological Science*, 37, 880–884.

Fumagalli, M., Pozzoli, U., Cagliani, R., Comi, G. P., Bresolin, N., Clerici, M. & Sironi, M. (2010). The landscape of human genes involved in the immune response to parasitic worms. *BMC Evolutionary Biology*, 10, 264.

GBD 2019 Diseases and Injuries Collaborators. (2020). Global burden of 369 diseases and injuries in 204 countries and territories, 1990–2019: a systematic analysis for the Global Burden of Disease Study 2019. *The Lancet*, 396(10258), 1204–1222.

Genovese, G., Friedman, D. J., Ross, M. D., Lecordier, L., Uzureau, P., Freedman, B. I., Bowden, D. W., Langefeld, C. D., Oleksyk, T. K., Uscinski Knob, A. L., Bernhardy, A. J., Hicks, P. J., Nelson, G. W., Vanhollebeke, B., Winkler, C. A., Kopp, J. B., Pays, E. & Pollak, M. R. (2010). Association of trypanolytic *ApoL1* variants with kidney disease in African Americans. *Science*, 329, 841–845.

Gonzalez-Quintela, A., Gude, F., Campos, J., Garea, M. T., Romero, P. A., Rey, J., Meijide, L. M., Fernandez-Merino, M. C. & Vidal, C. (2006). *Toxocara* infection seroprevalence and its relationship with atopic features in a general adult population. *International Archives of Allergy and Immunology*, 139, 317–324.

Graff, A., Bennion-Pedley, E., Jones, A. K., Ledger, M. L., Deforce, K., Degraeve, A., Byl, S. & Mitchell, P. D. (2020) A comparative study of parasites in three latrines from medieval and renaissance Brussels, Belgium (14th–17th centuries). *Parasitology*, 147(13), 1443–1451.

Gurven, M. D., Trumble, B. C., Stieglitz, J., Blackwell, A. D., Michalik, D. E., Finch, C. E. & Kaplan, H. S. (2016). Cardiovascular disease and type 2 diabetes in evolutionary perspective: a critical role for helminths? *Evolution, Medicine, and Public Health*, 2016, 338–357.

Gyorkos, T. W., Gilbert, N. L., Larocque, R., Casapía, M. & Montresor, A. (2012). Re-visiting *Trichuris trichiura* intensity thresholds based on anemia during pregnancy. *PLoS Neglected Tropical Diseases*, 6(9), e1783.

Harnett, M. M. & Harnett, W. (2017). Can parasitic worms cure the modern world's ills? *Trends in Parasitology*, 33, 694–705.

Harnett, W. (2014). Secretory products of helminth parasites as immunomodulators. *Molecular and Biochemical Parasitology*, 195, 130–136.

Harter-Lailheugue, S., Le Mort, F., Vigne, J-D., Guilaine, J., Le Brun, A. & Bouchet, F. (2005). Premiéres données parasitologiques sur les populations humaines précéramiques Chypriotes (VIIIe et VIIe millénaires av. J.-C.). *Paléorient*, 31, 43–54.

Hayes, K. S., Bancroft, A. J., Goldrick, M., Portsmouth, C., Roberts, I. S. & Grencis, R. K. (2010). Exploitation of the intestinal microflora by the parasitic nematode *Trichuris muris*. *Science*, 328(5984), 1391–1394.

Hoberg, E. P., Alkire, N. L., de Queiroz, A. & Jones, A. (2001). Out of Africa: origins of the *Taenia* tapeworms in humans. *Proceedings of the Royal Society B: Biological Sciences*, 268, 781–787.

Hotez, P. J., Bundy, D. A. P., Beegle, K., Brooker, S., Drake, L., de Silva, N., Montresor, A., Engels, D., Jukes, M., Chitsulo, L., Chow, J., Laxminarayan, R., Michaud, C., Bethony, J., Correa-Oliveira, R., Shuhua, X., Fenwick, A. & Savioli, L. (2006). 'Helminth infections: soil-transmitted helminth infections and schistosomiasis', in D. T. Jamison, J. G. Breman, A. R. Measham, G. Alleyne, M. Claeson, D. B. Evans, P. Jha, A. Mills & P. Musgrove (eds), *Disease control priorities in developing countries*, pp. 467–482. Oxford University Press, New York.

Hotez, P. J., Brindley, P. J., Bethony, J. M., King, C. H., Pearce, E. J. & Jacobson, J. (2008). Helminth infections: The great neglected tropical diseases. *The Journal of Clinical Investigation*, 118(4), 1311–1321.

Huang, X., Zeng, L. R., Chen, F. S., Zhu, J. P. & Zhu, M. H. (2018). *Trichuris suis* ova therapy in inflammatory bowel disease: A meta-analysis. *Medicine*, 97(34), e12087.

Hyde, J. E. (2005). Drug-resistant malaria. *Trends in Parasitology*, 21(11), 494–498.

Jackson, J. A., Friberg, I. M., Little, S. & Bradley, J. E. (2009). Review series on helminths, immune modulation and the hygiene hypothesis: immunity against helminths and immunological phenomena in modern human populations: coevolutionary legacies? *Immunology*, 126, 18–27.

Jourdan, P. M., Lamberton, P. H. L., Fenwick, A. & Addiss, D. G. (2018). Soil-transmitted helminth infections. *The Lancet*, 391(101117), 252–265.

Kahl, J., Brattig, N. & Liebau, E. (2018). The untapped pharmacopeic potential of helminths. *Trends in Parasitology*, 34, 828–842.

Karlsson, E. K., Kwiatkowski, D. P. & Sabeti, P. C. (2014). Natural selection and infectious disease in human populations. *Nature Reviews: Genetics*, 15, 379–393.

Kim, M. J., Seo, M., Oh, C. S., Chai, J. Y., Lee, J., Kim, G. J., Ma, W. Y., Choi, S. J., Reinhard, K., Araujo, A. & Shin, D. H. (2016). Paleoparasitological study on the soil sediment samples from archaeological sites of ancient Silla Kingdom in Korean peninsula. *Quaternary International*, 405, 80–86.

King, I. L. & Li, Y. (2018). Host-parasite interactions promote disease tolerance to intestinal helminth infection. *Frontiers in Immunology*, 9(2128), 1–10.

King, K. C., Stelkens, R. B., Webster, J. P., Smith, D. F. & Brockhurst, M. A. (2015). Hybridization in parasites: consequences for adaptive evolution, pathogenesis, and public health in a changing world. *PLoS Pathogens*, 11(9), e1005098.

Kwiatkowski, D. P. (2005). How malaria has affected the human genome and what human genetics can teach us about malaria. *American Journal of Human Genetics*, 77, 171–192.

Larsen, C. S. (2018). The bioarchaeology of health crisis: infectious disease in the past. *Annual Review of Anthropology*, 47, 295–313.

Le Bailly, M., Leuzinger, U. & Bouchet, F. (2003). Dioctophymidae eggs in coprolites from Neolithic site of Arbon-Bleiche 3 (Switzerland). *The Journal of Parasitology*, 89, 1073–1076.

Ledger, M. L., Stock, F., Schwaiger, H., Knipping, M., Brückner, H., Ladstätter, S. & Mitchell, P. D. (2018). Intestinal parasites from public and private latrines and the harbour canal in Roman Period Ephesus, Turkey (1st c. BCE to 6th c. CE). *Journal of Archaeological Science: Reports*, 21, 289–297.

Ledger, M. L., Anastasiou, E., Shillito, L-M., Mackay, H., Bull, I. D., Haddow, S. D., Knüsel, C. J. & Mitchell, P. D. (2019a). Parasite infection at the early farming community of Çatalhöyük. *Antiquity*, 93(369), 573–587.

Ledger, M. L., Grimshaw, E., Fairey, M., Whelton, H. L., Bull, I. D., Ballantyne, R., Knight, M. & Mitchell, P. D. (2019b). Intestinal parasites at the Late Bronze Age settlement of Must Farm, in the fens of East Anglia, UK (9th century B.C.E.). *Parasitology*, 146(12), 1583–1594.

Ledger, M. L., Rowan, E., Marques, F. G., Sigmier, J. H., Šarkić, N., Redžić, S., Cahill, N. D. & Mitchell, P. D. (2020). Intestinal parasitic infection in the Eastern Roman Empire during the Imperial Period and Late Antiquity. *American Journal of Archaeology*, 124(4), 631–657.

Ledger, M. L. & Mitchell, P. D. (2019). Tracing zoonotic parasite infections throughout human evolution. *International Journal of Osteoarchaeology*, 1–12. DOI: 10.1002/oa.2786.

Liu, Q., Sundar, K., Mishra, P. K., Mousavi, G., Liu, Z., Gaydo, A., Alem, F., Lagunoff, D., Bleich, D. & Gause, W. C. (2009). Helminth infection can reduce insulitis and type 1 diabetes through CD25- and IL-10-independent mechanisms. *Infection and Immunity*, 77, 5347–5358.

Machado, L. R., Hardwick, R. J., Bowdrey, J., Bogle, H., Knowles, T. J., Sironi, M. & Hollox, E. J. (2012). Evolutionary history of copy-number-variable locus for the low-affinity Fcγ receptor: mutation rate, autoimmune disease, and the legacy of helminth infection. *American Journal of Human Genetics*, 90, 973–985.

Maicher, C., Bleicher, N., & Le Bailly, M. (2019). Spatializing data in paleoparasitology: Application to the study of the Neolithic lakeside settlement of Zürich-Parkhaus-Opéra, Switzerland. *Holocene*, 29(7), 1198–1205.

Maizels, R. M. (2020). Regulation of immunity and allergy by helminth parasites. *Allergy,* 75, 524–534.

Maizels, R. M. & McSorley, H. J. (2016). Regulation of the host immune system by helminth parasites. *The Journal of Allergy and Clinical Immunology,* 138, 666–675.

Maizels, R. M. & Yazdanbakhsh, M. (2003). Immune regulation by helminth parasites: cellular and molecular mechanisms. *Nature Reviews. Immunology,* 3(9), 733–744.

Mangano, V. D. & Modiano, D. (2014). Host genetics and parasitic infections. *Clinical Microbiology and Infection,* 20, 1265–1275.

McSorley, H. J. & Maizels, R. M. (2012). Helminth infections and host immune regulation. *Clinical Microbiology Reviews,* 25(4), 585–608.

Menard, D. & Dondorp, A. (2017). Antimalarial drug resistance: a threat to malaria elimination. *Cold Spring Harbor Perspectives in Medicine,* 7(7), a025619.

Michelet, L. & Dauga, C. (2012). Molecular evidence of host influences on the evolution and spread of human tapeworms. *Biological Reviews of the Cambridge Philosophical Society,* 87, 731–741.

Mitchell, P. D. (2013). The origins of human parasites: exploring the evidence for endoparasitism throughout human evolution. *International Journal of Paleopathology,* 3, 191–198.

Mitchell, P. D. (2015). Human parasites in medieval Europe: lifestyle, sanitation and medical treatment. *Advances in Parasitology,* 90, 389–420.

Mitchell, P. D. (2017). Human parasites in the Roman world: health consequences of conquering an empire. *Parasitology,* 144, 48–58.

Mitchell, P. D., Anastasiou, E. & Syon, D. (2011). Human intestinal parasites in crusader Acre: evidence for migration with disease in the medieval period. *International Journal of Paleopathology,* 1, 132–137.

Moser, W., Schindler, C. & Keiser, J. (2017). Efficacy of recommended drugs against soil transmitted helminths: systematic review and network meta-analysis. *British Medical Journal,* 358, j4307.

Mpairwe, H., Webb, E. L., Muhangi, L., Ndibazza, J., Akishule, D., Nampijja, M., Ngom-wegi, S., Tumusime, J., Jones, F. M., Fitzsimmons, C., Dunne, D. W., Muwanga, M., Rodrigues, L. C. & Elliott, A. M. (2011). Anthelminthic treatment during pregnancy is associated with increased risk of infantile eczema: randomised-controlled trial results. *Pediatric Allergy and Immunology,* 22(3), 305–312.

Murray, P. R., Rosenthal, K. S. & Pfaller, M. A. (2009). *Medical microbiology.* Mosby, Philadelphia.

Nadkarni, G. N., Gignoux, C. R., Sorokin, E. P., Daya, M., Rahman, R., Barnes, K. C., Wassel, C. L. & Kenny, E. E. (2018). Worldwide frequencies of *APOL1* renal risk variants. *The New England Journal of Medicine,* 379(26), 2571–2572.

Namara, B., Nash, S., Lule, S. A., Akurut, H., Mpairwe, H., Akello, F., Tumusiime, J., Kizza, M., Kabagenyi, J., Nkurunungi, G., Muhangi, L., Webb, E. L., Muwanga, M. & Elliott, A. M. (2017). Effects of treating helminths during pregnancy and early childhood on risk of allergy-related outcomes: follow-up of a randomized controlled trial. *Pediatric Allergy and Immunology,* 28(8), 784–792.

Nyan, O. A., Walraven, G. E., Banya, W. A., Milligan, P., Van Der Sande, M., Ceesay, S. M., Del Prete, G. & McAdam, K. P. (2001). Atopy, intestinal helminth infection and total serum IgE in rural and urban adult Gambian communities. *Clinical and Experimental Allergy,* 31, 1672–1678.

Oleaga, A., Rey, O., Polack, B., Grech-Angelini, S., Quilichini, Y., Pérez-Sánchez, R., Boireau, P., Mulero, S., Brunet, A., Rognon, A., Vallée, I., Kincaid-Smith, J., Allienne, J. F. & Boissier, J. (2019). Epidemiological surveillance of schistosomiasis outbreak in Corsica (France): Are animal reservoir hosts implicated in local transmission? *PLoS Neglected Tropical Diseases,* 13(6), e0007543.

Owada, K., Nielsen, M., Lau, C. L., Clements, A. C. A., Yakob, L. & Soares Magalhães, R. J. (2017). Measuring the effect of soil-transmitted helminth infections on cognitive function in children: systematic review and critical appraisal of evidence. *Advances in Parasitology,* 98, 1–37.

Pabalan, N., Singian, E., Tabangay, L., Jarjanazi, H., Boivin, M. J. & Ezeamama, A. E. (2018). Soil-transmitted helminth infection, loss of education and cognitive impairment in school-aged children: a systematic review and meta-analysis. *PLoS Neglected Tropical Diseases,* 12(1), e0005523.

Perri, A. R., Power, R. C., Stuijts, I., Heinrich, S., Talamo, S., Hamilton-Dyer, S. & Roberts, C. (2018). Detecting hidden diets and disease: zoonotic parasites and fish consumption in Mesolithic Ireland. *Journal of Archaeological Science,* 97, 137–146.

Perry, G. H. (2014). Parasites and human evolution. *Evolutionary Anthropology,* 23, 218–228.

Pittman, K. J., Glover, L. C., Wang, L. & Ko, D. C. (2016). The legacy of past pandemics: common human mutations that protect against infectious disease. *PLoS Pathogens,* 12, e1005680.

Pritchard, D. I., Falcone, F. H. & Mitchell, P. D. (2020). The evolution of IgE-mediated type I hypersensitivity and its immunological value. *Allergy,* 1–17. doi: 10.1111/all.14570

Pullan, R. L. & Brooker, S. J. (2012). The global limits and population at risk of soil-transmitted helminth infections in 2010. *Parasites & Vectors,* 5, 81.

Rassi, A., Rassi, A. & Marin-Neto, J. A. (2010). Chagas disease. *The Lancet,* 375(9723), 1388–1402.

Reinhard, K. J., Confalonieri, U. E., Herrmann, B., Ferreira, L. F. & Araújo, A. J. G. (1986). Recovery of parasite remains from coprolites and latrines: aspects of paleoparasitological technique. *Homo*, 37, 217–239.

Reinhard, K. J., Ferreira, L. F., Bouchet, F., Sianto, L., Dutra, J. M. F., Iniguez, A., Leles, D., Le Bailly, M., Fugassa, M., Pucu, E. & Araújo, A. (2013). Food, parasites, and epidemiological transitions: a broad perspective. *International Journal of Paleopathology*, 3, 150–157.

Reinhard, K. J. & Araújo, A. (2015). Prehistoric earth oven facilities and the pathoecology of Chagas disease in the Lower Pecos Canyonlands. *Journal of Archaeological Science*, 53, 227–234.

Rodrigues, P. T., Valdivia, H. O., de Oliveira, T. C., Alves, J. M. P., Duarte, A. M. R. C., Cerutti-Junior, C., Buery, J. C., Brito, C. F. A., de Souza, J. C. Jr., Hirano, Z. M. B., Bueno, M. G., Catão-Dias, J. L., Malafronte, R. S., Ladeia-Andrade, S., Mita, T., Santamaria, A. M., Calzada, J. E., Tantular, I. S., Kawamoto, F., Raijmakers, L. R. J., Mueller, I., Pacheco, M. A., Escalante, A. A., Felger, I. & Ferreira, M. U. (2018). Human migration and the spread of malaria parasites to the New World. *Scientific Reports*, 8(1), 1993.

Romagnani, S. (2004). Immunologic influences on allergy and the TH1/TH2 balance. *The Journal of Allergy and Clinical Immunology*, 113(3), 395–400.

Rook, G. A. W. (2009). Review series on helminths, immune modulation and the hygiene hypothesis: the broader implications of the hygiene hypothesis. *Immunology*, 126, 3–11.

Rook, G. A. W. (2010). 99th Dahlem conference on infection, inflammation and chronic inflammatory disorders: Darwinian medicine and the 'hygiene' or 'old friends' hypothesis. *Clinical and Experimental Immunology*, 160(1), 70–79.

Sabin, S., Yeh, H. Y., Pluskowski, A., Clamer, C., Mitchell, P. D. & Bos, K. I. (2020). Estimating molecular preservation of the intestinal microbiome via metagenomic analyses of latrine sediments from two medieval cities. *Philosophical Transactions of the Royal Society B: Biological Sciences*, 375, 20190576.

Sandborn, W. J., Elliott, D. E., Weinstock, J., Summers, R. W., Landry-Wheeler, A., Silver, N., Harnett, M. D. & Hanauer, S. B. (2013). Randomised clinical trial: the safety and tolerability of *Trichuris suis* ova in patients with Crohn's disease. *Alimentary Pharmacology & Therapeutics*, 38(3), 255–263.

Saunders, K. A., Raine, T., Cooke, A. & Lawrence, C. E. (2007). Inhibition of autoimmune type 1 diabetes by gastrointestinal helminth infection. *Infection and Immunity*, 75, 397–407.

Schölmerich, J., Fellermann, K., Seibold, F. W., Rogler, G., Langhorst, J., Howaldt, S., Novacek, G., Petersen, A. M., Bachmann, O., Matthes, H., Hesselbarth, N., Teich, N., Wehkamp, J., Klaus, J., Ott, C., Dilger, K., Greinwald, R., Mueller, R.; International TRUST-2 Study Group. (2017). A randomised, double-blind, placebo-controlled trial of *Trichuris suis* ova in active Crohn's disease. *Journal of Crohn's & Colitis*, 11(4), 390–399.

Seo, M. & Shin, D. H. (2015). 'Parasitism, cesspits and sanitation in East Asian countries prior to modernisation', in P. D. Mitchell (ed), *Sanitation, latrines and intestinal parasites in past populations*, pp. 149–165. Ashgate, Farnham.

Sianto, L., Chame, M., S. P. Silva, C., L. C. Gonçalves, M., Reinhard, K., Fugassa, M. & Araújo, A. (2009). Animal helminths in human archaeological remains: a review of zoonoses in the past. *Revista do Instituto de Medicina Tropical de Sao Paulo*, 51, 119–130.

Sianto, L., Teixeira-Santos, I., Chame, M., Chaves, S. M., Souza, S. M., Ferreira, L. F., Reinhard, K. & Araujo, A. (2012). Eating lizards: a millenary habit evidenced by paleoparasitology. *BMC Research Notes*, 5, 586.

Sironi, M. & Clerici, M. (2010). The hygiene hypothesis: an evolutionary perspective. *Microbes and Infection*, 12, 421–427.

Smallwood, T. B., Giacomin, P. R., Loukas, A., Mulvenna, J. P., Clark, R. J. & Miles, J. J. (2017). Helminth immunomodulation in autoimmune disease. *Frontiers in Immunology*, 8, 453.

Smits, H. H., Everts, B., Hartgers, F. C. & Yazdanbakhsh, M. (2010). Chronic helminth infections protect against allergic diseases by active regulatory processes. *Current Allergy and Asthma Reports*, 10(1), 3–12.

Søe, M. J., Nejsum, P., Seersholm, F. V., Fredensborg, B. L., Habraken, R., Haase, K., Hald, M. M., Simonsen, R., Højlund, F., Blanke, L., Merkyte, I., Willerslev, E. & Kapel, C. M. O. (2018). Ancient DNA from latrines in Northern Europe and the Middle East (500 BC–1700 AD) reveals past parasites and diet. *PLoS One*, 13, e0195481.

Spyrou, M. A., Bos, K. I., Herbig, A. & Krause, J. (2019). Ancient pathogen genomics as an emerging tool for infectious disease research. *Nature Reviews: Genetics*, 20, 323–340.

Stearns, S. C. (2012). Evolutionary medicine: its scope, interest and potential. *Proceedings of The Royal Society B: Biological Sciences*, 279, 4305–4321.

Stephenson, L. S., Holland, C. V. & Cooper, E. S. (2000). The public health significance of *Trichuris trichiura*. *Parasitology*, 121(S1), S73–S95.

Strachan, D. P. (1989). Hay fever, hygiene, and household size. *British Medical Journal*, 299, 1259–1260.

Strunz, E. C., Addiss, D. G., Stocks, M. E., Ogden, S., Utzinger, J. & Freeman, M. C. (2014). Water, sanitation, hygiene, and soil-transmitted helminth infection: A systematic review and meta-analysis. *PLoS Medicine*, 11(3), e1001620.

Summers, R. W., Elliott, D. E., Qadir, K., Urban, J. F. Jr., Thompson, R. & Weinstock, J. V. (2003). *Trichuris suis* seems to be safe and possibly effective in the treatment of inflammatory bowel disease. *The American Journal of Gastroenterology*, 98(9), 2034–2041.

Summers, R. W., Elliott, D. E., Urban, J. F. Jr., Thompson, R. A. & Weinstock, J. V. (2005). *Trichuris suis* therapy for active ulcerative colitis: a randomized controlled trial. *Gastroenterology*, 128(4), 825–832.

Sutherland, R. E., Xu, X., Kim, S. S., Seeley, E. J., Caughey, G. H. & Wolters, P. J. (2011). Parasitic infection improves survival from septic peritonitis by enhancing mast cell responses to bacteria in mice. *PloS One*, 6(11), e27564.

Taghipour, A., Ghodsian, S., Jabbari, M., Olfatifar, M., Abdoli, A. & Ghaffarifar, F. (2020). Global prevalence of intestinal parasitic infections and associated risk factors in pregnant women: a systematic review and meta-analysis. *Transactions of the Royal Society of Tropical Medicine and Hygiene*, 1–14. doi: 10.1093/trstmh/traa101

Taylor, C. (2015). 'A tale of two cities: the efficacy of ancient and medieval sanitation methods', in P. D. Mitchell (ed), *Sanitation, latrines and intestinal parasites in past populations*, pp. 69–97. Ashgate, Farnham.

Taylor-Robinson, D. C., Maayan, N., Donegan, S., Chaplin, M. & Garner, P. (2019). Public health deworming programmes for soil-transmitted helminths in children living in endemic areas. *Cochrane Database of Systematic Reviews*, (9). doi: 10.1002/14651858.CD000371.pub7

Terefe, Y., Hailemariam, Z., Menkir, S., Nakao, M., Lavikainen, A., Haukisalmi, V., Iwaki, T., Okamoto, M. & Ito, A. (2014). Phylogenetic characterisation of *Taenia* tapeworms in spotted hyenas and reconsideration of the 'Out of Africa' hypothesis of *Taenia* in humans. *International Journal for Parasitology*, 44(8), 533–541.

van den Biggelaar, A. H., van Ree, R., Rodrigues, L. C., Lell, B., Deelder, A. M., Kremsner, P. G. & Yazdanbakhsh, M. (2000). Decreased atopy in children infected with *Schistosoma haematobium*: a role for parasite-induced interleukin-10. *The Lancet*, 356, 1723–1727.

Vaz Nery, S., Pickering, A. J., Abate, E., Asmare, A., Barrett, L., Benjamin-Chung, J., Bundy, D. A. P., Clasen, T., Clements, A. C. A., Colford, J. M. Jr., Ercumen, A., Crowley, S., Cumming, O., Freeman, M. C., Haque, R., Mengistu, B., Oswald, W. E., Pullan, R. L., Oliveira, R. G., Einterz Owen, K., Walson, J. L., Youya, A. & Brooker, S. J. (2019). The role of water, sanitation and hygiene interventions in reducing soil-transmitted helminths: interpreting the evidence and identifying next steps. *Parasites & Vectors*, 12(273), 1–8.

Webster, J. P., Molyneux, D. H., Hotez, P. J. & Fenwick, A. (2014). The contribution of mass drug administration to global health: past, present and future. *Philosophical Transactions of the Royal Society of London. Series B: Biological Sciences*, 369(1645), 20130434.

Weinstock, J. V. & Elliott, D. E. (2013). Translatability of helminth therapy in inflammatory bowel diseases. *International Journal for Parasitology*, 43(3–4), 245–251.

World Health Organization (WHO). (2012). *Eliminating soil-transmitted helminthiases as a public health problem in children: Progress report 2001–2010 and strategic plan 2011–2020*. World Health Organization, Geneva.

Willcocks, L. C., Carr, E. J., Niederer, H. A., Rayner, T. F., Williams, T. N., Yang, W., Scott, J. A., Urban, B. C., Peshu, N., Vyse, T. J., Lau, Y. L., Lyons, P. A. & Smith, K. G. (2010). A defunctioning polymorphism in FCGR2B is associated with protection against malaria but susceptibility to systemic lupus erythematosus. *Proceedings of the National Academy of Sciences of the United States of America*, 107, 7881–7885.

Williams, F. S., Arnold-Foster, T., Yeh, H-Y., Ledger, M. L., Baeten, J., Poblome, J. & Mitchell, P. D. (2017). Intestinal parasites from the 2nd–5th century AD latrine in the Roman baths at Sagalassos (Turkey). *International Journal of Paleopathology*, 19, 37–42.

Yazdanbakhsh, M., Kremsner, P. G. & van Ree, R. (2002). Allergy, parasites, and the hygiene hypothesis. *Science*, 296, 490–494.

Yeh, H.-Y., Mao, R., Wang, H., Qi, W. & Mitchell, P. (2016). Early evidence for travel with infectious diseases along the Silk Road: intestinal parasites from 2000-year-old personal hygiene sticks in a latrine at Xuanquanzhi Relay Station in China. *Journal of Archaeological Science: Reports*, 9, 758–764.

Yeh, H.-Y. & Mitchell, P. D. (2016). Ancient human parasites in ethnic Chinese populations. *Korean Journal of Parasitology*, 54, 565–572.

Zhan, X., Yeh, H. Y., Shin, D. H., Chai, J. Y., Seo, M. & Mitchell, P. D. (2019). Differential change in the prevalence of the *Ascaris*, *Trichuris*, and *Clonorchis* infection among past east Asian populations. *Korean Journal of Parasitology*, 57(6), 601–605.

Zhou, L., Chong, M. M. W. & Littman, D. R. (2009). Plasticity of CD4+ T cell lineage differentiation. *Immunity*, 30(5), 646–655.

Zhu, H.-R., (2015). Ecological model to predict potential habitats of *Oncomelania hupensis*, the intermediate host of *Schistosoma japonicum* in the mountainous regions, China. *PLoS Neglected Tropical Diseases*, 9, e0004028.

CHAPTER 12

Cardiovascular disease (CVD) in ancient people and contemporary implications.

Randall C. Thompson, Chris J. Rowan, Nicholas W. Weis, M. Linda Sutherland, Caleb E. Finch, Michaela Binder, Charlotte A. Roberts, and Gregory S. Thomas*

12.1 Introduction

Cardiovascular diseases (CVD), principally coronary artery disease (CAD) and stroke, are the leading cause of death throughout the globe (Roth et al., 2020). While decreasing over the last approximately fifty years in most high-income countries, CVD continues to increase in low- and middle-income countries (Roth et al., 2020; Thomas, 2007). Atherosclerosis (hardening and narrowing) of the coronary arteries (Box 12.1) results in blockages that cause myocardial infarction (i.e. necrosis of heart muscle) (Kannel, 1998). The majority of strokes are also caused by obstructive atherosclerosis in the arteries leading to and within the brain (Kannel, 1998). In 1990, 12.1 million people died of CVD, increasing to 18.6 million in 2019 (Roth et al., 2020). Today, the major risk factors for atherosclerosis, CAD, and strokes are cigarette smoking, elevated low-density lipoprotein (LDL) cholesterol, hypertension, diabetes, obesity

and physical inactivity, in addition to genetic predisposition (Virani et al., 2021).

* HORUS authors of this chapter, Thompson, Rowan, M.L. Sutherland, Finch and Thomas, thank the dedicated members of the HORUS team, Adel H. Allam, Michael A. Eskander, Caleb E. Finch, Klaus O. Fritsch, Bruno Fröhlich, Guido P. Lombardi, Crystal A. Medina, David E. Michalik, Matthew K. Miyamoto, Michael I. Miyamoto, Sallam Lotfy Mohamed, Jagat Narula, Navneet Narula, Sean J. Reardon, James D. Sutherland, Ian G. Thomas, Emily M. Venable, L. Samuel Wann, and Sir Magdi H. Yacoub.

Box 12.1   Atherosclerosis

Atherosclerosis is now accepted as an inherently inflammatory disease (Libby, 2021), first elucidated by Rudolf Virchow in 1856 (Libby, 2019; Mayerl et al., 2006). Over one and a half centuries later, our understanding of the fundamental causes and pathophysiology of atherosclerosis continues to evolve. Our current understanding is that lipid accumulation in major arteries may begin early in life (D'Armiento et al., 2001; de Negris et al., 2018), with atherosclerosis initiated by either dysfunction or injury of the arterial endothelium, the single cell layer that represents the innermost component of an artery (Furie & Mitchell, 2012) (Figure 12.1). This allows LDL and remnants of very low-density lipoproteins and chylomicrons to penetrate the endothelium into the next layer, the intima (Linton er al., 2019). Once in the intima, LDL becomes oxidised which causes it to become pro-inflammatory and to activate endothelial cells (Zalar et al., 2019). Activated endothelial cells induce the expression of adhesion molecules and pro-inflammatory cytokines. These attract inflammatory cells, including monocytes and T cells, which then enter the intima (Zalar et al., 2019; Libby, 2021). Once in the intima, monocytes transform into macrophages, which then engulf the oxidised LDL, eventually becoming foam cells. Foam cells and endothelial

Randall C. Thompson et al., *Cardiovascular disease (CVD) in ancient people and contemporary implications.* In: *Palaeopathology and Evolutionary Medicine.*
Edited by Kimberly A. Plomp et al., Oxford University Press. © Oxford University Press (2022). DOI: 10.1093/oso/9780198849711.003.0012

STAGES OF ATHEROSCLEROSIS

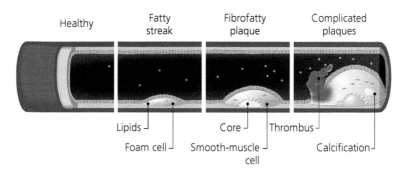

Figure 12.1 Stages of atherosclerosis.
Atherosclerosis begins with a fatty streak of foam cells containing low density lipoproteins (LDL) and remnants of very low-density lipoproteins and chylomicrons (lipids) beneath the inner cell layer (endothelium) of an artery. These cells secrete molecules that initiate an inflammatory response. As foam and inflammatory cells accumulate, a fibrous cap surrounds the plaque forming a fibrofatty plaque. If the lipid core continues to increase in size, haemorrhage within the plaque may occur or the fibrous cap may rupture or erode. Plaque erosion or rupture elicits clotting (thrombosis) at the site of the plaque injury, which contributes to the calcification that occurs with cell death.
Reprinted with permission from Shutterstock.

cells produce further proinflammatory molecules to recruit even more inflammatory cells, including B cells, into the arterial wall (Libby, 2021; Zalar et al., 2019). Once initiated, this becomes a chronic inflammatory focus (Linton et al., 2019; Zalar et al., 2019).

Over decades this now atherosclerotic plaque may remain quiescent or may increase in size enough to obstruct the artery. If a coronary artery that supplies the heart is affected, a myocardial infarction may occur. In sum, immuno-inflammatory mechanisms are directly involved in the initiation and progression of atherosclerosis from the incipient asymptomatic lesion to the advanced atherosclerotic plaque (Zalar et al., 2019). Interestingly, three recent clinical trials using oral or subcutaneously administered medications that block aspects of the inflammatory response in the artery have been shown to decrease myocardial infarction and/or stroke (Nidorf et al., 2020; Ridker et al., 2017; Tardif et al., 2019).

In contrast to contemporary populations, our ancestors are frequently portrayed in evolutionary medicine (EM) as immune to CVDs given their active lifestyles and healthy diets. Eaton and Eaton (1999b), for example, have discussed the lifestyle and health of contemporary hunter-gatherers as an analogy to understand the lives of ancient foragers. Hunter-gatherers generally have a diet consisting

of lean meat from wild animals (and fish, depending on location), and wild plant foods with a high fibre content (e.g. fruits, tubers, nuts, seeds) (Lee & Daly, 1999). However, the contribution of meat (including fat) at higher latitudes is proportionately greater than plant-based foods. In addition, some data suggest that native Arctic peoples consume less saturated fat in their meat—the fat that is linked to increased serum cholesterol and CVDs (Eaton & Eaton, 1999a). Today, we appreciate that there is no 'one size fits all' for the lifestyles of hunter-gatherers and this, undoubtably, was the case in the past (Kelly, 1995); the environment would drive what foods were available (Elton, 2008). Over time, climate change and environmental conditions, for example, would have affected the supply of wild animals and plants and, for some foraging societies today, traditional diets and lifestyles from neighbouring, agriculturally dependent communities can lead to change. We should also remember that different people within a group may not all have eaten the same combination of foodstuffs across their life course. Even so, blood pressure overall tends to be lower for hunter-gatherers than westernised groups, as does serum cholesterol.

A prime example of the healthy control of risk factors is provided by the Tsimané, a group who live at the base of the Andes in Bolivia (Kaplan et al., 2017). This population, numbering about 16,000, has lived a subsistence lifestyle until the last five to ten years, with little access to market goods. They fish, hunt and forage, and almost all are farmers, vital activities requiring high physical activity. Male and female adults are estimated to average the equivalent of 17,000 and 16,000 steps per day, respectively (adapted from Gurven et al., 2013). Their average lifetime LDL is about 71 mg/dL (Kaplan et al., 2017; Vasunilashorn et al., 2010).

In conjunction with the long-running Tsimané Health and Life History Project (THLHP), the HORUS team (described further later) performed non-contrast chest computed tomography (CT) scans in 2014–2016 on 705 Tsimané individuals ≥ 40 years of age (mean 58 years) looking for arterial calcification of the coronary arteries using the modern-day, coronary artery calcium scoring (CAC) method. Eighty-five per cent of them had no calcified coronary arteries, 13% had CAC scores of 1–99 and only 3% had CAC scores ≥ 100, characterised as moderate coronary artery calcification. Three per cent represents only about a tenth of the prevalence of this calcification score of individuals living in industrialised populations. The Tsimané thus represent a population with the least amount of CAC, and thus CAD, ever reported (Kaplan et al., 2017).

Evidence for atherosclerosis in past populations can be found in the bioarchaeological record. As plaques mature, cell death within the plaque occurs that results in calcification of some of the plaque in the form of calcium hydroxyapatite deposits. These mineral-like deposits are remarkably persistent and can survive in the archaeological record long after soft tissues have disintegrated post-mortem (Figure 12.2). In CT scans of mummies, calcification can also be seen either within the arterial wall or, if the wall has disintegrated over time, along the expected anatomic course of the artery (Allam et al., 2018). Calcification is visible during life as well and coronary calcium scoring is a test commonly ordered by physicians to determine if people have calcified plaque in their coronary arteries (Thompson et al., 2005). Calcification in other major arteries of the neck, chest, abdomen and legs can also be seen with routine CT scanning in living humans and in CT scans of mummies. If vascular

Figure 12.2 Calcified femoral artery in the grave of a female estimated to be 36 to 50 years old at death in a cemetery in Amara West in ancient Nubia.
Reprinted with permission from Binder and Roberts (2014).

calcification is present in living humans or mummies, it is pathognomonic (definitive for) atherosclerotic plaque (Stary et al., 1995).

As CT scanning of mummies is non-destructive, and ancient mummies can be transported from a museum location to a local medical facility for imaging, such scanning has been widely applied as a form of virtual autopsy. The CT scans have shown that arterial calcifications are identical in appearance to atherosclerotic calcification seen on CT scans of contemporary, living patients. They confirm the earlier descriptions of atherosclerosis made by Johan Czermak (1852) using microscopy and autopsy of atherosclerosis in the thoracic aorta of a female Egyptian mummy. They also confirm and expand the seminal research of Sir Mark Armand Ruffer (1910, 1911) who also used microscopy and autopsy to determine that atherosclerosis was common in ancient Egyptian mummies. This research has already shown the value of palaeopathology in offering a unique perspective and new understanding of CVDs. It also pushes back the history of CVDs to well before the present day and shows that non-industrialised people in the past could develop these diseases, contrary to previous views. Section 12.2 outlines some of the accumulating evidence for CVD in mummified and skeletal remains.

12.2 The evidence for ancient CVD

12.2.1 CT scans of ancient mummies

The largest collection of Egyptian mummies in the world is housed at the Egyptian National Museum of Antiquities in Cairo, Egypt. Thanks to the generosity of National Geographic and Siemens, a six-slice Siemens CT scanner (Florsheim, Germany) is available in the grounds of the museum and has been used by a number of Egyptian and international teams. In February 2009, a multi-disciplinary, international team called HORUS (named after the ancient Egyptian deity) was given permission by the Egyptian Supreme Council of Antiquities to image mummies and to look for evidence of atherosclerosis. On a visit to the Museum a year earlier, members of the HORUS team had already observed a placard concerning the mummy of Mernenptah, a pharaoh of the New Kingdom of the 19th Dynasty (reign ~ 1213–1203 BCE) noting he had arteriosclerosis (a form of atherosclerosis). Given that ancient lifestyles were presumed to be healthier than today, physicians on the HORUS team doubted this could be true.

The results of the CT scans, however, surprised the team. Sixteen of the twenty-two mummies imaged had identifiable vascular tissue. Of these, five had definite atherosclerosis with calcification seen in the arterial walls, while four had probable atherosclerosis, defined as calcification in the expected anatomic path of an artery that had disintegrated over the ages (Allam et al., 2009). Estimated age of death was determined by integrative assessment of architectural changes in the clavicle, humerus, femur and cranial suture closures (Buikstra & Ubelaker, 1994; Walker & Lovejoy, 1985). Among those who died after 45 years of age, calcification was present in seven of eight (87%) mummies, compared with only two of eight (25%) who died younger, aged < 45 years (Allam et al., 2009).

The mummies imaged in the current study were either priests or priestesses or members of the royal court (Allam et al., 2009), which raises the question of whether they had atherosclerosis because of their relative wealth and high status with potentially damaging diets and lifestyles. Over the next several years, the team collaborated with investigators across the globe, imaging and obtaining whole body CT scans of 137 mummies from Egypt, Peru, southwestern North America and Alaska, spanning more than 4000 years of human history (Thompson et al. 2013). The additional mummies that were scanned comprised seventy-six ancient Egyptians who lived between 3100 BCE to the

end of the Roman era (364 CE), fifty-one ancient Peruvians who lived between ~ 200 and 1500 CE, five Ancestral Puebloans who lived in the Four Corners region of southwestern North America (~ 200–600 CE) and five Aleutian Islanders in this sample who lived on the ring of volcanic islands that previously constituted the lower edge of the Bering land bridge from Asia to present-day Alaska (~ 1756–1930 CE).

Diets and living conditions varied greatly across the cultures represented by these mummies. Many ancient Egyptians lived in cities where jobs were differentiated, with some requiring strenuous physical activity such as farming, while members of the royal court likely had fewer physical demands, although potentially were subject to more social stress (Ikram, 2010). The Peruvians were primarily farmers living in stone houses who still used stone tools (de Mayolo, 1981), the Ancestral Puebloans farmed corn and squash and hunted small animals (Downum, 2012) while the Aleutian Islanders lived primarily as hunter-gatherers, eating fish, seals, sea otters, sea urchins, shellfish, ducks and fox (Fröhlich et al., 2002; Hrdlicka, 1945).

Atherosclerosis was visible in the people from all of these cultures, with its presence in the hunter-gatherers being a particular surprise (Brown, 1998). Earlier autopsy studies by Zimmerman and colleagues (1971, 1981), however, had clearly shown atherosclerosis in the mummies of two ancient hunter-gatherer Aleutian Islanders, a man in his 30s–40s and a female of roughly 50 years old. Another autopsy by Zimmerman and colleagues (1975) documented atherosclerosis, including CAD, in a female who lived on St Lawrence Island in the Bering Sea in ~ 400 CE, while other ancient Alaskan samples confirmed these findings (Zimmerman & Aufderheide, 1984).

In January 2018, the HORUS team also performed CT imaging of four Greenlandic, Inuit adult mummies and one infant in conjunction with the Harvard Peabody Museum of Archaeology and Ethnology and the Brigham and Women's Hospital in Boston (Wann et al., 2019). In 1929, Martin Luther of Harvard had removed these mummies from two caves adjacent to the ancient settlement of Qerrortut on the now uninhabited island of Uunartaq in southwestern Greenland (Hooton,

1930; Matthiassen, 1936). Danish anthropologist, Therkel Mathiassen, visited these caves in 1934 and identified grave goods dating the individuals to the 1500s (Matthiassen, 1936). Atherosclerosis was diagnosed in two of these four Inuit mummies and probable atherosclerosis in a third (Wann et al., 2019). A female, estimated to have died aged 18–22 years, had definite atherosclerosis in the thoracic and abdominal aorta and probable atherosclerosis of the left superficial femoral artery. A male of approximately 25 years of age had definite atherosclerosis in the thoracic and abdominal aorta. A 25–30-year-old male had probable left carotid atherosclerosis. Visually, the degree of atherosclerosis was much less than that observed in the ancient Egyptians who were older at the time of their deaths. The diet of the Inuit living in Southwestern Greenland at this time consisted predominately of marine animals and fish, but they also relied on animals from the interior. Marine resources included seals, walruses, whales, halibut, cod, birds and bird eggs of the guillemot, eider and kittiwake, but they also ate meat from caribou, reindeer and polar bears (Hansen et al., 1991; Kleivan, 1984).

12.2.2 Other evidence of atherosclerosis found in the remains of ancient people

The oldest mummy, the 'Iceman' or Ötzi, who has been confirmed to have atherosclerosis was found by two hikers in the Italian Alps in 1991. CT scanning of Ötzi demonstrated atherosclerosis of the carotid and iliac arteries and the aorta (Murphy et al., 2003). Ötzi was dated to about 3300 years BCE and was estimated to be 40–50 years old at the time of death from a flint-tipped arrow that pierced his back near his left shoulder blade (Murphy et al., 2003). Examination of his preserved gastrointestinal tract demonstrated that he was an omnivore, consuming deer and ibex (wild goat) meat, cereal crops, legumes, fruits and berries (Maixner et al., 2018; Rollo et al., 2002; Thompson et al., 2014). It might be assumed that living to his 40s or 50s, having an active lifestyle in the mountains, and eating a healthy, relatively unprocessed, mixed diet of carbohydrates, proteins and lipids, should have prevented him from vulnerability to atherosclerosis, yet he still had this condition.

Skeletal remains from Amara West, in ancient Nubia (modern-day Sudan), have also demonstrated atherosclerosis. Established in 1300 BC by the Egyptians during the New Kingdom Period (~ 1520–1075 BCE), Amara West was intended to be the capital of the Upper Nubia Province of Egypt (Binder & Roberts, 2014), and is located on the western side of the Nile River, about 100 miles into what is now modern-day Sudan. Two cemeteries in the town have been excavated, leading to the discovery of 180 skeletons. It is estimated that thirty-six of these individuals dated to the New Kingdom Period and 144 lived after it. Five of the excavated 180 individuals were discovered to have calcified atherosclerotic plaques within their graves associated with their skeletons, even in the absence of soft tissue (Binder & Roberts, 2014). Two were New Kingdom males and three were post-New Kingdom females. All five are believed to have died between 36 and 50 years of age. One male (Sk244-4) had a large, yellow calcified plaque in his chest cavity. The other male (Sk244-6) had five calcified plaques of various sizes. For both males, the artery where the plaque originated was likely the thoracic aorta. Sk243-3, one of three females, also had a plaque in their chest cavity, likely aortic in origin. Sk237 had eight unique plaques, all found in their abdominal area and likely related to the iliac artery. The third female, Sk305-4, had multiple calcifications running alongside the right femur consistent with femoral artery calcifications (Figure 12.2) (Binder & Roberts, 2014).

The people of Amara West are thought to have had a diet and lifestyle that may have promoted atherosclerosis to some degree, including access to domesticated animals for consumption. Dates were also popular foods, providing a high sugar content (Ryan et al., 2012), while another potential risk factor was smoke inhalation from internal pollution. All five skeletons had new bone formation on the visceral side of their ribs (Binder & Roberts, 2014), evidence of chronic lung inflammation, and possible infection of the lower respiratory tract (Roberts et al., 1994; Santos & Roberts, 2006). This inflammation may well have been exacerbated by smoke from the charcoal, wood and dung used in Amara West to fuel open fires for cooking and heating (Binder & Roberts, 2014). Upper respiratory tract inflammation (maxillary sinusitis) was also record-ed at a high level of 76% of all adults from the ceme-tery (Binder, 2014). Household smoke from cooking fires and landmass wildfires may also be athero-genic (Chen et al., 2021; Painschab et al., 2013). The World Health Organization attributed 3.8 million excess deaths in 2018 to indoor air pollution, of which nearly half were due to ischaemic heart dis-ease (27%) and stroke (18%) (WHO, 2018), while indoor air pollution ranks as one of the leading glob-al causes of mortality even today (GBD 2017 Risk Factor Collaborators, 2018).

Other skeletons from a documented forensic collection in Italy have revealed calcifications in individuals with recorded arterial disease (Biehler-Gomez et al., 2018). Supporting these more recent skeletal related findings, Sir Marc Armand Ruf-fer (1911, 455) commented in his autopsy study of ancient Egyptian mummies that, 'The femoral and profunda [arteries] were dissected out. Both were converted into rigid calcareous tubes.' The recent reports of calcified arteries associated with skele-tons show the potential of this type of evidence for contributing more to our knowledge about the ori-gin, evolution and history of CVD. Furthermore, as skeletons are the most commonly excavated human remains, they provide substantial opportunities for producing more evidence. Educating excavators to look for calcified blood vessels should be part of the induction of personnel who excavate cemeter-ies/burials.

12.2.3 Unangan female of the Aleutian Islands

It has been suggested that a marine-based diet with a high consumption of fish oils would be protec-tive against atherosclerosis (Dyerberg et al., 1978), and the HORUS team was keen to evaluate the sta-tus of ancient people from a culture with such a diet. They collaborated with Bruno Fröhlich of the National Museum of Natural History, Smithsoni-an Institution, Washington DC, USA (SI-NMNH) to review CT images of five Aleutian Islanders that he and Fulbright Scholar Birna Jonsdottir had pre-viously scanned in 2001 (Jonsdottir, 2002). With further investigation, and with Fröhlich soon join-ing the HORUS team, it was discovered that the five mummified individuals were of the Unangan

hunter-gatherer culture who lived between 1756 and 1930 CE. Their history is fascinating. The mum-mies had been collected from the 'Warm Cave' (so named because of steam vents from the volcano near its entrance) by Captain Ernest Hennig in 1874 and Aleš Hrdlička in 1936 on the Kagamil Island volcano of the Aleutian Island chain (Dall, 1878; Hrdlička, 1945) Hrdlička, while curator of the SI-NMNH, led a 1936 arctic expedition that included collecting mummies from the Warm and Cold Caves of Kagamil Island (Hrdlička, 1945). Figure 12.3 shows an image of Hrdlicka's expedition to this foreboding island.

Of these Kagamil Island mummies imaged and reported by Thompson and colleagues (2013), three of the five had calcified arteries on their CT scans consistent with atherosclerosis. One of the individ-uals (USNM-386,381) had been collected during the Hrdlička expedition from Warm Cave. This was a female, estimated to have died aged 47–51 years, who had atherosclerosis of the coronary, iliac and popliteal arteries, as well as of her aorta. Figure 12.4 shows an image of her right coronary artery and aortic arch disease and Figure 12.5 shows her iliac artery disease.

The severity of atherosclerosis in the Aleutian Islanders was another surprise to the HORUS team, as they were hunter-gatherers without access to a diet of domesticated plants or animals. As described earlier, their diet came from the sea, seashore and island interior, where they could find birds, bird eggs, berries and, at times, fox (Hrdlicka 1945; Jons-dottir 2002). From the sea and seashore they could fish/hunt/collect salmon, whales, eared and earless seals, otters, octopus, shellfish and sea urchins. One explanation for the degree of atherosclerosis found in the mummies of the Aleutian Islanders could be exposure to air pollution due to smoke inhalation, as previously discussed in relation to the skeletal remains from Amara West. The mean annual tem-perature in the Aleutians is 5°C (40 F°), higher in the summer and colder in the winter, with January tem-perature lows frequently reaching −12°C (10 F°). High winds and heavy fog are incessant (US Coast Pilot, 1947). The Aleutian Islanders lived in multi-family *barabaras*, earth-lodges in which inhabitants entered by a hole in the roof via a slatted timber lad-der. Shallow, bowl-shaped stone lamps were filled

Figure 12.3 Two members of Hrdlicka's Kagamil Island expedition, which included mummy retrieval in 1936. Steam can be seen coming from within the volcano. Reprinted with permission from The Wistar Institute, Philadelphia, PA.

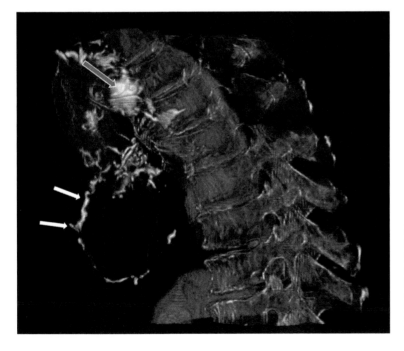

Figure 12.4 In this 3D volume reconstruction CT image, white arrows point to the calcifications of the right coronary artery in this Aleutian female hunter-gatherer who died at ~ 47–51 years old. Above the right coronary artery is the heavily calcified aortic arch identified by the blue arrow.

with seal and whale oil to light and heat their subterranean sod homes. Laughlin (1980) described the soot inside as so heavy at times that children's faces needed to be wiped of the accumulated soot in the morning.

Of relevance to the mummy studies, indoor air pollution was evaluated by Roberts (2007) in a study of maxillary sinusitis in skeletons from a range of locations representing different environments and subsistence, including the Aleutian Islands. Using

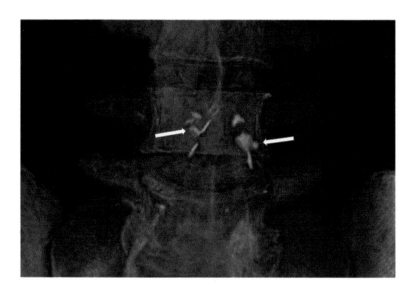

Figure 12.5 In this 3D volume reconstruction CT image, arrows point to the bilateral calcifications of the iliac arteries in this Aleutian female hunter-gatherer who died at ~47–51 years old.

this condition as indicative of indoor (and/or out-door) air pollution, she found maxillary sinusitis to be present in the skeletons of 43% of thirty-five Aleutian Islanders who lived 1500–1600 CE. Females were affected more than males, potentially reflecting tasks for the females, such as cooking, that were more likely to be undertaken inside the *barabaras*. It is also notable that, in volcanic areas today, sulphurous air pollution can lead to respiratory disease in residents of those areas (e.g. Longo & Yang, 2008). Roberts' (2007) findings are consistent with the studies of Zimmerman and colleagues (Zimmerman & Aufderheide, 1984; Zimmerman et al., 1971, 1981), who consistently found anthracosis (a pneumoconiosis caused by the accumulation of carbon in the lungs) in the Aleutian mummies they autopsied. Overall, these studies show that the living conditions within the *barabaras* may not only have contributed to the upper respiratory tract infections (Roberts, 2007), but also to the CVD identified in the Aleutian Island mummies.

Several teams other than HORUS have documented atherosclerosis in ancient mummies by radiographic imaging. These include investigators who used CT (Martina et al., 2005; Kim et al., 2015; Shin et al., 2017; Gaeta et al., 2019) and plain film radiography (Eiken, 1989; Fornaciari et al., 1989; Piombino-Mascali et al., 2017).

12.3 Other CVDs seen in mummies

Although CT evidence for at least pre-clinical atherosclerosis is common in mummies, evidence for end-organ damage from atherosclerotic disease, such as myocardial infarction, is rare. One possible exception is the Egyptian mummy of Lady Rai. Lady Rai lived during the reign of Amenhotep I during the 18th Dynasty, ~ 1570–1530 BCE (Allam et al. 2009). During her adult life, Lady Rai attended to Queen Ahmose-Nefertari, the mother of Amenhotep I, as a nursemaid, while the Queen ruled as the de facto monarch until Amenhotep came of age (Abdelfattah et al., 2013). In 2009, the mummy of Lady Rai was CT scanned by the HORUS team and found to have evidence of atherosclerosis with calcified plaques in her thoracic aorta. She also had very heavy calcification in the remnants of her heart's posterior wall suggesting that she may have suffered a myocardial infarction prior to her death (Abdelfattah et al., 2013).

Along with atherosclerosis, other cardiovascular pathologies can also be identified in mummies, including valvular and congenital heart disease. In the first fifty-two Egyptian mummies studied by the HORUS team (Allam et al., 2011), thirty-one had intact or remnant hearts and, of these, two had mitral annular calcification on CT imaging, appearing identical to mitral annular calcification seen in modern-day patients. This is a degenerative condition associated with atherosclerosis, which can lead to mitral valve dysfunction (Allison et al., 2006).

Congenital heart disease has been also found in at least one ancient mummy. The State Museum of Detmold, Germany houses the mummy of an infant child originally discovered in western South America. This mummy, nicknamed 'the Detmold Child,' was brought to Germany where radiocarbon dating revealed the body to be approximately 6500 years old (Haas et al. 2015). CT imaging of the teeth was consistent with an infant 6 months of age. Imaging was also consistent with hypoplastic left heart syndrome (HLHS), a congenital defect where the left side of the heart is underdeveloped. In the case of the Detmold Child, the left ventricle and ascending aorta were both hypoplastic. The average infant with HLHS lives an average of 4.5 days without medical care (Fruitman, 2000). The Detmold Child may have lived longer than expected because the high altitude and low oxygen tension of where he/she was located in the Andes, which would have increased the likelihood of continued patency of the ductus arteriosus and oxygenated blood being delivered to the body through this shunt (Haas et al., 2015).

12.4 CVD in great apes

Reviews of the causes of death of great apes include those by Vastesaeger and Delcourt (1961), Varki et al., (2009), Finch (2010) and Lowenstein et al. (2016). The most common cause of cardiac death in chimpanzees, gorillas and orangutans is diffuse interstitial myocardial fibrosis of unknown cause (Varki et al., 2009). Their most likely cause of death in the wild is infection (Finch, 2010; Varki et al., 2009). While atherosclerosis has been documented by autopsy, described by Manning (1942) and others (Vastesaeger & Delcourt, 1961), it has been an infrequent finding in great ape species.

The HORUS team evaluated thirty great apes (twenty adults, 10 juveniles) curated by the Division of Mammals, SI-NMNH and maintained post-mortem in alcohol tanks at the Smithsonian Museum Support Center in Suitland, Maryland (Thompson et al., 2020). The apes were collected

between 1908 and 2000 and included eight gorillas (*Gorilla gorilla gorilla*, *Gorilla beringei graueri*), seven chimpanzees (*Pan troglodytes*), thirteen orangutans (*Pongo pygmaeus*, *Pongo abelii*), and two bonobos (*Pan paniscus*). Almost all of the apes had spent some time in captivity. CT scanning could be performed on twenty-eight, and arterial vasculature was preserved well enough in twenty-two of these to evaluate using CT. Of these, thirteen apes showed evidence of mild arterial calcification on their CT scans with histologic confirmation in selected apes by CT guided biopsy (Thompson et al., 2020). The authors concluded that preclinical atherosclerosis may be more common in apes than previously thought. As apes share more than 95% of the human genome, some *tendency* to develop atherosclerosis has likely been present in humans since the beginning of our species.

12.5 Discussion

The CT and autopsy findings described above confirm that atherosclerosis was common prior to industrialisation. The findings have encompassed varied cultures consuming different diets and included people who were physically active in subsistence practices such as Peruvian and Ancient Pueblan farmers, and hunter-gatherers of the Aleutian Islands and present-day Greenland. In sum, palaeopathological analyses performed on mummies and skeletons challenge long-held scientific assumptions in EM that our ancient ancestors were free from CVD, encouraging researchers to re-examine their beliefs about the causes of atherosclerosis and to explore new environmental risk factors as well as the genetic underpinnings of this disease as discussed below.

In industrialised populations, atherosclerosis usually does not cause morbidity and death until after the prime reproductive years. In the USA, for example, the average age of first heart attack is 65.6 years for men and 72.0 years for women (Virani et al., 2021). Thus, natural selection would not be expected to act against atherosclerosis due to the 'declining force' of selection with age (Fisher, 1930). In 1956, George Williams (1966) presented his antagonistic pleiotropy hypothesis, whereby genes can benefit early life survival and reproduction, yet have harmful effects later in life when natural selection weakens. This hypothesis raises important questions related to EM. Which genes, and the mutations that resulted in their existence, might benefit humans early in life but be harmful by facilitating atherosclerosis later in life? Based on recent research, it appears these may be ones that mediate a robust human inflammatory response consistent with antagonistic pleiotropy, as explained further later (Finch, 2010; Trumble & Finch, 2019).

Until the twentieth century, the main cause of death in humans was infectious diseases (Finch, 2007, 2010; Finch & Crimmins, 2004; Trumble & Finch, 2019). In fact, the three leading causes of death in the USA in 1900 were, in order, pneumonia/influenza, tuberculosis and gastrointestinal infections (CDC, n.d.). Heart disease did not become the leading cause of death in the USA until 1930 (CDC, n.d.). Similarly, according to the limited record keeping at the time, there are data from 1750–54 from London showing that infections were then the leading cause of death (Matossian, 1985), and data from 1850–54 in both England and Wales tells the same story (UK Data Archive, 2021).

The recent evidence we present in this chapter linking atherosclerosis and inflammatory responses illuminates a potential ultimate cause of CVD in humans. Specifically, in the 200,000 years of human evolution, *Homo sapiens* has evolved a robust inflammatory/immune system to combat infectious diseases. However, this system can be counterproductive when LDL enters the arterial wall and initiates a robust inflammatory response that may cause and/or exacerbate atherosclerosis. If so, the prevalence of atherosclerosis in humans today could represent an evolutionary trade-off between fighting infectious diseases at younger ages and developing CVDs at older ages.

What conclusions can we draw then from studies of atherosclerosis in mummies? What are the implications for humans living today and for EM, especially given earlier assumptions concerning the low risk for CVD in ancestral humans? Prior to publication of studies of mummies and, in some cases, skeletal remains in which atherosclerosis was found to be common among ancient cultures, it was considered a contemporary disease

and a result of modern lifestyles. CVD has figured prominently in descriptions of 'mismatch' of humans that apparently evolved with presumably healthier lifestyles but then found themselves living in sedentary environments with atherogenic diets (Danziger 2016; Eaton & Eaton, 1999b; Nesse & Williams, 1994; Pollard, 2008; Swynghedauw, 2016). In contrast, the findings described here demonstrate that humans are inherently at risk for atherosclerosis and have been since early times. Specifically, in all the cultures with mummies that the HORUS team were able to evaluate, all had atherosclerosis. We conclude that *H. sapiens* has had an intrinsic risk for developing ischaemic grade atherosclerosis (Thompson et al., 2013).

When considering the severity of atherosclerosis in the four cultures represented in the study of 137 mummies by Thompson and colleagues (2013), there was a trend for the Egyptians to have more atherosclerosis than the other cultures, namely, the farmers of Peru, the farmer-foragers of the Four Corners region of the USA and the hunter-gatherers of the Aleutian Islands. By the time the Egyptians mummified their dead, ancient Egyptians lived in cities and job specialisation had already occurred. The Egyptian mummies examined were from royal courts and held specialised jobs that did not require farming, foraging or hunting for daily food needs, that is, less physical activity. The lifestyles of the other three cultures were closer to that of the Tsimané than the Egyptians, being either subsistence farmers or hunter-gatherers.

Since the late nineteenth century, researchers have had considerable interest in linking the development of atherosclerosis to chronic infections. Could infections themselves cause atherosclerosis? This association seems natural—chronic inflammatory pathways triggered by infections resemble the atherosclerotic processes in the arterial walls as described earlier. Circulating cytokines like interleukin-6 and acute phase reactants such as fibrinogen and plasminogen activating factor can be triggered by exotoxins from infection, and could play a role in the development of atherosclerosis and trigger endothelial cell responses (Rosenfeld et al., 2011). Several infectious organisms have been implicated in the development of atherosclerosis. These include, but are not limited to: human immunodeficiency virus (HIV), cytomegalovirus (CMV) and other herpes viruses, *Helicobacter pylori*, *Chlamydophila pneumoniae*, *Enterobacter hormaechei*, *Mycoplasma pneumoniae*, *Porphromonas gingivalis*, and other oral anaerobes (Reszka et al., 2008, Adiloglu et al., 2005, Ford et al., 2006). The association of different organisms with heart disease is based on several factors. These include the visual presence of organisms within atherosclerotic plaques detected via culture or polymerase chain reaction, atherogenesis epidemiology linked to positive serology, acceleration of plaque formation in animal models infected with viruses and bacteria and proatherogenic and prothrombotic virulence factors present in certain organisms. However, none of these have been proven to be causal.

Current research to determine if an infectious agent is causal are building on the challenges of previous studies and are also looking at new approaches. Specifically, with shared inflammatory pathways, could chronic infection with more than one organism collectively contribute to atherogenesis? There is momentum to the idea that, over time, chronic and cumulative infection risk may be linked to atherosclerosis (Espinosa-Klein et al., 2001). Additionally, there is interest in continuing to look at the beneficial effect of vaccines such as those against influenza on atherogenesis as well as a hypothetical beneficial effect of targeted antibiotic use (Fong et al. 2002; Muhlestein et al., 2000). While a clear connection linking infection to atherosclerosis has evaded researchers, that does not mean that one will not eventually be found with future investigations.

The findings that each population group studied, including hunter-gatherers, had atherosclerosis is also consistent with a genetic predisposition to atherosclerosis and thus myocardial infarction and stroke. One widely shared genetic risk factor of atherosclerosis is the *APOE4* allele, associated with elevated cholesterol and increased risk of myocardial infarction (Sing & Davignon, 1985; Yassine & Finch, 2020). APOE is a multifunctional protein in lipid metabolism, immunity and brain synapses. It was first known as a key transporter of cholesterol in the blood to the liver and, in the brain, to neurons (Trumble & Finch, 2019). The allele *APOE4* is the ancestral human gene (Hanlon & Rubenstein, 1995) from which the alleles *APOE3* and then *APOE2* evolved in the last 250,000 years (Fullerton et al., 2000). In all populations, *APOE3* is the most prevalent allele, followed by *APOE4*, and *APOE2*. While increasing morbidity from CVD and Alzheimer's disease, *APOE4* benefits hepatitis C infections by decreasing hepatic fibrosis (Wozniak et al., 2002). Moreover, *APOE4* increases survival in highly infectious environments of rural Ghana (van Excel et al., 2017) and resistance to diarrhoea in Brazilian slums (Mitter et al., 2012). Associations of *APOE4* with elevated cholesterol are also found in Ghana (van Exel et al., 2017) and among the Tsimané (Garcia et al., 2021).

In the Tsimané, *APOE4* carriers with elevated eosinophilia (i.e. a type of white blood cell that increases with parasitism) had better cognitive scores than *APOE3* carriers (Trumble et al., 2017). Thus, *APOE4* may have been advantageous to humans living in highly infectious environments, but became disadvantageous in low-infection environments, especially in contemporary populations with lifestyles of low physical activity and food in excess of caloric needs (Trumble & Finch, 2019). The transition of *APOE4* as an adaptive allele to being maladaptive can be described as representing the shift from pathogen-driven inflammation to sterile inflammation, exacerbated by obesogenic lifestyles, air pollution and smoking (Trumble & Finch, 2019). These findings may be viewed as a special case of antagonistic pleiotropy (Williams, 1966), thus contributing to further understanding of the evolution of CVD.

Much more work is needed to clarify the evolutionary reasons for a human predilection for atherosclerosis and CVD. Nonetheless, many of the causes of atherosclerosis, otherwise known as risk factors, are known and can be managed or minimised by humans today to avoid or delay the onset of atherosclerosis and its consequences. The dearth of coronary atherosclerosis in the Tsimané is a striking example of how specific lifestyles can lower the risk for human populations.

References

Abdelfattah, A., Allam, A. H., Wann, S., Thompson, R. C., Abdel-Maksoud, G., Badr, I., Amer, H. A., el-Halim Nur el-Din, A., Finch, C. E., Miyamoto, M. I., Sutherland, L., Sutherland, J. D. & Thomas, G. S. (2013). Atherosclerotic cardiovascular disease in Egyptian women: 1570 BCE–2011 CE. *International Journal of Cardiology*, 167(2), 570–574.

Adiloglu, A. K., Ocal, A., Can, R., Duver, H., Yavuz, T. & Aridogan, B. C. (2005). Detection of *Helicobacter pylori* and *Chlamydia pneumoniae* DNA in human coronary arteries and evaluation of the results with serologic evidence of inflammation. *Saudi Medical Journal*, 26, 1068–1074.

Allam, A. H., Thompson, R. C. Eskander, M. A., Mandour Ali, M. A., Sadek, A., Rowan, C. J., Sutherland, M. L., Sutherland, J. D., Frohlich, B., Michalik, D. E.; HORUS research team, Finch, C. E., Narula, J., Thomas, G. S. & Wann, L. S. (2018). Is coronary calcium scoring too late? Total body arterial calcium burden in patients without known CAD and normal MPI. *Journal of Nuclear Cardiology*, 25(6), 1990–1998. doi: 10.1007/s12350-017-0925-9

Allam, A. H., Thompson, R. C., Wann, L. S., Miyamoto, M. I., el-Halim Nur el-Din, A., El-Maksoud, G. A., Al-Tohamy Soliman, M., Badr, I., El-Rahman Amer, H. A., Sutherland, M. L., Sutherland, J. D. & Thomas, G. S. (2011). Atherosclerosis in ancient Egyptian mummies: the HORUS study. *JACC Cardiovascular Imaging*, 4(4), 315–327.

Allam, A. H., Thompson, R. C., Wann, L. S., Miyamoto, M. I. & Thomas, G. S. (2009). Computed tomographic assessment of atherosclerosis in ancient Egyptian mummies. *Journal of the American Medical Association*, 302, 2091–2094.

Allison, M. A., Cheung, P., Criqui, M. H., Langer, R. D. & Wright, C. M. (2006). Mitral and aortic annular calcification are highly associated with systemic calcified atherosclerosis. *Circulation*, 113(6), 861–866.

Biehler-Gomez, L., Maderma, E., Brescia, G., Caruso, V., Rizzi, A. & Cattaneo, C. (2018). Distinguishing atherosclerotic calcifications in dry bone: implications for forensic identification. *Journal of Forensic Sciences*, 64, 839–844.

Binder, M. (2014). *Health and diet in Upper Nubia through climate and political change: a bioarchaeological investigation of health and living conditions at ancient Amara West between 1300 and 800BC.* Unpublished PhD Thesis, Durham University, Durham.

Binder, M. & Roberts, C. A. (2014). Calcified structures associated with human skeletal remains: Possible atherosclerosis affecting the population buried at Amara West, Sudan (1300–800 BC). *International Journal of Paleopathology*, 6, 20–29.

Brown, P. J. (1998). *Understanding and applying medical anthropology.* Mayfield Publishing, Mountain View.

Buikstra, J. E. & Ubelaker, D. H. (eds). (1994). Standards for data collection from human skeletal remains. Arkansas Archeological Survey Research Series, No. 44. Arkansas Archeological Survey, Fayetteville.

Centers for Disease Control (CDC). (n.d.) Leading Causes of Death, 1900–1998. https://www.cdc.gov/nchs/data/dvs/lead1900_98.pdf

Chen, H., Samet, J. M., Bromberg, P. A. & Tong, H. (2021). Cardiovascular health impacts of wildfire smoke exposure. *Particle and Fibre Toxicology*, 18(1), 2.

Czermak, J. (1852). Description and microscopic findings of two Egyptian mummies. *Meeting of the Academy of Science*, 9, 27–69.

D'Armiento, F. P., Bianchi, A., De Nigris, F., Capuzzi, D. M., Armiento, M. R., Crimi, G., Abete, P., Palinski, W., Condorelli, M. & Napoli, C. (2001). Age-related effects on atherogenesis and scavenger enzymes of intracranial and extracranial arteries in men without classic risk factors for atherosclerosis. *Stroke*, 32(11), 2472

Dall, W. H. (1878). *On the remains of later prehistoric man obtained from the Catherina archipelago, Alaskan territory and especially from the caves of the Aleutian Islands.* Smithsonian Institution, Washington, DC.

Danziger R. S. (2016). 'Evolutionary imprints on cardiovascular physiology', in A. Alvergne, C. Jenkinson & C. Faurie (eds), *Evolutionary thinking in medicine: from research to policy and practice*, pp. 137–54. Springer, Cham.

de Mayolo, A. (1981). *La nutrición en el antiguo Perú.* Central de Reserva del Perú, Lima.

De Nigris, F., Cacciatore, F., Mancini, F. P., Vitale, D. F., Mansueto, G., D'Armiento, F. P., Schiano, C., Soricelli, A. & Napoli, C. (2018). Epigenetic hallmarks of fetal early atherosclerotic lesions in humans. *JAMA Cardiology*, 3(12), 1184–1191.

Downum, C. E. (ed.) (2012). *Hisat'sinom. Ancient peoples in a land without water.* School for Advanced Research Press, Santa Fe.

Dyerberg, J., Bang, H., Stoffersen, E., Moncada, S. & Vane, J. (1978). Eicosapentaenoic acid and prevention of thrombosis and atherosclerosis? *The Lancet*, 312(8081), 117–119.

Eaton, S. B., Eaton III, S. B. (1999a). 'Hunter-gatherers and human health', in R. B. Lee & R. Daly (eds), *The Cambridge encyclopedia of hunters and gatherers*, 449–456. Cambridge University Press, Cambridge.

Eaton S. B. & Eaton S. B. (1999b). 'The evolutionary context of chronic degenerative diseases', in S. C. Stearns (ed), *Evolution in health and disease*, pp. 251–259. Oxford University Press, Oxford.

Eiken, M. (1989). X-ray examination of the Eskimo mummies from Qilakitsoq. *Meddelelser Om Grønland (Man and Society)*, 12, 58–68.

Elton, S. (2008). 'Environments, adaptation, and evolutionary medicine', in S. Elton & P. O'Higgins (eds), *Medicine and evolution: current applications, future prospects*, pp. 9–33. CRC Press, Boca Raton.

Espinola-Klein, C., Rupprecht, H. J., Blankenberg, S., Bickel, C., Kopp, H., Victor, A., Hafner, G., Prellwitz, W., Schlumberger, W. & Meyer, J. (2001). Impact of infectious burden on extent and long-term prognosis of atherosclerosis. *Circulation*, 105, 15–21.

Finch, C. E. (2007). *The biology of human longevity. Inflammation, nutrition, and aging in the evolution of lifespans*. Amsterdam, Netherlands: Academic Press.

Finch, C. E. (2010). Evolution of the human lifespan and diseases of aging: Roles of infection, inflammation, and nutrition. Sackler colloquium. *Proceedings of the National Academy of Sciences of the United States of America*, 107(Suppl 1), 1718–1724.

Finch, C. E. & Crimmins, E. M. (2004). Inflammatory exposure and historical changes in human life-spans. *Science*, 305(5691), 1736–1739.

Fisher, R. H. (1930). *The genetical theory of natural selection*. Clarendon Press, Oxford.

Fong, I. W., Chiu, B., Viira, E., Jang, D. & Mahony, J. B. (2002). Influence of clarithromycin on early atherosclerotic lesions after *Chlamydia pneumoniae* infection in a rabbit model. *Antimicrobial Agents and Chemotherapy*, 46, 2321–2326.

Ford, P. J., Gemmell, E., Chan, A., Carter, C. L., Walker, P. J., Bird. P. S., West, M. J., Cullinan, M. P. & Seymour, G. J. (2006). Inflammation, heat shock proteins and periodontal pathogens in atherosclerosis: an immunohistologic study. *Oral Microbiology and Immunology*, 21, 206–211.

Fornaciari, G., Tognetti, A., Tornaboni, D., Pollina, L. & Bruno, J. (1989). 'Etude histologique, histochimique et ultrastructurelle des momies de la basilique de S. Domenico Maggiore à Naples (XVe–XVIe siècles)', in L. Capasso (ed), *Advances in Paleopathology. Proceedings of the VII European Meeting of the Paleopathology Association (Lyon, September 1988)*, pp. 93–96. Journal of Paleopathology, Monograph Publication 1. Solfanelli, Chieti.

Fruitman, D. S. (2000). Hypoplastic left heart syndrome: Prognosis and management options. *Paediatrics and Child Health*, 5(4), 219–225.

Fröhlich B., Harper A. B. & Gilberg R. (eds). (2002). *To the Aleutians and beyond: the anthropology of William S. Laughlin*. Publications of the National Museum Ethnographic Series, Volume 20. The National Museum of Denmark, Copenhagen.

Fullerton, S. M., Clark, A. G., Weiss, K. M., Nickerson, D. A., Taylor, S. L., Stengård, J. H., Salomaa, V., Vartiainen, E., Perola, M., Boerwinkle, E. & Sing, C. F. (2000). Apolipoprotein E variation at the sequence haplotype level: implications for the origin and maintenance of a major human polymorphism. *American Journal of Human Genetics*, 67, 881–900.

Furie, M. B. & Mitchell, R. N. (2012). Plaque attack: one hundred years of atherosclerosis in The American Journal of Pathology. *The American Journal of Pathology*, 180(6), 2184–2187.

Gaeta, R., Fornaciari, A., Izzetti, R., Caramella, D. & Giuffra, V. (2019). Severe atherosclerosis in the natural mummy of Girolamo Macchi (1648–1734), 'major writer' of Santa Maria della Scala Hospital in Siena (Italy). *Atherosclerosis*, 280, 66–74.

Garcia, A. R., Finch, C. E., Gatz, M., Kraft, T. S., Cummings, D., Charifson, M., Buetow, K., Beheim, B. A., Allayee, H., Thomas, G. S., Stieglitz, J., Gurven, M. D., Kaplan, H. & Trumble, B. C. (2021). APOE4 is associated with elevated blood lipids and lower levels of innate immune biomarkers in a tropical Amerindian subsistence population. eLife, 10, e68231. doi: 10.7554/eLife.68231.

GBD 2017 Risk Factor Collaborators. (2018). Global, regional, and national comparative risk assessment of 84 behavioural, environmental and occupational, and metabolic risks or clusters of risks for 195 countries and territories, 1990–2017: a systematic analysis for the Global Burden of Disease Study 2017. *The Lancet*, 392, 1923–1994.

Gurven, M., Jaeggi, A. V., Kaplan, H. & Cummings, D. (2013). Physical activity and modernization among Bolivian Amerindians. *PloS One*, 8(1), e55679. https://doi.org/10.1371/journal.pone.0055679

Haas, N. A., Zelle, M., Rosendahl, W., Zink, A., Preuss, R., Laser, K. T., Gostner, P., Arens, S., Domik, G. & Burchert, W. (2015). Hypoplastic left heart in the 6500-year-old Detmold Child. *The Lancet*, 385(9985), 2432.

Hanlon, C. S. & Rubinsztein, D. C. (1995). Arginine residues at codons 112 and 158 in the apolipoprotein E gene correspond to the ancestral state in humans. *Atherosclerosis*, 112, 85–90

Hansen, J. P. H., Meldgaard, J. & Nordqvist, J. (1991). *The Greenland mummies*. Smithsonian Institution, Washington, DC.

Hooton, E. A. (1930). Finns, Lapps, Eskimos, and Martin Luther. *Harvard Alumni Bulletin*, 32, 545–553.

Hrdlicka, A. (1945). *The Aleutian and Commander Islands and their inhabitants*. Wistar Institute of Anatomy and Biology, Philadelphia.

Ikram, S. (2010). *Ancient Egypt: an introduction*. Cambridge University Press, New York.

Jonsdottir, B. (2002). 'CT scanning of Aleutian mummies', in B. Fröhlich, A. B. Harper & R. Gilberg (eds), *To the Aleutians and beyond: The anthropology of William S. Laughlin*, pp. 155–167. Publications of the National

Museum Ethnographic Series, Volume 20. The National Museum of Denmark, Copenhagen.

Kannel, W. B. (1998). Overview of atherosclerosis. *Clinical Therapeutics*, 20, B2–B17.

Kaplan, H., Thompson, R. C., Trumble, B. C., Wann, L. S., Allam, A. H., Beheim, B., Frohlich, B., Sutherland, M. L., Sutherland, J. D., Stieglitz, J., Rodriguez, D. E., Michalik, D. E., Rowan, C. J., Lombardi, G. P., Bedi, R., Garcia, A. R., Min, J. K., Narula, J., Finch, C. E., Gurven, M. & Thomas, G. S. (2017). Coronary atherosclerosis in indigenous South American Tsimane: a cross-sectional cohort study. *The Lancet*, 389(10080), 1730–1739.

Kelly, R. L. (1995). *The foraging spectrum: diversity in hunter-gatherer lifeways*. Smithsonian Institution, Washington, DC.

Kim, M. J., Kim, Y-S., Oh, C. S., Go, J-H., Lee, I. S., Park, W-K., Cho, S-M., Kim, S-K. & Shin, D-H. (2015). Anatomical confirmation of computed tomography-based diagnosis of the atherosclerosis discovered in 17th-century Korean mummy. *PloS One*, 10(3), e0119474.

Kleivan, I. (1984). 'West Greenland before 1950', in D. Damas (ed), *Handbook of North American Indians Arctic*, pp. 595–621. Smithsonian Institution, Washington, DC.

Laughlin, W. S. (1980). *Aleuts: survivors of the Bering land bridge*. Holt, Rinehart and Winston, New York.

Lee, R. B. & Daly, R. (1999). 'Introduction. Foragers and others', in R. B. Lee & R. Daly (eds), *The Cambridge encyclopedia of hunters and gatherers*, 1–19. Cambridge University Press, Cambridge.

Libby, P. (2021). Inflammation in atherosclerosis—no longer a theory. *Clinical Chemistry*, 67(1), 131–142.

Libby, P. & Hansson, G. K. (2019). From focal lipid storage to systemic inflammation: JACC review topic of the week. *Journal of the American College of Cardiology*, 274(12), 1594–1607.

Linton, M. F., Yancey, P. G., Davies, S. S., Jerome, W. G., Linton, E. F., Song, W. L., Doran A. C. & Vickers, K. C. (2019). *The role of lipids and lipoproteins in atherosclerosis* MDText.com, Inc., South Darmouth. https://www.ncbi.nlm.nih.gov/sites/books/NBK343489/

Longo, B. M. & Yang, W. (2008). Acute bronchitis and volcanic air pollution: a community-based cohort study at Kilauea Volcano, Hawai'i, USA. *Journal of Toxicology and Environmental Health, Part A*, 71, 1565–1571.

Lowenstine, L. J., McManamon, R. & Terio, K. A. (2016). Comparative pathology of aging great apes: bonobos, chimpanzees, gorillas, and orangutans. *Veterinary Pathology*, 53(2), 250–276.

Maixner, F., Turaev, D. & Cazenave-Gassiot, A. (2018). The Iceman's last meal consisted of fat, wild meat, and cereals. *Current Biology*, 28, 2348–2355.

Manning, G. W. (1942). Coronary disease in the ape. *American Heart Journal*, 23, 719–724.

Martina, M. C., Cesarani, F., Boano, R., Roveri, A. M. D., Ferraris, A., Grilletto, R. & Gandini, G. (2005). Kha and Merit: multidetector computed tomography and 3D reconstructions of two mummies from the Egyptian Museum of Turin. *Journal of Biological Research*, 80, 42–44.

Matossian, M. K. (1985). Death in London, 1750–1909. *The Journal of Interdisciplinary History*, 16(2), 183–197.

Matthiassen, T. (1936). *The Eskimo archaeology of Julianehaab District: with a brief summary of the prehistory of the Greenlanders*. Meddelelser om Grønland series, Vol 118. C.A. Reitzels Forlag, Copenhagen.

Mayerl, C., Lukasser, M., Sedivy, R., Niederegger, H., Seiler, R. & Wick, G. (2006). Atherosclerosis research from past to present: on the track of two pathologists with opposing views. *Carl von Rokitansky and Rudolf Virchow Virchows Arch*ives, 449, 96–103. doi: 10.1007/s00428-006-0176-7

Mitter, S. S., Oriá, R. B., Kvalsund, M. P., Pamplona, P., Joventino, E. S., Mota, R. M., Gonçalves, D. C., Patrick, P. D., Guerrant, R. L. & Lima, A. A. (2012). Apolipoprotein E4 influences growth and cognitive responses to micronutrient supplementation in shantytown children from northeast Brazil. *Clinics (Sao Paulo)*, 67, 11–18.

Muhlestein, J. B., Anderson, J. L., Carlquist, J. F., Salunkhe, K., Horne, B. D., Pearson, R. R., Bunch, T. J., Allen, A., Trehan, S. & Nielson, C. (2000). Randomized secondary prevention trial of azithromycin in patients with coronary artery disease: primary clinical results of the ACADEMIC study. *Circulation*, 12, 1755–1760.

Murphy, W. A. Jr., Nedden, Dz., Gostner, P., Knapp, R., Recheis, W. & Seidler, H. (2003). The Iceman: discovery and imaging. *Radiology*, 226(3), 614–629.

Nesse, R. M. & Williams, G. C. (1994). Why we get sick. The new science of Darwinian medicine. Vintage Books, New York.

Nidorf, S. M., Fiolet, A. T., Mosterd, A., Eikelboom, J. W., Schut, A., Opstal, T. S., The, S. H. K., Xu, X. F., Ireland, M. A., Lenderink, T., Latchem, D., Hoogslag, P., Jerzewski, A., Nierop, P., Whelan, A., Hendriks, R., Swart, H., Schaap, J., Kuijper, A. F. M., van Hessen, M. W. J., Saklani, P., Tan, I., Thompson, A. G., Morton, A., Judkins, C., Bax, W. A., Dirksen, M., Alings, M., Hankey, G. J., Budgeon, C. A., Tijssen, J. G. P., Cornel, J. H., Thompson, P. L.; LoDoCo2 Trial Investigators. (2020). Colchicine in patients with chronic coronary disease. *New England Journal of Medicine*, 383(19),1838–1847.

Painschab, M. S., Davila-Roman, V. G., Gilman, R. H., Vasquez-Villar, A. D., Pollard, S. L., Wise, R. A., Miranda, J. J. & Checkley, W. (2013). CRONICAS Cohort Study Group. Chronic exposure to biomass fuel is associated with increased carotid artery intima-media thickness and a higher prevalence of atherosclerotic plaque. *Heart*, 99(14), 984–991.

Piombino-Mascali, D., Zink, A.R. & Panzer, S. (2017). Paleopathology in the Piraino mummies as illustrated by X-rays. *Anthropological Science*, 125(1), 25–33.

Pollard, T. M. (2008) *Western diseases: an evolutionary perspective*. Cambridge University Press, Cambridge.

Reszka, E., Jegier, B., Wasowicz, W., Lelonek, M., Banach, M. & Jaszewski, R. (2008). Detection of infectious agents by polymerase chain reaction in human aortic wall. *Cardiovascular Pathology*, 17, 297–302.

Ridker, P. M., Everett, B. M., Thuren, T., MacFadyen, J. G., Chang, W. H., Ballantyne, C., Fonseca, F., Nicolau, J., Koenig, W., Anker, S. D., Kastelein, J. J. P., Cornel, J. H., Pais, P., Pella, D., Genest, J., Cifkova, R., Lorenzatti, A., Forster, T., Kobalava, Z., Vida-Simiti, L., Flather, M., Shimokawa, H., Ogawa, H., Dellborg, M., Rossi, P. R. F., Troquay, R. P. T., Libby, P., Glynn, R. J.; CANTOS Trial Group. (2017). Anti-inflammatory therapy with Canakinumab for atherosclerotic disease. *New England Journal of Medicine*, 377, 1119–1131.

Roberts, C. A. (2007). A bioarchaeological study of maxillary sinusitis. *American Journal of Physical Anthropology*, 133, 792–807.

Roberts, C. A., Lucy, D. & Manchester, K. (1994). Inflammatory lesions of ribs: an analysis of the Terry Collection. *American Journal of Physical Anthropology*, 95(2), 169–182.

Rollo, F., Ubaldi, M., Ermini, L. & Marota, I. (2002). Ötzi's last meals: DNA analysis of the intestinal content of the Neolithic glacier mummy from the Alps. *Proceedings of the National Academy of Science of the United States of America*, 99(20), 12594–12599.

Rosenfeld, M. E. & Campbell, L. A. (2011). Pathogens and atherosclerosis: update on the potential contribution of multiple infectious organisms to the pathogenesis of atherosclerosis. *Thrombosis and Haemostasis*, 106, 858–867.

Roth, G. A., Mensah, G. A., Johnson, C. O., Addolorato, G., Ammirati, E., Baddour, L. M., Barengo, N. C., Beaton, A. Z., Benjamin, E. J., Benziger, C. P., Bonny, A., Brauer, M., Brodmann, M., Cahill, T. J., Carapetis, J., Catapano, A. L., Chugh, S. S., Cooper, L. T., Coresh, J., Criqui, M., DeCleene, N., Eagle, K. A., Emmons-Bell, S., Feigin, V. L., Fernández-Solà, J., Fowkes, G., Gakidou, E., Grundy, S. M., He, F. J., Howard, G., Hu, F., Inker, L., Karthikeyan, G., Kassebaum, N., Koroshetz, W., Lavie, C., Lloyd-Jones, D., Lu, H. S., Mirijello, A., Temesgen, A. M., Mokdad, A., Moran, A. E., Muntner, P., Narula, J., Neal, B., Ntsekhe, M., Moraes de Oliveira, G., Otto, C., Owolabi, M., Pratt, M., Rajagopalan, S., Reitsma, M., Ribeiro, A. L. P., Rigotti, N., Rodgers, A., Sable, C., Shakil, S., Sliwa-Hahnle, K., Stark, B., Sundström, J., Timpel, P., Tleyjeh, I. M., Valgimigli, M., Vos, T., Whelton, P. K., Yacoub, M., Zuhlke, L., Murray, C., Fuster, V.; GBD-NHLBI-JACC Global Burden of Cardiovascular Diseases Writing Group. (2020). Global burden of cardiovascular diseases and risk factors 1990-2019: update from the GBD 2019 Study. *Journal of the American College of Cardiology*, 76(25), 2982–3021.

Ruffer, M. A. (1910). Remarks on the histology and pathological anatomy of Egyptian mummies. *Cairo Scientific Journal*, 4, 40.

Ruffer, M. A. (1911). On arterial lesions found in Egyptian mummies (1580 BC–535 AD). *Journal of Pathology and Bacteriology*, 16, 453–462.

Ryan, P., Cartwright, C. & Spencer, N. (2012). Archaeobotanical research in a pharaonic town in ancient Nubia. *British Museum Technical Research Bulletin*, 6, 97–107.

Santos, A. L. & Roberts, C. A. (2006). Anatomy of a serial killer: differential diagnosis of tuberculosis based on rib lesions of adult individuals from the Coimbra Identified Skeletal Collection, Portugal. *American Journal Physical Anthropology*, 130, 38–49.

Shin, D. H., Oh, C. S., Hong, J. H., Kim, Y., Lee, S. D. & Lee, E. (2017). Paleogenetic study on the 17th-century Korean mummy with atherosclerotic cardiovascular disease. *PloS One*, 12(8), e0183098.

Sing, C. F. & Davignon, J. (1985). Role of the apolipoprotein E polymorphism in determining normal plasma lipid and lipoprotein variation. *American Journal of Human Genetics*, 37, 268–285.

Stary, H. C., Chandler, A. B., Dinsmore, R. E., Fuster, V., Glagov, S., Insull, W. Jr., Rosenfeld, M. E., Schwartz, C. J., Wagner, W. D. & Wissler, R. W. (1995). A definition of advanced types of atherosclerotic lesions and a histological classification of atherosclerosis. A report from the Committee on Vascular Lesions of the Council on Arteriosclerosis, American Heart Association. *Circulation*, 92(5), 1355–1374.

Swynghedauw, B. (2016). 'Evolutionary paradigms in cardiology: the case of chronic heart failure', in A Alvergne, C Jenkinson and C Faurie (eds), *Evolutionary thinking in medicine: from research to policy and practice*, pp. 137–154. Springer, Cham.

Tardif, J. C., Kouz, S., Waters, D. D., Bertrand, O. F., Diaz, R., Maggioni, A. P., Pinto, F. J., Ibrahim, R., Gamra, H., Kiwan, G. S., Berry, C., López-Sendón, J., Ostadal, P., Koenig, W., Angoulvant, D., Grégoire, J. C., Lavoie, M. A., Dubé, M. P., Rhainds, D., Provencher, M., Blondeau, L., Orfanos, A., L'Allier, P. L., Guertin, M. C. & Roubille, F. (2019). Efficacy and safety of low-dose colchicine after myocardial infarction. *New England Journal of Medicine*, 381, 2497–2505.

Thomas, G. S. (2007). President's message: the global burden of cardiovascular disease. *Journal of Nuclear Cardiology*, 14(4), 621–622.

Thompson, R. C., Allam, A. H., Lombardi, G. P., Wann, L. S., Sutherland, M. L., Sutherland, J. D., Soliman, M. A., Frohlich, B., Mininberg, D. T., Monge, J. M., Vallodolid, C. M., Cox, S. L., Abd el-Maksoud, G., Badr, I., Miyamoto, M. I., el-Halim Nur el-Din, A., Narula, J., Finch, C. E. & Thomas, G. S. (2013). Atherosclerosis across 4000 years of human history: the HORUS study of four ancient populations. *The Lancet*, 381, 1211–1222.

Thompson, R. C., Allam, A. H., Zink, A., Wann, L. S., Lombardi, G. P., Cox, S. L., Frohlich, B., Sutherland, M. L., Sutherland, J. D., Frohlich, T. C., King, S. I., Miyamoto, M. I., Monge, J. M., Valladolid, C. M., El-Halim Nur El-Din, A., Narula, J., Thompson, A. M., Finch, C. E. & Thomas, G. S. (2014). Computed tomographic evidence of atherosclerosis in the mummified remains of humans from around the world. *Global Heart*, 9, 2.

Thompson, R. C., McGhie, A. I., Moser, K. W., O'Keefe, J. H. Jr., Stevens, T. L., House, J., Fritsch, N. & Bateman, T. M. (2005). Clinical utility of coronary calcium scoring after nonischemic myocardial perfusion imaging. *Journal of Nuclear Cardiology*, 12(4), 392–400.

Thompson, R. C., Thompson, E. C., Narula, N., Wann, L. S., Kwong, I., Sutherland, M. L., Sutherland, J. D., Narula, J., Allam, A. H., France, C. A. M., Michalik, D. E., Lunde, D., Finch, C. E., Thomas, G. S., Hunt, D. R. & Frohlich, B. (2020). Imaging atherosclerosis in great apes. *JACC Cardiovascular Imaging*, 14(6), 1275–1277.

Trumble, B. C. & Finch, C. E. (2019). The exposome in human evolution: from dust to diesel. *The Quarterly Review of Biology*, 94(4), 333–394.

Trumble B. C., Stieglitz, J., Blackwell, A. D., Allayee, H., Beheim, B., Finch, C. E., Gurven, M. & Kaplan, H. (2017). Apolipoprotein E4 is associated with improved cognitive function in Amazonian forager-horticulturalists with a high parasite burden. *Federation of American Societies for Experimental Biology*, 31, 1508–1515.

UK Data Archive. (2021). SN 3552 - Causes of death in England and Wales, 1851–60 to 1891–1900: the decennial supplements. http://doc.ukdataservice.ac.uk/doc/3552/mrdoc/pdf/guide.pdf

US Coast Pilot. (1947). *Alaska part II: Yakutat Bay to Arctic Ocean*, 5th edn. US Department of Commerce, Washington DC.

Van Exel, E. Koopman, J. J. E., van Bodegom, D., Meij, J. J., de Knijff, P., Ziem, J. B., Finch, C. E. & Westendorp, R. G. J. (2017). Effect of *APOE* e4 allele on survival and fertility in an adverse environment. *PLoS One*, 12, e0179497. doi: 10.1371/journal.pone.0179497

Varki, N. Anderson, D., Herndon, J. G., Pham, T., Gregg, C. J., Cheriyan, M., Murphy, J. Strobert, E., Fritz, J.,

Else, J. G. & Varki, A. (2009). Heart disease is common in humans and chimpanzees, but is caused by different pathological processes. *Evolutionary Applications*, 2, 101–112. doi: 10.1111/j.1752-4571.2008.00064.x

Vastesaeger, M. M. & Delcourt, R. (1961). Spontaneous atherosclerosis and diet in captive animals. *European Review of Nutrition and Dietetics*, 3, 174–188.

Vasunilashorn, S., Crimmins, E. M., Kim, J. K., Winking, J., Gurven, M., Kaplan, H. & Finch, C. E. (2010). Blood lipids, infection, and inflammatory markers in the Tsimane of Bolivia. *American Journal of Human Biology*, 22, 731–740.

Virani, S. S., Alonso, A., Aparicio, H. J., Benjamin, E. J., Bittencourt, M. S., Callaway, C. W., Carson, A. P., Chamberlain, A. M., Cheng, S., Delling, F. N., Elkind, M. S.V., Evenson, K. R., Ferguson, J. F., Gupta, D. K., Khan, S. S., Kissela, B. M., Knutson, K. L., Lee, C. D., Lewis, T. T., Liu, J., Loop, M. S., Lutsey, P. L., Ma, J., Mackey, J., Martin, S. S., Matchar, D. B., Mussolino, M. E., Navaneethan, S. D., Perak, A. M., Roth, G. A., Samad, Z., Satou, G. M., Schroeder, E. B., Shah, S. H., Shay, C. M., Stokes, A., Van Wagner, L. B., Wang, N. Y., Tsao, C. W.; American Heart Association Council on Epidemiology and Prevention Statistics Committee and Stroke Statistics Subcommittee. (2021). Heart disease and stroke statistics—2021 update: a report from the American Heart Association. *Circulation*, 143(8), e254–e743.

Walker, R. A. & Lovejoy, C. O. (1985). Radiographic changes in the clavicle and proximal femur and their use in the determination of skeletal age at death. *American Journal of Physical Anthropology*, 68(1), 67–78.

Wann, L. S., Narula, J., Blankstein, R., Thompson, R. C., Frohlich, B., Finch, C. E. & Thomas, G. S. (2019). Atherosclerosis in 16th-century Greenlandic Inuit mummies. *JAMA Network Open*, 2(12).

World Health Organization (WHO). (2018). Household air pollution and health. https://www.who.int/newsroom/fact-sheets/detail/household-air-pollution-and-health

Williams, G. C. (1966). *Adaption and natural selection*. Princeton University Press, Princeton.

Wozniak, M. A., Itzhaki, R. F., Faragher, E. B., James, M. W., Ryder, S. D., Irving, W.L.; Trent HCV Study Group. (2002). Apolipoprotein E-epsilon 4 protects against severe liver disease caused by hepatitis C virus. *Hepatology*, 36, 456–463.

Yassine, H. & Finch, C. E. (2020). APOE alleles and diet in brain aging and AD (review). *Frontiers in Neuroscience*, 12, 150.

Zalar, D-M., Pop, C., Buzdugan, E., Todea, D. & Mogosan, C. I. (2019). The atherosclerosis-inflammation relationship–a pathophysiological approach. *Farmacia*, 67(6), 941–947.

Zimmerman, M. & Aufderheide, A. (1984). The frozen family of Utqiagvik: the autopsy findings. *Arctic Anthropology*, 21, 53–64.

Zimmerman, M. R. & Smith, G. S. (1975). A probable case of accidental inhumation of 1,600 years ago. *Bulletin of the New York Academy of Medicine*, 51(7), 828–837.

Zimmerman, M. R., Trinkaus, E., LeMay, M., Aufderheide, A. C., Reyman, T. A., Marrocco, G. R., Ortel, R. W., Benitez, J. T., Laughlin, W. S., Horne, P. D., Schultes, R. E. & Coughlin, E. A. (1981). The paleopathology of an Aleutian mummy. *Archives of Pathology and Laboratory Medicine*, 105, 638–641.

Zimmerman, M. R., Yeatman, G. W., Sprinz, H. & Titterington, W. P. (1971). Examination of an Aleutian mummy. *Bulletin of the New York Academy of Medicine*, 47 (1), 80–103.

CHAPTER 13

Connecting palaeopathology and evolutionary medicine to cancer research: past and present

Carina Marques, Zachary Compton and Amy M. Boddy*

13.1 Introduction

Cancer is a global disease with an overwhelming effect at multiple levels of our contemporary society, including individual, sociocultural, public health, and economic impacts, reaching an estimated 18.1 million active cases and 9.6 million cancer-related deaths in 2018 (Bray & Soerjomataram, 2019a). Although today's pervasiveness and high mortality are experienced worldwide, human populations vary in the burden of oncological disorders from a geographic and socio-economic perspective (Bray & Soerjomataram, 2019a). For example, cancer mortality, but not incidence, in the United States has shown a downward trend with a decline of 27% in deaths since the 1990s, whereas a significant rise in new cancer cases and mortality has been noted in low- and middle-income countries, and in certain geographic regions. Asia (57%) leads the global burden of cancer deaths, followed by Europe (20%), the Americas (14%), Africa (7%) and Oceania (0.7%) (Bray & Soerjomataram, 2019a). There is a general assumption

that the cancer prevalence was relatively low in past human populations, suggesting a link between modern lifestyles and the significant rise in cancer cases (Capasso, 2005; David & Zimmerman, 2010). In the first part of this chapter, we provide a comprehensive overview on why humans are vulnerable to cancer from an evolutionary perspective, while in the latter sections we re-evaluate the relatively low estimates of cancer in past human populations and suggest cancer in the past may not be as rare as it is often described. The palaeopathological data and evidence herein presented aims to contribute to the discussion on the evolutionary and environmental mismatches hypothesis (Box 13.1). Section 13.2 reviews the contribution of evolutionary medicine (EM) and palaeopathology to the field of cancer research. This work highlights the overlapping features of both disciplines and aims to provide a new perspective on cancer. To fully understand cancer's evolutionary trajectory, we must look to the past.

Box 13.1 Not a single disease

Cancer is not a single disease and instead the term encompasses multiple and heterogeneous conditions (Gluckman et al., 2016; Weinberg, 2014), each with its own biological properties, risk factors and outcomes. The fact that cancer is not a single disease complicates how

---

* The authors express their gratitude to the editors and reviewers. CM was supported by the Research Center for Anthropology and Health (CIAS), University of Coimbra, Portugal. AMB and ZC were supported by the National Cancer Institute of the National Institutes of Health under Award Number U54CA217376. The content is solely the responsibility of the authors and does not necessarily represent the official views of the National Institutes of Health.

Carina Marques et al., *Connecting palaeopathology and evolutionary medicine to cancer research*. In: *Palaeopathology and Evolutionary Medicine*. Edited by Kimberly A. Plomp et al., Oxford University Press. © Oxford University Press (2022). DOI: 10.1093/oso/9780198849711.003.0013

we can quantify cancer as a disease burden. The multiplicity of disease phenotypes that cancer can assume implicates significant barriers in studying it across populations. Indeed, cancer rates and origin tissue can vary by socio-economic and geographic regions, with lung (18%), colorectal (9%), stomach (8%), liver (8%) and female breast (7%) cancers leading the mortality statistics globally (Bray & Soerjomataram, 2019a). There is also variation in cancer subtypes at the population level. For example, infectious-related cancers (e.g. cervical cancer) are predominant in sub-Saharan Africa (30–50% of all cases for this region), whereas infectious-related cancers are relatively minor in Europe and North America (3–5% of all cases for this region) (Bray & Soerjomataram, 2019b). Further, lifestyle factors may influence cancer rates across different populations, for example, reproductive cancers, such as breast and prostate cancer, having a much higher prevalence in high-income countries (54.4% and 37.5%, respectively) compared to low- or medium-income countries (31.1% and 11.4%, respectively) (Bray & Soerjomataram, 2019a). Multiple interrelated factors underlie these differences in cancer risk across populations, such as aging, diet, reproductive patterns, population growth, development of public health programmes of prevention and early detection, environmental exposures, as well as genetic and sociocultural factors (Bray & Soerjomataram, 2019a,b; Weinberg, 2014).

The last few decades of biomolecular and clinical research have produced an extraordinary body of knowledge on the biological pathways, genetic mutations, environmental risk factors and sociocultural dynamics that lead to carcinogenesis (Weinberg, 2014). Despite these major advances, cancer incidence, morbidity and, in some contexts, mortality, are rising in many contemporary populations, strongly indicating that oncological research may benefit from new perspectives (Thomas et al., 2013). Perspectives in EM and palaeopathology can contribute to this transdisciplinary endeavour by developing models that evaluate how biological, ecological and sociocultural dynamics have prompted the development of cancer across deep time.

13.2 Cancer in EM

Evolutionary theory provides an important theoretical foundation for cancer that is useful at different levels of biological organisation, including somatic evolution at the cellular level, individual/population specific exposures and

vulnerabilities to cancer in organismal evolution. Evolution can explain why we get cancer and why cancer is so difficult to treat (Aktipis & Nesse, 2013; Gluckman et al., 2016; Greaves & Aktipis, 2016; Hanahan & Weinberg, 2014). A common theme in EM is the distinction between proximate and ultimate causes of a disease/phenotype, originating from the work of Ernst Mayr (1961), who suggested the distinctions between 'how' (proximate) and 'why' (ultimate) in biology ask very different questions (Nesse, 2019). The 'how' question is the immediate cause, while the 'why' question emphasises adaptive function. After this, Niko Tinbergen (1963) expanded on the proximate/ultimate paradigm to encompass different time scales, including ontogeny and phylogeny, which is now commonly known as 'Tinbergen's Four Questions' (Tinbergen, 1963). Individuals working within EM find this framework useful, as it moves medicine beyond mechanisms and can help provide new perspectives to disease vulnerability. Exploring answers to all four of Tinbergen's questions can provide a fuller picture of the direct mechanisms, developmental processes, evolutionary history and adaptive significance of the pathology. We review Tinbergen's four questions in the context of cancer biology (Figure 13.1) to help us understand cancer from both a mechanistic and an evolutionary perspective. It should be noted that while we are using this distinction between form and function as an exercise to understand cancer, it is a heterogenous condition with multiple intrinsic and extrinsic risk factors.

13.2.1 Proximate and ultimate perspectives on cancer biology

When considering mechanism (causation), we must ask what the molecular components are that produce the phenotype in question (Figure 13.1). A neoplasm is a mass of over-proliferating cells that lack the normal cellular controls on replication. Neoplasms can be benign or transform to malignancy, that is, invade the local tissue site and/or metastasise (travel to distant sites from the primary origin). Cancer is another term for malignant neoplasms and is thought to be derived from mutation accumulation (Weinberg, 2014). Cells accumulate mutations

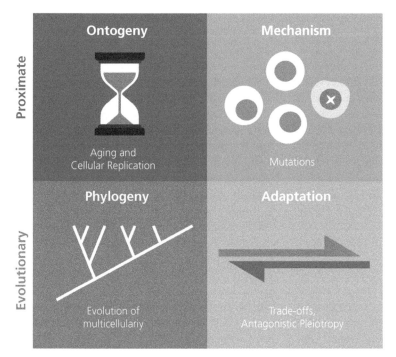

Figure 13.1 Tinbergen's Four Questions for Cancer Biology.
Illustration of the two proximate and two evolutionary perspectives for cancer biology. The proximate explanation describes cancer from a mechanistic point of view. Cancer is a result of genetic/epigenetic mutations that lead to over proliferation and disrupted cellular growth. Most species have cancer defense mechanisms, including DNA repair and immunosurveillance, and as such, cancer is a disease of aging populations. From an evolutionary perspective, cancer can be traced back to the evolution of multicellularity. There are beneficial adaptations, trade-offs and/or pleiotropic effects that may increase the risk for cancer.

(referred to as somatic mutations) during normal cellular replication and most of these mutations are harmless. However, the accumulation of mutations, both genetic and epigenetic, can eventually lead to dysregulation of normal cellular proliferation and invasion, especially when mutations disrupt genes important for suppressing cellular division (i.e. tumour suppressor genes) and promoting cellular division (i.e. oncogenes). Neoplasms can occur in any tissue type within the body and originate from different cell types. Importantly, cancer cells from different tissues and cell types all virtually lead to the same phenotypes, categorised as the 'hallmarks of cancer' (Hanahan & Weinberg, 2014; Weinberg, 2014). These characteristics include sustaining a proliferative signal, evading growth suppressors, avoiding immune destruction, inducing blood vessel growth (angiogenesis), genome instability and mutation, resisting cellular death and deregulating cellular energetics. Recently, there has been a significant debate in the field of cancer biology on what the driving mechanism for cancer risk is, including whether intrinsic (i.e. cellular division) or extrinsic (e.g. lifestyle, environmental exposures) play a bigger role in the somatic mutations that lead to

carcinogenesis. According to the 'Bad Luck' hypothesis, a correlation exists between cancer risk per tissue and lifetime number of stem cell divisions within each tissue. This hypothesis suggests cancer risk among tissue types can be explained by 'bad luck' mutations (i.e. intrinsic deleterious mutations accumulating during DNA replication) instead of lifestyle and hereditary genetic factors (Tomasetti & Vogelstein, 2015; Tomasetti et al., 2017). However, this hypothesis has been debated as a potential oversimplification because various tissue types may differ in the accumulation of cancer-causing mutations (Noble et al., 2015; Rozhok et al., 2015). Indeed, not all cells (or tissues) in the multicellular body are created equal and they differ in somatic mutation rates and cancer defences.

From an ontogenetic (developmental) perspective, we seek to know how the phenotype changes with age, and which developmental steps contribute to the phenotype (Figure 13.1). Our trillions of cells are constantly being exposed to DNA damaging agents within our natural environment, including radiation from the sun, free radicals from oxidative stress and even our own immune system causing DNA damage during chronic inflammatory conditions (Jaiswal et al., 2000). Yet, cancer is characterised as an age-related disease and cancer diagnoses in childhood are relatively rare. Indeed, our robust intrinsic defences protect us from developing cancer at younger ages. Along with the evolution of multicellularity, organisms have co-evolved robust mechanisms of cancer suppression and prevention (Aktipis et al., 2015), including redundant cell cycle control mechanisms, DNA repair mechanisms, surveillance of neoplastic cells by the immune system and even the physical structure of tissue within the body (reviewed in Harris et al., 2017). Most cancers affect individuals later in life due to the eventual breakdown of these cancer defence mechanisms, including the disruption of normal tissue homeostasis (Bissel & Hines, 2011). Cancer progression is a complex process that develops across many years of an individual's life. It is not aging, per say, that leads to cancer, but the relationship between aging, the accumulation of mutations and the tissue microenvironment that harbours potentially mutant cells. Many 'normal' tissues may harbour oncogenic mutations (Martincorena, 2019) but the surrounding microenvironment, or 'healthy' neighbour cells, may inhibit cell growth and suppress the malignant phenotype. As an individual ages, there is an increased likelihood for the accumulation of oncogenic mutations as well as disruptions in normal tissue homeostasis, eventually leading to clonal escape and expansion of the neoplastic cells (Weinberg, 2014).

Tinbergen's (1963) third question encompasses phylogeny (evolutionary history) and asks where the phenotype evolved on the tree of life (Figure 13.1). Evolutionary theory has been tremendously useful in understanding the initiation and progression of human neoplasms (Greaves & Maley, 2012; Merlo et al., 2006). Recently, the emergent field of comparative oncology seeks to implement evolution as an important framework for understanding cancer—not as a process within the individual, but in appreciating its implications and emergence across species (Nunney et al., 2015). Cancer is a threat to almost all multicellular species across the tree of life (Aktipis et al., 2015; Greaves, 2000), and neoplastic formation has been a persistent selective pressure since the dawn of multicellularity. The pervasiveness of cancer across the tree of

life is not limited to extant species, as fossilised neoplasia have been documented in two Jurassic period taxa, one being an *Apatosaurus*, a sauropod dinosaur (Capasso, 2005), and the other so far unidentified (Rothschild et al., 1999). However, the ubiquity of neoplastic pathologies is not limited to the vertebrate phylogeny (De Vico & Carella, 2014; Robert, 2010; Yevich & Barszcz, 1978). All species experience the selective pressure of neoplasia as the maintenance of healthy, functional and cooperative somatic cells was a central barrier to the evolutionary transition to multicellularity (Bertolaso & Dieli, 2017; Szathmáry & Smith, 1995). While there is a demonstrable risk of neoplasia across life, we can expect the relevance of such risk to scale with key life history factors. Life history theory, the slow–fast continuum of pace and scale of reproductive strategies, is a crucial foundation upon which hypotheses regarding the variance of cancer risk across species can be made. Positing the assumption that every somatic cell carries some risk of suffering a cancerous transformation, then lifetime cancer risk should increase steadily with an increase in species' body mass and longevity (Caulin & Maley, 2011). However, cancer prevalence rates across extant species demonstrates that this is not the case—a phenomenon known as Peto's paradox (Caulin & Maley, 2011; Ducasse et al., 2015). The surprisingly low cancer rates of the modern large, long-lived mammals remain as the most illuminating evidence of the force of selection against neoplastic growth.

While epidemiological cancer data in wild and zoological animals are limited, there have been a few studies to attempt to quantify cancer in non-human animals (Effron, 1977). Results from these observations show mammals appear to be more vulnerable to cancer than other vertebrates (Boddy et al., 2020a). Within mammals, species range in cancer prevalence, with estimates ranging from relatively cancer-free species, such as the naked-mole rat (Tian et al., 2013), to cancer-prone species, such as the Virginia opossum, with malignancy prevalence estimates as high as 53% (Boddy et al., 2020a). Within this limited dataset, current data confirm Peto's observation and life history predictions: bigger, long-lived animals do not get more cancer than smaller, short-lived animals, despite having more cells with more opportunities to accumulate cancer-causing mutations. Researchers propose that slow life history organisms must have evolved heightened cancer defence mechanisms to maintain the soma of large, long-lived multicellular bodies. An example of this is the discovery of extra copies of TP53, a critical tumour suppressor gene, in elephants, with functional studies identifying this gene expansion as a potential mechanism of cancer defence (Abegglen et al., 2015). Within a life history framework, the pace of human growth and reproduction places humans on the slow end of the slow–fast continuum. Reflecting on the cancer prevalence trends we see across vertebrates, humans are burdened with a higher cancer prevalence than species with a similar life history pace (Kokko & Hochberg, 2015). However, we argue that more data are needed to confirm these trends, as there is likely a bias in the diagnosis of human cancers compared to other animals. Indeed, it may be the case that the more carefully we screen for cancer, the more we are likely to find.

The lessons of life history theory bolster our understanding of human cancer biology even further when we shine the light of comparative oncology onto our closest living phylogenetic relatives, the non-human primates. With human's lifetime cancer risk of nearly 30% and 40% for females and males, respectively, if cancer risk is intrinsically derived, we expect a prevalence that mirrors that in the remainder of the extant primate species. Contrary to this assumption, a six-year study of post-mortem examinations in 1066 nonhuman primates at the Yerkes Primate Center yielded only six neoplasms (McClure, 1973). Although all the neoplasms were malignant, none of them aligned with the most common human malignancies. Upon examination, it was found that the primate neoplasms were primarily sarcomas or lymphomas, whereas cancers of the epithelia remain the lead actors in human neoplastic diseases (Varki & Varki, 2015). This decrease in prevalence compared to humans is even more astounding considering the dramatic increase in lifespan that modern veterinarian care affords to managed primate populations. The significant discrepancy in neoplastic type and frequency suggests that environmental exposure, rather than differences in evolutionary lineage, play the most significant role in cancer risk among primates. However,

given the absence of reliable data collected from the wild and the rarity of expansive necropsy studies like the one outlined here, additional non-human primate data are needed to draw conclusions on these patterns.

Last, Tinbergen's (1963) questions address adaptation (function): how the phenotype impacts survival and reproduction (i.e. fitness) (Figure 13.1). If cancer is evolutionarily ancient, why does the problem still persist today? Why haven't more species evolved more efficient DNA repair or higher sensitivity to DNA damage? The main hypothesis explaining this today is that many biological processes may contribute to the cancer phenotype due to trade-offs in function and energetics of the biological organism (Aktipis et al., 2015; Boddy et al., 2020b). In other words, cancer defences, such as immune surveillance, cell cycle control and DNA repair are costly and subject to trade-offs. These trade-offs in somatic maintenance (and cancer defences) depend on environmental availability of resources and extrinsic mortality conditions. In addition to the energetic trade-offs described, we may also be vulnerable to cancer due to constraints of natural selection operating on differential reproductive success. In other words, evolution will select traits that benefit reproduction over somatic maintenance and health. For example, within a given population, natural selection may favour traits important for early reproduction (such as fast growth). If this trait gives an advantage early in life, it will tend to be selected for, irrespective to the deleterious, late-onset effects. This term is referred to as 'antagonistic pleiotropy' in evolutionary biology and can explain why we see so many age-related diseases, including cancer (Byars & Voskarides, 2020). Indeed, the observation that mammals get more cancer than other vertebrates may be a case of antagonistic pleiotropy, where the reproductive benefits of internal gestation outweigh the potential cost to the mammalian body. These evolutionary explanations provide mechanisms for the persistence of cancer promoting genes in populations (Campisi, 2013; Crespi & Summers, 2006). Some examples of antagonistic pleiotropy in cancer biology include tumour suppressor variants that may also functionally contribute to increased fertility, such as BRCA 1/2 (Smith et al., 2012),

TP53 (Kang et al., 2009) and *KISS* (Mumtaz et al., 2017). For example, the human *BRCA* 1/2 genes are tumour suppressor genes involved in DNA damage repair. Mutations that lead to loss of function in *BRCA* increase the lifetime risk of breast and ovarian cancers (Easton et al., 1995; Gayther et al., 1995). One study, under natural fertility conditions in a population in Utah (~ 1930–1975), demonstrated that women who were *BRCA* mutation carriers had more children and a shorter interbirth window (Smith et al., 2012). Results from this study suggest some fertility advantage of *BRCA* mutation carriers and provide a potential explanation for why this variant may persist in the population, even if the variant increases the risk for cancer later in life. The findings of this relationship between *BRCA* mutations and fertility need to be replicated. Furthermore, other factors, such as founder events and genetic bottlenecks, may also contribute to the persistence and relatively high frequency of cancer-causing variants within in a population.

13.2.2 Evolutionary mismatch and cancer

Novel environments may contribute to cancer risk in human and nonhuman populations. In EM, this concept is called evolutionary mismatch and corresponds to when the environment/ecology changes faster than the population can adapt (Gluckman et al., 2016; Greaves & Aktipis, 2016; Hochberg & Noble, 2017). Human societies can reshape many environmental components at rapid rates, including (but not limited to) migration into new ecologies/landscapes with novel pathogens and UV exposure, introduction to novel foods, inhaling carcinogens through smoking, pollution from industrialisation, increase in energetic consumption and demographic transitions that impact fertility rates. Environmental mismatch cannot explain why we are ultimately susceptible to cancer (see Section 13.2.1), although novel environments can explain why certain cancers are increasing in prevalence in certain ecologies (Greaves, 2000). Environmental changes can shift the tissue-specific cancer risks through various mechanisms, such as directly increasing the mutation rate by various DNA damaging agents and indirectly increasing the

mutation rate by increased cellular proliferation (leading to more chances of mutations occurring). Section 13.4 describes how not all novel environmental shifts increase cancer risk, and some may indeed reduce cancer risk, although in the following section we highlight components of the rapidly changing environment that may influence cancer risk.

Common examples of environmental mismatch in human populations and cancer risk include sun exposure and skin cancer (Jablonski & Chaplin, 2010), demographic transitions and breast cancer (Aktipis et al., 2014; Eaton et al. 1994), lack of exposure to a diversity of pathogens and childhood acute lymphoblastic leukaemia (Greaves, 2018), as well as smoking and lung cancer (Bray & Soerjomataram, 2019a) (discussed in Section 13.4). Other examples are cancers originating from infectious diseases that have a higher burden in populations living in environments with high pathogen exposures (Ewald, 2018; Islami & Jemal, 2019). In the comparative oncology literature, there is evidence for high rates of cancer in domesticated species, including dogs, ferrets and chickens (Boddy et al., 2020b). While artificial selection is likely confounding these risks, without cancer prevalence data in non-captive/domestic animals for comparisons, it is difficult to understand if environmental mismatch or genetic bottlenecks are driving these diseases.

Environmental mismatch has been proposed to explain why there is a potential increase in cancer diagnoses in the recent past, but these theoretical predictions lack empirical data from the bioarchaeological record to quantify the temporal trends of cancer risk. Section 13.3 provides a comprehensive overview of cancer in the bioarchaeological literature in a temporal context. With the help of palaeo-oncologists, we can ask, and potentially answer, an important question: is cancer a disease of 'modernity'?

13.3 Cancer in palaeopathology

Cancer prevalence in past populations has been the focus of palaeo-oncologists, consisting of palaeopathologists or bioarchaeologists conducting research on oncological diseases in past peoples.

Palaeopathology can provide a novel perspective on the evolutionary history of cancer in human societies (Marques, 2019; Zuckerman et al., 2016). Evaluating the nature of past oncological diseases holds prospects for assessing the impact of cancer in past human societies, how widespread cancer was in our evolutionary lineage, how the evolution and natural history of cancer was modulated by major shifts in intrinsic and extrinsic risk factors (e.g. ecological or sociocultural conditions) and what tissue types were the most frequently affected by cancer in the past (Strouhal & Němečková, 2009; Brothwell, 2012; Hunt et al., 2018; Marques, 2019).

To our current knowledge, one of the oldest possible cancers in our evolutionary lineage is an osteosarcoma from 1.8–1.6 million year ago. The osteosarcoma was discovered in a fifth metatarsal (Sk 7923) of a hominin fossil (unknown specific taxon), from the site of Swartkrans in the Cradle of Humankind, South Africa (Odes et al., 2016). Within *Homo sapiens*, the oldest evidence of cancer is metastatic bone disease in a female skeleton from Indonesia who lived 16,000–18,000 years ago (Bulbeck, 2005). While the early Holocene seems to be sparse in cancer records, indisputable cancer cases on skeletal remains start to appear between 5000 BCE and 3000 BCE in Europe, Australia, Africa and America, representing at least eight high-confident diagnoses for this period (Brothwell & Sandison, 1967; de la Rúa et al., 1995; Hunt et al., 2018; Lieverse, 2005; Marques, 2019; Strouhal & Kritscher, 1990; Webb, 1995). Case reports of cancer in the palaeopathological literature begin to accrue after this time period, exceeding 250 cases, with the majority documented after 1000 CE (Hunt et al., 2018). Most of these reports refer to diagnosis of metastatic bone disease, specially from carcinomas, followed by bone changes associated with haematologic malignancies (multiple myeloma) and a smaller number of primary bone cancers (Hunt et al., 2018; Marques, 2019).

Despite the solid evidence of malignant neoplasms in past societies, the most widely accepted palaeo-oncological paradigm is the assumption that cancer in past humans was a relatively rare event (Capasso, 2005; David & Zimmerman, 2010). Many attribute this seeming paucity to 'modern' lifestyle risks; exposure to novel carcinogens; declining mortality due to infectious diseases, famine or violence; and extended life expectancy (Capasso, 2005; David & Zimmerman, 2010). Other scholars, in recognising the multiplicity of challenges to diagnose cancer in palaeopathology and the growing body of evidence, suggest that the past prevalence of neoplasms is much higher than previously thought (Hunt et al., 2018; Marques et al., 2018; Nerlich et al., 2006; Waldron, 1996; Zuckerman et al., 2016).

Some of the challenges in estimating cancer prevalence in past populations are intrinsic to the nature of palaeopathological research. The study of skeletal 'populations' excavated from archaeological sites represent a tiny and heterogenous proportion (in frailty and susceptibility to disease or demographic composition) of the once living (Wood et al., 1992). The sample sizes are generally small, human skeletal remains are often poorly preserved and fragmentary and they are seldom surveyed specifically for the identification of neoplastic conditions. Additionally, many cancers affect soft tissues and can be difficult—sometimes impossible—to identify unless they metastasise to the skeletal system. To further complicate matters, the lesions produced by cancer may be more prone to destruction, particularly for fossil and older archaeological sites. Lastly, there are limits in palaeopathological diagnoses, which most often rely only on the skeletal traces of cancer and absence of soft tissue and complementary examinations, that contribute to a likely underestimation of cancer prevalence in skeletal populations (Marques, 2019; Marques et al., 2018; Waldron, 1996).

Despite these challenges, palaeo-oncological research has made great progress in cancer recognition beyond the macroscopic identification in the skeleton and mummified tissues, including ancient DNA (aDNA) extraction, histological and immunohistochemical techniques. For example, aDNA samples extracted from dry bone and amplified via PCR have uncovered common cancer mutations that we see in tumours sequenced today. For example, a male adult skeleton exhumed from a 2500-year-old Scythian burial had evidence of widespread osteoblastic lesions. This individual was diagnosed with prostate cancer and ancient sequencing efforts uncovered hypermethylation of promoter sequences in the tumour suppressor gene, *p14ARF* (Schlott et al., 2007). These data are consistent with current knowledge of hypermethylated genes in prostate tumours (Park, 2010). In addition, *K-RAS* mutations were found in a colorectal adenocarcinoma sample in the mummy of Ferrante I of Aragon (1424–1494) (Forniaciari, 2018). In addition to genomic and histological techniques, studies on proteomics in skeletons with neoplastic lesions have shown great promise in case studies (e.g. Bona et al., 2014; Schultz et al., 2007). For example, Bona and colleagues (2014) identified protein biomarkers associated with osteosarcomas by analysing the osseous lesions observed in a humerus of a female that lived 2000 years ago in Hungary. Future technologies developed for sensitive samples, including aDNA sequencing and proteomics, provide an exciting path forward for palaeo-oncology and the field of evolution and cancer biology.

13.3.1 Case study: can palaeopathology provide information on temporal trends of cancer in ancient humans?

A recent compilation of case reports suggests malignant neoplasia may not be as rare as previously thought. A survey of the palaeopathological literature identified 272 possible cases of malignant neoplasm in skeletal and mummified remains from archaeological sites up to the 1900s (Hunt et al., 2018). For the meta-analysis in this chapter, we have expanded on this previous bioarchaeological survey of malignant neoplasms (Hunt et al., 2018) to include data published in the last three years in books, peer-reviewed articles, bioarchaeological reports, conference proceedings and dissertations. This resulted in 314 'probable' cases in the palaeopathological literature of malignant neoplasms diagnosed in skeletal populations, from more than 240 archaeological sites with diverse chronological and geographic origins. We implemented a meta-analysis with a palaeoepidemiological focus to evaluate time-trends of cancer in pre-1900s skeletal populations excavated from a wide range of archaeological sites and calculated crude and age-adjusted cancer rates. We addressed the question of whether malignant neoplasms are rare in

the palaeopathological record and tested if there are temporal trends as previously predicted. In other words, can palaeopathology provide the data and evidence needed to support the evolutionary mismatch hypotheses proposed in the field of EM?

Most reports of cancer in palaeopathology are published as case studies and lack detailed information required for a palaeoepidemiological approach (Hunt et al., 2018). Data on the total number of skeletons studied, age-at death, sex and time frame are all important measures needed for palaeoepidemiology and are often missing in case study reports. For the purpose of the meta-analysis in this chapter, we surveyed the aforementioned dataset and selected two cohorts for analyses following the criteria described in Figure 13.2a. The first cohort consists of skeletal populations for which it was only possible to calculate cancer crude prevalence due to lack of detailed biodemographic information. The second cohort consists of skeletal populations where the published data available allowed measures of standardisation of prevalence, namely age-adjusted prevalence. The geographic distribution of the first cohort is depicted in Figure 13.2b.

Using palaeoepidemiological methods (Waldron, 2007), we estimated cancer prevalence with the following calculations:

1) crude prevalence (%) of malignant neoplasms observed in the skeleton population/study-base ($n_{cancerskeleton}/n_{totalskeletons}$), herein reported as cancer crude prevalence (CCP);
2) the CCP (%) by sex and age-at-death classes ($n_{cancers}/n_{total}$ within each sex and age class); and
3) age-adjusted prevalence for comparative purposes, since disease prevalence among distinct groups cannot be estimated without accounting for differences in biodemographic composition.

We performed chi-squared tests to assess significance (set at $p \leq 0.05$). To address the small case numbers, 95% confidence intervals (CI) were calculated with the Clopper–Pearson exact method based on the beta distribution. Data analyses were performed using SPSS v. 26, Epitools (http://epitools.ausvet.com.au) and R statistical software (R Core Team, 2014).

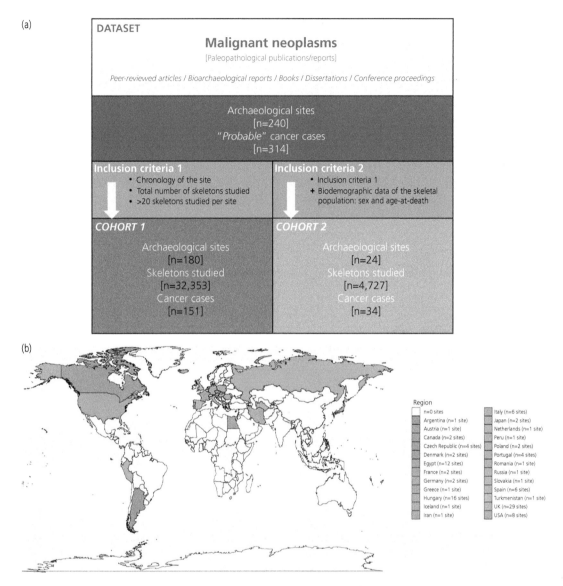

Figure 13.2 Selection criteria for the cohorts and geographic distribution of cases selected for the meta-analysis.
(a) First cohort selection criteria: in a first step of the analysis (cohort one) we selected all dataset entries of skeletal populations from archaeological sites where malignant neoplasms were diagnosed, for which there was also information on the chronology and total number of skeletons observed/studied in that population. We only included archaeological sites with more than twenty skeletons in the total sample. Second cohort selection criteria: for the second cohort (cohort two) we applied an additional filtering step and selected only reports with complete biodemographic baseline data (sex and age-at-death distribution). While this is a smaller dataset than the first cohort, it provides important biodemographic data that enables a palaeoepidemiological approach with age-standardization methods. (b) Geographic distribution of cancer cases surveyed. Selected skeletal populations from archaeological sites (n = 108) with cancer cases reported and information on the total number of skeletons observed by country for cohort one.

Out of the 108 skeletal populations selected for cohort one, we found that the CCP is relatively low per site, but mostly these values are variable, ranging from a minimum of 0.06% to a maximum of 4.6% (Figure 13.3). These results are not surprising due to the limitations of cancer diagnoses in the

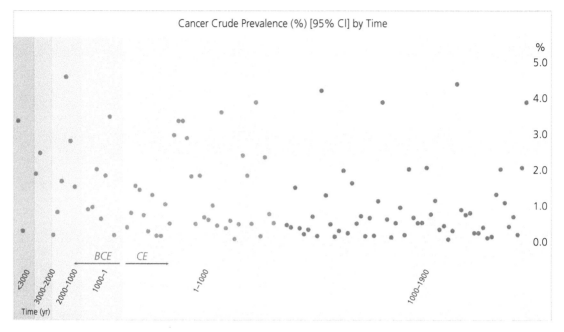

Figure 13.3 Time-trend assessment: cancer crude prevalence (CCP).

CCP ($n_{\sum cancer}/n_{\sum total}*100$) of 108 skeletal population distributed across time (between <3000 years BCE and 1900 CE). Mid-point values were used when there was an overlap of chronology with our defined intervals (< 3000 years BCE; 3000–2000 years BCE; > 2000–1000 years BCE; > 1000–1 BCE; 1–1000 CE; < 1000–1900 CE). The crude prevalence is highly dispersed and variable, ranging from a minimum of 0.06% to 4.6% (Mean = 1.2%, Median = 0.7%, SD = 1.1%, 95%CI = [0.96-1.4]). There is no increment across time.

BCE, Before Common Era; CE, Common Era.

bioarchaeological record (discussed in Section 13.4). The lowest estimate of cancer prevalence comes from a pre-contact Hawaiian site where one cancer case was detected among 1504 skeletons (Suzuki, 1987). The highest estimated prevalence (4.6%) corresponds to one malignant neoplasm detected in a sample of twenty-two skeletons from El-Assasif (Pw In Re, TT-39), Luxor, 1508–1425 BCE (Herrerín et al., 2014).

The distribution of CCP across time (Figure 13.3), starting in 5000 BCE (the earliest skeletal population in this review) until 1900 CE (the latest), does not show a trend of increasing prevalence as might be expected given the mismatch hypothesis and third epidemiological transition. The sum of cancer prevalence between these large temporal intervals (Figure 13.4) is similar (the 95% CI range overlaps) and shows a slight but significant decline for the last period ($\chi^2 = 13.040$; df = 5; $p = 0.023$). In other words, we find no support for an increase in CCP during

the last five millennia from current palaeopathology records.

The absence of a temporal trend in increasing cancer prevalence must be interpreted with caution because there are several limitations that can impede the ability of palaeopathologists to diagnose cancer in skeletal remains, such as methodological constraints, high variability of skeletal completeness and sample sizes across studies (i.e. the total number of skeletons observable) and the rather often poor representation of older time periods. Second, we must consider limitations of statistical estimates and inferences, since we are analysing small sample sizes in terms of the number of cancer cases (Figure 13.4), where the 95% confidence lower and upper limits are wide, indicating a low precision of the observed effect, particularly in older periods (Smith, 2020). Third, the comparison of crude prevalence rates between populations is unreliable when the age structure of the population

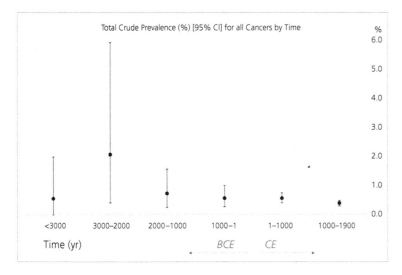

Figure 13.4 Time-trend assessment: total cancer crude prevalence by intervals. Computation of the total CCP ($n_{\sum cancer}/n_{\sum total}*100$) of 108 skeletal population by time intervals (between < 3000 years BCE and 1900 CE) with 95%CI. The total cancer rate is very similar across these time periods, with a slight decline for the latter one and a peak in the 3000–2000 BCE interval. However, these differences are not marked, as shown by the overlap of the 95% CI.

BCE, Before Common Era; CE, Common Era.

is not considered (Waldron,1996, 2007). The first two limitations can be resolved with increased sample sizes in future studies. To overcome the third pitfall, we analysed a second cohort, which is a subset of these skeletal populations where a basic demographic profile was available (n = 24 archaeological sites/skeletal populations, n = 34 cancers, n = 4727 skeletons studied) allowing the application of age-at-death standardisation (Figure 13.5).

We found no trend of increasing cancer prevalence after the age adjusted prevalence. These results show that the disparities in terms of demographic structure are not biasing the time-trend assessment. Current data estimates that childhood cancers (< 15 years) represent between 0.5% and 4.6% of the total number of cancer cases in a given country (Kumar et al., 2015). In support of these estimates, the palaeopathological literature (n = 314 neoplasms reported) reports that 5% (16/314) of malignant neoplasms recorded in skeletal remains are from non-adults (< 20 years), and these are mostly osteosarcomas and leukaemias. However, these values provide limited information if they are not considered within the proportion of adults and non-adults in a particular population. Using the subset of data filtered for known age/sex demographics (n = 24 sites), the age class 40+ years of age-at-death had a significantly higher proportion of cancer cases ($\chi^2 =$

24.377, df = 2, $p < 0.0001$) (Figure 13.6). The skeletons of non-adults with a diagnosis of cancer is 0.1%, whereas the proportion in the age-range of > 20–40 is 0.7% without significant differences in these cohorts. There were no significant differences between the male and female cohorts in the dataset (Figure 13.6).

13.4 Discussion: cancer is not new disease

It is undeniable that a marked increase in cancer lifetime risk, incidence, prevalence and mortality has characterised human populations in some parts of the world undergoing demographic, socioeconomic and epidemiological transitions after the nineteenth century. The first worldwide cancer mortality statistics provided in 1915 report an increasing temporal trend in cancer-related deaths (Hoffman, 1915). Cancer is often referred to as a 'disease of modernity', suggesting that recent lifestyle and environmental factors are mostly responsible for this disease burden. However, comparative data on cancer prevalence suggests that the disease is evolutionarily ancient and has been a health issue for almost all multicellular animals. Furthermore, despite difficulties in diagnosing cancer in archaeological skeletons, cancer is ubiquitous in the palaeopathological record, and

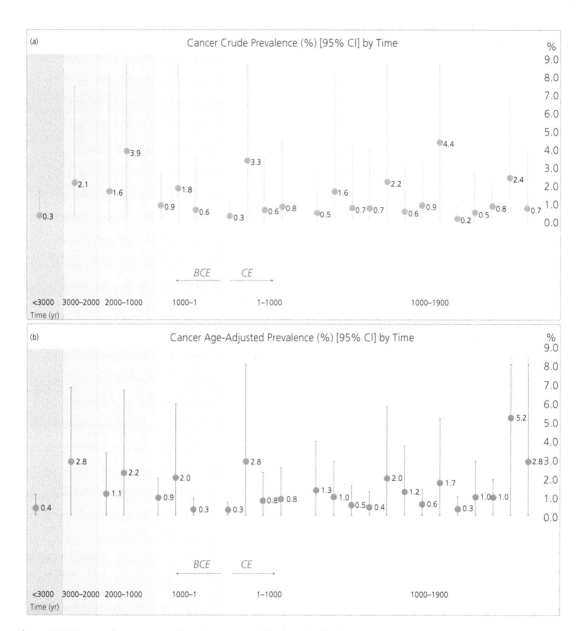

Figure 13.5 Time-trend assessment: crude prevalence *vs* age-at-death standardization.

Top image: CCP ($n_{\sum cancer}/n_{\sum total}$*100) with 95% CI (indicated by the vertical lines), of twenty-four skeletal populations with baseline biodemographic information, distributed across time (between < 3000 years BCE and 1900 CE). No significant differences between the crude prevalence were obtained ($\chi^2 = 33.082$, df = 24, $p = 0.102$). Bottom image: age-standardized rates with 95% CI interval (indicated by the vertical lines) for the same samples. Note that when we compare these skeletal populations taking into account differences in the age-at-death structures, the prevalence does not show marked differences along time. Age-adjusted rates estimated by the direct method of age-standardization (Silva, 1999), using the demographic distribution of the skeletal sample from the Global History of Health Project (GHHP): European Module (Steckel et al., 2019) as the reference population.

BCE, Before Common Era; CE, Common Era; . . . = the CI interval extends beyond the axis. Rates standardized by the direct method to the GHHP-European Module as standard population.

Cancer Crude Prevalence (%) by Sex

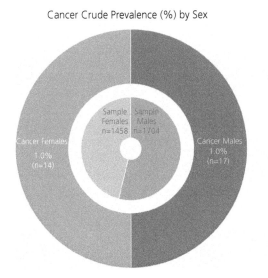

Cancer Crude Prevalence (%) by Age Class (years)

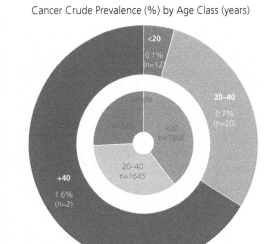

Figure 13.6 Biodemographic patterns.
Left: CCP ($n_{=\sum cancersex}/n_{=\sum totalsex} *100$) among males and females and sample size computed for the twenty-four skeletal populations with baseline biodemographic information beneath. Prevalence was not significantly different between males and females (Females: 1.0%, 95% CI = [0.5—1.5]; Males: 1.0%, 95% CI = [0.5–1.5]; χ^2 = 0.290, df = 1, p = 0.590). Right: CCP ($n_{\sum cancerageclass}/n_{\sum totalageclass}*100$) by age-at death intervals (years) and sample size, computed for the twenty-four skeletal populations with baseline biodemographic information beneath (< 20 years: 0.1%, 95% CI = [0.0–0.4]; > 20–40 years: 0.7%, 95% CI = [0.4–1.3]; +40 years: 1.6%, 95% CI = [1.0–2.5]; χ^2 = 24.377, df = 2, p < 0.000). Of note is that mid-point values to allocate individuals whose age-at-death range overlapped between the age classes herein established. For analyses of sex distribution, we excluded individuals with undetermined/unknown sex and non-adults.

its presence for the last millennia appears at rates higher than expected—a strong indication that cancer was a burden throughout human evolution.

13.4.1 Cancer prevalence is underestimated in palaeopathology

As discussed in Section 13.1, the burden of cancer today is considerable. Here, we review whether cancer is indeed as rare in past human populations as previous estimates suggest. A recent model proposes that a greater than ~ 5% lifetime risk of developing cancer, in a number of species (including humans), is largely attributable to novel environments through evolutionary mismatches (Hochberg & Noble, 2017). These estimates are based on a mathematical model of cancer risk drawn from data on modern forager populations and US cancer statistics. The use of extant forager data to infer the past introduces some assumptions and limitations. Cancer mortality in forager populations is

likely unknown due to lack of medical detection and diagnosis and extant environments cannot completely recapitulate the past (Pontzer et al., 2018). We suggest that the addition of palaeo-oncology records can address the current limitations to these estimates.

The seemingly low frequency of cancers detected in skeletal populations represents an underestimation of their true occurrence in past societies (Marques et al., 2018). Reports of cancer in pala-eopathology suffer from publication bias, with many cases only being published due to their peculiarities in terms of manifestation or chronological and/or geographic context. Additionally, current diagnostic measures also underestimate cancer prevalence. Palaeopathologists screen skeletal series through visual inspection of external bone surfaces. If an abnormal surface feature is identified, indicating possible cancer, then radiologic or histological techniques are performed to confirm diagnosis. Rothschild and Rothschild (1995) reported that the detection ability through radiological examination was nearly three times that of visual direct examination (26.5% and 8.6%, respectively), indicating that current palaeo-oncological prevalence is likely to be severely underestimated.

Furthermore, diagnoses of cancer in palaeopathology rely on soft tissue cancers that have metastasised to bone (some tissue-type cancers are more osteotropic, that is, likely to spread to bone, than others) and on the less-common primary bone malignancies. Marques and colleagues (2018) conducted a base-line study in two Portuguese identified skeletal collections from the nineteenth to the twentieth centuries composed of individuals with known causes of death (recorded in the death certificates). Cancer was the cause of death registered for 8.1% of individuals in this population. However, only 17.6% (95% CI: [11.0–24.1]) of the skeletons with an in vivo diagnosis of cancer (n = 131, mostly stomach, uterus, intestinal/colorectal and prostate cancers) showed bone lesions fully compatible with diagnosis, demonstrating that palaeo-oncological diagnostic sensitivity is low. Based on these diagnostic estimates, we can include this information to further refine cancer mortality estimations in past populations. For example, if the lifetime risk of dying from cancer was ~ 10%, then one would expect to identify between 1.1 and 2.4 cancer cases in an archaeological skeletal population of 100 well-preserved skeletons, based on the 95% CI interval of the 17.6% obtained by Marques and colleagues (2018). This value is not markedly divergent from what we are finding in most archaeological sites herein surveyed. Furthermore, a study from Nerlich and colleagues (2006) concluded that the frequency of cancer in skeletons from Egyptians that lived 3200–500 BCE did not differ markedly from British mortality data from the 1900s, when age-adjusted frequencies were considered. As such, the suggestion that cancer was rare in the past requires reassessment.

13.4.2 An overview of the evolutionary mismatch

What defines an environment as novel or ancestral? This question can be difficult to answer considering that anthropogenic transformations have been

a hallmark of our species' journey (Bovin et al., 2016). Some broadly defined characteristics of novel environments that contribute to increased cancer risk include survival to older ages, unhealthy lifestyles (i.e. diet/sedentariness), infections, mutagenic exposure (i.e. pollution/toxins) (Greaves & Aktipis, 2016; Hochberg & Noble, 2017) and demographic transitions. In the following sections, we attempt to provide more context to these broadly defined categories of mismatch, highlight some commonly perceived 'novel' environments that may not be so novel, and discuss the changes in lifestyle and environment that may have shifted distribution of cancer risk for humans across time.

13.4.3 The agricultural transition

With the first epidemiological transition, the shift from foraging to food production (~ 11,000 years ago), sedentism, an increase in population density, animal husbandry intensification leading to an increase in zoonotic infections and shifts to high-calorie diets and less dietary diversity are all associated with an overall decline in human health among Neolithic and later agricultural populations (Cohen & Armelagos, 1984; Harper & Armelagos, 2010; Larsen, 2015). In the palaeopathological record, both pre-Holocene and Holocene hunter-gatherer skeletal populations show evidence of cancer (Lieverse, 2005; Lieverse et al., 2014; Luna et al., 2008; Webb, 1995). In other words, current palaeo-oncological data provide no support for an increased risk of cancer with the agricultural transition. Future studies with larger sample sizes that aim to compare pre-agricultural and early agricultural populations may shed further light on the possibility of evolutionary mismatches.

13.4.4 Pathogen exposure

The transition to agriculture, with the spread of pathogens, the introduction of novel pathogens, and the subsequent phenomena of urbanisation and early industrialisation, may have influenced the risk for cancers originating from infection. This would not only have been due to the effects of oncogenic infectious agents, but also due to an overall diminished immune status and high levels of chronic inflammation associated with infectious diseases (Trumble & Finch, 2019).

Current estimates suggest ~ 20% of human cancers originate from pathogen-related infections (Ewald, 2018; Vandeven & Nghiem 2014). However, some suggest these estimates are low because it is difficult to determine the aetiology and driving factor for many cancer diagnoses (Ewald, 2018; Martel et al., 2012). Infectious pathogens responsible for oncogenic transformation in humans include several common viruses (e.g. HPVs, EBV, HHV-8, HTLV-1) and *Helicobacter pylori* (Martel et al., 2012). Cancer sites most commonly associated with known infectious agents include gastric, liver and uterine cervix cancers (Martel et al., 2012). There is a strong potential for infection-induced cancers in past populations and many of the commonly reported viruses have been detected in past human remains from a diverse time span (e.g. Castillo-Rojas et al., 2008; Gerszten et al., 2012). However, many of the infection-related cancers are also less likely to metastasise to the bones due to diverse factors (Marques et al., 2018), and hence are undetectable in past skeletal remains. We argue that the tissue-type vulnerability of cancer affecting human populations along our evolutionary journey underwent significant shifts, rather than simply being rare. If in general, agricultural and early urbanised societies had a higher burden of infectious-related cancer due to the introduction of novel pathogens, contemporary populations with reduced exposure to pathogens should have comparatively less infectious-related cancer risk.

The interaction between pathogen exposure, human evolution and cancer risk, however, is more complex than we have described. One of the most well-cited examples of evolutionary mismatch stems from the recent shift in exposure to microbes and parasites in populations living in industrialised, large-scale societies. Lifestyle and cultural practices of affluent individuals living in large-scale societies buffer infants and children from infections and exposures to pathogens, with socioeconomic status likely playing a role in the varying degrees of buffering. Lack of exposure to a diversity of pathogens in early childhood has consequently led to an imbalanced immune system and subsequent risk for allergies and autoimmune diseases (known as the 'hygiene hypothesis'; see Chapter 11) (Dunne & Cooke, 2005; Yazdanbakhsh et al., 2002). Building on the same underlying framework of the hygiene hypothesis, acute lymphoblastic leukaemia (ALL) is the most prevalent cancer diagnosed in contemporary children from industrialised societies. Epidemiological evidence suggests a recent increase in ALL, while cancer biologist Mel Greaves has proposed an equivalent evolutionary mismatch between immunological programming and ALL incidence in children (Greaves, 2018).

13.4.5 Diet

Although dietary diversity and flexibility characterises the Pleistocene and early Holocene hominins (Pontzer et al., 2018; also see Chapters 5 and 7), agriculturally driven dietary shifts could have impacted cancer occurrence. Further, lifestyle shifts in diet and the introduction of antibiotics have a large effect on the gut microbiome and its diversity (Sonnenburg & Sonnenburg, 2019). Recent research suggests a relationship between gut microbiota and gastrointestinal (GI) cancers, including colorectal cancer (Gagnière, 2016). Together, these data suggest cancers of the GI tract may be more common now than in past populations. However, high carbohydrate content, low dietary fibre, practices of preserving foods (salted or smoked) and mouldy stored food were also common in the past and a significant source of carcinogenic compounds (Friborg & Melbye, 2008). Mapping the introduction of different diets and their relevance in altering the prevalence of GI cancers could test these predictions; however, GI cancers are difficult to diagnose in the bioarchaeological record due to their lower osteotropic propensity.

13.4.6 Toxins and carcinogens

Throughout hominin evolution, humans have been exposed to various environmental carcinogens including plants toxins, dust, smoke, radiation, heavy metals and pathogens (Álvarez-Fernández et al., 2020; Ewald, 2018; Lanzirotti et al., 2014; Trumble & Finch, 2019; Whitley & Boyer, 2018;). While there is a link between smoking and lung cancer in contemporary populations, we suggest this is not a completely novel condition, as past humans were likely exposed to inhalation to at least some level of carcinogens from fire and cooking. As such, air pollution from biomass fuels (burned plant or animal material for cooking) is a cancer risk factor for both past and contemporary populations (Fullerton et al., 2008; Trumble & Finch, 2019).

While only a few studies address the impact of exogenous carcinogens on the prevalence of cancer in bioarchaeology, Whitley and Boyer (2018) detected high levels of naturally occurring radon in a pre-Columbian population from New Mexico, USA. The study of the skeletal remains of this population that lived around 1050–1320 CE also shows a particularly high prevalence of cancer. These results suggest that more data on exogenous carcinogen in relation to population specific cancer rates would be useful in understanding the role of environmental toxins and cancer risk. In addition to human cancers, we see links with toxins in the environments and the development of neoplasia in wildlife populations as well, including gastric papilloma in a population of Beluga whales living in polluted waters (Pesavento et al., 2018).

Skin cancer is another example in human populations of a neoplastic condition that may result from an environmental mismatch (Jablonski & Chaplin, 2010). For example, populations that live at higher, northern latitudes with less sun exposure have lower expression of the skin pigment melanin. This is thought to have evolved as an adaptation to ensure sufficient production of vitamin D (Greaves, 2014). However, when individuals with low skin pigmentation migrate to areas with high UV radiation exposure, they are more vulnerable to skin cancer compared to native populations with highly pigmented skin (Aktipis & Nesse, 2013). High melanoma rates in individuals of northern European ancestry are linked to intense UV exposure and burning, especially during childhood (Houghton & Polsky, 2002). Thus, movement of populations in the past to novel environments might have been an important factor on the occurrence of skin cancer.

13.4.7 Low birth rates

Breast cancer is one of the top diagnosed cancers in women worldwide (Bray et al., 2019a). Many have hypothesised the link between reproductive patterns and breast cancer risk (Aktipis et al., 2014; Greaves & Aktipis, 2016; Short, 1976). For 99% of human history, we lived in small-scale subsistence populations, hunted and gathered food sources and had no access to birth control. It has been estimated that women in subsistence populations have 100–150 ovulatory cycles over a lifetime (Eaton et al., 1994; Strassmann, 1999). In large-scale populations that have gone through epidemiological and demographic transitions, women tend to have a decreased age of menarche, delayed age of first birth, fewer children and a shorter duration of breastfeeding (Short, 1976). These shifts in reproductive demographics result in more menstrual cycles during the course of their lifetime (estimated 300–400 cycles) (Strassman, 1999). More menstrual cycles lead to higher cumulative exposure to reproductive hormones, such as oestrogen. Reproductive hormones are involved in many cellular pathways, including cell proliferation. It has been proposed that women in large-scale populations with delayed childbirth and fewer children may be at an increased risk for breast cancer from this chronic exposure to hormones (Short, 1976; Aktipis et al., 2014; Greaves and Aktipis, 2016). Similarly, one would also expect a lower frequency of reproductive cancers in past populations (Marques et al., 2018). Similar trends in mammary cancers can be observed in captive animals under managed breeding facilities such as zoos. In fact, one study reported a link between mammary carcinomas in captive tigers and lions and the administration of hormonal birth control (Mcaloose & Newton, 2009).

13.4.8 Increased adult survival

The current increase in cancer-related mortality may be related to the higher life expectancy in contemporary populations. While life expectancy (which is highly influenced by high infant mortality rates) in past populations and current forager populations is typically between the 30s and 40s, it is a common misconception to assume that these individuals did not live into older ages. Data from bioarchaeology and extant foragers showed that adult survivorship (if individuals survive to ~ 15 years of age) can be similar to early industrialised societies, with a modal age-at-death of ~ 72 years (68–78 years) (Chamberlain, 2006; Cave & Oxenham, 2016; Hill et al., 2007; Pontzer et al., 2018). It is likely that extended longevity has a long history in our lineage (Hill et al., 2007), and cancer at older ages was likely a reality in past populations as well, although they may have been different cancers than those that predominately affect industrialised societies today.

13.4.9 Culture as a buffer

Novel environments do not necessarily lead to poor health outcomes and increased cancer risk factors. Due to advancements in technology and medicine, many contemporary societies have better health care and preventive measures. Early recognition and detection of cancer, starting in the early 1900s, has increased cancer awareness and prevention. Additionally, populations living in large-scale industrialised environments are less impacted by infections and typically have improved health and immune status due to calorie-rich diets. Advances in medical access and technology, including vaccination programmes (e.g. HPV virus, hepatitis B), early screening measures (i.e. surgical removal of neoplasia prior to malignancy), sunscreen to prevent melanomas, as well as certain medications (Kostadinov et al., 2013), can buffer and mitigate some cancer risk factors. Controlling dietary quality and quantity, and exercise can also potentially lower risk for certain cancer subtypes.

13.5 Conclusion and future prospects

Common questions in cancer biology are how prevalent cancer was in the past, and how cancer prevalence has shifted throughout time. Many scholars suggest cancer is a disease of 'modernity'. However, a comparative and evolutionary approach to cancer demonstrates the disease has been a problem for animals since the dawn of multicellularity. If cancer is a common problem in our deep evolutionary past, why do we assume it was so rare in past human populations? Palaeopathological work has shown that cancer is ubiquitous in past human populations. Further, cancer prevalence in past populations are likely underestimated due to multiple biases addressed in this chapter, including small sample sizes, lack of fossilisation of soft-tissue tumours and diagnosis technologies. Despite these limitations, we provided a temporal overview of cancer prevalence estimates from current palaeopathological studies and found no temporal trend in cancer prevalence in the palaeopathology data, including that which encompasses major transitions in human societies.

As we highlighted through this chapter, cancer is not a disease of 'modern' times, but environmental changes can shift tissue-specific cancer risks. However, the overall consensus, in both fields of palaeo-oncology and EM, has been that cancer prevalence in human societies has increased significantly in the most recent period of our history, suggesting support for the evolutionary mismatch hypothesis. According to our analyses, we find no empirical support for evolutionary mismatch and increased cancer prevalence across these millennia. This may be due to various reasons. One reason could be due to the limited data, although we believe better palaeo-oncological data would support these trends, as cancer was likely underestimated in archaeological samples. Additionally, with recent advancements in cancer screening and diagnoses, it is difficult to say if humans are getting more cancer today or if we are instead getting better at detecting it. Another explanation may be that there is no significant increase for overall cancer prevalence per se, but there could be a substantial shift in cancer subtypes. For example, we may see less infection-related cancers but an increase in hormonally driven cancers after demographic transitions. Of note here is that many of the infection-related cancers do not always metastasise to the skeleton, hence these go unnoticed in palaeopathological studies. We also suggest that how we define environmental mismatch may need a more nuanced approach, as exposures to radiation, toxins and DNA-damaging agents have always been a part of our ecological niches. Thus, while it may be true that modern human biology has inherent mismatches with today's environments, the shift to these new environments is so recent that we were not able to capture trends in this current study.

Bridging the fields of palaeopathology and EM can provide insight into diseases in our evolutionary past. With our overlapping interests, we seek to answer the question of how evolutionary mismatches contribute to cancer risk in humans, but with the state of current data, we still may never reach a satisfying answer. However, we urge researchers to be more cautious in suggesting cancer was rare in our evolutionary past, because past and present populations have likely varied in their cancer risk depending on their ever-changing environment. All multicellular species are vulnerable to cancer and the most persuasive risk factors change both spatially and temporally. More detailed research on age-adjusted cancer prevalence, population-specific environmental exposures (including infections) and population genetics are required to begin to understand cancer risk in past and current human populations. Collaborations between cancer palaeopathologists, cancer biologists, epidemiologists, evolutionary biologists and historians are needed to address the complexities of cancer prevalence throughout human evolution.

References

Abegglen, L. M., Caulin, A. F., Chan, A., Lee, K., Robinson, R., Campbell, M. S., . . . & Schiffman, J. D. (2015). Potential mechanisms for cancer resistance in elephants and comparative cellular response to DNA damage in humans. *Journal of the American Medical Association*, 314(17), 1850–1860.

Aktipis, C. A., Boddy, A. M., Jansen, G., Hibner, U., Hochberg, M. E., Maley, C. C., et al. (2015). Cancer across the tree of life: cooperation and cheating in multicellularity. *Philosophical Transactions of the Royal Society of London Series B: Biological Sciences*, 370 (1673). doi:10.1098/rstb.2014.0219

Aktipis, C. A., Ellis, B. J., Nishimura, K. K. & Hiatt, R. A. (2014). Modern reproductive patterns associated with estrogen receptor positive but not negative breast cancer susceptibility. *Evolution, Medicine, and Public Health*, 2015 (1), 52–74. doi:10.1093/emph/eou028

Aktipis, C. A. & Nesse, R. M. (2013). Evolutionary foundations for cancer biology. *Evolutionary Applications*, 6(1), 144–159. doi:10.1111/eva.12034

Álvarez-Fernández, N., Martínez Cortizas, A. & López-Costas, O. (2020). Atmospheric mercury pollution deciphered through archaeological bones. *Journal of Archaeological Science*, 119, 105159. doi:https://doi.org/10.1016/j.jas.2020.105159

Bertolaso, M. & Dieli, A. M. (2017). Cancer and intercellular cooperation. *Royal Society Open Science*, 4(10), 170470. doi:10.1098/rsos.170470

Bissell, M. J. & Hines, W. C. (2011). Why don't we get more cancer? A proposed role of the microenvironment in restraining cancer progression. *Nature Medicine* 17(3),320–329.

Boddy, A. M., Abegglen, L. M., Pessier, A. P., Schiffman, J. D., Maley, C. C. & Witte, C. (2020a). Lifetime cancer prevalence and life history traits in mammals. *Evolution, Medicine, and Public Health*, 1, 187–195.

Boddy, A. M., Harrison, T. M. & Abegglen, L. M. (2020b). Comparative oncology: new insights into an ancient disease. *iScience*, 23(8), 101373.

Bona, A., Papai, Z., Maasz, G., Toth, G. A., Jambor, E., Schmidt, J., Toth, C., Farkas, C. & Mark, L. (2014). Mass spectrometric identification of ancient proteins as potential molecular biomarkers for a 2000-year-old osteogenic sarcoma. *PloS One*, 9(1), e87215. doi:10.1371/journal.pone.0087215

Bovin, N. L., Zeder, M. A., Fuller, D. Q., Crowther, A., Larson, G., Erlandson, J. M., Denham. T. & Petraglia, M. D. (2016). Ecological consequences of human niche construction: examining long-term anthropogenic shaping of global species distributions. *Proceedings of the National Academy of Sciences of the United States of America*, 113, 6388–6396.

Bray, F. & Soerjomataram, I. (2019a). 'The burden of cancer', in A. Jemal, L. Torre, I. Soerjomataram & F. Bray (eds), *The cancer atlas*, 3rd edn, pp. 36–37. American Cancer Society, Atlanta.

Bray, F. & Soerjomataram, I. (2019b). 'Overview of geographical diversity', in A. Jemal, L. Torre, I. Soerjomataram & F. Bray (eds), *The cancer atlas*, 3rd edn, pp. 48–49. American Cancer Society, Atlanta.

Brothwell, D. (2012). Tumours: problems of differential diagnosis in paleopathology. In A. L. Grauer (ed), *A companion to paleopathology*, pp. 420–433. Wiley-Blackwell, New York.

Brothwell, D. & Sandison, A. T. (1967). Disease in antiquity: a survey of the diseases, injuries and surgery of early populations. Charles C. Thomas, Springfield.

Bulbeck, F. D. (2005). 'The last glacial maximum human burial from Liang Lemdubu in Northern Sahulland', in S. O'Connor, M. Spriggs & P. Veth (eds), *The archeology of the Aru Islands, Eastern Indonesia*, pp. 255–294. Pandanus Books, Canberra.

Byars, S. G. & Voskarides, K. (2020). Antagonistic pleiotropy in human disease. *Journal of Molecular Evolution*, 88(1), 12–25. doi:10.1007/s00239-019-09923-2

Campisi, J. (2013). Aging, cellular senescence, and cancer. *Annual Review of Physiology*, 75, 685–705. doi:10.1146/annurev-physiol-030212-183653

Capasso, L. (2005). Antiquity of cancer. *International Journal of Cancer*, 113(1), 2–13. doi:10.1002/ijc.20610

Castillo-Rojas, G., Cerbon, M. A. & Lopez-Vidal, Y. (2008). Presence of *Helicobacter pylori* in a Mexican pre-Columbian mummy. *BMC Microbiology*, 8, 119. doi:10.1186/1471-2180-8-119

Caulin, A. F. & Maley, C. C. (2011). Peto's paradox: evolution's prescription for cancer prevention. *Trends in Ecology & Evolution*, 26(4), 175–182. doi:10.1016/j.tree.2011.01.002

Cave, C. & Oxenham, M. (2016). Identification of the archaeological 'invisible elderly': an approach illustrated with an Anglo-Saxon example. *International Journal of Osteoarchaeology*, 26(1), 163–175.

Chamberlain, A. T. (2006). *Demography in archaeology*. Cambridge University Press, Cambridge.

Cohen, M. N. & Armelagos, G. J. (1984). *Paleopathology at the origins of agriculture*. Academic Press, Inc, Orlando.

Crespi, B. J. & Summers, K. (2006). Positive selection in the evolution of cancer. *Biological Reviews of the Cambridge Philosophical Society*, 81(3), 407–424. doi:10.1017/s1464793106007056

David, A. R. & Zimmerman, M. R. (2010). Cancer: an old disease, a new disease or something in between? *Nature Reviews Cancer*, 10(10), 728–733. doi:10.1038/nrc2914

de La Rúa, C., Baraybar, J. & Etxeberria, F. (1995). Neolithic case of metastasizing carcinoma: multiple approaches to differential diagnosis. *International Journal of Osteoarchaeology*, 5(3), 254–264.

De Vico, G. & Carella, F. (2014). Tumors in invertebrates: molluscs as an emerging animal model for human cancer. *Invertebrate Survival Journal*, 1, 19–21.

Ducasse, H., Ujvari, B., Solary, E., Vittecoq, M., Arnal, A., Bernex, F., Pirot, N., Misse, D., Bonhomme, F., Renaud, F., Thomas, F. & Roche, B. (2015). Can Peto's paradox be used as the null hypothesis to identify the role of evolution in natural resistance to cancer? A critical review. *BMC Cancer*, 15, 792. doi:10.1186/s12885-015-1782-z

Dunne, D. W. & Cooke, A. (2005). A worm's eye view of the immune system: consequences for evolution of human autoimmune disease. *Nature Reviews Immunology*, 5(5), 420–426. doi:10.1038/nri1601

Easton, D. F., Ford, D. & Bishop, D. T. (1995). Breast and ovarian cancer incidence in BRCA1-mutation carriers. Breast Cancer Linkage Consortium. *American Journal of Human Genetics*, 56(1), 265–271.

Eaton, S. B., Pike, M. C., Short, R. V., Lee, N. C., Trussell, J., Hatcher, R. A., Wood, J. W., Worthman, C. M., Jones, N. G., Konner, M. J., Hill, K. R., Bailey, R. & Hurtado, A. M. (1994). Women's reproductive cancers in evolutionary context. *The Quarterly Review of Biology*, 69(3), 353–367. doi:10.1086/418650

Effron, M., Griner, L. & Benirschke, K. (1977). Nature and rate of neoplasia found in captive wild mammals, birds, and reptiles at necropsy. *Journal of the National Cancer Institute*, 59(1), 185–198.

Ewald, P. W. (2018). Ancient cancers and infection-induced oncogenesis. *International Journal of Paleopathology*, 21, 178–185. doi:10.1016/j.ijpp.2017.08.007

Forniaciari, G. (2018). Histology of ancient soft tissue tumors: a review. *International Journal of Paleopathology*, 21, 64–76.

Friborg, J. T. & Melbye, M. (2008). Cancer patterns in Inuit populations. *The Lancet Oncology*, 9(9), 892–900. doi:10.1016/S1470-2045(08)70231-6

Fullerton, D. G., Bruce, N. & Gordon, S. B. (2008). Indoor air pollution from biomass fuel smoke is a major health concern in the developing world. *Transactions of the Royal Society of Tropical Medicine and Hygiene*, 102(9), 843–851. doi:10.1016/j.trstmh.2008.05.028

Gagnière, J., Raisch, J., Veziant, J., Barnich, N., Bonnet, R., Buc, E., Bringer, M-A., Pezet, D. & Bonnet, M. (2016). Gut microbiota imbalance and colorectal cancer. *World Journal of Gastroenterology*, 22(2), 501–518. doi: 10.3748/wjg.v22.i2.501

Gayther, S. A., Warren, W., Mazoyer, S., Russell, P. A., Harrington, P. A., Chiano, M., Seal, S., Hamoudi, R., van Rensburg, E. J., Dunning, A. M., Love, R., Evans, G., Easton, D., Clayton, D., Stratton, M. R. & Ponder, B. A. J. (1995). Germline mutations of the BRCA1 gene in breast and ovarian cancer families provide evidence for a genotype-phenotype correlation. *Nature Genetics*, 11(4), 428–433. doi:10.1038/ng1295-428

Gerszten, E., Allison, M. J. & Maguire, B. (2012). Paleopathology in south American mummies: a review and new findings. *Pathobiology*, 79(5), 247–256.

Gluckman, P., Beedle, A., Buklijas, T., Low, F. & Hanson, M. (2016). *Principles of evolutionary medicine*, 2nd edn. Oxford University Press, Oxford.

Greaves, M. (2000). *Cancer: the evolutionary legacy*. Oxford University Press, Oxford.

Greaves, M. (2018). A causal mechanism for childhood acute lymphoblastic leukaemia. *Nature Reviews Cancer*, 18(8), 471–484. doi:10.1038/s41568-018-0015-6

Greaves, M. (2014). Was skin cancer a selective force for black pigmentation in early hominin evolution? *Proceedings of the Royal Society B: Biological Sciences*, 281, 20132955. http://doi.org/10.1098/rspb.2013.2955

Greaves, M. & Aktipis, A. (2016). 'Mismatches with our ancestral environments and cancer risk', in C. C. Maley & M. Greaves (eds), *Frontiers in cancer research: evolutionary foundations, revolutionary directions*, pp. 195-215. Springer, New York.

Greaves, M. & Maley, C. C. (2012). Clonal evolution in cancer. *Nature*, 481(7381), 306–313. doi:10.1038/nature10762

Hanahan, D. & Weinberg, R. (2014). 'Hallmarks of cancer: an organizing principle for cancer medicine', in V. T. DeVita, T. S. Lawrence & S. A. Rosenberg (eds), *DeVita, Hellman & Rosenberg's cancer: principles & practice of oncology*, pp. 24–44. Wolters Kluwer, Philadelphia.

Harper, K. & Armelagos, G. (2010). The changing disease-scape in the third epidemiological transition. *International Journal of Environmental Research and Public Health*, 7(2), 675–697.

Harris, V. K., Schiffman, J. D. & Boddy, A. M. (2017). 'Evolution of cancer defense mechanisms across species', in B. Ujvari, B. Roche & F. Thomas (eds), *Ecology and evolution of cancer*, pp. 99–110. Academic Press, London.

Herrerín, J., Baxarias, J., García-Guixe, E., Fointaine, V., Núñez, M. & Dinarés, R. (2014). 'Patologías de origen tumoral, infeccioso y traumático en cráneos procedentes de la tumba de Pw In Re (TT-39), Luxor (Egipto) (abstract)', in C. Marques, C. Lopes, I. Leandro, F. C. Silva, F. Curate, S. Assis & V. Matos (eds), *IV Jornadas Portuguesas de Paleopatologia: a saúde e a doença no passado. Abstract book*, p. 27. Universidade de Coimbra, Coimbra.

Hill, K., Hurtado, A. M. & Walker, R. S. (2007). High adult mortality among Hiwi hunter-gatherers: implications for human evolution. *Journal of Human Evolution*, 52, 443–454.

Hochberg, M. E. & Noble, R. J. (2017). A framework for how environment contributes to cancer risk. *Ecology Letters*, 2017, 1–18.

Hoffman, F. L. (1915). *The mortality from cancer throughout the world*. The Prudential Press, Newark.

Houghton, A. N. & Polsky, D. (2002). Focus on melanoma. *Cancer Cell*, 2(4), 275–278. doi:10.1016/s1535-6108(02)00161-7

Hunt, K. J., Roberts, C. & Kirkpatrick, C. (2018). Taking stock: a systematic review of archaeological evidence of cancers in human and early hominin remains. *International Journal of Paleopathology*, 21, 12–26. doi:10.1016/j.ijpp.2018.03.002

Islami, F. & Jemal, A. (2019). 'Overview of risk factors', in A. Jemal, L. Torre, I. Soerjomataram & F. Bray (eds), The cancer atlas, 3rd edn, pp. 18–19. American Cancer Society, Atlanta.

Jablonski, N. G. & Chaplin, G. (2010). Colloquium paper: human skin pigmentation as an adaptation to UV radiation. *Proceedings of the National Academy of Sciences of the United States of America*, 107(Suppl 2), 8962–8968. doi:10.1073/pnas.0914628107

Jaiswal, M., LaRusso, N. F., Burgart, L. J. & Gores, G. J. (2000). Inflammatory cytokines induce DNA damage and inhibit DNA repair in cholangiocarcinoma cells by a nitric oxide-dependent mechanism. *Cancer Research*, 60(1), 184–190.

Kang, H. J., Feng, Z., Sun, Y., Atwal, G., Murphy, M. E., Rebbeck, T. R., Rosenwaks, Z., Levine, A. J. & Hu, W. (2009). Single-nucleotide polymorphisms in the p53 pathway regulate fertility in humans. *Proceedings of the National Academy of Sciences of the United States of America*, 106(24), 9761–9766. doi:10.1073/pnas.0904280106

Kokko, H. & Hochberg, M. E. (2015). Towards cancer-aware life-history modelling. *Philosophical Transactions of the Royal Society of London B: Biological Sciences*, 370, 1673. doi:10.1098/rstb.2014.0234

Kostadinov, R. L., Kuhner, M. K., Li, X., Sanchez, C. A., Galipeau, P. C., Paulson, T. G., Sather, C. L., Srivastava, A., Odze, R. D., Blount, P. L., Vaughan, T. L., Reid, B. J. & Maley, C. C. (2013). NSAIDs modulate clonal evolution in Barrett's esophagus. *PLoS Genetics*, 9(6), e1003553. doi:10.1371/journal.pgen.1003553

Kumar, V., Abbas, A. & Aster, J. (2015). *Robbins and Cotran pathologic basis of disease*, 9th edn. Elsevier Saunders, Philadelphia.

Lanzirotti, A., Bianucci, R., LeGeros, R., Bromage, T. G., Giuffra, V., Ferroglio, E., Fornaciari, G. & Appenzeller, O. (2014). Assessing heavy metal exposure in renaissance Europe using synchrotron microbeam techniques. *Journal of Archaeological Science*, 52, 204–217.

Larsen, C. S. (2015). *Bioarchaeology: interpreting behavior from the human skeleton*, 2nd edn. Cambridge University Press, Cambridge.

Lieverse, A. R. (2005). *Bioarchaeology of the Cis-Baikal: biological indicators of Mid-Holocene hunter-gatherer adaptation and cultural change*. PhD Thesis. Cornell University, New York.

Lieverse, A. R., Temple, D. H. & Bazaliiski, V. I. (2014). Paleopathological description and diagnosis of metastatic carcinoma in an Early Bronze Age (4588+34 Cal. BP) forager from the Cis-Baikal region of Eastern Siberia. *PLoS One*, 9(12), e113919. doi:10.1371/journal.pone.0113919

Luna, L. H., Aranda, C. M., Bosio, L. A. & Beron, M. A. (2008). A case of multiple metastasis in Late Holocene hunter-gatherers from the Argentine Pampean region. *International Journal of Osteoarchaeology*, 18(5), 492–506.

Marques, C. (2019). 'Tumors of bone', in J. E. Buikstra (ed), Ortner's identification of pathological conditions in human skeletal remains, 3rd edn, pp. 639–718. Academic Press, London.

Marques, C., Matos, V., Costa, T., Zink, A. & Cunha, E. (2018). Absence of evidence or evidence of absence? A discussion on paleoepidemiology of neoplasms with contributions from two Portuguese human skeletal reference collections (19th-20th century). *International Journal of Paleopathology*, 21, 83–95. doi:10.1016/j.ijpp.2017.03.005

Martel, C., Ferlay, J., Franceschi, S., Vignat, J., Bray, F., Forman, D. & Plummer, M. (2012). Global burden of cancers attributable to infections in 2008: a review and synthetic analysis. *The Lancet Oncology*, 13(6), 607–615.

Martincorena, I. (2019). Somatic mutation and clonal expansions in human tissues. *Genome Medicine* 11(1), 35.

Mayr, E. (1961). Cause and effect in biology. *Science*, 134, 1501–1506.

McAloose, D. & Newton, A. L. (2009). Wildlife cancer: a conservation perspective. *Nature Reviews Cancer*, 9(7), 517–526.

McClure, H. M. (1973). Tumors in nonhuman primates: observations during a six-year period in the Yerkes Primate Center colony. *American Journal of Physical Anthropology*, 38(2), 425–429. doi:10.1002/ajpa.1330380243

Merlo, L. M., Pepper, J. W., Reid, B. J. & Maley, C. C. (2006). Cancer as an evolutionary and ecological process. *Nature Reviews Cancer*, 6(12), 924–935. doi:10.1038/nrc2013

Mumtaz, A., Khalid, A., Jamil, Z., Fatima, S. S., Arif, S. & Rehman, R. (2017). Kisspeptin: a potential factor for unexplained infertility and impaired embryo implantation. International Journal of Fertility & Sterility, 11(2), 99–104. doi:10.22074/ijfs.2017.4957

Nerlich, A. G., Rohrbach, H., Bachmeier, B. & Zink, A. (2006). Malignant tumors in two ancient populations: an approach to historical tumor epidemiology. *Oncology Reports*, 16(1), 197–202.

Nesse, R. M. (2019). Tinbergen's four questions: two proximate, two evolutionary. *Evolution, Medicine, and Public Health*, 2019(1), 2. doi:10.1093/emph/eoy035

Noble, R., Kaltz, O. & Hochberg, M. E. (2015). Peto's paradox and human cancers. *Philosophical Transactions of the Royal Society of London B: Biological Sciences*, 370(1673), 20150104. doi:10.1098/rstb.2015.0104

Nunney, L., Maley, C. C., Breen, M., Hochberg, M. E. & Schiffman, J. D. (2015). Peto's paradox and the promise of comparative oncology. *Philosophical Transactions of the Royal Society of London B: Biological Sciences*, 370(1673), 20140177. doi:10.1098/rstb.2014.0177

Odes, E. J., Randolph-Quinney, P. S., Steyn, M., Throckmorton, Z., Smilg, J. S., Zipfel, B., Augustine, T. N., de Beer, F., Hoffman, J. W., Franklin, R. D. & Berger, L. R. (2016). Earliest hominin cancer: 1.7-million-year-old osteosarcoma from Swartkrans Cave, South Africa. *South African Journal of Science*, 112(7/8), 5.

Park, J. Y. (2010). Promoter hypermethylation in prostate cancer. *Cancer Control*, 17(4), 245–255.

Pesavento, P. A., Agnew, D., Keel, M. K. & Woolard, K. D. (2018). Cancer in wildlife: patterns of emergence. *Nature Reviews Cancer*, 18, 646–661.

Pontzer, H., Wood, B. M. & Raichlen, D. A. (2018). Hunter-gatherers as models in public health. *Obesity Reviews*, 19(Suppl 1), 24–35. doi:10.1111/obr.12785

Robert, J. (2010). Comparative study of tumorigenesis and tumor immunity in invertebrates and nonmammalian vertebrates. *Developmental and Comparative Immunology*, 34(9), 915–925. doi:10.1016/j.dci.2010.05.011

Rothschild, B. M. & Rothschild, C. (1995). Comparison of radiologic and gross examination for detection of cancer in defleshed skeletons. *American Journal of Physical Anthropology*, 96(4), 357–363. doi:10.1002/ajpa.1330960404

Rothschild, B. M., Witzke, B. J. & Hershkovitz, I. (1999). Metastatic cancer in the Jurassic. *The Lancet*, 354(9176), 398. doi:10.1016/S0140-6736(99)01019-3

Rozhok, A. I., Wahl, G. M. & DeGregori, J. (2015). A critical examination of the 'Bad Luck' explanation of cancer risk. *Cancer Prevention Research*, 8(9), 762–764. doi:10.1158/1940-6207.Capr-15-0229

Schlott, T., Eiffert, H., Schmidt-Schultz, T., Gebhardt, M., Parzinger, H. & Schultz, M. (2007). Detection and analysis of cancer genes amplified from bone material of a Scythian royal burial in Arzhan near Tuva, Siberia. *Anticancer Research*, 27(6B), 4117–4119.

Schultz, M., Parzinger, H., Posdnjakov, D. V., Chikisheva, T. A. & Schmidt-Schultz, T. H. (2007). Oldest known case of metastasizing prostate carcinoma diagnosed in the skeleton of a 2,700-year-old Scythian king from Arzhan (Siberia, Russia). *International Journal of Cancer*, 121(12), 2591–2595. doi:10.1002/ijc.23073

Short, R. V. (1976). Definition of the problem-The evolution of human reproduction. *Proceedings of the Royal Society of London. Series B. Biological Sciences*, 195(1118), 3–24.

Smith, K. R., Hanson, H. A., Mineau, G. P. & Buys, S. S. (2012). Effects of *BRCA1* and *BRCA2* mutations on female fertility. *Proceedings of the Royal Society B: Biological Sciences*, 279(1732), 1389–1395. doi:10.1098/rspb.2011.1697

Smith, R. (2020). P > .05: the incorrect interpretation of 'not significant' results is a significant problem. *American Journal of Physical Anthropology*, 172(4), 521–527.

Sonnenburg, E. D. & Sonnenburg, J. L. (2019). The ancestral and industrialized gut microbiota and implications for human health. *Nature Reviews Microbiology*, 17(6), 383–390.

Steckel, R., Larsen, C. S., Roberts, C. & Baten, J. (2019). *The backbone of Europe: health, diet, work and violence over two millennia*. Cambridge University Press, Cambridge.

Strassmann, B. I. (1999). Menstrual cycling and breast cancer: an evolutionary perspective. *Journal of Women's Health*, 8(2), 193–202. doi:10.1089/jwh.1999.8.193

Strouhal, E. & Kritscher, H. (1990). Neolithic case of a multiple myeloma from Mauer (Vienna, Austria). *Anthropologie*, 28(1), 79–87.

Strouhal, E. & Němečková, A. (2009). History and palaeopathology of malignant tumours. *Anthropologie*, 47(3), 289–294.

Suzuki, T. (1987). Paleopathological study on a case of osteosarcoma. American Journal of Physical Anthropology, 74(3), 309–318. doi:10.1002/ajpa.1330740305

Szathmáry, E. & Smith, J. M. (1995). The major evolutionary transitions. *Nature*, 374(6519), 227–232. doi:10.1038/374227a0

Thomas, F., Fisher, D., Fort, P., Marie, J. P., Daoust, S., Roche, B., Grunau, C., Cosseau, C., Mitta, G., Baghdiguian, S., Rousset, F., Lassus, P., Assenat, E., Grégoire, D., Missé, D., Lorz, A., Billy, F., Vainchenker, W., Delhommeau, F., Koscielny, S., Itzykson, R., Tang, R., Fava, F., Ballesta, A., Lepoutre, T., Krasinska, L., Dulic, V., Raynaud, P., Blache, P., Quittau-Prevostel, C., Vignal, E., Trauchessec, H., Perthame, B., Clairambault, J., Volpert, V., Solary, E., Hibner, U. & Hochberg, M. E. (2013). Applying ecological and evolutionary theory to cancer: a long and winding road. *Evolutionary Applications*, 6(1), 1–10. doi:10.1111/eva.12021

Tian, X., Azpurua, J., Hine, C., Vaidya, A., Myakishev-Rempel, M., Ablaeva, J., Mao, Z., Nevo, E., Gorbunova, V. & Seluanov, A. (2013). High-molecular-mass hyaluronan mediates the cancer resistance of the naked mole rat. *Nature*, 499, 346–349. https://doi.org/10.1038/nature12234

Tinbergen, N. (1963) On the aims and methods of ethology. *Zeitschrift für Tierpsychologie*, 20, 410–433.

Tomasetti, C., Li, L. & Vogelstein, B. (2017). Stem cell divisions, somatic mutations, cancer etiology, and cancer prevention. *Science*, 355(6331), 1330–1334. doi:10.1126/science.aaf9011

Tomasetti, C. & Vogelstein, B. (2015). Cancer etiology. Variation in cancer risk among tissues can be explained by the number of stem cell divisions. *Science*, 347(6217), 78–81. doi:10.1126/science.1260825

Trumble, B. C. & Finch, C. E. (2019). The exposome in human evolution: from dust to diesel. *The Quarterly Review of Biology*, 94(4), 333–394. doi:10.1086/706768

Vandeven, N. & Nghiem, P. (2014). Pathogen-driven cancers and emerging immunetherapeutic strategies. *Cancer Immunology Research*, 2(1), 9–14.

Varki, N. M. & Varki, A. (2015). On the apparent rarity of epithelial cancers in captive chimpanzees. *Philosophical Transactions of the Royal Society of London Series B: Biological Sciences*, 370(1673), 20140225. doi:10.1098/rstb.2014.0225

Waldron, T. (1996). What was the prevalence of malignant disease in the past? *International Journal of Osteoarchaeology*, 6(5), 463–470.

Waldron, T. (2007). *Palaeoepidemiology: the measure of disease in the human past*. Left Coast Press Inc, Walnut Creek.

Webb, S. (1995). *Palaeopathology of aboriginal Australians: health and disease across a hunter-gatherer continent*. Cambridge University Press, Cambridge.

Weinberg, R. A. (2014). *The biology of cancer*, 2nd edn. Taylor & Francis, London.

Whitley, C. B. & Boyer, J. L. (2018). Assessing cancer risk factors faced by an Ancestral Puebloan population in the North American Southwest. *International Journal of Paleopathology*, 21, 166–177. doi:10.1016/j.ijpp.2017.06.004

Wood, J., Milner, G., Harpending, H. & Weiss, K. (1992). The osteological paradox. *Current Anthropology*, 33(4), 343–370.

Yazdanbakhsh, M., Kremsner, P. G. & van Ree, R. (2002). Allergy, parasites, and the hygiene hypothesis. *Science*, 296(5567), 490–494. doi:10.1126/science.296.5567.490

Yevich, P. P. & Barszcz, C. A. (1978). Neoplasia in soft-shell clams (*My a arenaria*) collected from oil-impacted sites. *Annals of the New York Academy of Sciences*, 298, 409–426. doi:10.1111/j.1749-6632.1977.tb19281.x

Zuckerman, M. K., Harper, K. N. & Armelagos, G. J. (2016). Adapt or die: three case studies in which the failure to adopt advances from other fields has compromised paleopathology. *International Journal of Osteoarchaeology*, 26(3), 375–383.

CHAPTER 14

Stress in bioarchaeology, epidemiology, and evolutionary medicine: an integrated conceptual model of shared history from the descriptive to the developmental

Daniel H. Temple and Ashley N. Edes*

14.1 Introduction

Stress has rapidly become one of the most compelling topics in both the lifestyle and scientific literature and lays claim to being one of the most referenced, yet misunderstood, terms in the anthropological lexicon (van der Kolk, 2014). The term is incorporated into studies of contemporary and past human populations with the goal of understanding the ways in which humans adapt to local environments and, though often defined according to fairly straightforward language (e.g. Goodman et al., 1988; Huss-Ashmore et al., 1982), causes great debate and tension surrounding appropriate usage (Ellison & Jasienska, 2007; Goodman et al., 1988; Jackson, 2014). For the purposes of this chapter, stress is defined 'as a disruption to physiological homeostasis produced by external agents' (Huss-Ashmore, 1982: 396). Despite complications surrounding terminology, the human stress response is now understood from a whole-organism perspective where interactions across the lifespan are observable. Since physiological disruption is embedded deep within the contexts of human-created cultural and ecological frameworks, the

study of stress has great resonance for anthropologists working in the past and present (Dressler, 1990, 2011; Larsen, 2015).

Bioarchaeological studies of stress rely upon theoretical developments in evolutionary medicine (EM) and evolutionary biology, although little evidence for cross-disciplinary interaction is documented (Reitsema & Kyle, 2014). However, the historical trajectories of both disciplines suggest continuity in research design, methods and results. Furthermore, greater clarity in defining, operationalising and, ultimately, contextualising stress experiences is realised using an integrated perspective (Reitsema & Kyle, 2014; Temple & Goodman, 2014). This chapter first provides a historical framework for the study of stress in epidemiology and EM as well as in bioarchaeological research. It then discusses how these shared histories may be integrated to arrive at a greater understanding of stress in the past and how such knowledge may better serve the contemporary human condition.

14.2 Starving animals and active glands: early models of stress in experimental settings

The concept of homeostasis, central to the definition of stress, was derived from the late 19th and early

* We thank Kimberly Plomp, Charlotte Roberts, Sarah Elton and Gillian Bentley for inviting this contribution. Jane Buikstra, Clark Larsen, two anonymous reviers, and the editors provided comments that greatly improved this manuscript.

Daniel H. Temple and Ashley N. Edes, *Stress in bioarchaeology, epidemiology, and evolutionary medicine*. In: *Palaeopathology and Evolutionary Medicine*. Edited by Kimberly A. Plomp et al., Oxford University Press. © Oxford University Press (2022). DOI: 10.1093/oso/9780198849711.003.0014 twentieth-century physiological work of Claude Bernard (1865) and Walter Cannon (1932). Bernard (1865) defined 'milieu interieur' (internal milieu) as the capacity of an organism to regulate internal conditions in a state of constancy that was independent of the surrounding environment, while Cannon (1915, 1932) emphasised a relatively constant state that shifts in response to environment.

Experimental work by Hans Selye (1936) introduced general adaptation syndrome, which includes three phases: alarm, resistance and exhaustion. The alarm phase engages the hypothalamic-pituitary-adrenal (HPA) axis and prompts the secretion of catecholamines (e.g. epinephrine) and glucocorticoids (cortisol in humans and most mammals, corticosterone in rodents and birds) (Figure 14.1). Resistance occurs when large quantities of cortisol are in circulation,

while the body increases blood pressure and glucose output to return to homeostasis. Long-term exposure to stress may elicit an exhaustion stage where mortality or permanent damage to organs systems is observed; at this stage, cortisol levels drop and the HPA axis is unable to mount a response to further challenges. General adaptation syndrome explained how organisms maintain homeostasis in response to external agents, thus promoting short-term survival as well as long-term consequences of stress exposure.

14.3 From the lab to the field: formalised studies of stress and environment

EM and epidemiological research surrounding stress existed in a descriptive state throughout the early twentieth century, although examples of

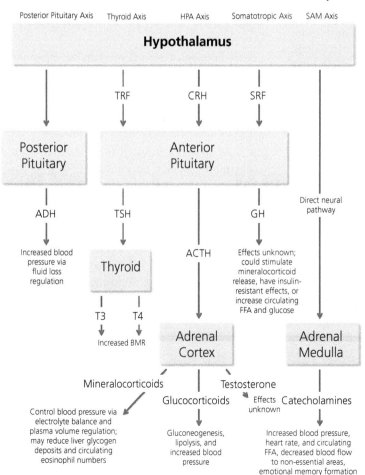

Figure 14.1 Primary axes activated by the hypothalamus during primate stress responses to an internal or external stressor. Reproduced with permission from Edes and Crews, 2017.

HPA, hypothalamic-pituitary-adrenal; SAM, sympathetic-adrenal-medullary; ACTH, adrenocorticotropin-releasing hormone; ADH, antidiuretic hormone; CRH, corticotropin-releasing hormone; GH, growth hormone (AKA somatotropin); SRF, somatotropic-releasing factor; TRF, thyrotropic-releasing factor; TSH, thyroid-stimulating hormone; T3, triiodothyronine; T4, thyroxine.

population approaches may be found (Dressler, 1990, 2011). Descriptive studies focused on disease aetiology and experimental treatments of animal subjects (Cannon, 1915, 1932; Seyle, 1936, 1938, 1946). By contrast, population approaches focused on stress in relation to psychosocial, socio-economic and ecological contexts (Donnison, 1929). Cassel and Tyroler (1961), in a critique of epidemiologically descriptive work, considered sociocultural and environmental contexts as primary agents of stress and disease. Anthropologists began exploring the adaptive limits of populations by recognising the adaptive aspects of the stress response and the interaction between human physiology with the built and natural environments (Baker, 1969). In this sense, the study of stress helps to explain human adaptation in terms of phenotypic traits that enhance survival and as a process that results in fitness-enhancing changes to human phenotypes (e.g., Dobzhansky, 1956, 1968). Action of the HPA axis was initially measured using growth, catecholamine and glucocorticoid metabolites found in urine, and body composition among populations from high altitudes (Baker, 1978; Frisancho & Baker, 1970; Hoon et al., 1976), those adjusting to new countries (Brown, 1981) or experiencing dietary insufficiency (Frisancho et al., 1973; Frisancho et al., 1980; Garn et al., 1964; Stini, 1981). Here, the goal was to understand the experience of human populations undergoing transition or placed within the context of environmental challenges, as well as to explore ways in which the human physiological response promoted survival.

In the late 1980s, the primacy of homeostasis came under question with the recognition that many physiological systems and parameters operate by making continual adjustments in opposition to assumptions of constancy (McEwen, 1998a, 1998b; McEwen & Wingfield, 2003; Schulkin, 2004). For example, blood pressure varies based on demand over the course of the day. If blood pressure were maintained through homeostasis, then this would inhibit optimal functioning during activity and rest. Instead, parameters such as blood pressure are regulated by the central nervous system, allowing for regular fluctuations in response to external challenges. Continual adjustments, as opposed to maintaining a constant state, was termed allostasis

(Box 14.1), or 'stability through change' (Sterling and Eyer 1988: 636).

Box 14.1   Allostasis

While not all organisms show the same responses to stressors (Johnson et al., 1992; Kemeny, 2003; Mason, 1975), the negative effects of stress on mental and physical health and longevity are well-documented (Charmandari et al., 2005; Guilliams & Edwards, 2010; Meaney et al., 1991). Homeostasis is such an integral concept in biology that medical conditions are often diagnosed and treated based on shifts of physiological parameters away from 'normal' ranges (Sterling, 2012). Many of the physiological markers used diagnostically respond to current demands allostatically. Allostatic responses range from daily mundane fluctuations to substantial shifts during different life history stages to dealing with unexpected challenges and circumstances. During such challenges, somatic systems enter an allostatic state characterised by fluctuations in physiological biomarkers. Typically, this allostatic activity is beneficial, allowing the organism to respond optimally in the current situation and then ending when the stressor is removed. However, when allostatic states are repeatedly or chronically activated, or when an individual is unable to habituate or shut off the response, damage gradually accumulates. This 'wear-and-tear' from excessive allostatic activity is termed 'allostatic load' (McEwen & Stellar, 1993). Allostatic load accumulates over the lifespan and can eventually result in disease, or allostatic overload (McEwen & Wingfield, 2003; Korte et al., 2005).

Moving beyond the traditional approach of characterising responses of individual biomarkers to stressors, researchers operationalised a method for estimating allostatic load by combining biomarkers from multiple physiological systems into a single score (Seeman et al., 1997). In just over two decades since the first measurement of allostatic load was published, upwards of five dozen biomarkers and a variety of methods to estimate allostatic load have been identified (Beckie, 2012; Edes & Crews, 2017; Juster et al., 2010). This research has shown that multi-biomarker approaches such as allostatic load often better reflect stress than individual biomarkers alone (Karlamangla et al., 2002, 2006; Seeman et al., 1997, 2001, 2004). These methods have been and continue to be applied to human populations across the globe to understand the effects of stress better, especially as it pertains to differential morbidity and mortality. For example, higher allostatic load has been associated with negative health outcomes such as cardiovascular disease, obesity, hypertension, diabetes, arthritis, periodontal disease and immunosuppression (Chen et al., 2014; Harding et al., 2016; Karlamangla et al., 2002, 2006; Logan & Barksdale, 2008; Mattei et al., 2010), as well as with increased mortality risk (Levine & Crimmins 2014; Robertson et al., 2017; Seeman et al., 1997). Higher allostatic load is even associated with mental health and cognition, including depression (Kobrosly et al., 2014), post-traumatic stress disorder (PTSD; Thayer et al., 2017) and cognitive declines (Karlamangla et al., 2002; Seeman et al. 1997, 2001).

14.4 Beyond cortisol: a multi-faceted approach to stress physiology

With Selye's (1946) explicit connections between cortisol levels and the stages of the general adaptation syndrome, there was a substantial increase in studies incorporating glucocorticoids as a measure of stress. Cortisol has been used to measure stress so ubiquitously that the words 'cortisol' and 'stress' are often (inaccurately) used interchangeably (MacDougall-Shackleton et al., 2019; McEwen, 2019). Research on cortisol has contributed invaluably to the collective knowledge surrounding stressors and stress responses as well as long-term effects. However, focusing so much effort on measuring cortisol can be misleading. Cortisol levels have been shown to vary widely during the stress response between individuals and even within the same individual at different points in time based on previous experience and perception of events (Breuner et al., 2008; Heim et al., 2000; Levine et al., 1989). Furthermore, cortisol and stress are not synonymous (MacDougall-Shackleton et al., 2019; McEwen, 2019). The term 'stress' is almost always used to refer to a negative stimulus but increases in cortisol also occur during positive and beneficial experiences, such as mating and exercise (Broom, 2017). Cortisol also serves other functions across the soma, including in energy metabolism. Therefore, accurately interpreting changes in cortisol levels requires knowledge of context, perception and activity levels.

Beyond the difficulty in accurately interpreting changes in circulating cortisol levels, approaches that focus strictly on cortisol may fail to take a whole-organism approach. Many other physiological systems and parameters fluctuate during stress, some in direct response to HPA axis activity and others far downstream or independent of it (Seaward, 2006; Everly & Lating, 2013; Figure 14.1). For example, the sympathetic-adrenal-medullary (SAM) axis releases adrenaline and noradrenaline within two to three seconds of a stressor being perceived, which function to prime the body to respond to potentially life-threatening situations (Everly & Lating, 2013). Beyond neuroendocrine axes, other biomarkers also have been observed to respond to stress, including albumin (Murata et al., 2004), C-reactive protein (CRP; Hostinar et al., 2015), dehydroepiandrosterone-sulfate (DHEA-S; Buford & Willoughby, 2008), inflammatory cytokines like interleukin-6 (IL-6) and tumour necrosis factor-α (TNF-α; Hostinar et al., 2015), luteinising hormone (Kalantaridou et al., 2010), oxytocin (Taylor et al., 2000), prolactin (Lennartsson & Jonsdottir, 2011) and testosterone (Charmandari et al., 2005; Kalantaridou et al., 2010). Even "stable" biomarkers change in response to some acute stressors, including lipids and lipoproteins (Niaura et al., 1992; van Doornen et al., 1998). Importantly, the physiological changes across somatic systems during stress responses are non-linear (McEwen, 2008, 2009) and may be dissociated (Hostinar et al., 2014), meaning activity within a single physiological axis cannot always accurately imply activity in other systems.

Other work has begun identifying factors which make some individuals more likely to have higher allostatic load than others, including age (Crimmins et al., 2009; Levine & Crimmins, 2014), psychosocial stress (Glei et al., 2007; Weinstein et al., 2003) and early-life adversity (Danese & McEwen, 2012; Horan & Widom, 2015). Moreover, factors such as social status (Crimmins et al., 2009; Seeman et al., 2004, 2010, 2014) and perceived social support help mediate allostatic load across the lifespan (Glei et al., 2007; Horan & Widom, 2015; Seeman et al., 2002, 2004; Weinstein et al., 2003). It has been estimated that allostatic load is approximately 30% heritable, with 70% of phenotypic variation accounted for by the environment (Petrovic et al., 2016).

14.5 Thrifty phenotypes, plasticity and constraint: the emergence of a life history perspective

While early work by Cannon (1915, 1932) and Selye (1936, 1938, 1946) characterised the endocrinological response to stress as adaptation by referencing alterations that promote short-term survival, later scholars argued that the study of stress evaluates failure of adaptation (Goodman et al., 1988). By contrast, experimental studies emphasise stress responses as an adaptive mechanism promoting short-term survival with long-term consequences (e.g. Crespi & Denver, 2005). The theory of allostasis, which is adaptive, contrasted with allostatic load and overload, which are the cumulative costs associated with these responses, epitomises this trade-off between immediate health and survival and later-life morbidity. Here, the study of stress may be contextualised as a discipline concerned with a combination of adaptive plasticity and physiological constraint.

Human life course approaches to stress have a long history in anthropological, clinical and sociological studies and demonstrate links between inequality, growth, and mortality (Bowditch, 1879; Banning, 1946; Boas, 1930; Cobb, 1936; Dubois, 1914; Forsdahl, 1977; Kimura & Kitano, 1959). A large-scale study of death records from more than 13,000 individuals from Great Britain revealed correlations between regions with high neonatal mortality and deaths due to heart disease and respiratory ailments (Barker & Osmond, 1986; Barker et al., 1989). Using records kept by midwife Ethel Margaret Burnside, the same research group found negative correlations between birth weight, placental weight and birth length with adult blood pressure (Barker, 1992; Barker et al., 1990). These findings contributed to the development of the thrifty phenotype hypothesis, which predicts that individuals survive neonatal stress through energy sparing, but are at a greater risk of metabolic disease at later stages of life (Bateson et al., 2004; Hales & Barker, 2001; see also Chapter 2).

Evolutionary mechanisms were further tethered to these findings as part of the predictive adaptive response (PAR) hypothesis. The PAR hypothesis predicts that organisms survive prenatal insults through energy sparing behaviour and physiologically prepare for birth in a resource deprived environment (Gluckman et al. 2005, 2007; Chapter 2). Furthermore, stressful in utero and early-life environments may also alter allostatic function and resilience to stressors, preparing developing phenotypes to be born into stressful environments (Danese & McEwen, 2012). Altered allostatic responses are characterised by either hypo- or hyper-reactivity to stressful stimuli, which promote higher allostatic load during adulthood and later life. However, individuals exposed to such stressors were at a greater risk of metabolic syndrome after being born into environments where obesogenic foods were readily available. Critics argue that neonatal body size has a stronger correlation with pre-maternal generations than foetal environment, while paleoclimatic modelling suggests that birth environment rarely reflects foetal experiences (Kuzawa, 2005; Wells, 2007, 2011). In addition, the PAR hypothesis failed to address the functional mechanisms promoting these conditions or the deeper trade-offs associated with selection acting on the trait (Worthmann & Kuzara, 2005).

Preliminary research indicates a relationship between allostatic load in women and adverse birth outcomes (Accortt et al., 2017; Barrett et al., 2018), as well as higher allostatic load in women who experienced adverse birth outcomes themselves (Barboza Solís et al., 2015; Christensen et al., 2018; Hux et al., 2014), suggesting allostatic load may be a mechanism underlying intergenerational effects of stress. Here, the study of stress and adaptation are best conceptualised using life history theory (LHT; Box 14.2). The use of LHT allows for a synthetic understanding of stress, where the mechanisms of host resistance may be understood in the context of adaptation, while the cost of survival is best explored through limits placed on organismal adaptability.

> **Box 14.2   Life history theory (LHT) and stress**
>
> LHT evaluates stress and survival mechanisms in association with the concepts of adaptive plasticity and

physiological constraint. Adaptive plasticity references the ability to produce multiple phenotypes through increased reaction norms. Under these circumstances, environmental perturbations may change the expression of a phenotype to promote survival and increase optimal values of a particular trait (DeWitt et al., 1998; Ghalambor et al., 2007). Careful evaluations of adaptive plasticity are required, as plasticity may be non-adaptive (Acuso-Rivero et al., 2019). Physiological constraints reflect costs and limits placed on adaptive plasticity through the modulation of energy to competing systems (Charnov, 1993; Futuyma, 1998; Stearns, 1992). Using the role of cortisol in the stress response as an example, Worthman & Kuzara (2005) demonstrate that the HPA axis promotes short-term survival for the individual, with long-term consequences in terms of reduced investment in growth, maintenance and reproduction. Similar response mechanisms across vertebrate species suggest that stress responses may represent a long-standing adaptation to adverse conditions by promoting reaction norms that move phenotypic traits closer to optimal values combined with reduced investment in later survival (Crespi & Denver, 2005; Thayer et al., 2017). These responses suggest a deeply evolved mechanism for the stress response that promotes short-term survival and is balanced by physiological constraint.

14.6 Stress in the past: foundational research in bioarchaeological approaches to physiological perturbation

For much of the twentieth century, the study of human skeletal remains mostly focused on descriptive methods that emphasised racial categorisation and differential diagnosis of skeletal lesions (Armelagos, 2003; Armelagos & Van Gerven, 2003). Bioarchaeology arose as a contextual challenge to this descriptive research, instead emphasising theoretically informed studies of human remains and mortuary practices (Buikstra, 1977; Armelagos, 2003). There exist numerous indicators of stress experiences in human skeletal remains that reveal evidence for growth disruption, dietary insufficiency, infectious disease and general episodic physiological perturbations (Figure 14.2) (Buikstra & Cook, 1980; Huss-Ashmore et al., 1982; Goodman et al., 1984; Larsen, 1987; Goodman & Martin, 2002). Skeletal indicators of stress were described and compared across cultural and ecological contexts

in a pre-formalised period of bioarchaeological research (Angel, 1946, 1947, 1954, 1966; Armelagos, 1969; Cohen, 1977; El-Najjar et al., 1976; Hooton, 1930; Johnston, 1962; Ubelaker, 1974). The integration of social and ecological conditions into these works were foundational in establishing the value of bioarchaeology to questions surrounding human adaptation in the past, although there existed little agreement on the definition for stress, nor consensus surrounding the cause of altered skeletal and dental phenotypes.

Skeletal and dental indicators of stress were described in a landmark paper that operationalised bioarchaeological research into a series of problem-oriented questions (Buikstra, 1977). Skeletal and dental indicators of stress were listed under a heading for diet and disease, then further divided into subsections for acute and chronic stress (Buikstra, 1977). Acute indicators of stress were episodic events where recovery is evident (e.g. linear enamel hypoplasia), while chronic stress was represented by disruptions that occur over extended durations and have long-term phenotypic consequences (e.g. long bone growth, stature). These skeletal and dental indicators of stress could not be assigned to a specific causative agent but did indicate deviations from normal physiological function. Cautious optimism for the use of these indicators was expressed, although the need for critical inquiry was made clear (Buikstra & Cook, 1980).

Formal bioarchaeological models defined stress as a '... physiological disruption of an organism resulting from environmental perturbation' (Huss-Ashmore et al., 1982, 395; also see Goodman & Armelagos, 1989; Goodman et al., 1984). The work built theoretical paradigms for comparing the stress experiences of past populations by drawing heavily on micro-economic models of subsistence change and detailed descriptions of skeletal and dental indicators of stress. Direct comparisons of lesion prevalence, growth trends and differences in body size were the most direct pathway to explore the process of human adaptation, and highlighted relationships between stress and cultural and ecological diversity (Goodman et al., 1984; Huss-Ashmore et al., 1982). The influential nature of this body of work is demonstrated by myriad bioarchaeological

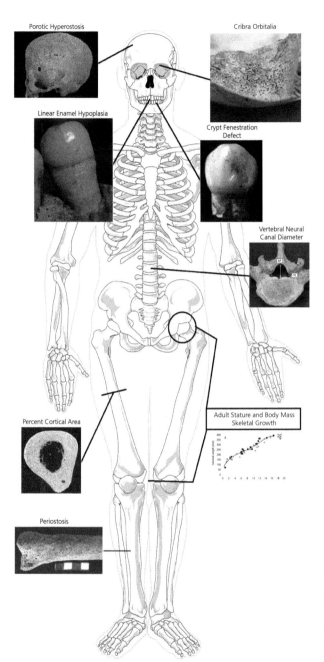

Figure 14.2 Commonly used skeletal indicators of stress.
Source: Modeled after work published by Steckel et al., 2002a. All images are courtesy of Daniel H. Temple, except the crypt fenestration defect, which is courtesy of Haagen D. Klaus. NB: Indicators such as Harris lines and fluctuating asymmetry may also be considered but are not pictured.

syntheses that explore stress experiences in relation to the agricultural transition, colonisation and population interaction (Cohen & Armelagos, 1984; Cohen & Crane-Kramer, 2012; Larsen, 1987, 1995, 2002, 2006; Larsen & Milner, 1994; Murphy & Klaus, 2012; Okazaki, 2004; Oxenham, 2006; Oxenham et al., 2005; Pechenkina & Oxenham, 2013; Pietrusewsky & Douglas, 2001; Pinhasi & Stock, 2011; Tayles & Oxenham, 2006; Temple, 2010; Temple & Larsen, 2013; Verano and Ubelaker, 1992). Caution in interpretation was suggested, however, due to associations between stress and contextual factors such as social structures (Buikstra, 1976, 1981; Cook, 1981, 1984; Powell, 1988).

These works illustrated a first wave of interaction in the study of stress between bioarchaeology and EM that portended similar trajectories. The studies sought explicit connections between the local environment and measures of stress within a sample while taking direction from the larger body of work in EM that explored stress in the context of cultural transition or environmental challenges in living populations. Similar to studies of living populations, bioarchaeological research emphasised adaptation as a process but envisioned skeletal indicators of stress in more dire terms. Goodman and colleagues (1988, 192) suggested that 'it is clear that many constraints such as poverty do not initiate adaptive processes. In this sense, stress is an important shadow image of the adaptive process . . . studying stress forces a consideration of the costs and limits of adaptation.'

That said, many skeletal indicators of stress are attributable to events where individuals survive. While these altered phenotypes indicate physiological disruption, this process remains part of a broader investment in short-term survival. In addition, these approaches only consider the relationship between stress and the HPA axis. Stress is a multi-system experience, and there exists great complexity in the contexts and feedback loops associated with this process. Further interaction with epidemiological modelling, evolutionary theory and the pathophysiology of phenotypic alteration was needed to facilitate greater synergies with EM.

14.7 Paradox, promise and the life history framework: contemporary studies of stress in bioarchaeology

14.7.1 Epidemiological modelling

Comparing stress indicators in human skeletal and dental remains is complicated by the fact that many of these conditions also suggest recovery (Buikstra, 1977; Ortner, 1991). These critical points culminated in a deeper discussion of hidden heterogeneity and selective mortality (Wood et al., 1992). Hidden heterogeneity refers to the concept that all individuals are heterogeneous in mortality risk owing to genetic, social and economic factors. Selective mortality occurs when survival is threatened by

specific underlying conditions. Bioarchaeologists may overcome these problems by engaging with statistical methods that help to demonstrate relationships between age at death and phenotypes altered by stress experience. Despite a lively debate at the onset of publication, very few studies interacted seriously with the problems of hidden heterogeneity and selective mortality prior to 2003 (DeWitte & Stojanowski, 2015; Wright & Yoder, 2003).

The early twenty-first century saw tremendous progress for the study of stress in bioarchaeological contexts, including use of epidemiological modelling, expansion of theoretical paradigms and development of technical methods for recording skeletal indicators of stress (Table 14.1). The multi-state model represents the first application of epidemiological methods to human skeletal remains from archaeological sites (Usher, 2000). The approach explores risk of death among individuals with (State 1) and without (State 2) indicators of stress (Usher, 2000). Samples are listed as covariates, and dying in States 1 and 2 is modelled as a proportional hazards specification where differences in mortality between samples results in a strongly negative or positive value. These models have been applied to samples derived from the Black Death and attritional cemeteries with the goal of understanding relationships between skeletal indicators of stress and mortality with sex and context surrounding death (DeWitte, 2010, 2017; DeWitte & Wood, 2008; Usher, 2000). Additional epidemiological methods include the use of survival and hazard analysis to compare mortality risk between samples in association with lesion presence and absence, lesion manifestation, timing of stress experience and skeletal indicators of growth (Boldsen, 2006; DeWitte, 2014a,b; DeWitte & Hughes-Morey, 2012; Garland, 2020; Saunders & Hoppa, 1993; Temple, 2014; Thomas et al., 2019; Watts, 2015; Wilson, 2014).

The application of epidemiological models to bioarchaeological settings was an important development in understanding relationships between stress indicators and survival, but greater articulation with theory was still needed (Temple & Goodman, 2014). In particular, the strength of epidemiological modelling includes the capacity

Table 14.1 Contemporary bioarchaeological approaches to the study of stress.

Approach	Methods	Advantages	Limitations
Epidemiological Modelling	One (1) or more indicator of stress; compared with indicators of morbidity and mortality	Explains relationship between lesion and morbidity/mortality; reveals evidence for heterogeneous frailty; contextual	Does not provide cumulative index; reduced emphasis on evolutionary theory
Skeletal Frailty Index	Multiple indicators of stress used as an index; estimates allostatic load	Provides cumulative indicator of stress over the life course; focuses on survival and mortality; contextual	Poor relationships with mortality; may reduce skeletal samples due to incomplete preservation
Life Course Life History	One (1) or more indicator of stress; compared with indicators of future morbidity and mortality; measured along ecological or social developmental gradients	Theorisation of adaptive plasticity and constraint; explores relationship between stress indicators over the lifespan; can be intergenerational contextual	Does not include cumulative index
Pathophysiology	One (1) or more indicators of stress; dense descriptions with emphasis on pathophysiology	Reveals physiological pathways for lesion formation; may be paired with cumulative or life history approaches; contextual	Does not include cumulative index; may not explore relationships with morbidity and mortality

to investigate hidden heterogeneity and selective mortality in the past. However, two drawbacks to this approach include a lack of explicit interaction with evolutionary theory and the incorporation of a cumulative framework for addressing questions of allostatic load. That said, interaction between stress indicators may be explored using epidemiological modelling (DeWitte & Wood, 2008; King et al., 2005; Temple, 2014; Yaussy, 2019). For example, relationships between enamel hypoplasia, periostosis and periodontal disease have been explored to understand interactions between early-life adversity and trade-offs in hyperinflammatory responses of the immune system (DeWitte & Bekvalac, 2011). While informative about interactions between stress and general physiological regulation, the epidemiological modeling approach does not include cumulative stress indices that are associated with studies of allostatic load.

14.7.2 Skeletal frailty index

A number of indices have been produced over the past decade that measure cumulative stress burdens in the past. These indices were calculated using presence or absence of skeletal indicators of stress, activity and disease that corresponded with an economically defined quality of life (Steckel et al.,

2002b). Later studies adopted a skeletal frailty index (SFI) around the concepts of allostasis and allostatic load (Marklein et al., 2016; Marklein & Crews, 2017). SFIs were developed using multiple biomarkers of stress, including skeletal and dental indicators of growth disruption, nutritional deficiency, infection, activity and trauma. The SFI may be compared between skeletal samples and incorporated into mortality models. While successful at measuring cumulative stress burden, mortality risks in relation to the SFI do not follow predicted patterns (Marklein & Crews, 2017). Senescent diseases such as osteoporosis, fractures and intervertebral disc disease were included and may have inflated SFIs among older individuals (e.g. Jurmain, 1999; Ortner, 2003). Another limitation to bioarchaeological studies of SFI is skeletal sampling. Estimates of SFI require the preservation of nearly complete skeletal remains. However, one challenge of bioarchaeological research is a near-universal reliance on fragmentary, incomplete assemblages, and as a result, sample sizes used to calculate SFI may be diminished. The SFI is an important contribution to exploring allostatic load in the past by including multiple biomarkers of stress and epidemiological models of mortality. Despite some limitations, the SFI carries great potential for the study of stress in mostly complete samples of human skeletal remains. Future approaches should pursue a critical perspective on the biomarkers included in the calculation of SFI and the relationship between these markers with mortality.

14.7.3 Life course and life history perspectives

Recent theoretical expansions of stress in bioarchaeology have been couched within evolutionary developmental biology, developmental systems theory, Developmental Origins of Health and Disease and LHT (Agarwal, 2016; Armelagos et al., 2009; Gowland, 2015; Temple, 2014, 2019a; Temple & Goodman, 2014; see Chapter 2). The life course paradigm evaluates skeletal indicators of stress within the context of trajectory and plasticity (Agarwal, 2016). Trajectory references the arc of development that may change in response to environmental and cultural experiences. Plasticity in reference to stress is defined in two contexts by bioarchaeological research. First, plasticity is generally defined as the capacity to alter phenotypes in response to environmental experience, and skeletal indicators of stress act as evidence for this malleability (Agarwal, 2016; Gowland, 2015; Temple, 2014, 2019b). Here, bioarchaeologists recognise that adaptive plasticity refers directly to the alteration of phenotypes that occur through physiological pathways that move populations towards fitness peaks by facilitating short-term survival (Agarwal, 2016; Temple, 2014, 2019b). An equally valuable approach emphasises intergenerational transmission of these phenotypes through the embodiment of maternal stress experiences during foetal development, infancy and childhood (Gowland, 2015; Meaney et al., 1991; 2007; Wells, 2016). Plasticity is observed through changes in skeletal phenotypes that result from the interconnected relationship between mothers and offspring, and in this sense may speak to transgenerational experiences.

Bioarchaeological studies of stress during the life course explore skeletal and dental phenotypes that may be altered by physiological disruptions, and often attempt to understand the duration, periodicity and timing of these events in relation to local environmental conditions. Enamel records evidence of growth disruptions that are possible to identify according to precise developmental timings

(Goodman & Rose, 1990; Hillson & Bond, 1997; Reid & Dean, 2006; King et al., 2002; Guatelli-Steinberg et al., 2012, 2014; Temple et al., 2012; Hillson, 2014; also see Chapter 6). Enamel does not remodel and allows these experiences to be understood across the life course, especially when placed in context with stable isotopes (Box 14.3; see also Chapter 5), historical record for diet and mortuary practices (Austin et al., 2013; Dirks et al., 2002, 2010; Garland et al., 2018; Guatelli-Steinberg et al., 2014; Humphrey, 2008; Sandberg et al., 2014; Temple, 2018, 2020). Stress experiences recorded in enamel may also be reconstructed in terms of periodicity and duration in relation to local environmental and cultural conditions (Guatelli-Steinberg et al., 2004; King et al., 2005; Merrett et al., 2016; Temple, 2016; Temple et al., 2013).

Box 14.3   Stable isotope analysis and intergenerational transmission of stress

Stable isotope analysis of carbon and nitrogen ratios derived from prenatal dentin may be compared to adult female means derived from bone collagen to gather a stronger indication of stress during pregnancy (Beaumont et al., 2015). Individuals who survive childhood have similar nitrogen profiles in prenatal enamel and bone collagen, which are also comparable to adult female mean nitrogen values. By contrast, the nitrogen profiles of non-survivors are variable, which suggests the possibility that stress in the prenatal and perinatal environment contributed to increased mortality risk. These results document the ways maternal stress may become embodied in offspring and the importance of developmental experiences in influencing life course trajectories (e.g. Gowland, 2015).

Physiological constraint was introduced as a theoretical construct from LHT to expand the life course approach to the context of death, and emphasises two important concepts: organisms do not experience limitless plasticity, and surviving early-life adversity often reduces investments in future growth and survival (Temple, 2014, 2019a). Bioarchaeology is, however, a contextual discipline. The use of context allows relationships between early-life adversity with morbidity and mortality to be explored in culturally and ecologically contingent frameworks (Temple, 2019a). In this sense, LHT unifies the adaptive plasticity component included in life course perspectives with trade-offs encountered at later stages of the life cycle. Physiological constraint is comparable to the trajectory concept but interacts directly with evolutionary life history and focuses specifically on limitations on adaptation. This approach is of great advantage to bioarchaeologists who use altered phenotypes as an indication of plasticity and compare surviving these episodes with skeletal evidence of morbidity and mortality risk.

The formation, timing and periodicity of stress indicators such as enamel hypoplasia or crypt fenestration enamel defects may be compared with factors such as mortality risk, diaphyseal growth and chronic infection at later stages of the life course (DeWitte & Bekvalac, 2011; Garland, 2020; Ham et al., 2021; Lorentz et al., 2019; Sciulli & Oberly, 2002; Temple, 2014; Thomas et al., 2019). In addition, the presence of enamel hypoplasia has also been used to explore relationships with somatic growth including adult stature (Floyd, 2007; Floyd & Littleton, 2006; Sciulli & Oberly, 2002) and vertebral neural canal stunting (Watts, 2013; see also Chapter 2). Relationships between enamel growth were associated with reduced adult stature in cases where defects were identified at early stages of the life cycle (Floyd, 2007; Floyd & Littleton, 2006), while vertebral neural canal dimensions show no relationship with these defects (Watts, 2013). However, vertebral neural canal dimensions develop at the very earliest stages of the lifespan and relationships between vertebral neural canal stunting and mortality have been reported, suggesting that surviving these growth disruptions may elicit trade-offs with mortality (Newman & Gowland, 2015; Watts, 2011, 2013, 2015). Surviving conditions that produce cribra orbitalia does not elicit mortality trade-offs in samples of adults (MacFadden & Oxenham, 2020), although mortality risk from ages 15 to 30 years was 3.9 times higher among individuals with cribra orbitalia in the same sample (Steckel, 2005). These findings suggest complex and contingent relationships between surviving stress in the early life environment and physiological trade-offs at later stages of the life cycle (see Chapter 2).

Stable isotope analysis has also been integrated into studies of enamel using life history perspectives. For example, quality of weaning foods is associated with mortality risk (Reitsema et al., 2016). Isotopic explorations suggest that individuals who experience weaning over extended periods of time experience elevated mortality risks compared to those who are weaned more quickly at younger ages (Sandberg et al., 2014). This differs from previous research that suggested elevated stress levels in a case of abrupt weaning (Austin et al., 2013). This may be associated with differences in the quality of weaning foods at Kulubnarti (Sandberg et al., 2014), and indeed, quality of weaning foods is associated with stress events and mortality (Reitsema et al., 2016; Garland et al., 2018). These studies imply that diet may be used to track stress experience in relation to mortality and life history trade-offs.

There is great value in life course and life history perspectives. First, these approaches draw upon evolutionary understandings of plasticity in relation to surviving stress events and resultant phenotypic alterations. Intergenerational explorations of plasticity are of paramount importance as these studies link stress experience across the entangled lives of parents and offspring. In addition, LHT focuses on the limits of adaptive plasticity through the exploration of physiological constraint. Finally, life course and LHT often incorporate epidemiological modelling to explore interactions with morbidity and mortality across the lifespan. One limitation of life course and life history perspectives must be referenced: the approach is not cumulative and mostly relies on non-specific indicators of stress experience. Greater interaction with pathophysiology may provide an important step forward in unlocking systemic conditions involved in the production of skeletal and dental indicators of stress (see also Chapter 2).

14.7.4 Pathophysiological approaches

EM and epidemiology look beyond cortisol as a singular stress regulator and consider the broad array of molecular factors involved in allostasis (Edes & Crews, 2017). Bioarchaeologists must be familiar with the molecular signalling pathways governing the appearance and distribution of lesions in the human skeleton (Gosman, 2012; Klaus, 2014; Ragsdale & Lehmer, 2012). For example, osteogenic signalling factors are triggered as downstream responses to inflammatory events, and this challenges assertions that periostosis does not reflect stress due to the general osteoclastogenic response of cortisol (Klaus, 2014). Stress hormones such as parathyroid hormone act as signalling factors for bone morphogenic proteins, while other osteogenic transcription factors are independently expressed in response to inflammation (Klaus, 2014). Similar findings are reported in cases of scurvy, where the depletion of ascorbic acid yields a strong influence over collagen formation, haemorrhaging, periosteal rupture and creation of new capillary networks via osteoclastic action on the cranial cortex (Klaus, 2017). Comparisons of lesions associated with scurvy have been useful for studies understanding the process of adaptation in relation to colonialism and changing climates (Klaus, 2017; Snoddy et al., 2020). These results suggest that the application of pathophysiology follows models established by EM and epidemiology by considering the multi-scalar pathways associated with stress.

Microscopic approaches to bone formation and resorption are another way forward in providing a more finite understanding of stress manifestation in the past. For example, the underlying aetiology of porotic hyperostosis and cribra orbitalia are consistently debated, with remarkably different interpretive value reflected in differential diagnoses (Angel, 1966; El-Najjar et al., 1976; Klaus, 2017; Oxenham & Cavill, 2010; Walker, 1985, 1986; Walker et al., 2009). The variety of conditions associated with these lesions include iron deficiency anaemia, rickets, scurvy and infectious disease, and these may be differentiated through careful evaluation of contributions to cranial porosity (Ortner, 2003; Schultz, 2001). Microscopic methods provide a more complete understanding surrounding the presence of new bone formation atop the external cortex, diploic expansion through the external cortex or resorption of the external cortex—skeletal phenotypes that provide near pathognomonic evidence of the underlying condition responsible for these lesions (Schultz, 2001; Wapler et al., 2004). These methods allow bioarchaeologists to move beyond

evaluations of general phenotypic alteration and examine the physiological pathways that later produce skeletal phenotypes.

Integrated studies of tooth microstructure, stable isotopes and proteins promise to shed light on deeper systemic processes involved in the stress experience. Microscopic striae are observable in human dentin as Anderson lines or peri-radicular bands (Smith & Reid, 2009; see also Chapter 6). These bands are comparable to striae of Retzius in periodicity and timing, while accentuated Anderson lines are associated with stress experience (Dean, 1999; Smith & Reid, 2009). Recent studies found elevated barium isotopes due to weight loss within accentuated Anderson lines (Austin et al., 2016). Heat-shock proteins that prevent denaturation of odontoblasts (HSP-70) were also recovered from dentin layers where accentuated lines were formed (Austin et al., 2016). Such results portend a bright future for the study of stress in the human past through the incorporation of novel methods that identify systemic pathways involved in the formation of skeletal and dental indicators of stress.

The pathophysiological approach reveals the systemic processes associated with skeletal phenotypes altered by stress. By contrast, life course, allostatic load and epidemiological modelling often discuss indicators of stress as non-specific responses to specific causative agents. In this sense, skeletal pathophysiology provides a more finite understanding of stress processes in the human past. In addition, skeletal pathophysiology has the capacity to understand stress as a multi-scalar event, rather than simply focusing on the HPA axis. There are, however, limitations to this approach. At present, skeletal pathophysiology does not include estimates of morbidity and mortality in association with these highly detailed descriptions of lesions. This research tends to focus on singular conditions, and therefore does not provide a cumulative index of stress within a sample. However, there exists great promise for integration, as combining detailed pathophysiological descriptions of lesions with morbidity and mortality estimates would answer challenges that have existed over decades within the palaeopathological community (i.e. Ortner, 1991). In addition, knowledge of pathophysiological pathways for lesion formation may be used to improve knowledge of biomarkers used in cumulative approaches to stress in the past.

14.8 Conclusions

Bioarchaeology and EM have followed comparable historical trajectories in stress research; both disciplines began as descriptive fields. A comparative approach that emphasised local, changing environments gained traction in the mid-twentieth century and was formalised under a broader conceptual framework that saw social and environmental conditions as paramount to stress experience. The comparative approach was later supplanted by research that found links between developmental and social environments as predictors of later-life morbidity and mortality, and this work interacted with evolutionary life history. While bioarchaeological research relied on comparative methods through the early twenty-first century, a new emphasis on theory prompted questions relating developmental environment to morbidity and mortality experienced at later ages. In this sense, skeletal indicators of stress are not simply variables to be compared in terms of prevalence, but instead as records of experience that have important connections to physiological outcomes across the lifespan.

Modern bioarchaeological research has evolved to rely upon EM and epidemiology in the development of methods, new theoretical paradigms and the application of laboratory technology to explore stress in human skeletal remains. Once a near-universally comparative discipline, bioarchaeologists have deployed epidemiological modelling to understand how stress in the past has contributed to morbidity and mortality. In addition, bioarchaeological research has benefited from theoretical paradigms centred on human life course and LHT. Bioarchaeologists are trained in microscopy, chemistry and advanced skeletal biology to provide greater clarity on stress experience. Some studies target specific indicators of stress experienced at early ages of development and track risks with morbidity and mortality experienced at later ages, while others engage in a cumulative systemic perspective. These choices are contingent and guided by the preservation of skeletal remains as well as the interests and expertise of the observer. In

this sense, it becomes impossible to advocate a singular bioarchaeological approach to the study of stress. Each has advantages and limitations that depend upon context (Table 14.1). The important point to emphasise is that interaction with EM, epidemiology, evolutionary biology and social theory have promoted approaches to stress that are gaining increasing resonance within the bioarchaeological community (e.g. Agarwal, 2016; Armelagos et al., 2009; Gowland, 2015; Temple, 2019a).

EM and epidemiological research have immediate ramifications for contemporary studies of stress, and many studies highlight the transformational nature of this work (Gillman et al., 2007; Nesse & Williams, 1994; Stearns et al., 2010; Trevathan et al., 1999; Wells, 2016). Bioarchaeologists are just beginning to leverage the deep time perspective to problems of contemporary import (Buikstra, 2019). EM and epidemiology interact with two of the most pressing challenges confronting human populations, including global climate change and systemic inequality. Global climate change increases uncertainty surrounding access to resources, and as local environments experience shifting climatic cycles, increased stress due to diminished access to food, water and disruptions to food production are predicted (Beniston, 2010; McMichael et al., 1996). Archaeological research finds that climate change elicits drastic transformation in rigid socioecological and cultural systems (Hegmon et al., 2008; Nelson et al., 2016). Increases in stress are noted following these transitions, suggesting that flexibility in socio-ecological and cultural systems may buffer against human suffering (Hegmon et al., 2008). Indeed, bioarchaeological studies of stress indicate that flexibility in socio-ecological and cultural systems may reduce precarity in response to climate change by facilitating resilience within them (Bartelink et al., 2019; Berger & Wang, 2020; Snoddy et al., 2020; Temple, 2019b). These findings carry transformational value by emphasising adaptability and flexibility in a world that is destabilised by inequality and rigidity in the face of a rapidly changing environment (see also Chapter 7).

EM and epidemiology find that the stressors introduced by colonialism, enslavement and systemic racism severely imperil community well-being (Benyshek et al., 2001; Gravlee et al., 2009;

Kuzawa & Sweet, 2009). Much of this violence is epigenomic, and these experiences can be transmitted to future generations (Jackson et al., 2018; Kuzawa & Sweet, 2009; see also Chapter 2). Bioarchaeological studies of stress also reveal the disproportionate impacts of colonialism, enslavement and systemic racism (Blakey et al., 1994, 2009; de la Cova, 2011, 2014; Larsen & Milner, 1994; Murphy & Klaus, 2017; Rankin-Hill, 1997; Verano and Ubelaker, 1992). Bioarchaeologists advocate a life history approach to these questions (Temple, 2019a). Results of these studies uncover evidence for inequality in mortality following survival of early-life stress in colonial environments (Ham et al., 2021; Thomas et al., 2019). As many institutions reconcile with these historically violent events, bioarchaeological studies of stress should be at the forefront of conversations that provide impact statements for descendant communities. This approach begins with training (and funding) a diverse generation of scholars (Gomez, 2015) and engaging with theoretical perspectives that emphasise marginalised viewpoints (Watkins, 2020). The integration of these perspectives suggests that bioarchaeological research on stress is poised to interact with questions of great relevance to the contemporary world.

References

Accortt, E. E., Mirocha, J., Dunkel Schetter, C. & Hobel, C. J. (2017). Adverse perinatal outcomes and postpartum multi-systemic dysregulation: adding Vitamin D deficiency to the allostatic load index. *Maternal and Child Health Journal*, 21(3), 398–406.

Acuso-Rivero, C., Murren, C. J., Schlichting, C. D., & Steiner, U. K. (2019). Adaptive phenotypic plasticity for life-history and less fitness-related traits. *Proceedings of the Royal Society B: Biological Sciences*, 286, 20190653.

Agarwal, S. C. (2016). Bone morphologies and life histories: life course approaches in bioarchaeology. *Yearbook of Physical Anthropology*, 159(S61), 130–149.

Angel, J. L. (1946). Skeletal change in ancient Greece. *American Journal of Physical Anthropology*, 4(1), 69–98.

Angel, J. L. (1947). Increase in length of life in ancient Greece. *American Journal of Physical Anthropology*, 5(2), 231.

Angel, J. L. (1954). Human biology, health, and history in Greece from first settlement until now. *Yearbook of the American Philosophical Society*, 98, 168–174.

Angel, J. L. (1966). Porotic hyperostosis, anemias, malarias, and marshes in the prehistoric Mediterranean. *Science*, 153(3737), 760–763.

Armelagos, G. J. (1969). Disease in ancient Nubia. *Science*, 163(3864), 255–259.

Armelagos, G. J. (2003). Bioarchaeology as anthropology. *Archaeological Papers of the American Anthropological Association*, 13, 27–41.

Armelagos, G. J., Goodman, A. H., Harper, K. N. & Blakely, M. L. (2009). Enamel hypoplasia and early mortality: bioarchaeological support for the Barker hypothesis. *Evolutionary Anthropology*, 18(6), 261–271.

Armelagos, G. J. & Van Gerven, D. (2003). A century of skeletal biology and paleopathology: contrasts, contradictions, and conflicts. *American Anthropologist*, 105(1), 53–64.

Austin, C., Smith, T. M., Bradman, A., Hinde, K., Joannes-Boyau, R., Bishop, D., Hare, D. J., Doble, P., Eskenazi, B. & Arora, M. (2013). Barium distributions in teeth reveal early-life dietary transitions in primates. *Nature*, 498, 216–219.

Austin, C., Smith, T. M., Farhani, R. M. Z., Hinde, K., Carter, E. A., Lee, J., Lay, P. A., Kennedy, B. J., Sarrafpour, B., Wright, R. J., Wright, R. O. & Arora, M. (2016). Uncovering system-specific stress signatures in primate teeth with multimodal imaging. *Science Reports*, 6, 18802.

Baker, P. T. (1969). The adaptive limits of human populations. *Man*, 19(1), 1–14.

Baker, P. T. (1978). *The biology of high-altitude peoples*. Cambridge University Press, New York.

Banning, C. (1946). Food shortages and public health, first half of 1945. *The Annals of the American Academy of Political and Social Sciences*, 245(1), 93–110.

Barboza Solís, C., Kelly-Irving, M., Fantin, R., Darnaudery, M, Torrisani, J., Lang, T. & Delpierre, C. (2015). Adverse childhood experiences and physiological wear-and-tear in midlife: findings from the 1958 British birth cohort. *Proceedings of the National Academy of Sciences of the United States of America*, 112(7), E738–E746.

Barker, D. J. (1992). *Fetal and infant origins of adult health and disease*. Tavistock, London.

Barker, D. J., Bull, A. R., Osmond, C. & Simmons, S. J. (1990). Fetal size and placental size in risk of hypertension in adult life. *British Medical Journal*, 301(6746), 259–262.

Barker, D. J. & Osmond, C. (1986). Infant mortality, childhood, nutrition, and ischaemic heart disease in England and Wales. *The Lancet*, 8489, 1077–1081.

Barker, D. J., Osmond, C. & Law, C. M. (1989). The intrauterine and early postnatal origins of cardiovascular disease and bronchitis. *Journal of Epidemiology and Community Health*, 43, 237–240.

Barrett, E. S., Vitek, W., Mbowe, O., Thurston, S. W., Legro, R. S., Alvero, R., Baker, V., Wright Bates, G., Casson, P., Coutifaris, C., Eisenberg, E., Hansen, K., Krawertz, K., Robinson, R., Rosen, M., Usadi, R., Zhang, H., Santoro, N. & Diamond, M. (2018). Allostatic load, a measure of chronic physiological stress, is associated with pregnancy outcomes, but not fertility, among women with unexplained infertility. *Human Reproduction*, 33(9), 1757–1766.

Bartelink, E. J., Bellifemine, V. I., Nechayev, I., Andrushko, V. A., Leventhal, A. & Jurmain, R. (2019). 'Biocultural perspectives on interpersonal violence in the prehistoric San Francisco Bay area', in D. H. Temple & C. J. Stojanowski (eds), *Hunter-gatherer adaptation and resilience: a bioarchaeological perspective*, pp. 274–302. Cambridge University Press, Cambridge.

Bateson, P., Barker, D., Clutton-Brock, T., Deb, D., D'Undine, B., Foley, R. A., Gluckman, P., Godfrey, K., Kirkwood, T., Mirazón Lahr, McNamara, J., Metcalfe, N. B., Monagan, P., Spencer, H. G. & Sultan, S. E. (2004). Developmental plasticity and human health. *Nature*, 430 (6998), 419–421.

Beaumont, J., Montgomery, J., Buckberry, J. & Jay, M. (2015). Infant mortality and isotopic complexity: new approaches to stress, maternal health, and weaning. *American Journal of Physical Anthropology*, 157(3), 441–557.

Beckie, T. M. (2012). A systematic review of allostatic load, health, and health disparities. *Biological Research for Nursing*, 14, 311–346.

Beniston, M. (2010). Climate change and its impacts: growing stress factors for human societies. *International Review of the Red Cross*, 92(879), 1–12.

Benyshek, D. C., Martin, J. F. & Johnston, C. F. (2001). A reconsideration of the origins of the type 2 diabetes epidemic among Native Americans and the implications for intervention policy. *Medical Anthropology*, 20, 25–64.

Berger, E. & Wang, H. (2020). 'Climate change and adaptive systems in Bronze Age, Gansu, China', in G. Robbins-Schug (ed.), *The Routledge handbook of the bioarchaeology of climate and environmental change*, pp. 83–102. Routledge, New York.

Bernard, C. (1865). *Introduction à l'étude de la médecine expérimentale*. J.-B. Baillière, Paris.

Blakey, M. L., Leslie, T. E. & Reidy, J. P. (1994). Frequency and chronological distribution of dental enamel hypoplasia in enslaved African Americans: a test of the weaning hypothesis. *American Journal of Physical Anthropology*, 95(4), 371–383.

Blakey, M. L., Mack, M. E., Barrett, A. R., Mahoney, S. S. & Goodman, A. H. (2009) 'Childhood health and dental development' in M. L. Blakey & L. M. Rankin-Hill (eds), *The New York African burial ground: unearthing the African presence in colonial New York, Volume 1: skeletal biology of the New York African burial ground*, pp. 143–156. Howard University Press, Washington.

Boas, F. (1930). Observations on the growth of children. *Science*, 72, 44–48.

Boldsen, J. L. (2006). Early childhood stress and adult mortality—a study of dental enamel hypoplasia in the medieval Danish village of Tirup. *American Journal of Physical of Anthropology*, 132(1), 59–66.

Bowditch, H. P. (1879). *The growth of children (a supplementary investigation): with suggestions in regard to method of research*. Kand, Abery, and Company, Boston.

Breuner, C. W., Patterson, S. H. & Hahn, T. P. (2008). In search of relationships between the acute adrenocortical response and fitness. *General and Comparative Endocrinology*, 157(3), 288–295.

Broom, D. (2017). Cortisol: often not the best indicator of stress and poor welfare. *Physiology News*, 107, 30–32.

Brown, D. E. (1981). General stress in anthropological fieldwork. *American Anthropologist*, 83, 74–92.

Buford, T. W. & Willoughby, D. S. (2008). Impact of DHEA(S) and cortisol on immune function in aging: a brief review. *Applied Physiology, Nutrition, and Metabolism*, 33, 429–433.

Buikstra, J. E. (1976). *Hopewell in the Lower Illinois River Valley: a regional approach to the study of human biological variability and prehistoric behavior*. Northwestern University Archaeological Program Scientific Papers, No. 2. Northwestern University, Evanston.

Buikstra, J. E. (1977). 'Biocultural dimensions of archaeological study: a regional perspective', in R. L. Blakely (ed), *Biocultural adaptation in prehistoric America*, pp. 67-84. University of Georgia Press, Athens.

Buikstra, J. E. (1981). 'Mortuary practices, paleodemography, and paleopathology', in R. Chapman, I. Kinnes & K. Randsborg (eds), *The archaeology of death*, pp. 123–132. Cambridge University Press, Cambridge.

Buikstra, J.E. (2019). *Bioarchaeologists speak out: deep time perspectives on contemporary issues*. Springer, New York.

Buikstra, J. E. & Cook, D. C. (1980). Paleopathology: an American account. *Annual Reviews in Anthropology*, 9, 433–470.

Cannon, W. B. (1915). *Bodily changes in pain, hunger, fear, and rage: an account of recent researches into the function of emotional excitement*. D Appleton and Company, Boston.

Cannon, W. B. (1932). *The wisdom of the body*. W. W. Norton and Company, New York.

Cassel, J. & Tyroler, H. A. (1961). Epidemiological studies of culture change: I. Health status and recency of industrialization. *Archives of Environmental Health*, 3, 25–33.

Charmandari, E., Tsigos, C. & Chrousos, G. (2005). Endocrinology of the stress response. *Annual Review of Physiology*, 67, 259–284.

Charnov, E. L. (1993). *Life history invariants: some explorations of symmetry in evolution and ecology*. Oxford University Press, Oxford.

Chen, X., Redline, S., Shields, A. E., Williams, D. R. & Williams, M. A. (2014). Associations of allostatic load with sleep apnea, insomnia, short sleep duration, and other sleep disturbances: findings from the National Health and Nutrition Examination Survey 2005 to 2008. *Annals of Epidemiology*, 24, 612–619.

Christensen, D. S., Flensborg-Madsen, T., Garde, E., Hanse, A. M., Masters Pederson, J. & Lykke Mortensen, E. (2018). Early life predictors of midlife allostatic load: a prospective cohort study. *PloS One*, 13(8), e0202395.

Cobb, W. M. (1936). *The laboratory of anatomy and physical anthropology at Howard University*. Howard University, Washington.

Cohen, M. N. (1977). *The food crisis in prehistory*. Yale University Press, New Haven.

Cohen, M. N. & Armelagos, G. J. (1984). *Paleopathology at the origins of agriculture*. Academic Press, Orlando.

Cohen, M. N. & Crane-Kramer, J. (2012). *Ancient health: skeletal indicators of agricultural and economic intensification*. University Press of Florida, Gainesville.

Cook, D. C. (1981), 'Mortality, age structure, and status in the interpretation of stress indicators in prehistoric skeletons: a dental example from the Lower Illinois River Valley', in R. Chapman, I. Kinnes, & K. Randsborg (eds), *The archaeology of death*, pp. 133–144. Cambridge University Press, Cambridge.

Cook, D. C. (1984). 'Subsistence and health in the Lower Illinois valley: osteological evidence' in M. N. Cohen & G. J. Armelagos (eds), *Paleopathology at the origins of agriculture*, pp. 235–269. Academic Press, Orlando.

Crespi, E. J. & Denver, R. J. (2005). Ancient origins of human developmental plasticity. *American Journal of Human Biology*, 17(1), 44–54.

Crimmins, E. M., Kim, J. K. & Seeman, T. E. (2009). Poverty and biological risk: the earlier 'aging' of the poor. *The Journals of Gerontology. Series A, Biological Sciences and Medical Sciences*, 64A, 286–292.

Danese, A. & McEwen, B. S. (2012). Adverse childhood experiences, allostasis, allostatic load, and age-related disease. *Physiology and Behavior*, 106, 29–39.

de La Cova, C. (2011). Race, health, and disease in 19th century born males. *American Journal of Physical Anthropology*, 144, 526–537.

de La Cova C. (2014). 'The biological effects of urbanization and in-migration on 19th-century-born African Americans and European Americans of low socioeconomic status: an anthropological and historical approach', in M. K. Zuckermann (ed), *Are modern environments bad for health? Revisiting the second epidemiological transition*, pp 243–264. Wiley Blackwell, New York.

Dean, M. C. (1999). 'Hominoid tooth growth: using incremental lines in dentine as markers of growth in modern human and fossil primate teeth', in R. D. Hoppa & C. M. Fitzgerald (eds), *Human growth in the past: studies from bones and teeth*, pp. 111–127. Cambridge University Press, Cambridge.

DeWitt, T. J., Sih, A. & Wilson, D. S. (1998). Costs and limits of phenotypic plasticity. *Trends in Ecology and Evolution*, 13, 77–81.

DeWitte, S. N. (2010). Sex differences in frailty in medieval England. *American Journal of Physical Anthropology*, 143(2), 285–297.

DeWitte, S. N. (2014a). Health in post-Black Death London (AD 1350–1538): age patterns of periosteal new bone formation in a post-epidemic population. *American Journal of Physical Anthropology*, 155(2), 260–267

DeWitte, S. N. (2014b). Differential survival among individuals with active and healed periosteal new bone formation. *International Journal of Paleopathology*, 7, 38–44.

DeWitte, S.N. (2017). 'Sex and frailty in medieval Europe: Patterns from catastrophic and attritional assemblages', in S. C. Agarwal & J. Wesp (eds), *Exploring sex and gender in bioarchaeology*, pp. 189–222. University of New Mexico Press, Albuquerque.

DeWitte, S.N. & Bekvalac, J. (2011). The association between periodontal disease and periosteal lesions. *American Journal of Physical Anthropology*, 145(4), 609–618.

DeWitte, S. N. & Hughes-Morey, G. (2012). Stature and frailty during the Black Death: the effect of stature on risks of epidemic mortality in London, A.D. 1348-1350. *Journal of Archaeological Science*, 39, 1412–1419.

DeWitte, S. N. & Stojanowski, C. M. (2015). The osteological paradox 20 years later: past perspectives and future directions. *Journal of Archaeological Research*, 23, 397–450.

DeWitte, S. N. & Wood, J. W. (2008). Selectivity of Black Death mortality with respect to pre-existing health. *Proceedings of the National Academy of Sciences of the United States of America*, 105, 1436–1441.

Dirks, W., Humphrey, L. T., Dean, M. C. & Jeffries, L. T. (2010). The relationship of accentuated lines in enamel to weaning stress in juvenile baboons. *Folia Primatologica*, 81(4), 207–223.

Dirks, W., Reid, D.J., Jolly, C. & Phillips-Conroy, J. (2002). Out of the mouths of baboons: stress, life history, and dental development in the Awash National Park hybrid zone, Ethiopia. *American Journal of Physical Anthropology*, 118, 239–252.

Dobzhansky, T. (1956). Genetics of natural populations. XXV. Genetic changes in populations of *Drosophila pseudoobscura* and *Drosophila persimilis* in some localities in California. *Evolution*, 10(1), 82–92.

Dobzhansky, T. (1968). 'On some fundamental concepts of Darwinian biology', in T. Dobzhansky, M. K. Hecht & W. C. Steere (eds), *Evolutionary biology*, vol. 2, pp. 1–34. Springer, New York.

Donnison, C. P. (1929). Blood pressure in the African native: its bearing upon the aetiology of hyperpiesia and arterio-sclerosis. *The Lancet*, 213(5497), 6–7.

Dressler, W. M. (1990). 'Culture, stress, and disease', in T. M. Johnson & C. F Sargent (eds), *Medical anthropology: a handbook of theory and method*, pp. 248–267. Greenwood Press, New York.

Dressler, W. M. (2011). 'Culture and the stress process', in M. Singer & P. Erickson (eds), *A companion to medical anthropology*, pp. 119–134. Wiley-Blackwell, New York.

Dubois, W. E. B. (1914). Our baby pictures. *The Crisis*, 8(6), 298–303.

Edes, A. N. & Crews, D. E. (2017). Allostatic load and biological anthropology. *American Journal of Physical Anthropology*, 162(S63), 44–70.

Ellison, P. W. & Jasienska, G. (2007). Adaptation, constraint, and pathology: how do we tell them apart? *American Journal of Human Biology*, 19(5), 1342–1348.

El-Najjar, M. Y., Ryan, D. J., Turner II, C. G. & Lozoff, B. (1976). The etiology of porotic hyperostosis among prehistoric and historic Anasazi Indians of Southwestern United States. *American Journal of Physical Anthropology*, 44(3), 477–487.

Everly, G. S. & Lating, J. M. (2013). 'The anatomy and physiology of the human stress response', in G. S. Everly & G. M. Lating (eds), *A clinical guide to the treatment of the human stress response*, 3rd edn, pp. 17–51. Springer, New York.

Floyd, B. (2007). Focused life history data and linear enamel hypoplasia to help explain intergenerational variation in leg length within Taiwanese families. *American Journal of Human Biology*, 19(3), 358–375.

Floyd, B. & Littleton, J. (2006). Linear enamel hypoplasia and growth in an Australian Aboriginal community: not so small, but not so healthy either. *Annals of Human Biology*, 33(4), 424–443.

Forsdahl, A. (1977). Are poor living conditions in childhood and adolescence an important risk factor for arteriosclerotic heart disease? *British Journal of Preventative Medicine*, 31(2), 91– 95.

Frisancho, A. R. & Baker, P. T. (1970). Altitude and growth: a study of the physical growth patterns of a high-altitude Peruvian Quechua population. *American Journal of Physical Anthropology*, 32(2), 279–292.

Frisancho, A. R., Guire, K., Babler, W., Borkan, G. & Way, A. (1980). Nutritional influence on childhood development and genetic control of adolescent growth of Quechuas and Mestizos from the Peruvian lowlands. *American Journal of Physical Anthropology*, 52(3), 367–375.

Frisancho, A. R., Sanchez, J., Pallardel, D. & Yanez, L. (1973). Adaptive significance of small body size under poor nutrient conditions in southern Peru. *American Journal of Physical Anthropology*, 39(2), 255–262.

Futuyma, D. J. (1998). *Evolutionary biology*. Sinauer and Associates, Sunderland.

Garland, C. J. (2020). Implications of accumulative stress burdens during critical periods of early postnatal life for mortality risk among Guale interred in a colonial era cemetery in Spanish Florida (ca. AD 1605–1680). *American Journal of Physical Anthropology*, 172(4), 621–637.

Garland, C. J., Reitsema, L. J., Larsen, C. S. & Thomas, D. H. (2018). Early life stress at Mission Santa Catalina de Guale: An integrative analysis of enamel defects and dentin incremental isotope variation in malnutrition. *Bioarchaeology International*, 2(2), 75–84.

Garn, S. M., Rohmann, C. G., Behar, M., Viteri, F. & Guzman, M.A. (1964).Compact bone deficiency in protein-calorie malnutrition. *Science*, 145(3639), 1444–1445.

Ghalambor, C. K., McKay, J. K., Carroll, S. P. & Reznick, D. N. (2007). Adaptive versus non- adaptive phenotypic plasticity and the potential for contemporary adaptation in new environments. *Functional Ecology*, 21(3), 394–407.

Gillman, M. W., Barker, D., Bier, D., Cagampang, F., Challis, J., Fall, C., Godfrey, K., Gluckman P., Hanson, M., Kuh, D., Nathanielsz, P., Nestel, P. & Thornburg, K. L. (2007). Meeting report on the 3rd international congress on developmental origins of health and disease (DOHaD). *Pediatric Research*, 61, 625–629.

Glei, D. A., Goldman, N., Chuang, Y-L. & Weinstein, M. (2007). Do chronic stressors lead to physiological dysregulation? Testing the theory of allostatic load. *Psychosomatic Medicine*, 69(8), 769–776.

Gluckman, P. D., Hanson, M. A., Morton, S. B. & Pinal, C. S. (2005). Life-long echoes—a critical reappraisal of the developmental origins of health and disease. *Biology of the Neonate*, 87(2), 127–139.

Gluckman, P. D., Hanson, M. A. & Spencer, H. G. (2007). Early life events and their consequences for later disease: a life history and evolutionary perspective. *American Journal of Human Biology*, 19(1), 1–19.

Gomez, M. (2015). Using the contents of the Cobb Research Laboratory to understand today's health disparities. *Backbone*, 1, 1–3.

Goodman, A.H. & Armelagos, G. J. (1989). Infant and childhood morbidity and mortality risks in archaeological populations. *World Archaeology*, 21, 225–243.

Goodman, A. H. & Martin, D. L. (2002). 'Reconstructing health profiles from human skeletal remains', in R. H. Steckel & J. C. Rose (eds), *The backbone of history: health and nutrition in the western hemisphere*, pp. 11–60. Cambridge University Press, Cambridge.

Goodman, A. H., Martin, D. L, Armelagos, G. J. & Clark, G. C. (1984). 'Indications of stress from bones and teeth', in M. N. Cohen & G. J. Armelagos (eds), *Paleopathology at the origins of agriculture*. pp. 13–49. Academic Press, Orlando.

Goodman, A. H. & Rose, J. C. (1990). Assessment of physiological perturbations from dental enamel hypoplasias and associated histological structures. *Yearbook of Physical Anthropology*, 33(S11), 59–110.

Goodman, A. H., Thomas, R. B., Swedlund, A. C. & Armelagos, G. J. (1988). Biocultural perspectives on stress in prehistoric, historic, and contemporary population research. *Yearbook of Physical Anthropology*, 31(S9), 169–202.

Gosman, J. H. (2012). 'The molecular biological approach in paleopathology', in A. L. Grauer (ed), *A companion to paleopathology*, pp. 76–96. Wiley-Blackwell, New York.

Gowland, R. L. (2015). Entangled lives: implications for the developmental origins of health and disease hypothesis. *American Journal of Physical Anthropology*, 158(4), 530–540.

Gravlee, C. (2009). How race becomes biology: Embodiment of social inequality. *American Journal of Physical Anthropology*, 139(1), 47–57.

Guatelli-Steinberg, D., Ferrell, R. & Spence, J. (2012). Linear enamel hypoplasia as an indicator of physiological stress in Great Apes: reviewing the evidence in light of enamel growth. *American Journal of Physical Anthropology*, 148(2), 191–204.

Guatelli-Steinberg, D., Larsen, C. S. & Hutchinson, D. L. (2004). Prevalence and duration of linear enamel hypoplasia: a comparative study of Neanderthals and Inuit foragers. *Journal of Human Evolution*, 47(1–2), 65–84.

Guatelli-Steinberg, D., Steinspring-Harris, T., Reid, D. J., Larsen, C. S., Hutchinson, D. L. & Smith, T. M. (2014). Chronology of linear enamel hypoplasia formation in the Krapina Neanderthals. *Paleoanthropology*, 2014, 431–445.

Guilliams, T. G. & Edwards, L. (2010). Chronic stress and the HPA axis. *Standard*, 9, 1–12.

Hales, C. N. & Barker, D. J. (2001). The thrifty phenotype hypothesis. *British Medical Bulletin*, 60(1), 5–20.

Ham, A. C., Temple, D. H., Klaus, H. D. & Hunt, D. R. (2021). Evaluating life history trade-offs through the presence of linear enamel hypoplasia at Pueblo Bonito and Hawikku: a biocultural study of early life stress and survival in the Ancestral Pueblo Southwest. *American Journal of Human Biology*, 33(2), e23506.

Harding, S., Silva, M. J., Molaodi, O. R., Enayat, Z. E., Cassidy, A., Karamano, A., Read, U. & Cruickshank, J. K. (2016). Longitudinal study of cardiometabolic risk from early adolescence to early adulthood in an ethnically diverse cohort. *BMJ Open*, 6, 1–10.

Hegmon, M., Peeples, M., Kinzig, A.P., Kulow, S., Meegan, C. & Nelson, M.C. (2008). Social transformation and its cost in the pre-Hispanic U.S. southwest. *American Anthropologist*, 110(3), 313–324.

Heim, C., Ehlert, U. & Hellhammer, D. H. (2000). The potential role of hypocortisolism in the pathophysiology of stress-related bodily disorders. *Psychoneuroendocrinology*, 25, 1–35.

Hillson, S. W. (2014). *Tooth development in human evolution and bioarchaeology*. Cambridge University Press, Cambridge.

Hillson, S. W. & Bond, S. (1997). The relationship of enamel hypoplasia to tooth crown growth: a discussion. *American Journal of Physical Anthropology*, 104(1), 89–103.

Hoon, R. S., Sharma, S. C., Balasubramanian, V., Chadha, K.S. & Mathew, O. P. (1976). Urinary catecholamine excretion on acute induction to high altitude (3,658 m). *Journal of Applied Physiology*, 41, 631–633.

Hooton, E. (1930). *The Indians of Pecos Pueblo: a study of their skeletal remains*. Phillips Academy, Andover.

Horan, J. M. & Widom, C. S. (2015), From childhood maltreatment to allostatic load in adulthood: the role of social support. *Child Maltreatment*, 20, 229–239.

Hostinar, C. E., Lachman, M. E., Mroczek, D. M., Seeman, T. E. & Miller, G. E. (2015). Additive contributions of childhood adversity and recent stressors to inflammation at midlife: findings from the MIDUS Study. *Developmental Psychology*, 51(11), 1630–1644.

Hostinar, C. E., Sullivan, R. M. & Gunnar, M. R. (2014). Psychobiological mechanisms underlying the social buffering of the hypothalamic-pituitary-adrenocortical axis: a review of animal models and human studies across development. *Psychological Bulletin*, 140(1), 256–282.

Humphrey, L. T., Dean, M. C., Jeffries, T. E. & Penn, M. (2008). Unlocking evidence of early diet from tooth enamel. *Proceedings of the National Academy of Sciences of the United States of America*, 105, 6834–6839. doi: 10.1073/pnas.0711513105

Huss-Ashmore, R., Goodman, A. H. & Armelagos, G. J. (1982). Nutritional inference from paleopathology. *Advances in Archaeological Method and Theory*, 5, 395–474.

Hux, V. J., Catov, J. M. & Roberts, J. M. (2014). Allostatic load in women with a history of low-birth-weight infants: the National Health and Nutrition Examination Survey. *Journal of Women's Health*, 23(12), pp. 1039–1045.

Jackson, L., Jackson, Z. & Jackson, F. (2018). Intergenerational resilience in response to the stress and trauma of enslavement and chronic exposure to institutionalized racism. *Journal of Clinical Epigenetics*, 4(15). doi: 10.21767/2472-1158.100100

Jackson, M. (2014). 'Evaluating the role of Hans Selye in the modern history of stress', in C. Ramsden (ed), *Stress, shock, and adaptation in the twentieth century*, pp. 21–48. University of Rochester Press, Rochester.

Johnson, E. O., Kamilaris, T. C., Chrousos, G. P. & Gold, P. W. (1992). Mechanisms of stress: a dynamic overview of hormonal and behavioral homeostasis. *Neuroscience and Biobehavioral Reviews*, 16, 115–130.

Johnston, F. E. (1962). Growth of the long bones of infants and children at Indian Knoll. *American Journal of Physical Anthropology*, 20(3), 249–253.

Jurmain, R. (1999). *Stories from the skeleton: behavioral reconstruction in human osteology*. Gordon and Breach, London.

Juster, R.-P., McEwen, B. S. & Lupien, S. J. (2010). Allostatic load biomarkers of chronic stress and impact on health and cognition. *Neuroscience and Biobehavioral Reviews*, 35(1), 2–16.

Kalantaridou, S. N., Zoumakis, E., Makrigiannakis, A., Lavasidis, L. G., Vrekoussis, T. & Chrousos, G. P. (2010). Corticotropin-releasing hormone, stress and human reproduction: an update. *Journal of Reproductive Immunology*, 85, 33–39.

Karlamangla, A. S., Singer, B. H., McEwen, B. S., Rowe, J. W. & Seeman, T. E. (2002). Allostatic load as a predictor of functional decline: MacArthur studies of successful aging. *Journal of Clinical Epidemiology*, 55(7), 696–710.

Karlamangla, A. S., Singer, B. H. & Seeman, T. E. (2006). Reduction in allostatic load in older adults is associated with lower all-cause mortality risk: MacArthur studies of successful aging. *Psychosomatic Medicine*, 68(3), 500–507.

Kemeny, M. E. (2003). The psychobiology of stress. *Current Directions in Psychological Science*, 12(4), 124–129.

Kimura, K. & Kitano, S. (1959). Growth of Japanese physiques in four successive decades after World War II. *Journal of the Anthropological Society of Nippon*, 67, 37–46.

King, T., Hillson, S. W. & Humphrey, L. T. (2002). A detailed study of enamel hypoplasia in a post-medieval adolescent of known age and sex. *Archives of Oral Biology*, 47(1), 29–39.

King, T., Humphrey, L. T. & Hillson, S. W. (2005). Linear enamel hypoplasias as indicators of systemic physiological stress: evidence from two known age-at-death and sex populations from post-medieval London. *American Journal of Physical Anthropology*, 128(3), 547–559.

Klaus, H. D. (2014). Frontiers in the bioarchaeology of stress and disease: cross-disciplinary perspectives from pathophysiology, human biology, and epidemiology. *American Journal of Physical Anthropology*, 155(2), 294–308.

Klaus, H. D. (2017). Paleopathological rigor and differential diagnosis: case studies involving terminology, description, and diagnostic frameworks for scurvy in skeletal remains. *International Journal of Paleopathology*, 19, 96–110.

Kobrosly, R. W., van Winjgaarden, E., Seplaki, C. L., Cory-Slechta, D. A. & Moynihan, J. (2014). Depressive symptoms are associated with allostatic load among community-dwelling older adults. *Physiology and Behavior*, 123, 223–230.

Korte, S. M., Koohaas, J. M., Wingfield, J. C. & McEwen, B. S. (2005). The Darwinian concept of stress: benefits of allostasis and costs of allostatic load and the trade-offs in health and disease. *Neuroscience and Biobehavioral Reviews*, 29(1), 3–38.

Kuzawa, C. (2005). Fetal origins of developmental plasticity: are fetal cues reliable predictors of future nutritional environments. *American Journal of Human Biology*, 17(1), 5–21.

Kuzawa, C. W. & Sweet, E. (2009). Epigenetics and the embodiment of race: developmental origins of US racial disparities in cardiovascular health. *American Journal of Human Biology*, 21(1), 2–15.

Larsen, C. S. (1987). Bioarchaeological interpretations of subsistence economy from human skeletal remains. *Advances in Archaeological Method and Theory*, 10, 339–445.

Larsen, C. S. (1995). Biological changes in human populations with agriculture. *Annual Reviews of Anthropology*, 24, 185–213.

Larsen, C. S. (2002). 'Post-Pleistocene human evolution: bioarchaeology of the agricultural transition', in P. S. Ungar & M. F. Teaford (eds), *Human diet: its origin and evolution*, pp. 19–35. Greenwood Publishing, Westport.

Larsen, C. S. (2006). The agricultural transition an environmental catastrophe: implications for health and lifestyle in the Holocene. *Quaternary International*, 150(1), 12–20.

Larsen, C. S. (2015). *Bioarchaeology: interpreting behavior from the human skeleton*. Cambridge University Press, Cambridge.

Larsen, C. S. & Milner, G. R. (1994). *In the wake of contact: biological responses to conquest*. Wiley-Liss, New York.

Lennartsson, A.-K. & Jonsdottir, I. (2011). Prolactin in response to acute psychosocial stress in healthy men and women. *Psychoneuroendocrinology*, 36, 1530–1539.

Levine, M. E. & Crimmins, E. M. (2014). A comparison of methods for assessing mortality risk. *American Journal of Human Biology*, 26(6), 768–776.

Levine, S., Coe, C. L. & Wiener, S. G. (1989). 'Psychoneu-roendocrinology of stress: a psychobiological perspective', in F. R. Brush & S. Levine (eds), *Psychoendocrinology*, pp. 341–377. Academic Press, Inc., San Diego.

Logan, J. G. & Barksdale, D. J. (2008). Allostasis and allostatic load: expanding the discourse on stress and cardiovascular disease. *Journal of Clinical Nursing*, 17, 201–208.

Lorentz, K. O., Lemmers, S. A. M., Chrysostomou, C. & Dirks, W. (2019). Use of dental microstructure to investigate the role of prenatal and early life physiological stress and death. *Journal of Archaeological Science*, 104, 85–96.

MacDougall-Shackleton, S. A., Bonier, F., Romero, L. M. & Moore, I. T. (2019). Glucocorticoids and 'stress' are not synonymous. *Integrative Organismal Biology*, 1(1), obz017.

MacFadden C. & Oxenham, M. F. (2020). A paleoepidemiological approach to the osteological paradox: Investigating stress, frailty, and resilience through cribra orbitalia. *American Journal of Physical Anthropology*, 173(2), 205–207.

Marklein, K. E. & Crews, D. E. (2017). Frail or hale: skeletal frailty indices in medieval London skeletons. *PLoS One*, 12(5), 1–28.

Marklein, K. E., Leahy, R. E. & Crews, D. E. (2016). In sickness and in death: assessing frailty in human skeletal remains. *American Journal of Physical Anthropology*, 161(2), 208–225.

Mason, J. W. (1975). A historical view of the stress field. *Journal of Human Stress*, 1, 6–12.

Mattei, J., Demissie, S., Falcon, L. M., Ordovas, J. M. & Tucker, K. (2010). Allostatic load is associated with chronic conditions in the Boston Puerto Rican Health Study. *Social Science and Medicine*, 70, 1988–1996.

McEwen, B. S. (1998a). Protective and damaging effects of stress mediators. *The New England Journal of Medicine*, 338, 171–179.

McEwen, B. S. (1998b). Stress, adaptation, and disease: allostasis and allostatic load. *Annals of the New York Academy of Sciences*, 840, 33–44.

McEwen, B. S. (2008).Central effects of stress hormones in health and disease: understanding the protective and damaging effects of stress and stress mediators. *European Journal of Pharmacology*, 583(2–3), 174–185.

McEwen, B. S. (2009). The brain is the central organ of stress and adaptation. *NeuroImage*, 47(3), 911–913.

McEwen, B. S. (2019). What is the confusion with cortisol? *Chronic Stress*, 3, 2470547019833647. doi: 10.1177/2470547019833647

McEwen, B. S. & Stellar, E. (1993). Stress and the individual: mechanisms leading to disease. *Archives of Internal Medicine*, 153, 2093–2101.

McEwen, B. S. & Wingfield, J. C. (2003). The concept of allostasis in biology and biomedicine. *Hormones and Behavior*, 43(1), 2–15.

McMichael, A. J., Haines, R., Sloof, S. & Kovaks, S. (1996). *Climate change and human health*. World Health Organization, Geneva.

Meaney, M. J., Mitchell, J. B., Aitken, D. H., Bhatnagar, S., Bodnoff, S. R., Iny, L.J. & Sarrieau, A. (1991). The effects of neonatal handling on the development of the adrenocortical response to stress: implications for neuropathology and cognitive deficits in later life. *Psychoneuroendocrinology*, 16(1–3), 85–103.

Meaney, M. J., Szyf, M. & Seckl, J. R. (2007). Epigenetic mechanisms of perinatal programming of hypothalamic-pituitary-adrenal function and health. *Trends in Molecular Medicine*, 13(7), 269–277.

Merrett, D. C., Zhang, H., Xiao, X., Wei, D., Wang, L., Zhu, H. & Yang, D. (2016). Enamel hypoplasia in Northeast China: evidence from Houtaomuga. *Quaternary International*, 405(16B), 11–21.

Murata, H., Shimada, N. & Yoshioka, M. (2004). Current research on acute phase proteins in veterinary diagnosis: an overview. *The Veterinary Journal*, 168, 28–40.

Murphy, M. S. & Klaus, H. D. (2012). *Colonized bodies, worlds transformed: towards a global bioarchaeology of contact and colonialism*. University Press of Florida, Gainesville.

Nelson, M. C., Ingram, S. E., Dugmore, A. J., Streeter, R., Peeples, M. A., McGovern, T. H., Hegmon, M., Arneborg, J., Kintigh, K. W., Brewington, S., Spielman, K. A., Simpson, I. A., Strawhacker, C., Comeau, L. E. L., Torvinen, A., Madsen, C., Hambrecht, G. & Smiarowski, C. (2016). Climate challenges, food security, and vulnerability. *Proceedings of the National Academy of Sciences of the United States of America*, 113(2), 298–303.

Nesse, R. & Williams G. C. (1994). *Why we get sick: the new science of Darwinian medicine*. Vintage Publishing, New York.

Newman, S. L. & Garland, R. L. (2015). The use of non-adult vertebral dimensions as indicators of growth disruption and non-specific health stress in skeletal populations. *American Journal of Physical Anthropology*, 158(1), 155–164.

Niaura, R., Stoney, C. M. & Herbert, P. N. (1992). Lipids in psychological research: the last decade. *Biological Psychology*, 34(1), 1–43.

Okazaki, K. (2004). A morphological study on the growth patterns of ancient people in Northern Kyushu, Yamaguchi region, Japan. *Anthropological Science*, 112(3), 219–234.

Ortner, D. J. (1991). 'Theoretical and methodological issues in paleopathology', in D. J Ortner & A. C. Aufderheide (eds), *Paleopathology: current synthesis and future options*, pp. 5–11. Smithsonian Institution, Washington, DC.

Ortner, D. J. (2003). *Identification of pathological conditions in human skeletal remains*. Academic Press, Amsterdam.

Oxenham, M. F. (2006). Biological responses to change in prehistoric Vietnam. *Asian Perspectives*, 45(2), 212–239.

Oxenham, M. F. & Cavill, I. (2010). Porotic hyperostosis and cribra orbitalia: the erythropoietic response to iron-deficiency anemia. *Anthropological Science*, 118(3), 199–200.

Oxenham, M. F., Nguyen, K. & Nguyen, N. L. (2005). Skeletal evidence for the emergence of infectious disease in Bronze Age and Iron Age Vietnam. *American Journal of Physical Anthropology*, 126(4), 359–376.

Pechenkina, E. & Oxenham, M. (2013). *Bioarchaeology of East Asia: movement, contact, health*. University Press of Florida, Gainesville.

Petrovic, D., Pivin, E., Ponte, B., Dhayat, N., Pruijm, M., Ehret, G., Ackermann, D., Guessos, I., Estoppey Younes, S., Pechere-Bertschi, A., Vogt, B., Mohaupt, M., Martin, P-Y., Paccaurd, F., Burnier, M., Bochud, M. & Stringhini, S. (2016). Sociodemographic, behavioral and genetic determinants of allostatic load in a Swiss population-based study. *Psychoneuroendocrinology*, 67, 76–85.

Pietrusewsky, M. P. & Douglas, M. T. (2001). Intensification of agriculture at Ban Chiang: is there evidence from the skeletons? *Asian Perspectives*, 40(2), 157–178.

Pinhasi, R. & Stock, J. T. (2011). *Human bioarchaeology of the agricultural transition*. Wiley-Blackwell, New York.

Powell, M. L. (1988). *Status and health in prehistory: a case study from Moundville*. Smithsonian Institution Press, Washington, DC.

Ragsdale, B. D. & Lehmer L. M. (2012). 'A knowledge of bone at the cellular (histological) level is essential to paleopathology', in A. L. Grauer (ed), *A companion to paleopathology*, pp. 227–249. Wiley-Liss, New York.

Rankin-Hill, L. (1997). *A biohistory of 19th century Afro-Americans: the burial remains of a Philadelphia cemetery*. Berin and Garvey, Westport.

Reid, D. J. & Dean, M. C. (2006). Variation in modern human enamel formation times. *Journal of Human Evolution*, 50(3), 329–346.

Reitsema, L. J. & Kyle Mcilvaine, B. (2014). Reconciling 'stress' and 'health' in physical anthropology: what can bioarchaeologists learn from other subdisciplines? *American Journal of Physical Anthropology*, 155(2), 181–185.

Reitsema, L. J., Vercellotti, G. & Boano, R. (2016). Subadult dietary variation at medieval Trino Vercellese, Italy and its relationship to mortality. *American Journal of Physical Anthropology*, 160(4), 653–654.

Robertson, T., Beveridge, G. & Bromley, C. (2017). Allostatic load as a predictor of all-cause and cause-specific mortality in the general population: Evidence from the Scottish health survey. *PloS One*, 12(8), e0183297.

Sandberg, P., Sponheimer, M., Lee-Thorp, J. & Van Gerven, D. (2014). Intra-tooth stable isotope analysis of dentine: A step towards addressing selective mortality in the reconstruction of life history in the archaeological record. *American Journal of Physical Anthropology*, 155(2), 281–293.

Saunders, S. R. & Hoppa R. D. (1993). Growth deficit in survivors and non-survivors: biological mortality bias in subadult skeletal samples. *Yearbook of Physical Anthropology*, 36(S17), 127–151.

Schulkin, J. (2004). *Allostasis, homeostasis, and the costs of physiological adaptation*. Cambridge University Press, New York.

Schultz, M. (2001). Paleohistopathology of bone: a new approach to the study of ancient diseases. *Yearbook of Physical Anthropology*, 116(S53), 106–147.

Sciulli, P.W. & Oberly, J. (2002). 'Native Americans in eastern North America', in R. H. Steckel, & J. C. Rose (eds), *The backbone of history: health and nutrition in the western hemisphere*, pp. 440–480. Cambridge University Press, Cambridge.

Seaward, B. L. (2006). *Managing stress: principles and strategies for health and well being*, 5th edn. Jones and Bartlett Publishers, Boston.

Seeman, M., Merkin, S., Karlamanga, A., Koretz, B. & Seeman, T. (2014). Social status and biological dysregulation: the 'status syndrome' and allostatic load. *Social Science and Medicine*, 118, 143–151.

Seeman, T. E., Eppel, E., Gruenewald, T., Karlamanga, A. & McEwen, B. S. (2010). Socio- economic differentials in peripheral biology: cumulative allostatic load. *Annals of the New York Academy of Sciences*, 1186(1), 223–239.

Seeman, T. E., Glei, D., Goldman, N., Weinstein, M., Singer, B. & Lin, Y-H. (2004). Social relationships and allostatic load in Taiwanese elderly and near elderly. *Social Science and Medicine*, 59, 2245–2257.

Seeman, T. E., McEwen, B. S., Rowe, J. W. & Singer, B. H. (2001). Allostatic load as a marker of cumulative biological risk: MacArthur studies of successful aging. *Proceedings of the National Academy of Sciences of the United States of America*, 98(8), 4770–4775.

Seeman, T. E., Singer, B. H., Rowe, J. W., Horwitz, R. I. & McEwen, B. S. (1997). Price of adaptation - allostatic load and its health consequences. *Archives of Internal Medicine*, 157, 2259–2268.

Seeman, T. E., Singer, B. H., Ryff, C. D., Love, G. D. & Levy-Storms, L. (2002). Social relationships, gender, and allostatic load across two age cohorts. *Psychosomatic Medicine*, 64(3), 395–406.

Selye, H. (1936). A syndrome produced by noxious agents. *Nature*, 138, 32.

Selye, H. (1938). Adaptation energy. *Nature*, 141, 926.

Seyle, H. (1946). The general adaptation syndrome and the diseases of adaptation. *Journal of Clinical Endocrinology*, 6(2), 119–131.

Smith, T. & Reid, D. J. (2009). 'Temporal nature of peri-radicular bands ("striae periradicales") on mammalian tooth roots', in T. Koppe, G. Meyer & K. W. Alt (eds), *Comparative dental morphology*, pp. 86–92. Karger, Basel.

Snoddy, A. M. E., King, C. L., Halcrow, S. E., Millard, A. R., Buckley, H. R., Standen, V. G., & Arriaza, B. T. (2020). 'Living on the edge: climate induced micronutrient famines in the Atacama Desert?', in G. Robbins-Schug, (ed), *The Routledge handbook of the bioarchaeology of climate and environmental change*, pp. 60–82. Routledge, New York.

Stearns, S. C. (1992). *Evolution of life histories*. Oxford University Press, Oxford.

Stearns, S. C., Nesse, R. M., Govindaraju, D. R. & Ellison, P. T. (2010). Evolutionary perspectives on health and medicine. *Proceedings of the National Academy of Sciences of the United States of America*, 107(S1), 1691–1695.

Steckel, R. H. (2005). Young adult mortality following severe physiological stress in childhood: skeletal evidence. *Economics and Human Biology*, 3(2), 314–328.

Steckel, R. H., Rose, J. C., Larsen, C. S. & Walker, P. L. (2002a). Skeletal health in the western hemisphere from 4000 B.C. to the present. *Evolutionary Anthropology*, 11(4), 142–155.

Steckel, R. H., Sciulli, P. W. & Rose, J. C. (2002b). 'A health index from skeletal remains', in R. H. Steckel & J. C. Rose (eds), *The backbone of history: health and nutrition in the western hemisphere*, pp. 61–93. Cambridge University Press, Cambridge.

Sterling, P. (2012). Allostasis: a model of predictive regulation. *Physiology and Behavior*, 106(1), 5–15.

Sterling, P. & Eyer, J. (1988). 'Allostasis: A new paradigm to explain arousal pathology', in S. Fisher & J. Reason (eds), *Handbook of life stress, cognition and health*, pp. 629–649. John Wiley and Sons, Chichester.

Stini, W. A. (1981). Association of early growth patterns with the process of aging. *Federation Proceedings*, 40, 2588–2594.

Tayles, N. & Oxenham, M. (2006). *Bioarchaeology of Southeast Asia*. Cambridge University Press, Cambridge.

Taylor, S. E., Klein, L. C., Lewis, B. P., Gruenewald, T. L., Gurung, R. A. & Updegraff, J. A. (2000). Biobehavioral responses to stress in females: tend-and-befriend, not fight-or-flight. *Psychological Review*, 107, 411–429.

Temple, D. H. (2019a). Bioarchaeological evidence for adaptive plasticity and constraint: exploring life history trade-offs in the human past. *Evolutionary Anthropology*, 28(1), 34–46.

Temple, D. H. (2016). Chronology and periodicity of linear enamel hypoplasia among Late/Final Jomon period foragers: evidence from incremental microstructures of enamel. *Quaternary International*, 405(16B), 3–10.

Temple, D. H. (2018). 'Exploring linear enamel hypoplasia as an embodied product of childhood stress in Late/Final Jomon period foragers using incremental microstructures of enamel', in P. Beauchesne & S. C. Agarwal (eds), *Children and childhood in bioarchaeology*, pp. 239–261. University Press of Florida, Gainesville.

Temple, D. H. (2020). 'The mother-infant nexus revealed by linear enamel hypoplasia: chronological and contextual evaluation of developmental stress using incremental microstructures of enamel in Late/Final Jomon period hunter-gatherers', in R. Gowland & S. Halcrow (eds), *The mother-infant nexus in anthropology: small beginnings, significant outcomes*, pp. 65–84. Springer, Cham.

Temple, D. H. (2010). Patterns of systemic stress during the agricultural transition in prehistoric Japan. *American Journal of Physical Anthropology*, 142(1), 112–124.

Temple, D. H. (2019b). 'Persistence of time: resilience and adaptability in prehistoric Jomon hunter-gatherers from the Inland Sea Region of Southwestern Honshu, Japan', in D. H. Temple & C. M. Stojanowski (eds), *Hunter-gatherer adaptation and resilience: a bioarchaeological perspective*, pp. 85–109. Cambridge University Press, Cambridge.

Temple, D. H. (2014). Plasticity and constraint in response to early life stressors: among Late/Final Jomon period foragers from Japan: evidence for life history trade-offs from incremental microstructures of enamel. *American Journal of Physical Anthropology*, 155(4), 537–545.

Temple, D.H. & Goodman., A.H. (2014). Bioarchaeology has a health problem: conceptualizing 'stress' and 'health' in bioarchaeological research. *American Journal of Physical Anthropology*, 151(2), 186–191.

Temple, D. H. & Larsen, C. S. (2013). 'Bioarchaeological perspectives on systemic stress during the agricultural transition in prehistoric Japan', in E. Pechenkina & M. F. Oxenham (eds), *Bioarchaeology of East Asia: movement, contact, health*, pp. 344–367. University Press of Florida, Gainesville.

Temple, D. H., McGroarty, J. N., Guatelli-Steinberg, D., Nakatasukasa, M. & Matsumura, H. (2013). A comparative study of stress episode prevalence and duration among Jomon-period foragers from Hokkaido. *American Journal of Physical Anthropology*, 152(2), 230–238.

Temple, D. H., McGroarty, J. N. & Nakatsukasa, M. (2012). Reconstructing patterns of systemic stress in a Jomon period subadult using incremental microstructures of enamel. *Journal of Archaeological Science*, 39(5), 1634–1641.

Thayer, Z., Barbosa-Leiker, C., McDonell, M., Nelson, L., Buchwald, D. & Manson, S. (2017). Early life trauma, post-traumatic stress disorder, and allostatic load in a sample of American Indian adults. *American Journal of Human Biology*, 29(3), 1–10.

Thomas, J. A., Temple, D. H. & Klaus, H. D. (2019). Crypt fenestration enamel defects and early life stress: contextual explorations of growth and mortality in Colonial Peru. *American Journal of Physical Anthropology*, 168(3), 582–594.

Trevathan, W. R., Smith, E. O. & McKenna, J. (1999). *Evolutionary medicine and health: new perspectives.* Oxford University Press, Oxford.

Ubelaker, D. J. (1974). *The reconstruction of demographic profiles from ossuary samples: a case study from the Tidewater Potomac.* Smithsonian Contributions to Anthropology, Washington, DC.

Usher, B. M. (2000). *A multi-state model of health and mortality for paleodemography.* PhD Thesis, Pennsylvania State University, Philadelphia.

Verano, J. & Ubelaker, D. H. (1992). *Disease and demography in the Americas.* Smithsonian Institution, Washington, DC.

van der Kolk, B. (2014). *The body keeps the score: brain, mind, and body in the healing of trauma.* Penguin Books, New York.

Van Doornen, L., Snieder, H. & Boomsma, D. I. (1998). Serum lipids and cardiovascular reactivity to stress. *Biological Psychology*, 47(3), 279–297.

Walker, P. L. (1985). 'Anemia among prehistoric Indians of the American Southwest', in C. F. Merbs and R. J. Miller (eds), *Health and disease in the prehistoric southwest*, pp. 139–164. Arizona State University Anthropological Research Papers, Tempe.

Walker, P. L. (1986). Porotic hyperostosis in a marine-dependent California Indian population. *American Journal of Physical Anthropology*, 69(3), 345–354.

Walker, P. L., Bathurst, R. R., Richman, R., Gjerdrum, T. & Andrushko, V. A. (2009). The causes of porotic hyperostosis and cribra orbitalia: a reappraisal of the iron-deficiency anemia hypothesis. *American Journal of Physical Anthropology*, 139(2), 109–125.

Wapler, U., Crubézy, E. & Schultz, M. (2004). Is cribra orbitalia synonymous with anemia? Analysis and interpretation of cranial pathology in Sudan. *American Journal of Physical Anthropology*, 123(4), 333–339.

Watkins, R. (2020). An alter(ed)native perspective on historical bioarchaeology. *Historical Archaeology*, 54, 17–33.

Watts, R. (2015). The long-term impacts of developmental stress: evidence from later and post medieval London (AD 1117–1853). *American Journal of Physical Anthropology*, 158(4), 569–580.

Watts, R. (2013). Lumbar vertebral canal size in adults and children: observations from a skeletal sample from London, England. *Homo*, 64(2), 120–128.

Watts, R. (2011). Non-specific indicators of stress and their association with age at death in medieval York: using stature and vertebral neural canal size to examine the effects of stress occurring during different periods of development. *International Journal of Osteoarchaeology*, 21(5), 568–576.

Weinstein, M., Goldman, N., Hedley, A., Yu-Hsuan, L. & Seeman. T. (2003). Social linkages to biological markers of health among the elderly. *Journal of Biosocial Science*, 35(3), 433–453.

Wells, J. C. K. (2007). Flaws in the theory of predictive adaptive responses. *Trends in Endocrinology Metabolism*, 18(9), 331–337.

Wells, J. C. K. (2016). *The metabolic ghetto: an evolutionary perspective on nutrition, power relations, and chronic disease.* Cambridge University Press, Cambridge.

Wells, J. C. K. (2011). The thrifty phenotype: an adaptation in growth or metabolism? *American Journal of Human Biology*, 23(1), 65–76.

Wilson, J. J. (2014). Paradox and promise: research on the role of recent advances in paleodemography and paleoepidemiology to the study of 'health' in pre-Columbian societies. *American Journal of Physical Anthropology*, 155(2), 268–280.

Wood, J. W., Milner, G. W., Harpending H.C. & Weiss, K. M. (1992). The osteological paradox: problems of inferring health from skeletal samples. *Current Anthropology*, 33(4), 343–370.

Worthmann, C. M. & Kuzara, J. (2005). Life history and early origins of health differentials. *American Journal of Human Biology*, 17(1), 95–112.

Wright, L. R. & Yoder, C. J. (2003). Recent progress in bioarchaeology: approaches to the osteological paradox. *Journal Of Archaeological Research*, 11(1), 43–70.

Yaussy S. (2019). The intersections of industrialization: variation in skeletal indicators of frailty by age, sex, and socioeconomic status in 18th and 19th-century London. *American Journal of Physical Anthropology*, 170(1), 116–130.

CHAPTER 15

Metabolic diseases in bioarchaeology: an evolutionary medicine approach

Jonathan C. Wells, Nellissa Y. Ling, Jay T. Stock, Hallie Buckley and William R. Leonard

15.1 Introduction

Until recently, the primary cause of human ill-health derived from infectious diseases, and the contribution of non-communicable diseases (NCD) was modest (Omran, 1971). Ostensibly, this rapid global shift in mortality from infectious to non-infectious causes can be attributed to diverse aspects of 'modernisation'. In high-income countries, the last 200 years has seen profound shifts in living conditions, involving changes in sanitation and hygiene levels, occupation, transport modes and food supply, alongside overall but unequal improvements in social welfare and medical care (Wells, 2016). From the late nineteenth century onwards, modern medicine developed on the back of its capacity to combat infectious diseases, while economic development improved access to better housing and more nutritious diets. However, these processes were also embedded in broader aspects of globalisation and market economic development, increasingly exposing populations to economic imperatives that drive patterns of consumption (Wells, 2016).

Given this recent epidemiological transition (Omran, 1971) associated with the industrialisation of many aspects of life, it is easy for many, including researchers in evolutionary medicine (EM), to portray NCDs as representing something new in human biology—a scenario provoked by new ways of living associated with our modern economic order. The global surge in NCDs closely tracks the global epidemic of obesity, which in turn parallels systematic changes in diet and physical activity patterns. These correlated trends are often approached through the lens of this transition in nutrition (Popkin et al., 2012). Although the global epidemics of cardiovascular disease (CVD), stroke and diabetes are strong correlates of a nutrition transition, however, none of these diseases should be regarded as entirely modern. Careful evaluation of the available evidence suggests that both obesity and its co-morbidities occurred in past populations, although at lower rates (Conard, 2009; Thompson et al., 2013). Moreover, major nutrition transitions also occurred in the past, most notably with the emergence and spread of agriculture (Cohen & Armelagos, 1984; Omran, 1971; Wells & Stock, 2020), and the colonisation of many geographical regions by Europeans since the mediaeval period (Wells, 2016). Elsewhere in this volume and relevant to this chapter, Thompson and colleagues (Chapter 12) focus on atherosclerosis and consider palaeopathological evidence for its manifestation in ancestral populations.

Identifying metabolic diseases in past populations is an important part of palaeopathological inquiry because their prevalence (or lack thereof) can provide a wealth of information about the general health and well-bring of a population, as well as an insight into social and economic transitions within a population and/or differences between populations. Despite the wealth of information that

Jonathan C. Wells et al., *Metabolic diseases in bioarchaeology*. In: *Palaeopathology and Evolutionary Medicine*. Edited by Kimberly A. Plomp et al., Oxford University Press. © Oxford University Press (2022). DOI: 10.1093/oso/9780198849711.003.0015 can be gained from interpreting skeletal indicators of metabolic stress in the past, palaeopathology is somewhat limited in only being able to diagnose a handful of so-called metabolic conditions in individual skeletons from archaeological sites. These include scurvy (i.e. vitamin C deficiency), rickets and osteomalacia (i.e. vitamin D deficiency), anaemia (i.e. iron deficiency), gout, non-specific stress indicators such as linear enamel hypoplasia, diffused idiopathic skeletal hyperostosis (DISH) and Harris lines of arrested growth (Brickley and Ives, 2008).

Here, we address three metabolic diseases, gout, diffuse idiopathic skeletal hyperostosis (DISH) and type 2 diabetes mellitus (T2DM), the first and last of which have received significant focus in EM. In their seminal book *Why We Get Sick*, Nesse and Williams (1994) refer to gout in the context of trade-offs and ageing, while discussion of T2DM has appeared in many other volumes in the context of environmental mismatch and early-life development (Gluckman & Hanson, 2006; Lieberman, 2008; Pollard et al., 2008). We opted to focus on these three conditions here because both gout and DISH are identifiable in the palaeopathological record while T2DM can rarely be identified from skeletal material. Moreover, gout and DISH are co-morbid (Littlejohn and Hall, 1982), and may be associated not only with T2DM, but other metabolic-related conditions such as CVDs (see Chapter 12, this volume; Elfishawi et al., 2018; Fassio et al., 2021; Pariente-Rodrigo et al., 2017; Singh & Gaffo, 2020). Therefore, investigating gout and DISH in the archaeological record provides an avenue to investigate related metabolic disorders as well.

15.2 Background to gout, DISH and Type 2 Diabetes Mellitus (T2DM)

15.2.1 Gout

Gout is a form of arthritis with a long history and can be a crippling disease for those who suffer from it (Nuki & Simkin, 2006). Gout occurs when urate levels become oversaturated and monosodium urate (MSU) crystals form in the body. Hyperuricaemia is a pre-requisite for gout, though being hyperuricaemic does not necessarily lead to the

condition (Neogi et al., 2015; Zhang, 2006). In ancient times, particularly in Europe, people considered gout a disease of the affluent, a marker of social status tied to a lifestyle with greater access to higher caloric foods, such as meat and alcohol (Nuki & Simkin, 2006). Today, however, genetic predisposition and the environment, both social and the physical (Box 15.1), are factors recognised in gout development (Dalbeth & McLean, 2019; Merriman, 2017, 2016).

Box 15.1 Pathogenesis of gout

Gout can be primary or secondary, the latter being due to renal failure (Waldron, 2019). The MSU crystals that cause gout are made of elongated stacks of purine sheets (Mandel & Mandel, 1976; Perrin et al., 2011) that develop into a distinctive needle shape (visible under a light microscope) (Perez-Ruiz et al., 2014). At an advanced stage, gout becomes chronic and tophaceous (Schlesinger, 2019). Tophi are large aggregates of these MSU crystals that are visible without a microscope as a white chalky substance and can be discernible through the skin if large enough (Jansen, 2016). It is in this later stage of gout that the tophi containing crystals within or around the joint can cause bone erosions (punched out appearance, with overhanging margins which may be sclerotic) and structural damage, usually asymmetrically; the first metatarsophalangeal joint is most affected. Over time, this may result in severe, and potentially irreversible, joint destruction (Perez-Ruiz et al., 2009). Because erosive lesions from MSU crystals do not completely remodel/heal, they are a hallmark criterion for diagnosing gout in joints known to be affected among archaeological populations. The preservation of urate crystals has been reported in a skeleton, but this is a rare finding (e.g. Wells 1973).

15.2.2 Bioarchaeological evidence of gout

In the skeleton, gout manifests as erosive lesions around the margins of synovial joints and can be seen by gross examination or from radiographs (Ortner, 2003). In particular, gouty lesions appear round and 'scooped/punched-out' with overhanging edges and sclerotic margins (Brower & Flemming, 2012; Resnick & Niwayama, 1995). A sclerotic margin is a layer of dense bone that forms inside the lesion cavity which creates an internal wall that lines the trabeculae (a series of bony rods or struts that provide the tensile strength of bone). Radiographically, this feature appears as a dense layer inside the lesion. Some lesions have overhanging edges that extend prominently from the original bone surface over the cavity as a result of bone remodelling along the surface of the tophi (Resnick & Niwayama, 1995). Often, the first metatarsophalangeal joint is the first region to be affected, and this is followed by other joints of the feet, then the hands (Schlesinger, 2019). Other regions of the body, such as the shoulders, spine and pelvis, are less frequently involved—the lower body is more commonly affected than the upper body (Resnick & Niwayama, 1995). Together these lesion characteristics and their distribution in the skeleton help to differentiate gout from other erosive arthropathies, such as rheumatoid arthritis (Jadon, 2018) and the seronegative spondyloarthropathies (Duba & Mathew, 2018).

Skeletal evidence of gout has been documented in archaeological sites from various regions, including Europe (Rothschild et al., 2004), mainland Southeast Asia (Pietrusewsky & Douglas, 2002) and the Pacific region (Buckley, 2007; Buckley et al., 2010; Rothschild & Heathcote, 1995; Suzuki, 1993). Prevalence of the condition differed notably among these groups. Only a handful of individuals have demonstrated skeletal evidence for gout in assemblages from Europe (Rothschild et al., 2004), and these are often considered high-status individuals. For example, European elites, such as the Dukes of Urbino (1422–1482 CE) (Fornaciari et al., 2018) and Tuscany (1549–1609 CE) (Fornaciari et al., 2009) and the Holy Roman Emperor Charles V (1516–1556 CE) (Ordi et al., 2006) were afflicted with gout, according to historical documents and/or based on the analyses of their remains. Similarly, only a small number of individuals in mainland Southeast Asia and Hawaii have been found with skeletal evidence for gout. For example, at the archaeological site of Ban Chiang, Thailand (~ 4100–2900 BP [before present]), two out of seventy-one individuals had evidence for the condition (Pietrusewsky & Douglas, 2002), and in an assemblage from Hawaii, it was one out of 349 individuals (Suzuki, 1993).

Conversely, Buckley (2007) and Buckley and colleagues (2010) found a high prevalence of probable gout in archaeological populations from Vanuatu and New Zealand. Of note is the Lapita site from Teouma, Vanuatu, dating to approximately 3000 BP, which has a very high prevalence (35%), with seven out of twenty individuals having skeletal traits indicating gout (Buckley, 2007). This evidence, if accepted as truly representing gout, could determine the existence of the condition in small Pacific forager and horticulturalist groups (Kinaston et al., 2016), standing in contrast to the classic view that gout was associated with upper-class members of large-scale societies in, for example, Europe (Nuki & Simkin, 2006). The prevalence of probable gout found among these ancient Pacific populations mirrors the high rate for individuals in today's indigenous populations from the same region. While this may not mean that modern groups residing in the locality are descendants of these ancient groups, the similarities between ancient and modern populations could suggest people in this region have demonstrated a greater susceptibility for gout for a very long time (Buckley, 2011).

15.2.3 DISH

Much the same as gout, DISH is a health condition affecting the ligaments and entheses that stretches back to antiquity (Box 15.2) and is found in the archaeological record as far back as 3000–3500 years ago (Foster et al., 2018; Oxenham et al., 2006). Neanderthals have been found with skeletal evidence for DISH from sites in Iraq and Russia (Crubézy & Trinkaus, 1992; Trinkaus et al., 2008). This bone-forming condition has also been reported in other species such as dogs (Kranenburg et al., 2010) and horses (Marković et al., 2019). Moreover, ossification of ligaments along the spine that appears to resemble DISH has also been found among dinosaur bones (Rothschild, 1987).

Box 15.2   Clinical manifestations of DISH

The skeletal expression of DISH is very different to gout. DISH is clinically defined as the presence of a flowing ectopic bone growth along the antero-lateral aspect of the vertebral column (anterior longitudinal ligament), commonly in the thoracic region and on the right side, and accompanied by extraspinal new bone formation, often bilateral, at entheseal sites (Resnick & Niwayama, 1995). The left side of the spine is not usually affected because it is occupied by the aorta which 'prevents' abnormal bony growth from forming there. Entheseal sites are areas around a joint where tissues, such as tendons, ligaments and the joint capsule attach to the bones (Goldring & Goldring, 2017). Moreover, DISH is associated with the preservation of the intervertebral disc and apophyseal joint spaces, the absence of erosive lesions in the sacroiliac joint (Resnick & Niwayama, 1995), and can lead to cartilage ossification. The condition is predominantly asymptomatic, leading some researchers to consider it clinically irrelevant (Schlapbach et al., 1989). However, some secondary conditions, such as dysphagia (difficulty swallowing) and back stiffness may result (Kuperus et al., 2020; Littlejohn, 2019). One study by Rogers and colleagues (1997) supports a possible relationship between DISH and 'bone formers', who are individuals prone to excess bony growths on the skeleton.

DISH may have an inherited component. For example, Jun and Kim (2012) reported an association among a Korean cohort between DISH and two SNPs (single nucleotide polymorphisms) from the fibroblast growth factor receptor 2 (FGF2) gene. In addition, Tsukahara and colleagues (2005) found an association between DISH and an SNP of COL6A1 in a Japanese cohort, but not in a Czech one. These mutations demonstrate that genetics may be one driver in DISH development, at least among some Asian populations. Several health conditions involving the axial region may appear similar on the skeleton to DISH and therefore should be considered in differential diagnosis. Ankylosing spondylitis (AS), for example, is considered to have bony manifestations similar to DISH (Forestier & Rotes-Querol, 1950; Resnick & Niwayama, 1995, 1976). AS manifests as abnormal bone growth along the spine in a bilateral contiguous manner, and the sacroiliac joints are involved. In contrast, DISH appears as flowing and bulbous ('candle wax' bone growth) representing ossification of the anterior longitudinal ligament, often occurring on the right side, and is diagnosed if four or more thoracic vertebrae are involved (Waldron, 2020). Moreover, whereas DISH is solely a bone-forming condition (Resnick

& Niwayama, 1995), AS may cause erosive lesions in the superior and inferior surfaces of the vertebral bodies, as well as bilaterally on the sacroiliac joints, the latter of which may fuse (Waldron, 2020). DISH can also lead to new bone formation at entheses and ossification of cartilage (e.g. the neck cartilages). Unlike AS, in DISH the vertebral disc and apophyseal joint spaces are preserved.

Abnormal bone formation in the axial region may also occur in other erosive arthropathies, including psoriatic arthropathy (PsA), reactive spondyloarthropathy (or Reiter's syndrome) (ReA) and enteropathic arthropathy (EnA). For PsA, bone changes are less common in the axial region than the periphery but, if it does occur, the cervical region is most likely to be affected (Resnick & Niwayama, 1995). Axial involvement is an uncommon event for ReA but is similar to PsA (Duba & Mathew, 2018). For EnA, bone changes are bilateral and symmetrical, and have been compared to the axial changes associated with AS (Resnick & Niwayama, 1995).

The presence of vertebral osteophytes, such as from spondylosis deformans, is another common spinal abnormality. Osteophytosis occurs with the degeneration of the intervertebral disc, leading to bone formation along the margins of the vertebral body and joint space narrowing (Resnick & Niwayama, 1976), though not always the latter (Rowe & Resnick, 2001). Other less common conditions noted to have some overlapping features with DISH are acromegaly, fluorosis and hypervitaminosis A (Harris et al., 1974; Resnick et al., 1975).

15.2.4 Skeletal evidence of DISH in the archaeological record

Bioarchaeological studies of DISH have been conducted on skeletal assemblages and individuals from various regions. Currently, most studies are of European individuals or groups, where associations have been made between DISH and high-status people, similar to the pattern seen in gout. Skeletal evidence of DISH has also been reported in the Egyptian Pharaoh, Ramses II (thirteenth–twelfth century BCE) (Chem et al., 2004), Italian Grand Dukes Cosimo I and Ferdinand I (1519–1574 CE and 1549–1609 CE, respectively) (Giuffra et al.,

2010) and a European General Johann Sporck (1596–1679 CE) (Dobisikova et al., 1999). These individuals with DISH were clearly affluent people, and DISH has been associated with obesity and rich diets (see, for example, review by Pillai and Littlejohn, 2014). Among some ancient groups, DISH has also been found in in greater frequency among burials of members of the European clergy (Kim et al., 2011; Rogers & Waldron, 2001; Verlaan, Oner & Maat, 2007). The association between DISH and monastic communities in Europe has led researchers to consider DISH to be a particular affliction in the past of clergy (Rogers & Waldron, 2001). The monastic life was thought to have been very comfortable, with access to higher caloric foods compared to the ordinary person. Gluttony, in fact, is a term that has been used to describe past monastic living in Europe (Rogers & Waldron, 2001; Waldron, 1985). Findings from stable isotope analyses have supported DISH and its association with richer diets among Europeans. For example, Quintelier and colleagues (2014) found that males and higher-status individuals consumed more animal protein than females or lay people from a post-medieval friary in Belgium.

While DISH and its association with diet and lifestyle is often discussed in the bioarchaeological literature, some studies have not found this link. One example is a skeleton with evidence of DISH from a male forager from Lake Baikal, dating back approximately 7500–6500 years (Faccia et al., 2016). Oxenham and colleagues (2006) also reported two individuals with DISH in a Jomon hunter-gatherer group from Hokkaido, Japan (1500–300 BCE). Skeletal evidence of DISH has also been identified in a 3000-year-old Lapita period skeletal assemblage from Vanuatu (Foster et al., 2018). Stable isotope results from the Lapita study found no association between DISH and diet among this group, who were foragers and horticulturalists (Foster et al., 2018). Yet, this Lapita group had a high prevalence of DISH at 42% (nineteen out of forty-five individuals studied) (Foster et al., 2018). Together, the existence of DISH in these non-European groups with different subsistence strategies suggests its development is associated with more than just diet and lifestyle. These variations make it challenging to

understand the mechanisms that underlie the condition. The proposed risk factors, such as high status and a rich diet, or potentially early-life stress, do not have to be mutually exclusive. Further research is thus required to shed light on the aetiology of DISH among these varied populations.

15.2.5 T2DM

T2DM is one of the most ubiquitous metabolic disorders in contemporary, industrialised populations and is increasing rapidly in low- and middle-income countries as they become globalised and introduced to obesogenic diets (Wells, 2016; Box 15.3). As noted earlier, T2DM has also appeared prominently in many EM textbooks as an exemplar of mismatch between modern and ancient human environments (e.g. Gluckman & Hanson, 2006; Gluckman et al., 2009; Lieberman, 2008; Pollard et al., 2008).

Box 15.3   Aetiology of type 2 diabetes

The primary cause of diabetes is the inability of the hormone insulin to fulfil its function, namely the regulation of levels of blood glucose (Gale, 2013). Normally, sufficient insulin is produced by the pancreas, but when the body is chronically challenged by excessive food intake or foods that are high in glucose, insulin resistance develops in some tissues, so that they no longer respond effectively to the hormone (Bergman et al., 2002; Gale, 2013). Initially, this can be compensated through the production of extra insulin by the pancreas, so that what is called glycaemic control (the maintenance of normal blood sugar levels) can be maintained through higher insulin turnover. Over time, however, this compensation may prove to be inadequate.

In contemporary populations, the supply of glucose, the body's primary fuel, is irregular because of the tendency to consume meals rather than to graze continuously. The importance of meals in pre-agricultural populations is unclear; however, characteristics of the human gut suggest that it has long been exposed to foods 'pre-digested' by technology such as cooking (Milton, 2000), which may indicate a lengthy history of consuming food via meals. Richard Wrangham (2009) has suggested this may date back to the earliest *Homo* species, when the use of fire may have become ubiquitous. Glucose will accumulate in the blood after every meal, requiring that the body responds by shifting the fuel either directly to tissues and organs, or to 'fuel dumps', the most important of which comprise glycogen in muscle or lipid in adipose tissue (Wells, 2009a). Glycaemic control is thus a central component of homeostasis, the maintenance of optimal cellular conditions through physiological feedback mechanisms.

Following digestion, macronutrient molecules pass from the gut to the liver, which converts glucose into glycogen, and another sugar, fructose, into fat (Ganong, 1999). The liver also synthesises glycoprotein molecules that transport free fatty acids to adipose tissue. However, the liver cannot achieve complete glycaemic control, so the pancreas contributes an additional level of regulation, secreting the hormone insulin (Ganong, 1999). This hormone functions to promote glucose uptake by organs and tissues, in particular muscle or adipose tissue, thereby preventing excessive blood sugar levels. If the demands of organs and tissues for glucose increase, the pancreas responds by secreting a second hormone, glucagon, which induces the liver to release glucose from its glycogen stores (Ganong, 1999). Muscle tissue can also switch to metabolising free fatty acids, released into the circulation from adipose tissue, which occurs when glucose supply is insufficient. In a state of physical fitness, muscle switches readily between these fuels, but such 'metabolic flexibility' can be compromised if physical fitness declines (Corpeleijn et al., 2009).

T2DM can be considered as a 'two-hit' disease, that involves both insulin resistance in muscle tissue and, ultimately, the failure of the pancreatic beta-cells to produce enough insulin to compensate (Bergman et al., 2002). The primary signs and symptoms are the production of high volumes of urine (polyuria), excessive thirst leading to high fluid intake (polydipsia), high circulating levels of blood glucose (hyperglycaemia), loss of glucose in urine (glycosuria), high levels of fat in the bloodstream (ketosis) as the body draws on alternative fuel supplies, weight loss despite high levels of appetite and increased acidity of the blood (acidosis), which impairs many vital functions (Ganong, 1999). Overall, the body fails to regulate the supply of the primary metabolic fuel to the organs and tissues.

Despite its close associations with sedentary behaviour, ultra-processed diets and obesity in contemporary populations, there is nothing particularly modern or even necessarily 'human" about T2DM. The disease can develop spontaneously in primates and rodents kept in captivity (Wagner et al., 2006), reflecting the failure to maintain homeostasis of blood glucose levels. Indeed, the similarity of the symptoms across primate and humans indicates that selection has favoured a common protective response to high circulating blood sugar levels to reduce toxic effects on various organs and tissues.

15.2.6 Identifying T2DM in the archaeological record

Direct detection of diabetes currently falls outside what would conventionally be considered possible in palaeopathology. A few examples of diabetes appear to have been documented in ancient historical records. For example, excessive urination (polyuria) was described in a passage dating from around 3400 BC from the Ebers papyrus from ancient Egypt, while ancient Indian texts record that the urine of some individuals tasted sweet, and would attract ants (Watve, 2013). The disease, however, leaves neither specific skeletal lesions, unlike many infectious diseases, nor developmental markers, unlike some forms of metabolic disease. Although T2DM has been shown to impact bone at the microscopic level, associated lesions of mineralised tissue would be very difficult to recognise during macroscopic assessment (Brickley et al., 2020). Nevertheless, expanding our remit to look at potentially related conditions like gout and DISH is certainly helpful and recent work has attempted to overcome the challenges of identifying T2DM by using a composite approach that links gout and DISH to risk for T2DM (Brickley et al., 2020: Faccia et al., 2016; Waldron, 1985).

Specifically, gout and DISH have been linked to several inter-related metabolic conditions, which may lead to the development of CVD and T2DM (Littlejohn, 2019; Elfishawi et al., 2018), although it should be remembered that most people suffering from gout do not have T2DM, and vice versa. Given these connections, it is possible that in future palaeopathological work gout or DISH could provide visible evidence that an individual might have suffered from at least one form of metabolic derangement in their lifetime. This would at best be indirect evidence of the existence of T2DM in the past and associated contextual information (archaeological/historical) would also be necessary to help with these interpretations. As an example, Dupras and colleagues (2010) identified many skeletal and dental changes (non-specific to diabetes) in an adult male skeleton from the Egyptian Middle Kingdom site of Dayr al-Barsha (~ 2055–1650 BCE) (see also Brickley et al., 2020, their Figure 10). These changes included Charcot's joint, osteoporosis, osteoarthritis, DISH, adhesive capsulitis, dental caries, periodontal disease and antemortem tooth loss. Considering these pathological conditions, Dupras and colleagues (2010) concluded that chronic T2DM was a very likely diagnosis. This broader approach emphasising phenotypic correlates of T2DM risk helps us to make inferences about the origins and ecological context of T2DM in the bioarchaeological record.

Such potential skeletal evidence is rare. We should also note that individuals may also suffer from T2DM but with lifestyle change subsequently resolve their symptoms (Lean et al., 2018; O'Dea, 1984). Building on the framework set out here, it may therefore be valuable to draw on broader forms of evidence, relating both to markers of environmental risk, and also life-course traits that increase the likelihood of developing T2DM. Indeed, this approach has already been used to consider long-term evolutionary trends that may have generically increased risk of T2DM in humans compared to other primates. Relevant trends include reductions in body size and muscle mass, encephalisation and increased cerebral glucose demand, as well as an enhanced tendency to store body fat (Leonard et al., 2003; Muchlinski et al., 2012; Wells, 2010, 2017). For example, the high-energy demands of the brain during infancy and childhood appear to have played a critical role in shaping distinctive elements of human glucose metabolism. The brain's requirements for glucose peak at age 4–5 years, when the body's fat stores are minimal (Kuzawa et al., 2014). This pattern suggests enhanced physiological capacity to shunt glucose to 'feed the brain' during periods of nutritional stress, helping us to understand how early malnutrition undermines long-term homeostatic maintenance.

In the future, ancient DNA studies might be able to track the distribution and prevalence of known diabetes risk alleles. Recent comparative analyses of ancient and modern genetic variants are providing insights into the origins of risks of T2DM. For example, work by the SIGMA Type 2 Diabetes Consortium (Consortium et al., 2014) has shown that a genetic variant at the *SLC16A11* locus associated with increased risk of T2DM in Mexican and other Latin American populations likely arose through introgression of anatomically modern humans with Neanderthals. This approach might demonstrate broad population contrasts but is challenged by the fact that many individual genes appear to contribute, each with a small magnitude of effect (Wells et al., 2016a).

In many palaeopathological analyses, there is a tendency to interpret phenotypic variation as evidence for variation in diet, growth and health that is either independent from, or complementary to, disease itself. We suggest instead that these phenotypic traits should be seen as part of a system of life-history trade-offs whereby, for example, investment in immune function or inflammation necessitates energetic trade-offs against other functions such as growth. Two of us recently applied this model to explain some of the variable outcomes in health, growth and demography associated with the transition to an agricultural subsistence, arguing, for example, that there is evidence for greater investment in reproduction and defence at a cost to skeletal growth and markers of homeostatic maintenance (Wells & Stock, 2020). A study identifying probable DISH in a mid-Holocene forager skeleton from Siberia demonstrated that this particular skeleton had a relatively high estimated body mass relative to stature, suggesting that phenotypic variation within skeletal populations can predispose specific individuals to metabolic disease (Faccia et al., 2016). This demonstrates the utility of the investigation of skeletal phenotypic data within the context of palaeopathological analyses.

15.3 Models of the aetiology of metabolic disorders

15.3.1 Ecological models of metabolic disease risk

Ecological factors can shape many of the traits, including adult body mass index (BMI) and body composition, dietary composition and growth patterns, that represent proximate components of metabolic disease risk, especially T2DM. Among the key ecological variables are climate, pathogen burden and food supply, though all of these may also be altered by niche construction (how humans have modified their environment and altered selective pressures operating on them). Due to the difficulty in obtaining direct skeletal evidence for T2DM in the bioarchaeological record, we suggest that indirect sources of evidence for T2DM and other metabolic disorders also merit attention in past environments. An ecological approach allows us to study the potential impact of changes in ecosystem or socio-economic conditions and/or the consequences of migration. Ecological stresses may elicit genetic or plastic responses from humans, depending on the duration and consistency of exposure.

Numerous studies have demonstrated population-associations of adult BMI with climatic conditions. Consistent with two 'ecological laws' that predict that components of body size and shape vary to favour heat retention in cold settings, and heat loss in hot settings, BMI tends to scale inversely with average annual temperature (Katzmarzyk & Leonard, 1998). A recent analysis found that both fat and lean mass also scaled inversely with average annual temperature (Wells, 2012). A few populations appear to contradict these trends, such as Polynesians, who have a high BMI despite occupying environments with high annual average temperatures (Bindon & Baker, 1997). However, Polynesians may be exposed to low temperatures during ocean-going activities (Houghton, 1990).

Research among arctic populations of Siberia has shown that beyond BMI, other aspects of metabolism relevant to T2DM risk also show evidence of adaptation to cold conditions. Indigenous Siberian populations have basal metabolic rates (BMR) 15–20% higher than expected for their body size and composition—an adaptation for increased heat production (Leonard et al., 2005). These elevated rates of energy and fat metabolism appear to play an important protective role in reducing T2DM risk. Indeed, work among the Siberian Yakut has shown that while BMI and body fatness are increasing with ongoing dietary and lifestyle change, fasting glucose levels (and hence T2DM risk) are rising much more modestly (Snodgrass et al., 2010).

In contrast, among populations from warmer, more tropical climes, increasing rates of T2DM are often observed at relatively low BMIs and levels of body fatness. This pattern may partly reflect the reduced BMR levels widely documented among populations of the tropics (Froehle, 2008). Figure 15.1 highlights the contrasts in body mass/composition and blood glucose levels between two indigenous populations of the arctic (Yakut) and tropics (Tsimané of Bolivia), both undergoing dietary and lifestyle transitions. Despite being significantly lighter and having lower fatness than their Yakut counterparts, Tsimané men and women have significantly higher fasting glucose levels and T2DM risk. Thus, one line of future palaeopathology research might be to evaluate whether archaeological groups from higher latitudes show a lower prevalence of skeletal traits associated with metabolic disorders than seen in groups from warmer, more tropical environments.

Beyond average annual temperature, volatile climatic environments also favour higher adiposity, as well as lower lean mass (Wells et al., 2019). This indicates that when food supply is less predictable, selection may favour lower levels of tissues that are relatively expensive to accrete and maintain, and higher levels of fat reserves to buffer against shortfalls in energy supply (Figure 15.2). This may help to explain several components of phenotype among South Asian populations, who have long been exposed to irregular famines when the monsoon fails. South Asians have a number of traits associated with increased T2DM susceptibility, including low birth weight, short adult stature, low muscle mass, reduced organ size, high fat–lean ratio and a predisposition to central adiposity (Wells et al., 2016a).

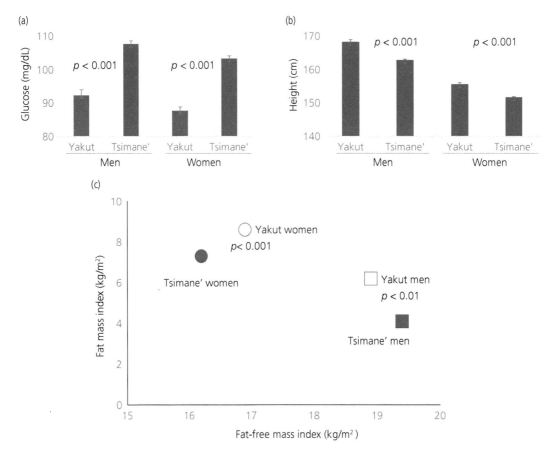

Figure 15.1 Values for (a) fasting blood glucose levels, (b) adult height and (c) body composition of men and women of the Yakut of Siberia and the Tsimané of Bolivia. All measurements are mean and standard deviation. Body composition data obtained from bioelectrical impedance analysis (BIA) and expressed as the Fat-free mass index (Fat-free mass/height^2) and Fat mass index (Fat mass/height^2) in the same kg/m^2 units as body mass index.

Although body fat is often assumed to have undergone selection in response to exposure to famines, the variability in fat distribution that characterises human populations is unlikely to be due to this stress, as the threat of death from starvation is relatively constant across settings (Wells, 2009a). An alternative hypothesis is that fat distribution is shaped by the local burden of infectious disease (Wells, 2009a). While peripheral fat is relatively inert, central abdominal fat is more metabolically active and is closely associated with immune function (Gabrielsson et al., 2003). An ecological study found that populations in environments with higher pathogen burdens had thinner subscapular skinfolds, indicating greater use of energy from this fat depot to fund immune function (Wells and Cortina-Borja, 2013). The anatomical distribution of fat may also be shaped by the types of pathogens in a given environment, as different pathogens impact the metabolism of different organs and tissues (Wells, 2009a).

Regarding dietary intake, the origins of agriculture were associated with an overall shift to greater carbohydrate content (Ulijaszek et al., 2012), particularly when larger-scale societies emerged with their focus on grain agriculture (Scott, 2017). In such societies, social hierarchies also exposed richer and poorer groups to very different diets, as represented in artefacts such as the Standard of Ur depicting a feast and its preparation in ancient Mesopotamia

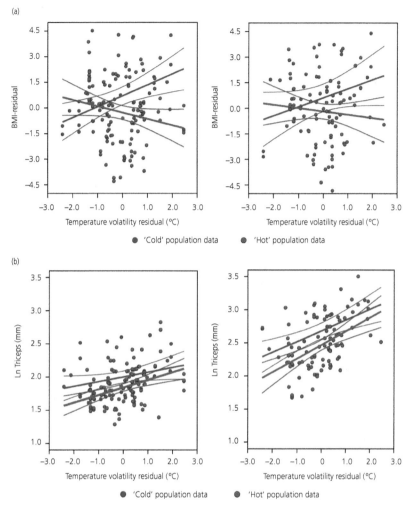

Figure 15.2 Associations of ecological uncertainty (proxied by inter-annual volatility in temperature) with markers of the two key components of body composition: (a) BMI residual (a marker of lean mass) and (b) Triceps skinfold, a marker of adiposity, stratified by hot and cold settings. In (a) and (b), males are in the left-hand panel, and females in the right-hand panel. In both sexes, greater ecological uncertainty is associated with lower lean mass in hot settings, but with higher lean mass in cold settings, and with higher adiposity in both hot and cold settings. *Source:* Reproduced with permission from Wells et al. (2019).

(Wells, 2016). However, pre-agricultural populations may also have regularly consumed high levels of carbohydrates, for example, honey (Crittenden, 2011) or fat by consuming the meat of animals that had gained large energy stores in preparation for winter survival (Hayden, 2001). During periods of scarcity, the consumption of foods with low protein content may also have led to the over-consumption of energy to satisfy protein requirements (Simpson and Raubenheimer, 2005). We assume that such

dietary changes played a key role in converting T2DM susceptibility into overt disease, but their effects would have differed according to their life-course impact. Undernutrition in early life would reduce metabolic capacity, whereas increased adiposity at later ages would increase metabolic load.

Using a life-course approach, we can consider how a variety of plastic responses at different stages of the life course might shape T2DM risk. From an ecological perspective, we can consider how different geographical and subsistence niches occupied by human populations have elicited adaptation (through either genetic or plastic mechanisms) that modify this risk. From an evolutionary perspective, we can apply life history theory to shed new light on why these developmental and ecological adaptations increase the risk of metabolic ill health. These diverse approaches may complement bioarchaeological analyses.

15.3.2 Life history approaches to understanding metabolic disease risk

We have argued that a more integrated approach to metabolic diseases could inform our understanding of their prehistoric origins. Life history traits (height, physique, maturation rate, fertility and longevity) can be considered as illustrating simultaneously life-course risk pathways, adaptation to local environments and variable investment across life history functions such as maintenance, reproduction or defence. Within any given setting, exposure to specific stimuli or stresses may either drive genetic change, or favour life-course adjustments in phenotype that are fitness-maximising. Examples include favouring offspring quantity over quality, early reproduction, reducing growth and expensive tissues and increasing energy stored in abdominal fat. Our key message is that, despite maximising fitness, these changes may also elevate the risk of metabolic disorders, while a basic principle of EM is that the tougher the environment, the more health (maintenance) is traded off against survival and reproduction (Nesse & Williams, 1994).

Recent research has highlighted growing opportunities to extract information about many of these risk traits from the archaeological record. Arguably the most difficult trait to detect directly in the skeletal record is adiposity, as this soft tissue leaves no material signal. However, most other traits can now be considered in more or less direct ways. Archaeologists have invested considerable effort in reconstructing past trends in stature, in particular through predictions based on long bones. There is widespread evidence that stature declined in many populations following the transition to agriculture (Mummert et al., 2011; Pomeroy et al., 2019a), though there is some evidence for temporal

and regional variation which includes increases in stature in some contexts (Wells & Stock, 2020). At the proximate level, declining stature indicates exposure to new stresses, such as more intense pathogen burdens, narrower diet ranges and greater food insecurity. From a life-history perspective, these trends indicate reducing energy allocation to growth and maintenance, in favour of survival/defence and reproduction (Figure 15.3) (Wells & Stock, 2020), thus indicating changes in T2DM risk.

Beyond stature, it would be particularly valuable to differentiate body mass into its key components—lean and fat mass—given that fat is a key marker of metabolic load, whereas lean mass is a marker of metabolic capacity and energy requirements. This relates to Wells's metabolic capacity load model, which addresses how development enables or constrains the ability of individuals to cope with metabolic demands across the life course (Wells, 2018). Until recently, it was difficult to know how trends in linear growth might have impacted tissue masses, but recent research is making inroads here. The most common approaches to body mass estimation from the skeleton use dimensions either of body breadth, or of weight-bearing joints such as the femoral head (Auerbach & Ruff, 2004; Kurki et al., 2010). While the diameter of the femoral head is correlated with weight at skeletal maturity, it is a poor predictor of the high body mass associated with obesity, which primarily represents adipose tissue accretion during adulthood (Young et al., 2018). At present, therefore, there are no direct ways to detect obesity from the skeleton, posing difficulty for the interpretation of past diabetes prevalence and aetiology. That obesity did occur in prehistoric populations is, however, strongly indicated by a number of corpulent 'Venus figurines' from Palaeolithic (Dixson & Dixson, 2011) and Neolithic (Trump & Cilia, 2002) Europe. Moreover, recent studies have shown that it is possible to improve our understanding of body mass components on the basis of skeletal evidence. Dual-energy X-ray absorptiometry (DXA) images of bone lengths can be used to generate region-specific equations to estimate stature (Pomeroy et al., 2018). These studies have shown that skeletal dimensions such as femoral head diameter are more strongly

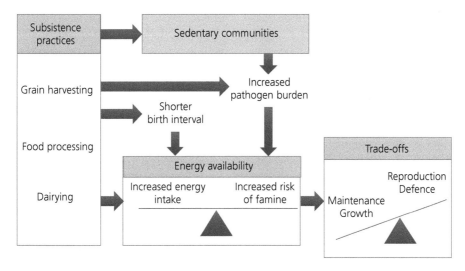

Figure 15.3 Summary of how the combination of changes in subsistence practices associated with the transition to agriculture may have increased dietary energy availability, but also changed diverse ecological stresses. These composite changes are proposed to have elicited life history trade-offs favouring reproduction and defence, over maintenance and growth.
Source: Reproduced with permission from Wells and Stock (2020).

correlated with lean body mass than total body mass (Pomeroy et al., 2018, 2019b).

This approach allows for more subtle integration of living and prehistoric human phenotypic data, and the exploration of temporal trends in components of phenotypic variation relevant to understanding metabolic diseases. An application of this method demonstrated that the 'thin' physique common in contemporary South Asia has a deep origin in the Mesolithic (Pomeroy et al., 2019a). Other work has used data on stature trends to simulate trends in neonatal size, predicting, for example, that birth weight in India would have been ~ 3440 g in early forager populations (8000 BP) and ~ 3200 g in early farming populations (4000 BP), compared to ~ 2800 g today (Wells, 2009b). These simulations suggest a 20% reduction in birth weight over the last 8000 years. The Developmental Origins of Health and Disease (DOHaD) model is discussed in more detail in Chapter 2. In brief, however, low birth weight is associated with higher risks of metabolic disorders in later life, particularly if individuals also gain weight quickly as infants and children (Wells et al., 2016a). While these findings do not demonstrate that T2DM was present in the past in India, they do demonstrate prehistoric origins of key risk factors and their likely

change over time and help us to understand the high susceptibility of South Asians to T2DM in contemporary obesogenic environments (Wells et al., 2016a).

Relating again to the metabolic capacity-load model (Wells, 2018), our understanding of past T2DM susceptibility would be greatly improved if we could identify subadults with low or high weight-for-height, which is a common component of analyses of growth variation and metabolic disease in contemporary populations. Stature estimation from subadult bone lengths is relatively common in forensic contexts (Smith, 2007), but is rarely employed in bioarchaeological studies due to the challenges in differentiating variability associated with age versus environmental conditions (Lewis, 2006). However, several recent studies have refined our ability to estimate both stature and body mass in subadult individuals (Cowgill, 2018; Robbins et al., 2010), which could be used to shed significant new light on prehistoric human health. A recent case study, where the integration of phenotypic data was employed, involved the identification of emaciation in prehistoric subadult remains from Inamgaon, India (Schug & Goldman, 2014). Small stature and low BMI were associated with reductions in bone mass and strength, and with greater pore volume relative to bone volume, indicating poor bone quality (Schug & Goldman, 2014).

Such creative approaches would also help integrate other phenotypic data within a broad life-history framework. For example, life-history theory predicts an inverse correlation between fertility and birth weights. Integrating estimates of fertility with perinatal and neonatal body mass estimates would therefore be informative. Similar work could examine links between longevity and maturation rate, which again would indicate the relative allocation of energy to maintenance versus defence and reproduction. Fertility and maturation are notoriously difficult to differentiate in bioarchaeological samples, but new opportunities are emerging (Lewis et al., 2016; Robbins, 2011). Likewise, improved understanding of dietary trends, for example, using stable isotope approaches, could identify periods of increased susceptibility to excess weight gain, and possibly which groups were most at risk. Whereas elites may have had plentiful access to calories, low protein intake can also drive weight gain (Simpson & Raubenheimer, 2005), suggesting that groups low in the hierarchy might also have been prone to overweight.

15.4 Conclusions

As yet, it is extremely rare for diabetes *sensu lato* to be directly identified in the bioarchaeological record. However, the broader approach we have set out here integrating knowledge of other metabolic conditions together with an understanding of human ecological adaptation and life-history theory, offers the opportunity to reconstruct T2DM risk profiles in past populations. While few prehistoric individuals may have had overt T2DM, some examples are indicated in the ancient historical record. One wonders whether bioarchaeologists have the possibility in their minds of finding T2DM in the archaeological record and/or have the tools and knowledge at their disposal to do so. Our logic is that the more risk traits demonstrated by any population, the higher the chances that T2DM manifested at some period in some individuals. The bioarchaeological record provides a broader picture of human skeletal variability than that offered by contemporary populations alone. In general, pre-agricultural

human populations were taller and more robust than those that followed (Ruff et al., 1997), suggesting a low generic risk of T2DM. However, there was also substantial regional variability, with a relatively slim physique evident on the Indian subcontinent well before the transition to agriculture (Pomeroy et al., 2019a). Understanding temporal and geographical variability in skeletal risk markers and elucidating the physical and social ecological factors that shaped this variability, will help reveal the processes of phenotypic adjustment through which T2DM risk emerged and changed.

In the last 12,000 years, major subsistence shifts such as the transition to agriculture and industrialisation and associated social stratification have 'reorganised' the life history strategies of most human populations (Wells & Stock, 2020), thereby changing their susceptibility to T2DM. In this context, it becomes clear that the ongoing nutrition transition is far from unique and is merely the latest driver of such life history transitions, orchestrated by new globalised forms of niche construction. Contemporary T2DM susceptibility emerged under the influence of long-term stresses such as local ecological conditions and disease burdens, the transition to agriculture, the emergence of sedentary lifestyles and the development of stratified societies. Human life history strategy responded to all of these stresses, and reshaped susceptibility to metabolic disorders in the process (Wells, 2016; Wells et al., 2016a). We suggest that substantial evidence on these transitions and associated risks could be gleaned from the bioarchaeological record, as shown by the various examples discussed in this chapter.

References

Auerbach, B. M. & Ruff, C. B. (2004). Human body mass estimation: a comparison of 'morphometric' and 'mechanical' methods. *American Journal of Physical Anthropology*, 125, 331–342.

Bergman, R. N., Ader, M., Huecking, K. & Van, C. G. (2002). Accurate assessment of beta-cell function: the hyperbolic correction. *Diabetes*, 51(Suppl 1), S212–S220.

Bindon, J. R. & Baker, P. T. (1997). Bergmann's rule and the thrifty genotype. *American Journal of Physical Anthropology*, 104(2), 201–210.

Brickley, M. B. & Ives, R. (2008). *The bioarchaeology of metabolic bone disease*. Academic Press, London.

Brickley, M. B., Ives, R. & Mays, S. (2020). *The bioarchaeology of metabolic bone disease*, 2nd edn. Academic Press, London.

Brower, A. C. & Flemming, D. J. (2012). 'Gout', in: *Arthritis in black and white*, 3rd edn, pp. 293–308. Elsevier/Saunders, Philadelphia.

Buckley, H. R. (2007). Possible gouty arthritis in Lapita-associated skeletons from Teouma, Efate Island, Central Vanuatu. *Current Anthropology*, 48(5), 741–749. https://doi.org/10.1086/520967.

Buckley, H. R. (2011). Epidemiology of gout: perspectives from the past. *Current Rheumatology Reviews*, 7, 106–113.

Buckley, H. R., Tayles, N., Halcrow, S. E., Robb, K. & Fyfe, R. (2010). View of the people of Wairau Bar: a re-examination. *Journal of Pacific Archaeology*, 1(1), 1–20.

Chhem, R. K., Schmit, P. & Fauré, C. (2004). Did Ramesses II really have ankylosing spondylitis? A reappraisal. *Canadian Association of Radiologists Journal*, 55(4), 211–7.

Cohen, M. N. & Armelagos, G. J. (1984). *Palaeopathology and the origins of agriculture*. Academic Press, Orlando.

Conard, N. J. (2009). A female figurine from the basal Aurignacian of Hohle Fels Cave in southwestern Germany. *Nature*, 459, 248–252.

SIGMA Type 2 Diabetes Consortium, Williams, A. L., Jacobs, S. B., Moreno-Macias, H., Huerta-Chagoya, A., Churchouse, C., Marquez-Luna, C., Garcia-Ortiz, H., Gomez-Vazquez, M. J., Burtt, N. P., Aguilar-Salinas, C. A., Gonzalez-Villapando, C., Florez, J. C., Orozco, L., Haiman, C. A., Tusie-Luna, T. & Altshuler, D. (2014). Sequence variants in SLC16A11 are a common risk factor for type 2 diabetes in Mexico. *Nature*, 506, 97–101.

Corpeleijn, E., Saris, W. H. & Blaak, E. E. (2009). Metabolic flexibility in the development of insulin resistance and type 2 diabetes: effects of lifestyle. *Obesity Reviews*, 10, 178–193.

Cowgill, L. (2018). Juvenile body mass estimation: a methodological evaluation. *Journal of Human Evolution*, 115, 78–84.

Crittenden, A. N. (2011). The importance of honey consumption in human evolution. *Foods and Foodways*, 19, 57–273.

Crubézy, E. & Trinkaus, E. (1992). Shanidar 1: A case of hyperostotic disease (DISH) in the middle Paleolithic. *American Journal of Physical Anthropology*, 89(4), 411–420. https://doi.org/10.1002/ajpa.1330890402.

Dalbeth, N. & McLean, L. (2019). 'Etiology and pathogenesis of gout', in M. C. Hochberg, E. M. Gravallese, A. J. Silman, J. S. Smolen, M. E. Weinblatt & M. H. Weisman (eds), *Rheumatology*, 7th edn, pp. 1555–1568. Elsevier, Philadelphia.

Dixson, A. F. & Dixson, B. J. (2011). Venus figurines of the European Paleolithic: symbols of fertility or attractiveness? *Journal of Anthropology*, 2011, 569120. doi.org/10.1155/2011/569120.

Dobisikova, M., Vyhnánek, L., Veleminsky, P., Kuzelka, V. & Cejkova, I. (1999). Palaeopathology and historical personality of General Johann Sporck (1595–1679). *International Journal of Osteoarchaeology*, 9(5), 297–301. https://doi.org/10.1002/(SICI)1099-1212(199909/10)9:5<297::AID-oa483>3.0.CO;2-C.

Duba, A. S. & Mathew, S. D. (2018). The seronegative spondyloarthropathies. *Primary Care: Clinics in Office Practice*, 45(2), 271–287. https://doi.org/10.1016/j.pop.2018.02.005.

Dupras, T. L., Williams, L. J., Willems, H. & Peeters, C. (2010). Pathological skeletal remains from ancient Egypt: the earliest case of diabetes mellitus? *Practical Diabetes International*, 27, 358-363a.

Elfishawi, M. M., Zleik, N., Kvrgic, Z., Michet, C. J., Crowson, C. S., Matteson, E. L. & Bongartz, T. (2018). The rising incidence of gout and the increasing burden of comorbidities: a population-based study over 20 years. *The Journal of Rheumatology*, 45(4), 574–579. https://doi.org/10.3899/jrheum.170806.

Faccia, K., Waters-Rist, A., Lieverse, A. R., Bazaliiskii, V. I., Stock, J. T. & Katzenberg, M. A. (2016). Diffuse idiopathic skeletal hyperostosis (DISH) in a middle Holocene forager from Lake Baikal, Russia: potential causes and the effect on quality of life. *Quaternary International*, 405, 66–79. https://doi.org/10.1016/j.quaint.2015.10.011.

Fassio, A., Adami, G., Idolazzi, L., Giollo, A., Viapiana, O., Bosco, E., Negrelli, R., Sani, E., Sandri, D., Mantovani, A., Targher, G., Rossini, M. & Gatti, D. (2021). Diffuse idiopathic skeletal hyperostosis (DISH) in type 2 diabetes: a new imaging possibility and a new biomarker. *Calcified Tissue International*, 108(2), 231–239. https://doi.org/10.1007/s00223-020-00768-2.

Forestier, J. & Rotés-Querol, J. (1950). Senile ankylosing hyperostosis of the spine. *Annals of the Rheumatic Diseases*, 9(4), 321.

Fornaciari, A., Giuffra, V., Armocida, E., Caramella, D., Rühli, F. J. & Galassi, F. M. (2018). Gout in Duke Federico of Montefeltro (1422–1482): a new pearl of the Italian renaissance. *Clinical and Experimental Rheumatology*, 36(1), 15–20.

Fornaciari, G., Giuffra, V., Giusiani, S., Fornaciari, A., Villari, N. & Vitiello, A. (2009). The 'gout' of the Medici, Grand Dukes of Florence: a palaeopathological study. *Rheumatology*, 48(4), 375–377.

Foster, A., Kinaston, R., Spriggs, M., Bedford, S., Gray, A. & Buckley, H. (2018). Possible diffuse idiopathic skeletal hyperostosis (DISH) in a 3000-year-old Pacific Island skeletal assemblage. *Journal of Archaeological Science: Reports*, 18, 408–419.

Froehle, A. W. (2008). Climate variables as predictors of basal metabolic rate: new equations. *American Journal of Human Biology*, 20(5), 510–529.

Gabrielsson, B. G., Johansson, J. M., Lönn, M., Jernås, M., Olbers, T., Peltonen, M., Larsson, I., Lönn, L., Sjöström, L., Carlsson, B. & Carlsson, L. M. (2003). High expression of complement components in omental adipose tissue in obese men. *Obesity Research*, 11(6), 699–708.

Gale, E. A. (2013). Is type 2 diabetes a category error? *Lancet*, 381, 1956–1957.

Ganong, W. F. (1999). *Review of medical physiology*, 19th edn. Appleton and Lange, Stamford.

Giuffra, V., Giusiani, S., Fornaciari, A., Villari, N., Vitiello, A. & Fornaciari, G., (2010). Diffuse idiopathic skeletal hyperostosis in the Medici, Grand Dukes of Florence (XVI century). *European Spine Journal*, 19(2), 103–107. https://doi.org/10.1007/s00586-009-1125-3.

Gluckman, P. & Hanson, M. (2006). *Mismatch: why our world no longer fits our bodies*. Oxford University Press, Oxford.

Gluckman, P., Hanson, M., Buklijas, T., Low, F. M. & Beedle, A. S. (2009). Epigenetic mechanisms that underpin metabolic and cardiovascular diseases. *Nature Reviews Endocrinology*, 5, 401–408. https://doi.org/10.1038/nrendo.2009.102.

Goldring, S. R. and Goldring, M. B. (2017). 'Biology of the normal joint', in G. S. Firestein, R. C. Budd, S. E. Gabriel, I. B. McInnes & J. R. O'Dell (eds), *Kelley and Firestein's textbook of rheumatology*, 10th edn, pp. 1–19. Elsevier, Philadelphia.

Harris, J., Carter, A. R., Glick, E. N. & Storey, G. O. (1974). Ankylosing hyperostosis. I. Clinical and radiological features. *Annals of the Rheumatic Diseases*, 33(3), 210.

Hayden, B. 2001. 'Richman, poorman, beggarman, chief: the dynamics of social inequality', in G. Feinman & T. D. Price (eds), *Archaeology at the millennium: a sourcebook*, pp. 231–272. Springer, New York.

Houghton, P. (1990). The adaptive significance of Polynesian body form. *Annals of Human Biology*, 17(1), 19–32.

Jadon, D. R. (2018). Psoriatic arthritis and seronegative spondyloarthropathies. *Medicine*, 46(4), 237–242. https://doi.org/10.1016/j.mpmed.2018.01.003.

Jansen, T. L. (2016). 'Clinical presentation of gout', in M. Doherty, D. J. Hunter, H. Bijlsma, N. Arden & N. Dalbeth (eds), *Oxford textbook of osteoarthritis and crystal arthropathy*, pp. 415–423. Oxford University Press, Oxford. https://doi.org/10.1093/med/9780199668847.001.0001.

Jun, J. K. & Kim, S. M. (2012). Association study of fibroblast growth factor 2 and fibroblast growth factor receptors gene polymorphism in Korean ossification of the posterior longitudinal ligament patients. *Journal of Korean Neurosurgical Society*, 52(1), 7.

Katzmarzyk, P. T. & Leonard, W. R. (1998). Climatic influences on human body size and proportions: ecological adaptations and secular trends. *American Journal of Physical Anthropology*, 106(4), 483–503.

Kim, D. K., Lee, I. S., Kim, W.-L., Lee, J. S., Koh, B. J., Kim, M. J., Youn, M. Y., Shin, M. H., Kim, Y.-S., Lee, S.-S., Oh, C. S. & Shin, D. H. (2011). Possible rheumatoid arthritis found in the human skeleton collected from the tomb of Joseon Dynasty, Korea, dating back to the 1700s AD. *International Journal of Osteoarchaeology*, 21(2), 136–149. https://doi.org/10.1002/oa.1112.

Kinaston, R. L., Bedford, S., Spriggs, M., Anson, D., & Buckley, H. R. (2016). Is there a 'Lapita diet'? A comparison of Lapita and post-Lapita skeletal samples from four Pacific Island archaeological sites', in M. Oxenham & H. Buckley (eds), *The Routledge handbook of bioarchaeology in Southeast Asia and the Pacific Islands*, 427–461. Routledge, Abingdon.

Kranenburg, H. C., Westerveld, L. A., Verlaan, J. J., Oner, F. C., Dhert, W. J. A., Voorhout, G., Hazewinkel, H. A. W. & Meij, B. P. (2010). The dog as an animal model for DISH? *European Spine Journal*, 19(8), 1325–1329. https://doi.org/10.1007/s00586-010-1280-6.

Kuperus, J. S., Mohamed Hoesein, F. A. A., de Jong, P. A. & Verlaan, J. J. (2020). Diffuse idiopathic skeletal hyperostosis: etiology and clinical relevance. *Best Practice & Research Clinical Rheumatology*, 34(3), 101527. https://doi.org/10.1016/j.berh.2020.101527.

Kurki, H. K., Ginter, J. K., Stock, J. T. & Pfeiffer, S. (2010). Body size estimation of small-bodied humans: applicability of current methods. *American Journal of Physical Anthropology*, 141, 169–180.

Kuzawa, C. W., Chugani, H. T., Grossman, L. I., Lipovich, L., Muzik, O., Hof, P. R., Wildman, D. E., Sherwood, C. C., Leonard, W. R. & Lange, N. (2014). Metabolic costs and evolutionary implications of human brain development. *Proceedings of the National Academy of Sciences of the United States of America*, 111, 13010–13015.

Lean, M. E., Leslie, W. S., Barnes, A. C., Brosnahan, N., Thom, G., McCombie, L., Peters, C., Zhyzhneuskaya, S., Al-Mrabeh, A., Hollingsworth, K. G., Rodrigues, A. M., Rehackova, L., Adamson, A. J., Sniehotta, F. F., Mathers, J. C., Ross, H. M., McIlvenna, Y., Stefanetti, R., Trenell, M., Welsh, P., Kean, S., Ford, I., McConnachie, A., Sattar, N. & Taylor, R. (2018). Primary care-led weight management for remission of type 2 diabetes (DiRECT): an open-label, cluster-randomised trial. *The Lancet*, 391(10120), 541–551.

Leonard, W. R., Robertson, M. L., Snodgrass, J. J. & Kuzawa, C. W. (2003). Metabolic correlates of hominid brain evolution. *Comparative Biochemistry & Physiology Part A: Molecular and Integrative Physiology*, 136, 5–15.

Leonard, W. R., Snodgrass, J. J. & Sorensen, M. V. (2005). Metabolic adaptation in indigenous Siberian populations. *Annual Review of Anthropology*, 34, 451–471.

Lewis, M., Shapland, F. & Watts, R. (2016). On the threshold of adulthood: A new approach for the use of maturation indicators to assess puberty in adolescents from medieval England. *American Journal of Human Biology*, 28, 48–56.

Lewis, M. E. (2006). 'The bioarchaeology of children', in *The bioarchaeology of children*, pp. 1–19. Cambridge University Press, Cambridge.

Lieberman, L. S. (2008). 'Diabesity and Darwinian medicine: the evolution of an epidemic', in W. R. Trevathan, E. O. Smith & J. J. McKenna (eds), *Evolutionary medicine and health: new perspectives*, pp.72–95. Oxford University Press, Oxford.

Littlejohn, G. O. (2019). 'Diffuse idiopathic skeletal hyperostosis', in M. C. Hochberg, E. M. Gravallese, A. J. Silman, J. S. Smolen, M. E. Weinblatt & M. H. Weisman (eds), *Rheumatology*, 7th edn, pp. 1801–1806. Elsevier, Philadelphia.

Littlejohn, G. O. & Hall, S. (1982). Diffuse idiopathic skeletal hyperostosis and new bone formation in male gouty subjects. *Rheumatology International*, 2(2), 83–86. https://doi.org/10.1007/BF00541250.

Mandel, N. S. & Mandel, G. S. (1976). Monosodium urate monohydrate, the gout culprit. *Journal of the American Chemical Society*, 98(8), 2319–2323. https://doi.org/10.1021/ja00424a054.

Marković, N., Stevanović, O., Krstić, N., Marinković, D. & Buckley, M., 2019. A case study of vertebral fusion in a 19th-century horse from Serbia. *International Journal of Paleopathology*, 27, 17–23. https://doi.org/10.1016/j.ijpp.2019.07.007.

Merriman, T. (2016). 'The genetic basis of gout', in M. Doherty, D. J. Hunter, H. Bijlsma, N. Arden & N. Dalbeth (eds), *Oxford textbook of osteoarthritis and crystal arthropathy*, pp. 415–423. Oxford University Press, Oxford. https://doi.org/10.1093/med/9780199668847.001.0001.

Merriman, T. (2017). Genomic influences on hyperuricemia and gout. *Rheumatic Disease Clinics of North America*, 43(3), 389–399. https://doi.org/10.1016/j.rdc.2017.04.004.

Milton, K. (2000). Hunter-gatherer diets: a different perspective. *American Journal of Clinical Nutrition*, 71, 665–667.

Muchlinski, M. N., Snodgrass, J. J. & Terranova, C. J. (2012). Muscle mass scaling in primates: an energetic and ecological perspective. *American Journal of Primatology*, 74(5), 395–407.

Mummert, A., Esche, E., Robinson, J. & Armelagos, G. J. (2011). Stature and robusticity during the agricultural transition: evidence from the bioarchaeological record. *Economics & Human Biology*, 9(3), 284–301.

Neogi, T., Jansen, T. L. T. A., Dalbeth, N., Fransen, J., Schumacher, H. R., Berendsen, D., Brown, M., Choi, H., Edwards, N. L., Janssens, H. J. E. M., Lioté, F.,

Naden, R. P., Nuki, G., Ogdie, A., Perez-Ruiz, F., Saag, K., Singh, J. A., Sundy, J. S., Tausche, A.-K., Vaquez-Mellado, J., Yarows, S. A. & Taylor, W. J. 2015. 2015 gout classification criteria: An American College of Rheumatology/European League Against Rheumatism collaborative initiative. *Arthritis & Rheumatology*, 67(10), 2557–2568. https://doi.org/10.1002/art.39254.

Nesse, R. M. & Williams, G. C. (1994). *Why we get sick: the new science of Darwinian medicine*. Times Books, New York.

Nuki, G. & Simkin, P. A. (2006). A concise history of gout and hyperuricemia and their treatment. *Arthritis Research & Therapy*, 8(Suppl 1), S1. https://doi.org/10.1186/ar1906.

O'Dea, K. (1984). Marked improvement in carbohydrate and lipid metabolism in diabetic Australian aborigines after temporary reversion to traditional lifestyle. *Diabetes*, 33, 596–603.

Omran, A. R. (1971). The epidemiologic transition: a theory of the epidemiology of population change. *Millbank Memorial Fund Quarterly*, 49, 509–538.

Ordi, J., Alonso, P. L., de Zulueta, J., Esteban, J., Velasco, M., Mas, E., Campo, E. & Fernández, P. L. (2006). The severe gout of Holy Roman Emperor Charles V. *New England Journal of Medicine*, 355(5), 516–520. doi:10.1056/NEJMon060780.

Ortner, D. J. (2003). *Identification of pathological conditions in human skeletal remains*, 2nd edn. Academic Press, London.

Oxenham, M. F., Matsumura, H. & Nishimoto, T. (2006). Diffuse idiopathic skeletal hyperostosis in Late Jomon Hokkaido, Japan. *International Journal of Osteoarchaeology*, 16(1), 34–46. https://doi.org/10.1002/oa.803.

Pariente-Rodrigo, E., Sgaramella, G. A., Olmos-Martínez, J. M., Pini-Valdivieso, S. F., Landeras-Alvaro, R. & Hernández-Hernández, J. L. (2017). Relationship between diffuse idiopathic skeletal hyperostosis, abdominal aortic calcification and associated metabolic disorders: data from the Camargo Cohort. *Medicina Clínica (English Edition)*, 149(5), 196–202. https://doi.org/10.1016/j.medcle.2017.07.020.

Perez-Ruiz, F., Castillo, E., Chinchilla, S. P. & Herrero-Beites, A. M. (2014). Clinical manifestations and diagnosis of gout. *Rheumatic Diseases Clinics of North America*, 40(2),193–206.

Perez-Ruiz, F., Dalbeth, N., Urresola, A., de Miguel, E. & Schlesinger, N. (2009). Gout. Imaging of gout: findings and utility. *Arthritis Research & Therapy*, 11(3), 1–8. https://doi.org/10.1186/ar2687.

Perrin, C. M., Dobish, M. A., Van Keuren, E. & Swift, J. A. (2011). Monosodium urate monohydrate crystallization. *CrystEngComm*, 13(4), 1111. https://doi.org/10.1039/c0ce00737d.

Pietrusewsky, M. & Douglas, M. T. (2002). Ban Chiang, a *prehistoric village site in northe*ast Thailand, *volume 1: the human skeletal remains.* UPenn Museum of Archaeology, Philadelphia.

Pillai, S. & Littlejohn, G. (2014). Metabolic factors in diffuse idiopathic skeletal hyperostosis – a review of clinical data. *Open Rheumatology Journal,* 8, 116–128. doi: 10.2174/1874312901408010116.

Pollard, T. M., Nunez-de la Mora, A. & Unwin, N. (2008). 'Evolutionary perspectives on Type 2 diabetes in Asia', in S. Elton and P. O'Higgins (eds), *Medicine and evolution: current applications, future prospects,* pp. 55–76. CRC Press, Boca Raton.

Pomeroy, E., Mushrif-Tripathy, V., Cole, T. J., Wells, J. C. K. & Stock, J. T. (2019a). Ancient origins of low lean mass among South Asians and implications for modern type 2 diabetes susceptibility. *Science Reports,* 9, 10515.

Pomeroy, E., Mushrif-Tripathy, V., Kulkarini, B., Kinra, S., Stock, J. T., Cole, T. J., Shirley, M. K. & Wells, J. C. K. (2019b). Estimating body mass and composition from proximal femur dimensions using dual energy X-ray absorptiometry. *Archaeological and Anthropological Sciences,* 11, 2167–2179.

Pomeroy, E., Mushrif-Tripathy, V., Wells, J. C. K., Kulkarini, B., Kinra, S. & Stock, J. T. (2018). Stature estimation equations for South Asian skeletons based on DXA scans of contemporary adults. *American Journal of Physical Anthropology,* 167, 20–31.

Popkin, B. M., Adair, L. S. & Ng, S. W. (2012). Global nutrition transition and the pandemic of obesity in developing countries. *Nutrition Reviews,* 70, 3–21.

Quintelier, K., Ervynck, A., Müldner, G., Neer, W.V., Richards, M. P. & Fuller, B.T. (2014). Isotopic examination of links between diet, social differentiation, and DISH at the post-medieval Carmelite Friary of Aalst, Belgium. *American Journal of Physical Anthropology,* 153(2), 203–213. https://doi.org/10.1002/ajpa. 22420.

Resnick, D. & Niwayama, G. (1976). Radiographic and pathologic features of spinal involvement in diffuse idiopathic skeletal hyperostosis (DISH). *Radiology,* 119(3), 559–568.

Resnick, D. & Niwayama, G. (1995). *Diagnosis of bone and joint disorders,* 3rd edn. Saunders, Philadelphia.

Resnick, D., Shaul, S. R. & Robins, J. M. (1975). Diffuse idiopathic skeletal hyperostosis (DISH): Forestier's disease with extraspinal manifestations. *Radiology,* 115(3), 513–524.

Robbins, G. (2011). Don't throw out the baby with the bathwater: estimating fertility from subadult skeletons. *International Journal of Osteoarchaeology,* 21, 717–722.

Robbins G., Sciulli, P. W. & Blatt S. H. (2010). Estimating body mass in subadult human skeletons. *American Journal of Physical Anthropology,* 143, 146–150.

Rogers, J., Shepstone, L. & Dieppe, P. (1997). Bone formers: osteophyte and enthesophyte formation are positively associated. *Annals of the Rheumatic Diseases,* 56(2), 85–90.

Rogers, J. & Waldron, T. (2001). DISH and the monastic way of life. *International Journal of Osteoarchaeology,* 11(5), 357–365. https://doi.org/10.1002/oa.574.

Rothschild, B. M. (1987). Diffuse idiopathic skeletal hyperostosis as reflected in the paleontologic record: dinosaurs and early mammals. *Seminars in Arthritis and Rheumatism,* 17(2), 119–125. https://doi.org/10.1016/0049-0172(87)90034-5.

Rothschild, B. M., Coppa, A. & Petrone, P. P. (2004). 'Like a virgin': absence of rheumatoid arthritis and treponematosis, good sanitation and only rare gout in Italy prior to the 15th century. *Reumatismo,* 56(1), 61–66. https://doi.org/10.4081/reumatismo.2004.61.

Rothschild, B. M. & Heathcote, G. M. (1995). Characterization of gout in a skeletal population sample: presumptive diagnosis in a Micronesian population. *American Journal of Physical Anthropology,* 98(4), 519–525. https://doi.org/10.1002/ajpa.1330980411.

Rowe, L. J. & Resnick, D. (2001). Degenerative diseases of the vertebral column. *Contemporary Spine Surgery,* 2(5), 29–34.

Ruff, C. B., Trinkaus, E. & Holliday, T. W. (1997). Body mass and encephalization in Pleistocene Homo. *Nature,* 387, 173–176.

Schlapbach, P., Beyeler, Ch., Gerber, N. J., Van Der Linden, S. J., Bürgi, U., Fuchs, W. A. & Ehrengruber, H. (1989). Diffuse idiopathic skeletal hyperostosis (DISH) of the spine: a cause of back pain? A controlled study. *Rheumatology,* 28(4), 299–303. https://doi.org/10.1093/rheumatology/28.4.299.

Schlesinger, N. (2019). 'Clinical features of gout', in M. C. Hochberg, E. M. Gravallese, A. J. Silman, J. S. Smolen, M. E. Weinblatt & M. H. Weisman (eds), *Rheumatology,* 7th edn, pp.1610–1612. Elsevier, Philadelphia.

Schug, G. R. & Goldman, H. M. (2014). Birth is but our death begun: a bioarchaeological assessment of skeletal emaciation in immature human skeletons in the context of environmental, social, and subsistence transition. *American Journal of Physical Anthropology,* 155, 243–259.

Scott, J. C. (2017). *Against the grain: a deep history of the earliest states.* Yale University Press, New Haven.

Simpson, S. J. & Raubenheimer, D. (2005). Obesity: the protein leverage hypothesis. *Obesity Reviews,* 6, 133–142.

Singh, J. A. & Gaffo, A. (2020). Gout epidemiology and comorbidities. *Seminars in Arthritis and Rheumatism,*

50(3), S11–S16. https://doi.org/10.1016/j.semarthrit.2020.04.008.

Smith, S. L. (2007). Stature estimation of 3–10-year-old children from long bone lengths. *Journal of Forensic Science*, 52, 538–546.

Snodgrass, J. J., Leonard, W. R., Tarskala, L. A., Egorova, A. G., Maharova, N. V., Pinigina, I. A., Halyev, S. D., Matveeva, N. P. & Romanova, A. N. (2010). Impaired fasting glucose and metabolic syndrome in an indigenous Siberian population. *International Journal of Circumpolar Health*, 69(1), 87–98.

Suzuki, T. (1993). Paleopathological and paleoepidemiological investigation of human skeletal remains of early Hawaiians from Mokapu Site, Oahu Island, Hawaii. *Japan Review*, 4, 83–128.

Thompson, R. C., Allam, A. H., Lombardi, G. P., Wann, L. S., Sutherland, M. L., Sutherland, J. D., Soliman, M. A., Frohlich, B., Miniberg, D. T., Monge, J. M., Vallodolid, C. M., Cox, S. L., Abd El-Maksoud, G., Badr, I., Miyamoto, M. I., El-Halism Nur El-Din, A., Narula, J., Finch, C. E. & Thomas, G. S. (2013). Atherosclerosis across 4000 years of human history: the HORUS study of four ancient populations. *The Lancet*, 381, 1211–1222.

Trinkaus, E., Maley, B. & Buzhilova, A.P. (2008). Brief communication: paleopathology of the Kiik-Koba 1 Neandertal. *American Journal of Physical Anthropology*, 137(1), 106–112. https://doi.org/10.1002/ajpa.20833.

Trump, D. & Cilia, D. (2002). *Malta: prehistory and temples*, Midsea Books Ltd, Malta.

Tsukahara, S., Miyazawa, N., Akagawa, H., Forejtova, S., Pavelka, K., Tanaka, T., Toh, S., Tajima, A., Akiyama, I. & Inoue, I. (2005). *COL6A1*, the candidate gene for ossification of the posterior longitudinal ligament, is associated with diffuse idiopathic skeletal hyperostosis in Japanese. *Spine*, 30(20), 2321–2324.

Ulijaszek, S. J., Mann, N. & Elton, S. (2012). *Evolving human nutrition: implications for public health*. Cambridge University Press, Cambridge.

Verlaan, J. J., Oner, F. C. & Maat, G. J. R. (2007). Diffuse idiopathic skeletal hyperostosis in ancient clergymen. *European Spine Journal*, 16(8), 1129–1135. https://doi.org/10.1007/s00586-007-0342-x.

Wagner, J. E., Kavanagh, K., Ward, G. M., Auerbach, B. J., Harwood, H. J. & Kaplan, J. R. (2006). Old world non-human primate models of type 2 diabetes mellitus. *ILAR Journal*, 47, 259–271.

Waldron, T. (1985). DISH at Merton Priory: evidence for a 'new' occupational disease? *British Medical Journal*, 291, 1762–1763.

Waldron, T. (2019). 'Joint diseases', in J. E. Buikstra (ed), *Ortner's identification of pathological conditions in human skeletal remains*, 2nd edn, pp. 719–748. Academic Press, London.

Waldron, T. (2020). *Palaeopathology*, 2nd edn. Cambridge University Press, Cambridge.

Watve, M. (2013). *Doves, diplomats and diabetes: a Darwinian interpretation of type 2 diabetes and related disorders*. Springer, New York.

Wells, C. (1973). A palaeopathological rarity in a skeleton of Roman date. *Medical History*, 17(4), 399–400.

Wells, J. C. (2009a). Ethnic variability in adiposity and cardiovascular risk: the variable disease selection hypothesis. *International Journal of Epidemiology*, 38, 63–71.

Wells, J. C. (2010). *The evolutionary biology of human body fatness: thrift and control*. Cambridge University Press, Cambridge.

Wells, J. C. (2016). *The metabolic ghetto: an evolutionary perspective on nutrition, power relations and chronic disease*. Cambridge University Press, Cambridge.

Wells, J. (2018). Optimising the balance between metabolic capacity and metabolic load for lifelong health. International Symposium on Understanding the Double Burden of Malnutrition for Effective Interventions. Vienna, Austria. 10–13 December.

Wells, J. C. (2012). Sexual dimorphism in body composition across human populations: associations with climate and proxies for short-and long-term energy supply. *American Journal of Human Biology*, 24(4), 411–419.

Wells, J. C. (2009b). What was human birth weight in the past? Simulations based on stature from the palaeolithic to the present. *Journal of Life Sciences*, 1, 115–120.

Wells, J. C. & Cortin-Borja, M. (2013). Different associations of subscapular and triceps skinfold thicknesses with pathogen load: an ecogeographical analysis. *American Journal of Human Biology*, 25(5), 594–605.

Wells, J. C., Pomeroy, E., Walimbe, S. R., Popkin, B. M. & Yajnik, C. S. (2016a). The elevated susceptibility to diabetes in India: an evolutionary perspective. *Front Public Health*, 4, 145.

Wells, J. C., Saunders, M. A., Lea, A. S., Cortina-Borja, M., & Shirley, M. K. (2019). Beyond Bergmann's rule: global variability in human body composition is associated with annual average precipitation and annual temperature volatility. *American Journal of Physical Anthropology*, 170(1), 75–87.

Wells, J. C. & Stock, J. T. (2020). Life history transitions at the origins of agriculture: a model for analysing how niche construction impacts human growth, demography and health. *Frontiers in Endocrinology* [online]. https://doi.org/10.3389/fendo.2020.00325.

Wells, J. C., Yao, P., Williams, J. E. & Gayner, R. (2016b). Maternal investment, life-history strategy of the offspring and adult chronic disease risk in South Asian women in the UK. *Evolution, Medicine, and Public Health*, 2016(1), 133–145.

Wells, J. C. K. (2017). Body composition and suscepti- bility to type 2 diabetes: an evolutionary perspective. *European Journal of Clinical Nutrition*, 71, 881–889.

Wrangham, R. (2009). *Catching fire: how cooking made us human*. Basic Books, New York.

Young, M., Johannesdottir, F., Poole, K., Shaw, C. & Stock, J. T. (2018). Assessing the accuracy of body mass esti- mation equations from pelvic and femoral variables among modern British women of known mass. *Journal of Human Evolution*, 115, 130–139.

Zhang, W. (2006). EULAR evidence-based recommenda- tions for gout. Part I: Diagnosis. Report of a task force of the standing committee for international clinical studies including therapeutics (ESCISIT). *Annals of the Rheumat- ic Diseases*, 65(10), 1301–1311. https://doi.org/10.1136/ ard.2006.055251.

CHAPTER 16

The palaeopathology of traumatic injuries: an evolutionary medicine perspective

Ryan P. Harrod and Anna J. Osterholtz

16.1 An evolutionary medicine approach to traumatic injury in the past

In modern society, people can be injured in a number of ways, including through accidents associated with slips, trips and falls, and trauma linked to specific activities like work-related hazards or sports; they may be victims of violence. Regardless of the specific mechanism, the injuries people suffer can, and often do, have a lasting impact on their lives. Traumatic injuries are characterised by 'sudden onset' that 'require immediate medical attention' according to medical professionals (Ahn et al., 2014, n.p.). According to the World Health Organization (WHO), trauma is a major global concern with millions of injuries occurring each year (WHO, 2014, 2). In fact, 'traumatic injury is responsible for 11% of global mortality and contributes to a significant amount of physical and psychological morbidity for all age groups' (Wiseman et al., 2015, 1). In the US, the National Center for Health Statistics (NCHS), a division of the Centers for Disease Control and Prevention (CDC), found that, in 2016, traumatic injuries were also a major medical concern for the nation. It was estimated that 39.5 million people in the US alone had visited the doctor's surgery for unintentional injuries, including 1.6 million who visited the emergency room for injuries from assaults (Rui et al., 2016).

Throughout our deep history, humans have been vulnerable to traumatic injuries related to daily activities including hunting, foraging, the risk of predation, using fire, making stone tools and other similar tasks, not to mention the intensification of intergroup conflict as human population densities increased. As such, humans have evolved mechanisms for dealing with and making sense of trauma, such as treatment technologies and neurological mechanisms to deal with pain. Furthermore, our closest primate relatives (chimpanzees, but not bonobos) frequently experience trauma from interpersonal attacks by conspecifics, while there is plenty of evidence from the bioarchaeological record for traumatic injuries as we outline below (Wrangham & Glowacki, 2012, Wrangham, Wilson & Muller, 2006, Wrangham & Peterson, 1996).

Nesse and Schilkin (2019) suggested that humans have developed neurological mechanisms for dealing with pain, and a greater susceptibility or resilience to chronic pain for some people might be a result of physical pain being part of our human heritage. Pain is an evolutionary adaptation that is instructive in that it alerts us to injury, and without pain feedback, humans would likely be incapable of surviving. Following studies of the Shiwiar culture in the Ecuadorian Amazon, Sugiyama (2004) suggests that traumatic injury and illness is common and that part of our evolutionary history as a species involved treating these injuries. As humans have evolved, we have also developed technological ways of caring for an injury, such as reducing fractures and splinting them using conservative and

Ryan P. Harrod and Anna J. Osterholtz, *The palaeopathology of traumatic injuries*. In: *Palaeopathology and Evolutionary Medicine.*
Edited by Kimberly A. Plomp et al., Oxford University Press. © Oxford University Press (2022). DOI: 10.1093/oso/9780198849711.003.0016

surgical methods, while our physiology may also have allowed us to adapt to life-threatening injuries (Stanzani et al., 2020). While some of the approaches to treating traumatic injuries are relatively new (e.g. orthopaedic surgery in the 1700s by Nicholas Andry), as a species, we have learned to adapt to trauma, a fact that is reflected in the low mortality rate of Shiwiar individuals who have survived severe injuries (Sugiyama, 2004).

Understanding the evolutionary and biocultural significance of violence and injury-related trauma in human history requires the deep time perspective that palaeopathological inquiry allows (Walker, 2001). Specifically, bioarchaeological evidence can provide valuable insight into the social, economic, biological and environmental factors that interplay to influence how humans prevent, treat and experience trauma. In fact, traumatic lesions are the among the most identified pathologies in archaeological remains (Walker, 2001). Thus, palaeopathology offers a unique perspective on trauma throughout human history that can help us develop new approaches in clinical research and modern medicine.

This chapter focuses on trauma throughout human history using an evolutionary perspective. We will also discuss the physical and mental consequences of trauma, such as the loss of mobility and function, daily pain, physical disfigurement, cognitive deficiency and fear of repeat injury. Traumatic injuries can leave permanent markers on the body and this evidence can provide insight into the lives of past peoples in a number of ways (e.g. socially sanctioned intimate partner violence, fighting for status, occupational injuries). Taking an evolutionary approach, we can understand how trauma and associated pain co-evolved, the ways that injuries impact an individual's lived experience and the development of community responses and medical interventions over time.

16.2 Understanding biomechanics and biology

Bioarchaeological studies of trauma focus on skeletal injuries and the identification of any bone remodelling that represents healing. One of the first (and most important) aspects of identifying

traumatic injury in skeletal remains is the determination of whether the trauma occurred peri- or antemortem. This can provide important information about whether the person survived the injury and the if/how it was treated. In addition, differential diagnoses can be conducted to determine possible causes of the injury, such as, for example, if a fracture is related to an underlying health condition like osteoporosis (Ortner, 2008). Also, palaeopathological evidence is not always limited to skeletal changes directly related to the traumatic event, since failure of other tissues can also cause affect the skeleton. For example, failure of ligamentous attachments surrounding joints, such as a torn anterior cruciate ligament (ACL) of the knee, which can ultimately lead to secondary arthritic change and pseudoarthrosis (the creation of a false joint between the two bone ends at the fracture location creating a new joint). The outcome of such injuries can sometimes be seen in archaeological remains, providing evidence for soft tissue trauma in the past. Considering this, an understanding of the biomechanics of the body and potential external and internal risk factors for trauma are necessary to identify and interpret injuries in human skeletal remains.

16.2.1 Biomechanics and tissue type

To understand what causes tissue to fail, it is necessary to attempt to reconstruct the direction of the applied force, mechanical loads placed on the tissue, and the cellular structure of the tissue itself (Galloway, 1999; Zephro & Galloway, 2014; Kieser, 2012; Schmitt et al., 2014; Wescott, 2013; Hannon, 2006a). There are kinetic forces (e.g. gravity, momentum, friction) constantly acting on tissues that dictate growth and development, distribution, movement and the progress of healing. As force is applied to a particular bone it causes mechanical loading that is associated with stress in the form of compression, shear, tension and torsion (see Hannon, 2006b, 80; Zephro & Galloway, 2014, 35–36). Stress occurs when loads are placed on tissue, and strain results from too much stress that begins to cause deformation (Hannon, 2006a; Zephro & Galloway, 2014).

The cellular structure and density of a bone determines if and how mechanical loading will result in deformation and/or trauma. The load necessary to exceed the ratio of stress to strain of a particular bone and the resulting fracture type will depend on specific qualities such as the amount, density and anisotropic (i.e. directionally dependent) characteristics of the bone's collagen (Hannon, 2006b), all of which may be influenced by age and overall health of the individual (Agarwal, 2019). According to Zephro and Galloway (2014, 34), deformation is reflective of the strain applied to the bone, as well as its rigidity or elasticity.

Mild-to-moderate stress and strain are important for maintaining proper bone health as they force bone to continually remodel. With increased muscular development, muscle attachment and skeletal insertion sites must correspondingly increase in rugosity to avoid traumatic-related evulsions of the muscle from the bone. Bone density (Fung, 2010, 503), direction of force (Yeni et al., 2004) and muscle development will all have an impact on how fractures occur. Often, a loaded bone will bend before breaking, with elastic deformation occurring below the yield point and resulting in temporary changes that may not visible after the event. However, if the loading extends beyond the bone's yield point, the elastic deformation can result in fracture. Plastic deformation, the shape changes that occur between the yield point and the failure point, results in permanent shape changes to the bone (Christensen et al., 2014). Overall, the amount of stress needed to exceed the yield and failure points decreases with increasing age, poor health and/or muscle atrophy and bone loss (e.g. osteoporosis).

There are a number of different types of tissue in the human body (connective, muscular, epithelial and nervous (Fung, 2010; Tortora & Anagnostakos, 1984). We focus primarily on the connective (Box 16.1) and muscular tissues, although damage to all tissue types from trauma may lead to skeletal lesions.

> ### Box 16.1   Connective tissues
>
> The connective tissue group is the most abundant in the body (Tortora & Anagnostakos, 1984) and includes bone, cartilage, ligaments and tendons, along with fascial, vascular and adipose tissues. Tendons are a connective tissue that attach bone to muscle and function to transmit tension from muscles to bone for movement. Ligaments are a connective tissue that attaches bone to bone and function to prevent excessive motion of joints (Benjamin et al., 2006; Coleman & Rodeo, 2003). Hyaline cartilage (i.e. the smooth cartilage covering the surfaces of synovial joints) serves to reduce friction between bone surfaces and reduce wear and tear (see for example, Triantafillopoulos et al., 2002). Muscular tissues are responsible for movement, either through active or passive means; strains and tears of muscle can leave lasting marks on the bones through the avulsion of tendons and ligaments from bone (Tortora & Anagnostakos, 1984; Benjamin et al., 2006).

Thus, to interpret trauma in palaeopathology, we must understand the stress–strain curve of particular bones and their associated articular cartilage, muscles, tendons or ligaments, as well as consider how damage to other tissues in the body can affect the skeleton (e.g. atrophy of the bones of one side of the body due to damage to the nervous system (Whiting & Zernicke, 2008).

16.3 The physiological impact of traumatic injuries

When bone is injured, there is a disruption of nociceptors (Mitchell et al., 2018, 326). Nociceptors produce pain signals that encourage the disuse or reduction of use of the injured bone. However, when the neural tissues are damaged, the result is a loss of sensation, which can lead to continued use of the broken bone. This may impact healing or result in malunion or pseudarthrosis (i.e. the creation of a false joint between the two bone ends at the fracture location). For example, if the nerve supply is disrupted due to injury of the vertebral column, the spinal cord or spinal nerves may be affected, leading to paralysis. Trauma-induced joint fusion and ossification of soft tissues may also occur, particularly in comminuted fractures (i.e. when the bone breaks in three or more pieces) because multiple bone fragments are involved. Essentially, the initial callus formation in the first stage of healing forms a 'natural splint', joining together two or more bones, preventing movement and thus helping the fracture to heal more effectively. This occurs commonly in the vertebral column when fractured vertebrae may fuse together, in paired bones (e.g. radius and ulna) where the non-fractured bone acts as a splint, as well as in the metapodials (hands and feet), where fractures may involve multiple small bones.

Injuries that affect both bone and soft tissue (Box: 16.2) to a comparable degree include amputations, decapitations and avulsion (i.e. tear off) of soft tissue (sometimes called degloving). Muhr (1984) highlights the complications associated with avulsion injuries via the example of a finger with a wedding ring stuck in industrial equipment—when under the pressure of entrapment in the equipment, the ring will tear off the soft tissue of the finger, not only removing the skin but also often causing skin necrosis that can affect the bone. Compound fractures can expose the individual to infectious agents due to their nature of puncturing the skin and result in more displacement of the bone ends, leading to poorer outcomes if medical intervention is not received (Muhr, 1984). Finally, those injuries confined to the bones of the joints can affect ligaments, tendons and muscles, and may be linked to dislocations, strains and sprains (Hannon & Knapp, 2006; Kieser et al., 2012; Box: 16.3).

Box 16.2   Soft tissue injuries

Soft tissue injuries also can affect the bone and include simple abrasions, more serious injuries like lacerations and puncture wounds and the fatal injury of evisceration. Any of these injuries can occasionally leave traces on the bones, especially if the injury introduces infectious agents into the wound that can ultimately cause a bony response (see, for example, Sharkey et al., 2011). For example, muscle trauma can lead to ossification of muscle tissue via the production of a haematoma (collection of blood). When it does not resolve itself, bone may be produced directly within the tissue itself, leading to a condition known as myositis ossificans traumatica (Ortner, 2003, 159; Lovell & Grauer, 2019, 338).

Box 16.3   Healing and remodelling

Regardless of the tissue, when a traumatic injury occurs, the body reacts. The process of healing the bone involves remodelling if the individual survives, which relies on the interplay between numerous types of cells that all must be functioning adequately for healing to occur (Ortner & Turner-Walker, 2003; Lynnerup & Klaus, 2019; White et al., 2012). This process of healing bone tissue, called remodelling, follows a standard pattern (Tsiridis et al., 2006; White et al., 2012). First, blood flows into the fractured area, forming a haematoma, which is then infiltrated by fibrous tissue, helping to unite the fractured ends of the bone with organic scaffolding (a fibrous callous). This fibrous callous provides the framework for the primary callous to form about one week after injury, which is then replaced with woven bone. Next, a secondary callous is formed when the poorly organised woven bone is replaced with lamellar bone, providing a stronger union across the fracture. The callous is then remodelled and integrated into the surrounding lamellar bone.

Osteogenic response to a fracture is typically expected to have begun as early as five days and up to around six weeks after the injury (Galloway et al., 2014, 48–49), but in older individuals or individuals who are immunosuppressed or fighting infectious processes, the healing time of fractures may be extended (Galloway et al., 2014).

16.4 Reconstructing an individual's lived experience

Given that bone is dynamic and remodels throughout an individual's lifetime, it can tell us a great deal about their lived experience (Agarwal, 2019). To reconstruct this lived experience, Harrod (2017) suggests that biological and cultural aspects of identity or 'biocultural identity' involves assessing as many skeletal indicators as possible with an emphasis placed on contextualisation. This approach follows a model first proposed in the 1960s by Frank Saul, who promoted the importance of understanding each individual by building what he called an 'osteobiography' (Saul, 1976, 1972; Saul & Saul, 1989; Agarwal, 2016). Bones provide clues for bioarchaeologists to reconstruct an individual's unique experiences and offer insights into the physical consequences of a particular person's life. Our goal is to attempt to understand aspects of that individual's identity, such as social position within their social group, lifestyle, nutritional status, health and well-being and even physical activity levels.

Bone fractures can result from numerous processes: acute traumatic events, repeated stress and an underlying weakness of the bone (Rodriguez-Martin, 2005). Because of the multiple pathways through which fractures can occur, understanding the ante-, peri- and post-mortem nature of the fracture is important, including its appearance (Sauer, 1998; Ubelaker & Adams, 1995; Spencer, 2012; Wescott, 2019). For bioarchaeological analysis of trauma, the category of *perimortem* may include individuals who survive fractures for a short period of time, but where the bone does not yet show an osteogenic response. This category may also include individuals who died at or around the time of fracture and those who experience trauma after death, as long as the bone has a 'green' response, meaning it was 'alive' when subjected to trauma, and will bend before breaking. Essentially, the perimortem category may include injuries that occurred as part of a mortuary process when organic materials have not yet sufficiently decomposed to cause 'dry' breaks. Post-mortem breaks will have a different appearance due to the loss of collagen in the bone, namely with edges that lack curves for the most part, consisting of straight, jagged lines.

All of this information can be gleaned from archaeological skeletons and pieced together to enable interpretation of the biocultural identify, including activity. For example, Osterholtz and colleagues (2019, 298) described the interaction between the articular cartilage, tendons and the bones of the knee including the femur, patella and tibia in the formation of chondral and osteochondral fractures or lesions (osteochondroses). As different tissues interact, they modify the stress–strain relationship as one tissue works to relieve stress from another (e.g. the muscle stress–strain range can be extended by the passive resistance of ligaments and tendons (Wescott, 2013; Kieser, 2012). The stress–strain relationship between these different anatomical structures was key to understanding the possible aetiology of changes to the articular surfaces of hundreds of patellae recovered from the Bronze Age (2100–2000 BCE) site of Tell Abraq in the United Arab Emirates. Based on comparison to identified traumas in modern groups with known activities, the authors concluded that the trauma to the knee found among the people at Tell Abraq was likely the result of rotational injuries associated with activities like throwing nets to catch fish

(Osterholtz et al., 2019, 300). This case study highlights how palaeopathologists can potentially identify and diagnose trauma in the past by 'reading the bones' of the dead (Armelagos, 2013) and reconstructing identity.

Obviously, some traumatic injuries can lead to death (i.e. lethal trauma) and even healed trauma can dramatically impact the lives of individuals in the past. However, it is possible that many traumatic injuries in the past were not recognised by the individuals or the people treating them as causing severe disabilities or impairments (Byrnes & Muller, 2017). In palaeopathology, the identification of a trauma or condition as an impairment is mediated based on the cultural and historical context. The term 'impairment', like 'disability', is complex and each has a different definition based on the social context (Barnes & Mercer, 2005). Barnes and Mercer (2005, 2–3) note that a medical definition of impairment would identify a loss of function, but that the culture we live in and the activities we engage in dictate if we identify and define something as an impairment. One example the authors provide is dyslexia, which would not be considered an impairment in a society where reading was not considered an essential function (Barnes & Mercer 2005, 3).

16.5 Using palaeopathology to understand trauma in the past

A growing number of today's bioarchaeologists advocate using a biocultural approach for trauma analysis (Zuckerman & Martin, 2016; Goodman & Leatherman, 1998), which attempts to contextualise trauma within its specific culture (e.g. understand the proximate cause) (see for example Tung, 2008; Redfern, 2008; Estabrook & Frayer, 2014; Robbins Schug et al., 2012; Šlaus et al., 2012; Scott & Buckley, 2010; Klaus, 2012; Meyer et al., 2009; Osterholtz, 2012; Martin & Harrod, 2015; Redfern & Fibiger, 2019), but as mentioned in the introduction, we also consider the evolutionary nature of injuries (e.g. the ultimate cause). Combining a biocultural approach with an evolutionary perspective allows us to think about how traumatic injuries are an intimate part of the human experience and realise that the explanation for, and treatment of, injuries has changed through time. We know that primates can survive traumatic injuries based on healed fractures found in wild populations of monkeys and apes (Nakai, 2003; Buikstra, 1975; Jurmain, 1997). However, with increased brain size and an increase in prosociality (Rosenberg, 1992; Hrdy, 2009; Fuentes, 2013), the human ability to care for severe injuries is a fundamental aspect of our biocultural evolution. Additionally, we must consider how the impact of trauma has changed as we live longer—thanks to modern medical interventions—and pass down knowledge to each generation with the development of more complex socio-political relations.

As with any health issue, there are many variables that impact and influence how an individual and their community experience trauma, especially when the trauma is inflicted by another person. The group experience of trauma is more than just a combination of all the individual experiences of trauma, making group-level analyses of trauma and patterns of trauma within different demographic categories, important considerations for understanding trauma. For example, how do males and females experience trauma differently? Are certain age categories more likely to be exposed to specific types of violence than others? Since interpersonal violence can be seen as a form of social communication and a way to negotiate social relationships and social standing, considering these variables can help palaeopathologists interpret trauma in the past.

In exploring interpersonal violence, we must distinguish between intra- and inter-group violence, since these two forms have different cultural meanings and expressions, and present different bioarchaeological signatures (Walker, 2001; Martin & Harrod, 2015). Singer (2009, 175) highlights how, during time of war, people are not only physically injured, but also succumb to infection and malnourishment, as well as lasting 'social suffering' in terms of the socio-political issues people face both during and after conflict (e.g. discrimination, marginalisation).

Walker's 1989 study is a prime example of the use of palaeopathological evidence of interpersonal violence to help interpret socio-political landscape of past populations. He analysed evidence for cranial trauma in prehistoric populations from the Channel Island area of Southern California. Historic documents written by Spanish colonisers had described the inhabitants of the Channel Island area to be peaceful and non-violent. However, Walker (1989) was able to use palaeopathological evidence to indicate that healed traumatic head injuries, likely from person-on-person violence, was not rare in these populations. Specifically, 24% of males and 10% of females sustained cranial depression fractures, many of which affected the facial or frontal bones, suggesting face-to-face combat. Those involved may have included high-status males and females based on the patterning of the traumatic injuries. Walker's findings contradict the historical writings describing these populations, and thus, demonstrate the wealth of biocultural information that can be obtained through archaeological evidence of trauma.

Walker (1989) also found that beginning in the Late Middle period (around 900 CE), the frequency of fractures increased through time and seemed to be linked to population growth and environmental instability. In particular, his findings demonstrate that trauma, at least in the prehistoric Channel Island area populations, was not something that only poor, unfortunate individuals experienced. This trend fits within a generally accepted framework of interpersonal violence in human populations. Similarly, Lambert (1997) investigated interpersonal violence in prehistoric populations from Southern California and, based on her findings, argues that conflict was designed to be non-lethal conflict with the purpose of diffusing tensions within the community, particularly in times of resource stress. She further noted that in this population, males and females experienced different patterns of interpersonal trauma, suggesting that the use of violence as a 'release valve' was culturally sanctioned differently for men and women. Trauma associated with inter-group warfare also increased during through time in prehistoric California, which can be linked with an increase in archaeological finds of weapons designed to kill, such as projectile points. For example, Lambert (1997) found a number of individuals with projectile points lodged in their vertebrae, particularly among young males between the ages of 15 and 40 at death, while females in the same age group were less than half as likely to be the victim of this type of violence. This demographic trend in weapon-based injuries supports Lambert's assertion that warfare increased in time, and not just violence.

Aside from interpersonal violence, palaeopathological evidence of trauma can also provide information on an individual's occupation, because certain occupations can increase risk of specific fractures. For example, farming activities, particularly those using animal labour, tend to be associated with higher rates of fractures. In a study of individuals from a medieval farming village in rural England, Judd and Roberts (1999) found that males had higher rates of fractures, likely due to different occupational-related labour practices. They compared their findings to the frequency of trauma found in contemporary farming communities that use 'traditional' methods (e.g. Amish farmers who eschew mechanical assistance) and concluded that, based on patterning of trauma in the skeletons, the injuries were likely due to physical activities performed by males, including tending to the animals, falls from horses, kicks and goring of people by cattle, etc.

In another example (Vietnamese Da But period skeletons c. 6800–6200 cal BP), Scott and colleagues (2019) examined fractures and related the pattern of fractures to the evidence of domestication and care of cattle. They found that males and females exhibited fractures in the axial skeleton, including the vertebrae and pelvis, as well as long bone fractures likely related to trauma experienced while caring for animals. Comparing their bioarchaeological findings to clinical literature, Scott and colleagues (2019) found that the fracture patterning identified was consistent with high-impact trauma as a result of animal husbandry based on modern case studies (Bjornstig et al., 1991; Busch et al., 1986; Murphy et al., 2010; Norwood et al., 2000; Shriyan et al., 2020). Essentially, the analysis of modern bovid-related trauma literature can be used to interpret patterns of trauma in bioarchaeological analyses, particularly in exploring trauma patterns by sex and age at death.

Taking an integrated approach, we can understand trauma through both the individual's lived experience and the way that their culture would have potentially viewed and treated the injuries. Violence (and the traumatic lesions resulting from violence) is historically and culturally contingent.

Exposure to violence may vary based on age, sex, gender, social class or occupation. Meanings and social uses of violence vary with cultural or group identity. Treatments of traumatic injury will be dependent upon cultural understandings of treatment and caregivers, as well as technology available. Violence, trauma and healthcare are also historically contingent, with ideas about what should constitute proper medical care shaped again by technological capabilities as well as social factors such as age, sex, gender and social class.

16.6 Trauma as part of syndemics and the importance of the One Health approach

Personal identity is an important consideration for understanding traumatic injury because the size and complexity of a group, their interactions with other cultures in a particular region and an individual's place in that group all impact the risk of traumatic injuries. For example, bioarchaeologists have assessed patterns of traumatic injuries and the demographic distribution of injured individuals to identify the presence of captives or slaves (Harrod & Martin, 2014; Corruccini et al., 1982; Terranova et al., 2000; Blakey & Rankin-Hill, 2004; Tung, 2012). Other researchers have also used those patterns to highlight socio-economic and gender-based differences in a particular culture and their impact on episodes of trauma (Mayes & Barber 2008; Powell, 1991; Storey, 2005; Redfern, 2005). If we can understand the social relationships between individuals and other groups, we are better able to determine how patterns of injury might shift based on subsistence strategy, regional conflict or major technological changes like industrialisation.

All traumatic injuries potentially require some degree of remediation. It could be that an individual simply avoids putting pressure on the injured bone or using the affected part of the body until it heals or, depending on the society, the injury might receive medical or surgical care. Understanding injuries, potential disabilities and possible treatments and acts of care are of particular interest to bioarchaeologists (Tilley & Schrenk, 2017; Tilley, 2012; Roberts, 2000; Waldron, 2000; Hawkey, 1998; Byrnes & Muller, 2017). Tilley and Cameron (2014) introduced the Index of Care, an online tool for recording injuries and assessing needed and actual care. This tool can be applied to individual skeletons or mummies using knowledge of potential care-practices and treatments available for the time and place.

While this does not necessarily indicate that the person received the care and treatment, it does allow for more objective assessments of possible care and enables scholars to suggest how trauma recovery and survival is reflected in a community's approach to care (Grauer & Roberts, 1996; Tilley, 2012). Depending on the degree of injury and level of disability, injured individuals can affect the way societies construct their 'built environment' (Battles, 2011, 111) and the importance the community places on the incorporation of traditional healers (Levey, 1963; Crockett, 1971; Warren, 1974; Hyatt, 1978; Izugbara & Duru, 2006; Bing & Hongcai, 2010; Asefzadeh & Sameefar, 2001). Evidence for a lack of intervention can also be very revealing, as it may indicate the social status, and thereby the treatment, of the impaired individual, which would also be an important aspect of their lived experience.

Just like trauma has a long evolutionary history, we suggest that the treatment of the resulting injuries is also a part of our evolution. Depending on the type of fracture and the severity of the break, individuals can take measures to promote healing by crafting a sling or splint and reducing activity to allow time for healing. The practice of caring for oneself is common today, and evolutionary approaches suggest self-care may even pre-date anatomically modern humans who evolved over 150,000 years ago (Hardy et al., 2013).

In addition to self-care, some injuries may also receive interventions from traditional healers including herbalists, spiritualists and specialists who manually manipulated the musculoskeletal system. Willett and Harrod (2017, 73) investigated skeletal remains from Pueblo societies in the Southwest US dating to 1125–1250 CE and highlighted how different types of traditional medicine (e.g. spiritual, emotional, physical) can be, and often are, part of the healing process.

When analysing human skeletons, it is not possible to identify most traditional healing efforts because they leave little to no trace on the bones. However, there is one practice that we can identify in the archaeological record—the act of manipulating the musculoskeletal system, or bone-setting, that may affect how the injury healed or not when the person was alive. The term 'bone-setter' is used to describe healers that were tasked with fixing alignments of the skeletal anatomy and mobility issues (Bennett, 1884; Oths & Hinojos, 2004a,b), and the value of this practice continues to be debated by practitioners of western medicine (Buchan, 1769; Bennett, 1884; Paget, 1867; Onuminya et al., 1999; Eshete, 2005; Gölge et al., 2015).

Traditional healing is not necessarily restricted to the deep past. For example, research by Mant (2016) found that people admitted to the Royal London Hospital in the late-eighteenth to early-nineteenth centuries often had fractured phalanges and ribs, which were treated outside of the medical system. In addition, Anderson (2004) conducted ethnographic interviews with traditional healers in modern-day Denmark, including bone-setters who helped heal fractures.

Despite debates about the effectiveness of methods of traditional healing, we highlight the practice because it has a long evolutionary history that continues today. If we are to understand how humans have adapted to trauma, we must consider if, and how, the individual and other members of the society treated traumatic injuries (see, for example, chapters in Tilley, 2012; Tilley & Schrenk, 2017; Roberts, 2011; Grauer & Roberts, 1996). Evidence for both lethal and non-lethal trauma may reveal signs of self-care and interventions from other individuals, including the realignment of fractured bones (Willett & Harrod, 2017), the use of a prosthetic (Eitler et al., 2016), or other signs indicating the treatment of an injury.

16.7 Conclusion

This chapter aimed to highlight ways that individuals may vary in how they accrue injuries and illustrate the need for understanding how humans evolved to live with and occasionally treat traumatic injuries. In archaeological settings researchers are typically focused on differentiating between accidental/occupational-related injuries versus those that result from interpersonal violence

(Harrod, 2018; Judd & Redfern, 2012; Wedel & Galloway, 2014). Yet, the most common injuries are those that occur as an accident or occupational hazard. These injuries include slips, trips, falls, collisions and industrial accidents, depending on the time and context. However, injuries as a result of violence have been explored in much great detail (see, for example, Schulting & Fibiger, 2012; Knüsel & Smith, 2014; Martin et al., 2012; Redfern, 2017).

Considering pathological conditions (e.g. genetic, nutritional, disease-related changes to the bone) is important for understanding the aetiology of trauma (Ubelaker, 2014). Pathological conditions, as well as changes related to aging, can weaken the body and result in a pre-existing condition that makes an individual more susceptible to future injury. The types of injuries associated with this category are hip dislocations, fractures due to osteoporosis and osteoarthritis causing pain in the lower back (Plomp, 2017). Poor overall health and senescence can lead to injuries, but so can repetitive, prolonged and strenuous activity (as discussed in the case of knee injuries at the UAE site of Tell Abraq). Examples of this type of injury are a slipped disc, torn ACL, carpal tunnel syndrome and numerous other activity-related injuries. However, we need to be cautious in our interpretations (Jurmain et al., 2012). It is important to realise that many of these mechanisms of injury can occur simultaneously within the same individual. The value of an evolutionary medicine perspective to understanding traumatic injury is that contextual reconstructions are essential to understanding injury aetiology and, combined with a biocultural approach bioarchaeologists can reveal patterns of trauma in the past that reveal information about both the individuals and their communities.

References

Agarwal, S. C. (2016). Bone morphologies and histories: life course approaches in bioarchaeology. *Yearbook of Physical Anthropology*, 159(S61), 130–149.

Agarwal, S. C. (2019). 'Understanding bone aging, loss, and osteoporosis in the past', in A. M. Katzenberg & A. L. Grauer (eds), *Biological anthropology of the human skeleton*, 3rd edn, pp. 385–414. John Wiley and Sons, Hoboken.

Ahn, J, Nana, A. D., Mirick, G. & Miller, A. N. (2014). *United States Bone and Joint Initiative: The burden of musculoskeletal diseases in the United States (BMUS)*, 3rd edn. United States Bone and Joint Initiative, Rosemont. http://www.boneandjointburden.org

Anderson, R. (2004) 'Indigenous bonesetters in contemporary Denmark', in K. S. Oths & S. Z. Hinojos (eds), *Healing by hand: manual medicine and bonesetting in global perspective*, pp. 5–22. Alta Mira Press, Walnut Creek.

Armelagos, G. J. (2013). Reading the bones. *Science*, 342(6164), 1291.

Asefzadeh, S. & Sameefar, F. (2001). Traditional healers in the Qazvin region of the Islamic Republic of Iran: a qualitative study', *Eastern Mediterranean Health Journal*, 7(3), 544–550.

Barnes, C. & Mercer, G. (2005). 'Understanding impairment and disability: towards an international perspective', in C. Barnes & G. Mercer (eds), *The social model of disability and the majority world*, pp. 1–16. The Disability Press, Leeds.

Battles, H. (2011). Toward engagement: exploring the prospects for an integrated anthropology of disability. *Vis-à-Vis: Explorations in Anthropology*, 11(1), 107–124.

Benjamin, M., Toumi, H., Ralphs, J. R., Bydder, G. M., Best, T. M. & Milz, S. (2006). Where tendons and ligaments meet bone: attachment sites ('entheses') in relation to exercise and/or mechanical load. *Journal of Anatomy*, 208(4), 471–490. http://www.ncbi.nlm.nih.gov/pmc/articles/PMC2100202/pdf/joa0208-0471.pdf

Bennett, G. M. (1884). *The art of the bone-setter: a testimony and a vindication*. Thomas Murby, London.

Bing, Z. & Hongcai, W. (2010). *Basic theories of traditional Chinese medicine*. Singing Dragon, London.

Bjornstig, U., Eriksson, A. & Ornehult, L. (1991). Injuries caused by animals. *Injury*, 22(4), 295–298. https://doi.org/10.1016/0020-1383(91)90009-4

Blakey, M. L. & Rankin-Hill, L. M. (eds). (2004). *Skeletal biology final report: The New York African burial ground, Vol. 1 and 2*. Howard University, Washington, DC.

Buchan, W. (1769). *Domestic medicine; or, the family physician: being an attempt to render the medical art more generally useful, by showing people what is in their own power both with respect to the prevention and cure of diseases*. Balfour, Auld, and Smellie, Edinburgh.

Buikstra, J. E. (1975). Healed fractures in *Macaca mulatta*: age, sex, and symmetry. *Folia Primatologica*, 23, 140–148.

Busch, H. M., Cogbill, T. H., Landercasper, J. & Landercasper, B. O. (1986). Blunt bovine and equine trauma. *Journal of Trauma*, 26 (6), 559–560. https://doi.org/10.1097/00005373-198606000-00013

Byrnes, J. F. & Muller, J. L. (eds). (2017). *Bioarchaeology of impairment and disability: theoretical, ethnohistorical, and methodological perspectives*. Springer, New York.

Christensen, A. M., Passalacqua, N. V. & Bartelink, E. J. (2014). *Forensic anthropology: current methods and practices.* Academic Press, Oxford.

Coleman, S.H., & Rodeo, S.A. (2003). 'Tendons and ligaments around the knee: the biology of injury and repair', in J. J. Callaghan, A. G. Rosenberg, H. E. Rubash, P. T. Simonian & T. L. Wickiewicz (eds), *The adult knee,* pp. 213–224. Lippincott Williams and Wilkins, Philadelphia.

Corruccini, R. S., Handler, J. S., Mutaw, R. J. & Lange, F. W. (1982). Osteology of a slave burial population from Barbados, West Indies. *American Journal of Physical Anthropology,* 59, 443–459.

Crockett, D. C. (1971)/ Medicine among the American Indians', *HSMHA Health Reports,* 86(5), 399–407.

Eitler, J., Deutschmann, J., Ladstätter, S., Glaser, F. & Fiedler, D. (2016). Prosthetics in Antiquity—An early medieval wearer of a foot prosthesis (6th century AD) from Hemmaberg/Austria. *International Journal of Paleopathology,* 12, 29–40.

Eshete, M. (2005). The prevention of traditional bone setter's gangrene. *The Journal of Bone and Joint Surgery,* 87(1), 102–103.

Estabrook, V. H. & Frayer, D. W. (2014). 'Trauma in the Krapina Neandertals: Violence in the Middle Paleolithic?', in C. Knüsel & M. J. Smith (eds), *The Routledge handbook of the bioarchaeology of human conflict,* pp. 67–89. Routledge, Abingdon.

Fuentes, A. (2013). 'Cooperation, Conflict and Niche Construction in the Genus Homo', in D. P. Fry, (ed), *War, peace, and human nature: the convergence of evolutionary and cultural views,* pp. 78–94. Oxford University Press, Oxford.

Fung, Y. C. (2010). *Biomechanics: mechanical properties of living tissues,* 2nd edn. Springer-Verlag, New York.

Galloway, A. (1999). 'The biomechanics of fracture production', in V. L. Wedel & A. Galloway (eds), *Broken bones: anthropological analysis of blunt force trauma,* pp. 35–62. Charles C. Thomas, Springfield.

Galloway, A., Zephro, L. & Wedel, V. L. (2014). 'Diagnostic criteria for the determination of timing and fracture mechanism', in V. L. Wedel & A. Galloway (eds), *Broken bones: anthropological analysis of blunt force trauma,* pp. 47–58. Charles C. Thomas, Springfield.

Gölge, U. H., Kaymaz, B., Kömürcü, E., Eroğlu, M., Göksel, F. & Nusran, G. (2015). Consultation of traditional bone setters instead of doctors: is it a sociocultural and educational or social insurance problem? *Tropical Doctor,* 45(2), 91–95.

Goodman, A.H. & Leatherman, T. L. (eds). (1998). *Building a new biocultural synthesis: political-economic perspectives on human biology.* The University of Michigan Press, Ann Arbor.

Grauer, A. L. & Roberts, C. A. (1996). Paleoepidemiology, healing, and possible treatment of traumas in the medieval cemetery population of St. Helen-on-the-Walls, York, England. *American Journal of Physical Anthropology,* 100, 531–544.

Hannon, P. (2006a). 'The basic principles of biomechanics', in P. Hannon & K. Knapp (eds), *Forensic biomechanics,* pp. 39–56. Lawyers & Judges Pub. Co, Tucson.

Hannon, P. (2006b). 'Biomechanics of Bone Tissue', in P. Hannon & K. Knapp (eds), *Forensic biomechanics,* pp. 73–92. Lawyers & Judges Pub. Co, Tucson.

Hannon, P. & Knapp, K. (eds). (2006). *Forensic biomechanics.* Lawyers & Judges Pub. Co, Tucson.

Hardy, K., Buckley, S. & Huffman, M. (2013). Neanderthal self-medication in context. *Antiquity,* 87(337), 873–878.

Harrod, R. P. (2017). *Bioarchaeology of social control.* Springer, New York.

Harrod, R.P. (2018) 'Trauma', in S. L. López Varela (ed), *The encyclopedia of archaeological sciences,* vol. 4. Wiley-Blackwell, Malden.

Harrod, R. P. & Martin, D. L. (2014). 'Signatures of captivity and subordination on skeletonized human remains: a bioarchaeological case study from the ancient southwest', in D. L. Martin & C. P. Anderson (eds), *Bioarchaeological and forensic perspectives on violence: how violent death is interpreted from skeletal remains,* pp. 103–119. Cambridge University Press, Cambridge.

Hawkey, D. E. (1998). Disability, Compassion and the Skeletal Record: Using musculoskeletal stress markers (MSM) to construct an osteobiography from early New Mexico'. *International Journal of Osteoarchaeology,* 8(5), 326–340.

Hrdy, S. B. (2009). *Mothers and others: the evolutionary origins of mutual understanding.* Harvard University Press, Cambridge.

Hyatt, R. (1978). *Chinese herbal medicine: ancient art and modern science.* Wildwood House, London.

Izugbara, C. O., & Duru, E. C. J. (2006). Transethnic sojourns for ethnomedical knowledge among Igbo traditional healers in Nigeria: preliminary observations. *Journal of World Anthropology: Occasional Papers,* 2(2), 27–48.

Judd, M. A. & Redfern, R. (2012). 'Trauma', in A. L. Grauer (ed), *A companion to paleopathology,* pp. 259–279. Blackwell Publishing Ltd, Malden.

Judd, M. A. & Roberts, C. A. (1999). Fracture trauma in a medieval British farming village. *American Journal of Physical Anthropology,* 109, pp. 229–243.

Jurmain, R. (1997). Skeletal evidence of trauma in African apes, with Special Reference to the Gombe Chimpanzees. *Primates,* 38, 1–14.

Jurmain, R., Alves Cardoso, F., Henderson, C. & Villotte, S. (2012) 'Bioarchaeology's holy grail: the reconstruction of activity', in A. L. Grauer (ed), *A companion to pale-opathology*, pp. 531–552. Blackwell Publishing Ltd, Malden.

Kieser, J. (2012). Biomechanics of bone and bony trauma', in J. Kieser, M. Taylor & D. Carr (eds), *Forensic Biomechanics*, pp. 35–70. John Wiley and Sons, Chichester.

Kieser, J., Taylor, M. & Carr, D. (eds). (2012). *Forensic Biomechanics*. John Wiley and Sons, Chichester.

Klaus, H. D. (2012). 'The bioarchaeology of structural violence: a theoretical model and a case study', in D. L. Martin, R. P. Harrod & P. V. (eds), *The bioarchaeology of violence*, pp. 29–62. University of Florida Press, Gainesville.

Knüsel, C. & Smith, M. J. (2014). *The Routledge handbook of the bioarchaeology of human conflict*. Routledge, Abingdon.

Lambert, P. M. (1997). 'Patterns of violence in prehistoric hunter-gatherer societies of coastal Southern California', in D. L. Martin (ed), *Troubled times: violence and warfare in the past*, pp 77–109. Gordon and Breach, Amsterdam.

Levey, M. (1963). Fourteenth century Muslim medicine and the Hisba. *Medical History*, 7(2), 176–182.

Lovell, N. C. & Grauer, A. L. (2019). 'Analysis and interpretation of trauma in skeletal remains', in M. A. Katzenberg & A. L. Grauer (eds), *Biological anthropology of the human skeleton*, 3rd edn, pp. 335–383. John Wiley and Sons, Hoboken.

Lynnerup, N. & Klaus, H. D. (2019). 'Fundamentals of human bone and dental biology: structure, function, and development', in J. E. Buikstra (ed), *Ortner's identification of pathological conditions in human skeletal remains*, 3rd edn, pp. 35–58. Academic Press, London.

Mant, M. (2016). '"Readmitted under urgent circumstance": uniting archives and bioarchaeology at the Royal London Hospital', in M. Mant & A. Holland (eds), *Beyond the bones: engaging with disparate datasets*, pp. 37–59. Academic Press, London.

Martin, D. L. & Harrod, R. P. (2015). Bioarchaeological contributions to the study of violence. *Yearbook of Physical Anthropology*, 156(S59), 116–145.

Martin, D. L., Harrod, R. P. & Pérez, V. R. (2012). *The bioarchaeology of violence*. University Press of Florida, Gainesville.

Mayes, A. T. & Barber, S. S. (2008). Osteobiography of a high-status burial from the lower Río Verde Valley of Oaxaca, Mexico. *International Journal of Osteoarchaeology*, 18(6), 573–588. https://doi.org/10.1002/oa.1011. http://dx.doi.org/10.1002/oa.1011

Meyer, C., Brandt, G., Haak, W., Ganslmeier, R. A., Meller, H. & Alt, K. W. (2009). The Eulau eulogy: bioarchaeological interpretation of lethal violence in Corded Ware multiple burials from Saxony-Anhalt, Germany. *Journal of Anthropological Archaeology*, 28(4), 412–423.

Mitchell, S. A. T., Majuta, L. A. & Mantyh, P. W. (2018). New insights in understanding and treating bone fracture pain. *Current Osteoporosis Reports*, 16(4), 325–332.

Muhr, G. (1984) 'Early Complications of fractures with Soft Tissue Injuries', in H. Tescherne & L. Gotzen (eds), *Fractures with soft tissue injuries*, pp. 131–138. Springer-Verlag, Berlin.

Murphy, C. G., McGuire, C. M., O'Malley, N. & Harrington, P. (2010). Cow-related trauma: a 10-year review of injuries admitted to a single institution. *Injury*, 41(5), 548–50. https://doi.org/10.1016/j.injury.2009.08.006

Nakai, M. (2003). Bone and joint disorders in wild Japanese macaques from Nagano Prefecture, Japan. *International Journal of Primatology*, 24(1), 179–195.

Nesse, R.M. & Schulkin, J. (2019). An evolutionary medicine perspective on pain and its disorders. *Philosophical Transactions of the Royal Society B: Biological Sciences*, 374(1785), 20190288.

Norwood, S., McAuley, C., Vallina, V. L., Fernandez, L. G., McLarty, J. W. & Goodfried, G. (2000). Mechanisms and patterns of injuries related to large animals. *Journal of Trauma*, 48(4), 740–4. https://doi.org/10.1097/00005373-200004000-00025

Onuminya, J. E., Onabowale, B. O., Obekpa, P. O. & Ihezue, C. H. (1999). Traditional bone setter's gangrene. *International Orthopaedics (SICOT)*, 23(2), 111–112.

Ortner, D. J. (2008) 'Differential Diagnosis of Skeletal Injuries', in E. H. Kimmerle & J. P. Baraybar (eds), *Skeletal trauma: identification of injuries resulting from human remains abuse and armed conflict*, pp. 21–93. CRC Press, Boca Raton.

Ortner, D. J. (2003). 'Trauma', in D. J. Ortner (ed.), *Identification of pathological conditions in human skeletal remains*, pp. 119–177. Academic Press, London.

Ortner, D. J. & Turner-Walker, G. (2003). 'The biology of skeletal tissues', in D. J. Ortner (ed.), *Identification of pathological conditions in human skeletal remains*, pp. 11–35. Academic Press, London.

Osterholtz, A. J. (2012).The social role of hobbling and torture: violence in the prehistoric Southwest. *International Journal of Paleopathology*, 2(2–3), 148–155.

Osterholtz, A. J., Harrod, R. P. & Miller, D.S. (2019). Analysis of pathology and activity-related changes to the patellae of individuals from Tell Abraq. *International Journal of Osteoarchaeology*, 29(2), 294–302.

Oths, K. S. & Hinojos, S. J. (eds). (2004a). *Healing by hand: manual medicine and bonesetting in global perspective*. Alta Mira Press, Walnut Creek.

Oths, K. S. & Hinojos, S. J. (2004b). 'Introduction', in K. S. Oths & S. J. Hinojos (eds), *Healing by hand:*

manual medicine and bonesetting in global perspective, pp. xiii–xxiii. Alta Mira Press, Walnut Creek.

Paget, J. (1867). Clinical lecture on cases that bone-setters cure. *British Medical Journal*, 1(314), 1–4.

Plomp, K. A. (2017). 'The bioarchaeology of back pain', in J. F. Byrnes & J. L. Muller (eds), *Bioarchaeology of impairment and disability: theoretical, ethnohistorical, and methodological perspectives*, pp. 141–158. Springer, New York.

Powell, M. L. (1991). 'Ranked status and health in the Mississippian chiefdom at Moundville', in M. L. Powell, P. S. Bridges & A. N. Wagner Mires (eds), *What mean these bones? Studies in southeastern bioarchaeology*, pp. 22–51. The University of Alabama Press, Tuscaloosa.

Redfern, R. (2005). *A gendered analysis of health from the Iron Age to the end of the Romano-British period in Dorset, England (mid to late 8th century B.C. to the end of the 4th century A.D.)*. Unpublished PhD Thesis, Institute of Archaeology and Antiquity, The University of Birmingham, Birmingham.

Redfern, R. (2008). 'A bioarchaeological analysis of violence in Iron Age females: a perspective from Dorset, England (fourth century BC to the first century AD)', In: O. Davis, N. M. Sharples & K. Waddington (eds), *Changing perspectives on the first millennium BC: Proceedings of the Iron Age research student seminar 2006*, pp. 139–160. Oxbow Books, Oxford.

Redfern, R. (2017). *Injury and trauma in bioarchaeology: interpreting violence in past lives*. Cambridge University Press, Cambridge.

Redfern, R. & Fibiger, L. (2019). 'Bioarchaeological evidence for prehistoric violence: use and misuse in the popular media', in J. E. Buikstra (ed), *Bioarchaeologists speak out*, pp. 59–77. Springer, New York.

Robbins Schug, G., Gray, K., Sankhyan, V. & Mushrif-Tripathy, A.R. (2012). A peaceful realm? Trauma and social differentiation at Harappa. *International Journal of Paleopathology*, 2(2–3), 136–147.

Roberts, C. A. (2000). 'Did they take sugar? The use of skeletal evidence in the study of disability in past populations', in J. Hubert (ed), *Madness, disability and social exclusion: the archaeology and anthropology of 'difference'*, pp. 46–59. Routledge, New York.

Roberts, C. A. (2011). 'The bioarchaeology of leprosy and tuberculosis: a comparative study of perceptions, stigma, diagnosis, and treatment', in S. C. Agarwal & B. A. Glencross (eds.), *Social bioarchaeology*, pp. 252–281. Wiley-Blackwell, Chichester.

Rodriguez-Martin, C. (2005). 'Identification and differential diagnosis of traumatic lesions of the skeleton', in A. Schmitt, E. Cunha & J. Pinheiro (eds), *Forensic anthropology and medicine: complementary sciences from recovery to cause of death*, pp. 197–221. Humana Press Inc, Totowa.

Rosenberg, K. G. (1992). The Evolution of Modern Human Childbirth. *Yearbook of Physical Anthropology*, 35, 89–124.

Rui, P., Kang, K. & Ashman, J. J. (2016). *National Hospital Ambulatory Medical Care Survey: 2016 emergency department summary tables*. National Center for Health Statistics, Hyattsville.

Sauer, N.J. (1998). 'The timing of injuries and manner of death: distinguishing among antemortem, perimortem, and postmortem trauma', in K. J. Reichs (ed), *Forensic osteology: advances in the identification of human remains*, 2nd edn, pp. 321–332. Charles C. Thomas, Springfield.

Saul, F. P. (1972). The human skeletal remains of altar de sacrificios: an osteobiographic analysis. *Papers of the Peabody Museum of American Archaeology and Ethnology, Harvard University*, 63(2).

Saul, F. P. (1976). 'Osteobiography: Life history recorded in bone', in E. Giles & J. S. Friedlaender (eds), *The measures of man: methodologies in biological anthropology*, pp. 372–382. Peabody Museum Press, Cambridge.

Saul, F. P. & Saul, J. M. (1989). 'Osteobiography: a Maya example', in *Reconstruction of life from the skeleton*, edited by M. Y. İşcan and Kenneth A. R. Kennedy. New York: Alan R. Liss, pp. 287–302.

Schmitt, K. U., Niederer, P. F., Cronin, D. S., Muser, M. H. & Walz, F. (2014). *Trauma Biomechanics: an introduction to injury biomechanics*, 4th edn. Springer, Heidelberg.

Schulting, R. J. & Fibiger, L. (2012). *Sticks, stones, and broken bones: Neolithic violence in a European perspective*. Oxford University Press, Oxford.

Scott, R. M. & Buckley, H. R. (2010). Biocultural interpretations of trauma in two prehistoric Pacific Island populations from Papua New Guinea and the Solomon Islands. *American Journal of Physical Anthropology*, 142(4), 509–518.

Scott, R. M., Buckley, H. R., Domett, K., Tromp, M., Trinh, H. H., Willis, A., Matsumura, H. & Oxenham, M. (2019). Domestication and large animal interactions: skeletal trauma in northern Vietnam during the hunter-gatherer Da But period. *PLoS One*, 14(9), e0218777. https://doi.org/https://doi.org/10.1371/journal.pone.0218777

Sharkey, E. J., Cassidy, M., Brady, J., Gilchrist, M. D. and NicDaeid, N. (2011). Investigation of the force associated with the formation of lacerations and skull fractures. *International Journal of Legal Medicine*, 126(6), 835–844.

Shriyan, S. V., Mani, U. A., Bhot, F. B., Sada, E. C., Ursekar, R. & Adake, D. (2020). Animal injuries; a case series of bull-induced injuries in India', *Advanced Journal of Emergency Medicine*, 4(1), e5. https://doi.org/10.22114/ajem.v0i0.244

Singer, M. (2009). *Introduction to syndemics: A critical systems approach to public and community health*. Jossey-Bass, San Francisco.

Šlaus, M., Novak, M., Bedić, Ž. & Strinović, D. (2012). Bone fractures as indicators of intentional violence in the eastern Adriatic from the antique to the late medieval period (2nd–16th century AD). *American Journal of Physical Anthropology*, 149(1), 26–38. https://doi.org/10.1002/ajpa.22083. http://dx.doi.org/10.1002/ajpa.22083

Spencer, S. D. (2012). detecting violence in the archaeological record: clarifying the timing of trauma and manner of death in cases of cranial blunt force trauma among pre-Columbian Amerindians of West-Central Illinois', *International Journal of Paleopathology*, 2(2–3), 112–122.

Stanzani, G., Tidswell, R. & Singer, M. (2020). Do critical care patients hibernate? Theoretical support for less is more. *Intensive Care Medicine*, 46, 495–497. https://doi.org/10.1007/s00134-019-05813-9

Storey, R. (2005). 'Health and lifestyle (before and after death) among the Copán elite', in E. Wyllys & W. L. Fash (eds), *Copán: The history of an ancient Maya kingdom*, pp. 315–343. School of American Research, Santa Fe.

Sugiyama, L. S. (2004). Illness, injury, and disability among Shiwiar forager-horticulturalists: implications of health-risk buffering for the evolution of human life history. *American Journal of Physical Anthropology*, 123(4), 371–389.

Terranova, C. J., Null, C. & Shujaa, K. (2000). Musculoskeletal indicators of work stress in enslaved Africans in Colonial New York: functional anatomy of the axial appendicular skeleton. *American Journal of Physical Anthropology Supplemental*, 30, 301.

Tilley, L. (2012). The bioarchaeology of care. *The SAA Archaeological Record*, 12(3), 39–41.

Tilley, L. & Cameron, T. (2014). Introducing the Index of Care: A web-based application supporting archaeological research into health-related care. *International Journal of Paleopathology*, 6(1), 5–9.

Tilley, L. & Schrenk, A. A. (eds). (2017). *New developments in the bioarchaeology of care*. Springer, New York.

Tortora, G. J. & Anagnostakos, N. P. (1984). *Principles of anatomy and physiology*. Vol. 4. Harper & Row, New York.

Triantafillopoulos, I. K., Papagelopoulos, P. J., Politi, P. K. & Nikiforidis., P. A. (2002). Articular changes in experimentally induced patellar trauma. *Knee Surgery, Sports Traumatology, Arthroscopy*, 10, 144–153.

Tsiridis, E., Sanjeev K. & Einhorn, T. A. (2006). 'Fracture healing and bone grafting', in J. A. Elstrom, W. W. Virkus & A. M. Pankovich (eds), *Handbook of fractures*, 3rd edn, pp. 49–61. McGraw-Hill, New York.

Tung, T. A. (2008). Violence after imperial collapse: A study of cranial trauma among Late Intermediate Period burials from the former Huari capital, Ayacucho, Peru. *Ñawpa Pacha*, 29, 101–118.

Tung, T. A. (2012). 'Violence against women: differential treatment of local and foreign females in the heartland of the Wari Empire, Peru', In D. L. Martin, R. P. Harrod and P. V. (eds), *The bioarchaeology of violence*, pp. 180–198. University of Florida Press, Gainesville.

Ubelaker, D. H. (2014). Contributions of pathological alterations to forensic anthropology interpretation. *Jangwa Pana*, 13, 140–151.

Ubelaker, D. H. & Adams, B. J. (1995). Differentiation of perimortem and postmortem trauma using taphonomic indicators. *Journal of Forensic Sciences*, 40(3), 509–512.

Waldron, T. (2000). 'Hidden or overlooked? Where are the disadvantaged in the skeletal record?', in J. Hubert (ed), *Madness, disability and social exclusion: the archaeology and anthropology of 'difference'*, pp. 29–45. Routledge, New York.

Walker, P. L. (1989). Cranial injuries as evidence of violence in prehistoric Southern California. *American Journal of Physical Anthropology*, 80(3), 313–323.

Walker, P. L. (2001). A bioarchaeological perspective on the history of violence. *Annual Review of Anthropology*, 30, 573–596.

Warren, D. M. (1974). Bono traditional healers. *Rural Africana*, 26, 25–39.

Wedel, V. L. and Galloway, A. (eds). (2014). *Broken bones: anthropological analysis of blunt force trauma*. Charles C. Thomas, Springfield.

Wescott, D. J. (2013). 'Biomechanics of bone trauma', in J. A. Siegel, P. J. Saukko & M. M. Houck (eds), *Encyclopedia of forensic sciences*, pp. 83–88. Academic Press, London.

Wescott, D. J. (2019). Postmortem change in bone biochemical properties: loss of plasticity. *Forensic Science International*, 300, 164–169.

White, T. D., Folkens, P. A. & Black, M. T. (2012). *Human osteology*, 3rd edn. Academic Press, Burlington.

Whiting, W. C. & Zernicke, R. F. (2008). *Biomechanics of musculoskeletal injury*. Human Kinetics, Champaign.

World Health Organization (WHO). (2014). *Injuries and Violence: The Facts 2014* [online]. https://apps.who.int/iris/handle/10665/149798. World Health Organization, Geneva.

Willett, A. Y. & Harrod, R. P. (2017). 'Cared for or outcasts: a case for continuous care in the precontact U.S. southwest', in L. Tilley & A. A. Schrenk (eds), *New developments in the bioarchaeology of care*, pp. 65–84. Springer, New York.

Wiseman, T.A., Curtis, K., Lam, M. and Forster, K. (2015). Incidence of depression, anxiety and stress following traumatic injury: a longitudinal study. *Scandinavian Journal of Trauma, Resuscitation and Emergency Medicine*, 23(29), 29. doi.org/10.1186/s13049-015-0109-z

Wrangham, R.W. & Glowacki, L. (2012). Intergroup aggression in chimpanzees and war in nomadic hunter-gatherers: evaluating the chimpanzee model', *Human Nature*, 23(1), 5–29.

Wrangham, R.W. & Peterson, D. (1996). *Demonic males: apes and the origins of human violence*. Houghton Mifflin Company, Boston.

Wrangham, R. W., Wilson, M. L. & Muller, M. N. (2006). Comparative rates of violence in chimpanzees and humans', *Primates*, 47, 14–26.

Yeni, Y. N., Dong, X. N., Fyhrie, D. P. & Les, C. M. (2004). The dependence between strength and stiffness of cancellous and cortical bone tissue for tension and compression: extension of a unifying principle. *Bio-Medical Materials and Engineering*, 14(3), 303–310.

Zephro, L. & Galloway, A. (2014). 'The biomechanics of fracture production', in V. L. Wedel & A. Galloway (eds.), *Broken bones: anthropological analysis of blunt force trauma*, pp. 33–45. Charles C. Thomas, Springfield.

Zuckerman, M. K. & Martin, D. L. (eds). (2016). *New directions in biocultural anthropology*. John Wiley and Sons, Hoboken.

CHAPTER 17

Uncovering tales of transmission: an integrated palaeopathological perspective on the evolution of shared human and animal pathogens

Elizabeth W. Uhl and Richard Thomas*

17.1 Overview

This chapter uses an integrated evolutionary and palaeopathological approach to explore the impact of shared human and animal pathogens. Traditionally, the focus has been on pathogens transmitted from animals to humans (zoonoses); however, this narrow perspective does not accurately reflect the complex interactions of pathogens with their hosts. In fact, many pathogens arose in environments shaped by humans, were shared across species and were spread by human activity around the world. Often they were subsequently introduced into wild animal populations (Bradley & Altizer, 2007; Miller & Olea-Popelka, 2013; Miller et al., 2013; Moreno, 2014; Viana et al., 2015). Our primary focus is identifying factors that have driven the transmission, distribution and virulence of specific pathogens across human environments. Since susceptibility to infection and disease is enhanced by environmental and genetic factors, understanding shared pathogens requires an evolutionary multidisciplinary approach that includes palaeopathological, historical archaeological, clinical, epidemiological and molecular evidence.

* The authors thank Dr. Mary Uhl for her invaluable organisational editorial assistance with this chapter.

The cross-species transfer of pathogens is best understood by investigating factors driving species susceptibility and microbial evolution, since these determine a pathogen's ability to cause disease and death. This virulence is influenced by factors impacting a pathogen's fitness (i.e. its ability to be transmitted to a new host), in particular the trade-off between the benefits and costs of rapid replication (Woolhouse et al., 2005). Even if a pathogen induces severe disease, there is an evolutionary advantage in rapid replication if it facilitates transmission before the host dies (Alizon & Michalakis, 2015; Ewald, 1996, 2004). In contrast, a pathogen that replicates slowly and does not kill the host has an advantage where few hosts are available and more time is needed for successful transmission (Figure 17.1). Although this predicted trade-off between transmission and virulence has been observed in natural disease outbreaks, it is difficult to study empirically. This is because the factors impacting it vary with the pathogen, host and environment (Bull & Lauring, 2014; Kerr, 2012; Mackinnon & Read, 2004; Rafaluk et al., 2015). The issue is further complicated by the difficulty in measuring virulence, as some but not all of the lesions induced by a pathogen will enhance transmission. For example, mycobacteria are usually transmitted from pulmonary lesions;

Elizabeth W. Uhl and Richard Thomas, *Uncovering tales of transmission.* In: *Palaeopathology and Evolutionary Medicine.* Edited by Kimberly A. Plomp et al., Oxford University Press. © Oxford University Press (2022). DOI: 10.1093/oso/9780198849711.003.0017

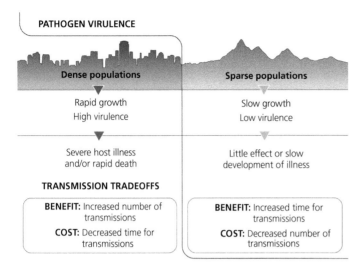

Figure 17.1 Pathogen virulence and transmission trade-offs. In sparse populations selection pressure is for slow growth and low virulence. This is because the benefit of increasing the time for transmission outweighs the cost of decreased numbers transmitted. In contrast, dense populations drive pathogens toward rapid growth and high virulence; the benefit of increased numbers transmitted is more than the cost in terms of decreased time for transmission.
Illustration by Ali Ennis Educational Resources, University of Georgia.

however, the same pathogenic mechanisms inducing lung pathology can also severely damage organs that do not facilitate transmission, such as bone (Maclean et al., 2013; Moule & Cirillo, 2020; Sakamoto, 2012).

The dramatic changes that occurred in the relationships between humans and animals as societies transitioned from hunting and foraging to herding and farming, and from living in rural to urban and industrialised environments, had major impacts on pathogen evolution (Table 17.1). The development, spread, and intensification of agriculture, coupled with urbanisation and increased connectivity through trade, enhanced opportunities for pathogens to be shared. Populations of potential hosts, human and animal, were increasingly crowded together often in unsanitary environments (Scott, 2017). These conditions drove pathogen transmission and virulence because they facilitated transmission from heavily infected and very sick individuals (Ewald, 2011; Scott, 2017).

This chapter does not focus on individual cases but rather on the conditions that allowed these pathogens to flourish in various human environments and to 'go global', as all eventually did. However, adopting a multidisciplinary evolutionary perspective is also necessary for placing individual cases and local outbreaks in context. Such a context is critical for palaeopathologists, as few lesions are truly pathognomonic, especially when found in individuals who lived in very different times and places.

Table 17.1 Human/animal environments and pathogen transmission

Human/animal environment	Factors affecting pathogen transmission
Hunter/gatherers	People/animals moving, animals in 'natural' environments, sparse populations
Herding	People/animals moving, 'natural' environments, crowding
Farming	Anthropogenic and 'natural' environments, crowding, species mixing
Urban centres: pre-mechanisation	People/animals in anthropogenic environments, moving, mixing, crowding
Urban centres: post-mechanisation	Anthropogenic and increasingly industrialised/artificial environments. People moving and crowded, large domestic animals moving less
War, trade, colonisation	People/animals moving across territories and in varied environments, crowding, species mixing

17.2 Shared pathogens in sparse populations: tapeworms (*Taeniid* sp.) in hunter-gatherers

Tapeworms (*Taeniid* sp.) are obligatory parasites (see Chapter 11) that are transmitted between predators and their prey (Koziol, 2017). The sexual stages develop in the intestines of the definitive meat-eating hosts (Haag et al., 2008). Eggs released into carnivore faeces are infectious to an intermediate host, usually an herbivore, where they develop into larvae and encyst in tissues. The cycle is completed when the encysted stages are consumed by a meat-eating host and develop into adult worms (Haag et al., 2008). That tapeworms can be maintained in both hosts for long periods of time has allowed them to persist in the relatively sparse populations typical of most predators and human hunter-gatherer societies. However, the domestication of animals provided new opportunities for transmission and has driven tapeworm evolution (Hoberg, 2006; Hoberg et al., 2001; Yanagida et al., 2014).

The greatest diversity of tapeworms occurs in canids (ten species) and felids (twelve species), and phylogenetic/divergence date analysis reveals they were the original definitive hosts (Hoberg et al., 2001; Hoberg, 2006). While it is difficult to distinguish between *Taenia* species from fossils, phylogenetic analysis indicates human infection with taenid tapeworms likely originated from two separate species jumps: one from a felid and the other from a hyaenid (Côté & Le Bailly, 2018; Hoberg, 2006; Hoberg et al., 2001; Terefe et al., 2014). Hunter-gatherers likely acquired *Taenia* infections when they competed with these predators for prey. The switch to humans as a definitive host significantly predates agriculture, as both transfers appear to have occurred in Africa during the Pleistocene (Hoberg, 2006, Hoberg et al., 2001, Michelet & Dauga, 2012). This was a time of substantial dietary change for hominins, shifting from a largely herbivorous to a more omnivorous diet (Milton, 2003). The increased exposure to *Taenia* parasites encysted in meat facilitated a species jump in which the parasites adapted to complete their life cycle in humans.

Once human hunter-gatherers became a new definitive host for *Taenia*, they in turn were the source of infection for the pigs and cattle they later domesticated (Harper & Armelagos, 2013). Domestication of animals facilitated tapeworm transmission by concentrating intermediate and definitive hosts together and increasing their exposure to the parasites. Direct evidence for this transmission is provided in the paleoparasitological record, which has documented *Taenia* spp. in archaeological sites dating from 10,500 to 2000 years ago (Bruschi et al.,

2006; Flammer et al., 2018; Sianto et al., 2009; Tams et al., 2018; Yeh & Mitchell, 2016; Yeh et al., 2019). While they have been found in sites all around the world, *Taenia* eggs found in human coprolites from prehistoric North America (6500 years ago and 1300 CE) likely represent colonisation by a different carnivore tapeworm, as there were no domesticated cattle or pigs in North America before the arrival of Europeans (Bruschi et al., 2006; Flammer et al., 2018; Reinhard et al., 2013; Sianto et al., 2009; Tams et al., 2018; Yeh & Mitchell, 2016; Yeh et al., 2019).

Three species of tapeworms: *Taenia solium* (intermediate host: pigs, also humans), *Taenia saginata* (cattle) and *Taenia asiatica* (pigs) cause human taeniasis (Hoberg, 2006). Although humans are the only definitive host for these parasites, increased prevalence of *T. solium* in domestic environments enhanced the likelihood of humans and dogs, which are not hosts for the adult worms, ingesting the parasite eggs and becoming infected with the intermediate stage (cysticercosis) (Blancou, 2003; Del Brutto & García, 2015; Ito et al., 2002). Indeed, in 1775 Paulet noted wild boar were not infected 'because they never live in fifth and rubbish' (Paulet, 1775 cited in Blancou, 2003, 320). The intermediate stage of *T. solium* can infect the brain (neurocysticercosis) and is a leading cause of adult onset seizures in areas where the parasite is endemic, although the infection may have occurred years before the seizures started (Bruschi, 2011; Del Brutto & García, 2015; Garcia et al., 2014). Thus, historical descriptions of adult-onset epilepsy, especially in the case of outbreaks, are potential indicators of *T. solium* infection. For example, during the 1930s, over 400 cases of neurocysticercosis-induced epilepsy were documented in British soldiers who had previously been stationed in India (Singh & Sander, 2019). Furthermore, it has even been proposed that Julius Caesar's epilepsy, first described when he was 54 years of age and mentioned by Shakespeare (*Julius Caesar* Act 1, Scene 2), was due to neurocysticercosis (Bruschi, 2011).

Domestication and human activity greatly expanded the distribution of tapeworms. While *T. solium* and *T. saginata* are found throughout the world, there is little genetic variability between isolates from geographically distant areas (Michelet & Dauga, 2012). For example, there are only two different *T. solium* mitochondrial genotypes:

African/American and Asian, which indicates that parasites from South and Central America share a common ancestor with those in Africa (Michelet & Dauga, 2012). This genetic similarity can be explained by the colonisation of both South America and West Africa by Europeans in the fifteenth century, and by the European origin of their swine populations. The Portuguese and Spanish trade routes that connected Europe and Asia to the Americas can explain the Far Eastern origin of some of the South American pigs and thus the distribution of the Asian *T. solium* genotypes (Michelet & Dauga, 2012).

In addition to farming practices and colonisation, human cultural and dietary preferences have also played a role in the distribution and incidence of taeniasis. In a variety of cultures, a preference for dishes made with raw meat has been identified as facilitating tapeworm transmission (Michelet & Dauga, 2012). In some cases, the taste for raw or undercooked meat may have developed in populations that were not originally exposed to *Taenia* parasites, leaving them susceptible to infection when the parasite was brought in through human activity. This happened in 1972 when a gift of *T. solium*-infected pigs from the Indonesian government led to an outbreak of seizures caused by neurocysticercosis in West New Guinea (Del Brutto & García, 2015). In contrast, there is a long history of concern about the consumption of pigs in areas where the infection was established early. This concern may have been due to the unpleasant 'measled' appearance of meat infected with *T. solium* cysts (Del Brutto & García, 2015). Pigs were considered unclean in ancient Greece, and eating pork was banned in both Jewish and Islamic traditions (Michelet & Dauga, 2012). Non-consumption of pork in Jewish and Islamic communities has been documented in the zooarchaeological record (Daróczi-Szabó, 2003; Grau-Sologestoa, 2017; Ijzereef et al., 1989; Redding, 2015; Valenzuela-Lamas, 2014). There is intriguing evidence that *T. solium* has adapted to being cooked, as compared to non-human adapted tapeworms, *T. solium* has an expanded number of heat shock protein 70 (HSP 70) genes (Perry, 2014). Expression of HSP 70 from duplicated genes protects against heat stress in oysters, but its role in *T. solium* needs to be investigated further (Perry, 2014; Tsai et al., 2013).

In summary, an evolutionary perspective indicates that while the two-host life cycle of *Taenia* parasites is well adapted for sparse predator and hunter-gatherer populations, opportunities for parasite transmission were greatly enhanced as infected humans first concentrated intermediate hosts on farms and moved them around the world. The opportunities provided by agricultural and cultural practices also facilitated the ability of the asexual stages of *T. solium* to infect humans as well as pigs, causing human cysticercosis, which is still an important if neglected tropical disease (Del Brutto & García, 2015; Djurković-Djaković et al., 2013; Garcia et al., 2014). As we will see, other pathogens would take similar advantage of transmission opportunities provided by human activities.

17.3 From sparse to crowded populations and from humans to animals to humans: the *Mycobacterium tuberculosis* complex

Few diseases have impacted human health as seriously as tuberculosis (see Chapter 10). It drove advances in medicine including the development of germ theory as well as diagnostic techniques (auscultation) and antibiotics (Webb, 1936). It also greatly affected history, culture and the arts (Chalke, 1962; Webb, 1936). For example Shakespeare makes references to consumption, 'wasting disease', 'phtisick' (phthisis), 'rotten lungs', 'wheezing lungs' and 'lethargies' in many of his plays (Chalke, 1962).

In addition to humans, tuberculosis also occurs in fish, reptiles, birds and mammals, although the mycobacteria inducing disease generally varies with the host. *Mycobacterium tuberculosis* complex (MTBC) is the most significant in terms of causing disease in mammals and is composed of the clonal progeny of a single ancestor. It includes the obligate human pathogen, *M. tuberculosis,* as well as *Mycobacterium canettii* (humans, possibly the ancestral strain), *Mycobacterium africanum* (humans), *Mycobacterium bovis* (cattle, humans and other animals), *Mycobacterium microti* (voles), *Mycobacterium caprae* (goats) and *Mycobacterium pinnipedii* (seals, sea lions and humans) (Gutierrez et al., 2005). *M.*

tuberculosis complex members have an extremely high sequence similarity and are basically one genetic species with lineages that have adapted to different hosts, often through genetic deletions (Riojas et al., 2018). These lineages are clonal as there is very little evidence of transfer of genetic material between them (Smith, 2012; Smith et al., 2006).

Evidence supporting an African origin for human tuberculosis includes the presence of all seven lineages of *M. tuberculosis* and the geographical restriction of *M. canettii* and *M. africanum* to Africa (Galagan, 2014). Phylogenetic analysis suggests the MTBC originated at least 70,000 years ago, but it is possible that human ancestors in eastern Africa may have been infected with ancestral mycobacteria as early as 3 million years ago (Cardona et al., 2020; Comas et al., 2013). After the tubercle bacilli emerged as a human pathogen in Africa, it diversified and spread around the world, an expansion facilitated by human migration, trade, warfare, conquest, and colonisation (Galagan, 2014; Gutierrez et al., 2005). Even though *M. tuberculosis* originated in Africa, tuberculosis was uncommon there prior to the introduction of more virulent strains from Europe (Comas et al., 2015).

Humans are the only known reservoir for *M. tuberculosis*. Although it can cause disease in many warm-blooded species, including birds, it is not easily transmitted except from humans (Montali et al. 2001). One possible explanation is indicated in the pathology. Although *M. tuberculosis* induces similar primary granulomatous lesions in animals, the disease is not usually characterised by an exudative pulmonary phase, which in humans currently accounts for 80% of the clinical disease and 100% of the transmission (Basaraba & Hunter, 2017; Hunter, 2011). The exudative lesions are generally in the upper lung lobes and consist of thin-walled cavities that release large numbers of bacteria into the airways (Basaraba & Hunter, 2017). Pierre C. A. Louis (1787–1872) noted the prevalence of pulmonary lesions: 'after the age of fifteen one could not have tubercle in any organ in the body without at the same time having tubercle in the lung' (Webb, 1936, 83). Pulmonary lesions allow *M. tuberculosis* to be transmitted by coughing, sneezing or speaking, and facilitate the shedding of infectious bacteria for long periods of time (Hunter, 2011). The capacity to induce both primary and chronic disease has adapted *M. tuberculosis* to human lifestyles and made it an extremely successful human pathogen; it is estimated that one quarter of the world's population is currently infected with latent TB (Houben & Dodd, 2016).

In sparse populations, *M. tuberculosis* is maintained by its ability to persist and be transmitted over a long period of time, either in an asymptomatic infection or associated with chronic disease (Figure 17.1). However, in dense populations, because *M. tuberculosis* is transmitted through aerosols, it can quickly disseminate to new hosts in an epidemic form. Crowded conditions also facilitate the transmission of more virulent forms, since transmission is primarily dependent upon having active lesions in the lungs (Bates & Stead, 1993; Comas et al., 2013). Amplified transmission from sick individuals in crowded conditions would explain why the virulence, as well as the incidence of *M. tuberculosis*, increased with the development and spread of agriculture, urbanisation and industrialisation (Comas et al., 2013; Hershberg et al., 2008). The effects of crowding on the incidence of tuberculosis can also be localised. In 1868 French military surgeon Jean-Antoine Villemin observed that tuberculosis was more common in soldiers stationed for long periods in barracks than among those in the field (Villemin, 1868, cited in Webb, 1936, 44).

Genetic resistance to tuberculosis is present in long-urbanised human populations (Barnes et al., 2011). This observation and the functional diversity of *M. tuberculosis* emphasise that humans and the pathogen have been co-evolving (Brites & Gagneux, 2012; Hershberg et al., 2008). Phylogenetic and palaeopathological evidence indicates human tuberculosis predates animal domestication and confirms its increase in agricultural societies resulted from evolution of the human adapted strain rather than a species jump from animals (Baker et al., 2015; Galagan, 2014).

While transmission is primarily from pulmonary infections, during the proliferative phase infection tubercular lesions can develop in many organs. Bone lesions occur in about 2–5% of untreated *M. tuberculosis* infections and have been commonly recognised in human skeletal remains from numerous archaeological sites. They have been documented in human remains from Egypt, the Mediterranean, Asia, pre- and post-Columbian America and from Europe stretching back to the early Neolithic (8800–8300 BCE cal.) (Donoghue, 2016; Pigrau-Serrallach & Rodríguez-Pardo, 2013). The most characteristic lesions are in the spine (Pott's disease), but tuberculosis can affect any part of the skeleton, in particular the knee, hip, and elbow joints, tarsals, metatarsals, carpals and metacarpals (Pigrau-Serrallach & Rodríguez-Pardo, 2013). In palaeopathological work, researchers often depend on the spinal changes for a diagnosis (Roberts & Buikstra, 2019). However, other skeletal lesions include chronic osteomyelitis, periosteal reactions, bone lysis, pathological fractures and hypertrophic pulmonary osteoarthropathy (Assis et al., 2011; Dangvard Pedersen et al., 2019; Horacio et al., 2015; Kelley & Micozzi, 1984; Pigrau-Serrallach & Rodríguez-Pardo, 2013). Advances in molecular techniques have made it easier to detect evidence of the pathogen in human remains, even when bone lesions are not present, and to track the emergence of lineages (Donoghue, 2016; Harkins et al., 2015).

Taken together, palaeopathological, historical and genetic evidence indicate more virulent lineages of *M. tuberculosis* arose after the development of agriculture and were subsequently spread around the world, often replacing older, less virulent strains (Galagan, 2014). This scenario explains why native Africans and Americans were initially considered to have been free of tuberculosis before European colonisation, even though the disease was present (Bates & Stead, 1993; Hershberg et al., 2008; Salo et al., 1994). Unfortunately, the imported lineages caused major outbreaks of tuberculosis in American and African native populations, especially in areas where they were in close contact with Europeans and crowded together (Bates & Stead 1993).

Historically, symptoms of tuberculosis were described much earlier in humans than in other animals. In addition, clinical, and genetic evidence indicates tuberculosis in animals is often caused by the human pathogen (Adesokan et al., 2019; Erwin et al., 2004; Koro Koro et al., 2015; Lanteri et al., 2011; Montali et al., 2001; Pesciaroli et al., 2014;

Zachariah et al., 2017). The earliest description of animal tuberculosis was in domesticated Indian elephants during the fifth century BCE (Gutierrez et al., 2006). Tuberculosis is still common in domesticated elephants and has spilled over into those in the wild. The primary cause is the human pathogen, *M. tuberculosis* (Maslow & Mikota, 2015; Zachariah et al., 2017). *M. tuberculosis* was also a major cause of great ape morbidity in early zoos and is considered one of the greatest health threats to wild ape populations living close to humans (Gilardi 2015; Snyder 1978).

M. bovis, best known as a cattle pathogen, has the broadest host range within the MTBC (Mostowy et al. 2005). Because an increase in human tuberculosis occurred around the time animals were domesticated, *M. tuberculosis* was initially considered to have originated from zoonotic transmission of *M. bovis*. However, genetic sequencing has revealed that *M. bovis* contains genetic deletions that are not present in the human mycobacterial strains—an indication that the original species jump was from humans to animals (Mostowy et al., 2005). Subsequent diversifications of *M. bovis* allowed it to infect other species including humans, a jump that was facilitated by the domestication of cattle and the development of dairying (Patané et al., 2017; Galagan, 2014). Phylogenetic analysis of almost 2000 modern *M. bovis* isolates from twenty-three countries and twenty-four host species has identified four distinct global lineages that are dispersed based upon geographic location rather than host species (Zimpel et al., 2020). Divergence analysis supports the role of human movement/trade in animals, as it indicates that while the current linages of *M. bovis* originated between 715 and 3556 years BP, they emerged later in the Americas and Oceania (Zimpel et al. 2020).

M. bovis is spread in cattle through inhalation/ingestion of infectious bacteria and induces 'pearl like' granulomas (Hunter, 2011). The disease is usually chronic, but unlike *M. tuberculosis* in humans, which can become dormant before inducing a cavity in the lung and being shed for decades, *M. bovis* is usually disseminated and can be shed from multiple orifices (Hunter, 2011). In particular, *M. bovis* can induce a mastitis that results in tubercle bacilli being excreted in the milk (Cousins, 2001).

UNCOVERING TALES OF TRANSMISSION **323**

Domestication has variably impacted the prevalence of cattle tuberculosis, the incidence of infection being lower in cattle grazing on pasture than where they are farmed intensively (Firdessa et al., 2012; Tschopp et al., 2010; Vordermeier et al., 2012). In free-ranging populations of cattle, as well as wild cervids, the prevalence of individuals infected with tuberculosis is approximately 1–5%, compared with 25–50% in dairy cattle and farmed deer that are housed or penned in small paddocks (O'Reilly & Daborn, 1995). During the Victorian period, a city's milk supply usually came from dairies in which cows were crowded together, conditions that enhanced transmission (Davies, 2006). Susceptibility to bovine tuberculosis also varies with subspecies: South Asian zebu cattle (*Bos taurus indicus*) are less susceptible than European breeds (*Bos taurus taurus*) (Vordermeier et al., 2012).

Historical and phylogeographical analysis indicate the exportation of infected cattle was important in the spread of *M. bovis* around the world (Smith, 2012; Webb, 1936; Zimpel et al., 2020). In particular Hereford cattle, the most widely distributed beef cattle in the world, were exported from England starting in the 1800s (Smith, 2012; Webb, 1936). Compatible with the role of infected imported cattle is that estimates of the most recent common ancestor for current *M. bovis* isolates from Northern Ireland range from 53–283 BP even though native cattle breeds have been farmed there since the Neolithic period (Zimpel et al., 2020). Phylogeographic analysis also ties the introduction of *M. bovis* with the importation of cattle to the Americas during the colonial period, and to New Zealand in the nineteenth century (Zimpel et al., 2020).

M. bovis has infected a variety of domesticated and wild animals, but many, including those in cats, dogs, horses and pigs, were spill overs in that they did not cause infections that were subsequently transmitted (Cousins, 2001). Most occurred where there was a high incidence of *M. bovis* in cattle, were often extra-pulmonary, and disappeared when the cattle infections were controlled. However, under conditions allowing an infected individual to be in close contact with another potential host (i.e. crowded herds, penning, supplemental feeding, fencing), wild animals including the Eurasian badger (*Meles meles*), the European wild boar (*Sus*

scrofa) in the Iberian Peninsula, the North American white tailed deer (*Odocoileus virginianus*), the African buffalo (*Syncerus caffer*) and the Australian brushtail possum (*Trichosurus vulpecula*) have maintained the infection within their populations and are currently potential sources of re-infection of cattle and humans (Cousins, 2001; Martín-Hernando et al., 2007; Naranjo et al., 2008; Palmer, 2013).

In addition to jumping from humans to animals, several members of the MTBC have been transmitted from animals to humans. *M. bovis*, the most common form of zoonotic tuberculosis, can be transmitted to humans through an aerosol or the ingestion of infected milk and meat. As in cattle and other species, the human disease is characterised by granulomas similar to those induced in the primary phase of *M. tuberculosis*; however, *M. bovis* does not induce a pulmonary exudative phase in humans or cattle (Hunter, 2011). Until pasteurisation of milk became widespread, *M. bovis* was a common cause of gastrointestinal tuberculosis particularly in children, but also in domesticated pigs fed dairy products and offal from infected cows (Cousins, 2001). Because human infection has been primarily through ingestion, *M. bovis* is more likely to cause extra-pulmonary infections and is less likely to be transmitted, although rare cases of human-to-human transmission have been reported (Evans et al., 2007). To date, the only documentation of *M. bovis* in archaeological human remains is as the cause of spinal tuberculosis in semi-nomadic Iron Age pastoralists from South Siberia (Taylor et al., 2007).

In contrast to the relative abundant evidence of tuberculosis in palaeopathological human remains, there are very few reported/potential cases of tuberculosis in zooarchaeology (Bartosiewicz, 2013, 2018; Bathurst, 2004; Rothschild et al., 2001; Wooding, 2010, 2019). This is compatible with the low frequency of skeletal involvement in animals infected with *M. bovis*: 0.5–1% for cattle, 8–9.5% for pigs and 1.7% for the Eurasian badger (*M. meles*) (Cohrs, 1967, Gallagher & Clifton-Hadley, 2000, Nieberle, 1938). However, it also reflects the fragmentary nature of most zooarchaeological assemblages as well as an inconsistent approach to recording ribs and vertebrae. The lack of clinical data on skeletal lesions in non-human tuberculosis complicates interpretation of zooarchaeological pathology, but what is available suggests a superficial similarity to the lesions in humans, as both axial and appendicular elements are affected in areas with abundant haemopoietic marrow (Lignereux, 1999; Mays, 2005; Wooding, 2010). There are species-specific differences in predilection sites (e.g. the pelvis in pigs) and in the character of the lesions (Cohrs, 1967). *M. bovis* infection in cats and dogs is more proliferative and potentially capable of inducing hypertrophic osteopathy (known as hypertrophic pulmonary osteoarthropathy in human medicine), although pulmonary neoplasia is currently a more common cause and should be considered as a differential diagnosis (Bathurst, 2004; Snider, 1971; Wooding, 2010). In contrast, in ruminants as in humans, bone resorption (e.g. osteomyelitis, osteoporosis, cavitation, destruction of articular surfaces) predominates (Bartosiewicz, 2013; Wooding, 2010). Rothschild and colleagues (2001, 306) have suggested that an undermined subchondral articular surface is 'relatively specific for the diagnosis of tuberculosis' in bovids; however this lesion is observed in cases of haematogenously seeded osteomyelitis caused by other types of bacteria and has not been confirmed as specific for mycobacterial infection using robust clinical evidence (Craig et al., 2016; Rothschild et al., 2001). The identification of *M. tuberculosis* complex DNA in an extinct bison (*Bison* cf. *antiquus*) metacarpal exhibiting this lesion from Natural Trap Cave, Wyoming and dated to ~ 17,000 BP, is interesting, but is a single example and is difficult to explain given the phylogenetic evidence that *M. tuberculosis* originated from human infection in Africa. The possibility of environmental mycobacteria—either ancient or modern—must also be entertained (Müller et al., 2016).

Mycobacteria of animal origin have been documented as causing human tuberculosis in the Americas prior to European contact (Bos et al., 2014). Phylogenetic analysis of mycobacteria isolated from diseased individuals who lived in Peru supports a single zoonotic transfer from pinnipeds to humans between 700 and 1000 CE (Bos et al., 2014). This species jump may have been followed by a re-adaptation to humans (Bos et al., 2014). *M. pinnipedii* has also induced tuberculosis in New Zealand beef cattle, where there was no evidence

of cattle-to-cattle transmission, and in zookeepers in contact with seals (Kiers et al., 2008; Loeffler et al., 2014).

Mycobacteria are examples of ancient human and animal pathogens that have co-evolved with their hosts by adapting to transmission opportunities and, unlike the 'crowd' disease pathogens (i.e. morbilliviruses), are able to persist in both low and high host densities. Human activity has spread both *M. tuberculosis* and *M. bovis* around the world and driven their evolution by providing new opportunities for transmission. Although it originated from a single clone, the evolution of the MTBC is characterised by numerous species jumps as its members have infected humans, non-human primates, cattle, goats, pigs, voles, elephants, marine mammals, dogs, cats, horses, deer, lions, rhinoceroses, parrots, tapirs, mongooses and badgers, among many others (Ghodbane & Drancourt, 2013). The MTBC exemplifies the fact that pathogens show no regard for the human versus animal distinction that has historically shaped our anthropocentric worldview.

17.4 Infection transmission by a vector: farming, urbanisation and the rise of *Plasmodium falciparum* as a human pathogen

Malaria is caused by over 450 different protozoan parasites that evolved from an ancient photosynthetic ancestor to infect a wide range of animals (Kalanon & McFadden, 2010). Primates have the most parasite species in mammals and five cause disease in humans (Garnham, 1996). To complete their life cycle, *Plasmodium* species require two hosts, one insect and one animal, and their evolution is constrained by their need to adapt to both. Natural transmission between vertebrates is through mosquitos. While *Plasmodium* species are generally host specific in their vertebrate hosts, they are less so in their invertebrate hosts and can infect many different species of anopheline mosquitoes (Molina-Cruz et al., 2016).

Malaria parasites alternate between asexual replication in the bloodstream of their vertebrate hosts and sexual reproduction in their mosquito vectors. Asexual replication through infection and rupture of red blood cells can induce disease in the vertebrate host. Gametes are produced with the asexual stages and are transmitted to the mosquito where sexual reproduction occurs. As might be predicted, malaria parasites adapt their replication versus reproductive cycles to take best advantage of opportunities for transmission (Rono et al., 2018). For example, it has been confirmed that in environments with low transmission opportunities, *P. falciparum* invests more in transmission to new hosts by producing more reproductive forms (gametes) than in asexual replication within the human host (Rono et al., 2018).

P. falciparum, the cause of the most virulent form of malaria in humans, is a good example of how human-induced environmental impacts can enhance the transmission and virulence of a vector borne pathogen. It is the only species of the *Laverania* subgenus that infects humans; all others infect African apes (Proto et al., 2019). Genetic evidence indicates the closest relative to *P. falciparum* is *Plasmodium praefalciparum*, a parasite infecting western lowland gorillas, and that the jump to humans occurred only once, possibly as recently as 10,000 years ago (Liu et al., 2010; Loy et al., 2017). The parasite subsequently lost the ape-specific erythrocyte invasion ligand, and adapted to infect humans (Loy et al., 2017; Proto et al., 2019). This adaptation has been highly beneficial to the parasite. Its human hosts have enabled it to spread worldwide and infect new vectors; it is now transmitted by over seventy different species of *Anopheles* mosquitos (Molina-Cruz et al., 2016).

That only one successful species transfer from gorillas to humans occurred implies there are obstacles to cross-species jumps for *Plasmodium* parasites (Liu et al., 2010; Loy et al., 2017). One of these is that *Plasmodium* parasites are constrained by the need to remain adapted to two very different hosts. This means factors driving evolution of the vector are as important as those driving the evolution of the parasite. The feeding and breeding preferences of the insect vectors have critical effects on transmission. In most parts of the world the anthropophilic index, which is the probability of a mosquito's blood meal being from a human, is often well below 50% (Bruce-Chwatt et al., 1966). However, in sub-Saharan Africa, where *P. falciparum* is common, it is between 80–100% (Bruce-Chwatt et al., 1966). A mosquito's taste for human blood is considered the most important single factor for the intensity of malaria transmission, and it can be enhanced by adaptation to breeding in human environments (Carter & Mendis, 2002; Coluzzi, 1999; Takken & Knols, 1999). Along with mosquito numbers, host preference is thus an important factor in maintaining the specificity of *Plasmodium* parasites and can explain malaria's historical and modern distribution (Coluzzi, 1999).

A change in a vector can expand a pathogen's host range. Genetic analysis indicates two human malaria parasites, *Plasmodium malariae* and *Plasmodium vivax* jumped into New World monkeys after the European colonisation of Central and South America, where they infected mosquitos adapted to feeding on both humans and monkeys (Carter & Mendis, 2002). These species jumps were likely facilitated by *P. malariae* and *P. vivax* having dormant stages in the liver, which allowed them to travel with their human host and remain infectious to mosquitos for a prolonged period (Hulden & Hulden, 2011).

Transmission of malaria parasites is critically affected by the density of their two host populations, both of which became more concentrated with the development of agriculture (Hulden & Hulden, 2011; Hume et al., 2003). Clearing land and the development of irrigation increased standing water and provided breeding sites for mosquitos, while the rise of towns and cities concentrated human populations (Carter & Mendis, 2002). The impact of farming as an amplifier of malarial infection is supported by a study in Borneo, which found that farming populations had a significantly higher prevalence of *P. falciparum* parasites than hunter-gatherers (Hume et al., 2003).

In contrast to *P. falciparum*, other forms of human malaria chronically infect the liver, which allows them to be maintained in sparse as well as dense populations. For example *P. vivax* can persist in populations even when vectors are not present (i.e. winter) and is the most widespread form of human malaria (Hulden & Hulden, 2011). Lacking a chronic stage, *P. falciparum* is dependent on a constant supply of mosquito vectors and susceptible hosts (Carter & Mendis, 2002; Hulden & Hulden, 2011).

However, increased transmission in dense human populations has also made *P. falciparum* more virulent.

The evolutionary pressure on vector borne pathogens is usually toward benign infection in the vector, but heavy parasitism and thus high virulence in the vertebrate host. *P. falciparum* has adapted to greater opportunities for transmission by increasing its replicative forms, which induces more severe disease in its host (Ewald, 1987; Mackinnon & Read, 2004; Rono et al., 2018). Evidence for a connection between transmission and virulence comes from a study that found *P. falciparum* from a high-transmission area was more virulent, more persistent and had more transmission to mosquitos than an isolate from a low-transmission area (Jeffery & Eyles, 1955; Mackinnon & Read, 2004). Further support of an evolutionary connection between transmission and virulence comes from the observation that repeated passage of *P. vivax* through humans (used to treat syphilis) produced a parasite strain that induced higher parasitaemia, fever, mortality and transmission to mosquitos than its ancestral form (James et al., 1936). Finally, the host immune response can also impact pathogen virulence. Since immunity to malaria parasites can have protective effects but is usually only partial, it can compensate for the transmission of more virulent forms within a population, which means that naïve, non-immune hosts like children or immigrants will develop more severe disease (Curtin, 1994; Mackinnon & Read, 2004).

Evidence of an increase in the virulence of malarial parasites comes from the appearance of protective human genetic changes around 5000 years ago (Hedrick, 2012; Loy et al., 2017). These multiple genetic changes arose independently in different areas and have a variety of detrimental disease trade-offs: sickle cell disease, thalassemia, glucose-6-phosphate dehydrogenase deficiency, haemoglobins C and E, ovalocytosis and the loss of Duffy antigen (protective against *P. vivax*). Together, they indicate malarial parasites have put a strong selection pressure on the human genome (Carter & Mendis, 2002). A β-thalassemia gene mutation that is relatively common in Sardinia has

been documented in an individual who lived there 2000 years ago, which indicates malaria was likely endemic on the island by the Roman period (Viganó et al., 2017). In contrast, the pre-Columbian populations of the Americas do not have genetic traits conferring protection from malaria (Cavalli-Sforza, 1994; Rodrigues et al., 2018).

Descriptions of malaria-like illnesses, with tertian and quartan fevers and enlargement of the spleen, occur in ancient texts from the Middle East, Africa, India and China (Carter & Mendis, 2002; Sabbatani et al., 2010). Malaria is a major cause of anaemia, due to the destruction of red blood cells, but is difficult to diagnose from human skeletal remains (Carter & Mendis, 2002; Håkan, 2003; Nerlich, 2016; Setzer, 2014). While some skeletal lesions, including cribra orbitalia, porotic hyperostosis and reduced trabecular and cortical bone mass, have been associated with anaemia, interpretation is complicated (Grauer, 2019). First, malaria is only one of many causes of anaemia; others include primary or secondary nutritional deficiencies, inherited disorders, infectious diseases, injury, menstruation and childbirth (Camaschella, 2015). Second, the primary causes of anaemia cannot be determined from skeletal lesions. More definitive documentation has come from DNA analysis (*P. falciparum* chloroquine-resistance transporter gene) and immunological assays (*P. falciparum* histidine rich protein-2 antigen and lactated dehydrogenase), which have confirmed the presence of malaria parasites in human remains from at least 4800 years ago, a finding compatible with the human genetic evidence indicating increased selection pressure at that time (Bianucci et al., 2015). *P. falciparum* DNA/protein has been detected in Egyptian mummies, fifth-century Roman remains and individuals from the Renaissance (Bianucci et al., 2015; Fornaciari et al., 2010a, 2010b; Nerlich, 1993, 2016; Nerlich et al., 2008). Advances in molecular techniques should continue to confirm the presence of *Plasmodium*, especially in cases without bone lesions, and will provide a more complete documentation of malaria from ancient to modern times. The presence/absence of protective alleles induced by selection pressure from malaria parasites can also be used as indirect evidence to identify areas where malaria was likely endemic (Anastasiou & Mitchell, 2013; Nerlich, 2016; Sabbatani et al., 2010).

From Africa, *P. falciparum* was spread by human migrations into the Mediterranean, the Middle East and beyond. While its expansion was ultimately limited to the geographic ranges of its mosquito vectors, transmission within these ranges varied and was enhanced by human population density and land/water management (Molina-Cruz et al., 2016; Sabbatani et al., 2010). Currently, the historical and palaeopathological evidence combined with the appearance of protective gene mutations indicate malaria reached southern Europe from both North Africa, where it was likely brought in by the Phoenicians, and the eastern Mediterranean, where it arrived via Greece (Sallares et al., 2004). This conclusion is supported by consideration of the mosquito vectors found along these two routes. The most important vector in central and southern Italy, *Anopheles labranchiae*, came from North Africa, while *Anopheles sacharovi* from the western Mediterranean became the dominant vector in north-eastern Italy following its arrival in the early medieval period (Sallares et al., 2004). Adaptations of the mosquitos likely affected the parasite's range. Whereas *A. sacharovi* has the ability to hibernate over winter, *A. labranchiae* does not and resided inside houses or barns to survive the winter weather (Sallares et al., 2004).

P. falciparum was brought to the New World with the slave trade through multiple independent introductions that occurred within a relatively short period (Molina-Cruz et al., 2016; Rodrigues et al., 2018). Once it arrived, the parasite readily adapted to new *Anopheles* vectors, even though the vectors were evolutionarily distant from those in Africa and spread with both its mosquito and human hosts (Molina-Cruz et al., 2016). Native Americans and most Europeans had not previously been exposed to this form of malaria and were very susceptible to severe disease.

Human migrations and environmental impacts also affected the prevalence and severity of malaria in other species. Birds native to Hawaii and other isolated islands, as well as penguins brought to zoological gardens, are highly susceptible to severe malarial disease when exposed to avian forms of *Plasmodium* because they evolved in environments without the pathogen (Grilo et al., 2016; Lapointe et al., 2012). Indeed, mosquitoes were introduced to Hawaii through the unintentional effects of European colonisation; there are no native mosquito species on the islands (Winchester & Kapan, 2013).

The historical impacts of malaria are too widespread to be adequately covered here, as it determined where and how people lived, affected the outcomes of wars and contributed to the economic factors driving slavery (Mann, 2011; McNeil, 2010). Malaria also had a great impact on the arts: it is mentioned by Dante (1265–1321), Geoffrey Chaucer (1342–1400) and in eight of Shakespeare's plays (1564–1616), as well as in numerous other well-known literary and historical works (Reiter, 2000). The biology of the vectors was especially critical to the parasite's impacts, as indicated by its range in Europe and the fact that the northern range of a major *P. falciparum* vector in the US (*Anopheles quadrimaculatus*) approximates the boundary between the Southern slave-holding and Northern free states (Mann, 2011; McNeil, 2010). Finally, the history of malaria also contains numerous warnings against complacency. For example, in 1914 the incidence of malaria was at a low point in Europe, and the consensus of opinion was that it was an anachronism in the 'developed' world (Hackett, 1937). However, it came roaring back with the disruption caused by the First World War and incapacitated armies on both sides (Hackett, 1937).

P. falciparum is a good example of the extra layer of complexity involved in the transmission of vector borne diseases. Because of this complexity, adopting an evolutionary perspective is essential for understanding the emergence of these pathogens while palaeopathological, historical, and genetic evidence is needed to assess their impacts. *P. falciparum* is a testament to how evolution driven by human facilitated opportunities enabled a small parasite passed to humans by a single horizontal transfer from gorillas could spread around the world and remain a major cause of human disease.

17.5 Herding, farming and warfare: transmission of shared intracellular bacterial pathogens: *Brucella* and *Burkholderia mallei*

The use of animals in agriculture, transport, trade and warfare brought them together with humans in increasingly confined and unsanitary spaces (Scott, 2017). These activities and their environmental impacts provided unique opportunities for pathogen transmission. The importance of human-manipulated environments in the rise of common zoonotic pathogens is supported by evidence that some infections (e.g. MTBC) were initially transmitted from humans to domestic animals and later spilled over to wild animal populations (Bengis et al., 2002; Harper & Armelagos, 2013).

For *Brucella* and *Burkholderia mallei*, transmission between animals and humans was facilitated by the use of cattle, pigs, sheep, and goats for meat and milk (*Brucella*) and horses for agriculture, transportation and warfare (*Burkholderia*) (Bendrey et al., 2020; Bendrey & Fournié, 2020; Wilkinson, 1981). Both pathogens have been transmitted between animals and humans for centuries, have had significant impacts on human history and are still considered potential bioterrorism agents (Anderson & Bokor 2012; Dance, 2009; Pappas et al., 2006).

The genus *Brucella* contains a monophyletic group of facultative intracellular bacteria that induce a chronic granulomatous disease in a variety of species. *Brucella* species are well adapted to their hosts, but those infecting domestic animals are more likely to be zoonotic and are more virulent than species infecting wild animals (Moreno, 2014). Currently, brucellosis is found all over the world, being especially common in the Mediterranean, Middle East, India, and Central and North America (Pappas et al., 2005; Rossetti et al., 2017). Transmission to humans is primarily from domestic animals, and four species cause human disease: *Brucella melitensis* (sheep and goats), *Brucella abortus* (cattle), *Brucella canis* (domesticated dogs) and *Brucella suis* (pigs) (Pappas et al., 2005). *B. melitensis*, the major cause of brucellosis in goats, is the most common and virulent species infecting humans (Rossetti et al., 2017).

A growth factor (erythritol) for *Brucella* bacteria is found in animal mammary glands and reproductive organs, and infection occurs through venereal transmission or contact with infected reproductive tissues (Akpinar, 2016; Essenberg et al., 2002; Petersen et al., 2013). Because *Brucella* causes abortions and is spread through reproductive activities, keeping animals in dense flocks/herds enhances transmission. Human infection occurs through consumption of infected meat and milk products, inhalation of infectious particles and contact with infected animal tissues (Akpinar, 2016; Miletski, 2009; Pappas et al., 2005). Brucellosis is often an occupation-related disease in people exposed to animal tissues (goat herders, dairy farmers, butchers, veterinarians). Bestiality, which is well documented throughout human history, is also a potential source of infection (Miletski, 2009).

In humans, brucellosis is generally more severe than in domestic animals and is characterised by fever, which can be relapsing (undulant) (Pappas et al., 2005; Moreno, 2014). Malodorous perspiration is almost pathognomonic, but other symptoms are variable and non-specific, as the disease can impact many different organs (Pappas et al., 2005). A common complication of brucellosis is bone and joint disease, which varies in presentation with age (Bendrey et al., 2020). It generally occurs in three distinct forms—peripheral arthritis (children and young adults), sacroiliitis and spondyloarthropathy (older adults) (Bravo et al., 2003; D'Anastasio et al., 2011; Esmaeilnejad-Ganji & Esmaeilnejad-Ganji, 2019). Non-erosive, peripheral arthritis in the knees, hips, ankles and wrists is common and occurs during acute infection (Bendrey et al. 2020; Esmaeilnejad-Ganji & Esmaeilnejad-Ganji, 2019). Sacroiliitis also occurs during acute infection, while spondyloarthropathy, usually in the lumbar spine, is more chronic and results from residual damage (Bendrey et al., 2020; Pappas et al., 2005). Because skeletal lesions are slow to develop, reactive bone forms and the vertebral lesions of brucellosis usually do not result in the vertebral body collapse commonly observed in tuberculosis (Bendrey et al., 2020; Sharif et al., 1989). Although lesions centred on the intervertebral discs are more typical of brucellosis, they are not unique, and other differential diagnoses should be considered in isolated cases, particularly from areas without domestic animals. Differential diagnoses for vertebral lesions similar to those induced by *Brucella* include tuberculosis, chronic spondylosis/osteomyelitis due to other haematogenously spread bacterial, parasitic or fungal pathogens and metastatic cancer (Chelli Bouaziz et al., 2008). Culture and/or detection of *Brucella* sp. DNA in combination with characteristic lesions provides the most definitive diagnosis. Consideration of the domestic environment in which the affected individual(s) lived can identify possible sources of infection, e.g. association with cattle, goats, sheep and pigs and in particular milk products.

Several lesions have been considered pathognomonic for brucellosis in animals (i.e. vertebral osteomyelitis in pigs, and fistulous withers and poll evil in horses); however, they are not unique to *Brucella* infection. The most typical are osteomyelitis of vertebrae and septic arthritis of joints, which are common manifestations of other forms of bacterial osteomyelitis in animals (Craig et al., 2016). Vertebral osteomyelitis is most common in young animals and results from hematogenous spread of bacteria (Craig et al., 2016). Bacteria from the environment commonly enter via the umbilicus, the respiratory and gastrointestinal tracks of young animals or through wounds (tail biting in piglets and tail docking in lambs) (Craig et al., 2016). Septic arthritis/osteomyelitis in the long bones commonly results from hematogenous infections and penetrating wounds (Firth et al., 1987; Verschooten et al., 2000). Bacterial infections may cause bursitis or tenosynovitis, which can also induce skeletal changes. While the lesions of vertebral osteomyelitis and septic arthritis are similar in animals, the most common causative pathogens vary with location and over time, which should be considered in the assessment of skeletal lesions in zooarchaeological remains. *Brucella* species have been documented as a cause of these lesions in environments where the pathogen was common, but the lesions by themselves, especially in individual cases, should be interpreted with caution. For example, *Trueperella* is currently the most common cause of bacterial vertebral osteomyelitis in large animals (Craig et al., 2016; Verschooten et al., 2000).

Domestic cattle and bison infected with *B. abortus* can develop arthritis with destruction of articular cartilage and eburnation/lysis of subchondral bone (Bracewell & Corbel, 1980; Tessaro et al., 1990).

Carpal and femoro-tibial joints are particularly susceptible, although all forms of arthritis commonly affect these sites (Lowbeer, 1959; Rohde et al., 2000). *B. abortus* infection in horses primarily occurs in animals pastured with infected cattle and causes fistulous withers and poll evil (Cohen et al., 1992). Both are chronic pyogenic infections that can extend into bone. Fistulous withers refers to an infection of the connective tissue overlying the second thoracic vertebra, while in poll evil, or cranial nuchal bursitis, the supra-atlantal bursa between the nuchal ligament and the dorsal arch of the atlas is affected (Cohen et al., 1992; Lowbeer, 1959). Both of these sites are under direct pressure from horse tack and the infections likely occur when areas of traumatised tissue are contaminated with bacteria (Gaughan et al., 1988). While *Brucella* infection has traditionally been associated with both fistulous withers and poll evil (Figure 17.2), it is not the only cause of these conditions and its prevalence varies with location and time period (Cohen et al., 1992; Duff, 1936; Gaughan et al., 1988; Roderick, 1948). Sterile cases of cranial nuchal bursitis and fistulous withers have also been reported, which supports the theory that tissue trauma from poorly fitting tack is a predisposing factor (García-López et al., 2010).

In pigs, osteomyelitis is a more common lesion of brucellosis than in other mammals, although the role of husbandry in its incidence needs to be considered (Lowbeer, 1959). Vertebral osteomyelitis and discospondylitis in the lumbar region are the most common lesions in *B. suis* infections, but arthritis, as well as chronic osteomyelitis of the long bones, ribs, sternum and phalanges have also been described (James, 1930; Lowbeer, 1959; Schlafer, 2007).

Historically, brucellosis in humans is an ancient disease that is directly connected to the domestication of animals (cattle, goats, pigs, sheep, camels and reindeer), and the consumption of animal milk (Moreno, 2014). The disease has been known by a variety of names reflecting both fever as a main feature of the disease and the geographic locations where it was common: Mediterranean Fever, Malta Fever, Gibraltar Fever, Rock Fever, Neapolitan Fever, Undulant Fever and Corp Disease (Vassallo, 1992). As these names imply, brucellosis has been

Figure 17.2 Chronic cranial nuchal bursitis ('poll evil') due to a possible Brucella infection in a Scythian horse from the Tuva Republic, Central Asia. Poll evil is often caused by *B. abortus* infection in horses that have been pastured with infected cattle. The infection can extend into bone in the wither ('fistulous withers') and poll areas, both of which are commonly subjected to trauma from tack. In this case, there is profuse enthesophyte formation and bone lysis at the site of the nuchal ligament attachment (the external occipital protuberance). *Brucella* and *M. bovis* DNA were extracted from human skeletons from Iron Age Tuva. The normal nuchal ligament attachment site is illustrated on the skull from a modern horse (left).
Right image, kindly provided by Robin Bendrey.

common around the Mediterranean and in Middle East, and it is also highly prevalent in Asia (Moreno, 2014). Brucellosis was spread to the Americas through the introduction of infected European animals and subsequently the pathogen jumped to the indigenous wildlife (deer, elk and buffalo) (Moreno, 2014).

Phylogenetic studies indicate *Brucella* emerged 86,000 to 296,000 years ago and began spreading in the Fertile Crescent about 6500 years ago (Foster et al., 2009; Pisarenko et al., 2018). Animal domestication facilitated the emergence of *Brucella* species adapted to specific hosts while increased transmission enhanced their virulence (Ficht, 2010; Moreno, 2014). A modelling study simulated the transmission of *B. melitensis* in Fertile Crescent Neolithic domestic goat populations, and confirmed that they would have been large enough to maintain the infection, even at low levels of transmission (Fournié et al., 2017). This indicates that human exposure likely increased even in the early stages of agricultural development (Fournié et al., 2017). As is the case today, management of the goat flocks was probably a factor in the spread of brucellosis in the past (Bendrey et al., 2020). Preferential retention of breeding and lactating females increases transmission, as infectious material from abortions and milk are common sources of infection.

Palaeopathological findings indicate human brucellosis is an ancient disease, even though it is not possible to definitively diagnose *Brucella* infection using bone lesions alone (Bendrey, 2008; Wooding, 2019). Typical skeletal lesions have been found at sites throughout the Mediterranean, Middle East and Europe from the Bronze Age to the Middle Ages, some of which have been confirmed as brucellosis using biomolecular analysis (Bendrey et al., 2020; Capasso, 1999; Curate, 2006; D'Anastasio et al., 2011; Etxeberria, 1994; Jones, 2019; Kay et al., 2014; Mays, 2007; Mutolo et al., 2012). A possible *B. melitensis* peptide has been identified in remnants of cheese found in an Egyptian tomb (1290–1213 BCE) and *Brucella*-like bacteria have been observed in cheese found in Herculaneum, Italy (79 CE) (Capasso, 2002; Greco et al., 2018). Lesions typical of brucellosis were also present in 17.4% of adult skeletons found at Herculaneum (Capasso, 2002; Sciubba et al., 2013).

In animals, evidence of 'poll evil' has been described in a Scythian horse (Figure 17.2) from the Tuva Republic, Central Asia (Bendrey et al., 2011). The occipital bone of this animal exhibited profuse ossification at ligament/tendon attachment sites (enthesophytes) and bone lysis at the nuchal ligament attachment to the external occipital protuberance, indicative of localised inflammation associated with bacterial infection (Bendrey et al., 2011). Notably, both *Brucella* and *M. bovis* DNA have been extracted from Iron Age human skeletons from this site (Bendrey et al., 2020; Taylor et al., 2007). Currently, brucellosis is one of the most common zoonotic diseases in developing countries and is continuing to evolve as new strains have been identified and human-to-human transmission has been recently documented (El-Sayed & Awad, 2018; Rossetti et al., 2017).

Glanders is another chronic granulomatous disease with important historical impacts (Sharrer, 1995; Wilkinson, 1981). *Burkholderia mallei* arose when a free-living strain of *Burkholderia pseudomallei* entered an animal host and evolved from an opportunistic to an obligate intracellular pathogen (Kettle & Wernery, 2016; Losada et al., 2010). *B. mallei* underwent genome reduction as it adapted to intracellular life, because the selection pressure to maintain the genes involved in survival outside a host was reduced (Losada et al., 2010; Song et al., 2010). This specialisation for intracellular environments, and being primarily a pathogen of equids, may have limited its opportunities for transmission. In the modern world, glanders is not as common as brucellosis, which is caused by a facultative intracellular pathogen that can survive outside a host.

Glanders is primarily a disease of equids and humans but has also been described in other species (Dvorak & Spickler, 2008). In equids, glanders is spread through contact and contaminated feed/water troughs. Animals with acute infections have high fevers with a thick nasal discharge and often die a few days after the appearance of symptoms (Dvorak & Spickler, 2008; Kettle & Wernery, 2016). In chronic infections, nasal, pulmonary and cutaneous pyogranulomas develop, and animals can shed bacteria for years (Kettle & Wernery, 2016). Glanders has been mistaken for tuberculosis, although tuberculosis is much less common in horses (Dvorak & Spickler, 2008; Wilkinson, 1981).

B. mallei is transmitted to humans through aerosols, ingestion and direct contact with mucosal surfaces or skin abrasions (Van Zandt et al., 2013). Like brucellosis, glanders is an occupational disease, with cavalry, grooms, veterinarians and others working closely with infected animals being the most susceptible (Van Zandt et al., 2013). Until the development of antibiotics, *B. mallei* infections were highly lethal in humans. Currently, glanders is endemic in Africa, Asia, the Middle East and Central and South America, but has been eradicated from North America, Australia and most of Europe. However, its high infectivity and ability to induce severe disease makes *B. mallei* a dangerous biological weapon (Akubek & Handelman, 1999; Lehavi et al., 2002; Wheelis, 1998). It was used in the American Civil War, the First and Second World Wars and by the Soviets in Afghanistan (Dance, 2009; Waag & Deshazer, 2004).

Glanders does not commonly affect the skeleton, making it difficult to document palaeopathologically (Grauer, 2019). In humans, haematogenous osteomyelitis, affecting the skull and tibia, can occur, but the lesions are not pathognomonic (Beitzke, 1934; Grauer, 2019). Nonetheless, because of the past importance of equids for agriculture, transportation and warfare, glanders is well documented historically, the first description being by Aristotle (384–322 BCE) (Wilkinson, 1981). Early authors considered glanders the worst disease horses could get, and it had a major impact, especially in wars. In the fourth century, Absyrtus, a Greek veterinarian in the army of Constantine the Great, described affected horses and recognised glanders was contagious (Wilkinson, 1981). Sollysel, the stable master to Louis XI, noted the importance of feed troughs and water buckets in spreading the disease in 1682 (Wilkinson, 1981). Tuberculosis was even compared to glanders to emphasise its contagious nature; in 1870 Villemin stated: 'The phthisical (tubercular) soldier is to his messmates what the glandered horse is to its yoke-fellow' (cited in Webb, 1936, 46).

In spite of the early recognition of its infectious nature and the identification of common routes of transmission, there was a reluctance to acknowledge glanders as a contagious disease. Resistance to making the diagnosis extended into the nineteenth century, since it meant valuable and often irreplaceable animals would have to be culled (Sharrer, 1995; Wilkinson, 1981). Glanders caused serious losses of horses in all the major European wars up to the Second World War. During the Crusades it also afflicted European knights who had ingested infected horsemeat, and it is mentioned in Shakespeare's *Taming*

Figure 17.3 This stereoscopic photograph shows the US Calvary depot at Giesboro (1864–1865), which could hold 30,000 horses, and where 17,000 horses were lost to disease, many to glanders. Crowding of large numbers of healthy and infected horses in confined spaces, with the constant addition of new individuals, communal feeding and shared water troughs was common when horses were critical for agriculture, transportation and warfare, and greatly facilitated the transmission of pathogens like *Burkholderia*. Such conditions do not characterize natural environments, where herds are much less dense, more stable in membership and constantly move to new food and water sources. US National Archives: photo 111-B-5544.

of the Shrew as one of the afflictions of Petruchio's horse (Act III, Scene 2) (Blancou, 2003).

In the US, glanders was a problem in the Second Seminole War (1834–1842), and was brought to Mexico by diseased US army horses during the Mexican American War (1846–1848) (Sharrer, 1995). In the American Civil War (1861–1865) glanders spread through the horses of both armies and from them into civilian equine populations (Sharrer, 1995; Bierer, 1939). Between 1864 and 1865, glanders caused devastating losses at the US cavalry depot in Giesboro, which could hold 30,000 horses (Figure 17.3); during this period 17,000 of them were lost to disease (Sharrer, 1995). Although this depot was unusually large, it exemplifies conditions common in both military and civilian settings where large numbers of healthy and infected animals were crowded together around communal feeding and watering stations. These cramped conditions and the constant introductions of new animals were especially conducive to the spread of *B. mallei*. Glanders continued to be a major problem into the early 1900s, especially in military and urban settings. Control efforts included culling infected animals and the removal of public watering troughs (Bierer, 1939). The rapid decline of glanders when mechanisation replaced horse-powered transport underscores the importance of husbandry factors in the transmission of *B. mallei*. Although there was still no effective treatment, it was eradicated in Britain in 1928 and from the US in 1934 (Wilkinson 1981).

17.6 Urbanisation, war and conquest: morbilliviruses—shared pathogens in crowded environments

The mixing of large human and animal populations increased opportunities for new pathogens, in particular those causing epidemic diseases. Many of these pathogens induce a persistent protective immune response and are dependent on large numbers of naïve individuals to maintain their transmission. Some of the most devastating of these 'crowd' diseases are caused by the morbilliviruses: human measles virus (HMV), rinderpest virus (RPV) of cattle (eradicated 2011) and canine distemper virus (CDV). These very closely related viruses are directly transmitted, have morbidity and mortality rates of 90–95% in naïve populations and are considered to be the most infectious viruses known (Barrett & Taylor, 2006; Cliff et al., 1993; Rima & Duprex, 2006). Morbilliviral diseases are generally similar across species, and infections vary in severity from subclinical to lethal, with clinical disease characterised by fever, respiratory and/or gastrointestinal signs and often a distinctive dermatitis (Beineke et al., 2009; Kirk, 1922; Morens & Taubenberger, 2015; Rima & Duprex, 2006).

Because morbilliviral infections either cause death or induce life-long protective immunity, population densities determine whether disease occurs periodically in epidemic form or persists in populations as an endemic or childhood disease. For HMV, it has been estimated that a population of at least 500,000 people is needed for the endemic form to be established (Black, 1966). Measles usually spreads from large cities to rural areas, with the endemic form persisting in large cities but fading in less populated areas, making them more susceptible to periodic epidemics (Cliff et al., 1993). A similar demographic pattern has been found for canine distemper. The disease is maintained as an endemic (puppy) disease in high-density canine populations but remains an epidemic disease in areas with smaller domestic dog populations (Cleaveland et al., 2000). Because large populations are needed to maintain the transmission of morbilliviruses, measles, rinderpest and canine distemper arose after the development of agriculture, when the large, dense, interacting populations of humans and animals required to maintain the viruses as endemic infections were established (Keeling & Grenfell, 1997; Pearce-Duvet, 2006).

HMV, RPV and CDV are some of the most closely related viruses known. They utilise the same two highly conserved cellular receptors and many of their genes are functionally interchangeable. The similarities of their proteins, nucleotide and amino acid sequences, codon usage and antigenic epitopes indicate these viruses share a single common progenitor (de Vries et al., 2014; Sheshberadaran et al., 1986; Sidhu et al., 1993; Stephenson & ter Meulen, 1979; Uhl et al., 2019; von Messling et al., 2001; White et al., 1961). High mutation rates, continued cross-species infections and changes in selection pressures have set up population bottlenecks, which have purged viral genetic variation and made determination of their original evolutionary relationships difficult to establish from sequence data. Supplementing this information with historical records and consideration of the environmental conditions that facilitate epidemics can provide a better understanding of the evolution of these important pathogens (Uhl et al., 2019).

Morbilliviruses are directly transmitted, thus their transfer between species is most likely to occur when large, infected populations of one species are in close contact with numerous naïve individuals of another. Cross-species infections are enabled because the viruses use highly conserved cell receptors. In addition, a very high infection rate facilitates transmission to new hosts. For HMV the average number of secondary cases arising from a single infected individual is six times higher than for influenza viruses (Biggerstaff et al., 2014; Guerra et al., 2017; Sakai et al., 2013). However, immunity is cross protective making species jumping less likely if a population has previously been infected, even sub-clinically, by a related morbillivirus. Immunity to HMV also protects against CDV infection and if it were to wane, CDV could jump to humans and induce a morbilliviral disease similar to measles (Cosby & Weir, 2018).

With the exception of CDV, which can cause dental enamel hypoplasia in puppies, morbilliviral infections do not induce definitive skeletal lesions (Dubielzig, 1979; Dubielzig et al., 1981). However, historical records describe highly fatal morbilliviral epidemics in humans and animals. Cattle plagues similar to rinderpest were reported in Asia around 3000 BCE and the disease was also described in Greek and Roman texts dating to ~ 376 BCE (Spinage, 2003). The Persian physician Rhazes gave the first definitive description of measles around 900 CE (Rhazes, 1848). This timeframe is compatible with phylogenetic analysis indicating RPV and HMV diverged between the eleventh and twelfth centuries CE (Furuse et al., 2010). Severe outbreaks of measles and rinderpest occurred in the Old World for the next millennium (Barrett & Taylor, 2006; Smallman-Raynor & Cliff, 2004; Cliff et al., 1993; Fleming, 1882; Kohn, 2001; McNeill, 1976; Newfield, 2009; Spinage, 2003), and were especially common during the pan-European wars of the seventeenth to nineteenth centuries when large numbers of people and cattle moved with the armies across the continent (Barrett & Taylor, 2006; Cliff et al., 1993; Fleming, 1882). Measles was introduced into the Americas soon after European contact and was one of the diseases that devastated indigenous people who had no prior immunity to morbilliviruses (Cook, 1998; Smallman-Raynor & Cliff, 2004).

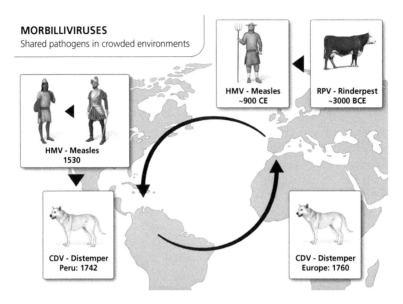

Figure 17.4 Morbilliviruses: shared pathogens in crowded environments.
The morbilliviruses were originally classified based upon the host from which they were isolated and the diseases they induced but are very closely related and are the most infectious viruses known, which allows them to cross species if opportunities for transmission arise. Historically, rinderpest was the first morbilliviral disease described (~ 3000 BCE) and was eliminated in 2011. Human measles virus (HMV) is believed to have separated from the rinderpest virus (RPV) around 900 CE. Canine distemper was first described in South America in 1742 CE, and subsequently in Europe (1761) as a new disease. Because of its close relationship to the Old World viruses, canine distemper virus (CDV) likely originated when HMV jumped into dogs during the massive measles epidemics that ravaged the indigenous populations of the Americas. Since wild dog packs are too small to maintain CDV, it likely became an endemic canine pathogen in large urban dog populations.
Illustration by Ali Ennis Educational Resources, University of Georgia.

The first descriptions of canine distemper were from the Americas (Blancou, 2004; Ulloa, 1806). Given HMV, RPV, and CDV are too similar to have been separated for thousands of years, the question of how CDV arose on a separate continent requires explanation. The groups of domestic dogs that moved to the Americas with migrating Pleistocene human populations were too small to maintain CDV as an endemic pathogen (Perri et al., 2019). Moreover, there were no reports of a fatal canine disease affecting imported European dogs. Considered together, palaeopathological, historical, molecular and epidemiological evidence indicates CDV originated as a pandemic pathogen in South America following the infection and adaptation of HMV to dogs (Figure 17.4) (Uhl et al., 2019). This cross-species transfer was facilitated by the severe measles epidemics that decimated indigenous American populations following the arrival of Europeans. Canine distemper was likely established as an endemic disease in the large domestic dog populations of

South American cities. It could not have become an endemic pathogen in the small populations characteristic of wild carnivore packs. After being established throughout South America, canine distemper was transported to Europe in 1760, where it initially induced widespread epidemics with the high morbidity and mortality typical of a new pathogen in a naïve population, before it eventually became endemic (Blancou, 2004; Uhl et al., 2019).

Canine distemper illustrates the advantages of a multi-disciplinary approach when interpreting palaeopathological evidence of disease, especially in animals. Animal diseases are not as well documented as human, and the veterinary literature tends to highlight the unusual over the common. Thus, given few lesions are pathognomonic, published reports of a lesion occurring in a particular disease do not mean the lesion is exclusive to, or even commonly caused by, the reported disease. For example, while it has been well documented that CDV infection in puppies results in dental enamel

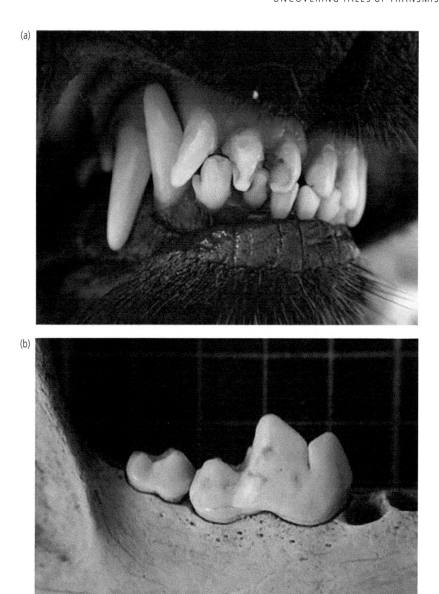

Figure 17.5 Palaeopathology in context: tooth enamel lesions in dogs.
Top: Enamel hypoplasia in a dog that was infected with CDV as a puppy Because CDV directly infects and kills the cells that make enamel, it induces plaque like lesions in multiple teeth. Bottom: enamel defects in an Iron Age dog (Balksbury Camp, Hampshire, UK, context F72/6). The lesions are more punctate and disconnected rather than plaque like. In addition, given the first historical description of canine distemper was in 1740 and large numbers of dogs are needed to maintain the infection CDV is extremely unlikely to have caused the lesions.
Top photograph curtesy of Dr. Rafael Fighera, Laboratório de Patologia Veterinária, Universidade Federal de Santa Maria, Santa Maria, RS, Brazil. Bottom photograph courtesy of Lauren Bellis and the Hampshire Cultural Trust.

hypoplasia (Figure 17.5), as in humans, there are many causes of tooth enamel damage in dogs, including trauma, fever, dietary factors, antibiotics and toxins, all of which need to be considered before making a diagnosis of canine distemper (Dubielzig et al., 1981; Hillson, 2005). In this case, as in many others, context and evidence from other sources is invaluable.

Animal diseases should be considered in the context of evolution-based susceptibilities, clinical features, pathogenesis and, for infectious diseases, pathogen history, prevalence and biology. This context requires a cross-disciplinary approach. For example, factors to consider when evaluating CDV as a potential cause of the tooth enamel lesions in an Iron Age dog from Britain (Figure 17.5) include:

1. time frame and geographic location (CDV was first described in 1740 in Peru);
2. population size (large canine populations are needed to maintain CDV as an endemic infection of puppies); and
3. lesion character and distribution (CDV causes plaque like enamel lesions by infecting ameloblasts before the permanent teeth erupt and thus multiple teeth are usually affected) (Blancou, 2004; Dubielzig et al., 1981; Uhl et al., 2019).

Given these considerations, it is extremely unlikely that CDV was the cause of the enamel lesions in this dog.

Cross-disciplinary investigations informed by evolution can enhance our understanding of the shared pathogens of the past. While evidence from different disciplines may not provide a definitive diagnosis for a particular specimen, such evidence can be both complimentary and compensatory to the pathological findings and thus can narrow down potential differential diagnoses and provide a context for their interpretation.

17.7 Conclusions

'We need to better understand the delicate balance between the host and pathogen in the context of the entire biological system and this requires a radical and transformational approach.' Anthony S. Fauci, MD, Director, National Institute of Allergy and Infectious Diseases (quoted in Basaraba & Hunter, 2017, 864)

The shared pathogens discussed are biologically diverse. They include cestode and protozoan parasites, bacteria and viruses that are transmitted in a variety of different ways: via aerosols, ingestion

and insect vectors. They cause diseases with manifestations ranging from acute epidemics to chronic debilitation and have infected a wide range of hosts over very long periods of time. For all the pathogens discussed, many of which were more benign in the sparse populations of their origin, human activities increased their transmission (Table 17.2). In particular, human migrations, trade, warfare, colonisation, the development and spread of agriculture and the rise of urbanism all provided new opportunities for pathogens to find hosts, and this in turn often drove an increase in virulence. The growth and crowding of settlements enabled the rise of especially virulent pathogens that are only maintained in high population densities and include some of the most devastating diseases in both humans and animals. In particular, the emergence of the morbilliviruses highlights how humans have provided opportunities that have driven pathogen evolution and facilitated species jumping.

The examples in this chapter illustrate how an evolutionary perspective informed by palaeopathology, history, medicine (human and veterinary), ecology, epidemiology and molecular analyses can often provide answers to questions about the origins of pathogens and why they caused disease in specific populations, times and locations. The factors facilitating the global emergence of pathogens shared between humans and animals are of particular importance because the diseases they induce have had major impacts on both human and animal health. This transfer of pathogens has occurred for thousands, if not millions, of years and continues today. Traditionally, these shared pathogens have been grouped together under 'zoonoses' and defined as diseases transmitted to humans from animals. The lessons from the past reveal that this classification is much too limited. It is informed by a biologically inaccurate distinction between humans and other animals that distorts our understanding of the factors affecting how pathogens interact with their host species.

In this chapter we showed that palaeopathology offers a crucial, direct line of evidence for tracking the evolutionary history of many pathogens. The detection of lesions in archaeological skeletal remains provides evidence that a disease

Table 17.2 Summary of selected shared pathogens

Pathogen/disease	Main hosts	Evolution	Palaeopa-thological evidence	Genetic/phylogenetic evidence	Historical descriptions	Factors enhancing transmission
Taenia saginata *Taenia solium* Taneasis Cystercercosis	From definitive hosts humans to intermediate hosts cattle (*T. saginata*) and pigs (*T. solium)* through faeces. Animals to humans from ingestion of meat.	Prolonged two-host cycle facilitates transfer in sparse populations.	*Taenia* eggs in preserved faeces from archaeological sites all over the world dating from 10,500 to 2000 years ago.	Switched from carnivores to humans as definitive host in the Pleistocene. Little genetic variability indicates were spread by humans.	Numerous: adult worms and cysticercosis ('measled' meat). Cause of adult epilepsy?	Domestication facilitated spread from humans to cattle and pigs, and to humans as intermediate host (cysticercosis). Trade/colonisation.
Mycobacterium tuberculosis Complex Tuberculosis	Humans, cattle and many other animals.	Prolonged course of infection. Direct transmission.	Humans: distinctive skeletal lesions from numerous sites in the Old World. Zoonotic form (*Mycobacterium pinipedii*) in the Americas before Europeans.	*M. tuberculosis* jumped from humans to cattle (*Mycobacterium bovis)* and other species. Has returned as zoonotic infections.	Numerous human and veterinary.	Crowding. Mixing. Urbanisation. Trade/colonisation.
Plasmodium falciparum Malaria	Human	Vector (mosquito), constraint from two host life cycle.	Skeletal evidence of chronic anaemia; *P. falciparu*m DNA and protein.	Recent species jump from gorilla. Protective gene mutations in humans.	Numerous: Africa, Middle East, Europe, India and China.	Land use that increased mosquitos. Crowding. Trade/colonisation.
Brucella melitensis Brucellosis	Sheep and goats: contact with infected reproductive tissues. Humans: contact with infected tissues, milk.	Facultative pathogen. Domestication led to emergence of species-adapted strains.	Human: distinctive skeletal lesions from several sites especially Mediterranean and Middle East, with *B. melitensis* confirmed in some cases.	Emerged 86,000–296,000 years ago. Spread began in the Fertile Crescent about 6500 years ago.	Numerous especially around Mediterranean.	Crowding. Herding. Farming/dairying. Trade/colonisation.
Burkholderia mallei Glanders	Horses and humans: close contact with infected animals, contaminated environment.	Evolved from soil dwelling *B. pseudomallei* (single jump). Obligate pathogen.	No distinctive lesions in skeletal remains.	Become an obligate intracellular pathogen through genome reduction.	Numerous: Aristotle (384-322 BCE) to modern times.	Environmental contamination. Crowding. Mixing. Horses used in transportation, farming, warfare.
Morbilliviruses Measles (HMV), Rinderpest (RPV), Canine Distemper (CDV)	Humans, cattle, domestic dogs Epidemic: high morbidity and mortality Endemic: dense populations	Extremely infectious. Species jumping: cattle to humans to dogs	HMV, RPV: No distinctive lesions in skeletal remains. CDV: Enamel hypoplasia in dogs infected as young puppies.	Are very closely related viruses	Numerous: Rinderpest 3000 BCE Measles ~ 900 CE Canine distemper 1743.	Crowding. Mixing species. Urbanisation. Warfare. Trade/colonisation.

process was present and identifies ones suitable for biomolecular analysis. Palaeopathology thus provides a critical context for the direct detection of causative agents, as pathogens can be present without causing disease. In addition to detecting infectious organisms, biomolecular analysis has the potential to unpick the genetic history of pathogens and their hosts. When findings from palaeopathology and biomolecular analysis are integrated with an evolutionary context and other lines of evidence, such as documentary sources, it is possible to generate new understandings of the factors driving pathogen evolution, producing disease outbreaks and facilitating cross species transmission, as we demonstrate in the case of canine distemper. This integration of multiple lines of evidence is important as palaeopathology does have limitations: there are many diseases that do not affect skeletal tissues, lesions are often limited to combinations of bone destruction and proliferation, few diseases have pathognomonic lesions, and differential diagnosis is highly dependent on skeletal completeness and preservation—a particular issue for zooarchaeological palaeopathology (Thomas, 2019). Nevertheless, securely dated palaeopathological evidence, grounded within a cultural, ecological, medical and biological context, enables the interplay between pathogens, their host(s) and environments in the past to be elucidated.

There are also lessons for the present and future in this interdisciplinary approach as the factors driving pathogen evolution, disease outbreaks and cross species infections are the same as those in the past. In particular, crowding and mass movements of people, animals and their products around the world has accelerated in modern times and continues to impact pathogen transmission. Indeed, the African swine fever epidemic that began in 2018 has similarities to historical plague epidemics, including the way it spread through trade routes and jumped into wildlife when it reached a new continent (Dixon et al., 2019). While it is currently the largest animal disease epidemic in history, it has recently been overshadowed in impact by the human coronavirus pandemic (Morens et al., 2020). This new human-adapted virus (SARS-CoV-2), like past pandemic pathogens, originated from a zoonotic jump to humans, who quickly spread it around the world (Sun et al., 2020; Ye et al., 2020). As has been the case with other pathogens, infected humans have transmitted the virus to a variety of animals including dogs, domestic cats, ferrets, mink, lions, gorillas and tigers at geographic locations far from its site of origin (Gollakner & Capua, 2020). In addition, as has happened with other pathogens in the past, strains of the virus with increased transmissibility in humans are rapidly evolving as it expands into new hosts across human environments. Understanding how human activities and the conditions in different domestic environments have driven the transmission and evolution of shared pathogens in the past can help inform effective responses to the current pandemic as well as facilitate the prevention of future ones.

HUMAN POPULATION: 8 BILLION

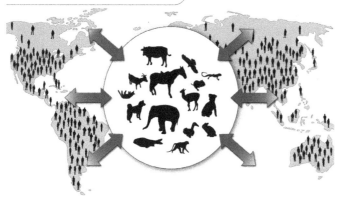

Figure 17.6 Today pathogen evolution continues to be driven by human activities, as there are 7.7 billion people on earth living in increasingly crowded conditions and rapidly expanding into new environments. The rapid spread of SARS-CoV-2 around the world illustrates that for a pathogen infecting humans, this large human population provides potentially innumerable opportunities for transmission both between human hosts and to other species. Illustration by Ali Ennis, Educational Resources, University of Georgia.

Humans have always lived surrounded by potential pathogens. Whether they co-exist relatively harmlessly or become a problem, cause acute or chronic disease and spread slowly or in epidemics has been, and still is, influenced by how we have impacted the environments we share with other animals. Pathogens are opportunists within these environments, capable and ready to take advantage of anything that promotes their transmission. Today there are almost eight billion people on earth, crowded together and travelling widely—this is 1300 times more than were present when the agricultural 'revolution' began around 10,000 years ago and facilitated the spread of many pathogens (Bocquet-Appel, 2011). A major concern, from an evolutionary standpoint, is the greatly enhanced transmission opportunities for pathogens that can infect humans (Figure 17.6).

References

Adesokan, H. K., Akinseye, V. O., Streicher, E. M., Van Helden, P., Warren, R. M. & Cadmus, S. I. (2019). Reverse zoonotic tuberculosis transmission from an emerging Uganda I strain between pastoralists and cattle in South-Eastern Nigeria. *BMC Veterinary Research*, 15, 437.

Akpinar, O. (2016). Historical perspective of brucellosis: a microbiological and epidemiological overview. *Le Infezioni Medicina*, 24, 77–86.

Akubek, K. & Handelman, S. (1999). *Biohazard: the chilling true story of the largest covert biological weapons program in the world*. Random House, New York.

Alizon, S. & Michalakis, Y. (2015). Adaptive virulence evolution: the good old fitness-based approach. *Trends in Ecology & Evolution*, 30, 248–254.

Anastasiou, E. & Mitchell, P. D. (2013). Palaeopathology and genes: investigating the genetics of infectious diseases in excavated human skeletal remains and mummies from past populations. *Gene*, 528, 33–40.

Anderson, P. D. & Bokor, G. (2012). Bioterrorism: pathogens as weapons. *Journal of Pharmacy Practice*, 25, 521–529.

Assis, S., Santos, A. L. & Roberts, C. A. (2011). Evidence of hypertrophic osteoarthropathy in individuals from the Coimbra Skeletal Identified Collection (Portugal). *International Journal of Paleopathology*, 1, 155–163.

Baker, O., Lee, O. Y., Wu, H. H., Besra, G. S., Minnikin, D. E., Llewellyn, G., Williams, C. M., Maixner, F., O'Sullivan, N., Zink, A., Chamel, B., Khawam, R., Coqueugniot, E., Helmer, D., Le Mort, F., Perrin, P.,

Gourichon, L., Dutailly, B., Pálfi, G., Coqueugniot, H. & Dutour, O. (2015). Human tuberculosis predates domestication in ancient Syria. *Tuberculosis (Edinb)*, 95 Suppl 1, S4–s12.

Barnes, I., Duda, A., Pybus, O. G. & Thomas, M. G. (2011). Ancient urbanization predicts genetic resistance to tuberculosis. *Evolution*, 65, 842–848.

Barrett T, P. R. & Taylor, W. P. (2006). *Rinderpest and peste des petits ruminants viral plagues of large and small ruminants*. Academic Press Elsevier, London.

Bartosiewicz, L. (2018). *Taphonomy and disease prevalence in animal palaeopathology: the proverbial 'veterinary horse'*. Oxbow Books, Oxford.

Bartosiewicz, L. G. & Gal, E. (2013). *Shuffling nags, lame ducks: the archaeology of animal disease*. Oxbow Books, Oxford.

Basaraba, R. J. & Hunter, R. L. (2017). Pathology of tuberculosis: how the pathology of human tuberculosis informs and directs animal models. *Microbiology Spectrum* [online] 5. doi: 10.1128/microbiolspec.TBTB2-0029-2016.

Bates, J. H. & Stead, W. W. (1993). The history of tuberculosis as a global epidemic. *Medical Clinics of North America*, 77, 1205–1217.

Bathurst, R. R. & Barta, J. L. (2004). Molecular evidence of tuberculosis-induced hypertrophic osteopathy in a 16th-century Iroquoian dog. *Journal of Archaeological Science* 31, 917–925.

Beineke, A., Puff, C., Seehusen, F. & Baumgartner, W. (2009). Pathogenesis and immunopathology of systemic and nervous canine distemper. *Veterinary Immunology & Immunopathology*, 127, 1–18.

Beitzke, H. (1934). Rotz der Knochen und Gelenke. *Hdb. spez. path. Anat. Hist*, 9, 583–593.

Bendrey, R., Cassidy, J. P., Bokovenko, N., Lepetz, S. & Zaitseva, G. I. (2011). A possible case of 'Poll Evil' in an early Scythian horse skull from Arzhan 1, Tuva Republic, Central Asia. *International Journal of Osteoarchaeology* 21, 111–118.

Bendrey, R., Cassidy, J. P., Fournié, G., Merrett, D. C., Oakes, R. H. A. & Taylor, G. M. (2020). Approaching ancient disease from a One Health perspective: interdisciplinary review for the investigation of zoonotic brucellosis. *International Journal of Osteoarchaeology* 30, 99–108.

Bendrey, R. & Fournié, G. (2020). Zoonotic brucellosis from the long view: can the past contribute to the present? *Infection Control & Hospital Epidemiology*, 42, 1–2.

Bendrey, R., Taylor, G. M., Bouwman, A. S. & Cassidy, J. P. (2008). Suspected bacterial disease in two archaeological horse skeletons from southern England: palaeopathological and biomolecular studies. *Journal of Archaeological Science* 35, 1581–1590.

Bengis, R. G., Kock, R. A. & Fischer, J. (2002). Infectious animal diseases: the wildlife/livestock interface. *Revue Scientifique et Technique (Paris)*, 21, 53–65.

Bianucci, R., Araujo, A., Pusch, C. M. & Nerlich, A. G. (2015). The identification of malaria in paleopathology: an in-depth assessment of the strategies to detect malaria in ancient remains. *ACTA Tropica (Amsterdam)*, 152, 176–180.

Bierer, B. W. (1939). *History of animal plagues of North America*. United States Department of Agriculture, Washington, DC.

Biggerstaff, M., Cauchemez, S., Reed, C., Gambhir, M. & Finelli, L. (2014). Estimates of the reproduction number for seasonal, pandemic, and zoonotic influenza: a systematic review of the literature. *BMC Infectious Diseases*, 14, 480.

Black, F. L. (1966). Measles endemicity in insular populations: critical community size and its evolutionary implication. *Journal of Theoretical Biology*, 11, 207–211.

Blancou, J. (2003). *History of the surveillance and control of transmissible animal diseases*. Office International des Epizooties, Paris.

Blancou, J. (2004). Dog distemper: imported into Europe from South America? *Historia Medicinae Veterinariae*, 29, 35–41.

Bocquet-Appel, J. P. (2011). When the world's population took off: the springboard of the Neolithic Demographic Transition. *Science*, 333, 560–561.

Bos, K. I., Harkins, K. M., Herbig, A., Coscolla, M., Weber, N., Comas, I., Forrest, S. A., Bryant, J. M., Harris, S. R., Schuenemann, V. J., Campbell, T. J., Majander, K., Wilbur, A. K., Guichon, R. A., Wolfe Steadman, D. L., Cook, D. C., Niemann, S., Behr, M. A., Zumarraga, M., Bastida, R., Huson, D., Nieselt, K., Young, D., Parkhill, J., Buikstra, J. E., Gagneux, S., Stone, A. C. & Krause, J. (2014). Pre-Columbian mycobacterial genomes reveal seals as a source of New World human tuberculosis. *Nature*, 514, 494–497.

Bracewell, C. D. & Corbel, M. J. (1980). An association between arthritis and persistent serological reactions to *Brucella abortus* in cattle from apparently brucellosis-free herds. *Veterinary Record*, 106, 99–101.

Bradley, C. A. & Altizer, S. (2007). Urbanization and the ecology of wildlife diseases. *Trends Ecology & Evolution*, 22, 95–102.

Bravo, M. J., de Dios Colmenero, J., Alonso, A. & Caballero, A. (2003). HLA-B*39 allele confers susceptibility to osteoarticular complications in human brucellosis. *Journal of Rheumatology*, 30, 1051–1053.

Brites, D. & Gagneux, S. (2012). Old and new selective pressures on Mycobacterium tuberculosis. *Infection, Genetics Evolution*, 12, 678–685.

Bruce-Chwatt, L. J., Garrett-Jones, C. & Weitz, B. (1966). Ten years' study (1955–64) of host selection by anopheline mosquitos. *Bulletin of the World Health Organization*, 35, 405–439.

Bruschi, F. (2011). Was Julius Caesar's epilepsy due to neurocysticercosis? *Trends in Parasitology*, 27, 373–374.

Bruschi, F., Masetti, M., Locci, M. T., Ciranni, R. & Fornaciari, G. (2006). Short report: cysticercosis in an Egyptian mummy of the late Ptolemaic period. *American Journal of Tropical Medicine & Hygiene*, 74, 598–599.

Bull, J. J. & Lauring, A. S. (2014). Theory and empiricism in virulence evolution. *PLoS Pathogens*, 10, e1004387.

Camaschella, C. (2015). Iron-deficiency anemia. *New England Journal of Medicine*, 372, 1832–1843.

Capasso, L. (1999). Brucellosis at Herculaneum (79 AD). *International Journal of Osteoarchaeology*, 9, 277–288.

Capasso, L. (2002). Bacteria in two-millennia-old cheese, and related epizoonoses in Roman populations. *Journal of Infection*, 45, 122–127.

Cardona, P. J., Català, M. & Prats, C. (2020). Origin of tuberculosis in the Paleolithic predicts unprecedented population growth and female resistance. *Scientific Reports*, 10, 42.

Carter, R. & Mendis, K. N. (2002). Evolutionary and historical aspects of the burden of malaria. *Clinical Microbiology Reviews*, 15, 564–594.

Cavalli-Sforza, L. L., Menozzi, P. & Piazza, A. (1994). The history and geography of human genes. Princeton University Press, Princeton.

Chalke, H. D. (1962). The impact of tuberculosis on history, literature and art. *Medical History*, 6, 301–318.

Chelli Bouaziz, M., Ladeb, M. F., Chakroun, M. & Chaabane, S. (2008). Spinal brucellosis: a review. *Skeletal Radiology*, 37, 785–790.

Cleaveland, S., Appel, M. G., Chalmers, W. S., Chillingworth, C., Kaare, M. & Dye, C. (2000). Serological and demographic evidence for domestic dogs as a source of canine distemper virus infection for Serengeti wildlife. *Veterinary Microbiology*, 72, 217–227.

Cliff, A., Haggett, P. & Smallman-Raynor, M. (1993). *Measles: an historical geography of a major human viral disease: from global expansion to local retreat, 1840–1990*. Blackwell Publishers, Cambridge.

Smallman-Raynor, M. R. & Cliff, A. (2004). *War epidemics: an historical geography of infectious diseases in military conflict and civil strife, 1850-2000*. Oxford University Press, Oxford.

Cohen, N. D., Carter, G. K. & McMullan, W. C. (1992). Fistulous withers in horses: 24 cases (1984–1990). *Journal American Veterinary Medical Association*, 201, 121–124.

Cohrs, P. (1967). *Nierberle and Cohrs textbook of the special pathological anatomy of domestic animals*. Pergamon Press, Oxford.

Coluzzi, M. (1999). The clay feet of the malaria giant and its African roots: hypotheses and inferences about origin, spread and control of Plasmodium falciparum. *Parassitologia*, 41, 277–283.

Comas, I., Coscolla, M., Luo, T., Borrell, S., Holt, K. E., Kato-Maeda, M., Parkhill, J., Malla, B., Berg, S., Thwaites, G., Yeboah-Manu, D., Bothamley, G., Mei, J., Wei, L., Bentley, S., Harris, S. R., Niemann, S., Diel, R., Aseffa, A., Gao, Q., Young, D. & Gagneux, S. (2013). Out-of-Africa migration and Neolithic coexpansion of *Mycobacterium tuberculosis* with modern humans. *Nature Genetics*, 45, 1176–1182.

Comas, I., Hailu, E., Kiros, T., Bekele, S., Mekonnen, W., Gumi, B., Tschopp, R., Ameni, G., Hewinson, R. G., Robertson, B. D., Goig, G. A., Stucki, D., Gagneux, S., Aseffa, A., Young, D. & Berg, S. (2015). Population genomics of *Mycobacterium tuberculosis* in Ethiopia contradicts the virgin soil hypothesis for human tuberculosis in sub-Saharan Africa. *Current Biology*, 25, 3260–3266.

Cook, N. D. (1998). *Born to die: disease and new world conquest*. Cambridge University Press, Cambridge.

Cosby, S. L. & Weir, L. (2018). Measles vaccination: threat from related veterinary viruses and need for continued vaccination post measles eradication. *Human Vaccines & Immunotherapy*, 14, 229–233.

Côté, N. M. & Le Bailly, M. (2018). Palaeoparasitology and palaeogenetics: review and perspectives for the study of ancient human parasites. *Parasitology*, 145, 656–664.

Cousins, D. V. (2001). *Mycobacterium bovis* infection and control in domestic livestock. *Revue Scientifique et Technique (Paris)*, 20, 71–85.

Craig L. E., Dittmer, K. E. & Thompson, K. G. 2016. 'Bones and joints', in M. G. Maxie (ed), *Jubb, Kennedy, and Palmer's pathology of domestic animals*, 6th edn, pp. 16–163. Elsevier, St. Louis.

Curate, F. (2006). Two possible cases of brucellosis from a Clarist monastery in Alcácer do Sal, southern Portugal. *International Journal of Osteoarchaeology*, 16, 453–458.

Curtin, P. D. (1994). Malarial immunities in nineteenth-century west Africa and the Caribbean. *Parassitologia*, 36, 69–82.

D'Anastasio, R., Staniscia, T., Milia, M. L., Manzoli, L. & Capasso, L. (2011). Origin, evolution and paleoepidemiology of brucellosis. *Epidemiology & Infection*, 139, 149–156.

Dance, D. A. B. (2009). 'Melioidosis and glanders as possible biological weapons', in I. W. Fong & K. Alibek (eds), Bioterrorism and infectious agents: a new dilemma for the 21st century, pp. 99–146. Springer, New York.

Dangvard Pedersen, D., Milner, G. R., Kolmos, H. J. & Boldsen, J. L. (2019). The association between skeletal lesions and tuberculosis diagnosis using a probabilistic approach. *International Journal of Paleopathology*, 27, 88–100.

Daróczi-Szabó, L. (2003). 'Animal bones as indicators of kosher food refuse from 14th century AD Buda, Hungary', in Behaviour *behind bones: the zooarchaeology of ritual, religion, status and identity*, pp. 252–261. Oxbow Books, Oxford.

Davies, P. D. (2006). Tuberculosis in humans and animals: are we a threat to each other? *Journal of the Royal Society of Medicine (London)*, 99, 539–540.

De Vries, R. D., Ludlow, M., Verburgh, R. J., Van Amerongen, G., Yuksel, S., Nguyen, D. T., Mcquaid, S., Osterhaus, A. D., Duprex, W. P. & De Swart, R. L. (2014). Measles vaccination of nonhuman primates provides partial protection against infection with canine distemper virus. *Journal of Virology*, 88, 4423–4433.

Del Brutto, O. H. & García, H. H. 2015. *Taenia solium* cysticercosis—The lessons of history. *Journal of the Neurological Sciences*, 359, 392–395.

Dixon, L. K., Sun, H. & Roberts, H. (2019). African swine fever. *Antiviral Res*, 165, 34–41.

Djurković-Djaković, O., Bobić, B., Nikolić, A., Klun, I. & Dupouy-Camet, J. (2013). Pork as a source of human parasitic infection. *Clinical Microbiology and Infection*, 19, 586–594.

Donoghue, H. D. (2016). Paleomicrobiology of human tuberculosis. *Microbiology Spectrum*, 4.

Dubielzig, R. R. (1979). The effect of canine distemper virus on the ameloblastic layer of the developing tooth. *Veterinary Pathology*, 16, 268–270.

Dubielzig, R. R., Higgins, R. J. & Krakowka, S. (1981). Lesions of the enamel organ of developing dog teeth following experimental inoculation of gnotobiotic puppies with canine distemper virus. *Veterinary Pathology*, 18, 684–689.

Duff, H. (1936). Fistulous withers and poll evil due to *Brucella abortus* in horses. *Veterinary Record*, 48, 175–177.

Dvorak, G. D. & Spickler, A. R. (2008). Glanders. *Journal of the American Veterinary Medical Association*, 233, 570–577.

El-Sayed, A. & Awad, W. (2018). Brucellosis: evolution and expected comeback. *International Journal of Veterinary Science and Medicine*, 6, S31–s35.

Erwin, P. C., Bemis, D. A., Mccombs, S. B., Sheeler, L. L., Himelright, I. M., Halford, S. K., Diem, L., Metchock, B., Jones, T. F., Schilling, M. G. & Thomsen, B. V. (2004).

Mycobacterium tuberculosis transmission from human to canine. *Emerging Infectious Diseases*, 10, 2258–10.

Esmaeilnejad-Ganji, S. M. & Esmaeilnejad-Ganji, S. M. R. (2019). Osteoarticular manifestations of human brucellosis: a review. *World Journal of Orthopedics*, 10, 54–62.

Essenberg, R. C., Seshadri, R., Nelson, K. & Paulsen, I. (2002). Sugar metabolism by Brucellae. *Veterinary Microbiology*, 90, 249–261.

Etxeberria, F. (1994). Vertebral epiphysitis: early signs of brucellar disease. *Journal of Paleopathology*, 6, 41–49.

Evans, J. T., Smith, E. G., Banerjee, A., Smith, R. M., Dale, J., Innes, J. A., Hunt, D., Tweddell, A., Wood, A., Anderson, C., Hewinson, R. G., Smith, N. H., Hawkey, P. M. & Sonnenberg, P. (2007). Cluster of human tuberculosis caused by *Mycobacterium bovis*: evidence for person-to-person transmission in the UK. *The Lancet*, 369, 1270–1276.

Ewald, P. W. (2004). Evolution of virulence. *Infectious Disease Clinics of North America*, 18, 1–15.

Ewald, P. W. (2011). Evolution of virulence, environmental change, and the threat posed by emerging and chronic diseases. *Ecological Research*, 26, 1017–1026.

Ewald, P. W. (1996). Guarding against the most dangerous emerging pathogens. *Emerging Infectious Diseases*, 2, 245–257.

Ewald, P. W. (1987). Transmission modes and evolution of the parasitism-mutualism continuum. *Annuals of the New York Academy of Sciences*, 503, 295–306.

Ficht, T. (2010). Brucella taxonomy and evolution. *Future Microbiology*, 5, 859–866.

Firdessa, R., Tschopp, R., Wubete, A., Sombo, M., Hailu, E., Erenso, G., Kiros, T., Yamuah, L., Vordermeier, M., Hewinson, R. G., Young, D., Gordon, S. V., Sahile, M., Aseffa, A. & Berg, S. (2012). High prevalence of bovine tuberculosis in dairy cattle in central Ethiopia: implications for the dairy industry and public health. *PLoS One*, 7, e52851.

Firth, E. C., Kersjes, A. W., Dik, K. J. & Hagens, F. M. (1987). Haematogenous osteomyelitis in cattle. *Veterinary Record*, 120, 148–152.

Flammer, P. G., Dellicour, S., Preston, S. G., Rieger, D., Warren, S., Tan, C. K. W., Nicholson, R., Přichystalová, R., Bleicher, N., Wahl, J., Faria, N. R., Pybus, O. G., Pollard, M. & Smith, A. L. (2018). Molecular archaeoparasitology identifies cultural changes in the Medieval Hanseatic trading centre of Lübeck. *Proceedings of the Royal Society B: Biological Sciences*, 285, 20180991.

Fleming, G. (1882). *A chronological history of animal plagues; their history, nature and prevention.* Chapman and Hall, London.

Fornaciari, G., Giuffra, V., Ferroglio, E. & Bianucci, R. (2010a). Malaria was 'the killer' of Francesco I de'

Medici (1531–1587). *American Journal of Medicine*, 123, 568–569.

Fornaciari, G., Giuffra, V., Ferroglio, E., Gino, S. & Bianucci, R. (2010b). Plasmodium falciparum immunodetection in bone remains of members of the renaissance Medici family (Florence, Italy, sixteenth century). *Transactions of the Royal Society of Tropical Medicine and Hygiene*, 104, 583–587.

Foster, J. T., Beckstrom-Sternberg, S. M., Pearson, T., Beckstrom-Sternberg, J. S., Chain, P. S., Roberto, F. F., Hnath, J., Brettin, T. & Keim, P. (2009). Whole-genome-based phylogeny and divergence of the genus Brucella. *Journal of Bacteriology*, 191, 2864–2870.

Fournié, G., Pfeiffer, D. U. & Bendrey, R. (2017). Early animal farming and zoonotic disease dynamics: modelling brucellosis transmission in Neolithic goat populations. *Royal Society Open Science*, 4, 160943.

Furuse, Y., Suzuki, A. & Oshitani, H. (2010). Origin of measles virus: divergence from rinderpest virus between the 11th and 12th centuries. *Virology Journal*, 7, 52.

Galagan, J. E. (2014). Genomic insights into tuberculosis. *Nature Reviews Genetics*, 15, 307–320.

Gallagher, J. & Clifton-Hadley, R. S. (2000). Tuberculosis in badgers: a review of the disease and its significance for other animals. *Research in Veterinary Science*, 69, 203–217.

Garcia, H. H., Nash, T. E. & Del Brutto, O. H. (2014). Clinical symptoms, diagnosis, and treatment of neurocysticercosis. *Lancet Neurology*, 13, 1202–1215.

García-López, J. M., Jenei, T., Chope, K. & Bubeck, K. A. (2010). Diagnosis and management of cranial and caudal nuchal bursitis in four horses. *Journal of the American Veterinary Medical Association*, 237, 823–829.

Garnham, P. C. C. (1996). *Malaria parasites and other haemosporidia.* Blackwell, Oxford.

Gaughan, E. M., Fubini, S. L. & Dietze, A. (1988). Fistulous withers in horses: 14 cases (1978–1987). *Journal of the American Veterinary Medical Association*, 193, 964–966.

Ghodbane, R. & Drancourt, M. (2013). Non-human sources of Mycobacterium tuberculosis. *Tuberculosis (Edinb)*, 93, 589–595.

Gilardi Kv, G. T., Leendertz, F. H., Macfie, E. J., Travis, D. A., Whittier, C. A. & Williamson, E. A. (2015). *Best practice guidelines for health monitoring and disease control in great ape populations.* GIUCN SSC Primate Specialist Group, Switzerland.

Gollakner, R. & Capua, I. (2020). Is COVID-19 the first pandemic that evolves into a panzootic? *Veterinaria Italiana*, 56, 7–8.

Grauer, A. L. (2019). 'Circulatory, reticuloendothelial, and hematopoietic disorders', in *Ortner's identification of pathological conditions in human skeletal remains*, 3rd edn, pp. 491–529. Academic Press, London.

Grau-Sologestoa, I. (2017). Socio-economic status and religious identity in medieval Iberia: the zooarchaeological evidence. *Environmental Archaeology*, 22, 189–199.

Greco, E., El-Aguizy, O., Ali, M. F., Foti, S., Cunsolo, V., Saletti, R. & Ciliberto, E. (2018). Proteomic analyses on an ancient Egyptian cheese and biomolecular evidence of brucellosis. *Analytical Chemistry*, 90, 9673–9676.

Grilo, M. L., Vanstreels, R. E., Wallace, R., García-Párraga, D., Braga É, M., Chitty, J., Catão-Dias, J. L. & Madeira De Carvalho, L. M. (2016). Malaria in penguins - current perceptions. *Avian Pathology*, 45, 393–407.

Guerra, F. M., Bolotin, S., Lim, G., Heffernan, J., Deeks, S. L., Li, Y. & Crowcroft, N. S. (2017). The basic reproduction number (R0) of measles: a systematic review. *Lancet Infectious Diseases*, 17, e420–e428.

Gutierrez, M. C., Ahmed, N., Willery, E., Narayanan, S., Hasnain, S. E., Chauhan, D. S., Katoch, V. M., Vincent, V., Locht, C. & Supply, P. (2006). Predominance of ancestral lineages of Mycobacterium tuberculosis in India. *Emerging Infectious Diseases*, 12, 1367–1374.

Gutierrez, M. C., Brisse, S., Brosch, R., Fabre, M., Omaïs, B., Marmiesse, M., Supply, P. & Vincent, V. (2005). Ancient origin and gene mosaicism of the progenitor of Mycobacterium tuberculosis. *PLoS Pathogens*, 1, e5.

Haag, K. L., Gottstein, B. & Ayala, F. J. (2008). Taeniid history, natural selection and antigenic diversity: evolutionary theory meets helminthology. *Trends in Parasitology*, 24, 96–102.

Hackett, L. W. (1937). *Malaria in Europe: an ecological study*. Oxford University Press, London.

Håkan, E. (2003). Malaria and anemia. *Current Opinion in Hematology*, 10, 108–114.

Harkins, K. M., Buikstra, J. E., Campbell, T., Bos, K. I., Johnson, E. D., Krause, J. & Stone, A. C. (2015). Screening ancient tuberculosis with qPCR: challenges and opportunities. *Proceedings of the Royal Society B: Biological Sciences*, 370, 20130622.

Harper, K. N. & Armelagos, G. J. (2013). Genomics, the origins of agriculture, and our changing microbe-scape: time to revisit some old tales and tell some new ones. *American Journal of Physical Anthropology*, 152(Suppl 57), 135–152.

Hedrick, P. W. (2012). Resistance to malaria in humans: the impact of strong, recent selection. *Malaria Journal*, 11, 349.

Hershberg, R., Lipatov, M., Small, P. M., Sheffer, H., Niemann, S., Homolka, S., Roach, J. C., Kremer, K., Petrov, D. A., Feldman, M. W. & Gagneux, S. (2008). High functional diversity in *Mycobacterium tuberculosis* driven by

genetic drift and human demography. *PLoS Biology*, 6, e311.

Hillson, S. (2005). *Teeth*. Cambridge University Press, Cambridge.

Hoberg, E. P. (2006). Phylogeny of *Taenia*: Species definitions and origins of human parasites. *Parasitology International*, 55(Suppl), S23–30.

Hoberg, E. P., Alkire, N. L., De Queiroz, A. & Jones, A. (2001). Out of Africa: origins of the *Taenia* tapeworms in humans. *Proceedings of the Royal Society B: Biological Sciences*, 268, 781–787.

Horacio, M. C., María, V. G. & Alonso, G. L. (2015). Hypertrophic osteoarthropathy as a complication of pulmonary tuberculosis. *Reumatologia Clinica*, 11, 255–257.

Houben, R. M. & Dodd, P. J. (2016). The global burden of latent tuberculosis infection: a re-estimation using mathematical modelling. *PLoS Medicine*, 13, e1002152.

Hulden, L. & Hulden, L. (2011). Activation of the hypnozoite: a part of *Plasmodium vivax* life cycle and survival. *Malaria Journal*, 10, 90.

Hume, J. C., Lyons, E. J. & Day, K. P. (2003). Human migration, mosquitoes and the evolution of Plasmodium falciparum. *Trends in Parasitology*, 19, 144–149.

Hunter, R. L. (2011). Pathology of post primary tuberculosis of the lung: an illustrated critical review. *Tuberculosis (Edinb)*, 91, 497–509.

Ijzereef, G. F., Serjeantson, D. & Waldron, T. (1989). *Social differentiation from animal bone studies*. ROB, Amersfoort.

Ito, A., Putra, M. I., Subahar, R., Sato, M. O., Okamoto, M., Sako, Y., Nakao, M., Yamasaki, H., Nakaya, K., Craig, P. S. & Margono, S. S. (2002). Dogs as alternative intermediate hosts of *Taenia solium* in Papua (Irian Jaya), Indonesia confirmed by highly specific ELISA and immunoblot using native and recombinant antigens and mitochondrial DNA analysis. *Journal of Helminthology*, 76, 311–314.

James, S. P., Nicol, W. D. & Shute, P. G. (1936). Clinical and parasitological observations on induced malaria: (section of tropical diseases and parasitology). *Proceedings of the Royal Society of Medicine*, 29, 879–894.

James, W. A., Graham, R. (1930). Porcine osteomyelitis, pyemic arthritis and pyemic bursitis associated with *Brucella suis* Traum. *Journal of the American Veterinary Medical Association*, 77, 774–782.

Jeffery, G. M. & Eyles, D. E. (1955). Infectivity to mosquitoes of *Plasmodium falciparum* as related to gametocyte density and duration of infection. *American Journal of Tropical Medicine & Hygiene*, 4, 781–789.

Jones, C. (2019). Brucellosis in an adult female from Fate Bell Rock Shelter, Lower Pecos, Texas (4000–1300 BP). *International Journal of Paleopathology*, 24, 252–264.

Kalanon, M. & McFadden, G. I. (2010). Malaria, *Plasmodium falciparum* and its apicoplast. *Biochemical Society Transactions*, 38, 775–782.

Kay, G. L., Sergeant, M. J., Giuffra, V., Bandiera, P., Milanese, M., Bramanti, B., Bianucci, R. & Pallen, M. J. (2014). Recovery of a medieval *Brucella melitensis* genome using shotgun metagenomics. *mBio*, 5, e01337–14.

Keeling, M. J. & Grenfell, B. T. (1997). Disease extinction and community size: modeling the persistence of measles. *Science*, 275, 65–67.

Kelley, M. A. & Micozzi, M. S. (1984). Rib lesions in chronic pulmonary tuberculosis. *American Journal of Physical Anthropology*, 65, 381–386.

Kerr, P. J. (2012). Myxomatosis in Australia and Europe: a model for emerging infectious diseases. *Antiviral Research*, 93, 387–415.

Kettle, A. N. & Wernery, U. (2016). Glanders and the risk for its introduction through the international movement of horses. *Equine Veterinary Journal*, 48, 654–658.

Kiers, A., Klarenbeek, A., Mendelts, B., Van Soolingen, D. & Koëter, G. (2008). Transmission of *Mycobacterium pinnipedii* to humans in a zoo with marine mammals. *International Journal of Tuberculosis and Lung Disease*, 12, 1469–1473.

Kirk, H. (1922). *Canine distemper: its complications, sequelae and treatment*. Bailliere, Tindall and Cox, London.

Kohn, G. (2001). *Encyclopedia of plague & pestilence from ancient times to the present*. Checkmark Books, New York.

Koro Koro, F., Ngatchou, A. F., Portal, J. L., Gutierrez, C., Etoa, F. X. & Eyangoh, S. I. (2015). The genetic population structure of *Mycobacterium bovis* strains isolated from cattle slaughtered at the Yaoundé and Douala abattoirs in Cameroon. *Revue Scientifique et Technique (Paris)*, 34, 1001–1010.

Koziol, U. (2017). Evolutionary developmental biology (evo-devo) of cestodes. *Experimental Parasitology*, 180, 84–100.

Lanteri, G., Marino, F., Reale, S., Vitale, F., Macrì, F. & Mazzullo, G. (2011). *Mycobacterium tuberculosis* in a red-crowned parakeet (*Cyanoramphus novaezelandiae*). *Journal of Avian Medicine and Surgery*, 25, 40–43.

Lapointe, D. A., Atkinson, C. T. & Samuel, M. D. (2012). Ecology and conservation biology of avian malaria. *Annuals of the New York Academy of Sciences*, 1249, 211–226.

Lehavi, O., Aizenstien, O., Katz, L. H. & Hourvitz, A. (2002). [Glanders—a potential disease for biological warfare in humans and animals]. *Harefuah*, 119, 88–91.

Lignereux, Y. & Peters, J. (1999). *Elements for the retrospective diagnosis of tuberculosis on animal bones from archaeological sites*. Golden Book and Tuberculosis Foundation, Szeged.

Liu, W., Li, Y., Learn, G. H., Rudicell, R. S., Robertson, J. D., Keele, B. F., Ndjango, J. B., Sanz, C. M., Morgan, D. B., Locatelli, S., Gonder, M. K., Kranzusch, P. J., Walsh, P. D., Delaporte, E., Mpoudi-Ngole, E., Georgiev, A. V., Muller, M. N., Shaw, G. M., Peeters, M., Sharp, P. M., Rayner, J. C. & Hahn, B. H. (2010). Origin of the human malaria parasite *Plasmodium falciparum* in gorillas. *Nature*, 467, 420–425.

Loeffler, S. H., De Lisle, G. W., Neill, M. A., Collins, D. M., Price-Carter, M., Paterson, B. & Crews, K. B. (2014). The seal tuberculosis agent, *Mycobacterium pinnipedii*, infects domestic cattle in New Zealand: epidemiologic factors and DNA strain typing. *Journal of Wildlife Diseases*, 50, 180–187.

Losada, L., Ronning, C. M., Deshazer, D., Woods, D., Fedorova, N., Kim, H. S., Shabalina, S. A., Pearson, T. R., Brinkac, L., Tan, P., Nandi, T., Crabtree, J., Badger, J., Beckstrom-Sternberg, S., Saqib, M., Schutzer, S. E., Keim, P. & Nierman, W. C. (2010). Continuing evolution of *Burkholderia mallei* through genome reduction and large-scale rearrangements. *Genome Biology Evolution*, 2, 102–116.

Lowbeer, L. (1959). Skeletal and articular involvement in brucellosis of animals. *Laboratory Investigations*, 8, 1448–1460.

Loy, D. E., Liu, W., Li, Y., Learn, G. H., Plenderleith, L. J., Sundararaman, S. A., Sharp, P. M. & Hahn, B. H. (2017). Out of Africa: origins and evolution of the human malaria parasites Plasmodium falciparum and Plasmodium vivax. *International Journal of Parasitology*, 47, 87–97.

Mackinnon, M. J. & Read, A. F. (2004). Virulence in malaria: an evolutionary viewpoint. *Philosophical Transactions of the Royal Society B: Biological Sciences*, 359, 965–986.

Maclean, K. A., Becker, A. K., Chang, S. D. & Harris, A. C. (2013). Extrapulmonary tuberculosis: imaging features beyond the chest. *Canadian Association of Radiologists Journal*, 64, 319–324.

Mann, C. C. (2011). *1493: Uncovering the new world Columbus created*. Vintage Books, New York.

Martín-Hernando, M. P., Höfle, U., Vicente, J., Ruiz-Fons, F., Vidal, D., Barral, M., Garrido, J. M., De La Fuente, J. & Gortazar, C. (2007). Lesions associated with *Mycobacterium tuberculosis* complex infection in the European wild boar. *Tuberculosis (Edinb)*, 87, 360–367.

Maslow, J. N. & Mikota, S. K. (2015). Tuberculosis in elephants—a reemergent disease: diagnostic dilemmas, the natural history of infection, and new immunological tools. *Veterinary Pathology*, 52, 437–440.

Mays, S. (2005). *Tuberculosis as a zoonotic disease in antiquity*. Oxbow Books, Oxford.

Mays, S. (2007). Lysis at the anterior vertebral body margin: evidence for brucellar spondylitis? *International Journal of Osteoarchaeology*, 17, 107–118.

McNeil, J. R. (2010). *Mosquito empires: ecology and war in the greater Caribbean, 1620–1914.* Cambridge University Press, New York.

McNeill, W. (1976). *Plagues and peoples.* Anchor Press/Doubleday, New York.

Michelet, L. & Dauga, C. (2012). Molecular evidence of host influences on the evolution and spread of human tapeworms. *Biological Reviews of the Cambridge Philosophical Society*, 87, 731–741.

Miletski, H. (2009). *Bestiality and zoophilia: sexual relations with animals.* Berg, Oxford.

Miller, M. & Olea-Popelka, F. (2013). One Health in the shrinking world: experiences with tuberculosis at the human-livestock-wildlife interface. *Comparative Immunology, Microbiology & Infectious Diseases*, 36, 263–268.

Miller, R. S., Farnsworth, M. L. & Malmberg, J. L. (2013). Diseases at the livestock-wildlife interface: status, challenges, and opportunities in the United States. *Preventative Veterinary Medicine*, 110, 119–132.

Milton, K. (2003). The critical role played by animal source foods in human (*Homo*) evolution. *Journal of Nutrition*, 133, 3886s–3892s.

Molina-Cruz, A., Zilversmit, M. M., Neafsey, D. E., Hartl, D. L. & Barillas-Mury, C. (2016). Mosquito vectors and the globalization of *Plasmodium falciparum* malaria. *Annual Review of Genetics*, 50, 447–465.

Montali, R. J., Mikota, S. K. & Cheng, L. I. (2001). *Mycobacterium tuberculosis* in zoo and wildlife species. *Revue Scientifique et Technique (Paris)*, 20, 291–303.

Moreno, E. (2014). Retrospective and prospective perspectives on zoonotic brucellosis. *Frontiers in Microbiology*, 5, 213.

Morens, D. M., Daszak, P., Markel, H. & Taubenberger, J. K. (2020). Pandemic COVID-19 joins history's pandemic legion. *mBio* [online] 11(3). doi:https://doi.org/10.1128/mBio.00812-20

Morens, D. M. & Taubenberger, J. K. (2015). A forgotten epidemic that changed medicine: measles in the US Army, 1917–18. *Lancet Infectious Diseases*, 15, 852–861.

Mostowy, S., Inwald, J., Gordon, S., Martin, C., Warren, R., Kremer, K., Cousins, D. & Behr, M. A. (2005). Revisiting the evolution of Mycobacterium bovis. *Journal of Bacteriology*, 187, 6386–6395.

Moule, M. G. & Cirillo, J. D. (2020). *Mycobacterium tuberculosis* dissemination plays a critical role in pathogenesis. *Frontiers in Cellular and Infection Microbiology*, 10, 65.

Müller, R., Roberts, C. A. & Brown, T. A. (2016). Complications in the study of ancient tuberculosis: presence of environmental bacteria in human archaeological remains. *Journal of Archaeological Science*, 68, 5–11.

Mutolo, M. J., Jenny, L. L., Buszek, A. R., Fenton, T. W. & Foran, D. R. (2012). Osteological and molecular identification of brucellosis in ancient Butrint, Albania. *American Journal of Physical Anthropology*, 147, 254–263.

Naranjo, V., Gortazar, C., Vicente, J. & De La Fuente, J. (2008). Evidence of the role of European wild boar as a reservoir of *Mycobacterium tuberculosis* complex. *Veterinary Microbiology*, 127, 1–9.

Nerlich, A. (2016). Paleopathology and paleomicrobiology of malaria. *Microbiology Spectrum* [online], 4. https://doi.org/10.1128/microbiolspec.PoH-0006-2015

Nerlich, A. G., Parsche, F., Von Den Driesch, A., And Löhrs, U. (1993). Osteopathological findings from mummified baboons from Ancient Egypt. *International Journal of Osteoarchaeology* 3, 189–198.

Nerlich, A. G., Schraut, B., Dittrich, S., Jelinek, T. & Zink, A. R. (2008). *Plasmodium falciparum* in ancient Egypt. *Emerging Infectious Diseases*, 14, 1317–1319.

Newfield, T. (2009). A cattle panzootic in the early fourteenth-century Europe. *Agricultureal History Review*, 57, 155–190(36).

Nieberle, K. (1938). *Tuberkulose und Fleischhygiene. Untersuchungen über die pathologische Anatomie und Pathogenese der Tuberkulose mit besonderer Berücksichtung für die Beurteilung des Fleisches der tuberkulösen Schlachtiere.* Fischer, Jena.

O'Reilly, L. M. & Daborn, C. J. (1995). The epidemiology of *Mycobacterium bovis* infections in animals and man: a review. *Tuberculosis and Lung Diseases*, 76(Suppl 1), 1–46.

Palmer, M. V. (2013). Mycobacterium bovis: characteristics of wildlife reservoir hosts. *Transboundary and Emerging Diseases*, 60(Suppl 1), 1–13.

Pappas, G., Akritidis, N., Bosilkovski, M. & Tsianos, E. (2005). Brucellosis. *New England Journal of Medicine*, 352, 2325–2336.

Pappas, G., Panagopoulou, P., Christou, L. & Akritidis, N. (2006). Brucella as a biological weapon. *Cellular and Molecular Life Sciences*, 63, 2229–2236.

Patané, J. S., Martins, J., Beatriz Castelão, A., Nishibe, C., Montera, L., Bigi, F., Zumárraga, M. J., Cataldi, A. A., Fonseca Junior, A., Roxo, E., Luiza, A., Osório, A. R., Jorge Ufms, K. S., Thacker, T. C., Almeida, N. F., Araújo, F. R. & Setubal, J. C. (2017). Patterns and processes of *Mycobacterium bovis* evolution revealed by phylogenomic analyses. *Genome Biology Evolution*, 9, 521–535.

Paulet, J. (1775). *Recherches historiques et physiques sur les maladies epizootiques avec les moyens d'y remedier, dans tous les cas.* Ruault, Paris.

Pearce-Duvet, J. M. (2006). The origin of human pathogens: evaluating the role of agriculture and domestic animals in the evolution of human disease. *Biological Reviews of the Cambridge Philosophical Society*, 81, 369–382.

Perri, A., Widga, C., Lawler, D., Martin, T., Loebel, T., Farnsworth, K., Kohn, L. & Buenger, B. (2019). New evidence of the earliest domestic dogs in the Americas. *American Antiquity*, 84, 68–87.

Perry, G. H. (2014). Parasites and human evolution. *Evolutionary Anthropology*, 23, 218–228.

Pesciaroli, M., Alvarez, J., Boniotti, M. B., Cagiola, M., Di Marco, V., Marianelli, C., Pacciarini, M. & Pasquali, P. (2014). Tuberculosis in domestic animal species. *Research in Veterinary Science*, 97(Suppl), S78–85.

Petersen, E., Rajashekara, G., Sanakkayala, N., Eskra, L., Harms, J. & Splitter, G. (2013). Erythritol triggers expression of virulence traits in *Brucella melitensis*. *Microbes and Infection*, 15, 440–449.

Pigrau-Serrallach, C. & Rodríguez-Pardo, D. (2013). Bone and joint tuberculosis. *European Spine Journal*, 22(Suppl 4), 556–566.

Pisarenko, S. V., Kovalev, D. A., Volynkina, A. S., Ponomarenko, D. G., Rusanova, D. V., Zharinova, N. V., Khachaturova, A. A., Tokareva, L. E., Khvoynova, I. G. & Kulichenko, A. N. (2018). Global evolution and phylogeography of *Brucella melitensis* strains. *BMC Genomics*, 19, 353.

Proto, W. R., Siegel, S. V., Dankwa, S., Liu, W., Kemp, A., Marsden, S., Zenonos, Z. A., Unwin, S., Sharp, P. M., Wright, G. J., Hahn, B. H., Duraisingh, M. T. & Rayner, J. C. (2019). Adaptation of *Plasmodium falciparum* to humans involved the loss of an ape-specific erythrocyte invasion ligand. *Nature Communications*, 10, 4512.

Rafaluk, C., Jansen, G., Schulenburg, H. & Joop, G. (2015). When experimental selection for virulence leads to loss of virulence. *Trends in Parasitology*, 31, 426–434.

Redding, R. W. (2015). The pig and the chicken in the Middle East: modeling human subsistence behavior in the archaeological record using historical and animal husbandry data. *Journal of Archaeological Research*, 23, 325–368.

Reinhard, K. J., Ferreira, L. F., Bouchet, F., Sianto, L., Dutra, J. M. F., Iniguez, A., Leles, D., Le Bailly, M., Fugassa, M., Pucu, E. & Araújo, A. (2013). Food, parasites, and epidemiological transitions: a broad perspective. *International Journal of Paleopathology*, 3, 150–157.

Reiter, P. (2000). From Shakespeare to Defoe: malaria in England in the Little Ice Age. *Emerging Infectious Diseases*, 6, 1–11.

Rhazes. (1848). *A treatise on the small pox and measles*. Sydenham Society, London.

Rima, B. K. & Duprex, W. P. (2006). Morbilliviruses and human disease. *Journal of Pathology*, 208, 199–214.

Riojas, M. A., McGough, K. J., Rider-Riojas, C. J., Rastogi, N. & Hazbón, M. H. (2018). Phylogenomic analysis of the species of the *Mycobacterium tuberculosis* complex demonstrates that *Mycobacterium africanum*, *Mycobacterium bovis*, *Mycobacterium caprae*, *Mycobacterium microti* and *Mycobacterium pinnipedii* are later heterotypic synonyms of *Mycobacterium tuberculosis*. *International Journal of Systematic and Evolutionary Microbiology*, 68, 324–332.

Roberts, C. A. & Buikstra, J. E. (2019). 'Bacterial infections', in *Ortner's identification of pathological conditions in human skeletal remains*, 3rd edn, pp. 321–439. Academic Press, London.

Roderick, L. (1948). *A study of equine fistulous withers and poll evil*. Kansas State College Agricultural Experimental Station Bulletin no. 63. Kansas State College of Agriculture and Applied Science, Manhattan.

Rodrigues, P. T., Valdivia, H. O., De Oliveira, T. C., Alves, J. M. P., Duarte, A., Cerutti-Junior, C., Buery, J. C., Brito, C. F. A., De Souza, J. C., Jr., Hirano, Z. M. B., Bueno, M. G., Catão-Dias, J. L., Malafronte, R. S., Ladeia-Andrade, S., Mita, T., Santamaria, A. M., Calzada, J. E., Tantular, I. S., Kawamoto, F., Raijmakers, L. R. J., Mueller, I., Pacheco, M. A., Escalante, A. A., Felger, I. & Ferreira, M. U. (). Human migration and the spread of malaria parasites to the New World. *Scientific Reports*, 8, 1993.

Rohde, C., Anderson, D. E., Desrochers, A., St-Jean, G., Hull, B. L. & Rings, D. M. (2000). Synovial fluid analysis in cattle: a review of 130 cases. *Veterinary Surgery*, 29, 341–346.

Rono, M. K., Nyonda, M. A., Simam, J. J., Ngoi, J. M., Mok, S., Kortok, M. M., Abdullah, A. S., Elfaki, M. M., Waitumbi, J. N., El-Hassan, I. M., Marsh, K., Bozdech, Z. & Mackinnon, M. J. (2018). Adaptation of *Plasmodium falciparum* to its transmission environment. *Nature Ecology & Evolution*, 2, 377–387.

Rossetti, C. A., Arenas-Gamboa, A. M. & Maurizio, E. (2017). Caprine brucellosis: a historically neglected disease with significant impact on public health. *PLoS Neglected Tropical Diseases*, 11, e0005692.

Rothschild, B. M., Martin, L. D., Lev, G., Bercovier, H., Bar-Gal, G. K., Greenblatt, C., Donoghue, H., Spigelman, M. & Brittain, D. (2001). *Mycobacterium tuberculosis* complex DNA from an extinct bison dated 17,000 years before the present. *Clinical Infectious Diseases*, 33, 305–311.

Sabbatani, S., Manfredi, R. & Fiorino, S. (2010). Malaria infection and human evolution. *Le Infezioni in Medicina*, 18, 56–74.

Sakai, K., Yoshikawa, T., Seki, F., Fukushi, S., Tahara, M., Nagata, N., Ami, Y., Mizutani, T., Kurane, I., Yamaguchi, R., Hasegawa, H., Saijo, M., Komase, K., Morikawa, S.

& Takeda, M. (2013). Canine distemper virus associated with a lethal outbreak in monkeys can readily adapt to use human receptors. *Journal of Virology*, 87, 7170–7175.

Sakamoto, K. (2012). The pathology of *Mycobacterium tuberculosis* infection. *Veterinary Pathology*, 49, 423–439.

Sallares, R., Bouwman, A. & Anderung, C. (2004). The spread of malaria to Southern Europe in antiquity: new approaches to old problems. *Medical History*, 48, 311–328.

Salo, W. L., Aufderheide, A. C., Buikstra, J. & Holcomb, T. A. (1994). Identification of *Mycobacterium tuberculosis* DNA in a pre-Columbian Peruvian mummy. *Proceedings of the National Academy of Sciences of the United States of America*, 91, 2091–2094.

Schlafer, D. H. & Miller, R. B. (2007). *Female genital system*, in M. G. Maxie (ed), *Kennedy and Palmer's pathology of domestic animals*, Vol. 3, pp. 429–564. Elsevier Saunders, Philadelphia.

Sciubba, M., Paolucci, A., D'Anastasio, R. & Capasso, L. (2013). [Paleopathology of Herculaneum's population (79 D.C.)]. *Medicina nei Secoli*, 25, 85–99.

Scott, J. C. (2017). *Against the grain*. Yale University Press, New Haven.

Setzer, T. J. (2014). Malaria detection in the field of paleopathology: a meta-analysis of the state of the art. *Acta Tropica*, 140, 97–104.

Sharif, H. S., Aideyan, O. A., Clark, D. C., Madkour, M. M., Aabed, M. Y., Mattsson, T. A., Al-Deeb, S. M. & Moutaery, K. R. (1989). Brucellar and tuberculous spondylitis: comparative imaging features. *Radiology*, 171, 419–425.

Sharrer, G. T. (1995). The great glanders epizootic, 1861–1866: a Civil War legacy. *Agricultural History*, 69, 79–97.

Sheshberadaran, H., Norrby, E., Mccullough, K. C., Carpenter, W. C. & Orvell, C. (1986). The antigenic relationship between measles, canine distemper and rinderpest viruses studied with monoclonal antibodies. *Journal of General Virology*, 67(Pt 7), 1381–1392.

Sianto, L., Chame, M., Silva, C. S., Gonçalves, M. L., Reinhard, K., Fugassa, M. & Araújo, A. (2009). Animal helminths in human archaeological remains: a review of zoonoses in the past. *Revista do Instituto de Medicina Tropical de Sao Paulo*, 51, 119–130.

Sidhu, M. S., Menonna, J. P., Cook, S. D., Dowling, P. C. & Udem, S. A. (1993). Canine distemper virus L gene: sequence and comparison with related viruses. *Virology*, 193, 50–65.

Singh, G. & Sander, J. W. (2019). Historical perspective: The British contribution to the understanding of neurocysticercosis. *Journal of the History of Neurosciences*, 28, 332–344.

Smith, N. H. (2012). The global distribution and phylogeography of *Mycobacterium bovis* clonal complexes. *Infection, Genetics and Evolution*, 12, 857–865.

Smith, N. H., Gordon, S. V., De La Rua-Domenech, R., Clifton-Hadley, R. S. & Hewinson, R. G. (2006). Bottlenecks and broomsticks: the molecular evolution of Mycobacterium bovis. *Nature Reviews Microbiology*, 4, 670–681.

Snider, W. R. (1971). Tuberculosis in canine and feline populations. *American Review of Respiratory Disease* 104, 877–887.

Snyder, R. (1978). *Historical aspects of tuberculosis in the Philadelphia Zoo*. Smithsonian Institution, Washington DC.

Song, H., Hwang, J., Yi, H., Ulrich, R. L., Yu, Y., Nierman, W. C. & Kim, H. S. (2010). The early stage of bacterial genome-reductive evolution in the host. *PLoS Pathogens*, 6, e1000922.

Spinage, C. (2003). *Cattle plague: a history*, Kluwer Academic/Plenum Publishers, New York.

Stephenson, J. R. & Ter Meulen, V. (1979). Antigenic relationships between measles and canine distemper viruses: comparison of immune response in animals and humans to individual virus-specific polypeptides. *Proceedings of the National Academy of Sciences of the United States of America*, 76, 6601–6605.

Sun, J., He, W. T., Wang, L., Lai, A., Ji, X., Zhai, X., Li, G., Suchard, M. A., Tian, J., Zhou, J., Veit, M. & Su, S. (2020). COVID-19: epidemiology, evolution, and cross-disciplinary perspectives. *Trends in Molecular Medicine*, 26, 483–495.

Takken, W. & Knols, B. G. (1999). Odor-mediated behavior of Afrotropical malaria mosquitoes. *Annual Reviews of Entomology*, 44, 131–157.

Tams, K. W., Jensen Søe, M., Merkyte, I., Valeur Seersholm, F., Henriksen, P. S., Klingenberg, S., Willerslev, E., Kjær, K. H., Hansen, A. J. & Kapel, C. M. O. (2018). Parasitic infections and resource economy of Danish Iron Age settlement through ancient DNA sequencing. *PLoS One*, 13, e0197399.

Taylor, G. M., Murphy, E., Hopkins, R., Rutland, P. & Chistov, Y. (2007). First report of *Mycobacterium bovis* DNA in human remains from the Iron Age. *Microbiology*, 153, 1243–1249.

Terefe, Y., Hailemariam, Z., Menkir, S., Nakao, M., Lavikainen, A., Haukisalmi, V., Iwaki, T., Okamoto, M. & Ito, A. (2014). Phylogenetic characterisation of *Taenia* tapeworms in spotted hyenas and reconsideration of the 'Out of Africa' hypothesis of *Taenia* in humans. *International Journal of Parasitology*, 44, 533–541.

Tessaro, S. V., Forbes, L. B. & Turcotte, C. (1990). A survey of brucellosis and tuberculosis in bison in and around

Wood Buffalo National Park, Canada. *Canadian Veterinary Journal*, 31, 174–180.

Thomas, R. (2019). 'Nonhuman animal paleopathology—are we so different?', in *Ortner's identification of pathological conditions in human skeletal remains*, 3rd edn, pp. 809–822. Academic Press, London.

Tsai, I. J., Zarowiecki, M., Holroyd, N., Garciarrubio, A., Sánchez-Flores, A., Brooks, K. L., Tracey, A., Bobes, R. J., Fragoso, G., Sciutto, E., Aslett, M., Beasley, H., Bennett, H. M., Cai, X., Camicia, F., Clark, R., Cucher, M., De Silva, N., Day, T. A., Deplazes, P., Estrada, K., Fernández, C., Holland, P. W. H., Hou, J., Hu, S., Huckvale, T., Hung, S. S., Kamenetzky, L., Keane, J. A., Kiss, F., Koziol, U., Lambert, O., Liu, K., Luo, X., Luo, Y., Macchiaroli, N., Nichol, S., Paps, J., Parkinson, J., Pouchkina-Stantcheva, N., Riddiford, N., Rosenzvit, M., Salinas, G., Wasmuth, J. D., Zamanian, M., Zheng, Y., Cai, J., Soberón, X., Olson, P. D., Laclette, J. P., Brehm, K. & Berriman, M. (2013). The genomes of four tapeworm species reveal adaptations to parasitism. *Nature*, 496, 57–63.

Tschopp, R., Schelling, E., Hattendorf, J., Young, D., Aseffa, A. & Zinsstag, J. (2010). Repeated cross-sectional skin testing for bovine tuberculosis in cattle kept in a traditional husbandry system in Ethiopia. *Veterinary Record*, 167, 250–256.

Uhl, E. W., Kelderhouse, C., Buikstra, J., Blick, J. P., Bolon, B. & Hogan, R. J. (2019). New world origin of canine distemper: interdisciplinary insights. *International Journal of Paleopathology*, 24, 266–278.

Ulloa, A. (1806). *A Voyage to South America*. Bond and Co, London.

Valenzuela-Lamas, S., Valenzuela-Suau, L., Saula, O., Colet, A., Mercadal, O., Subiranas, C. & Nadal. J. (2014). Shechita and Kashrut: Identifying Jewish populations through zooarchaeology and taphonomy. Two examples from medieval Catalonia (North-Eastern Spain). *Quaternary International*, 330, 109–117.

Van Zandt, K. E., Greer, M. T. & Gelhaus, H. C. (2013). Glanders: an overview of infection in humans. *Orphanet Journal of Rare Diseases*, 8, 131.

Vassallo, D. J. (1992). The corps disease: brucellosis and its historical association with the Royal Army Medical Corps. *Journal of the Royal Army Medical Corps*, 138, 140–150.

Verschooten, F., Vermeiren, D. & Devriese, L. (2000). Bone infection in the bovine appendicular skeleton: a clinical, radiographic, and experimental study. *Veterinary Radiology & Ultrasound*, 41, 250–260.

Viana, M., Cleaveland, S., Matthiopoulos, J., Halliday, J., Packer, C., Craft, M. E., Hampson, K., Czupryna, A., Dobson, A. P., Dubovi, E. J., Ernest, E., Fyumagwa, R., Hoare, R., Hopcraft, J. G., Horton, D. L., Kaare, M. T., Kanellos, T., Lankester, F., Mentzel, C., Mlengeya, T., Mzimbiri, I., Takahashi, E., Willett, B., Haydon, D. T. & Lembo, T. (2015). Dynamics of a morbillivirus at the domestic-wildlife interface: Canine distemper virus in domestic dogs and lions. *Proceeding of the National Academy of Sciences of the United States of America*, 112, 1464–1469.

Viganó, C., Haas, C., Rühli, F. J. & Bouwman, A. (2017). 2,000-year-old β-thalassemia case in Sardinia suggests malaria was endemic by the Roman period. *American Journal of Physical Anthropology*, 164, 362–370.

Villemin, J.-A. (1868). *Études sur la tuberculose*. JB Baillière et Fils, Paris.

Von Messling, V., Zimmer, G., Herrler, G., Haas, L. & Cattaneo, R. (2001). The hemagglutinin of canine distemper virus determines tropism and cytopathogenicity. *Journal of Virology*, 75, 6418–6427.

Vordermeier, M., Ameni, G., Berg, S., Bishop, R., Robertson, B. D., Aseffa, A., Hewinson, R. G. & Young, D. B. (2012). The influence of cattle breed on susceptibility to bovine tuberculosis in Ethiopia. *Comparative Immunology, Microbiology & Infectious Diseases*, 35, 227–232.

Waag, D. & Deshazer, D. (2004). 'Glanders: New insights into an old disease', in L. E. Lindler, F. J. Lebeda & G. W. Korch (eds), *Biological weapons defense: infectious diseases and counterbioterrorism*. Humana Press Inc, Totowa.

Webb, G. (1936). *Clio Medica: tuberculosis*. Paul B. Hoeber, Inc., New York.

Wheelis, M. (1998). First shots fired in biological warfare. *Nature*, 395, 213.

White, G., Simpson, R. M. & Scott, G. R. (1961). An antigenic relationship between the viruses of bovine rinderpest and canine distemper. *Immunology*, 4, 203–205.

Wilkinson, L. (1981). Glanders: medicine and veterinary medicine in common pursuit of a contagious disease. *Medical History*, 25, 363–384.

Winchester, J. C. & Kapan, D. D. (2013). History of *Aedes* mosquitoes in Hawaii. *Journal of the American Mosquito Control Association*, 29, 154–163.

Wooding, J. E. (2010). *The identification of bovine tuberculosis in zooarchaeological assemblages: working towards differential diagnostic criteria*. Unpublished PhD Thesis, University of Bradford, Bradford.

Wooding, J. E., King, S. S., Taylor, G. M., Knüsel, C. J., Bond, J. M. & Dent, J. S. (2019). Reviewing the palaeopathological evidence for bovine tuberculosis in the associated bone groups at Wetwang Slack, East Yorkshire. *International Journal of Osteoarchaeology* [online]. doi:10.1002/oa.2846

Woolhouse, M. E., Haydon, D. T. & Antia, R. (2005). Emerging pathogens: the epidemiology and evolution of species jumps. *Trends in Ecology & Evolution*, 20, 238–244.

Yanagida, T., Carod, J. F., Sako, Y., Nakao, M., Hoberg, E. P. & Ito, A. (2014). Genetics of the pig tapeworm in Madagascar reveal a history of human dispersal and colonization. *PLoS One*, 9, e109002.

Ye, Z. W., Yuan, S., Yuen, K. S., Fung, S. Y., Chan, C. P. & Jin, D. Y. (2020). Zoonotic origins of human coronaviruses. *International Journal of Biological Sciences*, 16, 1686–1697.

Yeh, H. Y. & Mitchell, P. D. (2016). Ancient human parasites in ethnic Chinese populations. *Korean Journal of Parasitology*, 54, 565–572.

Yeh, H. Y., Zhan, X. & Qi, W. (2019). A comparison of ancient parasites as seen from archeological contexts and early medical texts in China. *International Journal of Paleopathology*, 25, 30–38.

Zachariah, A., Pandiyan, J., Madhavilatha, G. K., Mundayoor, S., Chandramohan, B., Sajesh, P. K., Santhosh, S. & Mikota, S. K. (2017). *Mycobacterium tuberculosis* in Wild Asian Elephants, Southern India. *Emerging Infectious Diseases*, 23, 504–506.

Zimpel, C. K., Patané, J. S. L., Guedes, A. C. P., De Souza, R. F., Silva-Pereira, T. T., Camargo, N. C. S., De Souza Filho, A. F., Ikuta, C. Y., Neto, J. S. F., Setubal, J. C., Heinemann, M. B. & Guimaraes, A. M. S. (2020). Global Distribution and Evolution of *Mycobacterium bovis* Lineages. *Frontiers in Microbiology*, 11, 843.

CHAPTER 18

Now you have read the book, what next?

Gillian Bentley, Charlotte A. Roberts, Sarah Elton, and Kimberly A. Plomp

Prior to this volume, the principles and practices of evolutionary medicine (EM) and palaeopathology have rarely been considered together in research. This is surprising considering that both fields study evidence for the evolution of disease, as well as think about deep time and how long-term perspectives benefit interpretations of past and present health and well-being. While publications in both fields are prolific, they have typically remained separate in trying to address health problems, perhaps because there has been an enduring focus on addressing discipline-specific challenges, such as the identification and interpretation of disease in the skeletal record as exemplified by palaeopathology. In reality, researchers have worked for many years on successfully integrating the archaeological record with palaeopathological evidence, which has contributed considerably to the now-vibrant field of bioarchaeology. This focus may have distracted people from having productive debates and discussions about the integration of evolutionary theory within palaeopathology.

Why EM has not embraced the world of palaeopathology up until this point is more difficult to untangle—it may simply be an accident relating to the background and interests of the scholars who became immersed in the emerging discipline. George Williams, an evolutionary biologist, and Randy Nesse, a psychiatrist, wrote the pioneering article, the 'Dawn of Darwinian Medicine' in 1991. From that point, the field began to move in two ways as more and more researchers became

interested. First, a group primarily comprising biologists, including Stephen Stearns (Stearns, 1999; Stearns & Koella, 2008), wrote EM books focusing on host-parasite co-evolution, genetics, and cancer, while a second group (social scientists) began tackling topics that crossed more into anthropology and the social sciences, including obstetrics, sudden infant death syndrome, metabolic disorders, and other miscellaneous topics including sleep disorders, and myopia (Trevathan et al., 1999, 2008). It could be argued that a third strand also emerged with the Developmental Origins of Health and Disease (DOHaD) perspectives, spearheaded by Peter Gluckman, Mark Hanson, and others, which led to a series of publications on that particular theme (Gluckman & Hanson, 2005; Gluckman et al., 2016). Various position papers were often co-authored by combinations of people from these three different strands (Grunspan et al., 2018; Hidaka et al., 2015; Nesse, 2008, 2011; Nesse & Stearns, 2008; Nesse et al., 2010; Stearns et al., 2010). Palaeopathologists were, however, not among them. Other books on EM also did not feature articles on palaeopathology or bioarchaeology (Elton & O'Higgins, 2008; Alvergne et al., 2016).

Perhaps then, it is not so surprising that palaeopathology was slow to emerge into the domain of EM, given both the history and preoccupation of the newly emerging area with existing interests. Many proponents of EM were also fairly absorbed with the goal of achieving recognition in the medical world rather than expanding in

Gillian Bentley et al., *Now you have read the book, what next?* In: *Palaeopathology and Evolutionary Medicine.* Edited by Kimberly A. Plomp et al.,
Oxford University Press. © Oxford University Press (2022). DOI: 10.1093/oso/9780198849711.003.0018

new directions. The advantages to both EM and palaeopathology of working with each other should, however, be much clearer for those readers who have digested the contents of this book, or even skimmed lightly through its pages. There is much that each discipline can teach the other to the mutual benefit of both.

This one book, however, cannot create sufficient dynamic on its own, although it is a start. Other materials should—and must—follow to ensure that the two fields continue a productive dialogue. To begin with, future textbooks on EM geared towards students should make an effort to include at least one section on the value of bioarchaeology for reconstructing why humans are vulnerable to disease and how this manifested itself in the past. While bioarchaeologists who focus on palaeopathology have certainly had a presence at recent EM meetings, they need to be more visible and attend in greater numbers. In addition, plenaries and sessions at EM meetings should give more space to palaeopathologists to illuminate their work. Similarly, archaeology and palaeopathology meetings should include sessions on EM and encourage relevant evolutionary submissions. On top of this, journals that cover palaeopathology (e.g., *International Journal of Paleopathology*) should encourage submissions from authors that genuinely consider palaeopathological evidence within an evolutionary theoretical context. Conversely, the flagship journal (*Evolution, Medicine, and Public Health*) of the International Society for Evolution, Medicine, and Public Health (ISEMPH) has already advertised a special issue on palaeopathology, and the author of our Preface (Frank Rühli) will be the 2022–2024 President of the ISEMPH. Furthermore, journal editors should encourage research that explicitly links the two fields. Such activities create visibility that can only enhance the probability that scholars will continue to integrate palaeopathology and EM.

In terms of teaching, future and existing courses on EM should consider adding elements from bioarchaeology (including palaeopathology). This may be straightforward in US universities, where the four-field approach to anthropology (biological anthropology, cultural anthropology, linguistic anthropology, and archaeology) is common and

where scholars from different areas can contribute within a single department. It might not be as easy within an archaeology department that is a separate entity, as is the case in many European universities. When teaching bioarchaeology at all levels, from undergraduate to PhD, palaeopathology should always be included and taught in line with an evolutionary theoretical approach. If nothing else, inclusion of both fields in teaching will bring them more to the attention of academic scholars, as well as stakeholders beyond academia. Teaching curricula relevant to both fields could incorporate both approaches in showing the benefits of the synergies that clearly exist between EM and palaeopathology. This integration also has the potential to increase the relevancy of both EM and paleopathology to modern medicine and medical education.

Beyond publications and the classroom, public engagement activities could include both fields in discussing the benefits of taking deep time perspectives to understanding health and well-being today and planning for the future. Communication of the outcomes of co-productive research to relevant stakeholders would be very beneficial in highlighting the strengths of our research, including to the public. Bioarchaeological topics continue to fascinate the public (witness the enduring interest in mummies and other preserved bodies) and provide a promising platform for outreach. EM and palaeopathology could also work together in persuading governments to think outside the box in broadening their horizons with regards to managing health and taking a more interdisciplinary approach.

We propose that increased, multidisciplinary collaborations between bioarchaeologists and other scholars, such as evolutionary biologists, geneticists, biologists, biological anthropologists, and medical professionals, would enable more diverse and nuanced research projects that can investigate diseases with the deep-time perspective that is needed to plan for the future health of our species and our planet. The information and understanding that we would likely glean from these types of collaborations may even impact innovations in clinical prevention and/or treatments of health problems today. This would be especially true if future research in palaeopathology could continue to work towards more standardised ways of diagnosing, describing and reporting pathological lesions and other relevant variables to enable more effective sharing and combining of archaeological and modern data. Basic standard macroscopic methods for recording in palaeopathology are routinely used—a practice that has particularly gained attention since the publication of Buikstra and Ubelaker (1994) as a response to NAGPRA (Native American Graves Protection and Repatriation Act 1990). This Act prompted recognition of the need to record any culturally affiliated Native American human skeletal remains in federally funded institutions in a systematic and standardised manner prior to potential repatriation. By 2004, BABAO (British Association for Biological Anthropology and Osteoarchaeology) had published recording standards for Britain (Brickley & McKinley, 2004). Standard methods of recording are also used in the Global History of Health Project so that the large datasets generated can be reliably compared (Steckel et al., 2019), many of which have been adopted by scholars for their research.

These guidance documents have proven valuable for scholars in palaeopathology, such that pathological lesions are generally recorded using these 'standards'. Having such standardised methods for recording in palaeopathology is important because, ultimately, data comparability relies on scholars recording data in the same manner and using the same methods. This is especially relevant when comparing large datasets from around the world and across different time periods, and when thinking about frequency data. This is not to say that when baseline data have been documented using these methods, practitioners cannot add to the methods used if the research project/questions being addressed need to incorporate other methods. We therefore continue to advocate that standard recording methods should be implemented in any research project in palaeopathology.

Finally, we would encourage scholars in both EM and the palaeopathology of humans and other animals to continue to explore the One Health approach in their research. Our species is exposed to health risks in globally varying environments, and we share these with other animals. One Health

has become particularly prominent over 2020 and 2021 in discussions about Covid-19 (e.g., Ruckert et al., 2020), and in this volume we have several 'disease' chapters that are relevant (e.g., plague, leprosy, tuberculosis, parasites). Our species has co-evolved alongside many others to which we were in contact, especially through scavenging, hunting, and then animal husbandry, all of which would have exposed us to novel pathogens and zoonotic diseases. As a species, we had to adapt to these new pathogens without the benefits of modern medicine. The newly emerging diseases of recent decades, while novel in themselves, are but a repeat of patterns which humans have survived over several millennia. The past has much to teach us in this respect and we anticipate more work related to One Health.

We appreciate that some of our proposals will take time to mature but there is enormous promise with in these endeavours, as illuminated by the many chapters in this book. We hope they will have provided a flavour of how the two fields of EM and palaeopathology/bioarchaeology can be brought together in new and fruitful ways. We leave it to our readers to help advance the next frontiers in the continued integration of two fields that are so conceptually and temporally intertwined.

References

Alvergne, A., Jenkinson, C. & Faurie, C. (eds). (2016). *Evolutionary thinking in medicine: from research to policy and practice*. Springer, Cham.

Brickley, J. I. & McKinley, J. (2004). Recording guidelines for the standards for human remains. Institute of Field Archaeologists Paper 7. Institute of Field Archaeologists, Reading.

Buikstra, J. E. & Ubelaker, D. H. (1994). Standards for data collection from human skeletal remains. Arkansas Archeological Survey Research Series, No. 44. Arkansas Archeological Survey, Fayetteville.

Elton, S. & O'Higgins, P. (eds). (2008). *Medicine and evolution: current applications, future prospects*. CRC Press, Boca Raton.

Gluckman, P., Beedle, A., Buklijas, T., Low, F. & Hanson, M. (2016). *Principles of evolutionary medicine*, 2nd edn. Oxford University Press, Oxford.

Gluckman, P. & Hanson, M. (2005). *The fetal matrix: evolution, development and disease*. Cambridge University Press, Cambridge.

Grunspan, D. Z., Nesse, R. M., Barnes, M. E. & Brownell, S. E. (2018). Core principles of evolutionary medicine. *Evolutionary Medicine and Public Health*, 1, 13–23. doi: 10.1093/emph/eox025

Hidaka, B. H., Asghar, A., Aktipis, C. A., Nesse, R. M., Wolpaw, T. M., Skursky, N. K., Bennett, K. J., Beyrouty, M. W. & Schwartz, M. D. (2015). The status of evolutionary medicine education in North American medical schools. *BMC Medical Education*, 15, 38. doi: 10.1186/s12909-015-0322-5

Nesse R. (2008). What evolutionary biology offers public health. *Bulletin of the World Health Organization*, 86, 83. doi: 10.2471/BLT.07.049601

Nesse, R. M. (2011). Ten questions for evolutionary studies of disease vulnerability. *Evolutionary Applications* 4: 264–277. doi: 10.1111/j.1752-4571.2010.00181.x

Nesse, R. M., Bergstrom, C. T., Ellison, P. T., Flier, J. S., Gluckman, P., Govindaraju, D. R., Niethammer, D., Omenn, G. S., Perlman, R. L., Schwartz, M. D., Thomas, M. G., Stearns, S. & Valle, D. (2010). Making evolutionary biology a basic science for medicine. *Proceedings of the National Academy of Sciences of the United States of America*, 107, 1800–1807. doi: 10.1073/pnas.0906224106

Nesse, R.M. & Stearns, S. C. (2008). The great opportunity: evolutionary applications to medicine and public health. *Evolutionary Applications*, 1, 28–48. doi: 10.1111/j.1752-4571.2007.00006.x

Ruckert, A., Zinszer, K., Zarowsky, C., Labonté, R. & Carabin, H. (2020). What role for One Health in the COVID-19 pandemic? *Canadian Journal of Public Health*, 111, 641–644.

Stearns S. (ed). (1999). *Evolution in health and disease*. Oxford University Press, Oxford.

Stearns, S. & Koella, J. (eds). (2008). *Evolution in health and disease*, 2nd edn. Oxford University Press, Oxford.

Stearns, S. C., Nesse, R. M., Govindaraju, D. R. & Ellison, P. T. (2010). Evolutionary perspectives on health and medicine. *Proceedings of the National Academy of Sciences of the United States of America*, 107, 1691–1695. doi: 10.1073/pnas.0914475107.

Steckel RH, Larsen CS, Sciulli PW, Walker PL. 2019. 'Data collection codebook', in R. H. Steckel, C. S. Larsen, C. A. Roberts & J. Baten (eds), *The backbone of Europe. health, diet, work and violence over two millennia*, pp. 397–427. Cambridge University Press, Cambridge.

Trevathan, W., Smith, E. & McKenna, J. (eds). (1999). *Evolutionary medicine: new perspectives*. Oxford University Press, Oxford.

Trevathan, W., Smith, E. & McKenna, J. (eds). (2008). *Evolutionary medicine: new perspectives*, 2nd edn. Oxford University Press, Oxford.

Williams, G. C. & Nesse, R. M. (1991). The dawn of Darwinian medicine. *Quarterly Review of Biology*, 66, 1–22. doi: 10.1086/417048.

Afterword

Jane Buikstra

The rich tapestry of the human past presents scholars with a vast opportunity to explore the human condition under a variety of environmental, socio-economic, and political circumstances. A diverse, long-term history shows us, for example, that resilient groups are those who are flexible in the face of external challenges, such as epidemic disease and climate change. These are the ones who do not fall into the 'rigidity trap' of limited options, increasingly difficult for complex societies. This landscape of the past is thus a natural experiment where we may explore the ways in which individuals and groups adapted well to circumstances. Those who did not—the diseased, the disadvantaged, those dying young, those suffering from pain due to a combination of internal and external circumstances—along with the survivors and the healthy, all have a great deal to tell us about human evolution and hence, evolutionary medicine. Here we find deep time clinical trials with a time depth impossible in epidemiology today.

The term 'evolution', as defined by Merriam Webster (2021), includes the familiar 'descent with modification', as well as 'a process of change in a certain direction'. Synonyms include expansion, growth, and progress. Such might lead us to focus upon the valued well and the unfortunate ill as a dichotomy of faceless unnamed participants in the evolutionary process. Bioarchaeology, including the subfield of palaeopathology, is the study of people in the past, through the medium of human remains, archaeological and historical contexts. Thus, it is bioarchaeology that can explore such aspects of humankind's history as the interaction of morphological variation, activity, and physical degeneration. It can also provide unique perspectives on human diet in the past, the oral environment, and general health. Further, palaeopathology can explore the manner in which people have reacted to the ill and deformed—perhaps de-essentializing reactions that today are focused upon beauty, perfection, and perceived societal contributions that depend upon able-bodied status. We can 'people the past', giving faces to those who lived under different regimes of challenges and medical intervention.

A wise reviewer of any document first explores whether the authors and editors have met a stated goal. The goals of this volume are ambitious, clearly stated, and, to a large extent, achieved. This explicitly interdisciplinary effort purposely seeks to entangle the disciplines of palaeopathology and evolutionary medicine, primarily from the perspective of the former by demonstrating relevance to the latter. At the editors emphasise at outset, the two fields would seem to be closely aligned, though they have had surprisingly few true interactions. The reason for this may reflect the distinctive origins of the two fields, especially the social science and humanist background of bioarchaeology and most practitioners of the discipline. Physicians draw knowledge primarily from scientific disciplines; most focus upon *applying* knowledge, rather than *creating* it. To date, evolutionary medicine has focused effort within the biomedical community, convincingly arguing for this valued perspective in medical education through studies primarily (but not exclusively) rooted in contemporary medicine. Thus, while palaeopathologists and those engaged in evolutionary medicine would seem 'perfect partners' (as the editors maintain in their introduction), there are divides between disciplinary silos, applied versus theoretical knowledge and epistemologies that have limited knowledge transfer. This volume amply documents why these divisions should be overcome and how we should envision future collaborative ventures that will indeed advance medical practice, historical knowledge, and epidemiological perspectives.

Impressive examples can be drawn from any one of the chapters in this volume. Recognition of variations in spine morphology that predisposes to painful lower back conditions can lead to predictive models for clinicians advising about prospective sport or other activity regimes, as well as in rehabilitation. The long-term trajectories in the oral microbiome and human health are recorded in dental calculus, while histories of parasitic infections explore the beneficial nature of certain bacteria. Knowledge of the manner in which people with leprosy and other disfiguring conditions were stigmatised (or not) in the past leads us to question 'essentialised' universal reactions to such alterities.

As emphasised here, 'One Health' fits well structurally as the metaphorical 'third leg' to the stool already supported by evolutionary medicine and palaeopathology. The One Health movement was initiated by Charles Schwabe, a veterinarian trained in public health, who coined the term 'One Medicine' in 1984. One Medicine emphasised the similarities between diseases affecting humans and other animals. During the twenty-first century this concept has matured into a global public health initiative, focusing on the links between the environment, humans and non-human animals in promoting health worldwide (King et al., 2008). Inequalities in disease risk and conservation efforts characterise One Health initiatives today. One Health is then a logical partner for evolutionary medicine

and palaeopathology as we move into the middle portion of the twenty-first century.

In sum, it is important to think globally about health, interrogating the past in a framework that employs evolutionary principles. This volume represents a significant advance in drawing together the disciplines that should comprise the integrated anchor for this initiative.

References

King, L. J., Anderson, L. R., Blackmore, C. G, Blackwell, M. J., Lautner, E. A., Marcus, L. C., Meyer, T.E., Monath, T. P., Nave, J. E., Ohle, J., Pappaioanou, M. Sobata, J., Stokes, W. S., Davis, R. M., Glasser, J. H. & Mahr, R. K. (2008) Executive summary of the AVMA One Health Initiative Task Force report. *Journal of the American Veterinary Medical Association*, 233(2), 259–261.

Merriam-Webster (2021). 'Evolution'. https://www.merriam-webster.com/dictionary/evolution

Index

Please note that page references to Figures will be followed by the letter 'f', to Tables by the letter 't'